CALCULUS

CALCULUS

Produced by the Consortium based at Harvard and funded by a National Science Foundation Grant.

Deborah Hughes-Hallett
Harvard University

Andrew M. Gleason
Harvard University

Daniel E. Flath
University of South Alabama

Sheldon P. Gordon
Suffolk County Community College

David O. Lomen
University of Arizona

David Lovelock
University of Arizona

William G. McCallum
University of Arizona

Brad G. Osgood
Stanford University

Andrew Pasquale
Chelmsford High School

Jeff Tecosky-Feldman
Haverford College

Joe B. Thrash
University of Southern Mississippi

Karen R. Thrash
University of Southern Mississippi

Thomas W. Tucker
Colgate University

with the assistance of
Otto K. Bretscher
Harvard University

JOHN WILEY & SONS, INC.
New York Chichester Brisbane Toronto Singapore

Cover Photo: Eugene Gerhardt/FPG International

Acquisitions Editor	Ruth Baruth
Developmental Editor	Joan Carrafiello
Marketing Manager	Susan Elbe
Production Editor	Nancy Prinz
Designer	Laura Nicholls
Manufacturing Manager	Susan Stetzer

This book was set in Times Roman by the Consortium based at Harvard using TEX, Mathematica, and the package *Newcalcstyle*, which was written by Alex Kasman for this project. Special thanks also to S. Alex Mallozzi and Alice Wang for managing the process.

Recognizing the importance of preserving what has been written, it is a policy of John Wiley & Sons, Inc. to have books of enduring value published in the United States printed on acid-free paper, and we exert our best efforts to that end.

The paper in this book was manufactured by a mill whose forest management programs include sustained yield harvesting of its timberland. Sustained yield harvesting principles ensure that the number of trees cut each year does not exceed the amount of new growth.

Problems from *Calculus: The Analysis of Functions*, by Peter D. Taylor (Toronto: Wall & Emerson, Inc., 1992). Reprinted with permission of the publisher.

Library of Congress Cataloging-in-Publication Data

Calculus / Deborah Hughes-Hallett . . . [et al].
p. cm.
ISBN 0-471-58621-8 (cloth). — ISBN 0-471-31055-7 (pbk.)
1. Calculus. I. Hughes-Hallett, Deborah.
QA303.C155 1994
515—dc20 93-32103
CIP

Printed in the United States of America

10 9 8

Dedicated to Alex, Alice, Eric, and Alex for their resourcefulness, creativity, and endless good humor.

PREFACE

Calculus is one of the greatest achievements of the human intellect. Inspired by problems in astronomy, Newton and Leibniz developed the ideas of calculus 300 years ago. Since then, each century has demonstrated the power of calculus to illuminate questions in mathematics, the physical sciences, engineering, and the social and biological sciences.

Calculus has been so successful because of its extraordinary power to reduce complicated problems to simple rules and procedures. Therein lies the danger in teaching calculus: it is possible to teach the subject as nothing but the rules and procedures – thereby losing sight of both the mathematics and of its practical value. With the generous support of the National Science Foundation, our consortium set out to create a new calculus curriculum that would restore that insight. This book is part of that endeavor.

Basic Principles

Two principles guided our efforts. The first is our prescription for restoring the mathematical content to calculus:

The Rule of Three: *Every topic should be presented geometrically, numerically, and algebraically.*

We continually encourage students to think about the geometrical and numerical meaning of what they are doing. It is not our intention to undermine the purely algebraic aspect of calculus, but rather to reinforce it by giving meaning to the symbols. In the homework problems dealing with applications, we continually ask students to explain verbally what their answers mean in practical terms.

The second principle, inspired by Archimedes, is our prescription for restoring practical understanding:

The Way of Archimedes: *Formal definitions and procedures evolve from the investigation of practical problems.*

Archimedes believed that insight into mathematical problems is gained by first considering them from a mechanical or physical point of view.[1] For the same reason, our text is problem driven. Whenever possible, we start with a practical problem and derive the general results from it. By practical problems we usually, but not always, mean real world applications. These two principles have led to a dramatically new curriculum – more so than a cursory glance at the table of contents might indicate.

[1] . . . I thought fit to write out for you and explain in detail . . . the peculiarity of a certain method, by which it will be possible for you to get a start to enable you to investigate some of the problems in mathematics by means of mechanics. This procedure is, I am persuaded, no less useful even for the proof of the theorems themselves; for certain things first became clear to me by a mechanical method, although they had to be demonstrated by geometry afterwards because their investigation by the said method did not furnish an actual demonstration. But it is of course easier, when we have previously acquired, by the method, some knowledge of the questions, to supply the proof than it is to find it without any previous knowledge. From *The Method*, in *The Works of Archimedes* edited and translated by Sir Thomas L. Heath (Dover, NY)

Technology

We take advantage of computers and graphing calculators to help students learn to think mathematically. For example, using a graphing calculator to zoom in on functions is one of the best ways of seeing local linearity. Furthermore, the ability to use technology effectively as a tool is itself of the greatest importance. Students are expected to use their own judgement to determine where technology is useful.

However, the book does not require any specific software or technology. Test sites have used the materials with graphing calculators, graphing software, and computer algebra systems. Any technology with the ability to graph functions and perform numerical integration will suffice.

What Student Background is Expected?

We have found this curriculum to be thought-provoking for well-prepared students while still accessible to students with weak algebra backgrounds. Providing numerical and graphical approaches as well as the algebraic gives students several ways of mastering the material. This approach encourages students to persist, thereby lowering failure rates.

Content

When we designed this curriculum we started with a clean slate. We included some new topics, such as differential equations, and omitted some traditional topics whose inclusion we could not justify after discussions with mathematicians, engineers, physicists, chemists, biologists, and economists. In the process, we also changed the focus of certain topics. In order to meet individual needs or course requirements, topics can easily be added or deleted, or the order changed.

Chapter 1: A Library of Functions

Chapter 1 introduces all the elementary functions to be used in the book. Although the functions are probably familiar, the graphical, numerical, and modeling approach to them is fresh. Our purpose is to acquaint the student with each function's individuality: the shape of its graph, characteristic properties, comparative growth rates, and general uses. We expect to give the student the skill to read graphs and think graphically, to read tables and think numerically, and to apply these skills, along with their algebraic skills, to modeling the real world. We introduce exponential functions at the earliest possible stage, since they are fundamental to the understanding of real-world processes. Further attention is given to constructing new functions from old ones—how to shift, flip and stretch the graph of any basic function into a new, related function.

We encourage you to cover this chapter thoroughly, as the time spent on it will pay off when you get to the calculus.

Chapter 2: The Derivative

Chapter 2 presents the key concept of the derivative according to the Rule of Three. The purpose of this chapter is to give the student a practical understanding of the limit definition of the derivative and its interpretation as an instantaneous rate of change without complicating the discussion with differentiation rules. After finishing this chapter, a student will be able to find derivatives numerically (by taking arbitrarily fine difference quotients), visualize derivatives graphically as the slope of the graph, and interpret the meaning of first and second derivatives in various applications. The student will also understand local linearity and recognize the derivative as a function in its own right.

Chapter 3: The Definite Integral

Chapter 3 presents the key concept of the definite integral, along the same lines as Chapter 2. Some instructors using preliminary versions of the book have delayed covering Chapter 3 until after Chapter 5 without any difficulty.

The purpose of this chapter is to give the student a practical understanding of the definite integral as a limit of Riemann sums, and to bring out the connection between the derivative and the definite integral in the Fundamental Theorem of Calculus. We use the same method as in Chapter 2, introducing the fundamental concept in depth without going into technique. The motivating problem is computing the total distance traveled from the velocity function. The student will finish the chapter with a good grasp of the definite integral as a limit of Riemann sums, with the ability to compute it numerically, and with an understanding of how to interpret the definite integral in various contexts.

Chapter 4: Short-Cuts to Differentiation

Chapter 4 presents the symbolic approach to differentiation. The title is intended to remind the student that the basic methods of differentiation are not to be regarded as the definition of the derivative. The derivatives of all the functions in Chapter 1 are introduced as well as the rules for differentiating the combinations discussed in Chapter 1. Implicit differentiation is introduced and used to find derivatives of several basic functions. We give informal but mathematically sound justifications, introducing graphical and numerical reasoning where appropriate. The student will finish this chapter with basic proficiency in differentiation and an understanding of why the various rules are true.

Chapter 5: Using the Derivative

Chapter 5 presents applications of the derivative. It includes an investigation of parametrized families of functions according to the Way of Archimedes, using the graphing technology to observe basic properties and calculus to confirm them.

Our aim in this chapter is to enable the student to use the derivative in solving problems, rather than to learn a catalogue of application templates. It is not meant to be comprehensive, and you do not need to cover all the sections. The student should finish this chapter with the experience of having successfully tackled a few problems that required sustained thought over more than one session.

Chapter 6: Reconstructing a Function from Its Derivative

Chapter 6 focuses on "going backward" from a derivative to the original function, first graphically and numerically, then analytically. The chapter starts with the properties of the definite integral and its interpretation as an area (Sections 6.1 and 6.2) and ends with an analysis of motion under the influence of gravity. After finishing this chapter, students will understand how to "go backwards" from the derivative to the original function.

Chapter 7: The Integral

Chapter 7 investigates methods of finding integrals. We do not restrict our attention to functions that have closed-form antiderivatives. Instead, we emphasize the role of numerical integration as a basic tool. This chapter includes several techniques of integration; others are included in the table of integrals. While we do not specifically make use of computer algebra software, we certainly acknowledge that its existence changes the skills that students need to master.

This chapter threads practical skills with theoretical understanding. There are two groups of sections on computing definite integrals: Sections 7.1–7.5 on using the Fundamental Theorem, and Sections 7.6–7.7 on using numerical methods. Sections 7.8–7.9 are on improper integrals. The student will finish this chapter with proficiency in the basic methods of integration.

Chapter 8: Using the Definite Integral

Chapter 8 addresses applications of the definite integral. We emphasize the idea of subdividing a quantity to produce Riemann sums which, in the limit, yield a definite integral, and aim to show ways the integral is used without resorting to templates. The chapter starts with a discussion of how to set up definite integrals that represent given physical quantities, and then gives examples from geometry, physics, economics, and probability. You do not need to cover all the sections. The student will finish this chapter understanding how to form Riemann sums and knowing how they are used.

Chapter 9: Differential Equations

Chapter 9 introduces differential equations without too many technicalities. It is intended to show the power of the methods we have developed, using more realistic and complex applications. Slope fields are used to visualize the behavior of solutions of first-order differential equations. The emphasis is on qualitative solutions, modeling, and interpretation. We include applications to population models (exponential and logistic), the spread of disease, predator-prey equations, and competitive exclusion. There is also some material on second-order differential equations with constant coefficients: the spring equation, both damped and undamped, and solutions using complex numbers. The student will finish this chapter knowing what a differential equation is, how to approximate its solution graphically and numerically, and how to find some analytic solutions, all in the context of substantial applications.

Chapter 10: Approximations

Chapter 10 is an introduction to Taylor Series and Fourier series via the idea of approximating functions with simpler functions; the Taylor series is a local approximation, the Fourier series a global one. The primary focus is on Taylor polynomials and series. Geometric series and their applications are discussed. The notion of a convergent series is permitted to evolve naturally out of the investigation of Taylor polynomials. The graphical and numerical points of view are kept at the forefront throughout. The student will finish this chapter with a good grasp of Taylor approximations and understand how they differ from Fourier approximations.

Appendices

The appendices contain material on roots and accuracy, on continuity and bounds, on polar coordinates, and on complex numbers.

Our Experiences

In the process of developing the ideas incorporated in this book, we have been conscious of the need to test the materials thoroughly in a wide variety of institutions serving many different types of students. Consortium members have used previous versions of the book for several years at large and small liberal arts colleges, at large and small public universities, at a two-year institution, and at a high school. During the 1991-92 and 92-93 academic years, we were assisted by colleagues at over one hundred schools around the country who class-tested the book and reported their experiences and those of their students. This diverse group of schools used the book in semester and quarter systems, in large lecture sections and small classes, in computer labs, small groups, and traditional settings, and with a number of different technologies. We appreciate the valuable suggestions they made, which we have tried to incorporate into this first edition of the text.

Changes from the Preliminary Edition

Chapter 1. The number e is introduced later, at the same time as the natural logarithm. Compound interest has its own section, and the section on Roots, Continuity, and Accuracy has been moved to Appendix A and Appendix B.

Chapter 2. The dy/dx notation for the derivative is delayed until Section 2.4 where it is used to help students interpret the derivative in practical terms. Slightly more emphasis is put on limits in Section 2.1.

Chapter 3. The interpretation of the definite integral of a rate as total change has been moved to Section 3.4, where it is used to help students appreciate the Fundamental Theorem of Calculus.

Chapter 5. The material on antidifferentiation has been moved to the new Chapter 6: Reconstructing a Function from Its Derivative.

Chapter 6. This new chapter pulls together material on "going backward" from derivative to original function that was previously distributed through several chapters. It can be used to finish a study of Chapters 1–5 or as a starting point for Chapters 6–10.

Chapter 7. Introductory material has been moved to the new Chapter 6. A subsection on integration using partial fractions has been added to Section 7.5.

Chapter 8. The material on interpreting the definite integral as an area is now in Section 6.1. The section on probability and distributions has been split into two sections.

Chapter 9. This chapter has been rewritten to include more emphasis on applications. Section 9.7 is based on the growth of the US population since 1790. Section 9.8 contains the original material on two interacting populations as well as new material on the spread of a disease. Section 9.9 is entirely new and shows students how to investigate the phase plane using nullclines. The section on complex numbers and polar coordinates has been moved to Appendix C and Appendix D.

Chapter 10. The section on geometric series has been added. The section on estimating the error in a Taylor approximation has been rewritten, and now includes the Mean Value Theorem.

Appendices. The material in the appendices has been moved here for flexibility in scheduling.

Supplementary Materials

- **Instructor's Manual with Sample Exams** containing teaching tips, calculator programs, some overhead transparency masters and sample exams.
- **Instructor's Solution Manual** with complete solutions to all problems.
- **Answer Manual** with brief answers to all odd-numbered problems.
- **Student's Solution Manual** with complete solutions to half the odd-numbered problems.
- **Calculus Project Book** containing projects for students, and their solutions.

- **Orientation Video**, an orientation to teaching with the materials.
- **Workshop Video** to guide instructors conducting workshops on the materials in the textbook.
- **University of Arizona Software Manual** ties the University of Arizona software to specific problems in the text and includes data sets for working some problems.
- **Discovering Calculus with Derive**, a problems manual with brief instructions on the use of the software as well as additional problems which correspond to the text.

Acknowledgements

First and foremost, we want to express our appreciation to the National Science Foundation for their faith in our ability to produce a revitalized calculus curriculum and, in particular, to Louise Raphael, John Kenelly, John Bradley, and James Lightbourne. We also want to thank the members of our Advisory Board, Lida Barrett, Bob Davis, John Dossey, Ron Douglas, Seymour Parter and Steve Rodi for their ongoing guidance and advice.

In addition, a host of other people around the country and abroad deserve our thanks for all that they did to help our project succeed. They include: Wayne Anderson, Ruth Baruth, Maria Betkowski, Melkana Brakalova, Jackie Boyd-DeMarzio, Otto Bretscher, Morton Brown, Greg Brumfiel, Joan Carrafiello, Phil Cheifetz, Ralph Cohen, Bob Condon, Sterling G. Crossley, Ehud de Shalit, Bob Decker, Persi Diaconis, Tom Dick, Steve Doblin, Wade Ellis, Alice Essary, Sol Feferman, Hermann Flaschka, Patti Frazer Lock, Lynn Garner, Allan Gleason, Florence Gordon, Danny Goroff, Robin Gottlieb, JoEllen Hillyer, Luke Hunsberger, Richard Iltis, Rob Indik, Adrian Iovita, Jerry Johnson, Mille Johnson, Matthias Kawski, Gabriel Katz, David Kazhdan, Mike Klucznik, Donna Krawczyk, Robert Kuhn, Carl Leinbach, David Levermore, Don Lewis, John Lucas, Reginald Luke, Tom MacMahon, Dan Madden, Barry Mazur, Rafe Mazzeo, Dave Meredith, David Mumford, Alan Newell, Huriye Önder, Arnie Ostebee, Jose Padro, Mike Pavloff, Tony Phillips, John Prados, Amy Radunskaya, Wayne Raskind, Gabriella Ratay, Janet Ray, George Rublein, Wilfried Schmid, Marilyn Semrau, Pat Shure, Esther Silberstein, David Smith, Don Snow, Bob Speiser, Howard Stone, Steve Strogatz, "Suds" Sudholz, Cliff Taubes, Peter Taylor, Tom Timchek, Alan Tucker, Jerry Uhl, Bill Vélez, Gary Walls, Charles Walter, Mary Jean Winter, Debbie Yoklic, Lee Zia, Paul Zorn, and all the people in the Harvard mathematics department who shared their computers and their space with us.

Most of all, to the remarkable team that worked day and night to get the text into the computer (and out again), to get the solutions written and the pictures labeled: we greatly appreciate your ingenuity, energy and dedication. Thanks to: Stefan Bilbao, Ruvim Breydo, Will Brockman, Duff Campbell, Kenny Ching, Eric Connally, Radu Constantinescu, Radhika de Silva, Srdjan Divac, Patricia Hersh, Joseph Kanapka, Alex Kasman, Georgia Kamvosoulis, Dimitri Kountourogiannis, Alex Mallozzi, Mike Mitzenmacher, Ed Park, Jessica Polito, Sulian Tay, Alice Wang, Eric Wepsic, Gang Zhang.

Deborah Hughes-Hallett
Andrew M. Gleason
Daniel E. Flath
Sheldon P. Gordon
David O. Lomen
David Lovelock
William G. McCallum
Brad G. Osgood
Andrew Pasquale
Jeff Tecosky-Feldman
Joe B. Thrash
Karen R. Thrash
Thomas W. Tucker

To Students: How to Learn from this Book

- This book may be different from other math textbooks that you have used, so it may be helpful to know about some of the differences in advance. At every stage, this book emphasizes the *meaning* (in practical, graphical or numerical terms) of the symbols you are using. There is much less emphasis on "plug-and-chug" and using formulas, and much more emphasis on the interpretation of these formulas than you may expect. You will often be asked to explain your ideas in words or to explain an answer using graphs.
- The book contains the main ideas of calculus in plain English. Success in using this book will depend on reading, questioning, and thinking hard about the ideas presented. It will be helpful to read the text in detail, not just the worked examples.
- There are few examples in the text that are exactly like the homework problems, so homework problems can't be done by searching for similar–looking "worked out" examples. Success with the homework will come by grappling with the ideas of calculus.
- Many of the problems in the book are open-ended. This means that there is more than one correct approach and more than one correct solution. Sometimes, solving a problem relies on common sense ideas that are not stated in the problem explicitly but which you know from everyday life.
- This book assumes that you have access to a calculator or computer that can graph functions, find (approximate) roots of equations, and compute integrals numerically. There are many situations where you may not be able to find an exact solution to a problem, but can use a calculator or computer to get a reasonable approximation. An answer obtained this way is usually just as useful as an exact one. However, the problem does not always state that a calculator is required, so use your own judgement.

  If you mistrust technology, listen to this student, who started out the same way:

  > Using computers is strange, but surprisingly beneficial, and in my opinion is what leads to success in this class. I have difficulty visualizing graphs in my head, and this has always led to my downfall in calculus. With the assistance of the computers, that stress was no longer a factor, and I was able to concentrate on the concepts behind the shapes of the graphs, and since these became gradually more clear, I got increasingly better at picturing what the graphs should look like. It's the old story of not being able to get a job without previous experience, but not being able to get experience without a job. Relying on the computer to help me avoid graphing, I was tricked into focusing on what the graphs meant instead of how to make them look right, and what graphs symbolize is the fundamental basis of this class. By being able to see what I was trying to describe and learn from, I could understand a lot more about the concepts, because I could change the conditions and see the results. For the first time, I was able to see how everything works together . . . .

  That was a student at the University of Arizona who took calculus in Fall 1990, the first time we used the text. She was terrified of calculus, got a C on her first test, but finished with an A for the course.
- This book attempts to give equal weight to three methods for describing functions: graphical (a picture), numerical (a table of values) and algebraic (a formula). Sometimes it's easier to translate a problem given in one form into another. For example, you might replace the graph of a parabola with its equation, or plot a table of values to see its behavior. It is important to be flexible about your approach: if one way of looking at a problem doesn't work, try another.

- Students using this book have found discussing these problems in small groups helpful. There are a great many problems which are not cut-and-dried; it can help to attack them with the other perspectives your colleagues can provide. If group work is not feasible, see if your instructor can organize a discussion session in which additional problems can be worked on.
- You are probably wondering what you'll get from the book. The answer is, if you put in a solid effort, you will get a real understanding of one of the most important accomplishments of the millennium – calculus – as well as a real sense of how mathematics is used in the age of technology.

Deborah Hughes-Hallett
Andrew M. Gleason
Daniel E. Flath
Sheldon P. Gordon
David O. Lomen
David Lovelock
William G. McCallum
Brad G. Osgood
Andrew Pasquale
Jeff Tecosky-Feldman
Joe B. Thrash
Karen R. Thrash
Thomas W. Tucker

CONTENTS

CHAPTER ONE

A LIBRARY OF FUNCTIONS

Functions are truly fundamental to mathematics. For example, in everyday language we say, "The price of a ticket is a function of where you sit," or "The speed of a rocket is a function of its payload." In each case, the word *function* expresses the idea that knowledge of one fact tells us another. In mathematics, the most important functions are those in which knowledge of one number tells us another number. If we know the length of the side of a square, its area is determined. If the circumference of a circle is known, its radius is determined.

Calculus starts with the study of functions. This chapter will lay the foundation for calculus by surveying the behavior of the most common functions, including powers, exponentials, logarithms, and the trigonometric functions. Besides the behavior of these functions, we will explore ways of handling the graphs, tables, and formulas that represent them.

1.1 WHAT'S A FUNCTION?

Let's look at an example. In the summer of 1990, the temperatures in Arizona reached an all-time high (so high, in fact, that some airlines decided it might be unsafe to land their planes there). The daily high temperatures in Phoenix for June 19–29 are given in Table 1.1.

TABLE 1.1 *Temperature in Phoenix, Arizona, June 1990*

Date: June (1990)	19	20	21	22	23	24	25	26	27	28	29
Temperature (°F)	109	113	114	113	113	113	120	122	118	118	108

Although you may not have thought of something so unpredictable as temperature as being a function, the temperature *is* a function of date, because each day gives rise to one and only one high temperature. There is no formula for temperature (otherwise we would not need the weather bureau), but nevertheless the temperature does satisfy the definition of a function: Each date, t, has a unique high temperature, H, associated with it.

We define a function as follows:

One quantity, H, is a **function** of another, t, if each value of t has a unique value of H associated with it. We say H is the *value* of the function or the *dependent variable*, and t is the *argument* or *independent variable*. Alternatively, think of t as the *input* and H as the *output*. We write $H = f(t)$, where f is the name of the function.
The **domain** of a function is a set of possible values of the independent variable, and the **range** is the corresponding set of values of the dependent variable.

In the temperature example above, the independent variable is the date, and the dependent variable is the temperature. The domain is all possible dates, and the range is the high temperatures on those dates. The function assigns temperatures to dates.

Functions play an important role in science. Frequently, one observes that one quantity is a function of another and then tries to find a reasonable formula to express this function. For example, before about 1590 there was no quantitative idea of temperature. Of course, people understood relative notions like warmer and cooler, and some absolute notions like boiling hot, freezing cold, or body temperature, but there was no numerical measure of temperature. It took the genius of Galileo to realize that the expansion of fluids as they warmed was the key to the measurement of temperature. He was the first to think of temperature as a function of fluid volume.

Finding a function which represents a given situation is called making a *mathematical model.* Such a model can throw light on the relationship between the variables and can thereby help us make predictions.

Representation of Functions: Tables, Graphs, and Formulas

Functions can be represented in at least three different ways: by tables, by graphs, and by formulas. For example, the function giving the temperatures in Phoenix, Arizona, as a function of time can be represented by the graphs in Figure 1.1 as well as by a table.

Other functions arise naturally as graphs. Figure 1.2 contains electrocardiogram (EKG) pictures showing the heartbeat patterns of two patients, one normal and one not. Although it is possible to

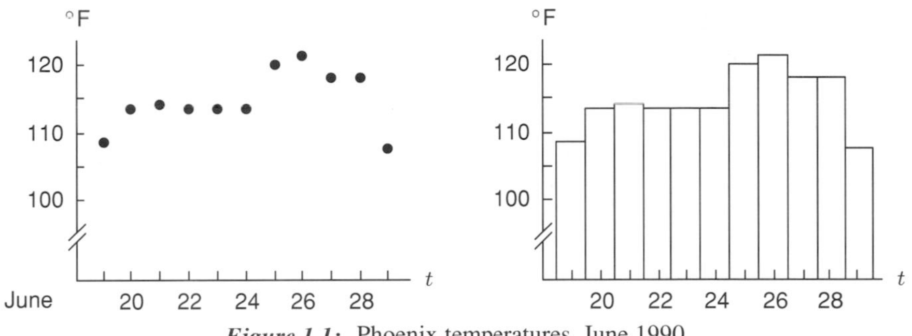

Figure 1.1: Phoenix temperatures, June 1990

construct a formula to approximate an EKG function, this is seldom done. The pattern of repetitions is what a doctor needs to know, and these are much more easily seen from a graph than from a formula. However, each EKG represents a function showing electrical activity as a function of time.

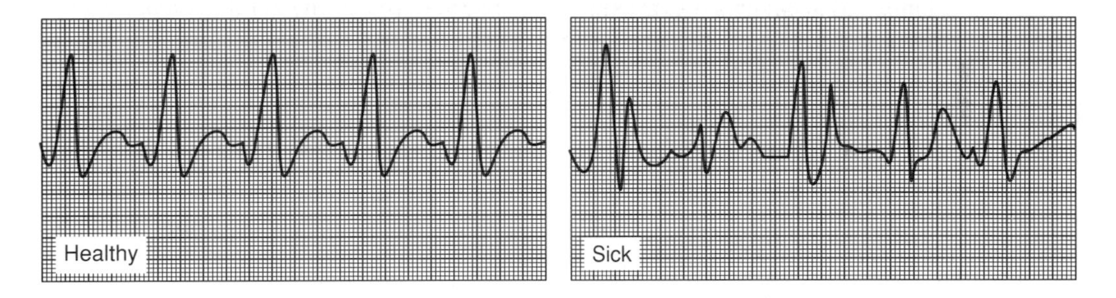

Figure 1.2: EKG readings on two patients

As another example of a function, consider the snow tree cricket. Surprisingly enough, all such crickets chirp at essentially the same rate if they are at the same temperature. That means that the chirp rate is a function of temperature. In other words, if we know the temperature, we can determine the chirp rate. Even more surprisingly, the chirp rate, C, in chirps per minute, increases steadily with the temperature, T, in degrees Fahrenheit, and to a high degree of accuracy can be computed by the formula

$$C = 4T - 160.$$

The formula for C is written $C = f(T)$ to express the fact that we are thinking of C as a function of T. The graph of this function is in Figure 1.3.

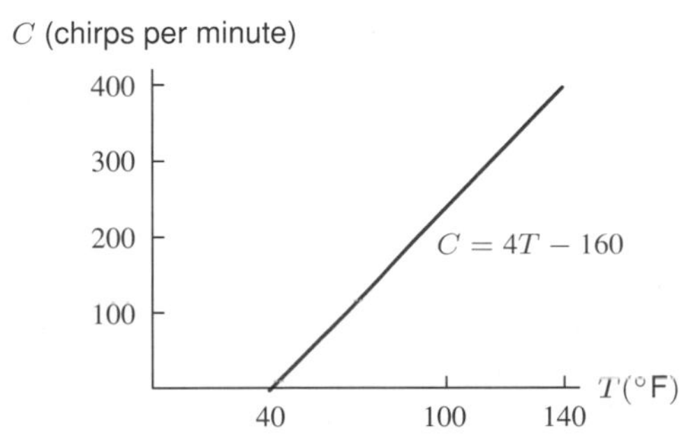

Figure 1.3: Cricket chirp rate versus temperature

Examples of Domain and Range

If the domain of a function is not specified, we will usually take it to be the largest possible set of real numbers. For example, we usually think of the domain of the function $f(x) = x^2$ as all real numbers, whereas the domain of the function $g(x) = 1/x$ is all real numbers except zero. Sometimes, however, we may specify, or restrict, the domain. For example, if the function $f(x) = x^2$ is used to represent the area of a square of side x, we consider only nonnegative values of x and restrict the domain to nonnegative numbers.

Example 1 Consider the function $C = f(T)$ giving chirp rate as a function of temperature. We assume that this equation holds for all temperatures for which the predicted chirp rate is positive, and up to the highest temperature ever recorded at a weather station, namely, 136°F. What is the domain of this function f?

Solution If we consider the equation

$$C = 4T - 160$$

simply as a mathematical relationship between C and T, any T value is possible. However, if we're thinking of it as a relationship between cricket chirps and temperature, then T cannot be less than 40°F, where C falls below the axis and becomes negative. (See Figure 1.3.) In addition, we are told that the formula doesn't hold at temperatures above 136°. Thus, for the function $C = f(T)$ we have

$$\begin{aligned}\text{Domain} &= \text{All } T \text{ values between } 40^\circ\text{F and } 136^\circ\text{F} \\ &= \text{All } T \text{ values with } 40 \le T \le 136.\end{aligned}$$

Therefore we say that the function $C = f(T)$ is represented by the formula

$$C = f(T) = 4T - 160 \quad \text{on the domain} \quad 40 \le T \le 136.$$

Example 2 Find the range of the function f, given the domain from Example 1. In other words, find all possible values of the chirp rate, C, in the equation $C = f(T)$.

Solution Again, if we consider $C = f(T)$ simply as a mathematical relationship, its range is all real C values. However, thinking of its meaning for crickets, the function will predict cricket chirps per minute between 0 (when $T = 40$°F) and 384 (when $T = 136$°F). Hence,

$$\begin{aligned}\text{Range} &= \text{All } C \text{ values from 0 to 384} \\ &= \text{All } C \text{ values with } 0 \le C \le 384.\end{aligned}$$

So far we have used the temperature to predict the chirp rate and thought of the temperature as the *independent variable* and the chirp rate as the *dependent variable*. However, we could do this backwards, and calculate the temperature from the chirp rate. From this point of view, the temperature is dependent on the chirp rate. Thus, which variable is dependent and which is independent may depend on your viewpoint.

Thinking of temperature as a function of chirp rate would enable us (in theory, at least) to use the chirp rate instead of a thermometer to measure temperature. The way we actually do measure temperature is based on another function: the relation between the height of the liquid in a thermometer and temperature. The height of the mercury is certainly a function of temperature; however, we always use this the other way around, and determine the temperature from the height of the mercury, as suggested by Galileo.

Proportionality

A common functional relationship occurs when one quantity is *proportional* to another. For example, if apples are 60 cents a pound, we say the price you pay, p cents, is proportional to the weight you buy, w pounds, because

$$p = f(w) = 60w.$$

As another example, the area, A, of a circle is proportional to the square of the radius, r:

$$A = f(r) = \pi r^2.$$

In general, y is (directly) **proportional** to x if there is a constant k such that

$$y = kx.$$

We also say that one quantity is *inversely proportional* to another if one is proportional to the reciprocal of the other. For example, the speed, v, at which you make a 50-mile trip is inversely proportional to the time, t, taken, because v is proportional to $1/t$:

$$v = \frac{50}{t} = 50\left(\frac{1}{t}\right).$$

Notice that if y is directly proportional to x, then the magnitude of one variable increases (decreases) when the magnitude of the other increases (decreases). If, however, y is inversely proportional to x, then the magnitude of one variable increases when the value of the other decreases.

Problems for Section 1.1

1. Which of the graphs in Figure 1.4 best match the following three stories?[1] Write a story for the remaining graph.
   (a) I had just left home when I realized I had forgotten my books, and so I went back to pick them up.
   (b) Things went fine until I had a flat tire.
   (c) I started out calmly but sped up when I realized I was going to be late.

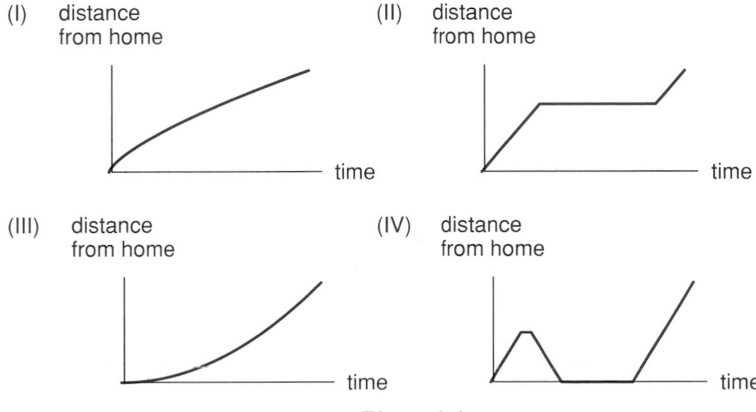

Figure 1.4

[1] Adapted from Jan Terwel. "Real Maths in Cooperative Groups in Secondary Education." In *Cooperative Learning in Mathematics*, edited by Neal Davidson, p 234. (Reading: Addison Wesley, 1990).

2. It warmed up throughout the morning, and then suddenly got much cooler around noon, when a storm came through. After the storm, it warmed up before cooling off at sunset. Sketch a possible graph of this day's temperature as a function of time.
3. Right after a certain drug is administered to a patient with a rapid heart rate, the heart rate plunges dramatically and then slowly rises again as the drug wears off. Sketch a possible graph of the heart rate against time from the moment the drug is administered.
4. Generally, the more fertilizer that is used, the better the yield of the crop. However, if too much fertilizer is applied, the crops become poisoned, and the yield goes down rapidly. Sketch a possible graph showing the yield of the crop as a function of the amount of fertilizer applied.
5. Describe what Figure 1.5 tells you about an assembly line whose productivity is represented as a function of the number of workers on the line.

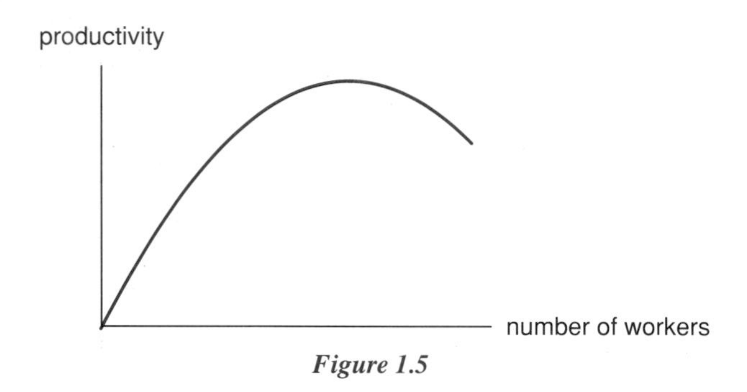

Figure 1.5

6. A flight from Dulles Airport in Washington, D.C., to LaGuardia Airport in New York City has to circle LaGuardia several times before being allowed to land. Plot a graph of distance of the plane from Washington against time, from the moment of takeoff until landing.
7. In her *Guide to Excruciatingly Correct Behavior*, Miss Manners states:

   > There are three possible parts to a date of which at least two must be offered: entertainment, food and affection. It is customary to begin a series of dates with a great deal of entertainment, a moderate amount of food and the merest suggestion of affection. As the amount of affection increases, the entertainment can be reduced proportionately. When the affection has replaced the entertainment, we no longer call it dating. Under no circumstances can the food be omitted.

   Based on this statement, sketch a graph showing entertainment as a function of affection, assuming the amount of food to be constant. Mark the point on the graph at which the relationship starts, as well as the point at which the relationship ceases to be called dating.

Problems 8 and 9 are about supply and demand curves. Economists are interested in how the quantity of an item which is manufactured and sold, q, depends on its price, p. They think of quantity as a function of price. However, for historical reasons[2] the economists put price (the independent variable) on the vertical axis and quantity (the dependent variable) on the horizontal axis. Since manufacturers and consumers react differently to changes in price, there are two functions relating p and q. The *supply curve* represents how the quantity of an item that manufacturers are willing to supply depends on the price for which the item can be sold. The *demand curve* represents how the quantity of an item demanded by consumers depends on its price.

[2]Originally, the economists thought of price as the dependent variable and put it on the vertical axis. Unfortunately, when their point of view changed, the axes did not.

(a) p (price per unit) p_0 q (quantity)

(b) p (price per unit) p_1 q_1 q (quantity)

Figure 1.6

8. One of the graphs in Figure 1.6 is a supply curve, and the other is a demand curve. Which is which? Why?
9. The price p_0 in Figure 1.6(a) represents the price below which the manufacturers are unwilling to produce any of the item. What do the price p_1 and the quantity q_1 in Figure 1.6(b) represent in practical economic terms?
10. Specify the domain and range of the function $y = f(x)$ whose graph is shown in Figure 1.7.

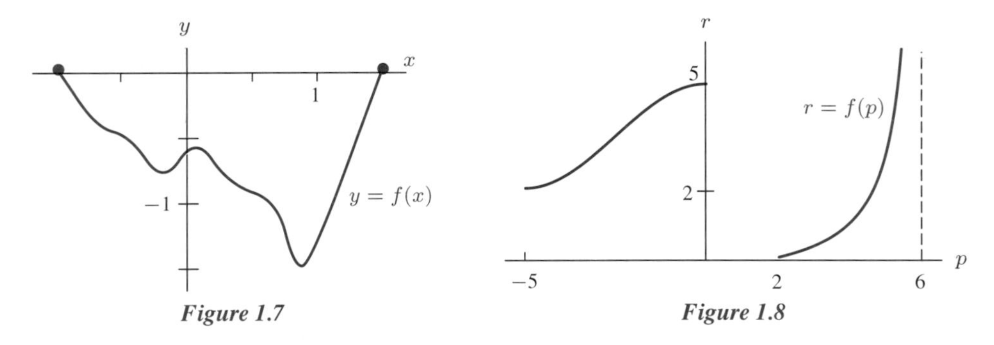

Figure 1.7

Figure 1.8

11. The graph of $r = f(p)$ is given in Figure 1.8.
    (a) What could be the domain of f?
    (b) What could be the range of f?
    (c) What values of r could correspond to exactly one value of p?
12. If $g(y) = 1/(y^2 - y)$, find all y values which do not determine a real value for $g(y)$. Solve $g(y) = 1/2$.
13. If $y = f(x) = 1/\sqrt{4 - x^2}$, what values of x do not determine a real value for y? Solve $f(x) = 5$.
14. When Galileo was formulating the laws of motion, he considered the motion of a body starting from rest and falling under gravity. He originally thought that the velocity of such a falling body was proportional to the distance it had fallen. What light does the data in Table 1.2 shed on Galileo's hypothesis? What alternative hypothesis is suggested by the two sets of data in Table 1.2 and Table 1.3?

TABLE 1.2

Distance (ft)	0	1	2	3	4
Velocity (ft/sec)	0	8	11.3	13.9	16

TABLE 1.3

Time (sec)	0	1	2	3	4
Velocity (ft/sec)	0	32	64	96	128

1.2 LINEAR FUNCTIONS

Probably the most commonly used functions are the *linear functions*. These are functions that represent a steady increase or a steady decrease. A function is linear if any change, or increment, in the independent variable causes a proportional change, or increment, in the dependent variable.

The Olympic Pole Vault

During the early years of the Olympics, the height of the winning pole vault increased approximately as shown in Table 1.4. Since the winning height increased regularly by 8 inches every four years, the height is a linear function of time over the period from 1900 to 1912. The height starts at 130 inches and increases by the equivalent of 2 inches every year, so if y is the height in inches and t is the number of years since 1900, we can write

$$y = f(t) = 130 + 2t.$$

The coefficient 2 tells us the rate at which the height increases and is the *slope* of the line $f(t) = 130 + 2t$.

TABLE 1.4 *Olympic pole vault records (approximate)*

Year	1900	1904	1908	1912
Height (inches)	130	138	146	154

You can visualize the slope in Figure 1.9 as the ratio

$$\text{Slope} = \frac{\text{Rise}}{\text{Run}} = \frac{8}{4} = 2.$$

Calculating the slope (rise/run) using any other two points on the line gives the same value. It is this fact—that the slope, or rate of change, is the same everywhere—that makes a line straight. For a function that is not linear, the rate of change will vary from point to point. Since $y = f(t)$ increases with t, we say that f is *an increasing function*. What about the constant 130? This represents the initial height in 1900, when $t = 0$. Geometrically, the 130 is the *intercept* on the vertical axis.

You may wonder whether the linear trend continues beyond 1912. Not surprisingly, it doesn't exactly. The formula $y = 130+2t$ predicts that the height in the 1988 Olympics would be 306 inches or 25 feet 6 inches, which is considerably higher than the actual value of 19 feet 9 inches. In fact, the height does increase at almost every session of the Olympics, but not at a constant rate. Thus, there is clearly a danger in *extrapolating* too far from the given data. You should also observe that the data in Table 1.4 is *discrete*, because it is given only at specific points (every four years). However, we have treated the variable t as though it were *continuous*, because the function $y = 130 + 2t$ makes sense for all values of t. The graph in Figure 1.9 is of the continuous function because it is a solid line, rather than four separate points representing the years in which the Olympics were held.

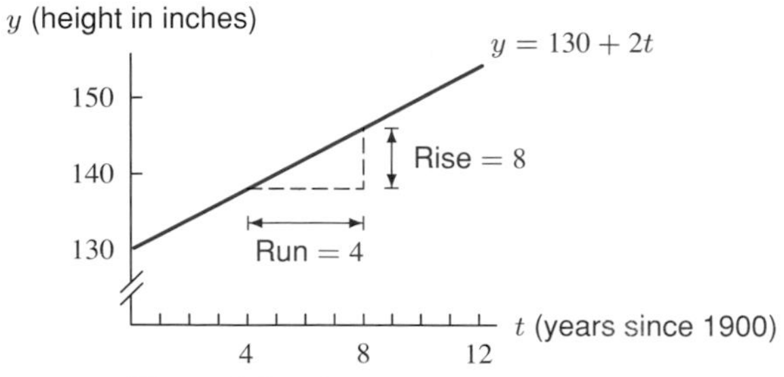

Figure 1.9: Olympic pole vault records

Linear Functions in General

A **linear function** has the form

$$y = f(x) = b + mx$$

Its graph is a line such that

- m is the slope, or rate of change of y with respect to x
- b is the vertical intercept, or value of y when x is zero.

Notice that if the slope is zero, $m = 0$, we have $y = b$, a horizontal line.

To recognize that a function $y = f(x)$ given by a table of data is linear, look for differences in y values that are constant for equal differences in x.

The slope of a linear function can be calculated from values of the function at two points, given by a and c, using the formula

$$m = \frac{\text{Rise}}{\text{Run}} = \frac{f(c) - f(a)}{c - a}.$$

The quantity $(f(c) - f(a))/(c - a)$ is called a *difference quotient* because it is the quotient of two differences. (See Figure 1.10). In Chapter 2, you will see that difference quotients play an important role in calculus.

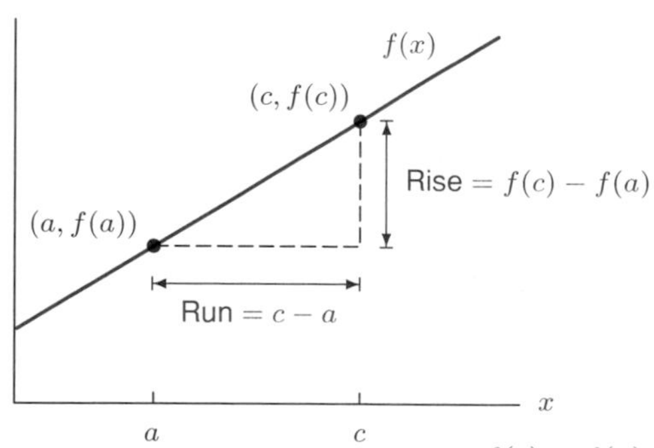

Figure 1.10: Difference quotient $= \dfrac{f(c) - f(a)}{c - a}$

The Success of Search and Rescue Teams

Consider the problem of the "search and rescue" teams working to find lost hikers in remote areas in the West. To search for an individual, members of the search team separate and walk parallel to one another through the area to be searched. Experience has shown that the team's chance of finding a lost individual is related to the distance, d, by which team members are separated. The percentage found[3] for various separations are recorded in Table 1.5.

[3]From *An Experimental Analysis of Grid Sweep Searching*, by J. Wartes (Explorer Search and Rescue, Western Region, 1974).

TABLE 1.5 *Separation of searchers versus success rate*

Separation distance d (ft)	Percent found, P
20	90
40	80
60	70
80	60
100	50

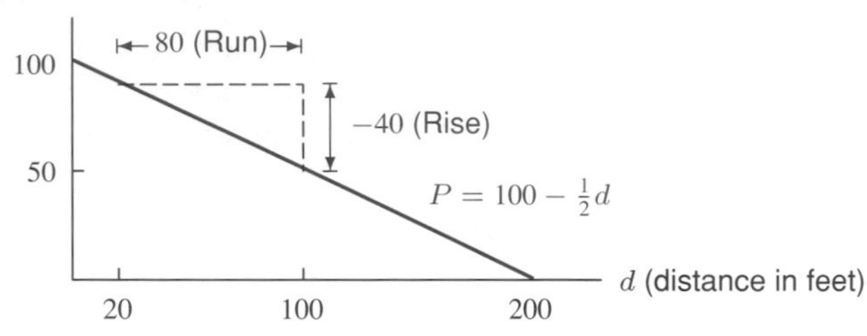

Figure 1.11: Separation of searchers versus success rate

From the data in the table, you can see that as the separation distance decreases, a larger percentage of the lost hikers is found, which makes sense. Since $P = f(d)$ decreases as d increases, we say that P is a *decreasing function* of d. You can also see that for the data given, each 20-foot increase in distance causes the percentage found to drop by 10. This constant decrease in P as d increases by a fixed amount is a clear indication that the graph of P against d is a line. (See Figure 1.11.) Notice that the slope is $-40/80 = -1/2$. The negative sign shows that P decreases as d increases. The slope is the rate at which P is increasing or decreasing, as d increases.

What about the vertical intercept? If $d = 0$, the searchers are walking shoulder to shoulder and you'd expect everyone to be found, so $P = 100$. This is exactly what you get if the line is continued to the vertical axis (a decrease of 20 in d causes an increase of 10 in P). Therefore, the equation of the line is

$$P = f(d) = 100 - \frac{1}{2}d.$$

What about the horizontal intercept? When $P = 0$, or $0 = 100 - \frac{1}{2}d$, then $d = 200$. The value $d = 200$ represents the separation distance at which, according to the model, no one is found. This is unreasonable, because even when the searchers are far apart, the search will sometimes be successful. This suggests that somewhere outside the given data, the linear relationship ceases to hold. As in the pole vault example, extrapolating too far beyond the given data may not give accurate answers.

Increasing versus Decreasing Functions

Let's summarize what we know about increasing and decreasing functions:

> A function f is **increasing** if the values of $y = f(x)$ increase as x increases.
> A function f is **decreasing** if the values of $y = f(x)$ decrease as x increases.
>
> The graph of an *increasing* function *climbs* as you move from left to right.
> The graph of a *decreasing* function *descends* as you move from left to right.

A Budget Constraint

An ongoing debate in the federal government concerns the allocation of money between defense and social programs. In general, the more that is spent on defense, the less that is available for social programs, and vice versa. Let's simplify the example to guns and butter. Assuming a constant budget, we will show that the relationship between the number of guns and the quantity of butter is linear. Suppose there is \$12,000 to be spent and that it is to be divided between guns, costing \$400 each, and butter, costing \$2000 a ton. Suppose the number of guns bought is g, and the number of tons of butter is b. Then the amount of money spent on guns is \$400g (because each one is \$400), and the amount spent on butter is \$2000b. Assuming all the money is spent,

$$\begin{array}{c}\text{Amount spent}\\ \text{on guns}\end{array} + \begin{array}{c}\text{Amount spent}\\ \text{on butter}\end{array} = \$12{,}000$$

or

$$400g + 2000b = 12{,}000$$

or

$$g + 5b = 30.$$

This equation is the budget constraint. Its graph is the line shown in Figure 1.12, which can be found by plotting points. We will calculate the points at which the graph crosses the axes. If $b = 0$, then

$$g + 5(0) = 30 \quad \text{so} \quad g = 30.$$

If $g = 0$, then

$$0 + 5b = 30 \quad \text{so} \quad b = 6.$$

Since the number of guns bought determines the amount of butter bought (because all the money which doesn't go to guns goes to butter), b is a function of g. Similarly, the amount of butter bought determines the number of guns, so g is a function of b. The budget constraint represents an *implicitly defined function*, because neither quantity is given explicitly in terms of the other. If we solve for g, giving

$$g = 30 - 5b$$

we have an *explicit* formula for g in terms of b. Similarly,

$$b = \frac{30 - g}{5} \quad \text{or} \quad b = 6 - 0.2g$$

gives b as an explicit function of g. Since the explicit functions

$$g = 30 - 5b \quad \text{and} \quad b = 6 - 0.2g$$

are linear, the graph of the budget constraint must be a line.

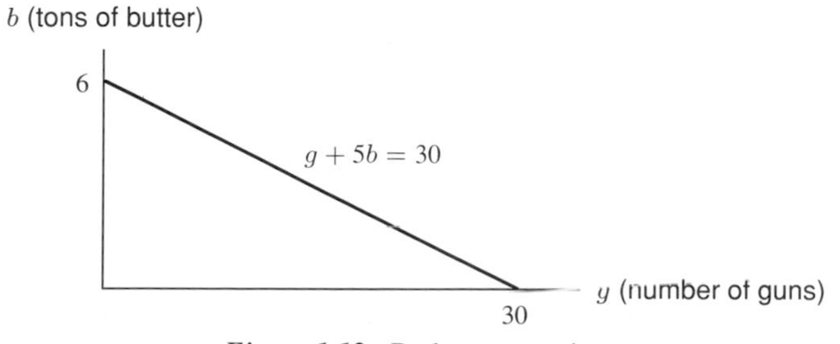

Figure 1.12: Budget constraint

Families of Linear Functions

Formulas such as $f(x) = mx$, and $f(x) = b + mx$, containing constants such as m and b which can take on various values, are said to define a *family of functions*. The constants m and b are called *parameters*. Each of the functions in this section belongs to the family $f(x) = b + mx$.

Grouping functions into families which share important features is particularly useful for mathematical modeling. We often choose a family to represent a given situation on theoretical grounds and then use data to determine the particular values of the parameters. The meaning of the parameters m and b in the family $f(x) = b + mx$ is shown in Figures 1.13 and 1.14.

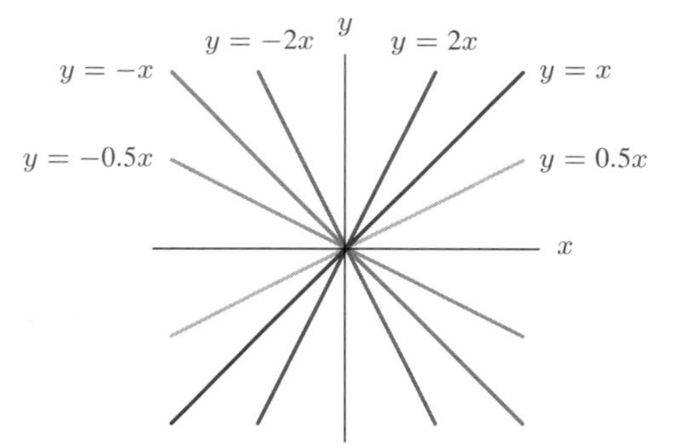

Figure 1.13: The family $y = mx$ (with $b = 0$)

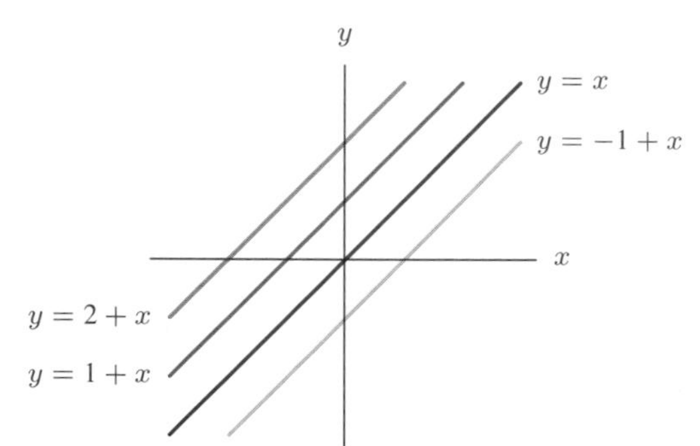

Figure 1.14: The family $y = b + x$ (with $m = 1$)

Problems for Section 1.2

1. Find the slope and vertical intercept of the line whose equation is $2y + 5x - 8 = 0$.
2. Find the equation of the line through the points $(-1, 0)$ and $(2, 6)$.
3. Find the equation of the line with slope m through the point (a, c).

For Problems 4–5, recall that parallel lines have equal slopes, and that two lines are perpendicular if their slopes are negative reciprocals.

4. Find the equation of the line through the point $(2, 1)$ which is perpendicular to the line $y = 5x - 3$.
5. Find the equations of the lines parallel to and perpendicular to the line $y + 4x = 7$ through the point $(1, 5)$.
6. Estimate the slope of the line shown in Figure 1.15 and use the slope to find an equation for that line. (Note that the x and y scales are unequal.)

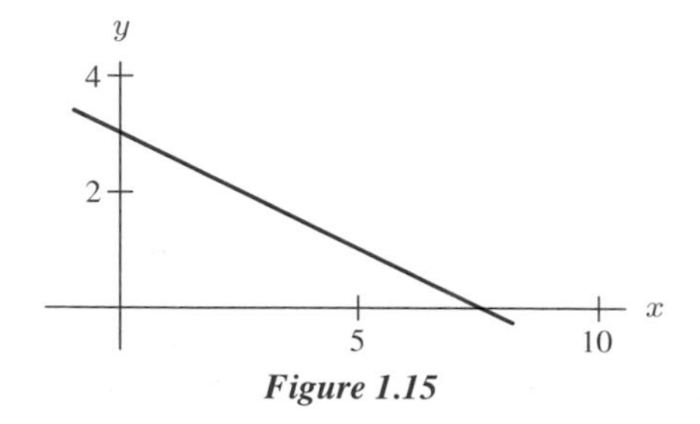

Figure 1.15

7. Match the graphs in Figure 1.16 with the equations below.

(a) $y = x - 5$	(c) $5 = y$	(e) $y = x + 6$
(b) $-3x + 4 = y$	(d) $y = -4x - 5$	(f) $y = x/2$

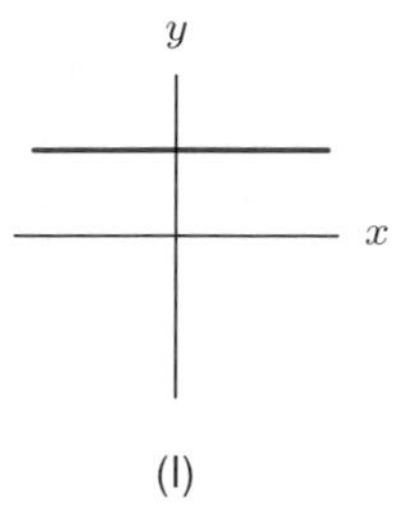

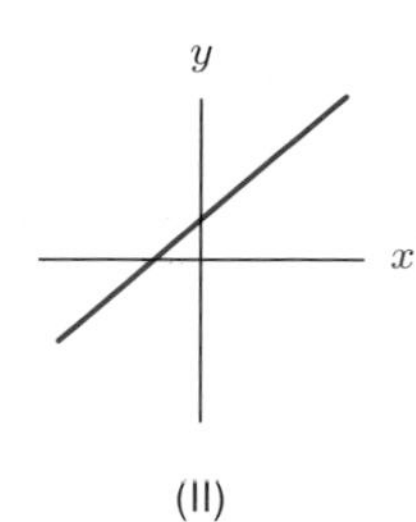

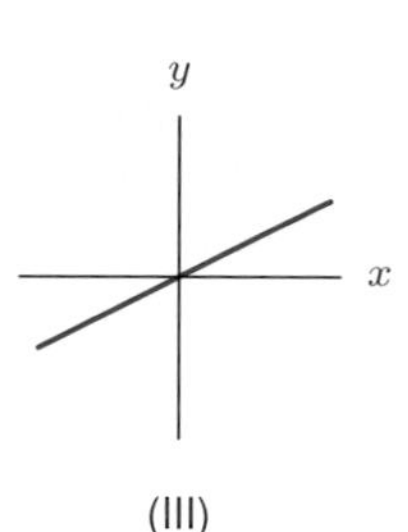

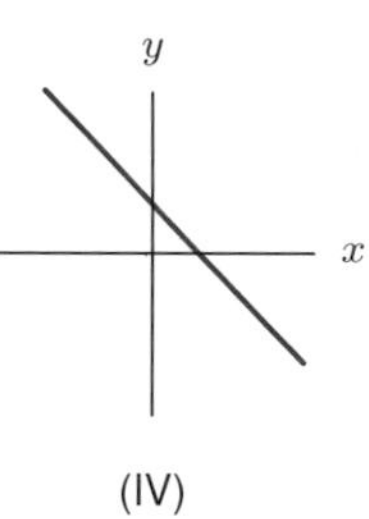

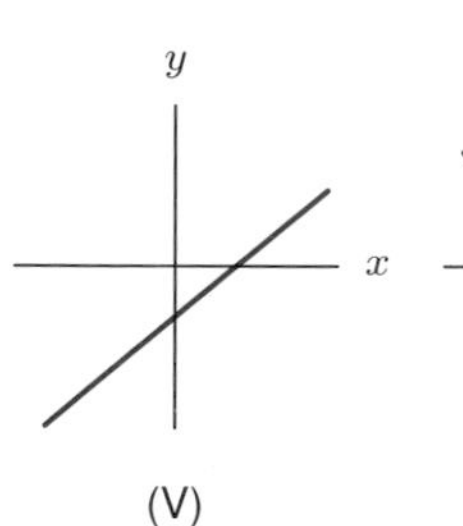

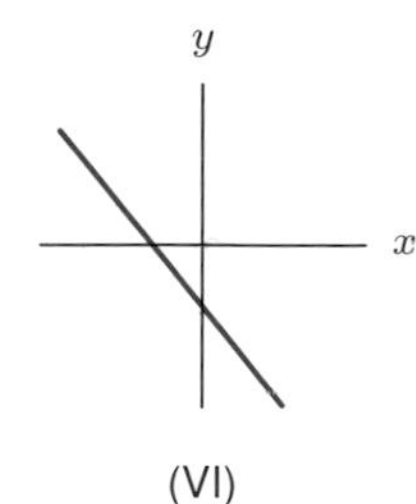

Figure 1.16

8. Match the graphs in Figure 1.17 with the equations below.

(a) $y = -2.72x$	(c) $y = 27.9 - 0.1x$	(e) $y = -5.7 - 200x$
(b) $y = 0.01 + 0.001x$	(d) $y = 0.1x - 27.9$	(f) $y = x/3.14$

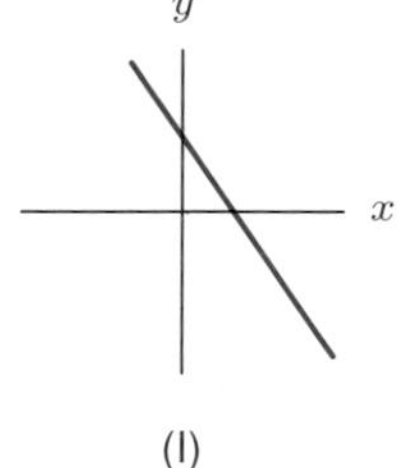

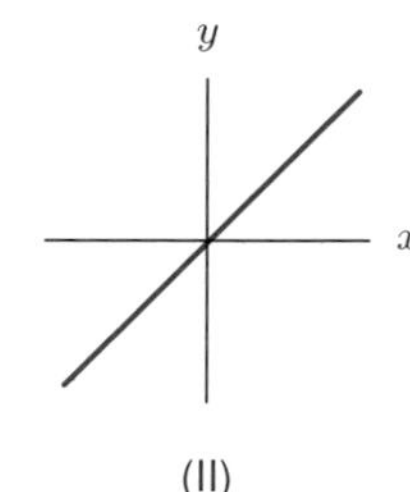

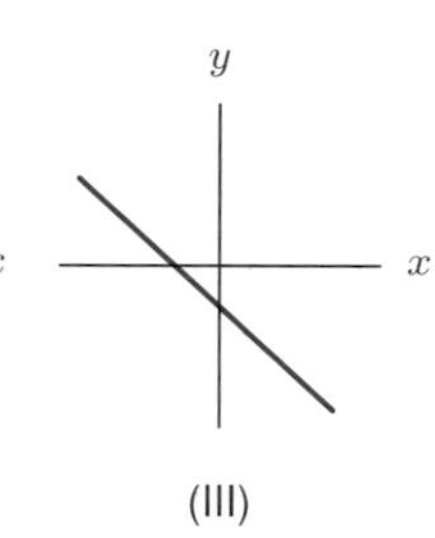

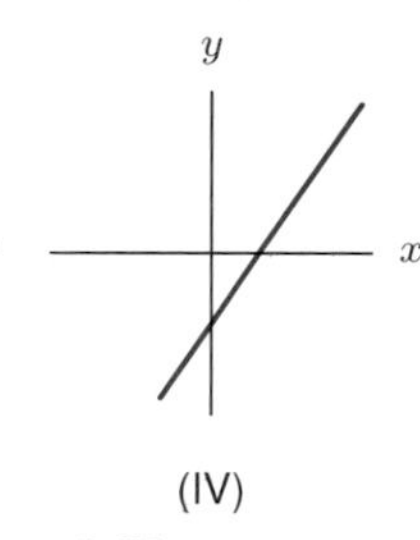

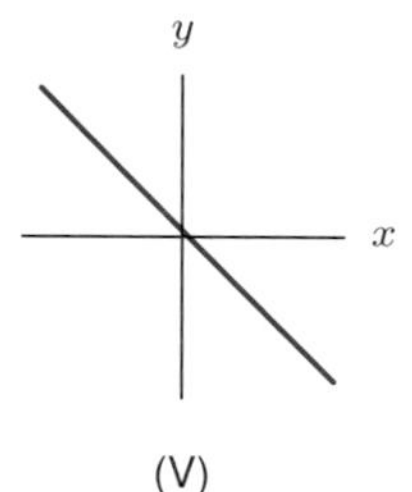

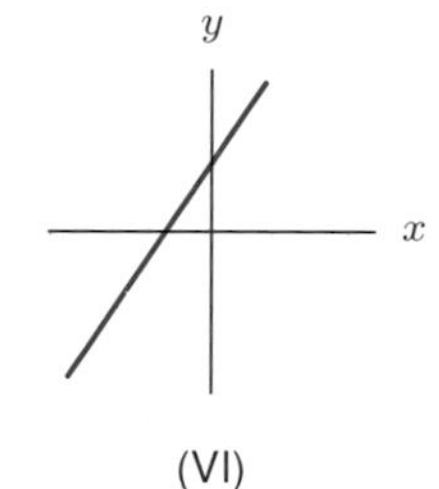

Figure 1.17

9. Corresponding values of p and q are given in the table below.
   (a) Find q as a linear function of p.
   (b) Find p as a linear function of q.

p	1	2	3	4
q	950	900	850	800

10. A linear equation was used to generate the values in the table below. Find that equation.

x	5.2	5.3	5.4	5.5	5.6
y	27.8	29.2	30.6	32.0	33.4

11. An equation of a line is $3x + 4y = -12$. Find the length of the portion of the line that lies between its x and y intercepts.

12. A car rental company offers cars at \$40 a day and 15 cents a mile. Its competitor's cars are \$50 a day and 10 cents a mile.
   (a) For each company, write a formula giving the cost of renting a car for a day as a function of the distance traveled.
   (b) On the same axes, sketch graphs of both functions.
   (c) How should you decide which company is cheaper?

13. Consider a graph of Fahrenheit temperature, °F, against Celsius temperature, °C, and assume that the graph is a line. You know that 212°F and 100°C both represent the temperature at which water boils. Similarly, 32°F and 0°C both represent water's freezing point.
    (a) What is the slope of the graph?
    (b) What is the equation of the line?
    (c) Use the equation to find what Fahrenheit temperature corresponds to 20°C.
    (d) What temperature is the same number of degrees in both Celsius and Fahrenheit?
14. The cost of planting seed is usually a function of the number of acres sown. The cost of the equipment is a *fixed cost* because it must be paid regardless of the number of acres planted. The cost of supplies and labor varies with the number of acres planted and are called *variable costs*. Suppose the fixed costs are \$10,000 and the variable costs are \$200 per acre. Let C be the total cost, measured in thousands of dollars, and let x be the number of acres planted.
    (a) Find a formula for C as a function of x.
    (b) Sketch a graph of C against x.
    (c) Explain how you can visualize the fixed and variable costs on the graph.
15. Suppose you are driving at a constant speed from Chicago to Detroit, about 275 miles away. About 120 miles from Chicago you pass through Kalamazoo, Michigan. Sketch a graph of your distance from Kalamazoo as a function of time.
16. Hot peppers have been rated according to Scoville units, with a maximum human tolerance level of 14,000 Scovilles per dish. The West Coast Restaurant, known for spicy dishes, promises a daily special to satisfy the most avid spicy-dish fans. The restaurant imports Indian peppers rated at 1200 Scovilles each and Mexican peppers with a Scoville rating of 900 each.
    (a) Determine the Scoville constraint equation relating the maximum number of Indian and Mexican peppers the restaurant should use for their specialty dish.
    (b) Solve the equation from part (a) to show explicitly the number of Indian peppers needed in the hottest dishes as a function of the number of Mexican peppers.
17. You have a fixed budget of $\$k$ to spend on soda and suntan oil, which cost $\$p_1$ per liter and $\$p_2$ per liter respectively.
    (a) Write an equation expressing the relationship between the number of liters of soda and the number of liters of suntan oil that you can buy if you exhaust your budget. This is your *budget constraint*.
    (b) Graph the budget constraint, assuming that you can buy fractions of a liter. Label the intercepts.
    (c) Suppose your budget is suddenly doubled. Graph the new budget constraint on the same axes.
    (d) With a budget of $\$k$, the price of suntan oil suddenly doubles. Sketch the new budget constraint on the same axes.
18. Since the opening up of the West, the US population has moved westward. To observe this, we look at the "population center" of the US, which is the point at which the country would balance if it were a flat plate with no weight, and every person had equal weight. In 1790 the population center was east of Baltimore, Maryland. It has been moving westward ever since, and in 1990 it crossed the Mississippi river to Steelville, Missouri (southwest of St. Louis). During the second half of this century, the population center has moved about 50 miles west every 10 years.
    (a) Express the approximate position of the population center as a function of time, measured in years from 1990. Measure position westward from Steelville, along the line running

through Baltimore.

(b) The distance from Baltimore to St. Louis is a bit over 700 miles. Could the population center have been moving at roughly the same rate for the last two centuries?

(c) Could the function in part (a) continue to apply for the next three centuries? Why or why not? [Hint: You may want to look at a map. Note that distances are in air miles and are not driving distances.]

19. For small changes in temperature, the formula for the expansion of a metal rod under a change in temperature is:

$$l - l_0 = al_0(t - t_0),$$

where l is the length of the object at temperature t, and l_0 is the initial length at temperature t_0, and a is a constant which depends on the type of metal.

(a) Express l as a linear function of t. Find the slope and y-intercept. [Hint: Treat the other quantities as constants.]

(b) Suppose you had a rod which was initially 100 cm long at 60°F and made of a metal with a equal to 10^{-5}. Write an equation giving the length of this rod at temperature t.

(c) What does the sign of the slope of the graph tell you about the expansion of a metal under a change in temperature?

20. When a cold yam is put into a hot oven to bake, the temperature of the yam rises. The rate, R (in degrees per minute), at which the temperature of the yam rises is governed by Newton's Law of Heating, which says that the rate is proportional to the temperature difference between the yam and the oven. If the oven is at 350°F and the temperature of the yam is H°F:

(a) Write a formula giving R as a function of H.

(b) Sketch the graph of R against H.

21. When a cup of hot coffee sits on the kitchen table, its temperature falls. The rate, R, at which its temperature changes is governed by Newton's Law of Cooling, which says that the rate is proportional to the temperature difference between the coffee and the surrounding air. Let's think of the rate, R, as a negative quantity because the temperature of the coffee is falling. If the temperature of the coffee is H°C and the temperature of the room is 20°C:

(a) Write a formula giving R as a function of H.

(b) Sketch a graph of R against H.

22. A body of mass m is falling downward with velocity v. Newton's Second Law of Motion, $F = ma$, says that the net downward force, F, on the body is proportional to its downward acceleration, a. The net force, F, consists of the force due to gravity, F_g, which acts downward, minus the air resistance, F_r, which acts upward. The force due to gravity is mg, where g is a constant. Assume the air resistance is proportional to the velocity of the body.

(a) Write an expression for the net force, F, as a function of the velocity, v.

(b) Write a formula giving a as a function of v.

(c) Sketch a against v.

1.3 EXPONENTIAL FUNCTIONS

Population Growth

Consider the data for the population of Mexico in the early 1980s in Table 1.6. To see how the population is growing, you might look at the increase in population from one year to the next, as shown in the third column. If the population had been growing linearly, all the numbers in the third column would be the same. But populations usually grow faster as they get bigger, because there are more people to have babies. So you shouldn't be surprised to see the numbers in the third column increasing.

TABLE 1.6 *Population of Mexico (estimated), 1980–1986*

Year	Population (millions)	Change in population (millions)
1980	67.38	
		1.75
1981	69.13	
		1.80
1982	70.93	
		1.84
1983	72.77	
		1.89
1984	74.66	
		1.94
1985	76.60	
		1.99
1986	78.59	

Suppose we divide each year's population by the previous year's population. We get, approximately,

$$\frac{\text{Population in 1981}}{\text{Population in 1980}} = \frac{69.13 \text{ million}}{67.38 \text{ million}} = 1.026$$

$$\frac{\text{Population in 1982}}{\text{Population in 1981}} = \frac{70.93 \text{ million}}{69.13 \text{ million}} = 1.026.$$

The fact that both calculations give 1.026 shows the population grew by about 2.6% between 1980 and 1981 *and* between 1981 and 1982. If you do similar calculations for other years, you will find that the population grew by a factor of about 1.026, or 2.6%, every year. Whenever you have a constant growth factor (here 1.026), you have *exponential growth*. If t is the number of years since 1980,

$$\begin{aligned}
\text{When } t = 0, &\quad \text{population} = 67.38 = 67.38(1.026)^0 \\
\text{When } t = 1, &\quad \text{population} = 69.13 = 67.38(1.026)^1 \\
\text{When } t = 2, &\quad \text{population} = 70.93 = 69.13(1.026) = 67.38(1.026)^2 \\
\text{When } t = 3, &\quad \text{population} = 72.77 = 70.93(1.026) = 67.38(1.026)^3
\end{aligned}$$

and so t years after 1980, the population is given by

$$P = 67.38(1.026)^t.$$

This is an *exponential function* with base 1.026. It is called exponential because the variable, t, is in the exponent. The base represents the factor by which the population grows each year.

If we assume that the same formula will hold for the next 50 years or so, the population will have the shape shown in Figure 1.18. Since the population is growing, the function is increasing. Notice also that the population grows faster and faster as time goes on. This behavior is typical of an exponential function. You should compare this with the behavior of a linear function, which climbs at the same rate everywhere and so has a straight-line graph. Because this graph is bending upward, we say it is *concave up*. Even exponential functions which climb slowly at first, such as this one, climb extremely quickly eventually. That is why exponential population growth is such a threat to the world.

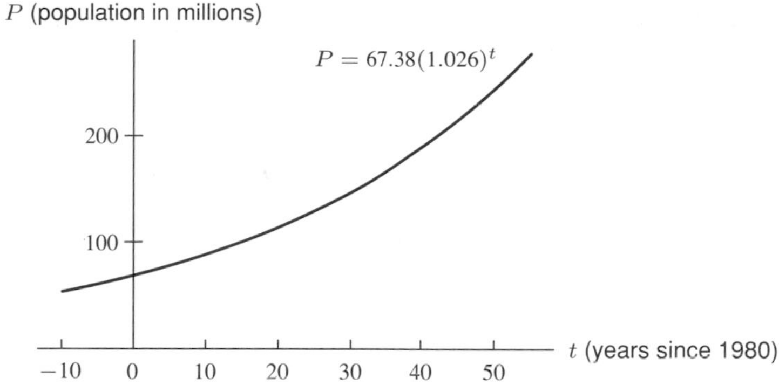

Figure 1.18: Population of Mexico (estimated): Exponential growth

Even if it represents reliable data, the smooth graph in Figure 1.18 is actually only an approximation to the true graph of the population of Mexico. Since we can't have fractions of people, the graph should really be jagged, jumping up or down by one each time someone is born or dies. However, with a population in the millions, the jumps are so small as to be invisible at the scale we are using. Therefore, the smooth graph is an extremely good approximation.

Example 1 Predict the population of Mexico in the year

(a) 2007 (when $t = 27$). (b) 2034 (when $t = 54$). (c) 2061 (when $t = 81$).

Solution Extrapolating so far into the future can be risky because it assumes that the population continues to grow exponentially with the same constant growth factor. (There could, for example, be a medical breakthrough that would increase the growth factor, or an epidemic that would decrease it.) Writing "$\approx$" to represent approximately equal, the model we are using predicts

(a) $P = 67.38(1.026)^{27} \approx 67.38(2) = 134.76$ million.

(b) $P = 67.38(1.026)^{54} \approx 67.38(4) = 269.52$ million.

(c) $P = 67.38(1.026)^{81} \approx 67.38(8) = 539.04$ million.

If you look at the answers to Example 1, you will see something that may surprise you. After 27 years the population has doubled; after another 27 years (at $t = 54$), it has doubled again. In another 27 years later (when $t = 81$), the population has doubled yet again. As a result, we say that the *doubling time* of the population of Mexico is 27 years.

Every exponentially growing population has a fixed doubling time. The world's population currently has a doubling time of about 38 years. Notice what this means: If you live to be 76, the world's population is expected to quadruple in your lifetime.

Concavity

We have used the term concave up to describe the graph in Figure 1.18. In Figure 1.21 on page 21, we will see a graph which is concave down. In general:

> The graph of any function is **concave up** if it bends upward, and it is **concave down** if it bends downward. A line is neither concave up nor concave down.

Musical Pitch

The pitch of a musical note is determined by the frequency of the vibration which causes it. Middle C on the piano, for example, corresponds to a vibration of 263 hertz (cycles per second). A note one octave above middle C vibrates at 526 hertz, and a note two octaves above middle C vibrates at 1052 hertz. (See Table 1.7.)

TABLE 1.7 *Pitch of notes above middle C*

Number, n, of octaves above middle C	Number of hertz $V = f(n)$
0	263
1	526
2	1052
3	2104
4	4208

TABLE 1.8 *Pitch of notes below middle C*

n	$V = 263 \cdot 2^n$
-3	$263 \cdot 2^{-3} = 263(1/2^3) = 32.875$
-2	$263 \cdot 2^{-2} = 263(1/2^2) = 65.75$
-1	$263 \cdot 2^{-1} = 263(1/2) = 131.5$
0	$263 \cdot 2^0 = 263$

Notice that

$$\frac{526}{263} = 2 \quad \text{and} \quad \frac{1052}{526} = 2 \quad \text{and} \quad \frac{2104}{1052} = 2$$

and so on. In other words, each value of V is twice the value before, so

$$f(1) = 526 = 263 \cdot 2 = 263 \cdot 2^1$$
$$f(2) = 1052 = 526 \cdot 2 = 263 \cdot 2^2$$
$$f(3) = 2104 = 1052 \cdot 2 = 263 \cdot 2^3.$$

In general

$$V = f(n) = 263 \cdot 2^n.$$

The base 2 represents the fact that as we go up an octave, the frequency of vibrations doubles. Indeed, our ears hear a note as one octave higher than another precisely because it vibrates twice as fast. For the negative values of n in Table 1.8, this function represents the octaves below middle C. The notes on a piano are represented by values of n between -3 and 4, and the human ear finds values of n between -4 and 7 audible.

Although $V = f(n) = 263 \cdot 2^n$ makes sense in musical terms only for certain values of n, values of the function $f(x) = 263 \cdot 2^x$ can be calculated for all real x, and its graph has the typical exponential shape, as can be seen in Figure 1.19. It is concave up, climbing faster and faster as x increases.

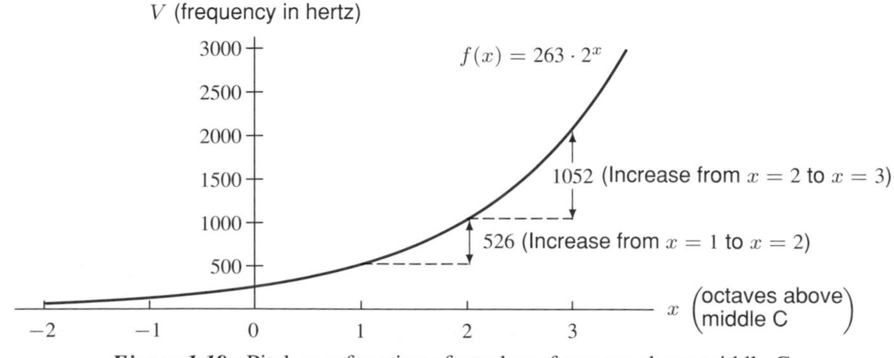

Figure 1.19: Pitch as a function of number of octaves above middle C

Removal of Pollutants from Jet Fuel

Now we will look at an example in which a quantity is decreasing instead of increasing. Before kerosene can be used as jet fuel, federal regulations require that the pollutants in it be removed by passing the kerosene through clay. We will suppose the clay is in a pipe and that each foot of the pipe removes 20% of the pollutants that enter it. Therefore each foot leaves 80% of the pollution. If P_0 is the initial quantity of pollutant and $P = f(n)$ is the quantity left after n feet of pipe:

$$\begin{aligned} f(0) &= P_0 \\ f(1) &= (0.8)P_0 \\ f(2) &= (0.8)(0.8)P_0 = (0.8)^2 P_0 \\ f(3) &= (0.8)(0.8)^2 P_0 = (0.8)^3 P_0 \end{aligned}$$

and so, after n feet,

$$P = f(n) = P_0(0.8)^n.$$

In this example, n must be non-negative. However, the *exponential decay function*

$$P = f(x) = P_0(0.8)^x$$

makes sense for any real x. We'll plot it with $P_0 = 1$ in Figure 1.20; some values of the function are in Table 1.9.

Notice the way the function in Figure 1.20 is decreasing: each downward step is smaller than the one before. This is because as the kerosene gets cleaner, there's less dirt to remove, and so each foot of clay takes out less pollutant than the previous one. Compare this to the exponential growth in Figures 1.18 and 1.19 on pages 17 and 18, where each step upward is larger than the one before. Notice, however, that all three graphs mentioned are concave up.

TABLE 1.9 *Values of decay function*

x	$P = (0.8)^x$
−2	1.56
−1	1.25
0	1
1	0.8
2	0.64
3	0.51
4	0.41

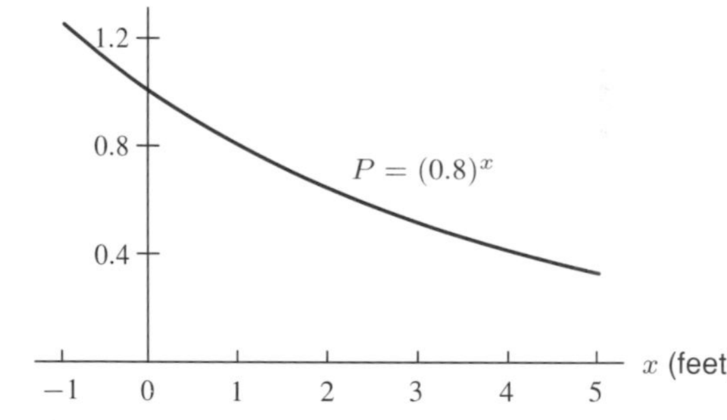

Figure 1.20: Pollutant removal: Exponential decay

The General Exponential Function

> P is an **exponential function** of t with base a if
>
> $$P = P_0 a^t$$
>
> where P_0 is the initial quantity (when $t = 0$) and a is the factor by which P changes when t increases by 1.
> If $a > 1$, we have exponential growth; if $0 < a < 1$, we have exponential decay.

The largest possible domain for the exponential function is all real numbers, provided $a > 0$. (Why do we not want $a \leq 0$?)

> To recognize that a function $P = f(t)$ given by a table of data is exponential, look for ratios of P values that are constant for equally spaced t values.

Radioactive Decay

Radioactive substances, such as uranium, decay by a certain percentage of their mass in a given unit of time. The most common way to express this rate of decay is to give the time period it takes for half the mass to decay. This period of time is called the *half-life* of the substance. The important thing to remember about radioactive decay is that two half-lives do not make a whole life! Rather, in the time period of two half-lives, the substance will decay to $(1/2) \cdot (1/2) = 1/4$ of its original mass.

One of the most well-known radioactive substances is carbon-14, which is used to date organic objects. When the object, such as a piece of wood or bone, was part of a living organism, it accumulated small amounts of radioactive carbon-14, so that a certain proportion of the carbon in the object was carbon-14. Once the organism dies, it no longer picks up carbon-14 through interaction with its environment (for example, through respiration).

By measuring the proportion of carbon-14 in the object and comparing that to the proportion in living material, we can estimate how much of the original carbon-14 has decayed. The half-life of carbon-14 is about 5730 years. Thus, after roughly 5000 years, we would find the object had about $1/2$ as much carbon-14 as when it was alive. After 10,000 years, we would find about $1/4$ as much, and after 15,000 years, about 1/8 as much. We can write an exponential function for the amount of carbon-14 left after a period of time t. First, suppose we measure time in units of 5730 years. Then if C_0 was the original amount of carbon-14, the amount, C, of carbon left after T "units" of time (namely, T half-lives), would be $C = C_0(1/2)^T$. However, we usually do not measure time in units of 5730 years, so if we let t be time measured in years (units of one year), then $T = t/5730$, and

$$C = C_0 \left(\frac{1}{2}\right)^{(t/5730)}.$$

In general, if a substance has a half-life of h years (or minutes or seconds), then the quantity, Q, of the substance left after t units of time, if there was Q_0 of the substance originally, is

$$Q = Q_0 \left(\frac{1}{2}\right)^{(t/h)}.$$

In summary, we use the following definitions:

> The **doubling time** of an exponentially increasing population is the time for it to double.
> The **half-life** of an exponentially decaying quantity is the time for it to be reduced to half.

Drug Buildup

Suppose that we want to model the amount of a certain drug in the body. Imagine that initially there is none, but that the quantity slowly starts to increase via a continuous intravenous injection. As the quantity in the body increases, so does the rate at which the body excretes the drug, so that eventually the quantity levels off at a saturation value, S. The graph of quantity against time will look something like that in Figure 1.21.

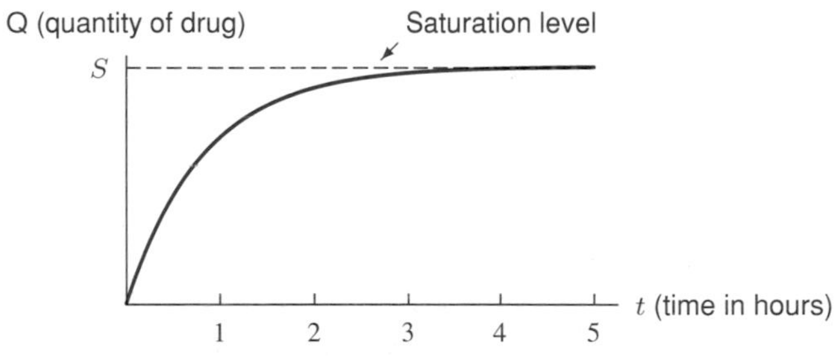

Figure 1.21: Buildup of drug in body

Notice that the quantity, Q, starts at zero and increases toward S. We say that the line representing the saturation level is a *horizontal asymptote*, because the graph gets closer and closer to it as time increases. Since the rate at which the quantity of the drug increases slows as it approaches S, this graph is bending downward; hence it is *concave down*.

Suppose we want to make a mathematical model of this situation; that is, suppose we want to find a formula giving the quantity, Q, in terms of time, t. Making a mathematical model often involves looking at a graph and deciding what kind of function has that shape. The graph in Figure 1.21 looks like an exponential decay function, upside down. What actually decays is the difference between the saturation level, S, and the quantity, Q, in the blood. Suppose the difference between the saturation level and the quantity in the body is given by the formula

$$\text{Difference} = (\text{Initial difference}) \cdot (0.3)^t$$

with t in hours. Since the difference is $S - Q$, and the initial value of this difference is $S - 0 = S$:

$$S - Q = S \cdot (0.3)^t.$$

Solving for Q as a function of t gives

$$Q = S - S \cdot (0.3)^t$$
$$Q = f(t) = S \cdot \left(1 - (0.3)^t\right).$$

Notice that the graph of this function is an upside-down exponential. As t gets larger, $(0.3)^t$ gets smaller, so Q gets closer to S. Using "$\rightarrow$" to mean "tends to," we can say $(0.3)^t \rightarrow 0$ as $t \rightarrow \infty$. This shows that

$$Q = S(1 - (0.3)^t) \rightarrow S(1 - 0) = S \quad \text{as} \quad t \rightarrow \infty$$

confirming that the graph of $Q = S(1 - (0.3)^t)$ has a horizontal asymptote at $Q = S$.

The Family of Exponential Functions

The formula $P = P_0 a^t$ gives a family of exponential functions with parameters P_0 (the initial quantity) and a (the base, or growth factor). The base is as important for an exponential function as the slope is for a linear function. We assume $a > 0$ and $a \neq 1$. Then the base tells you whether the function is increasing ($a > 1$) or decreasing ($0 < a < 1$). Since a is the factor by which P changes when t is increased by 1, large values of a mean fast growth; values of a near 0 mean fast decay. (See Figures 1.22 and 1.23.)

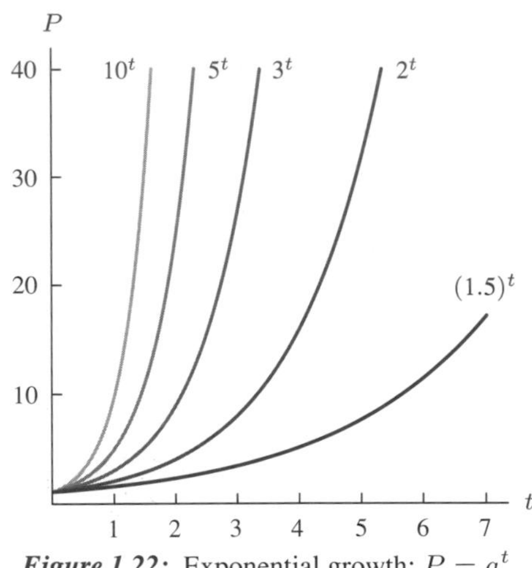

Figure 1.22: Exponential growth: $P = a^t$, $a > 1$

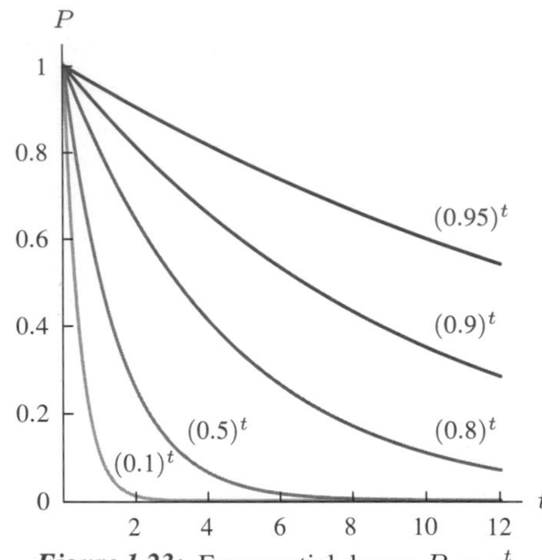

Figure 1.23: Exponential decay: $P = a^t$, $0 < a < 1$

Alternative Formula for the Exponential Function

Exponential growth is often described in terms of percentages. For example, the population of Mexico is growing at 2.6% per year; in other words, the growth factor is $a = 1.026$. Similarly, each foot of clay removes 20% of the pollution from jet fuel, so the decay factor is $a = 1 - 0.20 = 0.8$. In general, the following formulas apply.

> If r is the *growth* rate, then $a = 1 + r$, and
>
> $$P = P_0 a^t = P_0(1+r)^t.$$
>
> If r is the *decay* rate, then $a = 1 - r$, and
>
> $$P = P_0 a^t = P_0(1-r)^t.$$
>
> Note, for example, that $r = 0.05$ when the percentage growth rate is 5%.

Example 2 Suppose that $Q = f(t)$ is an exponential function of t. If $f(20.1) = 88.2$ and $f(20.3) = 91.4$:
(a) Find the base. (b) Find the percentage growth rate. (c) Evaluate $f(21.4)$.

Solution (a) Let

$$Q = Q_0 a^t.$$

Then

$$88.2 = Q_0 a^{20.1} \quad \text{and} \quad 91.4 = Q_0 a^{20.3}.$$

Dividing gives

$$\frac{91.4}{88.2} = \frac{Q_0 a^{20.3}}{Q_0 a^{20.1}} = a^{0.2}.$$

Solve for the base, a:

$$a = \left(\frac{91.4}{88.2}\right)^{1/0.2} = 1.195.$$

(b) Since $a = 1.195$, the growth rate is $r = 0.195 = 19.5\%$.

(c) We want to find $f(21.4) = Q_0 a^{21.4} = Q_0(1.195)^{21.4}$. First let's find Q_0 from the equation $91.4 = Q_0(1.195)^{20.3}$. Solving gives $Q_0 = 2.457$. Thus,

$$f(21.4) = 2.457(1.195)^{21.4} = 111.19.$$

Definition and Properties of Exponents

Below we list the definitions and rules that are used to manipulate exponents.

Definition of Zero, Negative, and Fractional Exponents

$$a^0 = 1, \quad a^{-1} = \frac{1}{a}, \quad \text{and, in general, } a^{-x} = \frac{1}{a^x}$$

$$a^{1/2} = \sqrt{a}, \quad a^{1/3} = \sqrt[3]{a}, \quad \text{and, in general, } a^{1/n} = \sqrt[n]{a}.$$

Rules for Computing Using Exponents

1. $a^x \cdot a^t = a^{x+t}$ For example, $2^4 \cdot 2^3 = (2 \cdot 2 \cdot 2 \cdot 2) \cdot (2 \cdot 2 \cdot 2) = 2^7$.
2. $\dfrac{a^x}{a^t} = a^{x-t}$ For example, $\dfrac{2^4}{2^3} = \dfrac{2 \cdot 2 \cdot 2 \cdot 2}{2 \cdot 2 \cdot 2} = 2^1$.
3. $(a^x)^t = a^{xt}$ For example, $(2^3)^2 = 2^3 \cdot 2^3 = 2^6$.

Problems for Section 1.3

1. The number of cancer cells grows slowly at first but then grows with increasing rapidity. Draw a possible graph of the number of cancer cells against time.
2. Each year the world's annual consumption of electricity rises. In addition, each year the increase in annual consumption also rises. Sketch a possible graph of the annual world consumption of electricity as a function of time.
3. A drug is injected into a patient's bloodstream over a five-minute interval. During this time, the quantity in the blood increases linearly. After five minutes the injection is discontinued, and the quantity then decays exponentially. Sketch a graph of the quantity versus time.

4. When there are no other steroid hormones (for example, estrogen) in a cell, the rate at which steroid hormones diffuse into the cell is fast. The rate slows down as the amount in the cell builds up. Sketch a possible graph of the quantity of steroid hormone in the cell against time, assuming that initially there are no steroid hormones in the cell.

5. Each of the functions in Table 1.10 is increasing, but each increases in a different way. Which of the graphs in Figure 1.24 below best fits each function?

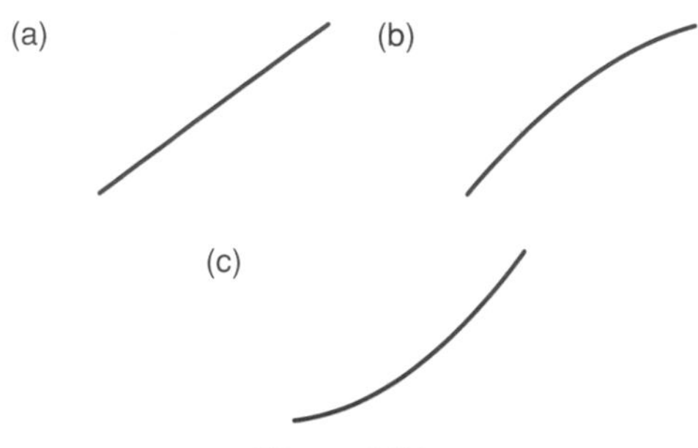

Figure 1.24

TABLE 1.10

t	$g(t)$	$h(t)$	$k(t)$
1	23	10	2.2
2	24	20	2.5
3	26	29	2.8
4	29	37	3.1
5	33	44	3.4
6	38	50	3.7

6. Each of the functions in Table 1.11 decreases, but each decreases in a different way. Which of the graphs in Figure 1.25 below best fits each function?

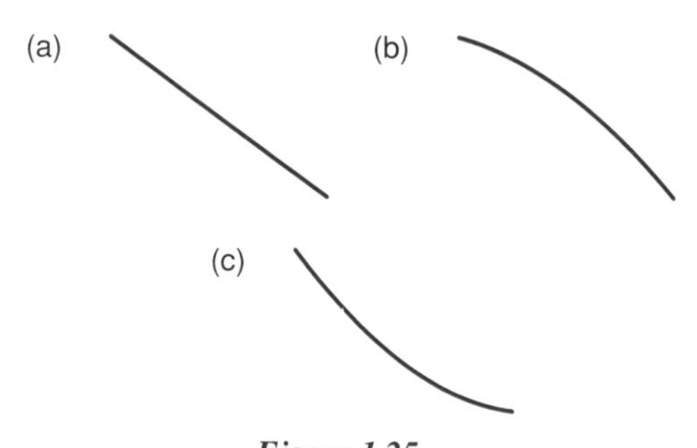

Figure 1.25

TABLE 1.11

x	$f(x)$	$g(x)$	$h(x)$
1	100	22.0	9.3
2	90	21.4	9.1
3	81	20.8	8.8
4	73	20.2	8.4
5	66	19.6	7.9
6	60	19.0	7.3

7. Match up the function values in Table 1.12 with the formulas

$$y = a(1.1)^s\ ,\quad y = b(1.05)^s\ ,\quad y = c(1.03)^s,$$

assuming a, b, and c are constants. Note that the function values have been rounded to two decimal places.

TABLE 1.12

s	$h(s)$	s	$f(s)$	s	$g(s)$
2	1.06	1	2.20	3	3.47
3	1.09	2	2.42	4	3.65
4	1.13	3	2.66	5	3.83
5	1.16	4	2.93	6	4.02
6	1.19	5	3.22	7	4.22

For Problems 8–9, find a possible formula for the functions represented by the data.

8.

x	0	1	2	3
$f(x)$	4.30	6.02	8.43	11.80

9.

t	0	1	2	3
$g(t)$	5.50	4.40	3.52	2.82

Find possible equations for the graphs in Problems 10–13.

10.

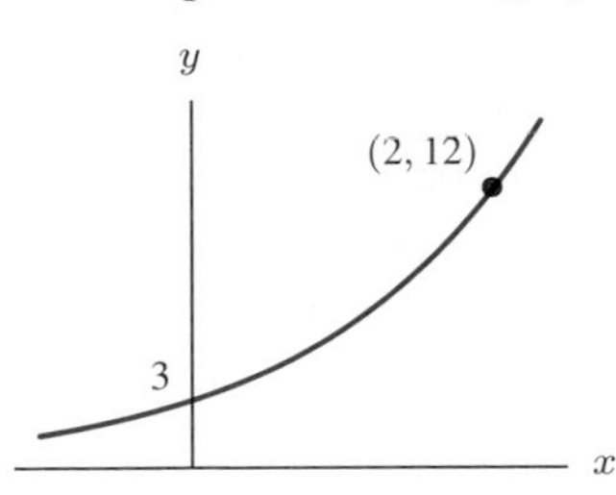

11.

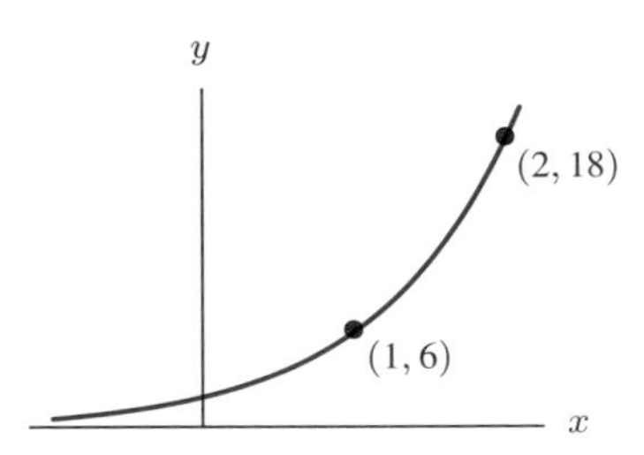

12.

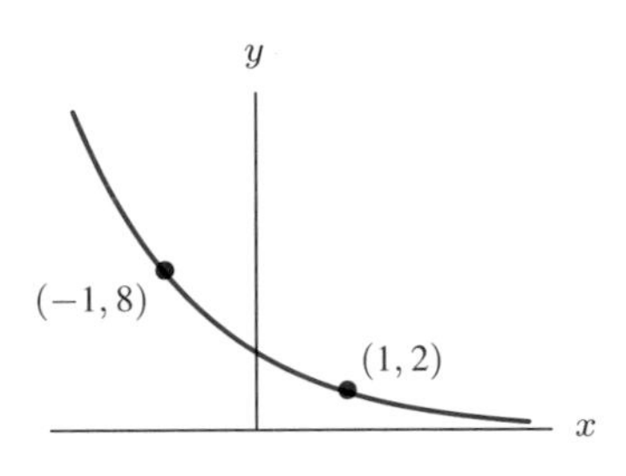

13.

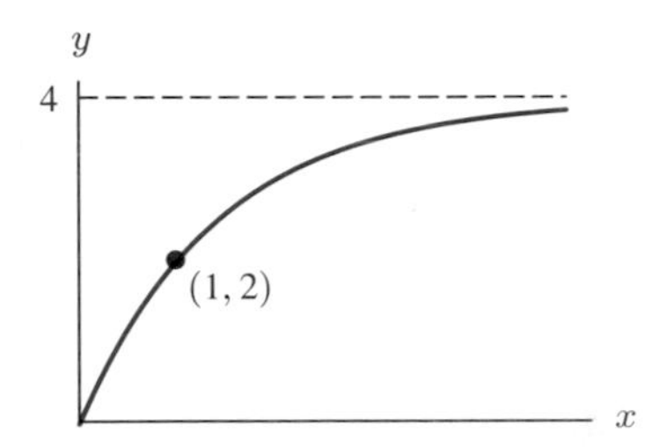

14. (a) The half-life of radium-226 is 1620 years. Write a formula for the quantity, Q, of radium left after t years, if the initial quantity is Q_0.
    (b) What percentage of an original amount of radium is left after 500 years?

15. In the early 1960s, radioactive strontium-90 was released during atmospheric testing of nuclear weapons and got into the bones of people alive at the time. If the half-life of strontium-90 is 29 years, what fraction of the strontium-90 absorbed in 1960 remained in people's bones in 1990?

16. When the Olympic Games were held outside Mexico City in 1968, there was much discussion about the effect the high altitude (7340 feet) would have on the athletes. Assuming air pressure decays exponentially at 0.4% every 100 feet, by what percentage is air pressure reduced by moving from sea level to Mexico City?

17. A population is known to be growing exponentially. Estimate the doubling time of the population shown by the graph in Figure 1.26, and verify graphically that the doubling time is independent of where you start on the graph.

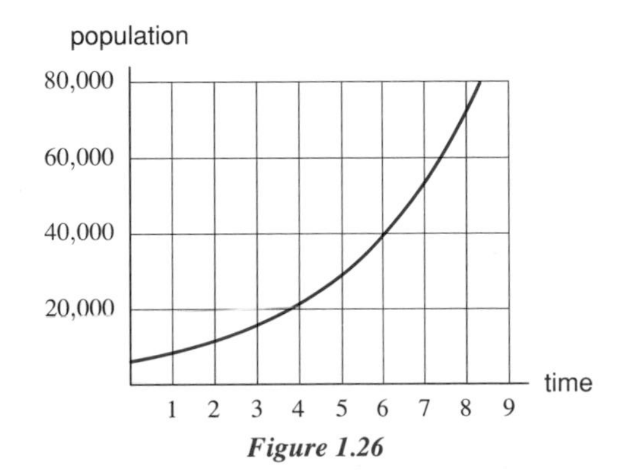

Figure 1.26

18. A certain region has a population of 10,000,000 and an annual growth rate of 2%. Estimate the doubling time by trial and error.

19. A certain radioactive substance decays exponentially in such a way that after 10 years, 70% of the initial amount remains. Find an expression for the quantity remaining after any number of years t. How much will be present after 50 years? What is the half-life? How long will it be before only 20% of the original amount is left? Before only 10% is left? (Use trial and error where necessary.)

20. Assume that the median price, P, of a home rose from $50,000 in 1970 to $100,000 in 1990. Let t be the number of years since 1970.

    (a) Assume the increase in housing prices has been linear. Find an equation for the line representing price, P, in terms of t. Use this equation to complete column (a) of Table 1.13. Work with the price in units of $1000.

    (b) If instead the housing prices have been rising exponentially, determine an equation of the form $P = P_0a^t$ which would represent the change in housing prices from 1970–1990, and complete column (b) of Table 1.13.

    (c) On the same set of axes, sketch the functions represented in column (a) and column (b) of Table 1.13.

TABLE 1.13

t	(a) Linear growth price in $1000 units	(b) Exponential growth price in $1000 units
0	50	50
10		
20	100	100
30		
40		

Countries with very high inflation rates often publish monthly rather than yearly inflation figures, because monthly figures are less alarming. Problems 21–22 involve such high rates, which are called *hyperinflation*.

21. In 1989, US inflation was 4.6% a year. In 1989 Argentina had an inflation rate of about 33% a month.

    (a) What is the yearly equivalent of Argentina's 33% monthly rate?

    (b) What is the monthly equivalent of the US 4.6% yearly rate?

22. Between December 1988 and December 1989, Brazil's inflation rate was 1290% a year. (This means that between 1988 and 1989, prices increased by a factor of $1 + 12.90 = 13.90$.)

    (a) What would an article which cost 1000 cruzeiros (the Brazilian currency unit) in 1988 cost in 1989?

    (b) What was Brazil's monthly inflation rate during this period?

1.4 POWER FUNCTIONS

Power functions are an important family of functions. A *power function* is one in which the dependent variable is proportional to a power of the independent variable. For example, the area, A, of a square of side s is given by

$$A = f(s) = s^2.$$

The volume, V, of a sphere of radius r is

$$V = f(r) = \frac{4}{3}\pi r^3.$$

Both of these are *power functions*. So is the function which describes how the gravitational attraction of the earth varies with distance. If g is the force of gravitational attraction on a unit mass at a distance r from the earth, Newton's Inverse Square Law of Gravitation says that

$$g = \frac{k}{r^2} \quad \text{or} \quad g = kr^{-2}$$

where k is a positive constant.

In general, a **power function** has the form

$$y = f(x) = kx^p$$

where k and p are any constants.

In this section we will compare various power functions with one another and with the exponential functions. Since $y = f(x) = mx$ (with m a constant) is also a power function (because $x = x^1$, the first power), linear functions are included in the comparison too.

Positive Integral Powers: $y = x$, $y = x^2$, $y = x^3$, . . .

First, we'll look at functions of the form $f(x) = x^n$, with n a positive integer. Figures 1.27 and 1.28 show that the graphs of these functions fall into two groups: the odd powers and the even powers. All the odd powers (x, x^3, x^5, and so on) are increasing everywhere and their graphs are symmetric about the origin. All odd powers above $n = 1$ have a bend, or "seat," at the origin. The even powers, on the other hand, are first decreasing and then increasing, making them $\bigcup$-shaped with symmetry about the y-axis. The even powers are concave up everywhere, whereas the odd ones (greater than 1) are concave down for negative x and concave up for positive x. All odd and even powers, however, go through the points $(0, 0)$ and $(1, 1)$.

Figure 1.29 shows that the higher the power of x, the faster the function climbs. For large values of x (in fact, for all $x > 1$), $y = x^5$ is above $y = x^4$, which is above $y = x^3$, and so on. Not only are the higher powers larger, but they are *much* larger. This is because if $x = 100$, for example, 100^5 is one hundred times as big as 100^4 which is one hundred times as big as 100^3. As x gets larger (written as $x \to \infty$), any positive power of x completely swamps all lower powers of x. We say that, as $x \to \infty$, higher powers of x *dominate* lower powers.

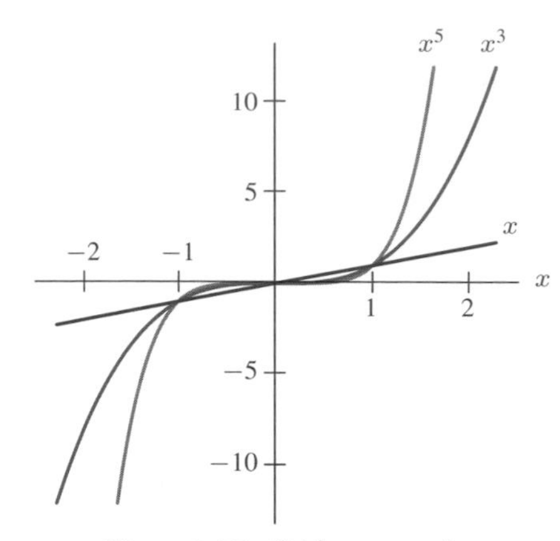

Figure 1.27: Odd powers of x

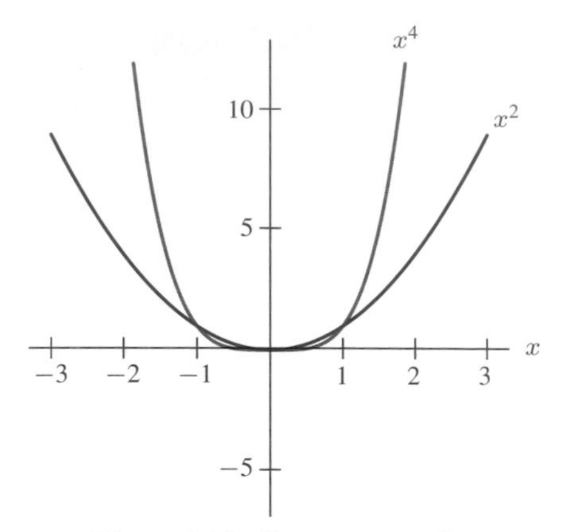

Figure 1.28: Even powers of x

The close-up view near the origin in Figure 1.30 shows an entirely different story. For x between 0 and 1, the order is reversed: x^3 is bigger than x^4, which is bigger than x^5. (Try $x = 0.1$ to confirm this.) The fact that higher powers of x climb faster is true for large values of x but not for small. For big values of x, the highest powers are largest; for values of x near zero, smaller powers dominate.

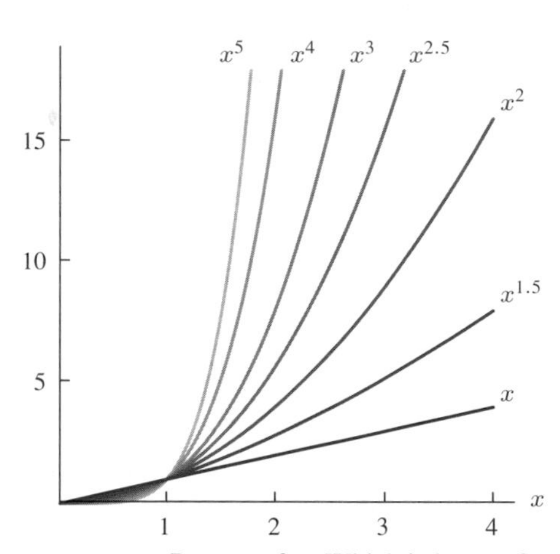

Figure 1.29: Powers of x: Which is largest for large values of x?

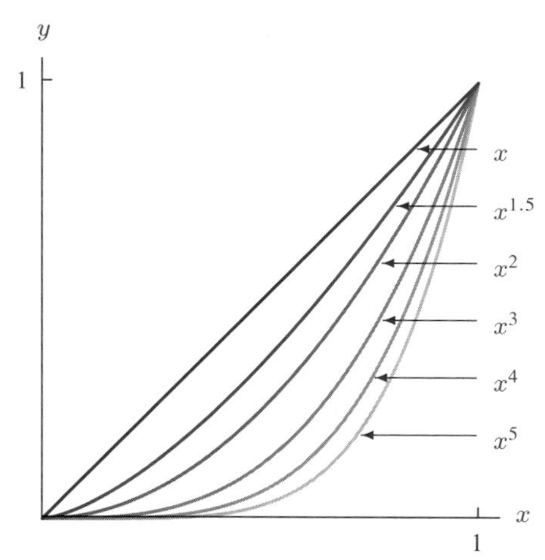

Figure 1.30: Between 0 and 1: Small powers of x dominate

Zero and Negative Integral Powers: $y = x^0$, $y = x^{-1}$, $y = x^{-2}, \ldots$

The function $y = x^0 = 1$ has a graph that is a horizontal line. Rewriting

$$y = x^{-1} = \frac{1}{x} \quad \text{and} \quad y = x^{-2} = \frac{1}{x^2}$$

makes it easier to see that as x increases, the denominators increase and the functions decrease. The graphs of $y = x^{-1}$ and $y = x^{-2}$ have both the x and y axes as asymptotes. (See Figure 1.31.) For $x > 1$, the graph of $y = x^{-2}$ is below that of $y = x^{-1}$, and both must stay below $y = x^0 = 1$.

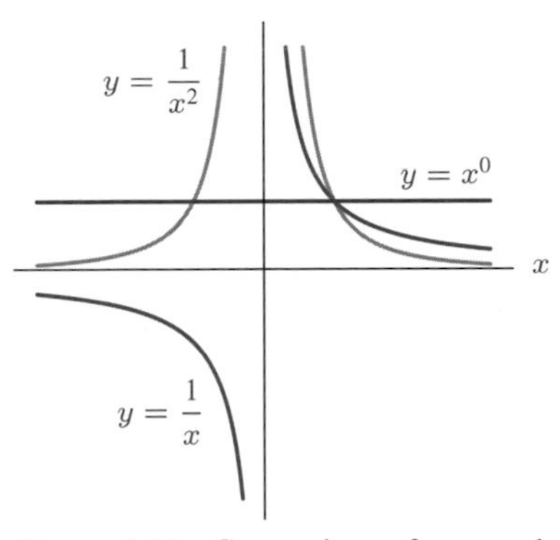

Figure 1.31: Comparison of zero and negative powers of x

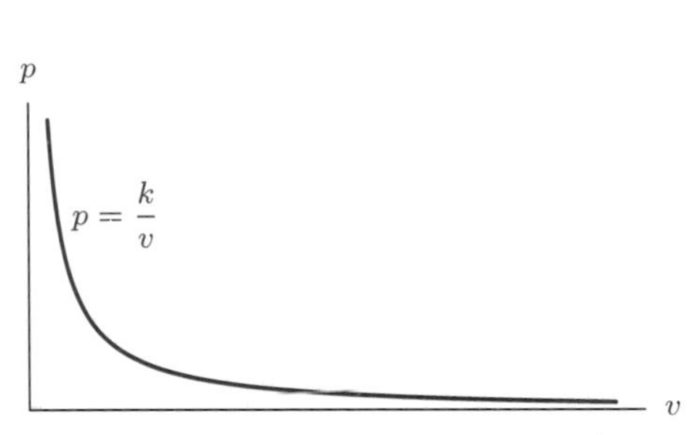

Figure 1.32: Graph of pressure, p, against volume, v, for Boyle's law

Example 1 Plot a graph of pressure against volume for a fixed quantity of gas at a constant temperature. Use the fact that pressure is inversely proportional to volume.

Solution Think of a fixed quantity of air — for example, inside a cylinder in a car engine. If the volume of the air is decreased (by moving the pistons), the pressure of the air will be increased. Conversely, if the volume is increased, the pressure will decrease. For an ideal gas, Boyle's law gives the exact relationship between pressure and volume, provided the temperature is constant. It says pressure times volume is constant, or

$$pv = k$$

with k positive. So

$$p = \frac{k}{v} = kv^{-1}.$$

The relationship $p = k/v$ is equivalent to saying p is inversely proportional to v. When v is large, p is small, and when v is small, p is large. For any (positive) value of k, the graph has the shape shown in Figure 1.32. Both axes are asymptotes, showing that as the volume tends to infinity, the pressure tends to zero, and vice versa. The shape is known as a rectangular *hyperbola*. The power function $p = k/v = kv^{-1}$ differs from exponential decay in that it is undefined for $v = 0$, so this graph does not cross the vertical axis. In addition, it approaches the horizontal axis more slowly than the exponential function.

Example 2 Quantum mechanics predicts that the force between two gas molecules has two components: an attractive force which is approximately proportional to r^{-7} (where r is the distance between the molecules) and a repulsive force approximately proportional to r^{-13}. How does the net force vary with r?

Solution A repulsive force is usually considered to be positive, whereas an attractive force is negative. Thus, we write $F = -ar^{-7} + br^{-13}$, where a, b are positive constants. If r is very small, r^{-13} is larger than r^{-7}, so $F \approx br^{-13}$ and the net force is repulsive. (This occurs when the molecules are so close together that the protons in the nuclei repel one another.) For larger r, r^{-7} is larger than r^{-13}, and $F \approx -ar^{-7}$, making the net force attractive. As $r \to \infty$, all the forces die away to zero.

Positive Fractional Powers: $y = x^{1/2}, y = x^{1/3}, y = x^{3/2}, \ldots$

The function giving the side of a square, s, in terms of its area, A, involves a root, or fractional power:

$$s = \sqrt{A} = A^{1/2}.$$

Similarly, the equation relating the average number of species found on an island and the size of the island involves a fractional power. If N is the number of species and A is the area of the island, observations have shown[4] that approximately

$$N = k\sqrt[3]{A} = kA^{1/3}$$

where k is a constant depending on the region of the world in which the island is found.

We will now look at functions of the form $y = x^{m/n} = \sqrt[n]{x^m}$. Since some fractional powers such as $x^{1/2}$ involve roots and are defined only for positive x and 0, we frequently restrict the domain of positive fractional powers of x to $x \geq 0$. Many calculators will not allow you to raise a negative number to a fractional power.

Figure 1.33 shows that for large x (in fact, all $x > 1$), the graph of $y = x^{1/2}$ is below the graph of $y = x$, and $y = x^{1/3}$ is below $y = x^{1/2}$. This is reasonable since, for example, $10^{1/2} = \sqrt{10} \approx 3.16$ and $10^{1/3} = \sqrt[3]{10} \approx 2.15$, so $10^{1/3} < 10^{1/2} < 10$. Between $x = 0$ and $x = 1$, the situation is reversed, and $y = x^{1/3}$ is on top. (Why?) Not surprisingly, $y = x^{3/2}$ is between $y = x$ and $y = x^2$ for all x.

The other important feature to notice about the graphs of $y = x^{1/2}$ and $y = x^{1/3}$ is that they bend in a direction opposite to that of the graphs of $y = x^2$ and x^3. For example, the graph of $y = x^2$ is climbing faster and faster as x increases; it is concave up. On the other hand, the graphs of $y = x^{1/2}$ and $y = x^{1/3}$ are climbing slower and slower; they are concave down. Despite this, all these functions do become infinitely large as x increases.

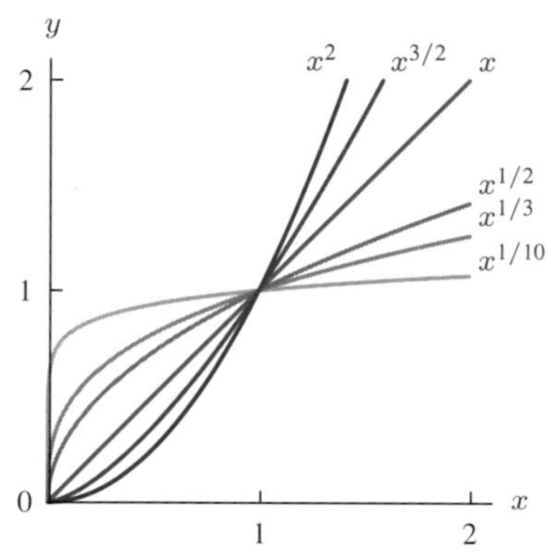

Figure 1.33: Comparison of some fractional powers of x

What Effect Do Coefficients Have?

We know that $x^2 < x^3$ for all $x > 1$. But which is larger, $50x^2$ or x^3? Eventually, $50x^2 < x^3$ too. In fact, $50x^2 < x^3$ for all $x > 50$. (See Figure 1.34.) The graphs of $y = x^2$ and $y = x^3$ cross at $x = 1$, whereas graphs of $y = 50x^2$ and $y = x^3$ cross at $x = 50$. Thus, the effect of the factor of 50 is to change the point at which the graphs cross. However, x^3 ends up on top in both cases: provided the coefficients are positive, as $x \to \infty$, the higher power is always larger eventually.

[4] *Scientific American*, September 1989, p. 112.

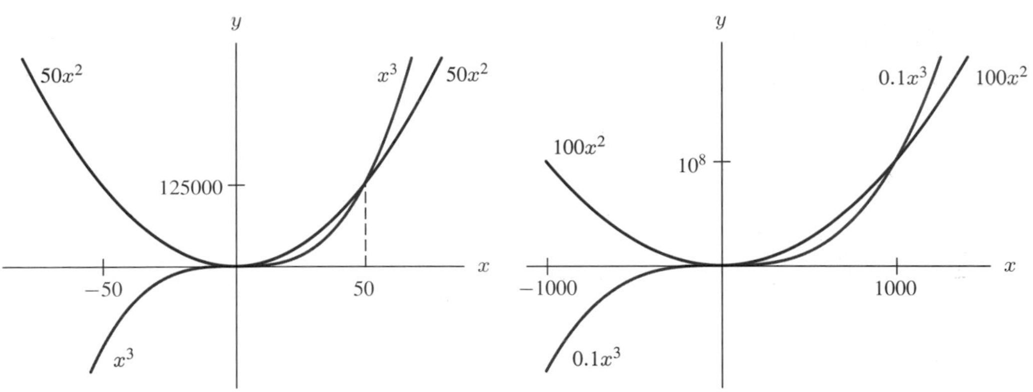

Figure 1.34: Graph of $y = x^3$ lies above graph of $y = 50x^2$ for large positive x

Figure 1.35: For large positive x, $y = 0.1x^3$ dominates $y = 100x^2$

Example 3 Which of $y = 100x^2$ and $y = 0.1x^3$ is larger as $x \to \infty$?

Solution Since $x \to \infty$, we are looking at large positive values of x, where the higher power will eventually be larger, or dominate. Thus, $y = 0.1x^3$ will be larger. (See Figure 1.35.)

Example 4 Sketch a global view of $f(x) = -x^3$, $g(x) = 40x^4$, and $h(x) = -0.1x^5$. Which function has the largest positive values as $x \to \infty$? Which function has the largest positive values as $x \to -\infty$ (i.e., as x gets more and more negative)?

Solution As $x \to \infty$, $g(x) = 40x^4$ is the only function which is positive. As $x \to -\infty$, the graph of $h(x) = -0.1x^5$ is eventually (for $x < -400$) above the graph of the other functions, so $h(x)$ has the largest positive values as $x \to -\infty$ (See Figure 1.36.) Notice that, for large x, the values of $f(x) = -x^3$ are so much smaller in magnitude than the values of the other functions that the graph of $f(x)$ cannot be seen in the far-away view.

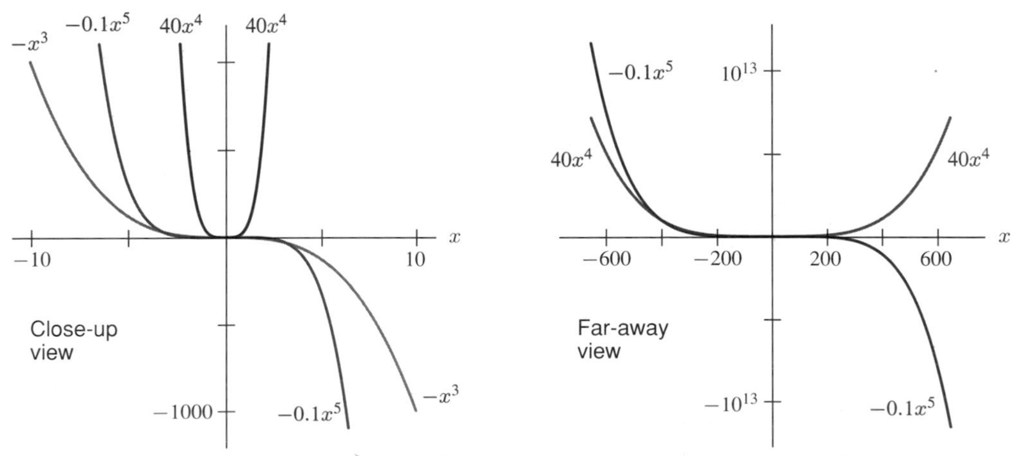

Figure 1.36: As $x \to \infty$, $g(x) = 40x^4$ dominates; as $x \to -\infty$, $h(x) = -0.1x^5$ dominates

Exponentials and Power Functions: Which Dominate?

In everyday language, *exponential* is often used to imply very fast growth. But do exponential functions always grow faster than power functions?

Let's consider $y = 2^x$ and $y = x^3$. The close-up, or local, view in Figure 1.37(a) shows that between $x = 2$ and $x = 5$, the graph of $y = 2^x$ lies below the graph of $y = x^3$. But the more global, or far–away, view in Figure 1.37(b) shows that the exponential function $y = 2^x$ eventually overtakes $y = x^3$. And Figure 1.37(c), which gives a very far–away, or global view, shows that, for large x, x^3 is insignificant compared to 2^x. Indeed, 2^x is growing so much faster than x^3 that its graph appears almost vertical—in comparison to the more leisurely climb of x^3.

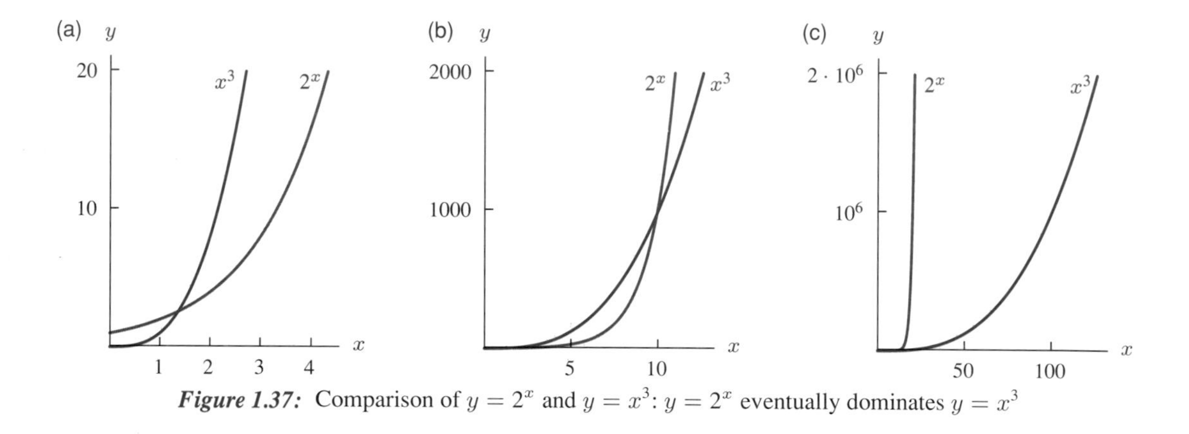

Figure 1.37: Comparison of $y = 2^x$ and $y = x^3$: $y = 2^x$ eventually dominates $y = x^3$

In fact, *every* exponential growth function eventually dominates *every* power function. Although an exponential function may be below a power function for some values of x, if you look at large enough x values, a^x (with $a > 1$) will eventually dominate x^n, no matter what n is. Two more examples are presented in Figure 1.38 and Table 1.14.

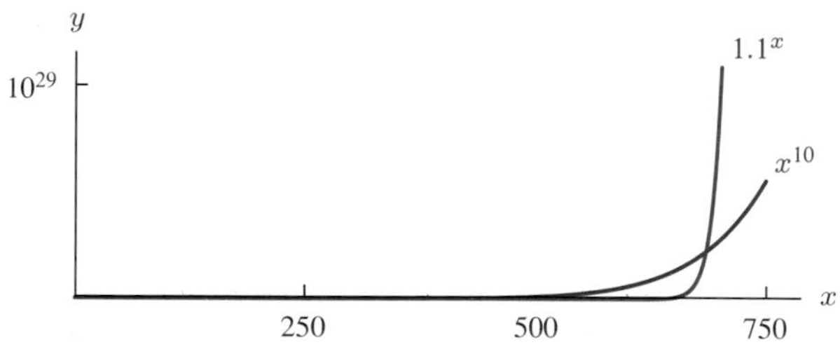

Figure 1.38: Exponential function eventually dominates power function

TABLE 1.14 *Comparison of x^{100} and 1.01^x*

x	x^{100}	1.01^x
10^4	10^{400}	$1.6 \cdot 10^{43}$
10^5	10^{500}	$1.4 \cdot 10^{432}$
10^6	10^{600}	$2.4 \cdot 10^{4321}$

Can you guess what happens in the case of negative powers and negative exponents? For example, consider $y = 2^{-x}$ and $y = x^{-2}$. Since $y = 2^{-x} = 1/2^x$ and $y = x^{-2} = 1/x^2$, knowing that 2^x is eventually larger than x^2 tells you that 2^{-x} is eventually smaller than x^{-2}. Hence $y = 2^{-x}$ is eventually below $y = x^{-2}$. (See Figure 1.39.) This behavior is also typical: Every exponential decay function will eventually approach 0 faster than every power function with a negative exponent.

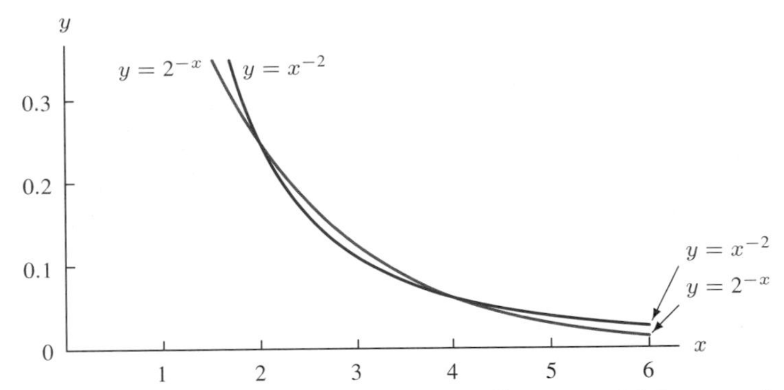

Figure 1.39: Comparison of $y = 2^{-x}$ and $y = x^{-2}$: Exponential dies away faster

Problems for Section 1.4

1. Simplify each of the following: (a) $8^{2/3}$ (b) $9^{-3/2}$
2. Sketch graphs of $y = x^{1/2}$ and $y = x^{2/3}$ on the same axes. Which function has larger values as $x \to \infty$?
3. What happens to the value of $y = x^4$ as $x \to \infty$? As $x \to -\infty$?
4. What happens to the value of $y = -x^7$ as $x \to \infty$? As $x \to -\infty$?
5. Sketch a graph of $y = x^{-4}$.
6. On a graphing calculator or computer, plot graphs of the following functions, first for $-5 \le x \le 5$, $-100 \le y \le 100$, and then for $-1.2 \le x \le 1.2$, $-2 \le y \le 2$.
   (a) $y = x, y = x^3, y = x^6, y = x^9$
   (b) $y = x, y = x^4, y = x^7, y = x^{10}$
   Observe the general shape of these functions: Do the odd powers have the same general shape? What about the even powers? Which function is largest in magnitude for big x? For x near 0? Is this what you expected?
7. Do some calculations using specific values of x to verify that $y = x^{1/3}$ is above $y = x^{1/2}$ and that $y = x^{1/2}$ is above $y = x$ for $0 < x < 1$.
8. (a) Use a graphing calculator (or a computer) to plot the graphs of x^3, x^4, and x^5 on the interval $-0.1 \le x \le 0.1$. Determine an appropriate range for y so that all powers will be distinguishable in the viewing rectangle.
   (b) Plot the same graphs for $-100 \le x \le 100$, and determine an appropriate range for y.
9. By hand, sketch global pictures of $f(x) = x^3$ and $g(x) = 20x^2$ on the same axes. Which function has larger values as $x \to \infty$?
10. By hand, sketch pictures of $f(x) = x^5$, $g(x) = -x^3$, and $h(x) = 5x^2$ on the same axes. Which has larger positive values as $x \to \infty$? As $x \to -\infty$?
11. By trial and error, use a calculator to find to two decimal places the point near $x = 10$ at which $y = 2^x$ and $y = x^3$ cross.
12. Use a graphing calculator to find the point(s) of intersection of the graphs of $y = (1.06)^x$ and $y = 1 + x$.
13. For what values of x is $4^x > x^4$?

14. For what values of x is $3^x > x^3$? (Note: You will need to think about how to deal with the fact that the graphs of 3^x and x^3 are relatively close together for values of x near 3.)

15. According to the April 1991 issue of *Car and Driver*, an Alfa Romeo going at 70 mph requires 177 feet to stop. Assuming that the stopping distance is proportional to the square of velocity, find the stopping distances required by an Alfa Romeo going at 35 mph and at 140 mph (its top speed).

16. Poiseuille's law gives the rate of flow, R, of a gas through a cylindrical pipe in terms of the radius of the pipe, r, for a fixed drop in pressure. Assume a constant drop in pressure throughout the remainder of this problem.
    (a) Determine a formula for Poiseuille's Law, given that the rate of flow is proportional to the fourth power of the radius.
    (b) If $R = 400$ cm^3/sec in a pipe of radius 3 cm for a certain gas, determine an explicit formula for the rate of flow of that gas through a pipe of radius r cm.
    (c) What is the rate of flow of the same gas through a pipe with a 5-cm radius?

17. The values of three functions are given in Table 1.15. One function is of the form $y = ab^t$, one is of the form $y = at^2$, and one is of the form $y = bt^3$. Which function is which?

TABLE 1.15

t	$f(t)$	t	$g(t)$	t	$h(t)$
2.0	4.40	1.0	3.00	0.0	2.04
2.2	5.32	1.2	5.18	1.0	3.06
2.4	6.34	1.4	8.23	2.0	4.59
2.6	7.44	1.6	12.29	3.0	6.89
2.8	8.62	1.8	17.50	4.0	10.33
3.0	9.90	2.0	24.00	5.0	15.49

18. Values of three functions are contained in Table 1.16. (The numbers have been rounded to two decimal places.) Two are power functions and one is an exponential. One of the power functions is a quadratic and one a cubic. Which one is exponential? Which one is quadratic? Which one is cubic?

TABLE 1.16

x	$f(x)$	x	$g(x)$	x	$k(x)$
8.4	5.93	5.0	3.12	0.6	3.24
9.0	7.29	5.5	3.74	1.0	9.01
9.6	8.85	6.0	4.49	1.4	17.66
10.2	10.61	6.5	5.39	1.8	29.19
10.8	12.60	7.0	6.47	2.2	43.61
11.4	14.82	7.5	7.76	2.6	60.91

19. Owing to improved seed types and new agricultural techniques, the grain production of a region has been increasing. Over a 20-year period, annual production (in millions of tons) was as follows:

1970	1975	1980	1985	1990
5.35	5.90	6.49	7.05	7.64

At the same time the population (in millions) was:

1970	1975	1980	1985	1990
53.2	56.9	60.9	65.2	69.7

(a) Find a linear or exponential function which approximately fits each set of data. (Pick whichever type of function fits better.)
(b) If this region was self-supporting in this grain in 1970, was it self-supporting between 1970 and 1990? (Being self-supporting means that each person has enough of the grain. How does the amount of grain each person has in later years compare?)
(c) What are your predictions for the future if the trends continue?

20. Use a graphing calculator (or a computer) to graph $y = x^4$ and $y = 3^x$. Determine the appropriate domains and ranges that will give each of the graphs in Figure 1.40.

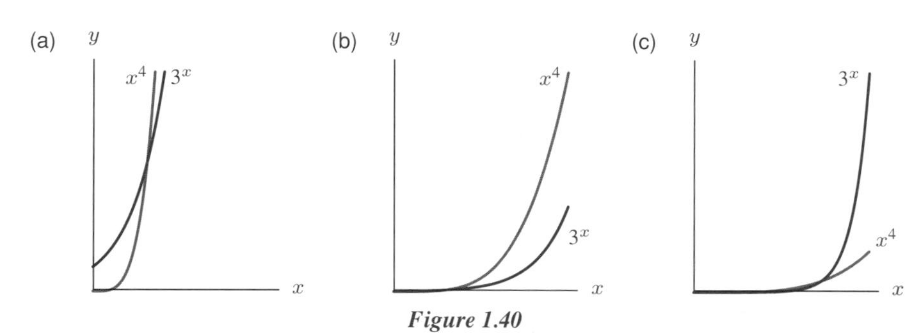

Figure 1.40

1.5 INVERSE FUNCTIONS

From Distance to Time and Back

On August 18, 1989, Arturo Barrios of Mexico set a world record in the 10,000-meter run with a time of 27 minutes and 8.23 seconds. His times, in seconds, at 2000-meter intervals are recorded in Table 1.17, where $f(d)$ is the number of seconds Barrios took to complete the first d meters of the race. For example, Barrios ran the first 4000 meters in 650.1 seconds, so $f(4000) = 650.1$. The function f is useful to athletes planning to compete with Barrios.

Let us now change our point of view and ask for distances rather than times. If we ask how far Barrios ran during the first 650.1 seconds of his race, the answer is clearly 4000 meters. Going backwards in this way from numbers of seconds to numbers of meters gives a function called the *inverse function* of f, denoted by f^{-1}. Thus, $f^{-1}(t)$ is the number of meters that Barrios ran during the first t seconds of his race. To find values of f^{-1}, we can either read Table 1.17 backwards, or use Table 1.18 which contains values of f^{-1}.

The two functions f and f^{-1} convey the same information, but they express it differently. For example, the fact that Barrios ran the first 6000 meters in 975.5 seconds can be written with either f or f^{-1}:

$$f(6000) = 975.5 \quad \text{or} \quad f^{-1}(975.5) = 6000.$$

TABLE 1.17 *Barrios's running time*

d (meters)	$f(d)$ (seconds)
0	0.00
2000	325.90
4000	650.10
6000	975.50
8000	1307.00
10000	1628.23

TABLE 1.18 *Distance run by Barrios*

t (seconds)	$f^{-1}(t)$ (meters)
0.00	0
325.90	2000
650.10	4000
975.50	6000
1307.00	8000
1628.23	10000

The independent variable for f is the dependent variable for f^{-1}, and vice versa. The domains and ranges of f and f^{-1} are also interchanged. The domain of f is all distances d such that $0 \leq d \leq 10000$, which is the range of f^{-1}. The range of f is all times t, such that $0 \leq t \leq 1628.23$, which is the domain of f^{-1}.

Definition of an Inverse Function

Not every function f has an inverse, a problem we will discuss later in this section. But when an inverse exists, it is defined as follows:

$$f^{-1}(t) = d \quad \text{means} \quad f(d) = t$$

The notation f^{-1} to denote an inverse function is perhaps unfortunate, as it is easy to confuse with a reciprocal, which the inverse function is not. However, there's no changing such well-established notation!

Example 1 The temperature at which water boils decreases as altitude increases, a fact important to cooks. Let $f(h)$ be the boiling point (°C) of water at an altitude of h meters above sea level during standard atmospheric conditions. What is the meaning in practical terms of $f^{-1}(90)$ and of $f^{-1}(90) = 3000$? Evaluate $f^{-1}(100)$.

Solution The function f goes from altitude to temperature, so f^{-1} goes back from temperature to altitude. Thus, $f^{-1}(90)$ is the altitude in meters at which the boiling point of water is 90°C. The equation $f^{-1}(90) = 3000$ means that the boiling point of water is 90°C at an altitude of 3000 meters. The equation $f(3000) = 90$ has the same meaning. Since the boiling point of water is 100°C at sea level (where the altitude is 0 meters), we must have $f^{-1}(100) = 0$.

Formulas for Inverse Functions

If a function is defined by a formula, it is sometimes possible to find a formula for the inverse function as well. In Section 1.1, we looked at the snow tree cricket, whose chirp rate, C, in chirps per minute, is approximated by the formula

$$C = f(T) = 4T - 160$$

where T is the temperature in degrees Fahrenheit. So far we have used this formula to predict the chirp rate from the temperature. But it is perfectly possible to use this formula backwards to calculate the temperature from the chirp rate.

Example 2 Find the formula for the function giving temperature in terms of the number of cricket chirps per minute; that is, find the inverse function f^{-1} such that

$$T = f^{-1}(C).$$

Solution We know $C = 4T - 160$. We solve for T, giving

$$T = \frac{C}{4} + 40$$

so

$$f^{-1}(C) = \frac{C}{4} + 40.$$

When Does a Function Have an Inverse?

If a function has an inverse, we say it is *invertible*. Not all functions have inverses, and the best way to understand which ones do have inverses is to look at an example which does not.

Consider the flight of the Mercury spacecraft *Freedom 7*, which carried the first American, Alan Shepard, Jr., into space in May 1961. The spacecraft rose to an altitude of 116 miles, and then came down into the sea for a total flight time of 15 minutes. The function $f(t)$ giving the altitude in miles t minutes after lift-off does not have an inverse. To see why not, try to decide on a value for $f^{-1}(100)$. Clearly, $f^{-1}(100)$ should tell us the time when the altitude of the spacecraft was 100 miles. However, there are two such times, one when the spacecraft was ascending and one when it was descending. (See Figure 1.41.)

Since functions must take single definite values, there is no inverse function f^{-1}. The reason why the altitude function $f(t)$ does not have an inverse is that the altitude first increases and then decreases, so each altitude corresponds to two times, t. The reason why the Barrios time and chirp rate functions do have inverses is that they both increase everywhere. Thus, each running time, t, corresponds to some unique distance, d, and any biologically reasonable number of chirps, C, corresponds to some particular temperature, T.

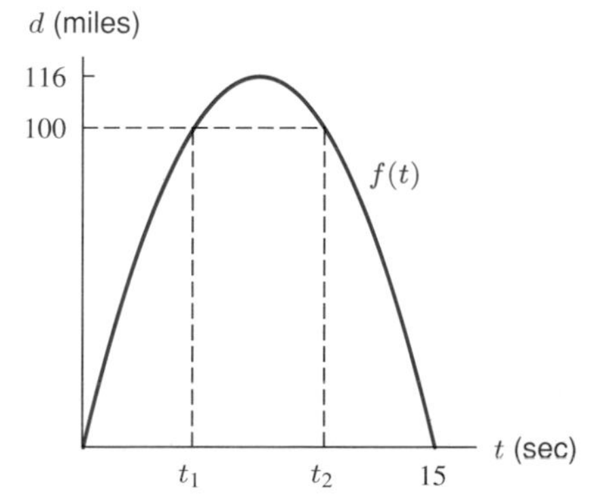

Figure 1.41: Two times, t_1 and t_2, at which altitude of spacecraft is 100 miles

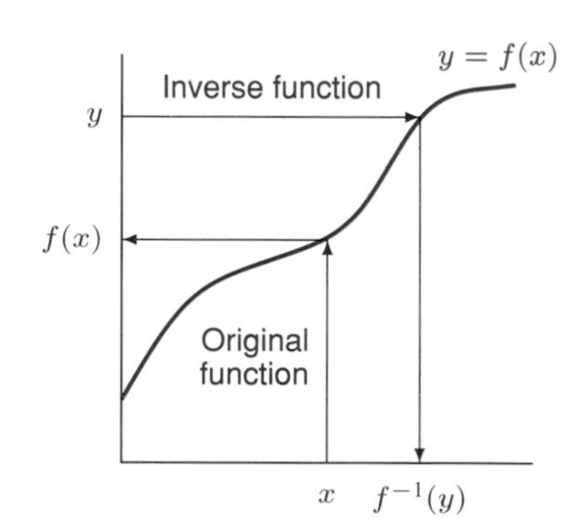

Figure 1.42: Why an increasing function has an inverse

A function does not have to be increasing everywhere to have an inverse, but the graph in Figure 1.42 does suggest when an inverse will exist. The original function, f, takes you from x to y, as shown in Figure 1.42. Since having an inverse means there is a function going from y to x, the crucial question is whether you can get back. In other words, does each y value correspond to a unique x value? If so, there's an inverse; if not, there isn't. This principle may be stated geometrically, as follows:

> A function has an inverse if (and only if) its graph intersects any horizontal line at most once.

Thus, for example, the function $f(x) = x^2$ does not have an inverse because many horizontal lines intersect the parabola twice. (See Figure 1.43.)

Graphs of Inverse Functions

The graph of $f(x) = x^3$ shows that this function is increasing everywhere and so has an inverse. To find the inverse, we solve

$$y = x^3$$

for x, giving

$$x = \sqrt[3]{y}.$$

Thus, the inverse function is

$$f^{-1}(y) = \sqrt[3]{y}$$

or, if we want to call the variable x,

$$f^{-1}(x) = \sqrt[3]{x}.$$

The graphs of $y = x^3$ and $y = x^{1/3}$ are shown in Figure 1.44. Notice that these graphs are the reflections of one another about the line $y = x$. For example, $(8, 2)$ is on the graph of $y = x^{1/3}$ because $2 = 8^{1/3}$, and $(2, 8)$ is on the graph of $y = x^3$ because $8 = 2^3$. The points $(8, 2)$ and $(2, 8)$ are reflections of one another about the line $y = x$. In general:

> The graph of f^{-1} is the reflection of the graph of f about the line $y = x$.

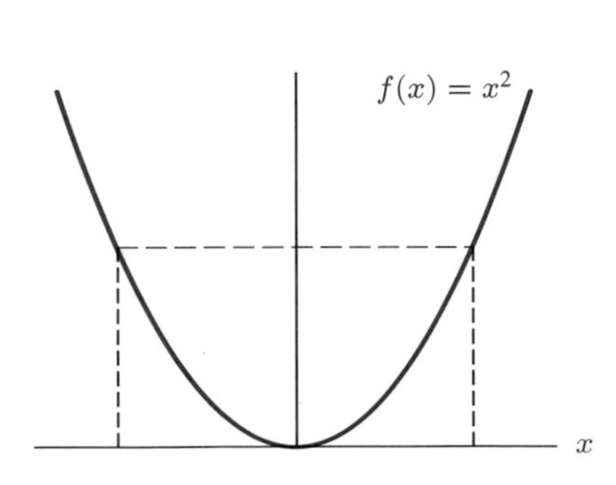

Figure 1.43: Why $f(x) = x^2$ has no inverse: Horizontal line intersects the graph twice

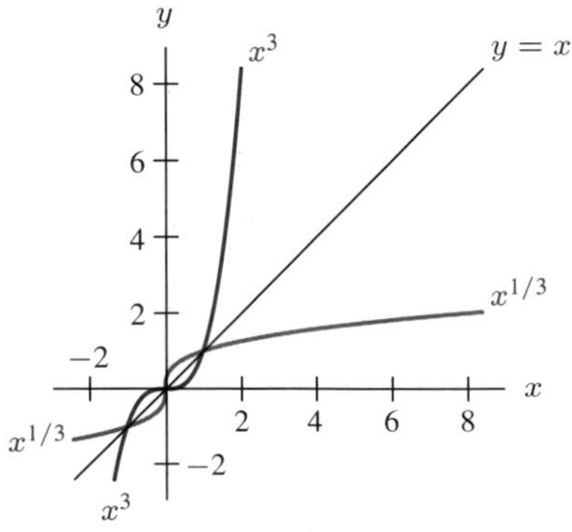

Figure 1.44: Graphs of inverse functions, $y = x^3$ and $y = x^{1/3}$, are reflections in the line $y = x$

Problems for Section 1.5

1. Let $f(x)$ equal the temperature (°F) when the column of mercury in a particular thermometer is x inches long. What is the meaning of $f^{-1}(75)$ in practical terms?

For Problems 2–6, decide whether the function f is invertible.

2. $f(d)$ is the total number of gallons of fuel an airplane has used by the end of d minutes of a particular flight.
3. $f(t)$ equals the number of customers in Macy's department store at t minutes past noon on December 18, 1993.
4. $f(x)$ is the volume in liters of x kilograms of water at 4°C.
5. $f(w)$ is the cost in cents of mailing a letter that weighs w grams.
6. $f(n)$ is the number of students in your calculus class whose birthday is on the nth day of the year.
7. Write a table of values for f^{-1}, where f is as given below. The domain of f is the integers from 1 to 7. State the domain of f^{-1}.

x	1	2	3	4	5	6	7
$f(x)$	3	−7	19	4	178	2	1

8. The function $f(x) = x^3 + x + 1$ is invertible. Use a graphing calculator to give an approximate value for $f^{-1}(20)$.

For Problems 9–11, use a graphing calculator or computer to sketch the graphs of the following functions, and decide whether or not they are invertible.

9. $f(x) = x^2 + 3x + 2$ 10. $f(x) = x^3 - 5x + 10$ 11. $f(x) = x^3 + 5x + 10$

12. The cost of producing q articles is given by the function

$$C = f(q) = 100 + 2q.$$

(a) Find a formula for the inverse function.
(b) Explain in practical terms what the inverse function tells you.

13. A kilogram weighs about 2.2 pounds.

(a) Write a formula for the function, f, which gives an object's mass in kilograms, k, as a function of its weight in pounds, p.
(b) Find a formula for the inverse function of f. What does this inverse function tell you, in practical terms?

14. Suppose f is invertible and increasing. What can you say about whether its inverse is increasing or decreasing?
15. If a function f is invertible and concave up, what can you say about the concavity of its inverse?

16. Figure 1.45 is the graph[5] of the function f, where $f(t)$ is the number (in millions) of motor vehicles registered in the world in the year t. (In 1988, one-third of the registered vehicles in the world were in the United States.)
    (a) Is f invertible? Explain.
    (b) What is the meaning of $f^{-1}(400)$ in practical terms? Evaluate $f^{-1}(400)$.
    (c) Sketch the graph of f^{-1}.

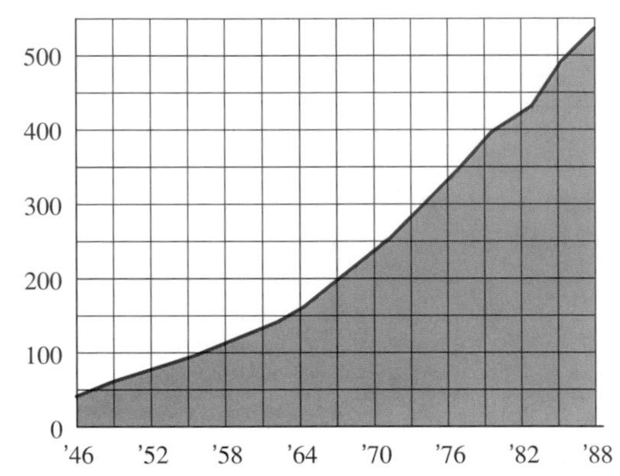

Figure 1.45: World population of motor vehicles

1.6 LOGARITHMS

In Section 1.3, we set up a function approximating the population of Mexico (in millions) as

$$P = f(t) = 67.38(1.026)^t$$

where t is the number of years since 1980. Writing the function this way shows that we are thinking of the population as a function of time, and that we believe the population to be 67.38 million in 1980 and to grow by 2.6% every year.

Now suppose that instead of calculating the population, we want to find when the population is expected to reach 100 million. This means we want to find the value of t for which

$$100 = f(t) = 67.38(1.026)^t.$$

Since the exponential function is always increasing and is eventually more than 100, there's exactly one value of t making $P = 100$. How should we find it? A reasonable way to start is trial and error. Taking $t = 10$ and $t = 20$ we get

$$P = f(10) = 67.38(1.026)^{10} = 87.1\ldots \quad \text{(so } t = 10 \text{ is too small)}$$
$$P = f(20) = 67.38(1.026)^{20} = 112.58\ldots \quad \text{(so } t = 20 \text{ is too large)}$$

Some more experimenting leads to

$$P = f(15) = 67.38(1.026)^{15} \approx 99.0$$
$$P = f(16) = 67.38(1.026)^{16} \approx 101.6$$

[5]From D. Blerics and P. Walzer, "Energy for Motor Vehicles," *Scientific American*, September 1990.

so t is between 15 and 16. In other words, the population is projected to reach 100 million sometime during 1995.

Although it is always possible to approximate t by trial and error like this, it would clearly be better to have a formula that gives t in terms of P. The *logarithm* function will enable us to do this.

Definition and Properties of Logs to Base 10:

We define the *logarithm* function, $\log_{10} x$, to be the inverse of the exponential function, 10^x. Thus we say

$$\log_{10} x = c \quad \text{means} \quad 10^c = x.$$

We call 10 the *base*. We will use $\log x$ to mean $\log_{10} x$, which is the notation your calculator uses. In other words,

> The **logarithm** to base 10 of x is the power of 10 you need to get x.

Thus, for example, $\log 1000 = \log 10^3 = 3$ since 3 is the power of 10 needed to get 1000. Similarly, $\log(0.1) = -1$ because $0.1 = 1/10 = 10^{-1}$. However, try to find $\log(-3)$ on your calculator. What happens? The reason you cannot find $\log(-3)$ is that no power of 10 is negative or 0. Therefore, in general

> $\log_{10} x$ is not defined if x is negative or 0.

In working with logarithms, you will need to use the following properties:

Rules for Computing Using Logarithms

1. $\log(AB) = \log A + \log B$
2. $\log\left(\frac{A}{B}\right) = \log A - \log B$
3. $\log\left(A^p\right) = p \log A$
4. $\log\left(10^x\right) = x$
5. $10^{\log x} = x$

In addition, $\log 1 = 0$ because $10^0 = 1$.

The fact that $\log x$ is the power of 10 giving x justifies the rules 4 and 5.

Solving Equations Using Logarithms

Logs are frequently useful when we have to solve for unknown exponents, as in the next examples.

Example 1 Find t such that $2^t = 7$.

Solution First, notice that we expect t to be between 2 and 3 (because $2^2 = 4$ and $2^3 = 8$). To find t exactly, take logs to base 10:

$$\log(2^t) = \log 7.$$

Then use the third log rule, which says $\log(2^t) = t \log 2$:

$$t \log 2 = \log 7.$$

Using a calculator to find the logs gives

$$t(0.301) = 0.845$$

so

$$t = \frac{0.845}{0.301} \approx 2.81.$$

Example 2 Solve $100 = 67.38(1.026)^t$ for t, using logs.

Solution Dividing both sides of the equation by 67.38, we get

$$\frac{100}{67.38} = 1.484 = (1.026)^t.$$

Now take logs of both sides, and use the fact that $\log(A^t) = t \log A$ to get

$$\log 1.484 = \log(1.026^t) = t \log(1.026).$$

Then, using a calculator to find the logs, we get

$$0.1714 = 0.0111t$$

Solving this equation gives

$$t \approx 15.4$$

which is between $t = 15$ and $t = 16$, as we found at the beginning of this section.

Example 3 Find a formula for the inverse of the function

$$P = f(t) = 67.38(1.026)^t.$$

Solution We must find a formula expressing t as a function of P. Take logs:

$$\log P = \log[67.38(1.026)^t].$$

Since $\log(AB) = \log A + \log B$,

$$\log P = \log 67.38 + \log[(1.026)^t].$$

Now use $\log(A^t) = t\log A$:

$$\log P = \log 67.38 + t\log 1.026.$$

Now solve for t in two steps, using a calculator at the final stage:

$$t\log 1.026 = \log P - \log 67.38$$

$$t = \frac{\log P}{\log 1.026} - \frac{\log 67.38}{\log 1.026} = 89.7\log P - 164.0$$

Thus,

$$f^{-1}(P) = 89.7\log P - 164.0.$$

Note that

$$f^{-1}(100) = 89.7(\log 100) - 164.0 = (89.7)(2) - 164.0 = 15.4,$$

which agrees with the result of Example 2.

Example 4 Find the half-life of the decaying exponential $P = P_0(0.8)^x$ that we used to model the removal of pollutants in jet fuel on page 19. What does your answer mean in practical terms?

Solution We seek the value of x such that

$$P = \frac{1}{2}P_0.$$

Thus, we must solve the equation

$$\frac{1}{2}P_0 = P_0(0.8)^x$$

Dividing both sides by P_0 leaves

$$(0.8)^x = \frac{1}{2}.$$

Taking logs of both sides gives

$$x(\log 0.8) = \log\left(\frac{1}{2}\right)$$
$$-0.097x = -0.30$$
$$x = 3.1$$

The half-life is 3.1. In practical terms, this tells us that forcing kerosene through 3.1 feet of clay pipe removes half of its impurities.

Logs can't always be used to find exponents. The next example shows equations which can only be solved numerically or by using a graph because they contain both linear and exponential terms.

Example 5 Solve: (a) $1 + x = 2^x$, (b) $1 + x = 5^x$.

Solution (a) Taking logs is of no help, because it gives an equation that is no easier to solve than the original:

$$\log(1 + x) = \log(2^x) = x\log 2 = (0.301)x$$

The next possibility is to guess, and you can check that $x = 0$ and $x = 1$ both satisfy the equation. The graphs of $y = 1 + x$ and $y = 2^x$ in Figure 1.46 show that there are no other solutions.

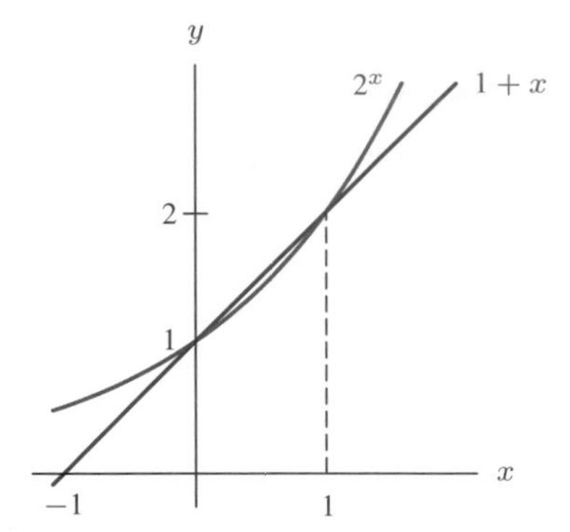

Figure 1.46: Graph showing solutions of $1 + x = 2^x$

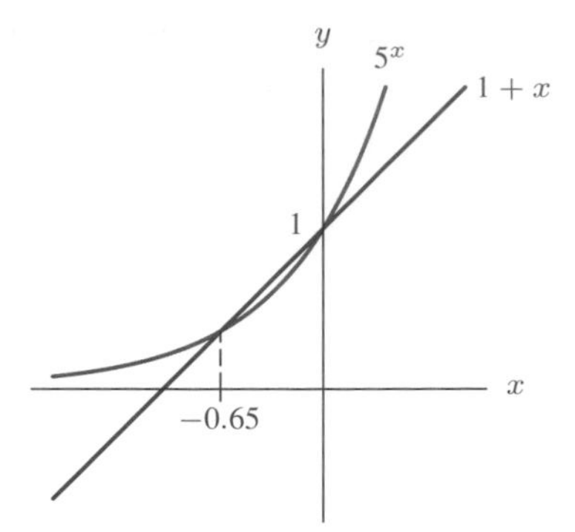

Figure 1.47: Graph showing solutions of $1 + x = 5^x$

(b) Logs are no help here either, and guessing isn't quite as easy. Again, $x = 0$ is a solution, and Figure 1.47 shows that there is another solution to the left of $x = 0$. By zooming in on the graph, or by trial and error, we find $x \approx -0.65$ is also a solution.

The Graph of log x

Example 6 Graph $f(x) = \log x$, and compare it to the graph of $g(x) = 10^x$.

Solution The values of $\log x$ in Table 1.19 were found using a calculator. Because no power of 10 gives 0, $\log 0$ is undefined. See Figure 1.48 for the graphs. You will notice that one of the big differences between $g(x) = 10^x$ and $f(x) = \log x$ is that the exponential function grows extremely quickly whereas the log function grows extremely slowly. Of course, these two facts are directly related. Since $\log x$ is the power of 10 you need to get x, and you only need a relatively small power of 10 to get a pretty large x, the log doesn't grow very fast. Indeed, to make $\log x$ large, you will need a gigantic value of x. However, albeit slowly, the log function does go to infinity as x increases.

The log function is not defined for $x \leq 0$ and has a vertical asymptote at $x = 0$. As x gets closer to 0 from the right (written $x \to 0^+$), the log graph drops toward $-\infty$. Check this on a calculator by finding log(0.1), log(0.01), log(0.001), and so on. In addition, you should notice that the log function crosses the x-axis at $x = 1$ because $10^0 = 1$, so $\log(1) = 0$.

TABLE 1.19 *Values for* $\log x$ *and* 10^x

x	$\log x$
0	undefined
1	0
2	0.3
3	0.5
4	0.6
⋮	⋮
10	1

x	10^x
0	1
1	10
2	100
3	10^3
4	10^4
⋮	⋮
10	10^{10}

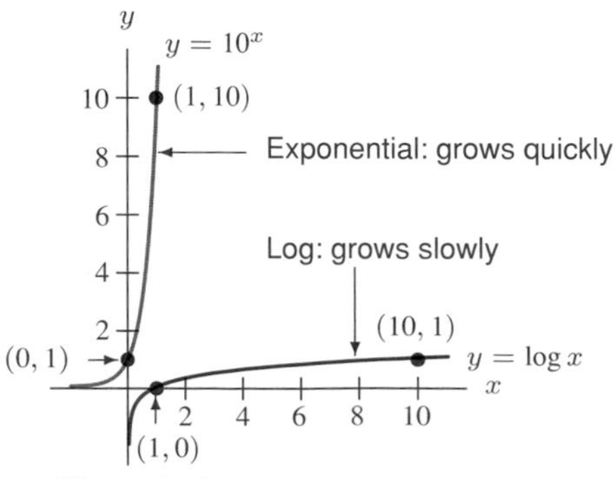

Figure 1.48: Graphs of $\log x$ and 10^x

Since $f(x) = \log x$ and $g(x) = 10^x$ are inverse functions, the graphs of the two functions are reflections of one another in the line $y = x$. (See Figure 1.48.) For example,

$$\log 10 = 1 \quad \text{so } (10, 1) \text{ is on the log graph}$$

and this means that

$$10^1 = 10 \quad \text{so } (1, 10) \text{ is on the exponential graph.}$$

The points $(10, 1)$ and $(1, 10)$ are reflections of one another in the line $y = x$. A similar argument shows that if (a, b) is on the exponential graph, (b, a) is on the log graph.

Comparison between Logarithms and Power Functions

There is a superficial resemblance between the graphs of $y = x^{1/3}$ and $y = \log x$, shown in Figure 1.49. But note the differences. The graph of $x^{1/3}$ includes the origin $(0, 0)$, whereas the graph of $\log x$ never reaches the vertical axis $x = 0$, which is an asymptote. Another difference is the rate of growth of the two functions for large x. The fact that $\log(1{,}000{,}000) = 6$ and $\log(10{,}000{,}000) = 7$ make it pretty clear that the logarithm climbs slowly as x increases beyond 1. You won't be surprised to discover, therefore, that the logarithm climbs slower than any positive power of x.

Figure 1.50 compares $x^{1/3}$ and $100 \log x$; Table 1.20 compares $x^{0.001}$ and $1000 \log x$. In fact, x^p dominates $A \log x$ for large x no matter what the values of $p > 0$ and $A > 0$.

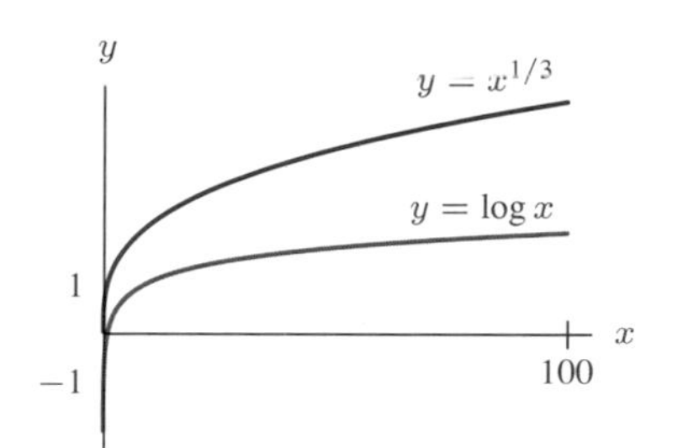

Figure 1.49: Comparison of $y = \log x$ and $y = x^{1/3}$

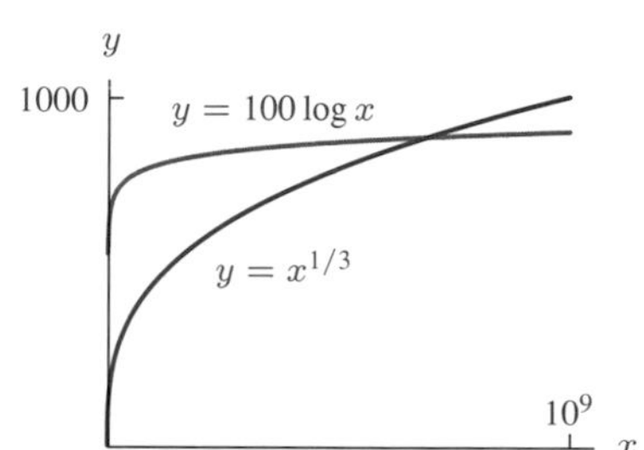

Figure 1.50: Comparison of $y = x^{1/3}$ and $y = 100 \log x$

TABLE 1.20 *Comparison of $x^{0.001}$ and $1000 \log x$*

x	$x^{0.001}$	$1000 \log x$
10^{5000}	10^5	$5 \cdot 10^6$
10^{6000}	10^6	$6 \cdot 10^6$
10^{7000}	10^7	$7 \cdot 10^6$

Problems for Section 1.6

1. Construct a table of values to compare the values of $f(x) = \log x$ and $g(x) = \sqrt{x}$ for $x = 1, 2, \ldots, 10$. Round to two decimal places. Use these values to graph both functions.

For Problems 2–3, plot a graph of the given function on a calculator or computer. Describe and explain what you see.

2. $y = \log 10^x$
3. $y = 10^{\log x}$

For Problems 4–11, solve for t using logs. (In Problems 4–11, you can check your answer by using a graphing calculator or computer if you like.)

4. $5^t = 7$
5. $2 = (1.02)^t$
6. $7 \cdot 3^t = 5 \cdot 2^t$
7. $P_0\, a^t = Q_0\, b^t$
8. $a = b^t$
9. $P = P_0\, a^t$
10. $Q = Q_0\, a^{nt}$
11. $5.02\,(1.04)^t = 12.01\,(1.03)^t$

For Problems 12–19, simplify the expression as much as possible.

12. $\log A^2 + \log B - \log A - \log B^2$

13. $\log(10^{x+7})$

14. $10^{\log A^2}$

15. $10^{2 \log Q}$

16. $10^{-\log P}$

17. $10^{-(\log B)/2}$

18. $\dfrac{\log A^2 - \log A}{\log B - \frac{1}{2}\log B}$

19. $2\log\alpha - 3\log B - \dfrac{\log\alpha}{2}$

20. Find the equation of the line l in Figure 1.51.

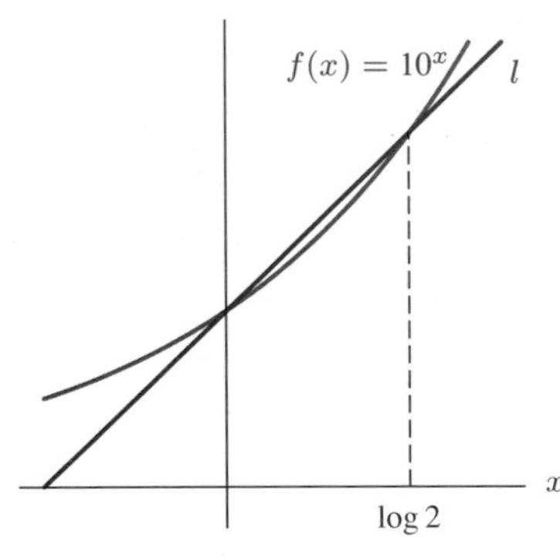

Figure 1.51

21. Find the inverse function of $p(t) = (1.04)^t$.
22. What is the doubling time of prices which are increasing by 5% a year?
23. In 1980, there were about 170 million vehicles (cars and trucks) and about 227 million people in the United States. If the number of vehicles was growing at 4% a year, while the population was growing at 1% a year, in what year was there, on average, one vehicle per person?
24. The population of a region is growing exponentially. If there were 40,000,000 people in 1980 ($t = 0$) and 56,000,000 in 1990, find an expression for the population at any time t. What would you predict for the year 2000? What is the doubling time?
25. (a) Find the doubling time, D, for annual growth rates, i%, of 2%, 3%, 4%, and 5%.
    (b) Since D decreases as i increases, we might guess that D is inversely proportional to i, that is, $D = k/i$. Use your answers to part (a) to confirm that, approximately, $D = 70/i$. This is the "Rule of 70" used by bankers. To compute the approximate doubling time of an investment, the banker divides 70 by the yearly interest rate.
26. The half-life of a certain radioactive substance is 12 days. If there are 10.32 grams initially:
    (a) Write an equation to determine the amount, A, of the substance as a function of time.
    (b) When will the substance be reduced to 1 gram?
27. Owing to an innovative rural public health program, infant mortality in Senegal, West Africa, is being reduced at a rate of 10% per year. How long will it take for infant mortality to be reduced by 50%?
28. Assume that the rate of inflation continues at the 1991 rate of 4.6% a year. If postage stamps go up, on average, at the rate of inflation, when will it cost $1 to mail a letter? [Note: That cost rose to 29 cents in 1990.]

1.7 THE NUMBER *e* AND NATURAL LOGARITHMS

All the calculations in the previous section were made with *common logarithms*, that is, logarithms to base 10. However, in science, the most frequently used base is the famous number $e = 2.71828\ldots$. In fact, this base is used so often that the logarithm to base e is called the *natural logarithm* and is denoted by "ln." You will find a ln button on most scientific calculators as well as an e^x button; that should be some indication of how important the base e is. At first glance, this is all somewhat mysterious. What can possibly be natural about using logarithms to the base 2.71828? The full answer to that question must wait until Chapter 4, where we show that many calculus formulas come out neater when e is used as the base rather than any other base.

An exponential function can be expressed either using base e or using any other base a (provided $a > 0$, $a \neq 1$). However, the base e is generally used whenever calculus is involved. Logarithms can be defined for any base a (with $a > 1$), although most calculators contain logs only to base 10 and e. We will only use logs to these two bases.

Definition and Properties of the Natural Logarithm

The natural logarithm of x, written $\ln x$, is defined to be the inverse function of e^x:

$$\ln x = \log_e x = c \quad \text{means} \quad e^c = x$$

so

$\ln x$ is the power of e needed to get x.

The rules for manipulating natural logs are similar to those for logs to base 10, and $\ln x$ is not defined when x is negative or 0.

Rules For Computing Using Natural Logarithms

1. $\ln(AB) = \ln A + \ln B$
2. $\ln\left(\frac{A}{B}\right) = \ln A - \ln B$
3. $\ln(A^p) = p\ln A$
4. $\ln e^x = x$
5. $e^{\ln x} = x$

In addition, $\ln 1 = 0$ because $e^0 = 1$.

Using the LN button on a calculator to plot a graph of $f(x) = \ln x$ for $0 < x \leq 10$, we get Figure 1.52. Observe that the graph of $y = \ln x$, like the graph of $y = \log x$, climbs very slowly as x increases.

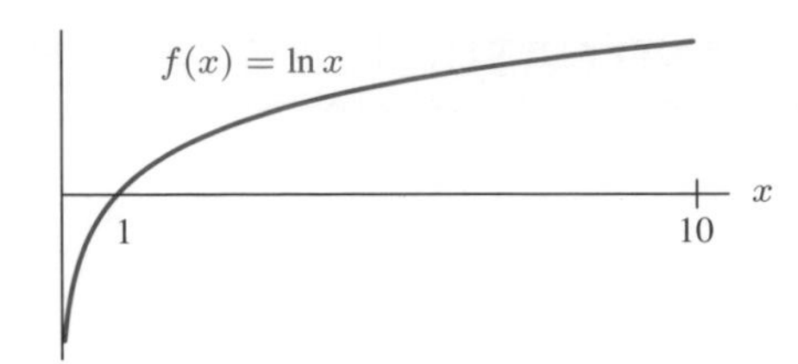

Figure 1.52: Graph of the natural logarithm

In Section 1.3, we saw the family of exponential functions

$$P = P_0 a^t,$$

where P_0 is the initial value of P and a is the growth factor. The case $a > 1$ represents exponential growth; $0 < a < 1$ represents exponential decay. For any positive number a, we can write

$$a = e^k$$

for $k = \ln a$. If $a > 1$, k is positive, and if $0 < a < 1$, k is negative. Thus, the function representing an exponentially growing population can be rewritten as

$$P = P_0 a^t = P_0 (e^k)^t = P_0 e^{kt}$$

with k positive. In the case when $0 < a < 1$, we can use another positive constant, k, and write

$$a = e^{-k}.$$

Thus if Q is a quantity that is decaying exponentially and Q_0 is the initial quantity, at time t we will have

$$Q = Q_0 a^t = Q_0 (e^{-k})^t = Q_0 e^{-kt} = \frac{Q_0}{e^{kt}}.$$

Since e^{kt} is now in the denominator, Q will decrease as time goes on—as you'd expect if Q is decaying.

> Any **exponential growth** function can be written in the form
>
> $$P = P_0 e^{kt}$$
>
> and any **exponential decay** function can be written as
>
> $$Q = Q_0 e^{-kt}$$
>
> where P_0 and Q_0 are the initial quantities and k is positive.
>
> We say that P and Q are growing or decaying at a *continuous rate* of k. (Note that, for example, $k = 0.02$ corresponds to a continuous growth rate of 2%.)

The reason that k is called the continuous rate will be explained in Section 1.8, and will be explored in detail in Chapter 9.

Example 1 Sketch the graphs of $P = e^{0.5t}$ and $Q = e^{-0.2t}$.

Solution

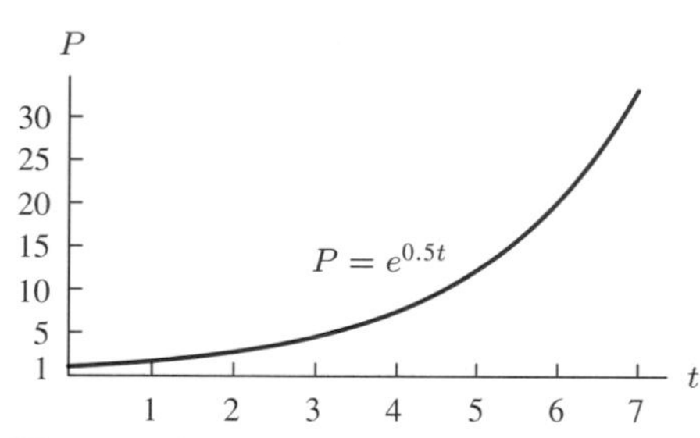

Figure 1.53: An exponential growth function

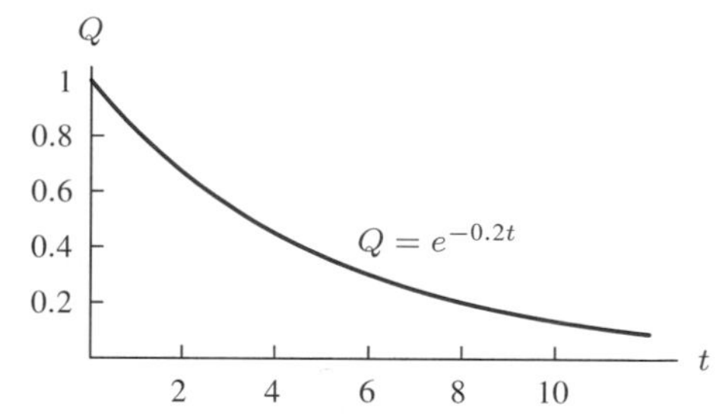

Figure 1.54: An exponential decay function

The graph of $P = e^{0.5t}$ is in Figure 1.53. Notice that the graph is the same shape as the previous exponential growth curves: increasing and concave up. Indeed, the families a^t, $a > 1$, and e^{kt}, $k > 0$, are one and the same. The graph of $Q = e^{-0.2t}$ is in Figure 1.54; it too has the same shape as other exponential decay functions. The families a^t, $0 < a < 1$, and e^{kt}, $k < 0$, are one and the same.

Example 2 The release of chlorofluorocarbons used in air conditioners and, to a lesser extent, in household sprays (hair spray, shaving cream, etc.) destroys the ozone in the upper atmosphere. At the present time, the amount of ozone, Q, is decaying exponentially at a continuous yearly rate of 0.25%. What is the half-life of ozone? In other words, at this rate, how long will it take for half the ozone to disappear?

Solution If Q_0 is the initial quantity of ozone, then

$$Q = Q_0 e^{-0.0025t}.$$

We want to find T, the value of t making $Q = Q_0/2$, so

$$\frac{Q_0}{2} = Q_0 e^{-0.0025T}.$$

Taking natural logs yields

$$-0.0025T = \ln\left(\frac{1}{2}\right) = -0.6931,$$

so

$$T \approx 277 \text{ years.}$$

In Example 2 the decay rate was given. However, in many situations where we expect to find exponential growth or decay, the rate is not given. To find it, we must know the quantity at two different times and then solve for the growth rate, as in the next example.

Example 3 The population of Kenya was 19.5 million in 1984 and 21.2 million in 1986. Assuming it increases exponentially, find a formula for the population of Kenya as a function of time.

Solution If we measure the population, P, in millions and time, t, in years since 1984, we can say

$$P = P_0 e^{kt} = 19.5e^{kt}$$

where $P_0 = 19.5$ is the initial value of P. We find k by using the fact that $P = 21.2$ when $t = 2$, so

$$21.2 = 19.5e^{k \cdot 2}.$$

To find k, we divide both sides by 19.5, giving

$$\frac{21.2}{19.5} = 1.087 = e^{2k}.$$

Now take natural logs of both sides:

$$\ln(1.087) = \ln(e^{2k}).$$

Using a calculator and the fact that $\ln(e^{2k}) = 2k$, this becomes

$$0.0834 = 2k.$$

So

$$k \approx 0.042,$$

and therefore,

$$P = 19.5e^{0.042t}.$$

Since $k = 0.042 = 4.2\%$, we say the population of Kenya was growing continuously at 4.2% a year.

Relationship between a^t and e^{kt}

In Example 3 we chose to use e for the base of the exponential function representing Kenya's population, making clear that the continuous growth rate was 4.2%. If, however, we had wanted to emphasize the annual growth rate, we could have expressed the exponential function in the form

$$P = P_0 a^t.$$

Since the population grew from 19.5 to 21.2 million in 2 years, we know that

$$21.2 = 19.5a^2$$

so

$$a = \left(\frac{21.2}{19.5}\right)^{(1/2)} \approx 1.043.$$

Thus,

$$P = P_0(1.043)^t.$$

Notice that $a \approx 1.043$ corresponds to an annual growth rate of about 4.3%, which is just slightly more than the continuous growth rate of 4.2% found in Example 3.

In general, an exponential function of the form $P_0 e^{kt}$ can always be written in the form $P_0 a^t$, and vice versa, simply by letting $e^k = a$ or $k = \ln a$. The two different formulas, $P = P_0 e^{kt}$ and $P = P_0 a^t$, have the same graph and represent the same function.

To convert between a^t and e^{kt} use:

$$a^t = (e^k)^t = e^{kt} \quad \text{where} \quad k = \ln a$$

If a comes from a percentage growth rate r, that is, $a = 1 + r$, then the continuous growth rate $k = \ln(1 + r)$ will be slightly less than, but very close to, r, provided r is small.

Example 4 In Section 1.3, we saw that radioactive decay of carbon-14 could be modeled by the function

$$C = C_0 \left(\frac{1}{2}\right)^{(t/5730)}$$

where C is the quantity of carbon-14 at time t, C_0 is the initial quantity, and the half-life of carbon-14 is 5730 years. Express this function in terms of e.

Solution We want to rewrite the function as

$$C = C_0 \left(\frac{1}{2}\right)^{(t/5730)} = C_0 e^{kt}.$$

Canceling C_0, we have

$$\left(\frac{1}{2}\right)^{(t/5730)} = e^{kt}.$$

Taking natural logs of both sides gives

$$\frac{t}{5730} \ln\left(\frac{1}{2}\right) = \ln(e^{kt}) = kt.$$

Canceling the t's gives

$$k = \frac{1}{5730} \ln\left(\frac{1}{2}\right) = -0.000121.$$

Thus,

$$C = C_0 e^{-0.000121t}.$$

Problems for Section 1.7

1. Construct a table of values to compare the values of $f(x) = \log x$ and $g(x) = \ln x$ for $x = 1, 2, \ldots, 10$. Round to two decimal places. Use these values to graph both functions.

For Problems 2–5, find the indicated quantities.

2. $3 \ln e + \ln(1/e)$
3. $\ln(e^2) + e^{-\ln e}$
4. $3 \ln(e \ln e) + \ln(\ln e)$
5. $e^{-\ln \sqrt{e}}$

Simplify the expressions in Problems 6–9 completely.

6. $2 \ln A - 3 \ln B + \ln(AB)$
7. $e^{2 \ln A - (\ln B)/2}$
8. $\ln(xe^{-\ln x})$
9. $\ln(e^2 \ln(e \ln e))$

Solve for x in Problems 10–13.

10. $2^x = e^{x+1}$

11. $2e^{3x} = 4e^{5x}$

12. $4e^{2x-3} - 5 = e$

13. $10^{x+3} = 5e^{7-x}$

Convert the functions in Problems 14–18 into the form $P = P_0 a^t$. Which represent exponential growth and which represent exponential decay?

14. $P = P_0 e^{0.2t}$

15. $P = 10e^{0.917t}$

16. $P = P_0 e^{-0.73t}$

17. $P = 79e^{-2.5t}$

18. $P = 7e^{-\pi t}$

Convert the functions in Problems 19–22 into the form $P = P_0 e^{kt}$.

19. $P = P_0 2^t$

20. $P = 10(1.7)^t$

21. $P = 5.23(0.2)^t$

22. $P = 174(0.9)^t$

For Problems 23–24, plot a graph of the given function on a calculator or computer. Describe and explain what you see.

23. $y = \ln e^x$

24. $y = e^{\ln x}$

25. (a) Evaluate the quantity $\ln x / \log x$ for several values of x (say $x = 0.5, 5, 10, 100$). What do you notice?
    (b) Plot a graph of $y = \ln x / \log x$ on a computer or calculator. What do you see? How does it relate to your results in part (a) ?

For Problems 26–29, solve for t.

26. $a = be^t$

27. $P = P_0\, e^{kt}$

28. $ae^{kt} = e^{bt}$, where $k \neq b$

29. $ce^{-\alpha t} = be^{-\gamma t/n}$, where $\alpha n \neq \gamma$

30. Find the inverse function of $f(t) = 50e^{0.1t}$.

31. Define $f(x) = \dfrac{1}{1+e^{-x}}$.
    (a) Is f increasing or decreasing?
    (b) Explain why f is invertible, and find a formula for $f^{-1}(x)$.
    (c) What is the domain of f^{-1}?
    (d) Sketch the graphs of f and f^{-1} on the same axes, and explain their relationship.

32. (a) A population grows according to the equation $P = P_0 e^{kt}$ (with P_0, k constants). Find the population as a function of time, t, if it grows at a continuous rate of 2% a year and starts at 1 million.
    (b) Plot a graph of the population you found in part (a) against time.

33. Air pressure, P, decreases exponentially with the height above the surface of the earth, h:

$$P = P_0 e^{-1.2\times 10^{-4} h}$$

where P_0 is the air pressure at sea level and h is in meters.

(a) If you go to the top of Mount McKinley, height 6,198 meters (about 20,320 feet), what is the air pressure, as a percent of the pressure at sea level?

(b) The maximum cruising altitude of an ordinary commercial jet is around 12,000 meters (about 40,000 feet). At that height, what is the air pressure, as a percent of the sea level value?

34. Under certain circumstances, the velocity, V, of a falling raindrop is given by

$$V = V_0(1 - e^{-t}),$$

where t is time and V_0 is a positive constant.

(a) Sketch a rough graph of V against t, for $t \geq 0$.
(b) What does V_0 represent?

35. The air in a factory is being filtered so that the quantity of a pollutant, P (measured in mg/liter), is decreasing according to the equation $P = P_0e^{-kt}$, where t represents time in hours . If 10% of the pollution is removed in the first five hours:

(a) What percentage of the pollution is left after 10 hours?
(b) How long will it take before the pollution is reduced by 50%?
(c) Plot a graph of pollution against time. Show the results of your calculations on the graph.
(d) Explain why the quantity of pollutant might decrease in this way.

36. The population, P, in millions, of Nicaragua was 3.6 million in 1990 and growing at 3.4% per year. Let t be time in years since 1990.

(a) Express P as a function in the form $P = P_0a^t$.
(b) Express P as an exponential function using the base e.
(c) Compare the annual and continuous growth rates.

37. One of the main contaminants of a nuclear accident, such as that at Chernobyl, is strontium-90, which decays exponentially at a continuous rate of approximately 2.47% per year. Preliminary estimates after the Chernobyl disaster suggested that it would be about 100 years before the region would again be safe for human habitation. What percent of the original strontium-90 would still remain by this time?

38. The quantity, Q, of radioactive carbon-14 remaining t years after an organism dies is given by the formula

$$Q = Q_0e^{-0.000121t},$$

where Q_0 is the initial quantity.

(a) A skull uncovered at an archeological dig has 15% of the original amount of carbon-14 present. Estimate its age.
(b) Show how you can calculate the half-life of carbon-14 from this equation.

39. A Vermeer (1632–1675) picture contains 99.5% of its carbon-14 (half-life 5730 years). From this information can you determine whether or not the picture is a fake? Explain your reasoning.

40. Geological dating of rocks is done using potassium-40 rather than carbon-14 because potassium has a longer half-life. The potassium decays to argon, which remains trapped in the rocks and can be measured; thus the original quantity of potassium can be calculated. The half-life of potassium-40 is $1.28 \cdot 10^9$ years. Find a formula giving the quantity, P, of potassium-40 remaining as a function of time in years, assuming the initial quantity is P_0:

(a) Using base 1/2
(b) Using base e

1.8 NOTES ON COMPOUND INTEREST

If you have some money, you may decide to invest it to earn interest. The interest can be paid in many different ways — for example, once a year or many times a year. If the interest is paid more frequently than once per year and the interest is not withdrawn, there is a benefit to the investor since the interest earns interest. This effect is called *compounding*. You may have noticed banks offering accounts that differ both in interest rates and in compounding methods. Some offer interest compounded annually, some quarterly, and others daily. Some even offer continuous compounding.

What is the difference between a bank account advertising 8% compounded annually (once per year) and one offering 8% compounded quarterly (four times per year)? In both cases 8% is an annual rate of interest. The expression 8% *compounded annually* means that at the end of each year, 8% of the current balance is added. This is equivalent to multiplying the current balance by 1.08. Thus, if \$100 is deposited, the balance, B, in dollars, will be

$$\begin{aligned} B &= 100(1.08) && \text{after one year,} \\ B &= 100(1.08)^2 && \text{after two years,} \\ B &= 100(1.08)^t && \text{after } t \text{ years.} \end{aligned}$$

The expression 8% *compounded quarterly* means that interest is added four times per year (every three months) and that $\frac{8}{4} = 2\%$ of the current balance is added each time. Thus, if \$100 is deposited, at the end of one year, four compoundings will have taken place and the account will contain $\$100(1.02)^4$. Thus, the balance, B, in dollars, will be

$$\begin{aligned} B &= 100(1.02)^4 && \text{after one year,} \\ B &= 100(1.02)^8 && \text{after two years,} \\ B &= 100(1.02)^{4t} && \text{after } t \text{ years.} \end{aligned}$$

Note that 8% is *not* the rate used for each three month period; the annual rate is divided into four 2% payments. Calculating the total balance after one year under each method shows that

$$\begin{aligned} \text{Annual compounding:} \quad & B = 100(1.08) = 108.00 \\ \text{Quarterly compounding:} \quad & B = 100(1.02)^4 = 108.24 \end{aligned}$$

Thus, more money is earned from quarterly compounding, because the interest earns interest as the year goes by. In general, the more often interest is compounded, the more money will be earned (although the increase may not be very large).

We can measure the effect of compounding by introducing the notion of *effective annual yield*. Since \$100 invested at 8% compounded quarterly grows to \$108.24 by the end of one year, we say that the *effective annual yield* in this case is 8.24%. We now have two interest rates which describe the same investment: the 8% compounded quarterly and the 8.24% effective annual yield. Banks call the 8% the *annual percentage rate*, or *APR*. We may also call the 8% the *nominal rate* (nominal means "in name only"). However, it is the effective yield which tells you exactly how much interest the investment really pays. Thus, to compare two bank accounts, simply compare the effective annual yields. The next time that you walk by a bank, look at the advertisements, which should (by law) include both the APR, or nominal rate, and the effective annual yield. We will often abbreviate *annual percentage rate* to *annual rate*.

Using the Effective Annual Yield

Example 1 Which is better: Bank X paying a 7% annual rate compounded monthly or Bank Y offering a 6.9% annual rate compounded daily?

Solution We will find the effective annual yield for each bank.

Bank X: There are 12 interest payments in a year, each payment being $0.07/12 = 0.005833$ times the current balance. If the initial deposit were \$100, then the balance B will be

$$
\begin{aligned}
B &= 100(1.005833) \quad \text{after one month,} \\
B &= 100(1.005833)^2 \quad \text{after two months,} \\
B &= 100(1.005833)^t \quad \text{after } t \text{ months.}
\end{aligned}
$$

To find the effective annual yield, we need to look at one year, or 12 months, which gives $B = 100(1.005833)^{12} = 100(1.072286)$, so the effective annual yield $\approx 7.23\%$.

Bank Y: There are 365 interest payments in a year (assuming it is not a leap year), each being $0.069/365 = 0.000189$ times the current balance. Then the balance B in the account is

$$
\begin{aligned}
B &= 100(1.000189) \quad \text{after one day,} \\
B &= 100(1.000189)^2 \quad \text{after two days,} \\
B &= 100(1.000189)^t \quad \text{after } t \text{ days.}
\end{aligned}
$$

so at the end of one year we have multiplied the initial deposit by

$$(1.000189)^{365} = 1.071413$$

so the effective annual yield for Bank Y $\approx 7.14\%$.

Comparing effective annual yields for the banks, we see that Bank X is offering a better investment, by a small margin.

Example 2 If \$1000 is invested in each bank in Example 1, write an expression for the balance in each bank after t years.

Solution For Bank X, the effective annual yield $\approx 7.23\%$, so after t years the balance, in dollars, will be

$$B = 1000(1.0723)^t.$$

For Bank Y, the effective annual yield $\approx 7.14\%$, so after t years the balance, in dollars, will be

$$B = 1000(1.0714)^t.$$

(Again, we are ignoring leap years.)

If interest at an annual rate of r is compounded n times a year, then r/n times the current balance is added n times a year. Thus, with an initial deposit of \$$P$, the balance t years later is

$$B = P\left(1 + \frac{r}{n}\right)^{nt}$$

Note that r is the nominal rate and that, for example, $r = 0.05$ when the annual rate is 5%.

Increasing the Frequency of Compounding: Continuous Compounding

---

Example 3 Find the effective annual yield for a 7% annual rate compounded
(a) 1000 times a year. (b) 10,000 times a year.

Solution (a)

$$\left(1+\frac{0.07}{1000}\right)^{1000} \approx 1.0725056$$

giving an effective annual yield of about 7.25056%.

(b)

$$\left(1+\frac{0.07}{10{,}000}\right)^{10{,}000} \approx 1.0725079$$

giving an effective annual yield of about 7.25079%.

You can see that there's not a great deal of difference between compounding 1000 times each year (about three times per day) and 10,000 times each year (about 30 times per day). What happens if we compound more often still? Every minute? Every second? You may be surprised that the effective annual yield does not increase indefinitely, but tends to a finite value. The benefit of increasing the frequency of compounding becomes negligible beyond a certain point.

For example, if you were to compute the effective annual yield on a 7% investment compounded n times per year for values of n larger than 100,000, you would find that

$$\left(1+\frac{0.07}{n}\right)^{n} \approx 1.0725082.$$

So the effective annual yield is about 7.25082%. Even if you take $n = 1{,}000{,}000$ or $n = 10^{10}$, the effective annual yield will not change appreciably. The value 7.25082% is an upper bound which is approached as the frequency of compounding increases.

When the effective annual yield is at this upper bound, we say that the interest is being *compounded continuously*. (The word *continuously* is used because the upper bound is approached by compounding more and more frequently.) Thus, when a 7% nominal annual rate is compounded so frequently that the effective annual yield is 7.25082%, we say that the 7% is compounded *continuously*. This represents the most one can get from a 7% nominal rate.

Where Does the Number e Fit In?

It turns out that e is intimately connected to continuous compounding. To see this, use your calculator to check that $e^{0.07} \approx 1.0725082$, which is the same number we obtained when we compounded 7% a large number of times. So you have discovered that for very large n

$$\left(1+\frac{0.07}{n}\right)^{n} \approx e^{0.07}.$$

As n gets larger, the approximation gets better and better, which we write as

$$\left(1+\frac{0.07}{n}\right)^{n} \longrightarrow e^{0.07}$$

meaning that as n increases, the value of $\left(1+0.07/n\right)^n$ gets closer and closer to $e^{0.07}$.

Thus, if $\$P$ is deposited at an annual rate of 7% compounded continuously, the balance, $\$B$, will be

$$\begin{aligned} B &= P(1.0725082) = Pe^{0.07} && \text{after one year,} \\ B &= P(1.0725082)^2 = P\left(e^{0.07}\right)^2 = Pe^{(0.07)2} && \text{after two years,} \\ B &= P(1.0725082)^t = P\left(e^{0.07}\right)^t = Pe^{0.07t} && \text{after } t \text{ years.} \end{aligned}$$

If interest on an initial deposit of $\$P$ is *compounded continuously* at an annual rate r, the balance t years later can be calculated using the formula

$$B = Pe^{rt}.$$

Again, r is the nominal rate, and, for example, $r = 0.05$ when the annual rate is 5%.

In solving a problem involving compound interest, it is important to be clear whether interest rates are nominal rates or effective yields, as well as whether compounding is continuous or not.

Example 4 Find the effective annual yield of a 6% annual rate, compounded continuously.

Solution In one year, an investment of P becomes $Pe^{0.06}$. Using a calculator, we see that

$$Pe^{0.06} = P(1.0618365)$$

So the effective annual yield is about 6.18%.

Example 5 Suppose you want to invest money in a certificate of deposit (CD) for your child's education. You need it to be worth \$12,000 in 10 years. How much should you invest if the CD pays interest at a 9% annual rate compounded quarterly? Continuously?

Solution Suppose you invest $\$P$ initially. A 9% annual rate compounded quarterly has an effective annual yield given by $(1+0.09/4)^4 = 1.0930833$, or 9.30833%. So after 10 years you will have

$$P(1.0930833)^{10} = 12000.$$

Thus, you should invest

$$P = \frac{12000}{(1.0930833)^{10}} = \frac{12000}{2.4351885} = 4927.75$$

On the other hand, if the CD pays 9% compounded continuously, after 10 years you will have

$$Pe^{(0.09)10} = 12000.$$

So you would need to invest

$$P = \frac{12000}{e^{(0.09)10}} = \frac{12000}{2.4596031} = 4878.84$$

Notice that to achieve the same result, continuous compounding requires a smaller initial investment than quarterly compounding. This is to be expected since the effective annual yield is higher for continuous than for quarterly compounding.

Problems for Section 1.8

1. Use a graph of $y = \left(1 + 0.07/x\right)^x$ to find the value of $\left(1 + 0.07/x\right)^x$ as $x \to \infty$. Confirm that the value you get is $e^{0.07}$.

2. If you deposit \$10,000 in an account earning interest at an 8% annual rate compounded continuously, how much money is in the account after five years?

3. (a) Find the effective annual yield for a 5% annual interest rate compounded
   (i) 1000 times/year, (ii) 10,000 times/year, (iii) 100,000 times/year.
   (b) Look at the sequence of answers in part (a), and deduce the effective annual yield for a 5% annual rate compounded continuously.
   (c) Compute $e^{0.05}$. How does this confirm your answer to part (b)?

4. (a) Find $\left(1 + 0.04/n\right)^n$ for $n = 10{,}000$, and 100,000, and 1,000,000. Use the results to deduce the effective annual yield of a 4% annual rate compounded continuously.
   (b) Confirm your answer by computing $e^{0.04}$.

5. Use the number e to find the effective annual yield of a 6% annual rate, compounded continuously.

6. A bank account is earning interest at 6% per year compounded continuously.
   (a) By what percentage has the bank balance in the account increased over one year? (This is the effective annual yield.)
   (b) How long does it take the balance to double?
   (c) Assuming now that the interest rate is i, find a formula giving the doubling time in terms of the interest rate.

7. Suppose \$1000 is invested at 6% annual interest compounded continuously.
   (a) How long will it take for the investment to double?
   (b) Use your answer to part (a) to express the value of the investment after t years in terms of a base 2 exponential function.

8. What is the effective annual yield of an investment paying at a 12% annual rate, compounded continuously?

9. (a) The Banque Nationale du Zaïre pays 100% nominal interest on deposits, compounded monthly. You invest 1 million zaïre. (The "zaïre" is the unit of currency of the Republic of Zaïre.) How much money do you have after one year?
   (b) How much money do you have after one year if you invest 1 million zaïre with interest compounded (i) Daily? (ii) Hourly? (iii) Each minute?
   (c) Does this amount increase without bound as interest is compounded more and more often, or does it level off? If it levels off, provide a close "upper" estimate for the total after one year.

10. Explain how you can match the interest rates (a)–(e) with the effective annual yields (I)–(V) without doing any calculations.

(a) 5.5% annual rate, compounded continuously.	(I) 5%
(b) 5.5% annual rate, compounded quarterly.	(II) 5.06%
(c) 5.5% annual rate, compounded weekly.	(III) 5.61%
(d) 5% annual rate, compounded yearly.	(IV) 5.651%
(e) 5% annual rate, compounded twice a year.	(V) 5.654%

11. When you rent an apartment, you are often required to give the landlord a security deposit which is returned if you leave the apartment undamaged. In Massachusetts the landlord is required to pay the tenant interest on the deposit once a year, at a 5% annual rate, compounded annually. The landlord, however, may invest the money at a higher (or lower) interest rate. Suppose the landlord invests a $1000 deposit at an annual rate of
(a) 6%, compounded continuously (b) 4%, compounded continuously.
In each case, determine the net gain or loss by the landlord at the end of the first year. (Give your answer to the nearest cent.)

12. The newspaper article below is from *The New York Times*, May 27, 1990. Fill in the three blanks. (For the last blank, assume the interest has been compounded yearly, and give your answer in dollars. Exclude the occurrence of leap years.)

213 Years After Loan, Uncle Sam Is Dunned

By LISA BELKIN

Special to The New York Times

SAN ANTONIO, May 26 — More than 200 years ago, a wealthy Pennsylvania merchant named Jacob DeHaven lent $450,000 to the Continental Congress to rescue the troops at Valley Forge. That loan was apparently never repaid.

So Mr. DeHaven's descendants are taking the United States Government to court to collect what they believe they are owed. The total: ____ in today's dollars if the interest is compounded daily at 6 percent, the going rate at the time. If compounded yearly, the bill is only ____.

Family Is Flexible

The descendants say that they are willing to be flexible about the amount of a settlement and that they might even accept a heartfelt thank you or perhaps a DeHaven statue. But they also note that interest is accumulating at ____ a second.

1.9 NEW FUNCTIONS FROM OLD

Shifts and Stretches

The graph of a constant multiple of a given function is easy to visualize: each y value is stretched or shrunk by that multiple. For example, consider the function $f(x)$ and its multiples $y = 3f(x)$ and $y = -2f(x)$, whose graphs are in Figure 1.55. The factor 3 in the function $y = 3f(x)$ stretches each $f(x)$ value by multiplying it by 3; the factor -2 in the function $y = -2f(x)$ stretches $f(x)$ by multiplying by 2 and reflecting it about the x-axis. You can think of the multiples of a given function as a family of functions.

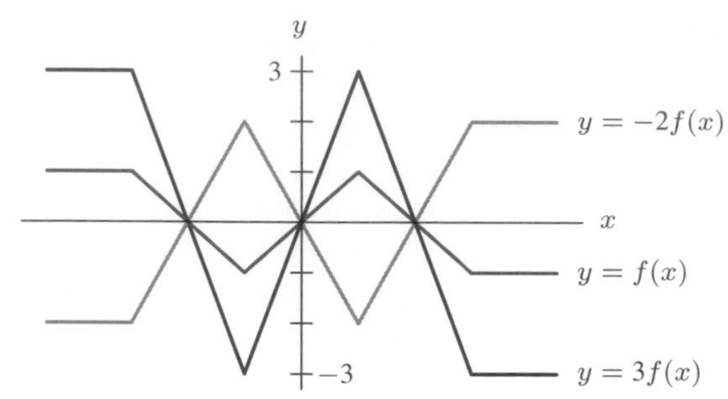

Figure 1.55: The family of multiples of the function $f(x)$

It is also easy to create families of functions by shifting graphs. For example, $y - 4 = x^2$ is the same as $y = x^2 + 4$, which is the graph of $y = x^2$ shifted up by 4, and $y = (x - 2)^2$ is the graph of $y = x^2$ shifted right by 2. (See Figure 1.56.)

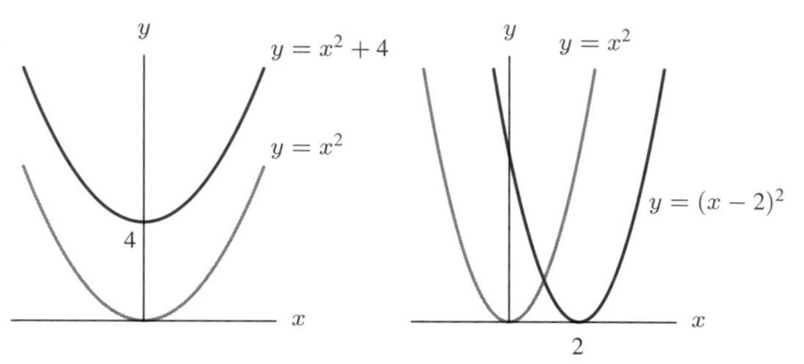

Figure 1.56: Graphs of $y = x^2$ with $y = x^2 + 4$ and $y = (x - 2)^2$

In general:

- Multiplying a function by a constant stretches or shrinks its graph vertically; a negative sign reflects the graph about the x-axis.
- Replacing y by $(y - k)$ moves a graph up by k (down if k is negative).
- Replacing x by $(x - h)$ moves a graph to the right by h (to the left if h is negative).

Sums of Functions

The right-hand graph in Figure 1.57 shows the number of US students majoring in science and engineering, broken down into men and women.[6] The top line represents the total; the upper curve represents the women, and the bottom curve represents the men. It is really a picture of three functions: the number of men majoring in science and engineering, $m(t)$, the number of women majoring in science and engineering, $w(t)$, and the total number of students, $n(t)$, where t is the year. The men's and women's graphs are shown separately to the left.

[6]Data from the National Science Foundation, Washington D.C.

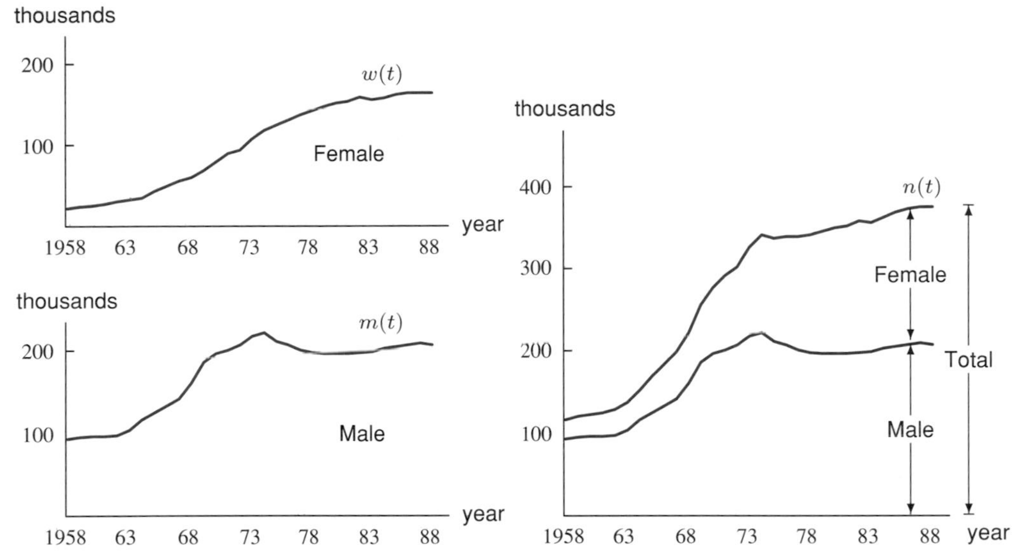

Figure 1.57: Numbers of US students majoring in science and engineering (1958–1988)

Now

$$\text{Total number} = \text{Number of men} + \text{Number of women}$$

so, in functional notation,

$$n(t) = m(t) + w(t).$$

Figure 1.57 makes it clear that the graph of the sum function is obtained by stacking the other graphs one on top of the other. Let's look at the same idea in a more mathematical setting.

Example 1 Graph the function $y = 2x^2 + 1/x$ for $x > 0$.

Solution We graph both of the functions $y = 2x^2$ and $y = 1/x$ separately first. Now imagine the corresponding y values stacked on top of one another, and you will get the graph of $y = 2x^2 + 1/x$. (See Figure 1.58.) Notice that, for x near 0, the graph of $y = 2x^2 + 1/x$ looks much like the graph of $y = 1/x$ (because $1/x$ is much larger than $2x^2$). For large x, the graph of $y = 2x^2 + 1/x$ looks like the graph of $y = 2x^2$.

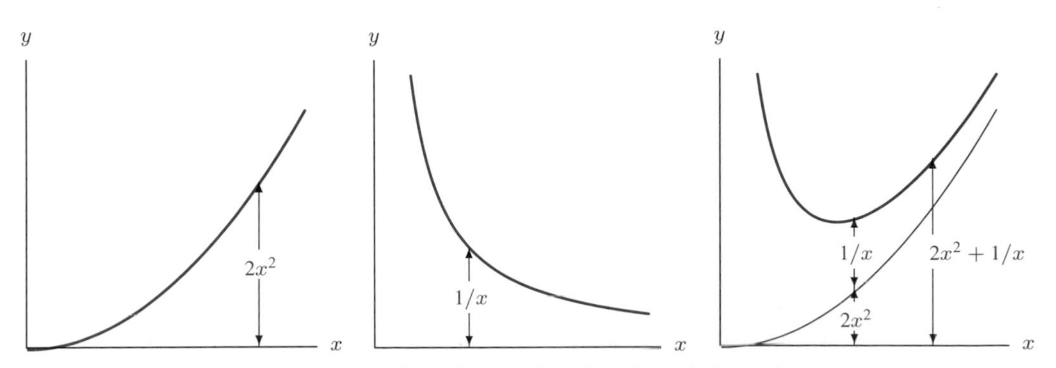

Figure 1.58: Summing two functions from their graphs

Composite Functions

If oil is spilled from a tanker, the area of the oil slick will grow with time. Suppose that the oil slick is always a perfect circle. (In practice, this does not happen because of winds and tides and the location of the coastline.) The area of the slick is a function of its radius

$$A = f(r) = \pi r^2.$$

The radius is also a function of time, because the radius increases as more oil spills. Thus, the area, being a function of the radius, is also a function of time. If, for example, the radius is given by

$$r = g(t) = 1 + t$$

then area is given as a function of time by substitution

$$A = \pi r^2 = \pi(1+t)^2.$$

We say that A is a *composite function* or a "function of a function," which is written

$$A = \underbrace{f(g(t))}_{\substack{\text{Composite function;} \\ f \text{ is outside function} \\ g \text{ is inside function}}} = \pi(g(t))^2 = \pi(1+t)^2.$$

Here the *inside function*, g, represents the calculation which is done first, and the *outside function*, f, represents the calculation done second. Look at the example, and think about how you would calculate A using the formula $\pi(1+t)^2$. For any given t, the first step is to find $1+t$, and the second step is to square and multiply by π. So the first step corresponds to the inside function $g(t) = 1+t$, and the second step corresponds to the outside function $f(r) = \pi r^2$.

Example 2 Express each of the following functions as a composition:
(a) $h(t) = (1+t^3)^{27}$ (b) $k(x) = \log x^2$ (c) $l(x) = \log^2 x$ (d) $n(y) = e^{-y^2}$

Solution In each case think about how you would calculate a value of the function. The first stage of the calculation will give you the inside function, and the second stage will give you the outside one.

(a) The first stage is cubing and adding 1, so the inside function is $g(t) = 1+t^3$. The second stage is taking the 27th power, so the outside function is $f(y) = y^{27}$. Then

$$f(g(t)) = f(1+t^3) = (1+t^3)^{27}.$$

In fact, there are lots of different answers: $g(t) = t^3$ and $f(y) = (1+y)^{27}$ is another possibility.

(b) By convention, $\log x^2$ means $\log(x^2)$, so evaluating this function involves squaring x first and then taking the log. So if $g(x) = x^2$ is the inside and $f(y) = \log y$ is the outside, then $f(g(x)) = \log x^2$.

(c) By convention, $\log^2 x$ means $(\log x)^2$, so this function involves taking the log first and then squaring. Using the same definitions of f and g as in part (b), namely $f(x) = \log x$ and $g(t) = t^2$, the composition is $g(f(x)) = (\log x)^2$. By evaluating the functions in parts (b) and (c) for $x = 2$, say, giving $\log(2^2) = 0.602$, and $\log^2(2) = 0.091$, you can see that the order in which you compose two functions certainly can make a difference.

(d) To calculate e^{-y^2} you square y, take its negative, and then take e to that power. So $g(y) = -y^2$ and $f(z) = e^z$. Then $f(g(y)) = e^{-y^2}$. Alternatively, you could take $g(y) = y^2$ and $f(z) = e^{-z}$.

Odd and Even Functions

There is a certain symmetry apparent in the graphs of $f(x) = x^2$ and $g(x) = x^3$ in Figure 1.59. Namely, for each point (x, x^2) on the graph of f, the point $(-x, x^2)$ is also on the graph; and for each point (x, x^3) on the graph of g, the point $(-x, -x^3)$ is also on the graph (see Figure 1.59). The graph of $f(x) = x^2$ is symmetric in the y-axis, whereas the graph of $g(x) = x^3$ is symmetric about the origin. We say that the squaring function is *even* and the cubing function is *odd*. The names come from the fact that the even powers are even functions and odd powers are odd functions.

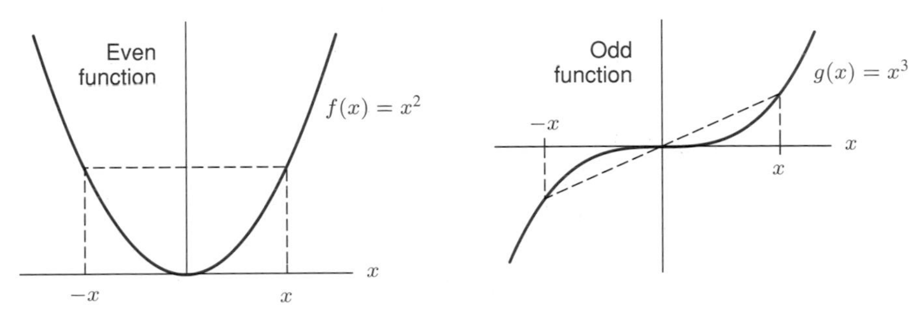

Figure 1.59: Symmetry of odd and even functions

In general, for any function f,

f is an **even** function if $f(-x) = f(x)$ for all x.

f is an **odd** function if $f(-x) = -f(x)$ for all x.

You should be aware that many functions do not have any symmetry and so are neither even nor odd.

Problems for Section 1.9

1. (a) Write an equation for a graph obtained by vertically stretching the graph of $y = x^2$ by a factor of 2, followed by a vertical upward shift of 1 unit. Sketch the graph.
   (b) What is the equation if the order of the transformations (stretching and shifting) in part (a) is interchanged?
   (c) Are the two graphs the same? Explain the effect of reversing the order of transformations.
2. If $f(x) = 2x + 3$ and $g(x) = \log x$, find
   (a) $g(f(x))$ (b) $f(g(x))$ (c) $f(f(x))$
3. What is the difference (if any) between $\ln[\ln(x)]$ and $\ln^2(x)$ $[= (\ln x)^2]$?
4. If $h(x) = x^3 + 1$ and $g(x) = \sqrt{x}$, find
   (a) $g(h(x))$ (b) $h(g(x))$ (c) $h(h(x))$ (d) $g(x) + 1$ (e) $g(x + 1)$
5. If $f(t) = (t + 7)^2$ and $g(t) = 1/(t + 1)$, find
   (a) $f(g(t))$ (b) $g(f(t))$ (c) $f(t^2)$ (d) $g(t - 1)$
6. If $f(z) = 10^z$ and $g(z) = \log(z)$, find
   (a) $f(g(100))$ (b) $g(f(3))$ (c) $f(g(x))$ (d) $g(f(x))$

For Problems 7–10, let $m(z) = z^2$. Find and simplify the following quantities:

7. $m(z+1) - m(z)$

8. $m(z+h) - m(z)$

9. $m(z) - m(z-h)$

10. $m(z+h) - m(z-h)$

For Problems 11–14 determine functions f and g such that $h(x) = f(g(x))$. [There is more than one correct answer. Do not choose $f(x) = x$ or $g(x) = x$.]

11. $h(x) = x^3 + 1$

12. $h(x) = (x+1)^3$

13. $h(x) = \ln^3 x$

14. $h(x) = \ln(x^3)$

15. The Heaviside step function, H, is graphed in Figure 1.60. Sketch graphs of the following functions.
   (a) $2H(x)$
   (b) $H(x) + 1$
   (c) $H(x+1)$
   (d) $-H(x)$
   (e) $H(-x)$

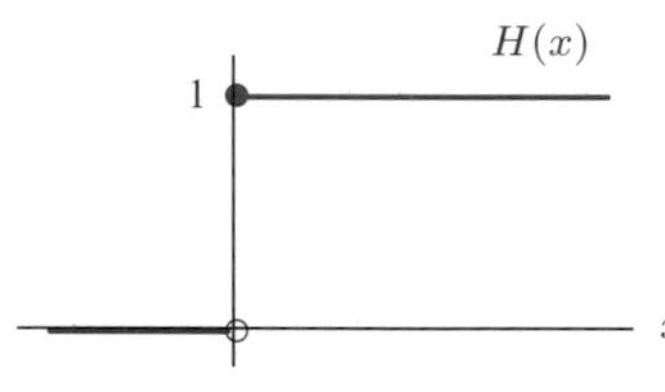

Figure 1.60

For each of the functions $y = f(x)$ whose graphs are in Problems 16 and 17, sketch graphs of
(a) $y = 2f(x)$ (b) $y = f(x+1)$ (c) $y = f(x) + 1$
Where possible, identify x and y intercepts and horizontal and vertical asymptotes for each graph.

16.

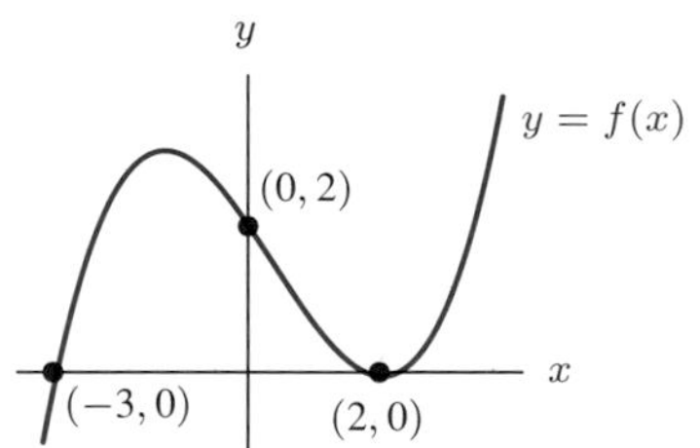

17.

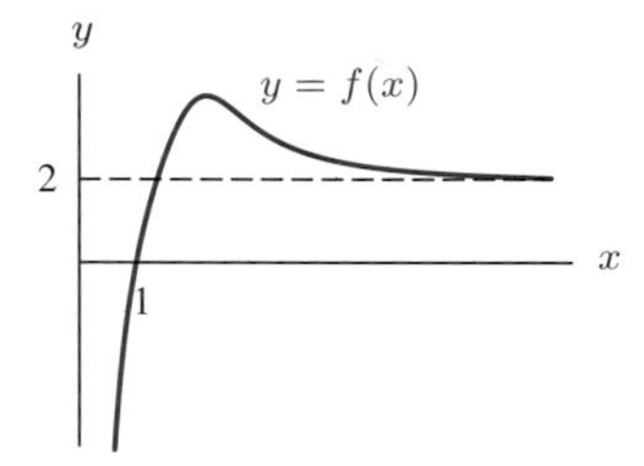

18. What symmetries do the graphs of even and odd functions have?

19. Are the functions $f(x) = 1/x$, $g(x) = \ln(x^2)$, $h(x) = e^x$ even, odd, or neither?

For Problems 20–22, use the graph of $y = f(x)$ in Figure 1.61 to sketch the graph indicated:

20. $y = 2f(x)$

21. $y = 2 - f(x)$

22. $y = \dfrac{1}{f(x)}$

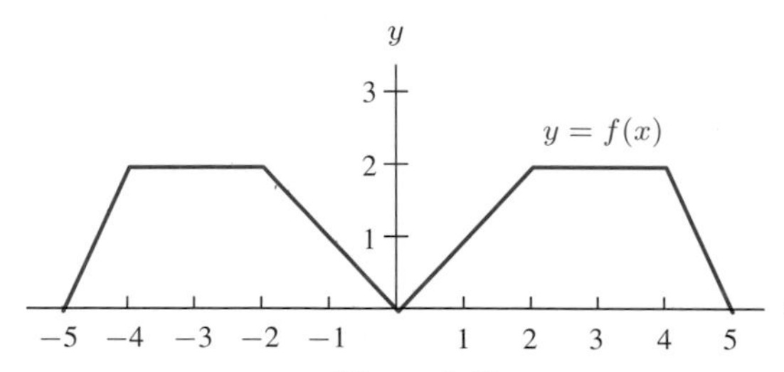

Figure 1.61

For Problems 23–28, suppose that f and g are given by the graphs in Figure 1.62:

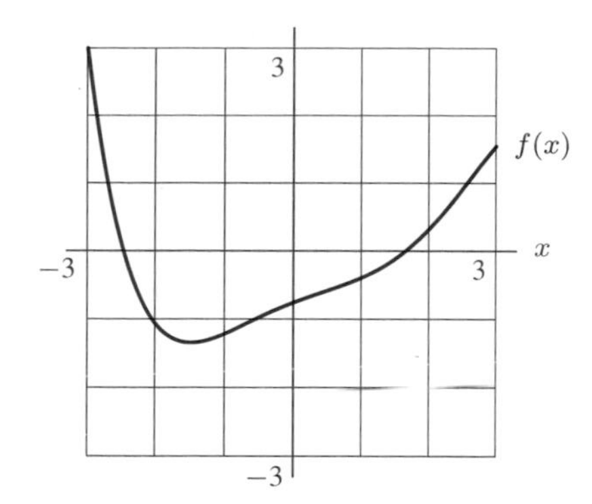

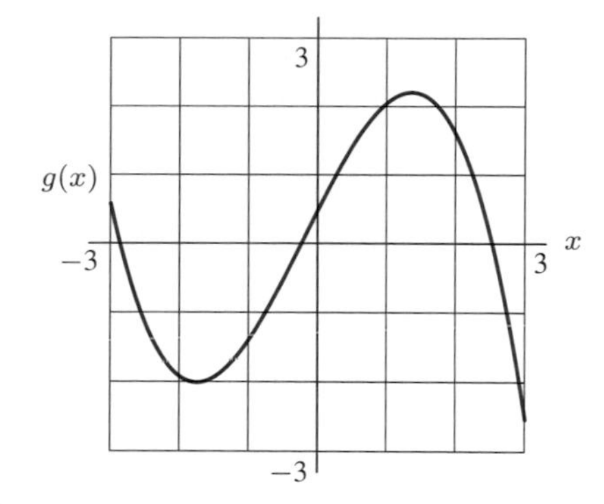

Figure 1.62

23. Find $f(g(1))$.
24. Find $g(f(2))$.
25. Find $f(f(1))$.
26. Sketch a graph of $f(g(x))$.
27. Sketch a graph of $g(f(x))$.
28. Sketch a graph of $f(f(x))$.

29. Complete Table 1.21 to show values for functions f, g, and h, given the following conditions:
(a) f is symmetric about the y-axis. (b) g is symmetric about the origin.
(c) h is the composition of f with g [that is, $h(x) = g(f(x))$].

TABLE 1.21

x	$f(x)$	$g(x)$	$h(x)$
−3	0	0	
−2	2	2	
−1	2	2	
0	0	0	
1			
2			
3			

30. Complete the graphs of $f(x)$ and $g(x)$ for $-10 \leq x \leq 10$ in Figure 1.63 given that $f(x)$ is even and $g(x)$ is odd.

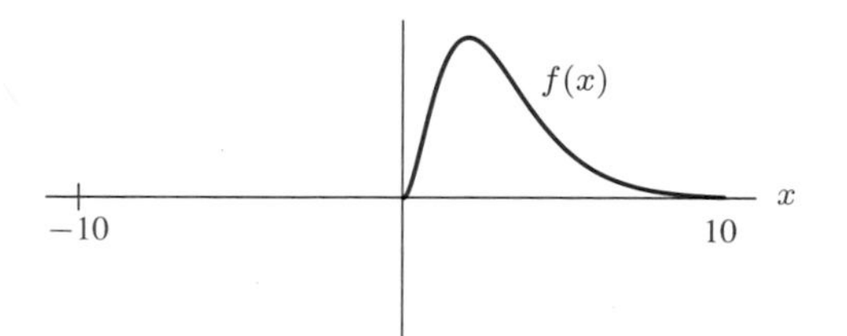

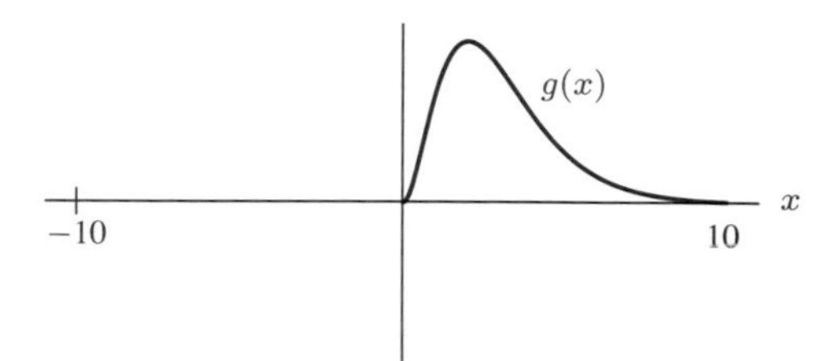

Figure 1.63

1.10 THE TRIGONOMETRIC FUNCTIONS

Trigonometry originated as part of the study of triangles. Indeed, the name *tri-gon-o-metry* means the measurement of three-cornered figures. Thus, the first definitions of the trigonometric functions were in terms of triangles. The trigonometric functions can also be defined using the unit circle, a definition that makes them periodic, or repeating. Many naturally occurring processes are also periodic. The water level in a tidal basin, the blood pressure in a heart, an alternating current, and the position of the air molecules transmitting a musical note all fluctuate regularly. Such phenomena can be represented by trigonometric functions.

We will use only the three trigonometric functions found on a calculator: the sine, the cosine, and the tangent.

Radians

When trigonometry is used to represent oscillations, the angles are usually measured in radians. The formulas of calculus, as you will see, come out much more neatly in radians than in degrees.

A **radian** is defined to be the angle at the center of a unit circle which cuts off an arc of length 1, measured counterclockwise. (See Figure 1.64(a).) A unit circle has radius 1.

An angle of 2 radians cuts off an arc of length 2. A negative angle, such as $-1/2$ radians, cuts off an arc of length $1/2$, but measured clockwise. (See Figure 1.64(b).)

It is useful to think of angles as rotations, since then we can make sense of angles of over 360°. Since one full rotation of 360° cuts off an arc of length 2π, the circumference of the unit circle, it follows that:

$$360^\circ = 2\pi \text{ radians} \quad \text{so} \quad 180^\circ = \pi \text{ radians}.$$

Since 1 radian $= 180^\circ/\pi$, one radian is about 60°. The word radians is often dropped, so if an angle or rotation is referred to without units, it is understood to be in radians.

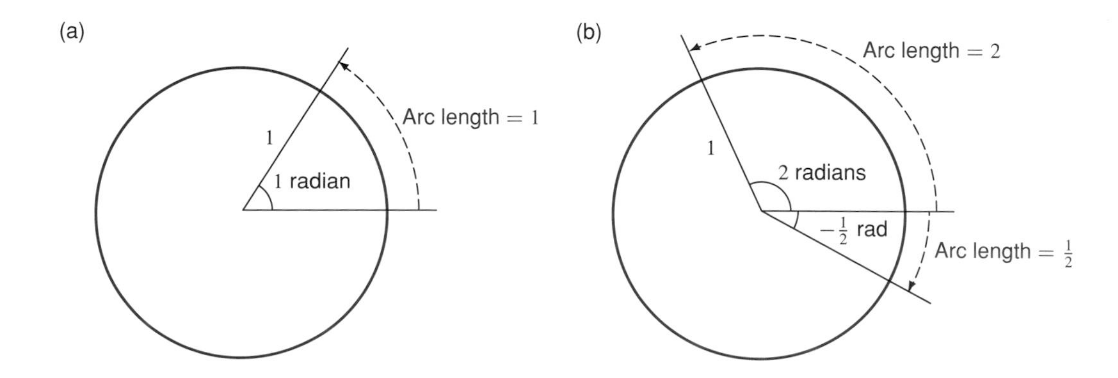

Figure 1.64: Radians defined using unit circle

Arc Length

Radians are also useful for computing the arc length of a sector of a circle of radius other than 1. If the circle has radius r and the sector has angle θ, as in Figure 1.65, then we have the following relation:

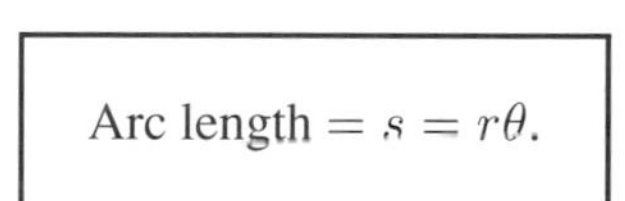

$$\text{Arc length} = s = r\theta.$$

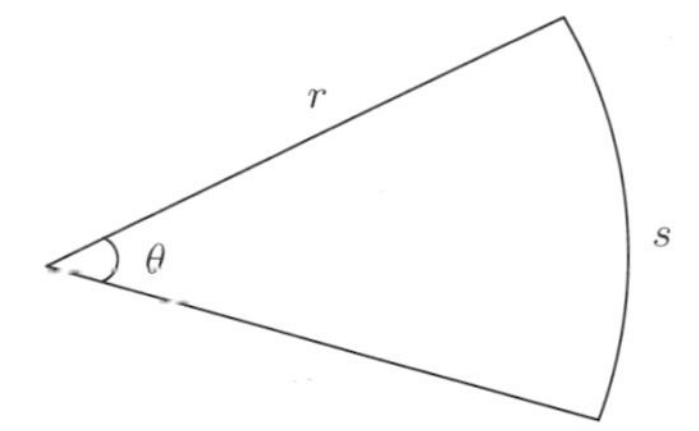

Figure 1.65: Arc length of a sector of a circle

The Sine and Cosine

The two basic trigonometric functions—the sine and cosine—are defined using a unit circle. In Figure 1.66, a length t is measured counterclockwise around the circle from the point $(1, 0)$ to the point P, giving an angle, or rotation, of t radians at the origin. If P has coordinates (x, y), we define

$$\cos t = x \quad \text{and} \quad \sin t = y.$$

Note that we will assume that the trigonometric functions are *always* in radians unless specified otherwise.

Since the equation of the unit circle is $x^2 + y^2 = 1$, we have the following fundamental identity

$$\cos^2 t + \sin^2 t = 1.$$

As t increases and P moves around the circle, the values of $\sin t$ and $\cos t$ oscillate between 1 and -1, and eventually repeat as P moves through points where it has been before. If t is negative, the length is measured clockwise around the circle.

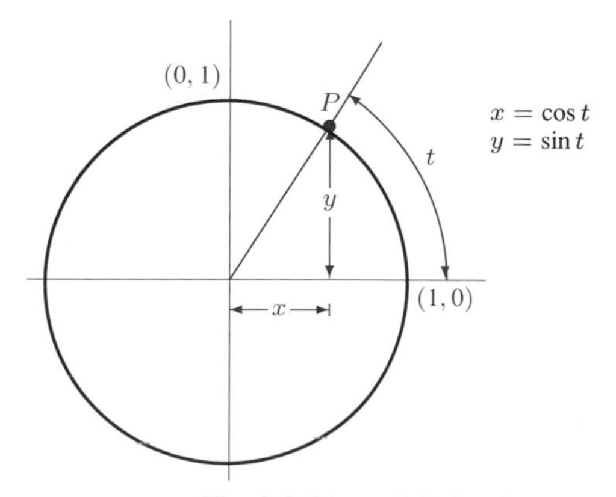

Figure 1.66: The definitions of $\sin t$ and $\cos t$

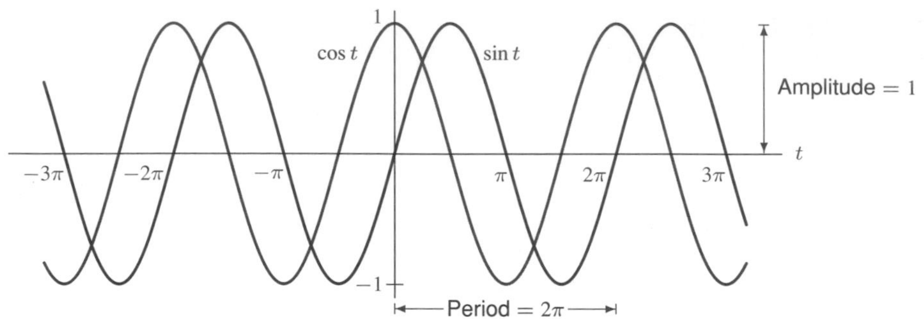

Figure 1.67: Graphs of $\cos t$ and $\sin t$

Amplitude, Period and Phase

The graphs of the sine and cosine are in Figure 1.67. Notice that the sine is an odd function, and the cosine is even. The maximum and minimum values of the sine and cosine are $+1$ and -1, because those are the maximum and minimum values of y and x on the unit circle. After the point P has moved around the complete circle once, the values of $\cos t$ and $\sin t$ start to repeat; we say the functions are *periodic*.

> The **amplitude** of an oscillation is half the distance between maximum and minimum values. The **period** of an oscillation is the time needed for the oscillation to execute one complete cycle.

The amplitude of $\cos t$ and $\sin t$ is 1, and the period is 2π. Why 2π? Because that's the value of t when the point P has gone exactly once around the circle. (The circle has radius 1 and circumference $2\pi \cdot 1 = 2\pi$.)

In looking at Figure 1.67 you can see that the sine and cosine graphs are exactly the same shape, only shifted horizontally. Since the cosine is the sine shifted $\pi/2$ to the left,

$$\cos t = \sin(t + \pi/2).$$

This says that the cosine of any number is the same as the sine of the number that is $\pi/2$ further to the right—which is exactly what the graphs show. Since the sine is the cosine shifted $\pi/2$ to the right,

$$\sin t = \cos(t - \pi/2).$$

We say that the *phase difference* between $\sin t$ and $\cos t$ is $\pi/2$. (Why does this make sense from the definitions of the sine and cosine on the unit circle?)

> To describe arbitrary amplitudes and periods, we use functions of the form
>
> $$f(t) = A \sin Bt \qquad \text{and} \qquad g(t) = A \cos Bt,$$
>
> where A is the amplitude and $2\pi/B$ is the period.
> To represent arbitrary phase differences, we shift a graph of the correct amplitude and period horizontally by replacing t with $t - h$ or $t + h$.

Example 1 Find and show on a graph the amplitude and period of the functions

(a) $y = 5\sin 2t$ (b) $y = -5\sin\left(\frac{t}{2}\right)$ (c) $y = 1 + 2\sin t$

Solution (a) From Figure 1.68, you can see that the amplitude of $y = 5\sin 2t$ is 5 because the factor of 5 stretches the oscillations up to 5 and down to -5. The period of $y = \sin 2t$ is π, because when t changes from 0 to π, the quantity $2t$ changes from 0 to 2π, so the sine function will have gone through one complete oscillation.

(b) Figure 1.69 shows that the amplitude of $y = -5\sin(t/2)$ is again 5, because the negative sign reflects the oscillations in the t-axis but does not change how far up or down they go. The period of $y = -5\sin(t/2)$ is 4π because when t changes from 0 to 4π, the quantity $t/2$ changes from 0 to 2π, so the sine function goes through one complete oscillation.

(c) The 1 shifts the graph $y = 2\sin t$ up by 1. Since $y = 2\sin t$ has an amplitude of 2 and a period of 2π, the graph of $y = 1 + 2\sin t$ goes up to 3 and down to -1, and has a period of 2π. (See Figure 1.70.) Thus, $y = 1 + 2\sin t$ also has amplitude 2 and period 2π.

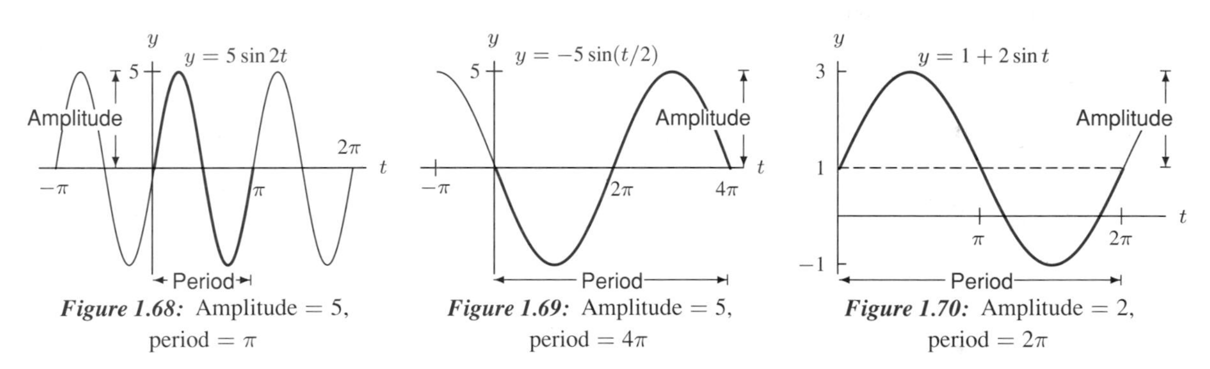

Figure 1.68: Amplitude = 5, period = π

Figure 1.69: Amplitude = 5, period = 4π

Figure 1.70: Amplitude = 2, period = 2π

Example 2 Find formulas for the functions describing the oscillations in Figures 1.71–1.73.

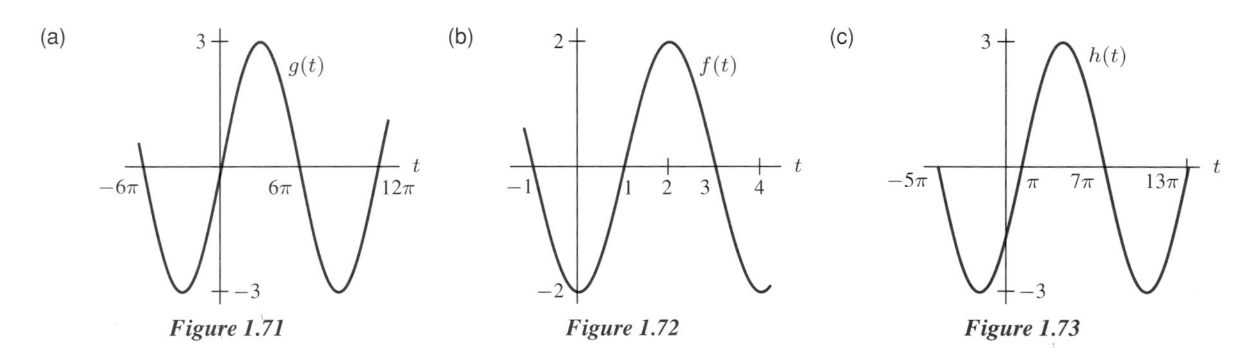

Figure 1.71

Figure 1.72

Figure 1.73

Solution (a) The function in Figure 1.71 looks like a sine function of amplitude 3, so $g(t) = A\sin Bt$ and $A = 3$. Since the function executes one full oscillation between $t = 0$ and $t = 12\pi$, when t changes by 12π, the quantity Bt changes by 2π. This means $B \cdot 12\pi = 2\pi$, so $B = 1/6$. Therefore, $g(t) = 3\sin(t/6)$ has the graph shown.

(b) The function in Figure 1.72 looks like a cosine function with amplitude 2, so $f(t) = 2\cos Bt$. The function completes one oscillation between $t = 0$ and $t = 4$. Thus, when t changes by 4, the quantity Bt changes by 2π, so $B \cdot 4 = 2\pi$, or $B = \pi/2$. The function $2\cos(\pi t/2)$ has the right amplitude and period but has a vertical intercept of 2 instead of -2. Therefore, $f(t) = -2\cos(\pi t/2)$ has the graph shown.

(c) The function in Figure 1.73 looks like the function $g(t)$ in Figure 1.71, but shifted a distance of π to the right. Since $g(t) = 3\sin(t/6)$, we replace t by $(t-\pi)$ to obtain $h(t) = 3\sin[(t-\pi)/6]$.

There are many possible equations for these graphs; you may be able to find others.

Example 3 On February 10, 1990, high tide in Boston was at midnight. The water level at high tide was 9.9 feet; later, at low tide, it was 0.1 feet. Assuming the next high tide is exactly 12 hours later and that the height of the water is given by a sine or cosine curve, find a formula for water level in Boston as a function of time.

Solution Suppose the height of the water level is y feet, and let t be the time measured in hours from midnight. Then the oscillations are to have amplitude 4.9 feet $(= (9.9 - 0.1)/2)$ and period 12, so $12B = 2\pi$ and $B = \pi/6$. Since the water is highest at midnight, when $t = 0$, the oscillations are best represented by a cosine, because the cosine is at its maximum at the beginning of the cycle. (See Figure 1.74.) Thus we can say

$$\text{Height above average} = 4.9\cos\left(\frac{\pi}{6}t\right).$$

Since the average depth of the water was 5 feet $(= (9.9 + 0.1)/2)$, we want a cosine curve shifted up by 5. We get this by adding 5:

$$y = 5 + 4.9\cos\left(\frac{\pi}{6}t\right).$$

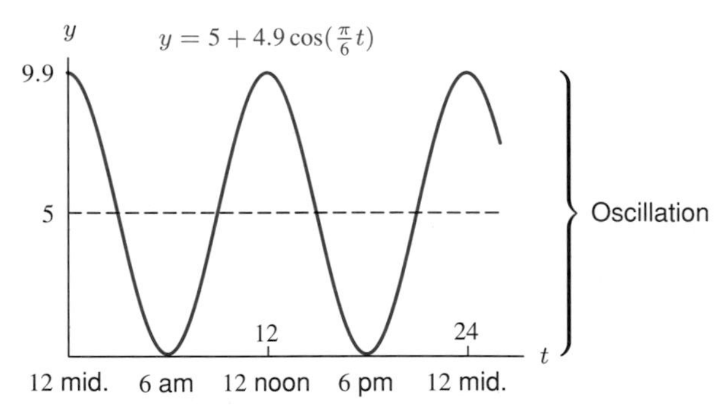

Figure 1.74: Function approximating the tide in Boston on February 10, 1990

Example 4 Of course, there's something wrong with the assumption in Example 3 that the next high tide will be at noon. If so, the high tide would always be at noon or midnight, instead of progressing slowly through the day, as in fact it does. The interval between successive high tides actually averages about 12 hours 24 minutes. Using this, give a more accurate formula for the height of the water as a function of time.

Solution The period is 12 hours 24 minutes = 12.4 hours, so $B = 2\pi/12.4$, giving

$$y = 5 + 4.9\cos\left(\frac{2\pi}{12.4}t\right) = 5 + 4.9\cos(0.507t).$$

Example 5 Use the information from Example 4 to write a formula for the water level in Boston on a day when the high tide is at 2 pm.

Solution When the high tide is at midnight

$$y = 5 + 4.9\cos(0.507t).$$

Since 2 pm is 14 hours after midnight, we replace t by $(t - 14)$. Thus, on a day when the high tide is at 2 pm,

$$y = 5 + 4.9\cos[0.507(t - 14)].$$

The Tangent Function

If t is any number with $\cos t \neq 0$, we define the tangent function by

$$\tan t = \frac{\sin t}{\cos t}.$$

Figure 1.66 on page 67 shows the geometrical meaning of the tangent function: $\tan t$ is the slope of the line through the origin $(0, 0)$ and the point $P = (\cos t, \sin t)$ on the unit circle.

The tangent function is undefined wherever $\cos t = 0$, namely, at $t = \pm\pi/2, \pm 3\pi/2, \ldots$, and it has a vertical asymptote at each of these points. Now think about where $\tan t$ will be positive and where it will be negative: it will be positive where $\sin t$ and $\cos t$ have the same sign. The graph of the tangent is in Figure 1.75.

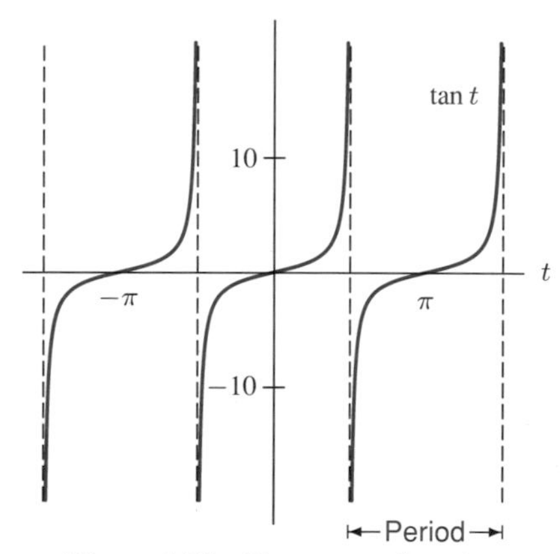

Figure 1.75: The tangent function

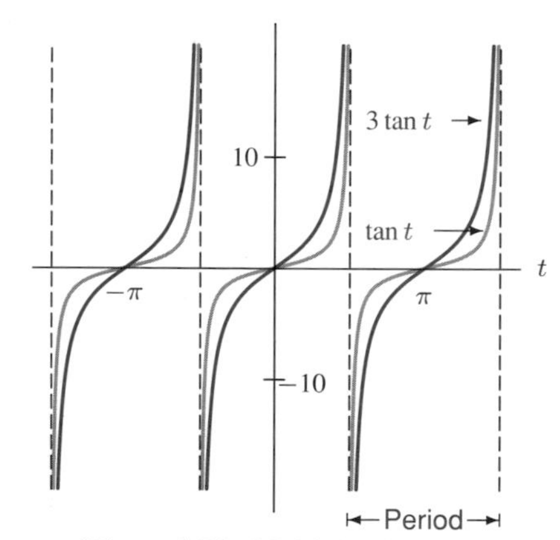

Figure 1.76: Multiple of tangent

Notice that the tangent function has period π, because it repeats every π units. Does it make sense to talk about the amplitude of the tangent function? Not if we're thinking of the amplitude as a measure of the size of the oscillation, because the tangent becomes infinitely large near each vertical asymptote. We can still multiply the tangent by a constant, but that constant no longer represents an amplitude. (See Figure 1.76.)

The Inverse Trigonometric Functions

On occasion, you may need to find a number with a given sine. For example, you might want to find x such that

$$\sin x = 0$$

or such that

$$\sin x = 0.3.$$

The first of these equations can be solved by inspection; the solution is $x = 0, \pm\pi, \pm 2\pi, \ldots$. Solving the second equation requires a calculator but it also has infinitely many solutions. For each equation, we pick out the only solution between $-\pi/2$ and $\pi/2$ as the preferred solution. For example, the preferred solution to $\sin x = 0$ is $x = 0$. We define the inverse sine, written "arcsin" or "$\sin^{-1}$," as the function which gives the preferred solution.

For $-1 \leq y \leq 1$,

$$\arcsin y = x$$
$$\text{means } \sin x = y \quad \text{with} \quad -\frac{\pi}{2} \leq x \leq \frac{\pi}{2}$$

Thus the arcsine is the inverse function to the piece of the sine function having domain $[-\pi/2, \pi/2]$. (See Table 1.22 and Figure 1.77.) On a calculator, the arcsine function is usually denoted by $\boxed{\sin^{-1}}$.

TABLE 1.22 *Values of* $\sin x$ *and* $\sin^{-1} x$

x	$\sin x$	x	$\sin^{-1} x$
$-\frac{\pi}{2}$	-1.000	-1.000	$-\frac{\pi}{2}$
-1.0	-0.841	-0.841	-1.0
-0.5	-0.479	-0.479	-0.5
0.0	0.000	0.000	0.0
0.5	0.479	0.479	0.5
1.0	0.841	0.841	1.0
$\frac{\pi}{2}$	1.000	1.000	$\frac{\pi}{2}$

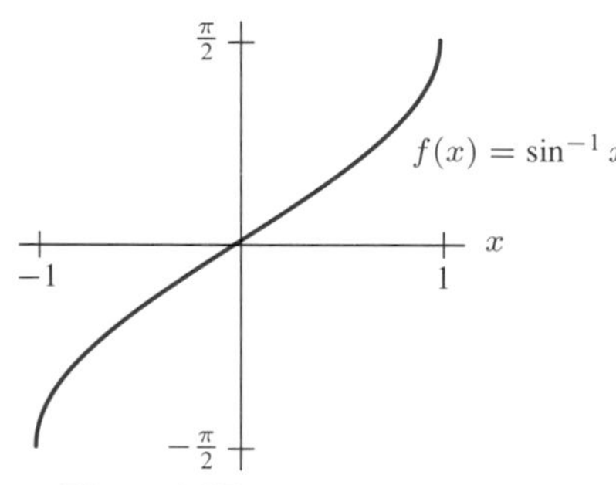

Figure 1.77: The arcsine function

The inverse tangent, written "arctan" or "$\tan^{-1}$," is the inverse function for the piece of the tangent function having domain $(-\pi/2, \pi/2)$.

For any y,

$$\arctan y = x$$
$$\text{means } \tan x = y \quad \text{with} \quad -\frac{\pi}{2} < x < \frac{\pi}{2}$$

On a calculator, the inverse tangent is usually denoted by $\boxed{\tan^{-1}}$. The graph of the arctangent is in Figure 1.78.

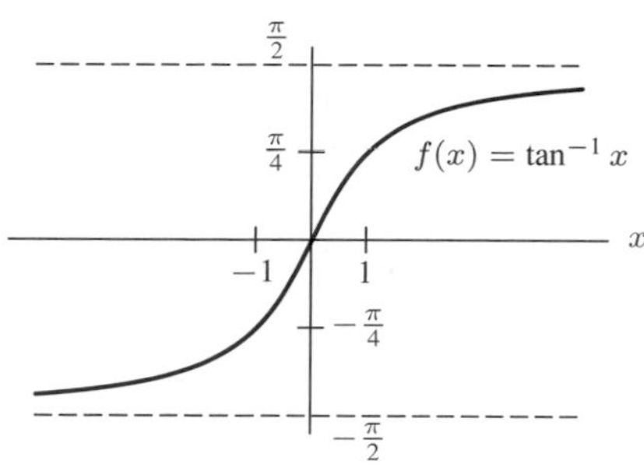

Figure 1.78: The arctangent function

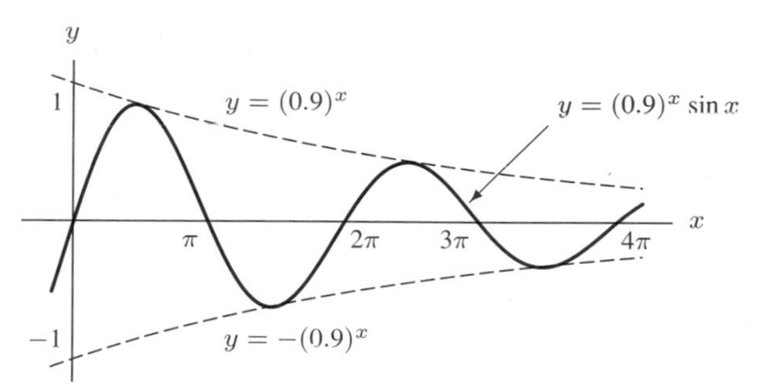

Figure 1.79: A damped oscillation

Further Examples with Trigonometric Functions

Example 6 Explain why the function

$$f(x) = (0.9)^x \sin x$$

represents an oscillation which is dying out, called a *damped oscillation*.

Solution You can think of $f(x)$ as a sine function with an amplitude of $(0.9)^x$, which decays as x increases. Figure 1.79 shows that $f(x)$ is an oscillation which dies away as x increases. This is the kind of motion you'd expect from a weight on a spring bobbing up and down, or any other kind of oscillation dying out because of friction.

Example 7 Plot the graph of the function $f(x) = x + \sin x$ and explain how it is related to the graphs of $y = x$ and $y = \sin x$.

Solution The graphs of $y = x$ and $y = \sin x$ are shown in Figure 1.80(a). Imagine the corresponding y values stacked vertically, and you will get the graph of $f(x) = x + \sin x$ in Figure 1.80(b).

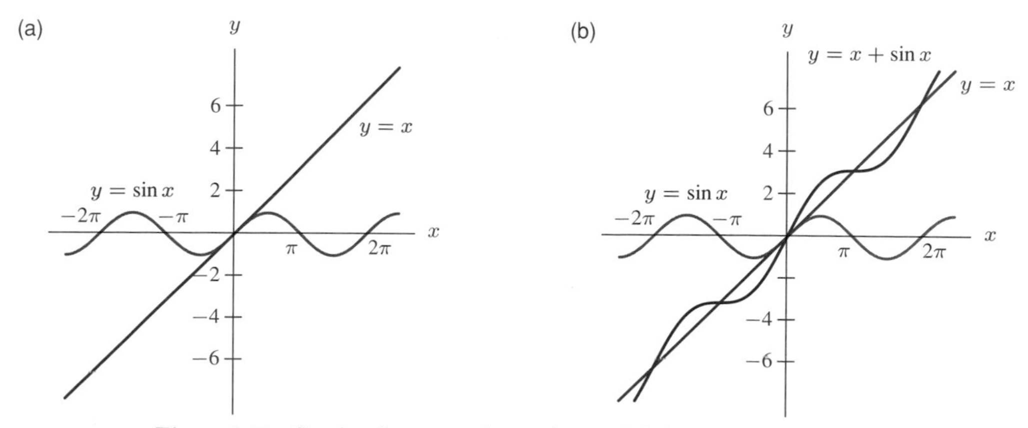

Figure 1.80: Graphs of $y = x$ and $y = \sin x$ and their sum, $y = x + \sin x$

Problems for Section 1.10

For Problems 1–8, draw the angle using a ray through the origin, and determine whether the sine, cosine, and tangent of that angle are positive, negative, zero, or undefined.

1. $\frac{\pi}{4}$
2. $\frac{\pi}{2}$
3. 3π
4. $\frac{4\pi}{3}$
5. $-\frac{\pi}{12}$
6. $-\frac{7\pi}{6}$
7. 4
8. -1

Given that $\sin(\pi/12) = 0.258$ and $\cos(\pi/5) = 0.809$, compute (without using the trigonometric functions on your calculator) the quantities in Problems 9–18. You may want to draw a picture showing the angles involved, and then check your answers on a calculator.

9. $\sin\frac{11\pi}{12}$
10. $\cos\left(-\frac{\pi}{5}\right)$
11. $\sin\frac{13\pi}{12}$
12. $\cos\frac{4\pi}{5}$
13. $\sin\frac{23\pi}{12}$
14. $\cos\left(-\frac{4\pi}{5}\right)$
15. $\sin\frac{37\pi}{12}$
16. $\cos\left(-\frac{11\pi}{5}\right)$
17. $\cos\frac{\pi}{12}$
18. $\sin\frac{\pi}{5}$

19. Use the solution to Example 3 on page 70 to estimate the water level in Boston Harbor at 3:00 am, 4:00 am, and 5:00 pm.
20. What is the period of the earth's revolution around the sun?
21. What is the approximate period of the moon's revolution around the earth?
22. What is the period of the motion of the minute hand of a clock?
23. An LP record rotates $33\frac{1}{3}$ times in a minute. What is the period of its motion?
24. What is the difference between $\sin x^2$, $\sin^2 x$, and $\sin(\sin x)$? Express each of the three as a composition. (Note: $\sin^2 x = (\sin x)^2$.)
25. (a) Match the functions f, g, h, k, whose values are given in the table, with the functions whose formulas are:
(i) $\omega = 1.5+\sin t$ (ii) $\omega = 0.5+\sin t$ (iii) $\omega = -0.5+\sin t$ (iv) $\omega = -1.5+\sin t$.

t	$\omega = f(t)$	t	$\omega = g(t)$	t	$\omega = h(t)$	t	$\omega = k(t)$
6.0	−0.78	3.0	1.64	5.0	−2.46	3.0	0.64
6.5	−0.28	3.5	1.15	5.1	−2.43	3.5	0.15
7.0	0.16	4.0	0.74	5.2	−2.38	4.0	−0.26
7.5	0.44	4.5	0.52	5.3	−2.33	4.5	−0.48
8.0	0.49	5.0	0.54	5.4	−2.27	5.0	−0.46

(b) Based on the table, what is the relationship between the values of $g(t)$ and $k(t)$? Explain this relationship using the formulas you chose for g and k.
(c) Using the formulas you chose for g and h, explain why all the values of g are positive, whereas all the values of h are negative.

For Problems 26–31, sketch graphs of the functions. What are their amplitudes and periods?

26. $y = 3\sin x$
27. $y = 3\sin 2x$
28. $y = -3\sin 2\theta$
29. $y = 4\cos 2x$
30. $y = 4\cos(\frac{1}{2}t)$
31. $y = 5 - \sin 2t$

32. Match the functions below with the graphs in Figure 1.81.
(a) $y = 2\cos(t - \pi/2)$ (b) $y = 2\cos t$ (c) $y = 2\cos(t + \pi/2)$

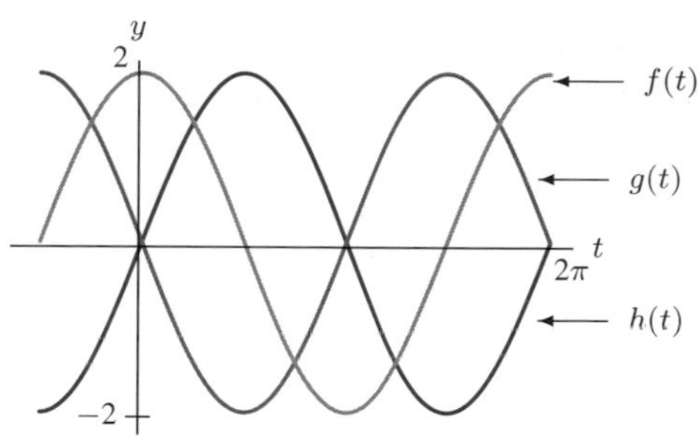

Figure 1.81

For Problems 33–40, find a possible formula for each graph.

33.

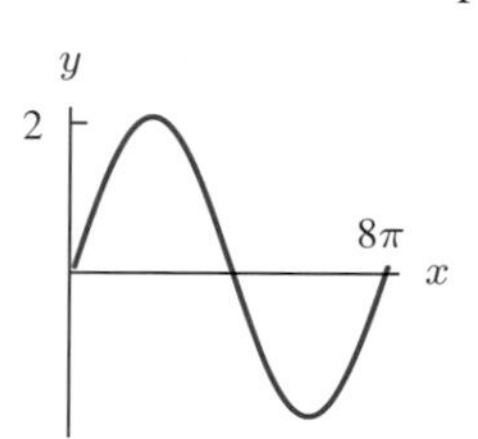

34.

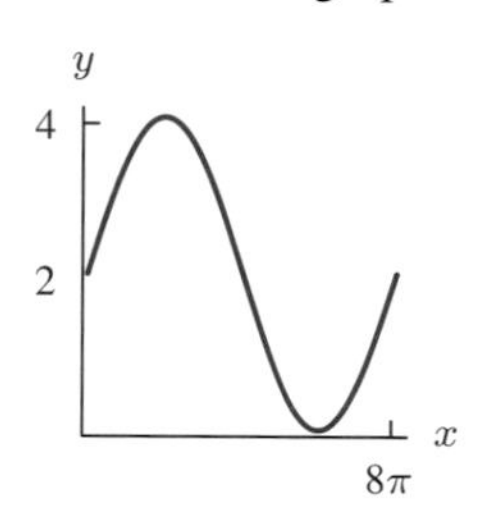

35.

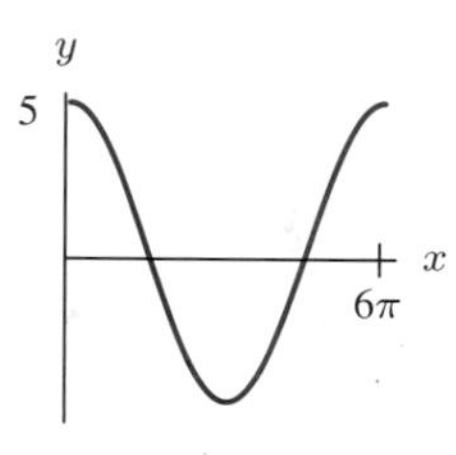

36.

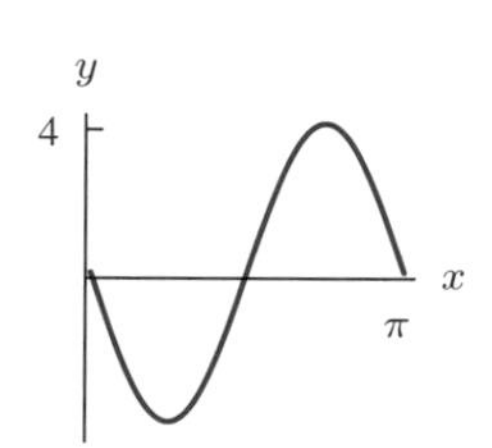

37.

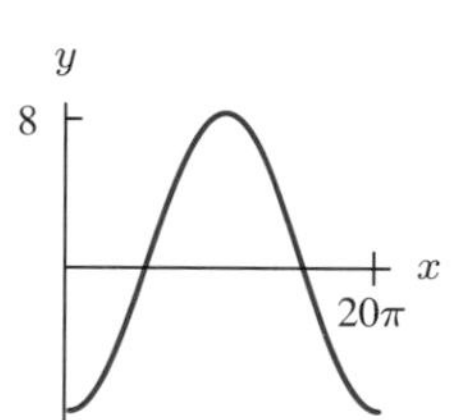

38.

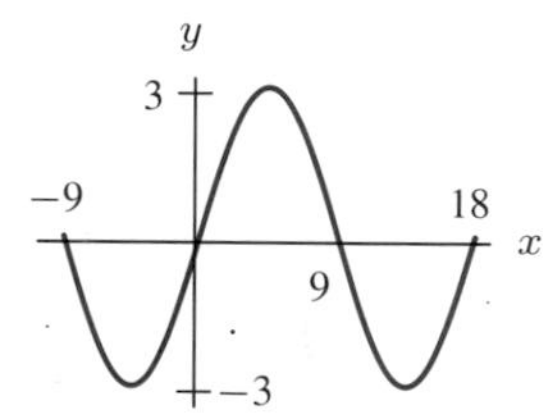

39.

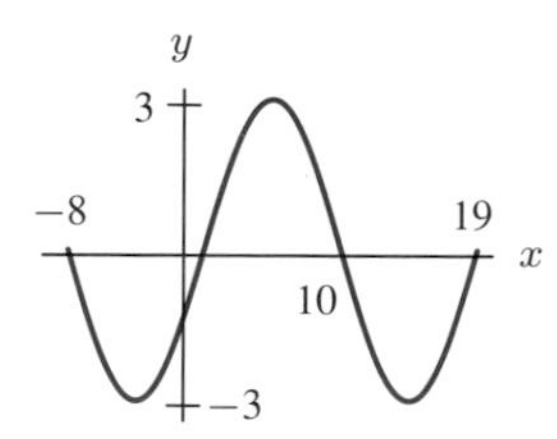

40.

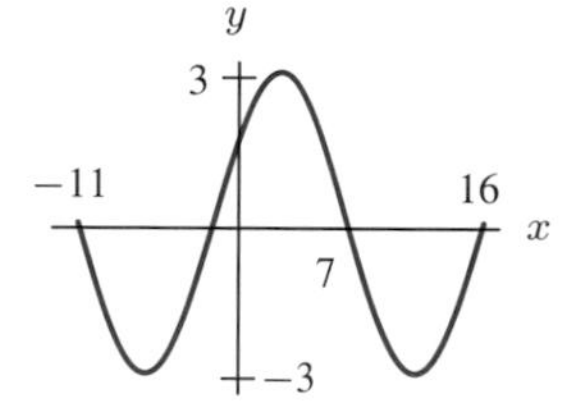

41. A population of animals varies sinusoidally between a low of 700 on January 1 and a high of 900 on July 1.
 (a) Graph the population against time.
 (b) Find a formula for the population as a function of time, t, measured in months since the start of the year.

42. The Bay of Fundy in Canada is reputed to have the largest tides in the world, with the difference between low and high water level being as much as 15 meters (nearly 50 feet). Suppose at a particular point in the Bay of Fundy, the depth of the water, y meters, as a function of time, t, in hours since midnight on January 1, 1994, is given by

$$y = y_0 + A\cos[B(t - t_0)].$$

(a) What is the physical meaning of y_0?
(b) What is the value of A?
(c) What is the value of B? Assume the time between successive high tides is $12\frac{1}{2}$ hours.
(d) What is the physical meaning of t_0?

43. The visitors' guide to St. Petersburg, Florida, contains the chart shown in Figure 1.82 to advertise their good weather. Fit a trigonometric function approximately to the data. The independent variable should be time in months. In order to do this, you will need to find the amplitude and period of the data, and when the maximum occurs. (There are many possible answers to this problem, depending on how you read the graph.)

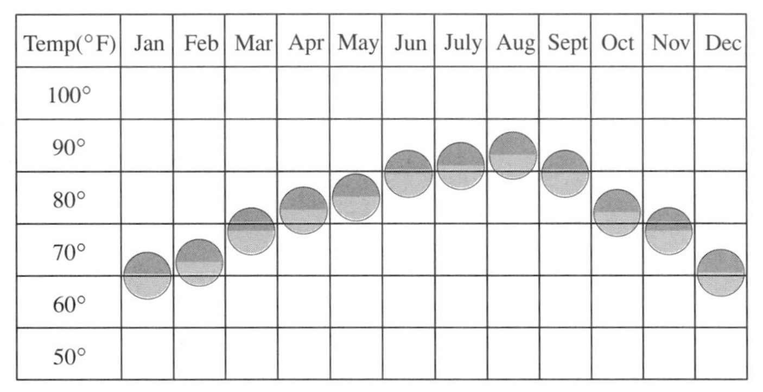

Figure 1.82: "St. Petersburg...where we're famous for our wonderful weather and year-round sunshine." (Reprinted with permission)

44. The voltage, V, of an electrical outlet in a home is given as a function of time, t (in seconds), by $V = V_0 \cos(120\pi t)$.
(a) What is the period of the oscillation?
(b) What does V_0 represent?
(c) Sketch the graph of V against t. Label the axes.

45. (a) Using a calculator set in radians, make a table of values, to two decimal places, of $f(x) = \arcsin x$, for $x = -1, -0.8, -0.6, \ldots, 0, \ldots, 0.8, 1$. (arcsin is denoted by $\boxed{\sin^{-1}}$ on most calculators).
(b) Sketch $f(x) = \arcsin x$. Mark on the graph the domain and range of f.

46. This problem introduces the arccosine function, or inverse cosine, denoted by $\boxed{\cos^{-1}}$ on most calculators.
(a) Using a calculator set in radians, make a table of values, to two decimal places, of $g(x) = \arccos x$, for $x = -1, -0.8, -0.6, \ldots, 0, \ldots, 0.8, 1$.
(b) Sketch the graph of $g(x) = \arccos x$.
(c) Based on your graph, what are the domain and range of the arccosine?
(d) Why is the domain of the arccosine the same as the domain of the arcsine?
(e) Why is the range of the arccosine *not* the same as the range of the arcsine? To answer this, look at how the domain of the original sine function was restricted to construct the arcsine. Why can't the domain of the cosine be restricted in exactly the same way to construct the arccosine? How should the domain of the cosine be restricted?

47. From the data in Table 1.23, determine a possible formula for each function. Explain your reasoning. (Note: Before checking the values of a trigonometric function, be certain your calculator is in radian mode.)

TABLE 1.23

x	$f(x)$	$g(x)$	$h(x)$	$F(x)$	$G(x)$	$H(x)$
−5	−10	20	25	0.958924	0.544021	2.958924
−4.5	−9	19	20.25	0.97753	−0.412118	2.97753
−4	−8	18	16	0.756802	−0.989358	2.756802
−3.5	−7	17	12.25	0.350783	−0.656987	2.350783
−3	−6	16	9	−0.14112	0.279415	1.85888
−2.5	−5	15	6.25	−0.598472	0.958924	1.401528
−2	−4	14	4	−0.909297	0.756802	1.090703
−1.5	−3	13	2.25	−0.997495	−0.14112	1.002505
−1	−2	12	1	−0.841471	−0.909297	1.158529
−0.5	−1	11	0.25	−0.479426	−0.841471	1.520574
0	0	10	0	0	0	2
0.5	1	9	0.25	0.479426	0.841471	2.479426
1	2	8	1	0.841471	0.909297	2.841471
1.5	3	7	2.25	0.997495	0.14112	2.997495
2	4	6	4	0.909297	−0.756802	2.909297
2.5	5	5	6.25	0.598472	−0.958924	2.598472
3	6	4	9	0.14112	−0.279415	2.14112
3.5	7	3	12.25	−0.350783	0.656987	1.649217
4	8	2	16	−0.756802	0.989358	1.243198
4.5	9	1	20.25	−0.97753	0.412118	1.02247
5	10	0	25	−0.958924	−0.544021	1.041076

48. Using a calculator or computer, estimate all points of intersection of the graphs of $f(x) = x + \sin x$ and $g(x) = x^3$. How do you know you have found all of them?

49. You are told that two trigonometric functions each have period π and that their graphs intersect at $x = 3.64$, but you are told nothing else about the functions.
    (a) Can you say whether these two graphs intersect at any smaller positive x value? If so, what is it?
    (b) Find an x value greater than 3.64 at which these graphs intersect.
    (c) Find a negative x value at which these graphs intersect.

50. (a) Use a graphing calculator or computer to find the period of $2\sin 3t + 3\cos t$.
    (b) What is the period of $\sin 3t$? Of $\cos t$?
    (c) Use your answers to part (b) to explain your answer to part (a).

1.11 POLYNOMIALS AND RATIONAL FUNCTIONS

Polynomials

Some of the best-known functions for which there are formulas are the polynomials:

$$y = p(x) = a_n x^n + a_{n-1}x^{n-1} + \cdots + a_1 x + a_0,$$

where n is a positive integer, called the *degree* of the polynomial (provided $a_n \neq 0$). The shape of the graph of a polynomial depends on its degree, as shown in Figure 1.83. These graphs correspond to a positive coefficient for x^n; a negative coefficient flips the graph over. Notice that the graph of the quadratic "turns around" once, the cubic "turns around" twice, and the quartic (fourth degree) "turns around" three times. An nth degree polynomial "turns around" at most $n - 1$ times (where n is a positive integer), but there may be fewer turns.

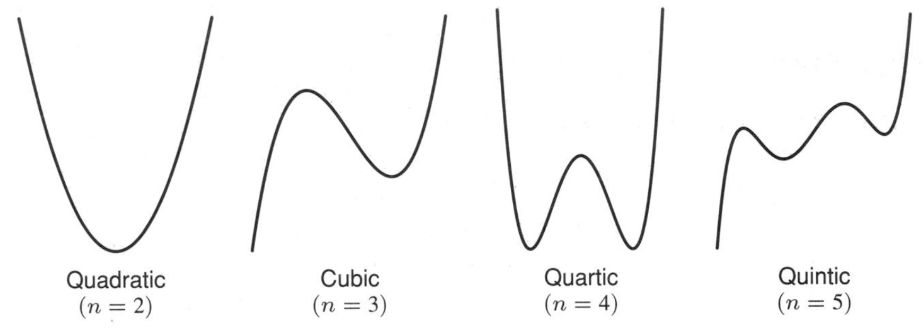

Figure 1.83: Graphs of typical polynomials of degree n

Example 1 Find possible formulas for the polynomials whose graphs are in Figure 1.84.

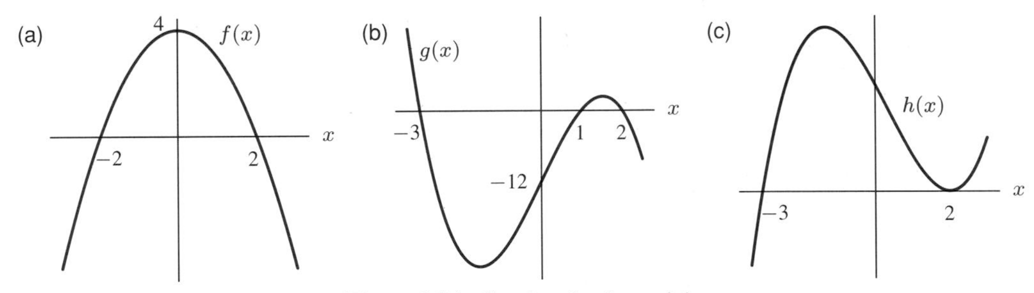

Figure 1.84: Graphs of polynomials

Solution (a) This graph appears to be a parabola, turned upside down and moved up by 4, so

$$f(x) = -x^2 + 4.$$

The minus sign turns the parabola upside down and the $+4$ moves it up by 4. You should notice that this formula does give the correct x-intercepts since $0 = -x^2 + 4$ has solutions $x = \pm 2$.

You can also solve this problem by looking at the x-intercepts first, which tell you $f(x)$ must have factors of $(x + 2)$ and $(x - 2)$, so

$$f(x) = k(x + 2)(x - 2).$$

To find k, use the fact that the graph has a y intercept of 4, so $f(0) = 4$, giving

$$4 = k(0 + 2)(0 - 2)$$

so $k = -1$. Therefore, $f(x) = -(x + 2)(x - 2)$, which multiplies out to $-x^2 + 4$. Note that

$$f(x) = 4 - \frac{x^4}{4}$$

also fits the requirements, but its "shoulders" are sharper. There are many possible answers to these questions.

(b) This looks like a cubic with factors $(x + 3)$, $(x - 1)$, and $(x - 2)$, one for each intercept:

$$g(x) = k(x + 3)(x - 1)(x - 2).$$

Since the y intercept is -12,

$$-12 = k(0 + 3)(0 - 1)(0 - 2)$$

so $k = -2$, and

$$g(x) = -2(x + 3)(x - 1)(x - 2).$$

(c) This also looks like a cubic with zeros at $x = 2$ and $x = -3$. Notice that at $x = 2$ the graph of $h(x)$ touches the x-axis but doesn't cross it, whereas at $x = -3$ the graph crosses the x-axis. We say that $x = 2$ is a *double zero*, but that $x = -3$ is a single zero.

To find a formula for $h(x)$, first imagine the graph of $h(x)$ to be slightly lower down, so that the graph has one x intercept near $x = -3$ and two near $x = 2$, say at $x = 1.9$ and $x = 2.1$. Then

$$h(x) \approx k(x + 3)(x - 1.9)(x - 2.1).$$

Now move the graph back to its original position. The zeros at $x = 1.9$ and $x = 2.1$ move toward $x = 2$, giving

$$h(x) = k(x + 3)(x - 2)^2.$$

Thus the double zero leads to a repeated factor, $(x - 2)^2$. Notice that when $x > 2$, the factor $(x - 2)^2$ is positive, and when $x < 2$, $(x - 2)^2$ is still positive. This reflects the fact that $h(x)$ doesn't change sign near $x = 2$. Compare this with the behavior near the single zero, where h does change sign.

You cannot find k, as no coordinates are given for points off of the x-axis. Inserting any positive value of k will stretch the graph but not change the zeros and therefore will still work.

Example 2 Using a calculator or computer, sketch graphs of $y = x^4$ and $y = x^4 - 15x^2 - 15x$ for $-4 \leq x \leq 4$ and for $-20 \leq x \leq 20$. Set the y range to $-100 \leq y \leq 100$ for the first domain, and to $-100 \leq y \leq 200{,}000$ for the second. What do you observe?

Solution

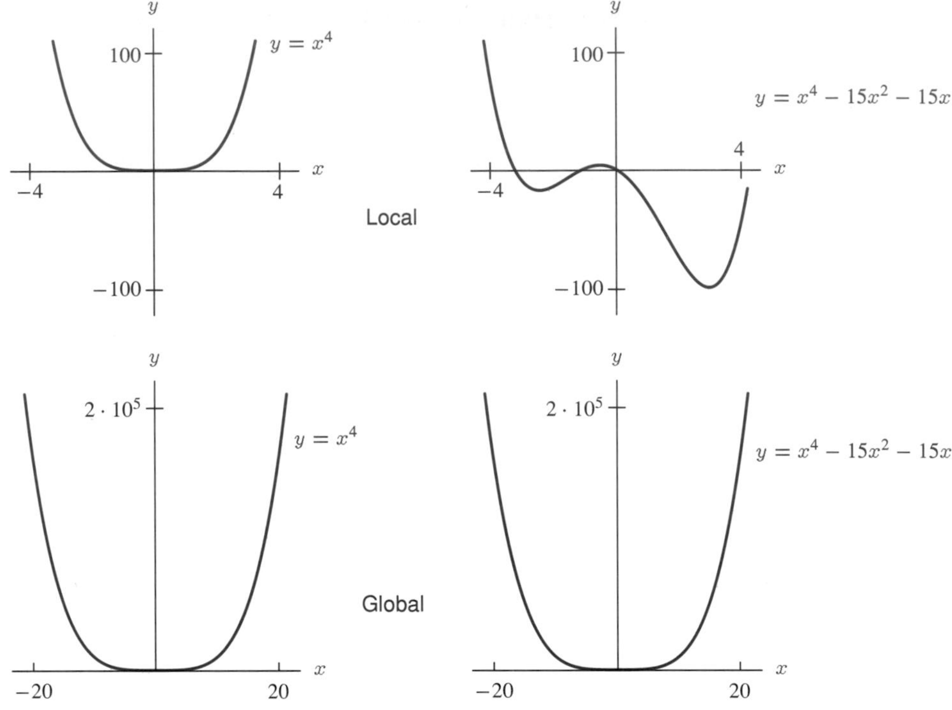

Figure 1.85: Local and global views of $y = x^4$ and $y = x^4 - 15x^2 - 15x$

From the graphs in Figure 1.85 you can see that close up (for $-4 \leq x \leq 4$) the graphs look different; from far away, however, they are almost indistinguishable. The reason is that the leading terms (those with the highest power of x) are the same, namely x^4, and for large values of x, the leading term dominates the other terms.

Looked at numerically in Table 1.24, the differences in the values of the two functions when $x = \pm 20$, although large, are tiny compared with the vertical scale (0 to 200,000) and so can't be seen on the graph.

TABLE 1.24 *Numerical values of $y = x^4$ and $y = x^4 - 15x^2 - 15x$*

x	$y = x^4$	$y = x^4 - 15x^2 - 15x$	Difference
−20	160,000	154,300	5700
−15	50,625	47,475	3150
15	50,625	47,025	3600
20	160,000	153,700	6300

Example 3 If a tomato is thrown vertically into the air at time $t = 0$ with velocity 48 feet per second, its distance, y, above the surface of the earth at time t (in seconds) is given by the equation

$$y = -16t^2 + 48t.$$

Sketch a graph of position against time, and mark on the graph the coordinates of the points corresponding to instants when the tomato (a) hits the ground and (b) reaches its highest point.

Solution This graph is a parabola opening downward. Its intercepts are given by

$$0 = -16t^2 + 48t$$

or

$$0 = 16t(t - 3) \quad \text{so } t = 0, 3 \text{ seconds.}$$

The tomato's journey occurs between $t = 0$ and $t = 3$. The point representing the instant when the tomato hits the ground is where $y = 0$ and $t = 3$. Its highest point occurs halfway through the journey, at $t = 1.5$, when $y = -16(1.5)^2 + 48(1.5) = 36$ feet. See Figure 1.86. You should notice that although the graph is arch shaped, the tomato in fact moves vertically up and down and will land at exactly the spot from which you threw it. The horizontal axis in Figure 1.86 represents time, not horizontal displacement in space.

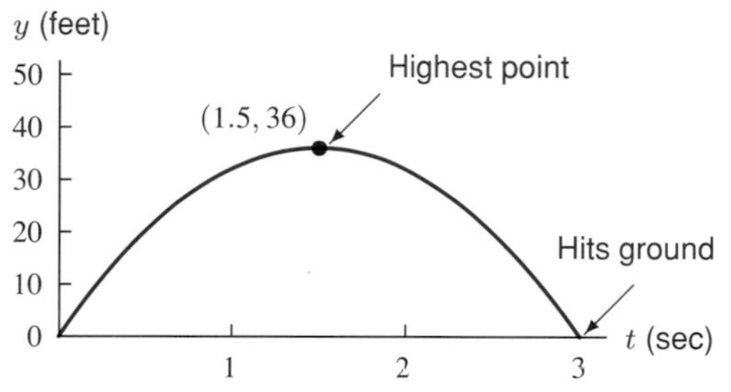

Figure 1.86: The height of the tomato

Rational Functions

Rational functions are those of the form

$$f(x) = \frac{p(x)}{q(x)}$$

where p and q are polynomials. Their graphs often have vertical asymptotes where the denominator is zero. If the denominator is nowhere zero, there are no vertical asymptotes. They may also have horizontal asymptotes which occur if $f(x)$ approaches a finite number as $x \to \infty$ or $x \to -\infty$. The behavior of a function as $x \to \pm\infty$ is called its *end behavior*.

Example 4 Plot and describe the graph of $y = \dfrac{x^2 - 4}{x^2 - 1}$, including end behavior.

Solution Factoring gives

$$y = \frac{x^2 - 4}{x^2 - 1} = \frac{(x + 2)(x - 2)}{(x + 1)(x - 1)}$$

so $x = \pm 1$ are vertical asymptotes. If $y = 0$, then $(x + 2)(x - 2) = 0$ or $x = \pm 2$; these are the x intercepts. Note that the vertical asymptotes arise from zeros of the denominator, while zeros of the numerator give rise to x-intercepts. Substituting $x = 0$ gives $y = 4$; this is the y-intercept. To see what happens as $x \to \pm\infty$, look at the y values in Table 1.25. Notice that, in this example, positive and negative x's give the same y value.

TABLE 1.25 *Values of* $y = \frac{x^2-4}{x^2-1}$

x	$y = \frac{x^2-4}{x^2-1}$
± 10	0.969697
± 100	0.999700
± 1000	0.999997

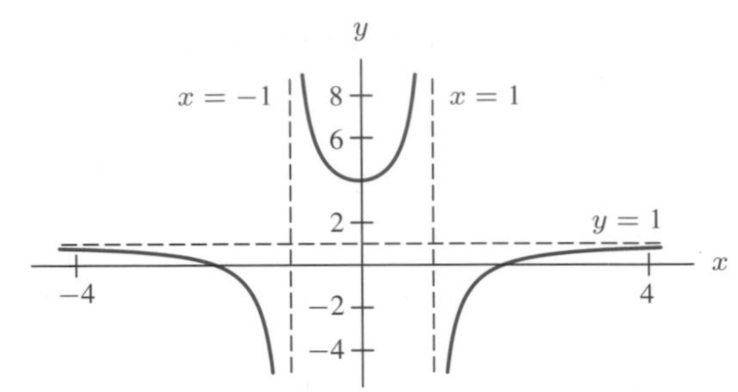

Figure 1.87: Graph of the function $y = \frac{x^2-4}{x^2-1}$

Clearly y is getting closer to 1 as x gets large positively or negatively. This can also be seen by realizing that as $x \to \pm\infty$, only the highest powers of x really matter. For large x, the 4 and the 1 are insignificant compared to x^2, so

$$y = \frac{x^2 - 4}{x^2 - 1} \approx \frac{x^2}{x^2} = 1.$$

Therefore the horizontal asymptote is $y = 1$. Since, for $x > 1$, the denominator is positive and the numerator is less than the denominator, the graph lies *below* its asymptote. (Why doesn't the graph lie below $y = 1$ when $-1 < x < 1$?) See Figure 1.87.

Problems for Section 1.11

1. Assume that each of the graphs in Figure 1.88 is of a polynomial. For each graph:
   (a) What is the minimum possible degree of the polynomial?
   (b) Is the leading coefficient of the polynomial positive or negative? (The leading coefficient is the coefficient of the highest power of x.)

(I) 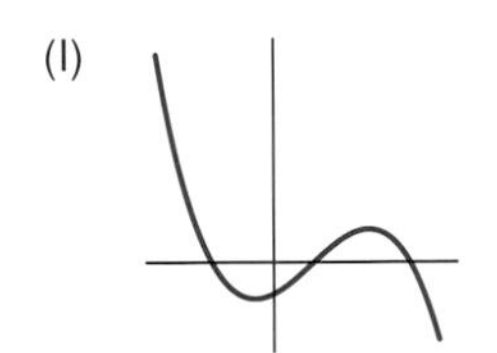
(II) 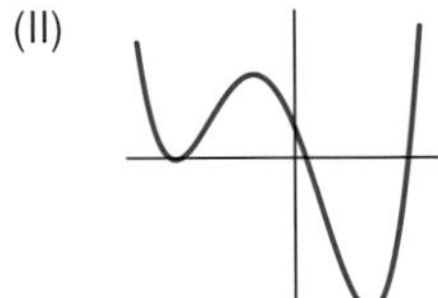
(III) 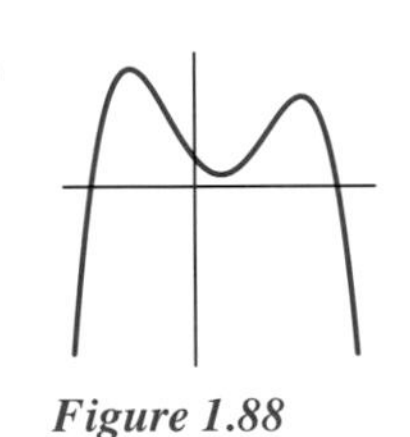
(IV) 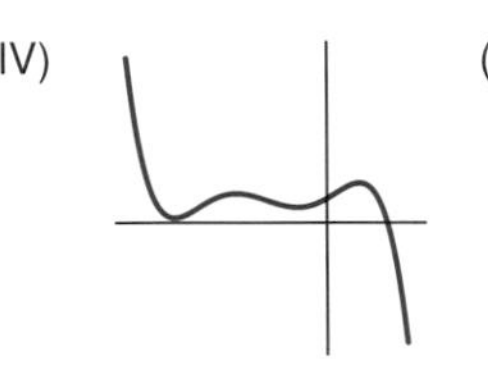
(V) 

Figure 1.88

2. Describe the end behavior of each of the functions below. Specifically:
   - As $x \to +\infty$, does $f(x) \to +\infty$ or $-\infty$? Why?
   - As $x \to -\infty$, does $f(x) \to +\infty$ or $-\infty$? Why?

   (a) $f(x) = -3x^3 + 70x^2 - 20$
   (b) $f(x) = 20x^4 + 3x^3 + x^2 + 1000$
   (c) $f(x) = -3x^4 + 20x^3 - 5x^2 + x - 20$
   (d) $f(x) = x^4 + x^5$
   (e) $f(x) = 4x^4 + 5x^5 - 6x^6$

For Problems 3–6, sketch graphs of the polynomials:

3. $f(x) = (x+2)(x-1)(x-3)$

4. $f(x) = 5(x^2-4)(x^2-25)$

5. $f(x) = -5(x^2-4)(25-x^2)$

6. $f(x) = 5(x-4)^2(x^2-25)$

For each of the rational functions in Problems 7–10, find asymptotes, the behavior of the function as $x \to \pm\infty$, and the behavior of the function near any vertical asymptotes. Then use this information to sketch a graph. Check your work by looking at the actual graph on a computer or graphing calculator.

7. $y = \dfrac{1-4x}{2x+2}$

8. $y = \dfrac{3x^3+2x^2-11}{x^3+8}$

9. $y = \dfrac{x^2+2x+1}{x^2-4}$

10. $y = \dfrac{1-x^2}{x-2}$

11. For which positive integers n is $f(x) = x^n$ even? Odd?

12. Which polynomials are even? Odd? Are there polynomials which are neither?

13. If $f(x) = ax^2 + bx + c$, what do you know about the values of a, b, and c if:
    (a) $(1, 1)$ is on the graph of $f(x)$?
    (b) $(1, 1)$ is the vertex of the graph of $f(x)$? (You may want to use the fact that the equation for the axis of symmetry of the parabola of $y = ax^2 + bx + c$ is $x = -b/2a$.)
    (c) The y intercept of the graph is $(0, 6)$?
    (d) Find a quadratic function that satisfies all three conditions.

14. A pomegranate is thrown from ground level straight up into the air at time $t = 0$ with velocity 64 feet per second. Its height at time t will be $f(t) = -16t^2 + 64t$. Find the time it hits the ground and the instant that it reaches its highest point. What is the maximum height?

15. The height of an object above the ground at time t is given by

$$s = v_0 t - \frac{g}{2}t^2$$

where v_0 represents the initial velocity and g is a constant called the acceleration due to gravity.
    (a) At what height is the object initially?
    (b) How long is the object in the air before it hits the ground?
    (c) When will the object reach its maximum height?
    (d) What is that maximum height?

For Problems 16–19:

(a) Find a possible formula for the graph.

(b) For each graph, read off approximate intervals over which the function is increasing and over which it is decreasing.

16.

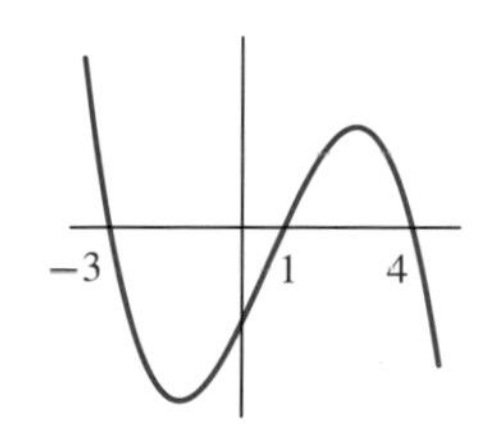

17.

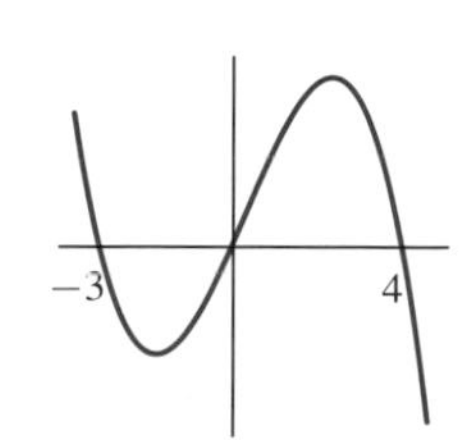

18.

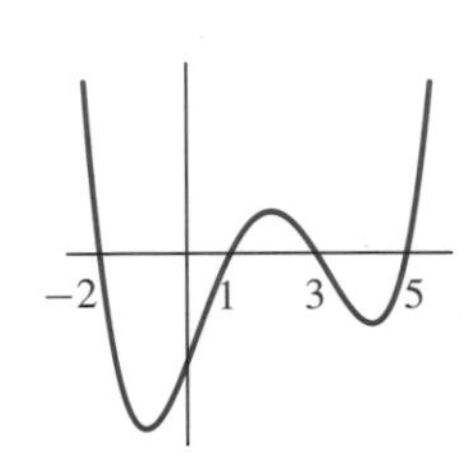

19.

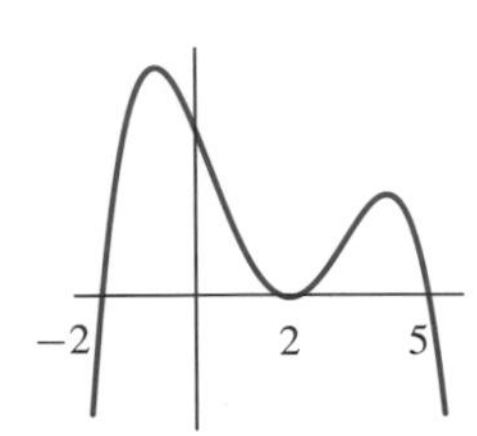

Determine cubic polynomials that represent each of the graphs in Problems 20–21.

20.

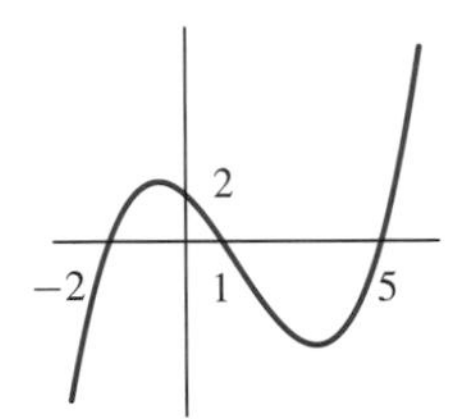

21.

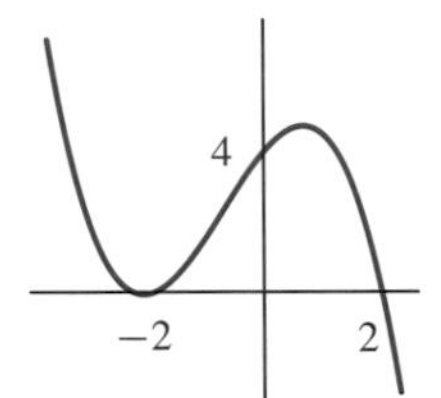

22. By shifting the graph of $y = x^3$, find a cubic polynomial with a graph similar to that in Figure 1.89.

23. The graph of a rational function $y = f(x)$ is given in Figure 1.90. If $f(x) = g(x)/h(x)$ with $g(x)$ and $h(x)$ both quadratic functions, give possible formulas for $g(x)$ and $h(x)$. (There are many possible answers.)

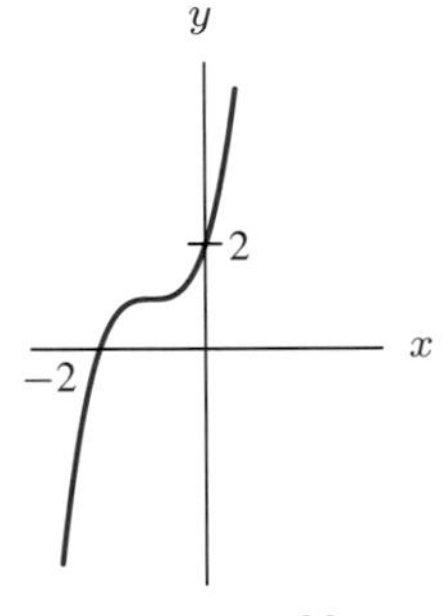

Figure 1.89

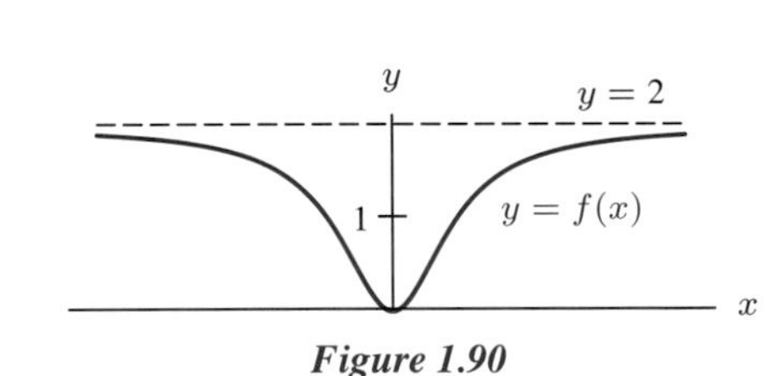

Figure 1.90

24. Newton's Second Law of Motion, $F = ma$, tells us that the net force, F, on a train of mass m is proportional to its acceleration, a. Suppose that the only forces are those of the engine, which exerts a constant force, F_E, in the direction of motion, and the wind resistance, which exerts a force proportional to the square of the train's velocity, v, but in the opposite direction.
    (a) Write a formula giving a as a function of v.
    (b) Sketch a graph of a against v.

25. The rate, R, at which a population in a confined space increases is proportional to the product of the current population, P, and the difference between the *carrying capacity*, L, and the current population. (The carrying capacity is the maximum population the environment can sustain.)
    (a) Write R as a function of P.
    (b) Sketch R as a function of P.

REVIEW PROBLEMS FOR CHAPTER ONE

1. A car starts out slowly and then goes faster and faster until a tire blows out. Sketch a possible graph of the distance the car has traveled as a function of time.
2. Having left home in a hurry, I'd only gone a short distance when I realized I hadn't turned off the washing machine, and so I went back to do so. I then set out again immediately. Sketch my distance from home as a function of time.
3. The graph in Figure 1.91 shows how the usage of household gas, e.g., for cooking, varies with the time of day in Ankara, the capital city of Turkey. Give a possible explanation for the shape of the graph.

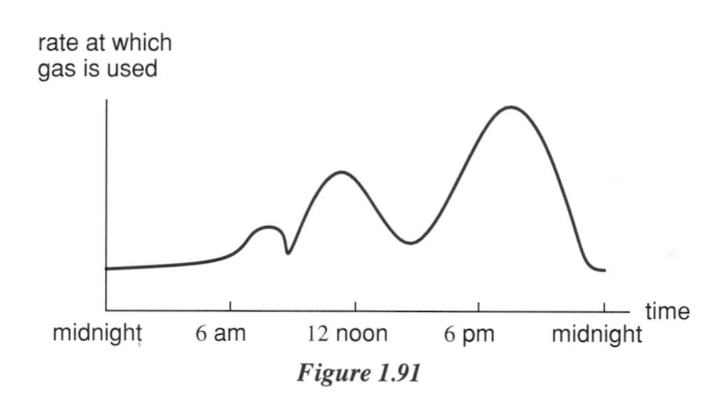

Figure 1.91

4. Sketch a possible graph for a function that is decreasing everywhere, concave up for negative x and concave down for positive x.
5. Consider the graph in Figure 1.92.

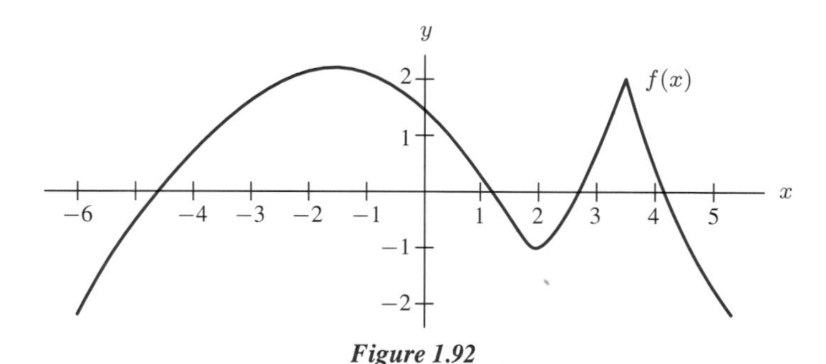

Figure 1.92

   (a) How many zeros does this function have? Approximately where are they?
   (b) Give approximate values for $f(2)$ and $f(4)$.
   (c) Is the function increasing or decreasing near $x = -1$? How about near $x = 3$?
   (d) Is the graph concave up or concave down near $x = 2$? How about near $x = -4$?
   (e) List all intervals (approximately) on which the function is increasing.
6. Table 1.26 gives the average temperature in Wallingford, Connecticut, for the first 10 days in March 1990.
   (a) Over which intervals was the average temperature increasing? Decreasing?

(b) Find a pair of consecutive intervals over which the average temperature was increasing at a decreasing rate. Find another pair of consecutive intervals over which the average temperature was increasing at an increasing rate.

TABLE 1.26

Date in March	1	2	3	4	5	6	7	8	9	10
Average temperature (°F)	42°	42°	34°	25°	22°	34°	38°	40°	49°	49°

7. For tax purposes, you may have to report the value of your assets, such as cars or refrigerators. The value you report depreciates, or drops, with time. The idea is that a car you originally paid $10,000 for may be worth only $5000 a few years later. The simplest way to calculate the value of your asset is using "straight-line depreciation," which assumes that the value is a linear function of time. If a $950 refrigerator depreciates completely in seven years, find a formula for its value as a function of time.

8. An airplane uses a fixed amount of fuel for takeoff, a (different) fixed amount for landing, and a fixed amount per mile when it is in the air. How does the total quantity of fuel required depend on the length of the trip? Write a formula for the function involved. Explain the meaning of the constants in your formula.

9. Sketch reasonable graphs for the following. Pay particular attention to the concavity of the graphs, and explain your reasoning.

(a) The total revenue generated by a car rental business, plotted against the amount spent on advertising.
(b) The temperature of a cup of hot coffee standing in a room, plotted as a function of time.

10. If $f(x) = x^2 + 1$, find:

(a) $f(t+1)$ (b) $f(t^2+1)$ (c) $f(2)$ (d) $2f(t)$ (e) $[f(t)]^2 + 1$

11. For $g(x) = x^2 + 2x + 3$, determine:

(a) $g(2+h)$ (b) $g(2)$ (c) $g(2+h) - g(2)$

12. For $f(n) = 3n^2 - 2$ and $g(n) = n + 1$, determine:

(a) $f(n) + g(n)$
(b) $f(n)g(n)$
(c) The domain of $f(n)/g(n)$.
(d) $f(g(n))$
(e) $g(f(n))$

13. Sketch the graph of a function defined for $x \geq 0$ with all of the following properties. (There are lots of possible answers.)

(a) $f(0) = 2$.
(b) $f(x)$ is increasing for $0 \leq x < 1$.
(c) $f(x)$ is decreasing for $1 < x < 3$.
(d) $f(x)$ is increasing for $x > 3$.
(e) $f(x) \to 5$ as $x \to \infty$.

14. When a new product is advertised, more and more people try it. However, the rate at which new people try it slows as time goes on.

(a) Sketch a graph of the total number of people who have tried such a product against time.
(b) What do you know about the concavity of the graph?

Convert the functions in Problems 15–16 into the form $P = P_0a^t$.

15. $P = 2.91e^{0.55t}$

16. $P = (5 \cdot 10^{-3})e^{-1.9 \cdot 10^{-2}t}$

17. (a) Use the data from Table 1.27 to determine a formula of the form

$$Q = Q_0e^{rt}$$

which would give the number of rabbits, Q, at time t (in months).

(b) What is the approximate doubling time for this population of rabbits?

(c) Use your equation to predict when the rabbit population will reach 1000.

TABLE 1.27

t	0	1	2	3	4	5
Q	25	43	75	130	226	391

18. If you need \$20,000 in your bank account in 6 years, how much must be deposited now? (Assume an annual interest rate of 10%, compounded continuously.)

19. What nominal annual interest rate has an effective annual yield of 5%?

20. What is the effective annual yield for a nominal annual interest rate of 8%?

21. Different kinds of the same element (called different *isotopes*) can have very different half-lives. The decay of plutonium-240 is described by the formula

$$Q = Q_0e^{-0.00011t}$$

whereas the decay of plutonium-242 is described by

$$Q = Q_0e^{-0.0000018t}.$$

Find the half-lives of plutonium-240 and plutonium-242.

22. An animal skull still has 20% of the carbon-14 that was present when the animal died. The half-life of carbon-14 is 5730 years. Find the approximate age of the skull.

23. Suppose prices are increasing by 0.1% a day.

(a) By what percent do prices increase a year?

(b) Looking at your answer to part (a), guess the approximate doubling time of prices increasing at this rate. Check your guess.

24. In the early 1920s, Germany had tremendously high inflation, called hyperinflation. Photographs of the time show people going to the store with wheelbarrows full of money. If a loaf of bread cost $1/4$ DM in 1919 and 2,400,000 DM in 1922, what was the average yearly inflation rate between 1919 and 1922?

25. Each planet moves around the sun in an elliptical orbit. The orbital period, T, of a planet is the time it takes the planet to go once around the sun. The semimajor axis of each planet's orbit is the average of the largest and the smallest distances between the planet and the sun. Kepler discovered that the period of a planet is proportional to the $\frac{3}{2}$ power of its semimajor axis. What is the orbiting period (in days) of Mercury, the closest planet to the sun, with a semimajor axis of 58 million km? What is the period (in years) of Pluto, the farthest planet, with a semimajor axis of 6000 million km? The semimajor axis of the earth is 150 million km. [Hint: What is the earth's period?]

26. Match the following formulas with the graphs in Figure 1.93:

(a) $y = 1 - 2^{-x}$ (d) $y = 1 - x^2$ (g) $y = 2^{-x}\sin x$

(b) $y = \log(x+1)$ (e) $y = \tan x$ (h) $y = 1 + \cos x$

(c) $y = 2\cos x$ (f) $y = x^3 - x^2 - x + 1$ (i) $y = \arctan x$

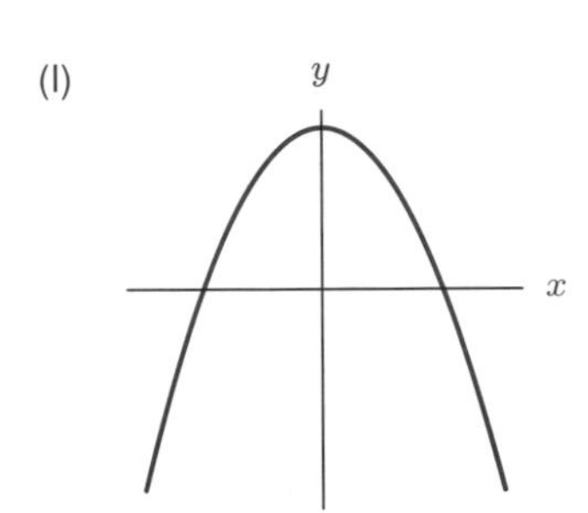

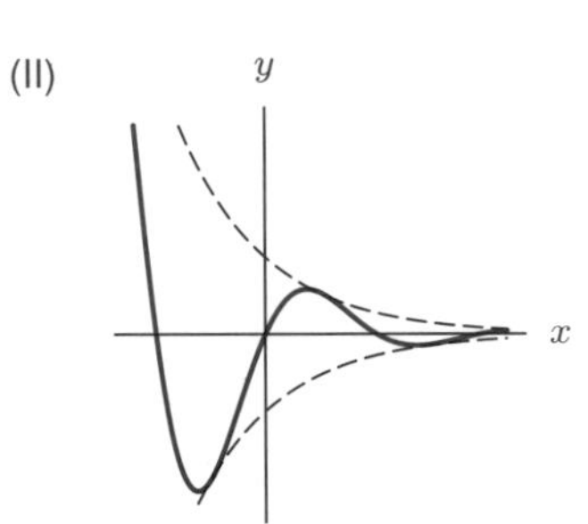

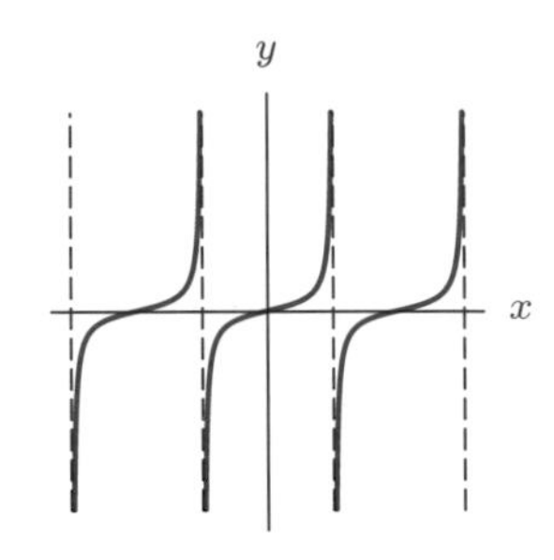

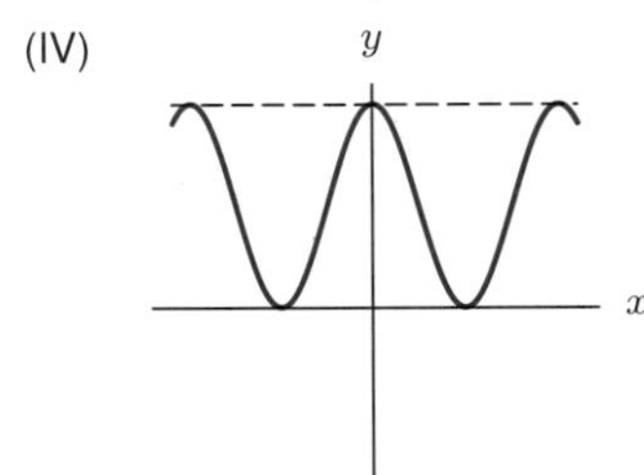

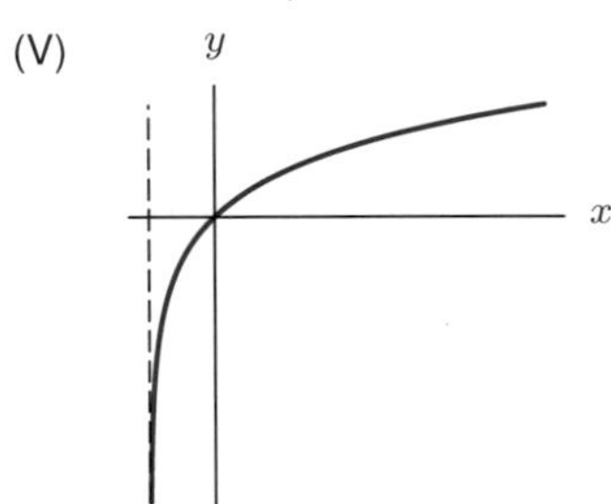

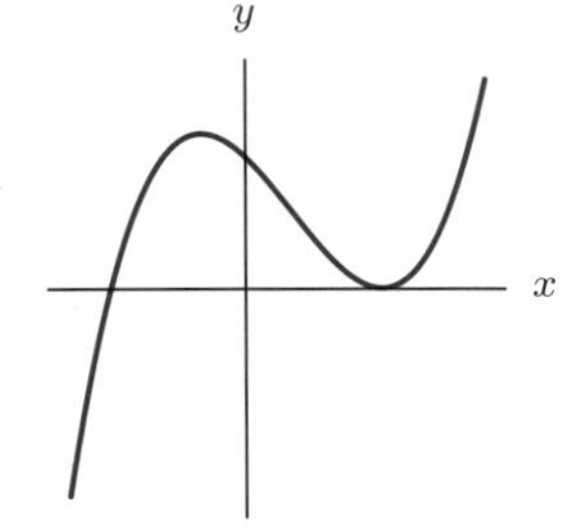

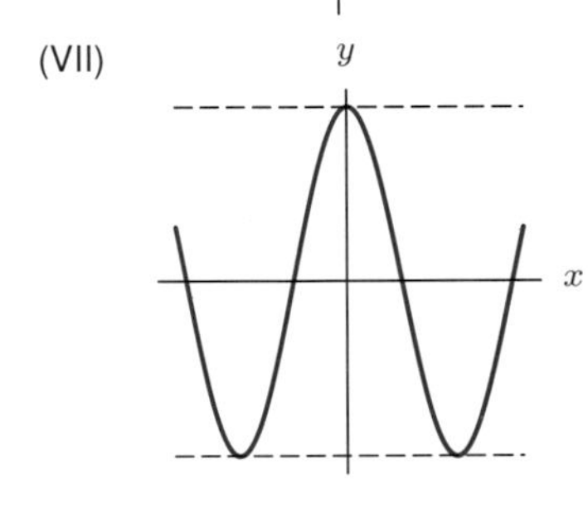

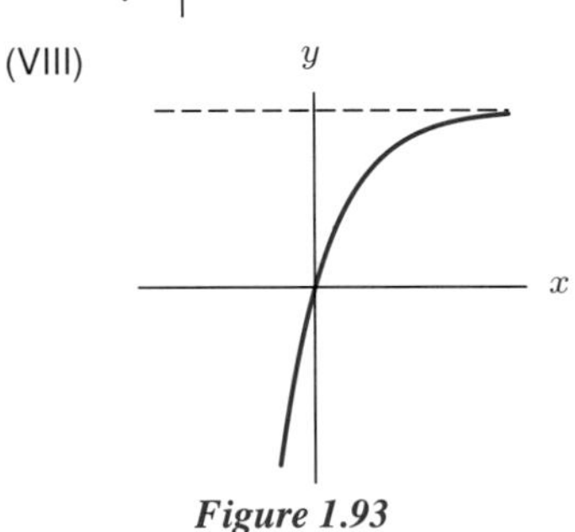

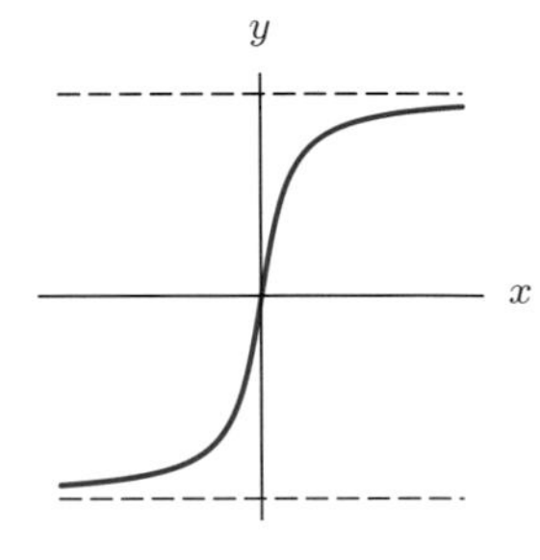

Figure 1.93

27. (a) Consider the functions shown in Figure 1.94(a). Find the coordinates of C.
(b) Consider the functions shown in Figure 1.94(b). Find the coordinates of C in terms of b.

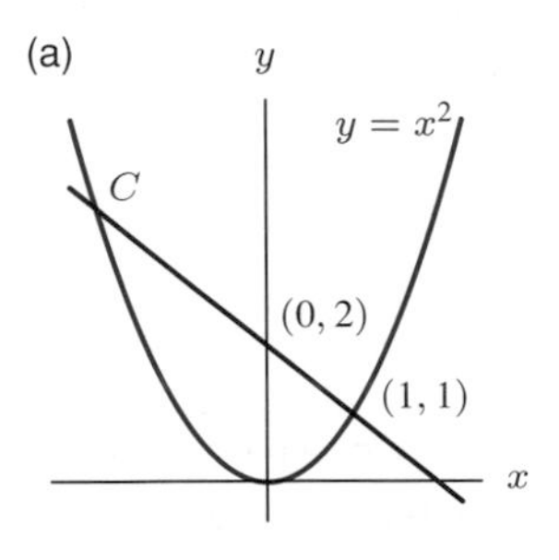

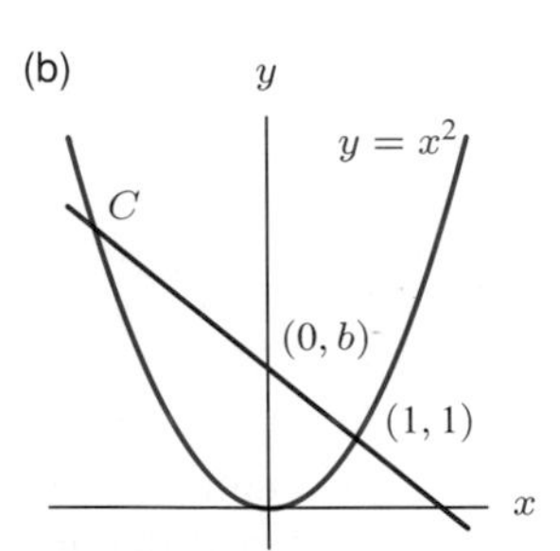

Figure 1.94

Find possible formulas for the functions graphed in Problems 28–36.

28.

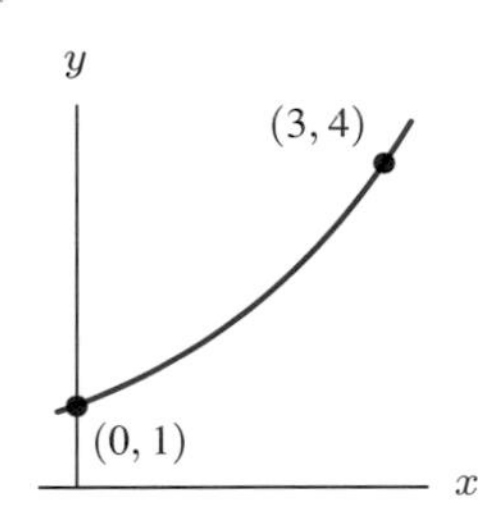

29.

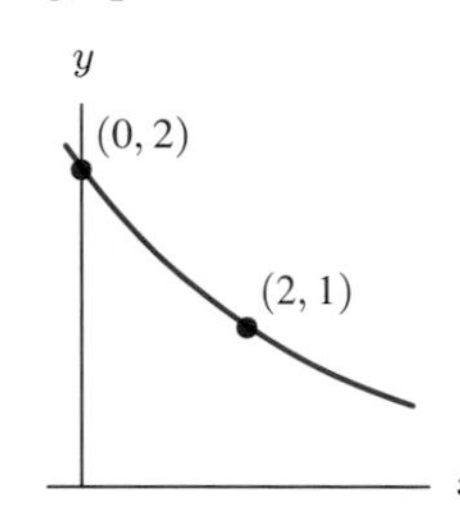

30.

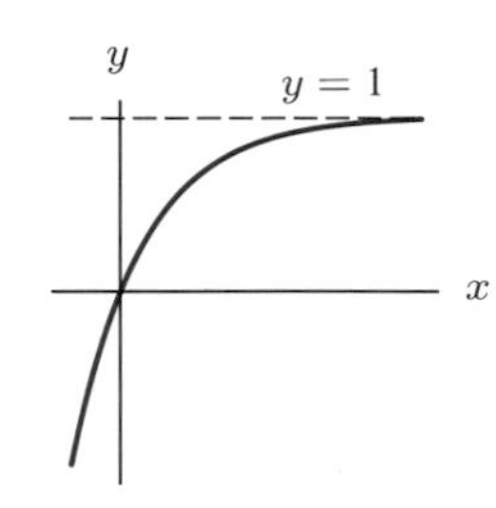

31.

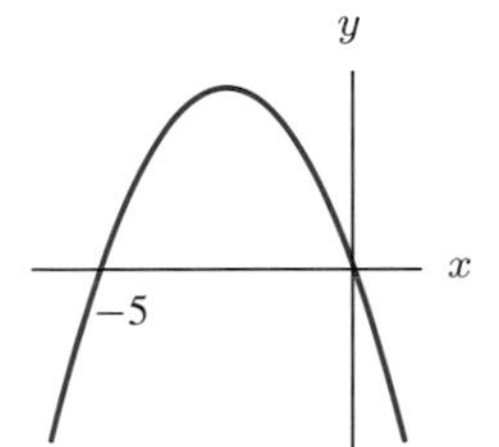

32.

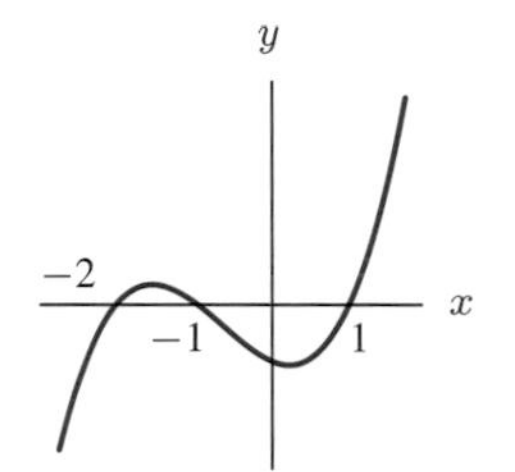

33.

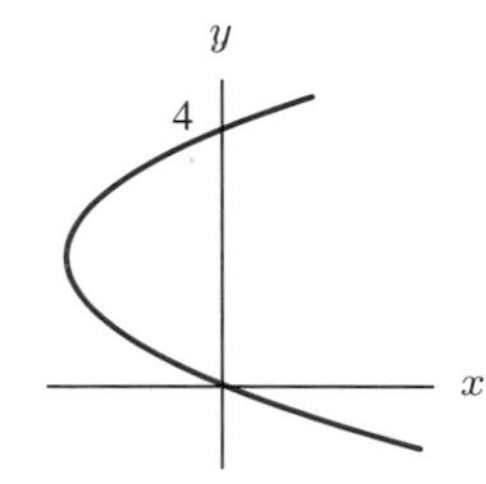

34.

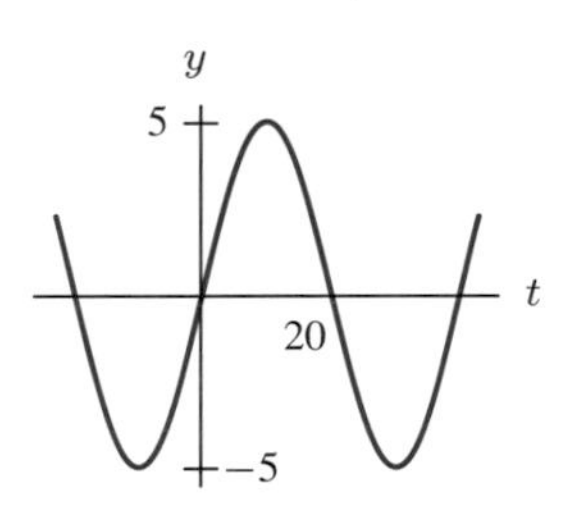

35.

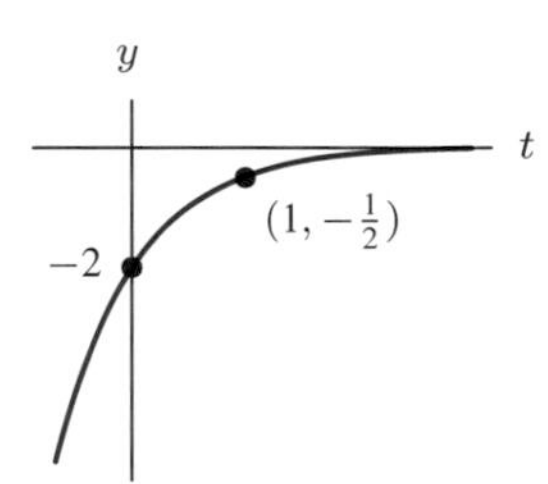

36.

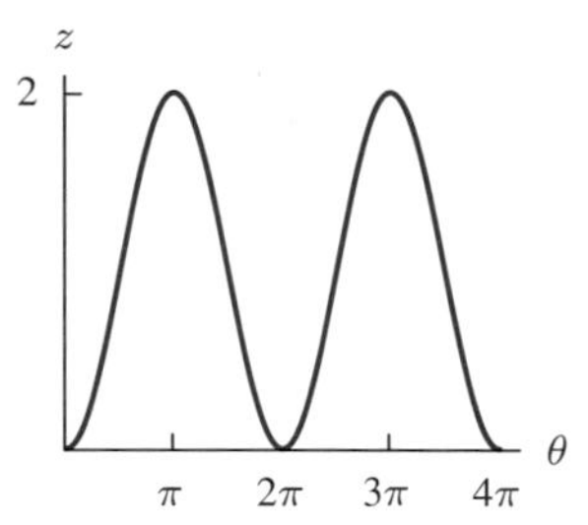

37. (a) What does Table 1.28 tell you about the number of roots of $\sin\left(t^2\right) = 0$ in the interval $0 \le t \le 4$? Approximately where do they lie?

TABLE 1.28

t	0	1	2	3	4
$\sin(t^2)$	0	0.84	−0.76	0.41	−0.29

(b) What does a graphing calculator or computer tell you about the number of roots the equation has in the interval $0 \le t \le 4$?

(c) Use the calculator or computer to estimate each of these roots to one decimal place.

(d) Explain why the smallest positive root is $\sqrt{\pi}$.

(e) Find an exact, symbolic expression (like $\sqrt{\pi}$) for each of the other positive roots you found.

38. (a) Define $\arcsin x$.

(b) Find $\arcsin(\sin 1)$, $\arcsin(\sin 2)$, $\sin(\arcsin 1)$, $\sin(\arcsin 2)$

(c) Explain your answers to part (b).

39. Using a calculator or computer, sketch a graph of $y = \sin x$, $y = 0.4$, and $y = -0.4$.

(a) From the graph, estimate to one decimal place all the solutions of $\sin x = 0.4$ with $-\pi \le x \le \pi$.

(b) Use a calculator to find arcsin(0.4). What is the relation between arcsin(0.4) and each of the solutions you found in part (a)?

(c) Estimate all the solutions to $\sin x = -0.4$ with $-\pi \leq x \leq \pi$ (again, to one decimal place).

(d) What is the relation between arcsin(0.4) and *each* of the solutions you found in part (c)?

40. The depth of water in a tank oscillates sinusoidally once every 6 hours around an average depth of 7 feet. If the smallest depth is 5.5 feet and the largest depth is 8.5 feet, find a formula for the depth in terms of time, measured in hours. (There are many possible answers.)

41. (a) Use a graphing calculator or computer to find the period of $2\sin 4x + 3\cos 2x$.

(b) Give your answer in an exact form (as a multiple of π).

(c) Find the periods of $\sin 4x$ and $\cos 2x$, and use them to explain your answer to part (a).

42. Each of the functions described by the data in Table 1.29 is increasing over its domain, but each increases in a different way. Which of the graphs in Figure 1.95 best fits each function?

TABLE 1.29

x	$f(x)$
1	1
2	2
4	3
7	4
11	5
16	6
22	7
29	8
37	9
47	10

x	$g(x)$
3.0	1
3.2	2
3.4	3
3.6	4
3.8	5
4.0	6
4.2	7
4.4	8
4.6	9
4.8	10

x	$h(x)$
10	1
20	2
28	3
34	4
39	5
43	6
46.5	7
49	8
51	9
52	10

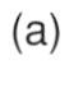

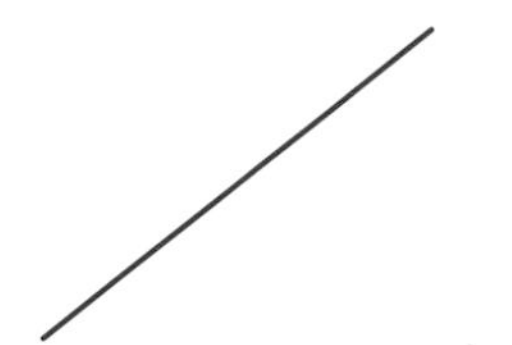

(b)

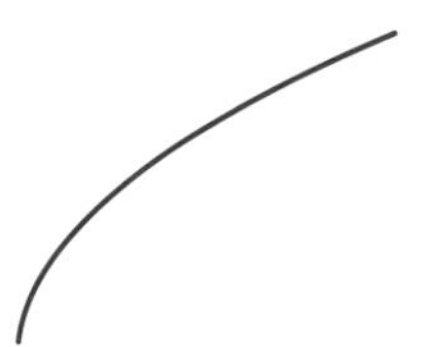

(c)

Figure 1.95

43. Given the graph of $y = h(x)$ in Figure 1.96:

(a) Sketch a graph of:

(i) $y = h^{-1}(x)$

(ii) $y = \dfrac{1}{h(x)}$

(b) What becomes of the asymptote when you sketch the inverse function?

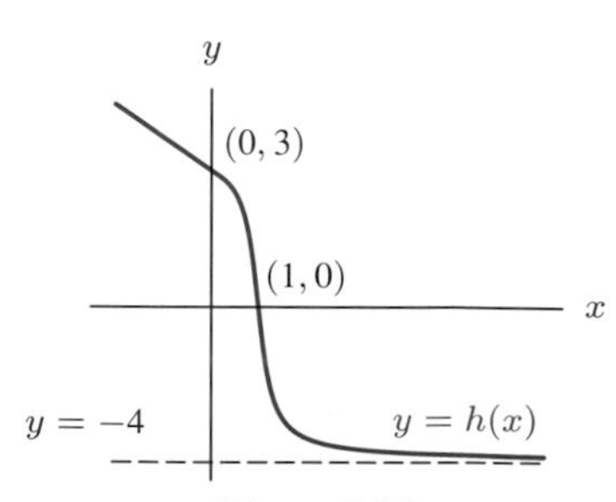

Figure 1.96

44. (a) What effect does the transformation

$$y = p(x) \qquad \text{to} \qquad y = p(1 + x)$$

have on the graph of $p(x)$?

(b) If p is a polynomial of degree ≤ 2 such that for all x

$$p(x) = p(1 + x),$$

what can you say about p?

45. For each of the following two conditions, find all polynomials, p, of degree ≤ 2 which satisfy the condition for all x. (a) $p(x) = p(-x)$ (b) $p(2x) = 2p(x)$

46. A *catalyst* in a chemical reaction is a substance which speeds up the reaction but which does not itself change. If the product of a reaction is itself a catalyst, the reaction is said to be *autocatalytic*. Suppose the rate, r, of a particular autocatalytic reaction is proportional to the quantity of the original material remaining times the quantity of product, p, produced. If the initial quantity of the original material is A and the amount remaining is $A - p$:

(a) Express r as a function of p.
(b) What is the value of p when the reaction is proceeding fastest?

47. Glucose is fed by intravenous injection at a constant rate, k, into a patient's bloodstream. Once there, the glucose is removed at a rate proportional to the amount of glucose present. If R is the net rate at which the quantity, G, of glucose in the blood is increasing:

(a) Write a formula giving R as a function of G.
(b) Sketch a graph of R against G.

48. A fish population is reproducing at an annual rate equal to 5% of the current population, P. Meanwhile, fish are being caught by fishermen at a constant rate, Y (measured in fish per year).

(a) Write a formula for the rate, R, at which the fish population is increasing as a function of P.
(b) Sketch a graph of R against P.

49. Let $S(x)$ be the number of sunlight hours on a cloudless June 21, as a function of the latitude x (measured in degrees).

(a) What is $S(0)$? [Hint: Latitude 0° is the equator.]
(b) Let x_0 be the latitude of the Arctic Circle ($x_0 \approx 66°30'$). In the Northern Hemisphere, $S(x)$ is given by:

$$S(x) = \begin{cases} a + b \arcsin\left(\dfrac{\tan x}{\tan x_0}\right) & \text{for } 0 \leq x \leq x_0 \\ 24 & \text{for } x_0 \leq x \leq 90 \end{cases}$$

for some constants a and b. Find a and b.

(c) Compute $S(x)$ for Tucson, Arizona ($x = 32°13'$) and Walla Walla, Washington ($46°4'$).
(d) Graph $S(x)$, for $-90° \leq x \leq 90°$, using a graphing calculator or computer. Is the graph smooth?

50. Figure 1.97 is the graph of the function f, where $f(t)$ is the depth in meters below the Atlantic Ocean floor where t million-year-old rock can be found.[7]

(a) Evaluate $f(15)$, and say what it means in practical terms.

(b) Is f invertible? Explain.

(c) Evaluate $f^{-1}(120)$, and say what it means in practical terms.

(d) Sketch a graph of f^{-1}.

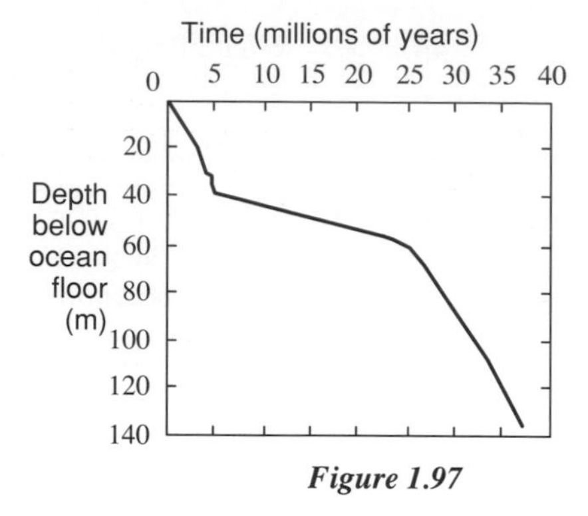

Figure 1.97

[7]Data of Dr. Murlene Clark based on core samples drilled by the research ship *Glomar Challenger*, taken from *Initial Reports of the Deep Sea Drilling Project*

CHAPTER TWO

KEY CONCEPT: THE DERIVATIVE

We begin this chapter by investigating the problem of speed: how can we measure the speed of a moving object at a given instant in time? Or, more fundamentally, what do we mean by the term *speed*? We'll come up with a definition of speed that has wide-ranging implications — not just for the speed problem, but for measuring any rate of change. Our journey will lead us to the key concept of *derivative*, which forms the basis for our study of calculus.

The derivative can be interpreted geometrically as the slope of a curve, and physically as a rate of change. Because derivatives can be used to represent everything from fluctuations in interest rates to the rates at which fish are dying and gas molecules are moving, they have applications throughout the sciences.

2.1 HOW DO WE MEASURE SPEED?

The notion of speed—and, in particular, the speed of an object at an instant in time—is surprisingly subtle and difficult to define precisely. Consider the statement "At the instant it crossed the finish line, the horse was traveling at 42 mph." How can such a claim be substantiated? A photograph taken at that instant will show the horse motionless—it is no help at all. There is some paradox in trying to quantify the property of motion at a particular instant in time, since by focusing on a single instant you stop the motion!

A similar difficulty arises whenever we attempt to measure the rate of change of anything—for example, oil leaking out of a damaged tanker. The statement "One hour after the ship's hull ruptured, oil was leaking at a rate of 200 barrels per second" seems not to make sense. You could argue that at any given instant *no* oil is leaking.

Problems of motion were of central concern to Zeno and other philosophers as early as the fifth century B.C. The modern approach, made famous by Newton's calculus, is to stop looking for a simple notion of speed at an instant, and instead to look at speed over small intervals containing the instant. This method sidesteps the philosophical problems mentioned earlier but brings new ones of its own.

We shall illustrate the ideas discussed above by an idealized example, called a thought experiment. It is idealized in the sense that we assume that we can make measurements of distance and time as accurately as we wish. In fact, the numbers we shall use come from a mathematical formula, not from real measurements, but that doesn't matter for our purposes.

A Thought Experiment: Average and Instantaneous Velocity

We shall look at the speed of a small object (say, a grapefruit) that is thrown high in the air at $t = 0$ seconds. The behavior of the grapefruit is not surprising: it goes up, slows down, reverses direction, falls down, and finally, "*Splat!*" Informally, we can say that the grapefruit leaves the thrower's hand at high speed, slows down until it reaches its maximum height, and then gradually speeds up in the downward direction. (See Figure 2.1.)

But suppose that we would like to be more precise in our determination of the speed, say, at $t = 1$ second. We'll assume that we can measure the height of the grapefruit above the ground at any time t; we'll think of the height, y, as a function of time. (See Table 2.1.) "*Splat!*" comes sometime between 6 and 7 seconds. The numbers show the behavior noted above: in the first second the grapefruit travels $90 - 6 = 84$ feet, and in the second second it travels only $142 - 90 = 52$ feet. Hence the grapefruit traveled faster over the first interval, $0 \leq t \leq 1$, than the second interval, $1 \leq t \leq 2$. Since speed = distance/time, we say that the average speed over the interval $0 \leq t \leq 1$ is 84 ft/sec, and the average speed over the next interval is 52 ft/sec.

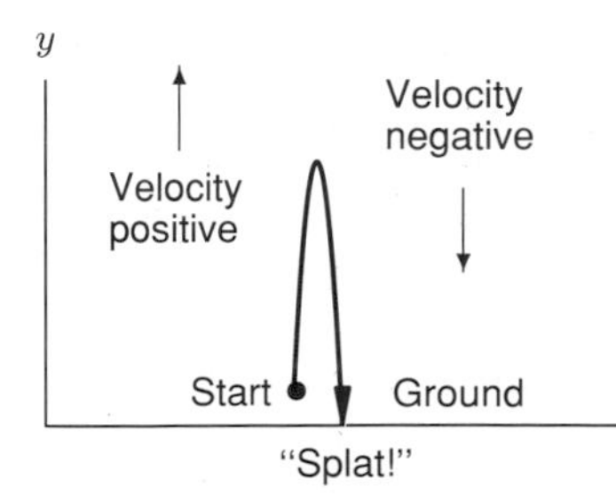

Figure 2.1: The grapefruit's path

TABLE 2.1 *Height of the grapefruit above the ground*

t (sec)	0	1	2	3	4	5	6
y (feet)	6	90	142	162	150	106	30

Velocity versus Speed

From now on, we will make a distinction between velocity and speed. Suppose an object moves along a line. If we pick one direction to be positive, the *velocity* is positive if it is in the same direction, and negative if it is in the opposite direction. For the grapefruit, upward is positive and downward is negative. (See Figure 2.1.) *Speed* is the magnitude of the velocity and so is always positive or zero.

> The **average velocity** of an object over the interval $a \leq t \leq b$ is the net change in position during the interval divided by the change in time, $b - a$.

Example 1 Compute the average velocity of the grapefruit over the interval $4 \leq t \leq 5$. What is the significance of the sign of your answer?

Solution During this interval, the grapefruit moves $(106 - 150) = -44$ feet. Therefore the average velocity is -44 ft/sec. The negative sign means the height is decreasing and the grapefruit is moving downward.

Example 2 Compute the average velocity of the grapefruit over the interval $1 \leq t \leq 3$.

Solution Average velocity $= (162 - 90)/(3 - 1) = 72/2 = 36$ ft/sec.

The average velocity is a useful concept since it gives a rough idea of the behavior of the grapefruit: if two identical grapefruits are hurled into the air, and one has an average velocity of 10 ft/sec over the interval $0 \leq t \leq 1$ while the second has an average velocity of 100 ft/sec over the same interval, clearly the second one was thrown harder.

But average velocity over an interval doesn't solve the problem of measuring the velocity of the grapefruit at *exactly* $t = 1$ second. To get closer to an answer to that question, we'll have to look at what happens near $t = 1$ in more detail. Suppose we are given the data[1] in Figure 2.2, showing the average velocity over small intervals on either side of $t = 1$.

Notice that the average velocity before $t = 1$ is slightly more than the average velocity after $t = 1$. We would expect to define the velocity *at* $t = 1$ to be between these two average velocities. As the size of the interval shrinks, the values of the velocity before $t = 1$ and the velocity after $t = 1$ get closer together. By the smallest interval in Figure 2.2, both velocities are 68.0 ft/sec (to one decimal place), so we will define the velocity at $t = 1$ to be 68.0 ft/sec (to one decimal place).

Of course, if we showed more decimal places, average velocities before and after $t = 1$ would no longer agree even in the smallest interval. To calculate the velocity at $t = 1$ to more decimal places, we would have to take smaller and smaller intervals on either side of $t = 1$ until the average velocities agree to the number of decimal places we wanted. The velocity at $t = 1$ is then defined to be this common average velocity.

[1] The data is in fact calculated from the formula $y = 6 + 100t - 16t^2$.

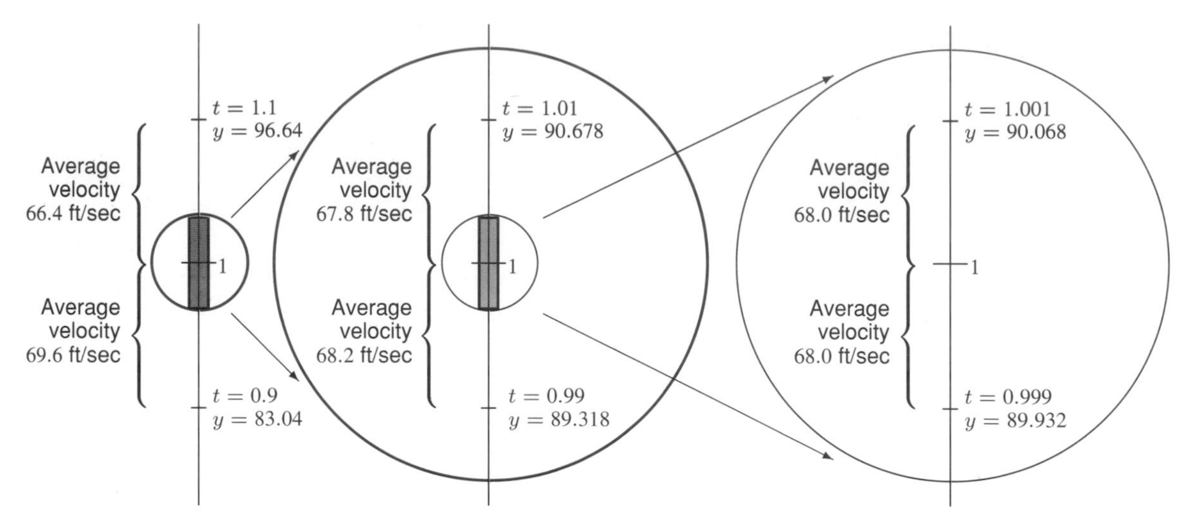

Figure 2.2: Average velocities over intervals on either side of $t = 1$: Showing successively smaller intervals

Defining Instantaneous Velocity Using the Idea of a Limit

When you take smaller and smaller intervals, it turns out that the average velocities are always just above or just below 68 ft/sec. It seems natural, then, to define velocity at the instant $t = 1$ to be 68 ft/sec. This is called the *instantaneous velocity* at this point, and its definition depends on our being adequately convinced that smaller and smaller intervals will provide average speeds that come arbitrarily close to 68. Modern mathematics has a name for this process: it is called *taking the limit*.

> The **instantaneous velocity** of an object at time t is given by the limit of the average velocity over the interval as the interval shrinks around t.

Make sure that you see how we have replaced the original difficulty of computing velocity at a point by a search for an argument to convince ourselves that the average velocities do approach a number as the time intervals shrink in size. In a sense, we have traded one hard question for another, since we don't yet have any idea how to be certain what number the average velocities are approaching. In our thought experiment, the number seems to be exactly 68, but what if it were 68.000001? How can we be sure that we have taken small enough intervals?

For most practical purposes, it is not likely to be important whether the velocity is exactly 68 or 68.000001. Showing that the limit is exactly 68 requires more precise knowledge of how the velocities were calculated and of the limiting process; we will see this in Chapter 4.

Visualizing Velocity: Slope of Curve

We have just seen how to estimate velocity numerically. Now we will see how to visualize velocity using a graph of height. Let's go back to the grapefruit. Suppose that Figure 2.3 shows the height of the grapefruit plotted against time. (Note that this is not a picture of the grapefruit's path, which is almost straight up and down.)

How can we visualize the average velocity on this graph? Suppose $y = s(t)$. Let's consider the interval $1 \le t \le 2$ and the expression

$$\text{Average velocity} = \frac{\text{Distance moved}}{\text{Time elapsed}} = \frac{s(2) - s(1)}{2 - 1} = \frac{142 - 90}{1} = 52 \text{ ft/sec}.$$

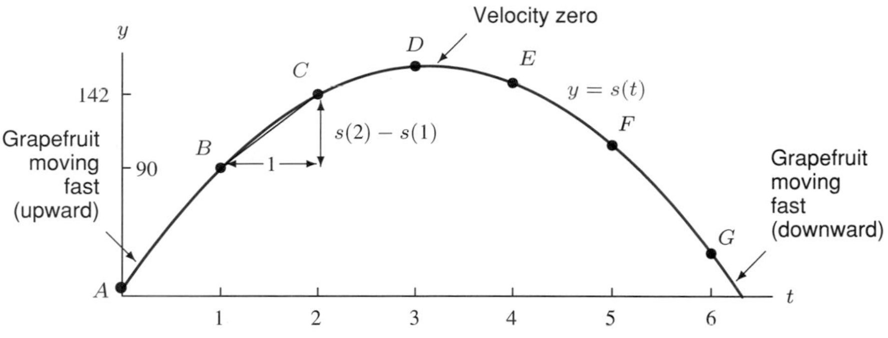

Figure 2.3: The height, y, of the grapefruit at time t

Now $s(2) - s(1)$ is the change in height over the interval, or the distance moved, and it is marked vertically in Figure 2.3. The 1 in the denominator is the time elapsed and is marked horizontally in Figure 2.3. Therefore,

$$\text{Average velocity} = \frac{\text{Distance moved}}{\text{Time elapsed}} = \text{Slope of line joining } BC.$$

(See Figure 2.3.) A similar argument shows the following:

> The **average velocity** over any interval is the slope of the line joining the points on the graph of $s(t)$ corresponding to the endpoints of the interval.

The next question is how to visualize the velocity at an instant. Let's think about how we found the instantaneous velocity. We took average velocities across smaller and smaller intervals ending at the point $t = 1$. Two such velocities are represented by the slopes of the lines in Figure 2.4. As the length of the interval shrinks, the slope of the line gets closer to the slope of the curve at $t = 1$.

The cornerstone of the idea is the fact that, on a very small scale, most functions look almost like straight lines. Imagine taking the graph of a function near a point and "zooming in" to get a close-up view. (See Figure 2.5.) The more you zoom in, the more the curve will appear to be a straight line. In other words, if you repeatedly zoom in on a section of the curve centered at a point of interest, the section of curve will eventually look like a straight line. We call the slope of this line the *slope of the curve* at the point. Therefore, the slope of the magnified line is the instantaneous velocity. Thus, we can say:

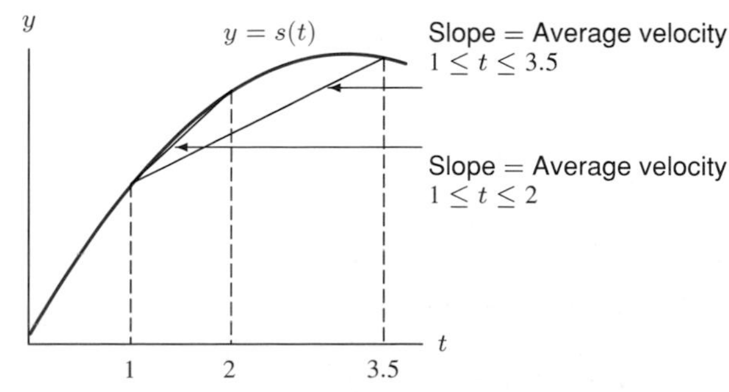

Figure 2.4: Average velocities over small intervals

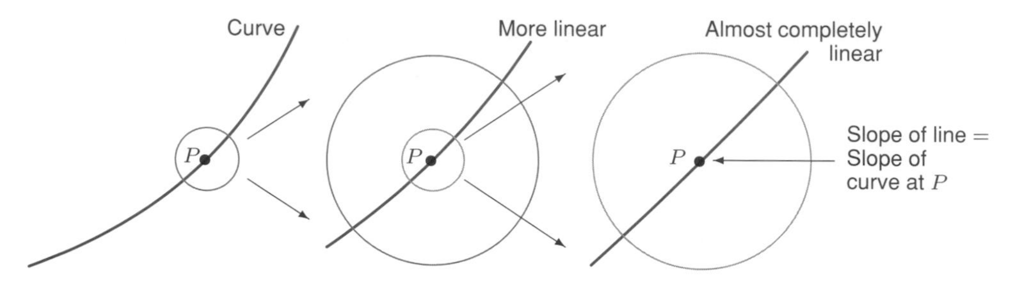

Figure 2.5: Finding the slope of the curve at the point by "zooming in"

> The **instantaneous velocity** is the slope of the curve at a point.

Look back at the graph of the grapefruit's height as a function of time in Figure 2.3. If you think of the velocity at any point as the slope of the curve there, you can see how the grapefruit's velocity varies during its journey. At points A and B the curve has a large positive slope, indicating that the grapefruit is traveling up rapidly. Point D is almost at the top: the grapefruit is slowing as it reaches the peak. At the peak, the slope of the curve is zero: the fruit has slowed to a stop in preparation for its return to earth. At point E the curve has a small negative slope, indicating a slow velocity of descent. Finally, the slope of the curve at point G is large and negative, indicating a large downward velocity that is responsible for the *"Splat."*

Example 3 At the top of the grapefruit's path, its velocity is zero. Does this mean the grapefruit has stopped?

Solution At the top of its path the grapefruit changes direction—from moving upward to moving downward—and at the moment it makes this change, the grapefruit's instantaneous velocity is zero. Deciding whether the grapefruit has stopped depends on what you think "stopped" means. Most people think an object has stopped only if it has zero velocity for some interval of time; with this definition, our grapefruit has not stopped, but only had zero instantaneous velocity for one instant of time.

The Precise Definition of Velocity and the Notation for Limit

To recap the results so far, we found that a natural way to define the velocity of an object at an instant t was to look at the average velocity over smaller and smaller intervals containing t and see to what number those average velocities converge. Let us introduce a formal expression for this idea. As before, let s be the name for the function of time that gives the position (or height) y of the grapefruit at time t, so $y = s(t)$. Then we need to write down an expression that gives the average velocity over an interval.

If $a \le t \le b$,

$$\text{Average velocity} = \frac{\text{Change in position}}{\text{Change in time}} = \frac{s(b) - s(a)}{b - a}.$$

If we're interested in looking for instantaneous velocity at $t = a$, then we want to look at smaller and smaller intervals near $t = a$. We will consider intervals of the form $a \le t \le a + h$, where h is the size of the interval. Then, over the interval $a \le t \le a + h$,

$$\text{Average velocity} = \frac{s(a+h) - s(a)}{h}.$$

We get a similar formula when $h < 0$. Now remember that in the previous section we agreed that the instantaneous velocity is the number that the average velocities approach as the intervals decrease in size, that is, as h becomes smaller. Thus, the instantaneous velocity at $t = a$ is

$$\text{The limit, as } h \text{ approaches 0, of } \quad \frac{s(a+h) - s(a)}{h}.$$

One final small change is for the sake of economy: Instead of writing the phrase "limit, as h approaches 0, of," we'll just use

$$\lim_{h \to 0}$$

Now we can exhibit our definition in all of its finery:

$$\text{Instantaneous velocity at } t = a = \lim_{h \to 0} \frac{s(a+h) - s(a)}{h}.$$

This expression forms the foundation of the rest of calculus. Be sure that you are not confused by the notation and recognize it for what it is: the number that the average velocities approach as the intervals shrink. To find that number, the limit, we look at intervals of smaller and smaller, but never zero, length. You should realize that we haven't introduced any new ideas in this definition; we have simply found a compact way to write the ideas developed previously.

What Is a Limit, and How Do We Find One?

Let's look at what it means to take a limit in more detail. Suppose, for example, that $a = 5$ and $s(t) = t^2$, and that we want to calculate the limit, L:

$$L = \lim_{h \to 0} \frac{s(5+h) - s(5)}{h}.$$

Then L is the *number* obtained by evaluating the expression $(s(5+h) - s(5))/h$ at values of h getting closer and closer to 0, and observing to what number these values converge.

TABLE 2.2 *Finding a limit numerically*

h	0.1	0.01	0.001	0.0001	−0.1	−0.01	−0.001	−0.0001
$(s(5+h) - s(5))/h$	10.1	10.01	10.001	10.0001	9.9	9.99	9.999	9.9999

The values of $(s(5+h) - s(5))/h$ in Table 2.2 seem to be converging to the number 10 as $h \to 0$. So it is a good guess that

$$L = \lim_{h \to 0} \frac{s(5+h) - s(5)}{h} = 10.$$

However, from Table 2.2 we can't be sure that the limit isn't, for example, 10.0000001 or 9.99999. Showing that the limit is *exactly* 10 requires algebra. Further discussion of the concept of limit is in Section 2.7.

Notice that by taking a small value of h, we get a good approximation to L, though we don't get L exactly. We generally estimate limits numerically by the method below.

If
$$L = \lim_{h \to 0} \frac{s(a+h) - s(a)}{h},$$
then we often approximate L by taking a small value of h, giving
$$L \approx \frac{s(a+h) - s(a)}{h}.$$

Problems for Section 2.1

1. A car is driven at a constant speed. Sketch a graph of the distance the car has traveled as a function of time.
2. A car is driven at an increasing speed. Sketch a graph of the distance the car has traveled as a function of time.
3. A car starts at a high speed, and its speed then decreases slowly. Sketch a graph of the distance the car has traveled as a function of time.
4. A bicyclist pedals at a fairly constant rate, with evenly spaced intervals of coasting. Sketch a graph of the distance she has traveled as a function of time.
5. Find the average velocity over the interval $0 \le t \le 0.2$, and estimate the velocity at $t = 0.2$ of a car whose position, s, is given by

t (sec)	0	0.2	0.4	0.6	0.8	1.0
s (ft)	0	0.5	1.8	3.8	6.5	9.6

6. A ball is tossed into the air from a bridge, and its height, y (in feet), above the ground t seconds after it is thrown is given by
$$y = f(t) = -16t^2 + 50t + 36.$$
   (a) How high above the ground is the bridge?
   (b) What is the average velocity of the ball for the first second?
   (c) Approximate the velocity of the ball at $t = 1$ second.
   (d) Graph the function f, and determine the maximum height the ball will reach. What should the velocity be at the time the ball is at the peak?
   (e) Use the graph to decide at what time, t, the ball reaches its maximum height.

7. Match the points labeled on the curve in Figure 2.6 with the given slopes.

Slope	Point
-3	
-1	
0	
$1/2$	
1	
2	

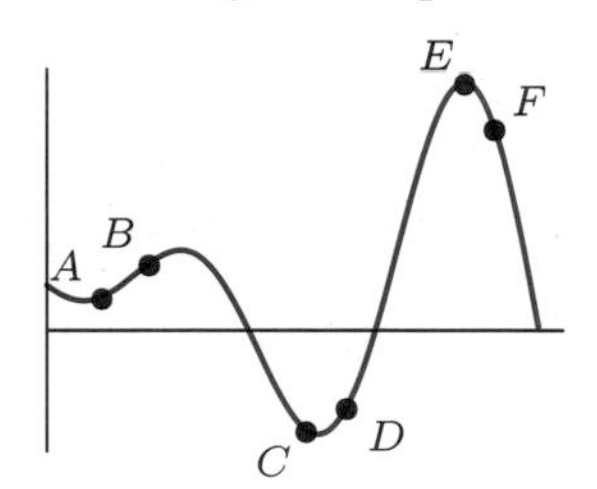

Figure 2.6

8. For the function shown in Figure 2.7, at what labeled points is the slope of the curve positive? Negative? Which labeled point has the greatest (i.e., most positive) slope? The least slope (i.e., negative and with the largest magnitude)?

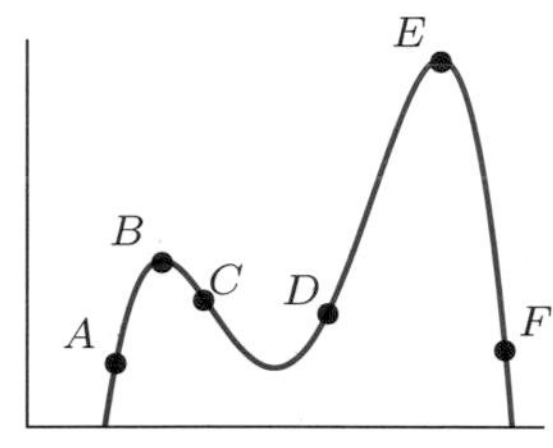

Figure 2.7

9. For the graph $y = f(x)$ shown in Figure 2.8, arrange the following numbers in ascending (i.e., smallest to largest) order:
   - The slope of the curve at A.
   - The slope of the curve at B.
   - The slope of the curve at C.
   - The slope of the line AB.
   - The number 0.
   - The number 1.

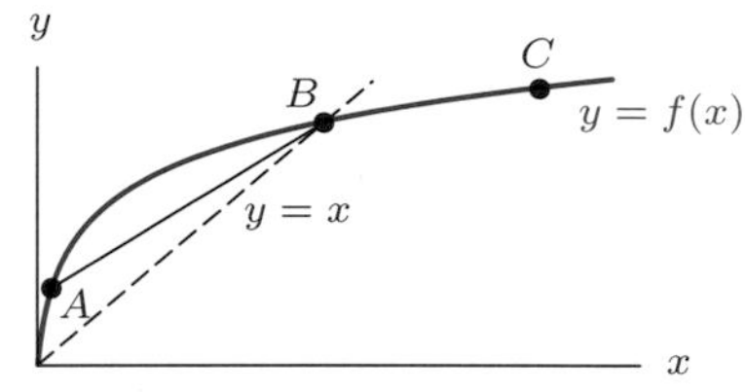

Figure 2.8

10. Suppose a particle is moving at varying velocity along a straight line and that $s = f(t)$ represents the distance of the particle from a point as a function of time, t. Sketch a possible graph for f if the average velocity of the particle between $t = 2$ and $t = 6$ is the same as the instantaneous velocity at $t = 5$.

Estimate the limits in Problems 11–14 by substituting smaller and smaller values of h. Give your answers to one decimal place.

11. $\displaystyle\lim_{h\to 0} \frac{(3+h)^2 - 9}{h}$

12. $\displaystyle\lim_{h\to 0} \frac{7^h - 1}{h}$

13. $\displaystyle\lim_{h\to 0} \frac{\cos h - 1}{h}$

14. $\displaystyle\lim_{h\to 0} \frac{e^{1+h} - e}{h}$

2.2 THE DERIVATIVE AT A POINT

Average Rate of Change

Now we will apply the analysis of Section 2.1 to any function $y = f(x)$, not just to height as a function of time. In the case of height, we looked at

$$\frac{s(a+h) - s(a)}{h},$$

which is the change in height divided by the length of a small time interval. Now we'll consider

$$\frac{f(a+h)-f(a)}{h}$$

for some function f. This ratio is called the *difference quotient*. What does it represent? The numerator, $f(a+h)-f(a)$, measures the change in the value of f during the interval from a to $a+h$. Thus, the difference quotient is the change in f divided by the change in x. Hence we can say:

$$\begin{array}{c}\text{Average rate of change}\\ \text{of } f \text{ over the interval}\\ \text{from } a \text{ to } a+h\end{array} = \frac{f(a+h)-f(a)}{h}$$

This ratio compares the change in the value of the function, $f(x)$, with h, the change in x. If a small change in x produces a large change in $f(x)$, this ratio will be large; conversely, if a small change in x produces an even smaller change in $f(x)$, the ratio will be small. Although the interval is no longer necessarily a time interval, we still talk about the *average rate of change* of f over the interval. If we want to emphasize the independent variable, we talk about the average rate of change of f *with respect to* x.

Blowing Up a Balloon

Consider the function which gives the radius of a sphere in terms of its volume. For example, think of blowing air into a balloon: As you change the volume of air, the radius changes. You've probably noticed that a balloon seems to blow up faster at the start and then slows down as you blow more air into it. What you're seeing is the rate of change of the radius with respect to change in volume. You can compute average rates of change of any function over any interval exactly as we did above. When the balloon is small, a small change in volume produces a relatively large change in radius. As the balloon grows, the same small change in volume (a puff of air) produces a much smaller change in the radius.

Example 1 The function r for the radius of a sphere in terms of its volume is given by the formula

$$r(V)=\left(\frac{3V}{4\pi}\right)^{1/3}.$$

Calculate the average rate of change of r with respect to V over the intervals $0.5 \le V \le 1$ and $1 \le V \le 1.5$.

Solution

$$\begin{array}{c}\text{Average rate of change}\\ \text{of radius for } 0.5 \le V \le 1\end{array} = \frac{r(1)-r(0.5)}{0.5} = 2\left(\left(\frac{3}{4\pi}\right)^{1/3}-\left(\frac{1.5}{4\pi}\right)^{1/3}\right) \approx 0.26$$

$$\begin{array}{c}\text{Average rate of change}\\ \text{of radius for } 1 \le V \le 1.5\end{array} = \frac{r(1.5)-r(1)}{0.5} = 2\left(\left(\frac{4.5}{4\pi}\right)^{1/3}-\left(\frac{3}{4\pi}\right)^{1/3}\right) \approx 0.18$$

As you can see, the rate does decrease as the volume increases.

Average Rate of Change versus Absolute Change

The average rate of change of a function during an interval is not at all the same as the absolute change. Absolute change is just the difference in the values of f at the ends of the interval, that is,

$$f(a+h) - f(a),$$

whereas the rate of change is the absolute change divided by the size of the interval,

$$(f(a+h) - f(a))/h.$$

The rate of change tells how quickly (or slowly) the function changes from one end of the interval to the other, relative to the size of the interval. It is often more useful to know the rate of change than the absolute change. For example, if someone offers you a \$150 return on a \$100 investment, you will want to know how long it is going to take to make that money. Just knowing the absolute change in your money, \$50, is not enough, but knowing the rate of change (i.e., \$50 divided by the time it takes to make it) will help you decide whether or not to make the investment.

Instantaneous Rate of Change: The Derivative

We can also define the *instantaneous rate of change* of a function at a point. We simply mimic what we did for velocity, namely, look at the average rate of change over smaller and smaller intervals. By taking the limit, we get:

$$\text{Rate of change of } f \text{ at } a = \lim_{h \to 0} \frac{f(a+h) - f(a)}{h}.$$

This number is so important that it is given its own name, the *derivative of f at a*, denoted by $f'(a)$. We define it as follows:

The derivative of f at a

$$f'(a) = \lim_{h \to 0} \frac{f(a+h) - f(a)}{h}.$$

If we want to emphasize that $f'(a)$ is the rate of change of $f(x)$ as the variable x changes, we call $f'(a)$ the derivative of f *with respect to x* at $x = a$.

Example 2 By choosing small values for h, estimate the instantaneous rate of change of the radius of a sphere with respect to change in volume at $V = 1$.

Solution With $h = 0.01$ and $h = -0.01$,

$$\frac{r(1.01) - r(1)}{0.01} \approx 0.2061 \quad \text{and} \quad \frac{r(0.99) - r(1)}{-0.01} \approx 0.2075.$$

With $h = 0.001$ and $h = -0.001$,

$$\frac{r(1.001) - r(1)}{0.001} \approx 0.2067 \quad \text{and} \quad \frac{r(0.999) - r(1)}{-0.001} \approx 0.2069.$$

The values of these difference quotients suggest that the limit is between 0.2067 and 0.2069. Rounding off suggests that the value is about 0.207; taking smaller h values confirms this. So we say

$$r'(1) = \begin{array}{c}\text{Instantaneous rate of change} \\ \text{of radius with respect to} \\ \text{change in volume at } V = 1\end{array} \approx 0.207.$$

In the previous example we got an approximation to the instantaneous rate of change, or derivative, by substituting in smaller and smaller values of h. In the next example we will compute the exact value of the derivative.

Example 3 Find the derivative of the function $f(x) = x^2$ at the point $x = 1$.

Solution We need to look at

$$f'(1) = \lim_{h\to 0} \frac{f(1+h) - f(1)}{h}.$$

This is the same as

$$\lim_{h\to 0} \frac{(1+h)^2 - 1^2}{h} = \lim_{h\to 0} \frac{(1 + 2h + h^2) - 1}{h} = \lim_{h\to 0} \frac{2h + h^2}{h}.$$

Now choose several small values for h (e.g. 0.1, 0.01, 0.001), compute $(2h+h^2)/h$ for these, and see if you can guess the limit. An alternative approach is to divide by h in the expression $(2h + h^2)/h$, a valid operation since the limit only examines values of h close to, but not equal to zero. We get

$$\lim_{h\to 0} \frac{h(2+h)}{h} = \lim_{h\to 0} (2 + h).$$

You can see that this limit is 2, so $f'(1) = 2$. Thus, at $x = 1$ the rate of change of x^2 is 2.

Graphically, what does it mean that the derivative is 2? Since the derivative is the rate of change, it means that for small changes in x, near $x = 1$, the change in $f(x) = x^2$ is about twice as big as the change in x. As an example, if x changes from 1 to 1.1, a net change of 0.1, $f(x)$ should change by about 0.2. Figure 2.9 shows this geometrically.

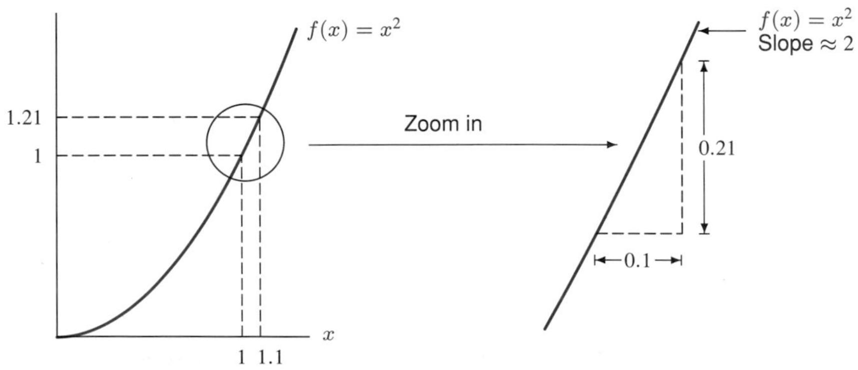

Figure 2.9: Graph of $f(x) = x^2$ near $x = 1$ has slope ≈ 2

Table 2.3 shows the derivative of $f(x) = x^2$ numerically. Notice that near $x = 1$, every time the value of x increases by 0.001, the value of x^2 increases by approximately 0.002. Thus near $x = 1$ the graph is approximately linear with slope $0.002/0.001 = 2$.

TABLE 2.3 *Values of $f(x) = x^2$ near $x = 1$*

x	x^2	Difference in successive x^2 values
0.998	0.996004	
		0.001997
0.999	0.998001	
		0.001999
1.000	1.000000	
		0.002001
1.001	1.002001	
		0.002003
1.002	1.004004	
↑ x increments 0.001		↑ all approximately 0.002

Visualizing the Derivative: Slope of Curve and Slope of Tangent

As with velocity, we can visualize the derivative $f'(a)$ as the slope of the graph of f at a. In addition, there is another way to think of $f'(a)$. Consider the difference quotient $(f(a+h) - f(a))/h$. The numerator, $f(a+h) - f(a)$, is the vertical distance marked in Figure 2.10 and h is the horizontal distance, so

$$\text{Average rate of change of } f = \frac{f(a+h) - f(a)}{h} = \text{Slope of line } AB.$$

As h becomes smaller, the line AB approaches the tangent line to the curve at A. (See Figure 2.11.) Thus,

$$\begin{matrix}\text{Instantaneous} \\ \text{rate of change of } f \\ \text{at } a\end{matrix} = \lim_{h \to 0} \frac{f(a+h) - f(a)}{h} = \text{Slope of tangent at } A.$$

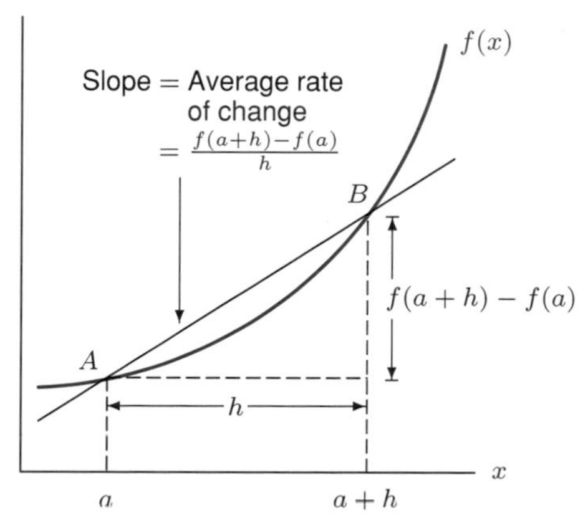

Figure 2.10: Visualizing the average rate of change of f

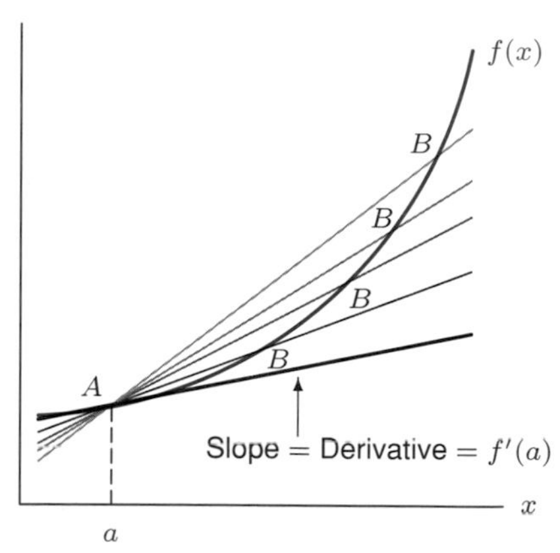

Figure 2.11: Visualizing the instantaneous rate of change of f

Therefore, the derivative at point A can be thought of as:

- The slope of the curve at A.
- The slope of the tangent line to the curve at A.

The slope interpretation is often useful in gaining rough information about the derivative, as the following examples show.

Example 4 Is the derivative of $\sin x$ at $x = \pi$ positive or negative?

Solution Looking at a graph of $\sin x$ in Figure 2.12, we see that a tangent line drawn at $x = \pi$ has negative slope, so the derivative at this point is negative. (Remember, x is in radians.)

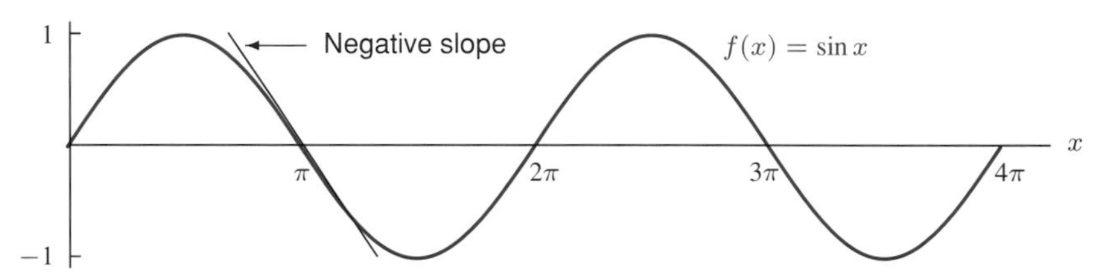

Figure 2.12: Tangent line to $\sin x$ at $x = \pi$

Remember that if you zoom in on the graph of a function $y = f(x)$ at the point where $x = a$, you will usually find that the graph looks more and more like a straight line with slope $f'(a)$.

Example 5 By zooming in on the point $(0, 0)$ on the graph of the sine function, estimate the value of the derivative of $\sin x$ at $x = 0$, with x in radians.

Solution Figure 2.13 shows successive graphs of $\sin x$, with smaller and smaller scales. On the interval $-0.1 \leq x \leq 0.1$, the graph looks like a straight line of slope 1. Thus, the derivative of $\sin x$ at $x = 0$ is about 1.

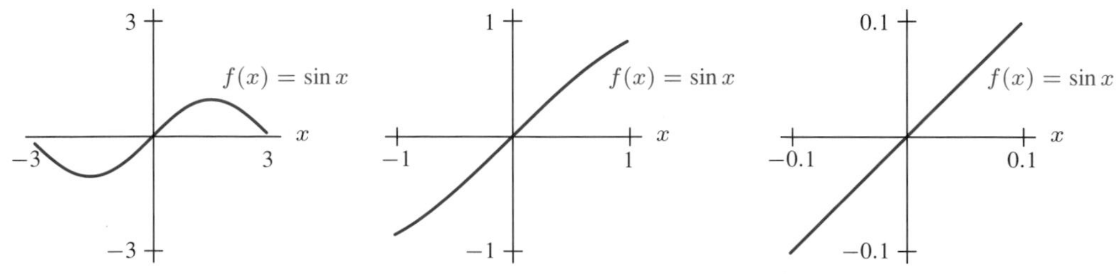

Figure 2.13: Zooming in on the graph of $\sin x$ near $x = 0$

Later we will show that the derivative of $\sin x$ at $x = 0$ is exactly 1. (See Problems 33 and 34 on page 224 in Section 4.6.) From now on we will assume that this is so.

Example 6 Use the tangent line at $x = 0$ to estimate values of $\sin x$ near $x = 0$.

Solution The previous example shows that near $x = 0$, the graph of $y = \sin x$ looks like the graph of the straight line $y = x$; we can use this to estimate values of $\sin x$ when x is close to 0. For example, the point on the straight line $y = x$ with x coordinate 0.32 is $(0.32, 0.32)$. Since the line is close to the graph of $y = \sin x$, we estimate that $\sin 0.32 \approx 0.32$. (See Figure 2.14.) Checking on the calculator, we find that $\sin 0.32 = 0.3146$, so our estimate is quite close. Notice that the graph leads you to expect that the real value of $\sin 0.32$ is slightly less than 0.32.

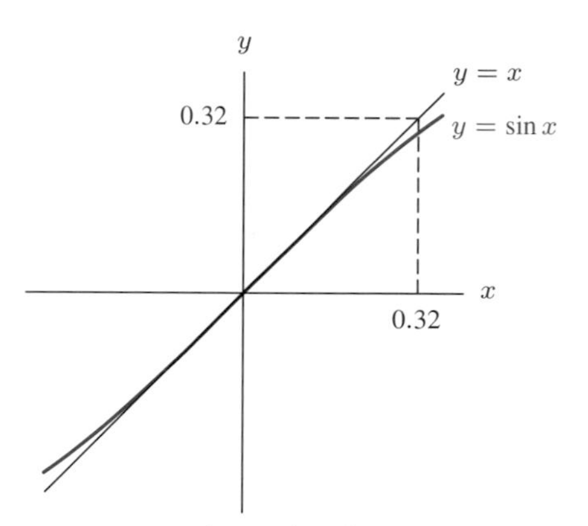

Figure 2.14: Approximating $y = \sin x$ by $y = x$

Why Do We Use Radians and Not Degrees?

In Example 5 we saw that the derivative of $\sin x$ at $x = 0$ is 1, when x is in radians. This is the reason we choose to use radians. If we had done the previous example in degrees, the derivative of $\sin x$ would have turned out to be a much messier number. (See Problem 16, page 110.)

Examples: Computing the Derivative at a Point

Example 7 Estimate the value of the derivative of $f(x) = 2^x$ at $x = 0$ graphically and numerically.

Solution Graphically: If you draw a tangent line at $x = 0$ to the exponential curve in Figure 2.15, you'll see that it has a positive slope. Since the slope of the line BA is $(2^0 - 2^{-1})/(0 - (-1)) = 1/2$ and the slope of the line AC is $(2^1 - 2^0)/(1 - 0) = 1$, we know that the derivative is between $1/2$ and 1.
Numerically: To find the derivative at $x = 0$, we need to look at values of the difference quotient

$$\frac{f(0+h) - f(0)}{h} = \frac{2^h - 2^0}{h} = \frac{2^h - 1}{h}$$

for small h. Table 2.4 shows some values of 2^h together with values of the difference quotients. (See Problem 25 on page 111 for what happens for very small values of h.)

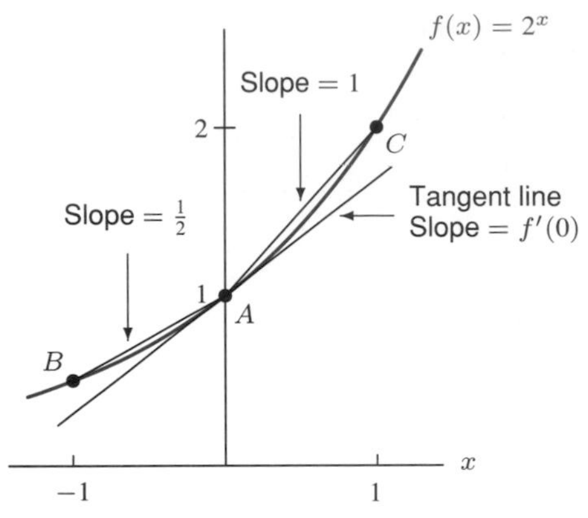

Figure 2.15: Graph of $y = 2^x$ showing the derivative at $x = 0$

TABLE 2.4 *Numerical values for difference quotient of 2^x near $x = 0$*

h	2^h	Difference quotient: $\frac{2^h-1}{h}$
−0.0003	0.999792077	0.693075
−0.0002	0.999861380	0.693099
−0.0001	0.999930688	0.693123
0	1	
0.0001	1.00006932	0.693171
0.0002	1.00013864	0.693195
0.0003	1.00020797	0.693219

Figure 2.15 suggests that difference quotients calculated with negative h's are smaller than the derivative, and those calculated with positive h's are larger. Thus, from Table 2.4 we see that the derivative is between 0.693123 and 0.693171. To three decimal places, $f'(0) = 0.693$.

Example 8 Find an approximate equation for the tangent line to $f(x) = 2^x$ at $x = 0$.

Solution From the previous example, we know the slope of the tangent line is about 0.693. Since we also know the line has y intercept 1, its equation is

$$y = 0.693x + 1.$$

Example 9 Find the value of the derivative of $f(x) = x^2 + 1$ at $x = 3$ algebraically. Find the equation of the tangent line to f at $x = 3$.

Solution We need to look at the difference quotient and take the limit as h approaches zero. The difference quotient is

$$\frac{f(3+h)-f(3)}{h} = \frac{(3+h)^2+1-10}{h} = \frac{9+6h+h^2-9}{h} = \frac{6h+h^2}{h} = \frac{h(6+h)}{h}.$$

Since $h \neq 0$, we can divide by h in the last expression to get $6 + h$. Now the limit as h goes to 0 of $6 + h$ is clearly 6, so,

$$f'(3) = \lim_{h\to 0} \frac{6h+h^2}{h} = \lim_{h\to 0}(6+h) = 6.$$

Thus, we know that the slope of the tangent line at $x = 3$ is 6. Since $f(3) = 10$, the tangent passes through (3, 10), so the equation of the tangent line is

$$y - 10 = 6(x-3) \quad \text{or} \quad y = 6x - 8.$$

Example 10 Table 2.5 shows values of $f(x) = x^3$ near $x = 2$. Use it to estimate $f'(2)$.

TABLE 2.5 *Values of x^3 (to three decimal places)*

x	1.998	1.999	2.000	2.001	2.002
x^3	7.976	7.988	8.000	8.012	8.024

Solution The derivative, $f'(2)$, is the rate of change of x^3 at $x = 2$. Notice that each time x changes by 0.001 in the table, the value of x^3 changes by 0.012. Therefore, we estimate

$$f'(2) = \begin{matrix}\text{Rate of change}\\ \text{of } f \text{ at } x = 2\end{matrix} \approx \frac{0.012}{0.001} = 12.$$

The function values in the table look exactly linear because they have been rounded. For example, the exact value of x^3 when $x = 2.001$ is 8.012006001, not 8.012. Thus, the table can tell us only that the derivative is approximately 12. You can show that the derivative is exactly 12 using the method of Example 3, page 104.

Problems for Section 2.2

1. (a) Make a table of values, rounded to two decimal places, for $f(x) = \log x$ (that is, log base 10) with $x = 1, 1.5, 2, 2.5, 3$. Then use this table to answer parts (b) and (c).
   (b) Find the average rate of change of $f(x)$ between $x = 1$ and $x = 3$.
   (c) Use average rates of change to approximate the instantaneous rate of change of $f(x)$ at $x = 2$.
2. Sketch a rough graph of $f(x) = \sin x$, and use the graph to decide whether the derivative of $f(x)$ at $x = 3\pi$ is positive or negative. Give reasons for your decision.
3. On a sketch of $y = f(x)$ similar to that in Figure 2.16, mark lengths that represent the quantities in parts (a) – (d). (Pick any convenient x, and assume $h > 0$.)
   (a) $f(x)$ (b) $f(x+h)$ (c) $f(x+h) - f(x)$ (d) h
   (e) Using your answers to parts (a)–(d), show how the quantity $\dfrac{f(x+h) - f(x)}{h}$ can be represented as the slope of a line on the graph.
4. On a sketch of $y = f(x)$ similar to that in Figure 2.17, mark lengths that represent the quantities in parts (a) – (d). (Pick any convenient x, and assume $h > 0$.)
   (a) $f(x)$ (b) $f(x+h)$ (c) $f(x+h) - f(x)$ (d) h
   (e) Using your answers to parts (a)–(d), show how the quantity $\dfrac{f(x+h) - f(x)}{h}$ can be represented as the slope of a line on the graph.
5. Show how you can represent the following on a sketch similar to that in Figure 2.18.
   (a) $f(4)$ (b) $f(4) - f(2)$ (c) $\dfrac{f(5) - f(2)}{5 - 2}$
6. Consider the function $y = f(x)$ shown in Figure 2.18. For each of the following pairs of numbers, decide which is larger. Explain your answer.
   (a) $f(3)$ or $f(4)$? (b) $f(3) - f(2)$ or $f(2) - f(1)$?
   (c) $\dfrac{f(2) - f(1)}{2 - 1}$ or $\dfrac{f(3) - f(1)}{3 - 1}$?

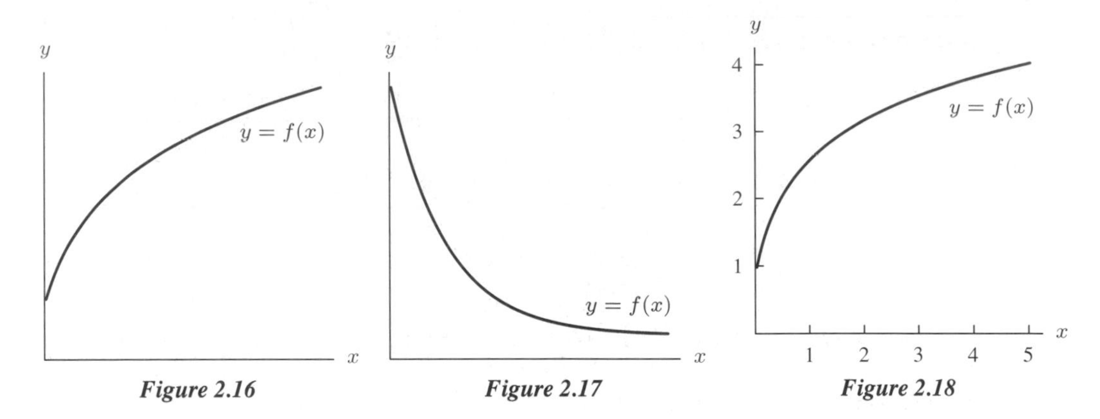

Figure 2.16 Figure 2.17 Figure 2.18

7. Suppose $y = f(x)$ graphed in Figure 2.18 represents the cost of manufacturing x kilograms of a chemical. Then $f(x)/x$ represents the average cost of producing 1 kilogram when x kilograms are made. This problem asks you to visualize these averages graphically.
   (a) Show how to represent $f(4)/4$ as the slope of a line.
   (b) Which is larger, $f(3)/3$ or $f(4)/4$?
8. With the function f given by Figure 2.18, arrange the following quantities in ascending order:
$$0, \quad 1, \quad f'(2), \quad f'(3), \quad f(3) - f(2)$$
9. Refer to the graph of the function k in Figure 2.19:
   (a) Between which pair of consecutive points is the average rate of change of k greatest?
   (b) Between which pair of consecutive points is the average rate of change of k closest to zero?
   (c) Between which two pairs of consecutive points are the average rates of change of k closest?

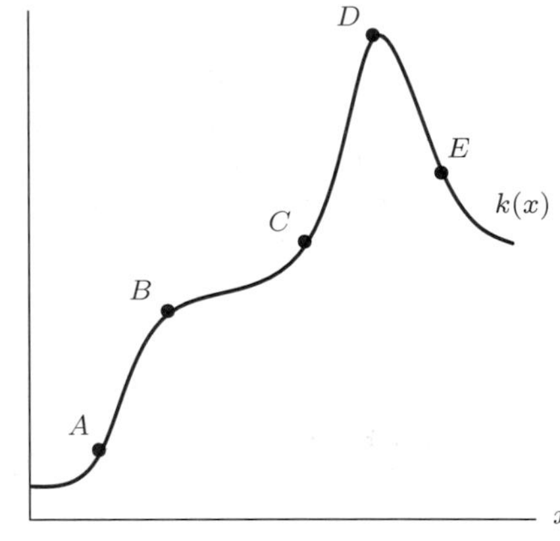

Figure 2.19

10. Find the derivative of $f(x) = x^3 + 5$ at $x = 1$ algebraically.
11. Find the derivative of $g(x) = 1/x$ at $x = 2$ algebraically.
12. If $g(t) = 3t^2 + 5t$, find $g'(-1)$ algebraically.
13. If $g(z) = z^{-2}$, find $g'(2)$ algebraically.
14. Find the equation of the tangent line to $f(x) = x^2 + x$ at $x = 3$. Sketch a graph of the function and this tangent line.
15. Find the equation of the tangent line to the graph of $f(x) = 1/x^2$ at the point $(1, 1)$.
16. (a) Estimate $f'(0)$ if $f(x) = \sin x$, with x in degrees.
    (b) In Example 5, page 106, you found that the derivative of $\sin x$ at $x = 0$ was 1. Why do you get a different result here?

17. If $f(x) = x^3 + 4x$, estimate $f'(3)$ using a table similar to that in Example 10 on page 109.

18. For $g(x) = x^5$, use tables similar to that in Example 10 on page 109 to estimate $g'(2)$ and $g'(-2)$. What relationship do you notice between $g'(2)$ and $g'(-2)$? Explain geometrically why this must occur.

19. Estimate the derivative of $f(x) = x^x$ at $x = 2$.

20. For $y = f(x) = 3x^{3/2} - x$, use your calculator to construct a graph of $y = f(x)$, for $0 \le x \le 2$. From your graph, estimate $f'(0)$ and $f'(1)$.

21. (a) Use your calculator to approximate the derivative of the hyperbolic sine function (written sinh) at the points 0, 0.3, 0.7, and 1.
    (b) Can you find a relation between the values of this derivative and the values of the hyperbolic cosine (written $\cosh x$)?

22. Let $f(x) = \ln(\cos x)$. Use your calculator to approximate the instantaneous rate of change of f at the point $x = 1$. Do the same thing for $x = \pi/4$. (Note: Be sure that your calculator is set in radians.)

23. The population, P, of China, in billions, can be approximated by the function

$$P = 1.15(1.014)^t$$

where t is the number of years since the start of 1993. According to this model, how fast is the population growing at the start of 1993 and at the start of 1995? Give your answers in millions of people per year.

24. (a) Sketch graphs of the functions $f(x) = \frac{1}{2}x^2$ and $g(x) = f(x) + 3$ on the same set of axes. What can you say about the slopes of the tangent lines to the two graphs at the point $x = 0$? $x = 2$? $x = x_0$?
    (b) Show that adding a constant value, C, to any function does not change the value of the slope of its graph at any point. [Hint: Let $g(x) = f(x) + C$, and calculate the difference quotients for f and g.]

25. Suppose Table 2.4 on page 108 is continued with smaller values of h, using a calculator. The results are in Table 2.6.

TABLE 2.6 *More difference quotients of 2^x near $x = 0$*

h	Difference quotient: $(2^h - 1)/h$
10^{-4}	0.6931496
10^{-6}	0.69315
10^{-8}	0.693
10^{-10}	0.7
10^{-12}	0

Comment on the values of the difference quotient in Table 2.6. In particular, why is the last value of $(2^h - 1)/h$ zero? What do you expect the value of $(2^h - 1)/h$ to be when $h = 10^{-20}$?

2.3 THE DERIVATIVE FUNCTION

In the last section we looked at the derivative of a function at a fixed point. Now we'll consider what happens at a variety of points and see that, in general, the derivative takes on different values at different points and is itself a function.

First, remember that the derivative of a function at a point tells you the rate at which the value of the function is changing at that point. Geometrically, if you "zoom in" on a point in the graph until it looks like a straight line, the slope of that line is the derivative at that point. Equivalently, you can think of the derivative as the slope of the tangent line at the point, because as you "zoom in," the curve and the tangent line become indistinguishable.

Example 1 Estimate the derivative of the function given by the graph in Figure 2.20 at $x = -2, -1, 0, 1, 2, 3, 4, 5$.

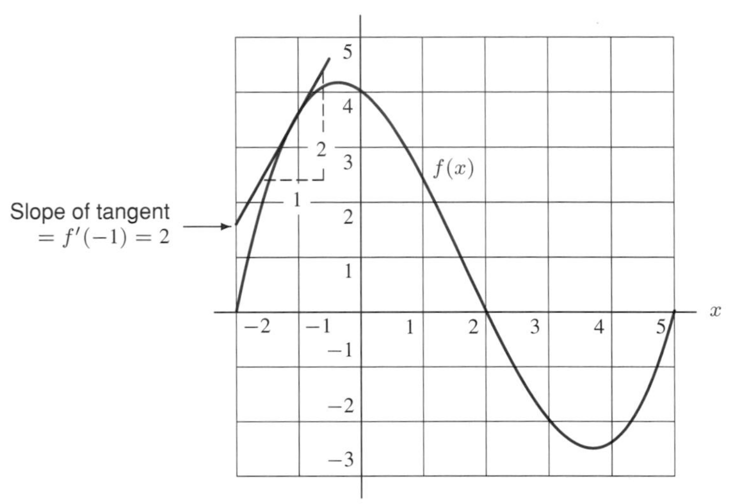

Figure 2.20: Estimating the derivative graphically as the slope of the tangent line

Solution From the graph you can estimate the derivative at any point by placing a straight edge so that it forms the tangent at that point, and then using the grid squares to estimate the slope of the straight edge. (For example, the tangent at $x = -1$ is drawn in Figure 2.20, and has a slope of about 2, so $f'(-1) \approx 2$.) Notice that the slope at $x = -2$ is positive and fairly large; the slope at $x = -1$ is positive but smaller. At $x = 0$, the slope is negative, by $x = 1$ it has become more negative, and so on. Some estimates of the derivative are listed in Table 2.7. You should check these values yourself. Are they reasonable? Is the derivative positive where you expect? Negative?

TABLE 2.7 *Estimated values of derivative of function in Figure 2.20*

x	-2	-1	0	1	2	3	4	5
Derivative at x	6	2	-1	-2	-2	-1	1	4

The important point to notice is that for every x value, there's a corresponding value of the derivative. The derivative, therefore, is itself a function of x.

For any function f, we define the **derivative function**, f', by

$$f'(x) = \text{Rate of change of } f \text{ at } x = \lim_{h\to 0} \frac{f(x+h) - f(x)}{h}.$$

For every x value for which this limit exists, we say f is *differentiable* at that x value. If the limit exists for all x in the domain of f, we say f is *differentiable everywhere*. The functions we shall deal with will be differentiable at every point in their domain, except perhaps for a few isolated points.

Finding the Derivative of a Function Given Graphically

Example 2 Sketch the graph of the derivative of the function shown in Figure 2.20.

Solution Table 2.7 gives some values of this derivative which we can plot. However, it is a good idea first to identify some of the key features of the derivative graph from the graph of the original function. For example, we can see from Figure 2.20 (repeated in Figure 2.21) that the function is increasing from $x = -2$ to about $x = -0.5$. Thus, the derivative is positive in this interval, and so we must draw the graph of f' above the x-axis from $x = -2$ to about $x = -0.5$. Between $x = -0.5$ and about $x = 3.7$ the function is decreasing, so the derivative is negative and its graph must be below the x-axis. Beyond $x = 3.7$ the function is increasing, so the derivative is positive again and the graph of f' is above the axis. Somewhere in the region where the derivative is negative, it is going to reach its lowest point; this will be at the point where the graph of the original function is decreasing most steeply. From Figure 2.20 we see that this occurs a little before $x = 2$, where the slope is slightly steeper than -2. Thus, our derivative graph should have a minimum value of slightly below -2 occurring slightly to the left of $x = 2$. With this in mind, and using the data in Table 2.7, we obtain Figure 2.21, which shows a graph of the derivative (the black curve), along with the original function (color).

You should check for yourself that this graph of f' makes sense. Notice that at the points where f has large upward slope, such as $x = -2$, the graph of the derivative is far above the x-axis, as it should be, since the value of the derivative should be large there. On the other hand, at points where the slope is gentle, the graph of f' is close to the x-axis, since the derivative is small.

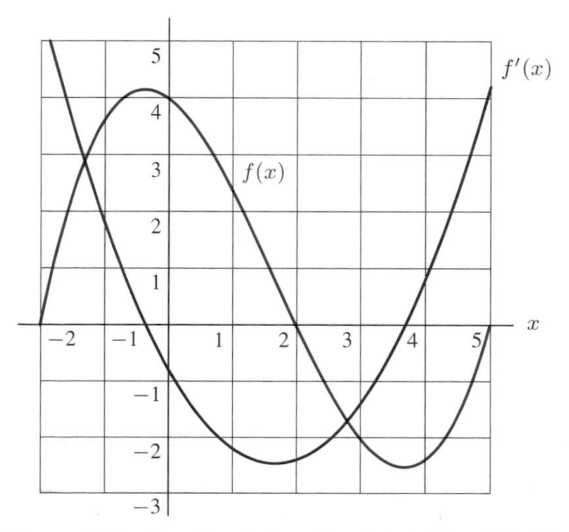

Figure 2.21: Function (colored) and derivative (black) from Example 2

Finding the Derivative of a Function Given Numerically

If we are given a table of function values instead of a graph, we can estimate values of its derivative.

Example 3 Suppose Table 2.8 gives values of $c(t)$, the concentration (mg/cc) of a drug in the bloodstream at time t (min). Construct a table of estimated values for $c'(t)$, the rate of change of $c(t)$ with respect to time.

TABLE 2.8 *Concentration as a function of time*

t (min)	0	0.1	0.2	0.3	0.4	0.5	0.6	0.7	0.8	0.9	1.0
$c(t)$ (mg/cc)	0.84	0.89	0.94	0.98	1.00	1.00	0.97	0.90	0.79	0.63	0.41

Solution We want to estimate the derivative of c using the values in the table. To do this, we have to assume that the data points are close enough together that the concentration doesn't change wildly between them. From the table, we can see that the concentration is increasing between $t = 0$ and $t = 0.4$, so we'd expect a positive derivative there. However, the increase is quite slow, so we would expect the derivative to be small. The concentration doesn't change between 0.4 and 0.5, so we expect the derivative to be 0 there. From $t = 0.5$ to $t = 1.0$, the concentration starts to decrease, and the rate of decrease gets larger and larger, so we would expect the derivative to be negative and of greater and greater magnitude.

Using the data in the table, we can estimate the derivative using the difference quotient:

$$c'(t) \approx \frac{c(t+h) - c(t)}{h}$$

with $h = 0.1$. For example,

$$c'(0) \approx \frac{c(0.1) - c(0)}{0.1} = \frac{0.89 - 0.84}{0.1} = 0.5 \text{ mg/cc/min.}$$

Thus, we estimate that

$$c'(0) \approx 0.5.$$

Similarly, we get the estimates

$$c'(0.1) \approx \frac{c(0.2) - c(0.1)}{0.1} = \frac{0.94 - 0.89}{0.1} = 0.5$$

$$c'(0.2) \approx \frac{c(0.3) - c(0.2)}{0.1} = \frac{0.98 - 0.94}{0.1} = 0.4$$

$$c'(0.3) \approx \frac{c(0.4) - c(0.3)}{0.1} = \frac{1.00 - 0.98}{0.1} = 0.2$$

$$c'(0.4) \approx \frac{c(0.5) - c(0.4)}{0.1} = \frac{1.00 - 1.00}{0.1} = 0.0$$

and so on. These values are tabulated in Table 2.9. Notice that the derivative has small positive values up until $t = 0.4$, and then it gets more and more negative, as we expected. The slopes are shown on the graph of $c(t)$ in Figure 2.22.

TABLE 2.9 *Derivative of concentration*

t	$c'(t)$
0	0.5
0.1	0.5
0.2	0.4
0.3	0.2
0.4	0.0
0.5	−0.3
0.6	−0.7
0.7	−1.1
0.8	−1.6
0.9	−2.2

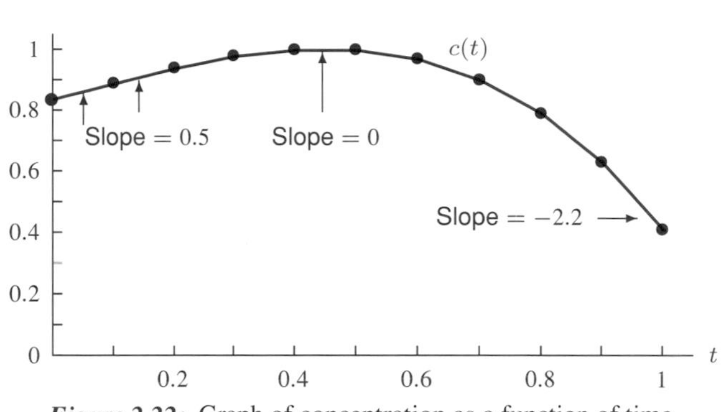

Figure 2.22: Graph of concentration as a function of time

Other Ways We Could Calculate the Derivative Numerically

In the previous example, our estimate for the derivative at 0.2 used the point to the right. We found the average rate of change between $t = 0.2$ and $t = 0.3$. However, we could equally well have gone to the left and used the rate of change between $t = 0.1$ and $t = 0.2$ to approximate the derivative at 0.2. For a more accurate result, we could average these slopes and say

$$c'(0.2) \approx \frac{1}{2}\left(\begin{matrix}\text{slope to left}\\ \text{of } 0.2\end{matrix} + \begin{matrix}\text{slope to right}\\ \text{of } 0.2\end{matrix}\right) = \frac{0.5 + 0.4}{2} = 0.45.$$

Each of these methods of approximating the derivative gives a reasonable answer. For convenience, unless there is a reason to do otherwise, we will estimate the derivative by going to the right.

Finding the Derivative of a Function Given by a Formula

If we are given a formula for f, can we come up with a formula for f'? Using the definition of the derivative, we often can, as shown in the next example. Indeed, much of the power of calculus depends on our ability to find formulas for the derivatives of all the functions in our library. This is done in detail in Chapter 4.

Example 4 Find a formula for the derivative of $f(x) = x^2$.

Solution Before computing the formula for $f'(x)$ algebraically, let's try to guess the formula by looking for a pattern in the values of $f'(x)$. Table 2.10 contains values of $f(x) = x^2$ (rounded to three decimals), which we can use to estimate the values of $f'(1)$, $f'(2)$, and $f'(3)$.

TABLE 2.10 *Values of $f(x) = x^2$ near $x = 1$, $x = 2$, $x = 3$ (rounded to three decimals)*

x	x^2 (approx)
0.999	0.998
1.000	1.000
1.001	1.002
1.002	1.004

x	x^2 (approx)
1.999	3.996
2.000	4.000
2.001	4.004
2.002	4.008

x	x^2 (approx)
2.999	8.994
3.000	9.000
3.001	9.006
3.002	9.012

Near $x = 1$, x^2 increases by about 0.002 each time x increases by 0.001, so

$$f'(1) \approx \frac{0.002}{0.001} = 2.$$

Similarly,

$$f'(2) \approx \frac{0.004}{0.001} = 4$$
$$f'(3) \approx \frac{0.006}{0.001} = 6.$$

Knowing the value of f' at specific points can never tell us the formula for f', but it certainly can be suggestive: knowing $f'(1) \approx 2$, $f'(2) \approx 4$, $f'(3) \approx 6$ certainly suggests that $f'(x) = 2x$.

Now we'll show that $f'(x) = 2x$ is the correct formula algebraically. The derivative is calculated by forming the difference quotient and taking the limit as h goes to zero. The difference quotient is

$$\frac{f(x+h) - f(x)}{h} = \frac{(x+h)^2 - x^2}{h} = \frac{x^2 + 2xh + h^2 - x^2}{h} = \frac{2xh + h^2}{h}.$$

Since h never actually reaches zero, we can divide it out in the last expression to get $2x + h$. The limit of this as h goes to zero is $2x$, so

$$f'(x) = \lim_{h \to 0} (2x + h) = 2x.$$

What Does the Derivative Tell Us Graphically?

When f' is positive, the tangent is sloping up; when f' is negative, the tangent is sloping down. If $f' = 0$ everywhere, then the tangent is horizontal everywhere and so f is constant. Thus, the sign of f' tells us whether f is increasing or decreasing.

> If $f' > 0$ on an interval, then f is *increasing* over that interval.
> If $f' < 0$ on an interval, then f is *decreasing* over that interval.
> If $f' = 0$ on an interval, then f is *constant* over that interval.

Moreover, the magnitude of the derivative gives us the magnitude of the rate of change; so if f' is large (positive or negative), then the graph of f will be steep (up or down), whereas if f' is small the graph of f will slope gently. With this in mind, you can deduce a lot about the behavior of a function from the behavior of its derivative.

Example 5 Suppose the derivative of f is the spike function illustrated in Figure 2.23. What can you say about the graph of f itself?

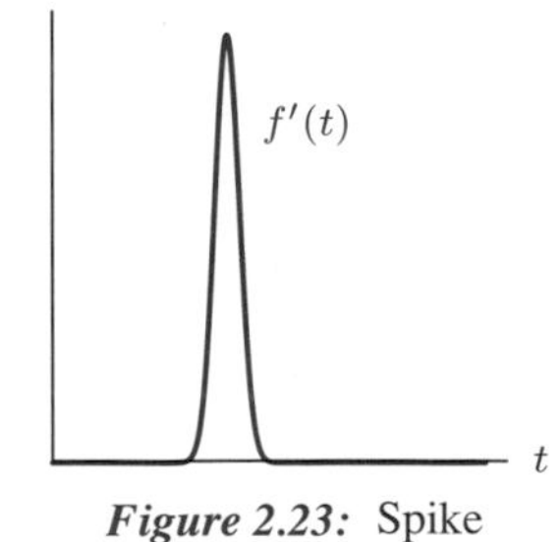

Figure 2.23: Spike derivative function

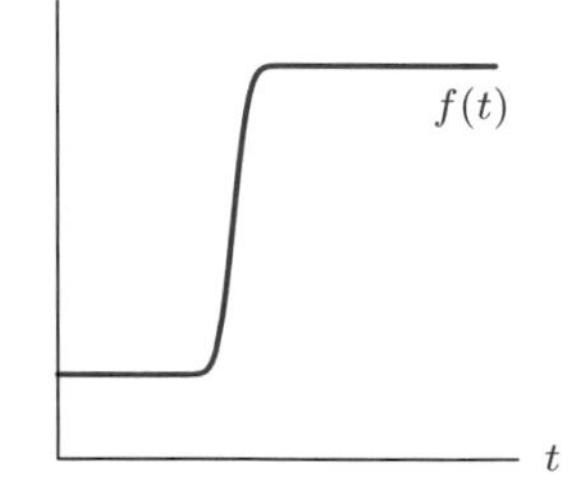

Figure 2.24: Step function

Solution On intervals where $f' = 0$, f is not changing at all, and is therefore constant. On the small interval where $f' > 0$, f is increasing; at the point where f' hits the top of its spike, f is increasing quite sharply. So f should be constant for a while, have a sudden increase, and then be constant again. A possible graph for f is shown in Figure 2.24.

Problems for Section 2.3

For Problems 1–6, sketch a graph of the derivative function of each of the given functions.

1.

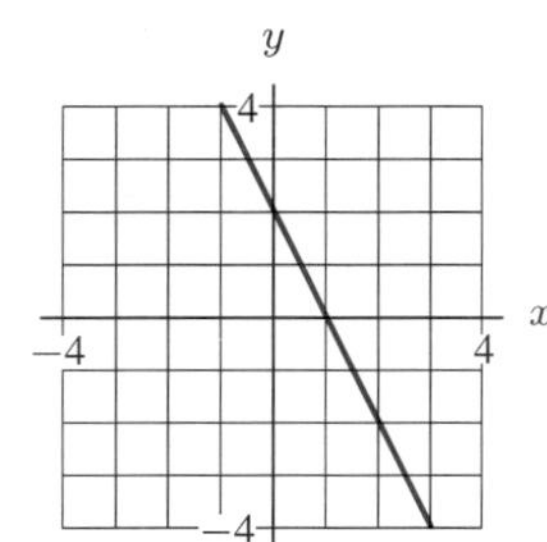

2.

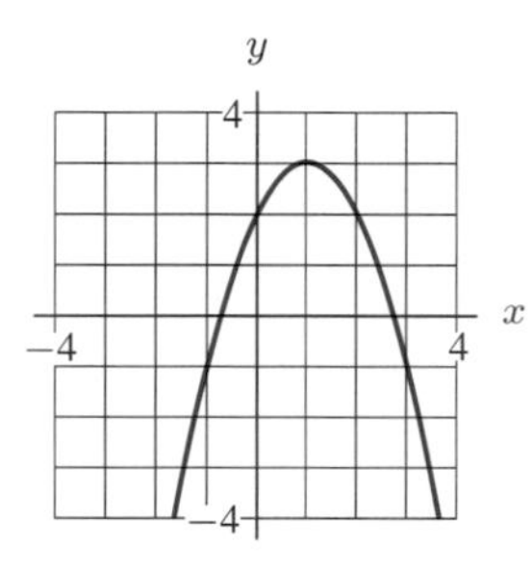

3.

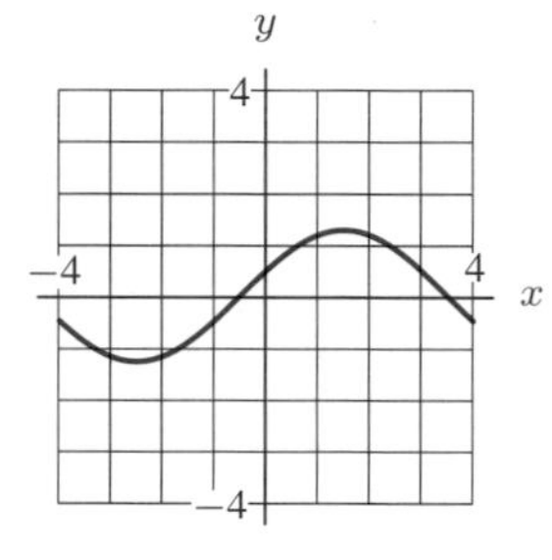

4.

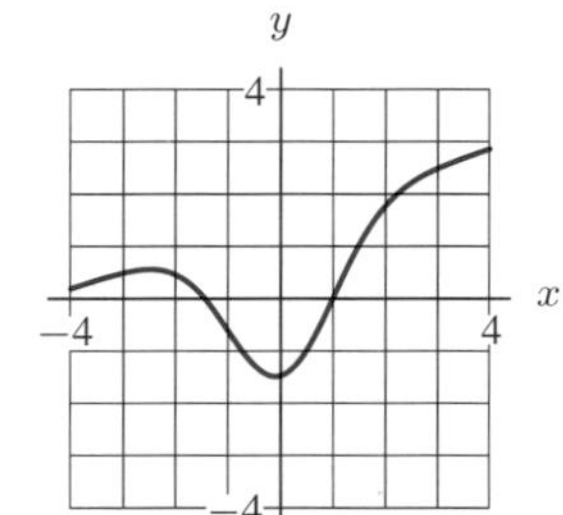

5.

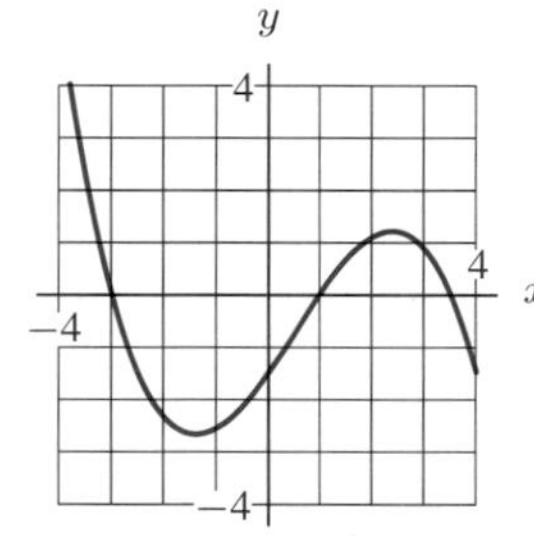

6.

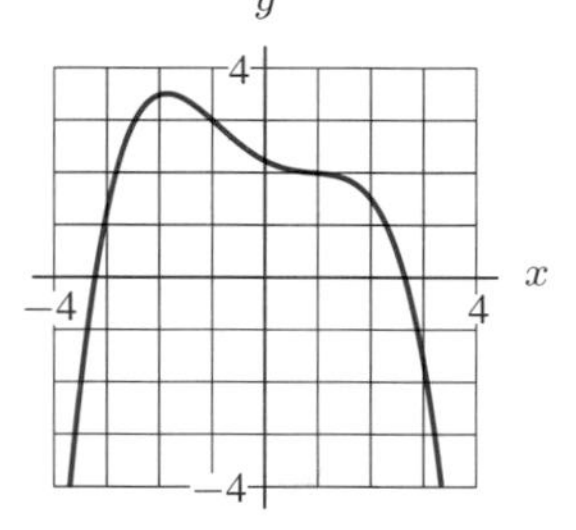

7. Sketch the derivative of the downward step function in Figure 2.25.

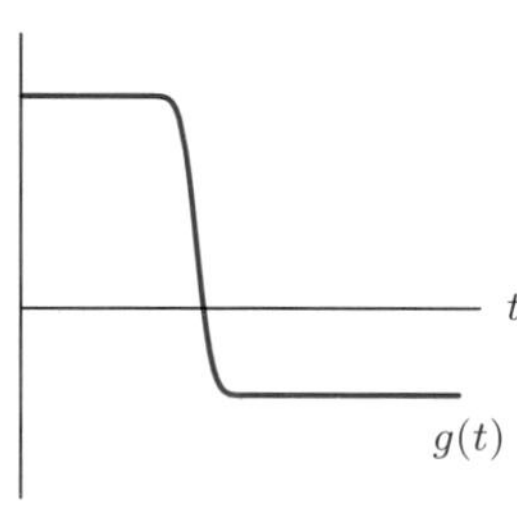

Figure 2.25: Downward step function

8. (a) Sketch a smooth curve whose slope is everywhere positive and increasing gradually.
   (b) Sketch a smooth curve whose slope is everywhere positive and decreasing gradually.
   (c) Sketch a smooth curve whose slope is everywhere negative and increasing gradually (i.e., becoming less and less negative).
   (d) Sketch a smooth curve whose slope is everywhere negative and decreasing gradually (i.e., becoming more and more negative).

9. Given the numerical values shown, find approximate values for the derivative of $f(x)$ at each of the x values given. Where is the rate of change of $f(x)$ positive? Where is it negative? Where does the rate of change of $f(x)$ seem to be greatest?

x	0	1	2	3	4	5	6	7	8
$f(x)$	18	13	10	9	9	11	15	21	30

10. Suppose $f(x) = \frac{1}{3}x^3$. Make tables similar to Table 2.10 on page 115 to estimate $f'(2)$, $f'(3)$, and $f'(4)$. What do you notice? Can you guess a formula for $f'(x)$?

11. If $g(t) = t^2 + t$, use tables similar to Table 2.10 on page 115 to estimate $g'(1)$, $g'(2)$, and $g'(3)$. Use these to guess a formula for $g'(t)$.

Find a formula for the derivatives of the functions in Problems 12–15 algebraically.

12. $f(x) = x^3$
13. $g(x) = 2x^2 - 3$
14. $k(x) = 1/x$
15. $l(x) = 1/x^2$

16. Draw a possible graph of $y = f(x)$ given the following information about its derivative.
- $f'(x) > 0$ on $1 < x < 3$
- $f'(x) < 0$ for $x < 1$ and $x > 3$
- $f'(x) = 0$ at $x = 1$ and $x = 3$

17. Draw a possible graph of $y = f(x)$ given the following information about its derivative.
- $f'(x) > 0$ for $x < -1$
- $f'(x) < 0$ for $x > -1$
- $f'(x) = 0$ at $x = -1$

18. In the graph of f in Figure 2.26, at which of the labeled x values is
(a) $f(x)$ greatest? (b) $f(x)$ least? (c) $f'(x)$ greatest? (d) $f'(x)$ least?

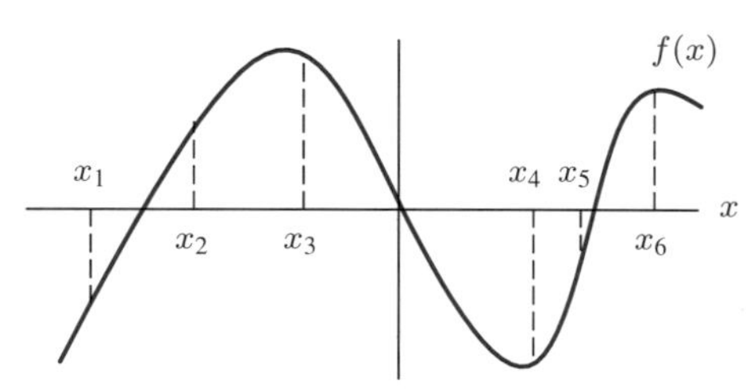

Figure 2.26

19. (a) On the same set of axes, graph $f(x) = \sin x$ and $g(x) = \sin 2x$ from $x = 0$ to $x = 2\pi$.
(b) On a second set of axes, sketch the graphs of $f'(x)$ and $g'(x)$ and compare them. (Be careful when comparing the slopes of $f(x)$ and $g(x)$ at each point.)

For Problems 20–25, sketch the graph of $f(x)$, and use this graph to sketch the graph of $f'(x)$.

20. $f(x) = x(x - 1)$
21. $f(x) = 5x$
22. $f(x) = \cos x$
23. $f(x) = e^x$
24. $f(x) = \log x$
25. $f(x) = \dfrac{1}{x^3}$

For Problems 26–31, sketch the graph of $y = f'(x)$ for the function given.

26.

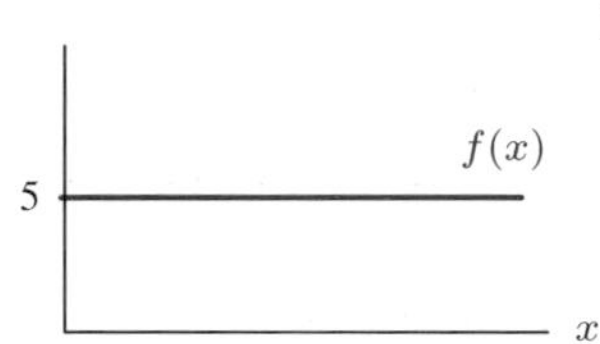

27.

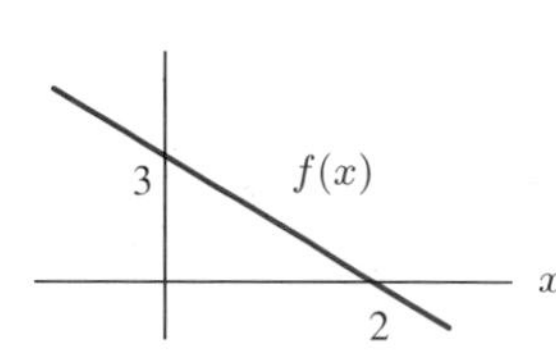

28.

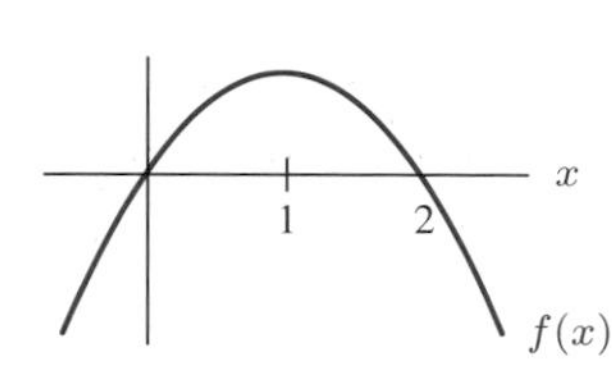

29.

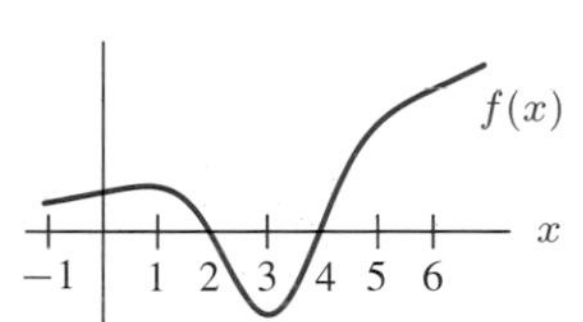

30.

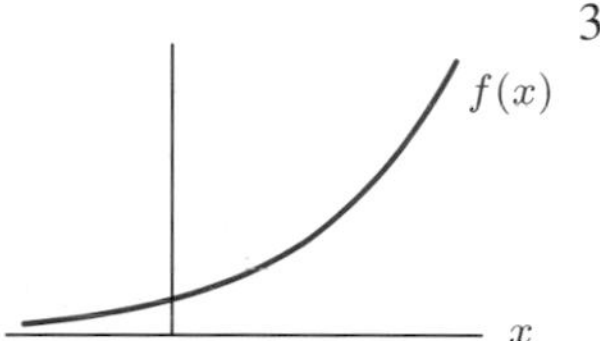

31.

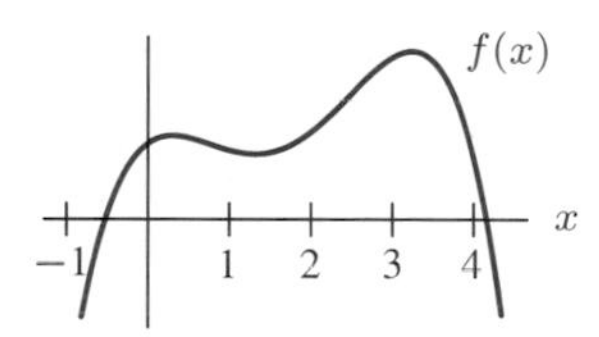

32. The population of a herd of deer is modeled by

$$P(t) = 4000 + 500 \sin\left(2\pi t - \frac{\pi}{2}\right)$$

where t is measured in years.

(a) How does this population vary with time? Sketch a graph of $P(t)$ for one year.

(b) Use the graph to decide when in the year the population is a maximum. What is that maximum? Is there a minimum? If so, when?

(c) Use the graph to decide when the population is growing fastest. When is it decreasing fastest?

(d) Estimate roughly how fast the population is changing on the first of July.

33. Show that if $f(x)$ is an even function, then $f'(x)$ is odd.

34. Show that if $g(x)$ is an odd function, then $g'(x)$ is even.

35. To give a patient an antibiotic slowly, the drug is injected into the muscle. (For example, penicillin for venereal disease is administered this way.) The quantity of the drug in the bloodstream starts out at zero, increases to a maximum, and then decays to zero again.

(a) Sketch a possible graph of the quantity of the drug in the bloodstream as a function of time. Mark the time at which the drug is at a maximum by t_0.

(b) Describe in words how the rate at which the drug is entering or leaving the blood changes over time. Sketch a graph of this rate against time, marking t_0 on the time axis.

2.4 INTERPRETATIONS OF THE DERIVATIVE

We have already seen how the derivative can be interpreted as a slope and as a rate of change. In this section, you will see examples of other interpretations. The point of these examples is not to make a catalogue of interpretations but to illustrate the process of obtaining them.

Acceleration

We started this chapter by showing how velocity could be calculated as a rate of change of position with respect to time. If $s(t)$ measures the distance an object has moved from a reference point along a straight line, and $v(t)$ is its velocity at time t, then:

$$\text{Instantaneous velocity} = s'(t) = \lim_{h \to 0} \frac{s(t+h) - s(t)}{h}.$$

Now, *acceleration*, $a(t)$, is the rate of change of velocity with respect to time, so we define

$$\begin{array}{c}\text{Average acceleration}\\ \text{from } t \text{ to } t+h\end{array} = \frac{v(t+h)-v(t)}{h}$$

and

$$\text{Instantaneous acceleration} = v'(t) = \lim_{h\to 0} \frac{v(t+h)-v(t)}{h}.$$

If the term velocity or acceleration is used alone, it is assumed to be instantaneous.

Example 1 An accelerating sports car goes from 0 mph to 60 mph in five seconds. Its velocity is given in Table 2.11, converted from miles per hour to feet per second, so that all time measurements are in seconds. (Note: 1 mph is 22/15 ft/sec.) Find the average acceleration of the car over each of the first two seconds.

TABLE 2.11 *Velocity of sports car*

Time, t (sec)	0	1	2	3	4	5
Velocity, v (ft/sec)	0	30	52	68	80	88

Solution To measure the average acceleration over an interval, we calculate the average rate of change of velocity over the interval. The units of acceleration are ft/sec per second, or (ft/sec)/sec, written ft/sec^2.

$$\begin{array}{c}\text{Average acceleration}\\ \text{for } 0 \le t \le 1\end{array} = \frac{\text{Change in velocity}}{\text{Time}} = \frac{30-0}{1} = 30 \text{ ft/sec}^2.$$

$$\begin{array}{c}\text{Average acceleration}\\ \text{for } 1 \le t \le 2\end{array} = \frac{52-30}{2-1} = 22 \text{ ft/sec}^2.$$

Example 2 The graph in Figure 2.27 shows the velocity of a racing car as a function of time. Estimate its acceleration when $t = 1$.

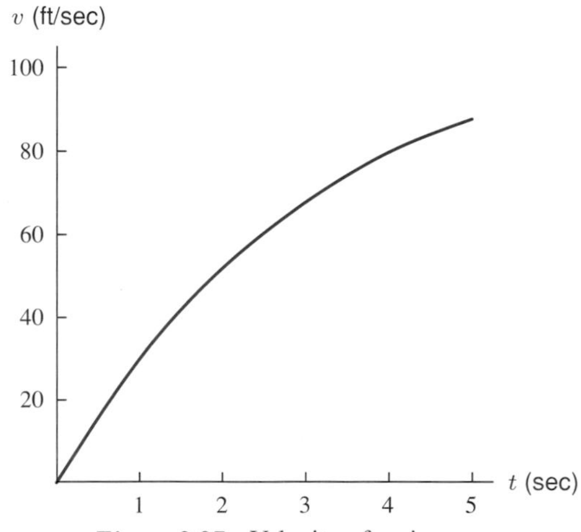

Figure 2.27: Velocity of racing car

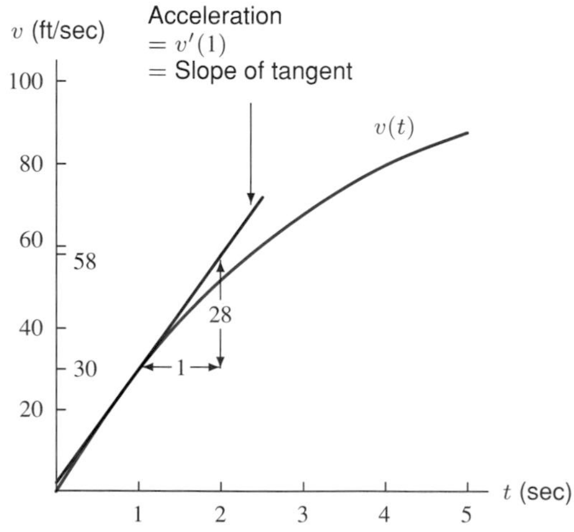

Figure 2.28: Estimating racing car's acceleration from velocity graph

Solution The acceleration at $t = 1$ is the derivative $v'(1)$, the slope of the tangent to the velocity curve at $t = 1$. (See Figure 2.28.) Thus, we estimate

$$\text{Acceleration at } t = 1 = v'(1) \approx \frac{28}{1} = 28 \text{ ft/sec}^2.$$

An Alternative Notation for the Derivative

So far we have used the notation f' to stand for the derivative of the function f. An alternative notation for derivatives was introduced by the German mathematician Gottfried Wilhelm Leibniz (1646–1716) when calculus was first being developed in the seventeenth century. If the variable y depends on the variable x, that is, $y = f(x)$, then we let dy/dx stand for the derivative $f'(x)$. In other words, if

$$y = f(x),$$

then we write

$$\frac{dy}{dx} = f'(x).$$

Leibniz's notation is quite suggestive, especially if you think of the letter d in dy/dx as standing for "small difference in . . . ." The notation dy/dx reminds us that the derivative is a limit of ratios of the form

$$\frac{\text{Difference in } y \text{ values}}{\text{Difference in } x \text{ values}}.$$

It is always a good idea to have a mathematical symbol let us know where it came from and what it means: dy/dx does this and $f'(x)$ does not. The notation dy/dx is useful for determining the units for the derivative: the units for dy/dx are the units for y divided by (or "per") the units for x. The d/dx notation can also be very convenient. For example, it is a lot easier to say

"$\frac{d}{dx}(x^2 + 3x) = 2x + 3$" than it is to say "if $f(x) = x^2 + 3x$, then $f'(x) = 2x + 3$."

The separate entities dy and dx officially have no independent meaning: they are all part of one notation. In fact, a good formal way to view the notation dy/dx is to think of d/dx as a single symbol meaning "the derivative with respect to x of . . .". Thus dy/dx could be viewed as

$$\frac{d}{dx}(y), \quad \text{meaning "the derivative with respect to } x \text{ of } y\text{."}$$

On the other hand, many scientists and mathematicians really do think of dy and dx as separate entities representing "infinitesimally" small differences in y and x, even though it is difficult to say exactly how small "infinitesimal" is. It may not be formally correct, but it is very helpful to the intuition to think of dy/dx as a very small change in y divided by a very small change in x.

For example, recall that if $s = f(t)$ is the position of a moving object at time t, then $v = f'(t)$ is the velocity of the object at time t. Writing

$$v = \frac{ds}{dt}$$

directly reminds you of this fact, since it suggests a distance, ds, over a time, dt, and we know that distance over time is velocity. Similarly, we recognize

$$\frac{dy}{dx} = f'(x)$$

as the slope of the graph of $y = f(x)$ by remembering that slope is vertical rise, dy, over horizontal run, dx.

The disadvantage of the Leibniz's notation is that it is rather awkward if you want to specify the x value at which you are evaluating the derivative. To specify $f'(2)$, for example, we have to write

$$\left.\frac{dy}{dx}\right|_{x=2}.$$

Using Units to Interpret the Derivative

The following examples illustrate the fact that if you want to interpret the meaning of a derivative in practical terms, it often helps to think about units of measurement.

For example, suppose $s = f(t)$ gives the position in meters of a body from a fixed point as a function of time, t, in seconds. Then knowing that

$$\frac{ds}{dt} = f'(2) = 10 \text{ meters/sec}$$

tells us that when $t = 2$ sec, the body is moving at a velocity of 10 meters/sec. This is an instantaneous velocity, meaning that if the body continued to move at this speed for a whole second, it would cover 10 meters. In practice, however, the velocity of the body is probably changing and so doesn't remain 10 meters/sec for long. Notice that the units of instantaneous velocity and of average velocity are the same.

Example 3 The cost C (in dollars) of building a house A square feet in area is given by the function $C = f(A)$. What is the practical interpretation of the function $f'(A)$?

Solution In the alternative notation,

$$f'(A) = \frac{dC}{dA}.$$

This is a cost divided by an area, so it is measured in dollars per square foot. You can think of dC as the extra cost of building an extra dA square feet of house. Thus, dC/dA is the additional cost per square foot. So if you are planning to build a house roughly A square feet in area, $f'(A)$ is the cost per square foot of the *extra* area involved in building a slightly larger house, and is called the *marginal cost*. The marginal cost is not necessarily the same thing as the average cost per square foot for the entire house, since once you are already set up to build a large house, the cost of adding a few square feet could be comparatively small.

Example 4 The cost of extracting T tons of ore from a copper mine is $C = f(T)$ dollars. What does it mean to say that $f'(2000) = 100$?

Solution In the alternative notation,

$$f'(2000) = \left.\frac{dC}{dT}\right|_{T=2000}.$$

Since C is measured in dollars and T is measured in tons, dC/dT must be measured in dollars per ton. So the statement

$$\left.\frac{dC}{dT}\right|_{T=2000} = 100$$

says that when 2000 tons of ore have already been extracted from the mine, the cost of extracting the next ton is approximately \$100. In other words, after 2000 tons have been removed, extraction costs are \$100 per ton. Another way of saying this is that it costs about \$100 to extract ton number 2000 or 2001. Note that this may well be different from the cost of extracting the tenth ton, which is likely to be more accessible.

Example 5 If $q = f(p)$ gives the number of pounds of sugar produced when the price per unit is p dollars, then what are the units and the meaning of

$$\frac{dq}{dp} = f'(3) = 50?$$

Solution Since $f'(3)$ is the limit as $h \to 0$ of

$$\frac{f(3+h) - f(3)}{h}$$

and $f(3 + h) - f(3)$ is in pounds, while h is in dollars, the units of the difference quotient is pounds/dollar. Since $f'(3)$ is the limit of the difference quotient, its units are also pounds/dollar. The statement

$$\frac{dq}{dp} = f'(3) = 50 \text{ pounds/dollar}$$

tells us that the rate of change of q with p is 50 when $p = 3$. Rephrasing, this means that when the price is \$3, the quantity produced is increasing at 50 pounds/dollar. This is an instantaneous rate of change, meaning that, if the rate were to remain 50 pounds/dollar, and if the price were to increase by a whole dollar, the quantity produced would increase by 50 pounds. In fact, the rate probably doesn't remain constant and so the quantity produced would probably not be exactly 50 pounds. Notice, however, that the units of the derivative and of the average rate of change are again the same. This is because the units of the instantaneous and the average rate of change are *always* the same.

Example 6 You are told that water is flowing through a pipe at a rate of 10 cubic feet per second. Interpret this rate as the derivative of some function.

Solution You might think at first that the statement has something to do with the velocity of the water, but in fact a flow rate of 10 cubic feet per second could be achieved either with very slowly moving water through a large pipe, or with very rapidly moving water through a narrow pipe. If we look at the units — cubic feet per second — we realize that we are being given the rate of change of a quantity measured in cubic feet. But a cubic foot is a measure of volume, so we are being told the rate of change of a volume. If you imagine all the water that is flowing through ending up in a tank somewhere and let $V(t)$ be the volume of the tank at time t, then we are being told that the rate of change of $V(t)$ is 10, or

$$V'(t) = \frac{dV}{dt} = 10.$$

Problems for Section 2.4

1. Consider the graph shown in Figure 2.29.
   (a) If $f(t)$ gives the position of a particle at time t, list the points at which the particle has zero velocity.
   (b) If we now suppose instead that $f(t)$ is the *velocity* of a particle at time t, what is the significance of the points listed in your answer to part (a)?

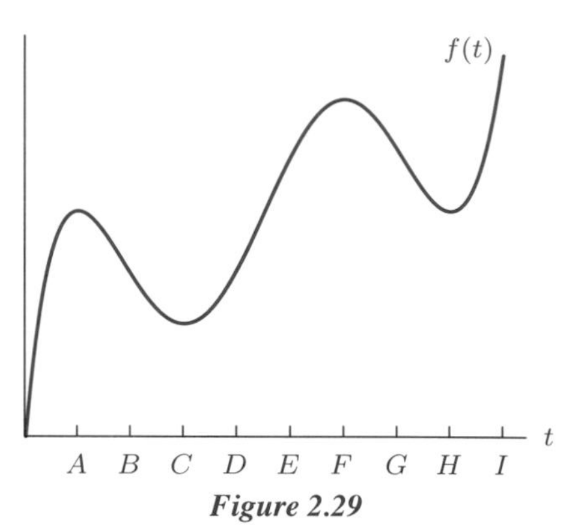

Figure 2.29

2. If $\lim_{x\to\infty} f(x) = 50$ and $f'(x)$ is positive for all x, what is $\lim_{x\to\infty} f'(x)$? (Assume this limit exists.) Explain your answer with a picture.
3. Let $f(x)$ be the elevation in feet of the Mississippi river x miles from its source. What are the units of $f'(x)$? What can you say about the sign of $f'(x)$? (Assume that $0 \le x \le$ length of the river.)
4. Let $g(t)$ be the height, in inches, of Amelia Earhart (one of the first woman airplane pilots) t years after her birth. What are the units of $g'(t)$? What can you say about the signs of $g'(10)$ and $g'(30)$? (Assume that $0 \le t < 39$, the age at which Amelia Earhart's plane disappeared.)
5. Suppose $C(r)$ is the total cost of paying off a car loan borrowed at an annual interest rate of $r\%$. What are the units of $C'(r)$? What is the practical meaning of $C'(r)$? What is its sign?
6. Suppose $P(t)$ is the monthly payment on a mortgage which will take t years to pay off. What are the units of $P'(t)$? What is the practical meaning of $P'(t)$? What is its sign?
7. After investing \$1000 at an annual interest rate of 7% compounded continuously for t years, your balance is $\$B$, where $B = f(t)$. What are the units of dB/dt? What is the financial interpretation of dB/dt?
8. An economist is interested in how the price of a certain commodity affects its sales. Suppose that at a price of $\$p$, a quantity, q, of the commodity is sold. If $q = f(p)$, explain in economic terms the meaning of the statements $f(10) = 240{,}000$ and $f'(10) = -29{,}000$.
9. The temperature, T, in degrees Fahrenheit, of a cold yam placed in a hot oven is given by $T = f(t)$, where t is the time in minutes since the yam was put in the oven.
   (a) What is the sign of $f'(t)$? Why?
   (b) What are the units of $f'(20)$? What is the practical meaning of the statement $f'(20) = 2$?
10. Investing \$1000 at an annual interest rate of $r\%$, compounded continuously, for 10 years gives you a balance of $\$B$, where $B = g(r)$. What is a financial interpretation of the statements
    (a) $g(5) \approx 1649$?
    (b) $g'(5) \approx 165$? What are the units of $g'(5)$?
11. If $g(v)$ is the fuel efficiency of a car going at v miles per hour (i.e., $g(v) =$ the number of miles per gallon at v mph), what are the units of $g'(55)$? What is the practical meaning of the statement $g'(55) = -0.54$?

12. Let P be the total petroleum reservoir on earth in the year t. (In other words, P represents the total quantity of petroleum, including what's not yet discovered, on earth at time t.) Assume that no new petroleum is being made and that P is measured in barrels. What are the units of dP/dt? What is the meaning of dP/dt? What is its sign? How would you set about estimating this derivative in practice? What would you need to know to make such an estimate?

13. (a) If you jump out of an airplane without a parachute, you will fall faster and faster until wind resistance causes you to approach a steady velocity, called a *terminal* velocity. Sketch a graph of your velocity against time.
    (b) Explain the concavity of your graph.
    (c) Assuming wind resistance to be negligible at $t = 0$, what natural phenomenon is represented by the slope of the graph at $t = 0$?

14. A company's revenue from car sales, C (measured in thousands of dollars), is a function of advertising expenditure, a, also measured in thousands of dollars. Suppose $C = f(a)$.
    (a) What does the company hope is true about the sign of f'?
    (b) What does the statement $f'(100) = 2$ mean in practical terms? How about $f'(100) = 0.5$?
    (c) Suppose the company plans to spend about $100,000 on advertising. If $f'(100) = 2$, should the company spend slightly more or slightly less than $100,000 on advertising? What if $f'(100) = 0.5$?

15. Let $P(x)$ = the number of people in the US of height $\leq x$ inches. What is the meaning of $P'(66)$? What are its units? Estimate $P'(66)$ (using common sense). Is $P'(x)$ ever negative? [Hint: You may want to approximate $P'(66)$ by a difference quotient, using $h = 1$. Also, you may use the fact that the US population is about 250 million, and that 66 inches = 5 feet 6 inches.]

16. Census figures for the US population (in millions) are listed in Table 2.12. Since population varies with time, there is a function, f, such that $P = f(t)$. Assume that f is increasing (as the values in the table suggest). Then f is invertible.
    (a) What is the meaning of $f^{-1}(100)$?
    (b) What does the derivative of $f^{-1}(P)$ at $P = 100$ represent? What are its units?
    (c) Estimate $f^{-1}(100)$.
    (d) Estimate the derivative of $f^{-1}(P)$ at $P = 100$.

TABLE 2.12 *US population (in millions), 1790–1990*

Year	Population	Year	Population	Year	Population	Year	Population
1790	3.9	1850	23.1	1910	92.0	1970	205.0
1800	5.3	1860	31.4	1920	105.7	1980	226.5
1810	7.2	1870	38.6	1930	122.8	1990	248.7
1820	9.6	1880	50.2	1940	131.7		
1830	12.9	1890	62.9	1950	150.7		
1840	17.1	1900	76.0	1960	179.0		

17. Table 2.12 gives the US population figures between 1790 and 1990.
    (a) Estimate the rate of change of the population for the years 1900, 1945, and 1990.
    (b) When, approximately, was the rate of change of the population greatest?
    (c) Estimate the US population in 1956.
    (d) Based on the data from the table, what would you predict for the census in the year 2000?

18. (a) In Problem 17, we thought of the US population as a smooth function of time. To what extent is this justified? What happens if we zoom in at a point of the graph? What about events such as the Louisiana Purchase? Or the moment of your birth?
(b) What do we in fact mean by the rate of change of the population for a particular time t?
(c) Give another example of a real-world function which is not smooth but is usually treated as such.

2.5 THE SECOND DERIVATIVE

Since the derivative is itself a function, we can often calculate its derivative. For a function f, the derivative of its derivative is called the *second derivative*, and written f'' (read "f double-prime"). If $y = f(x)$, the second derivative can also be written as $\frac{d^2y}{dx^2}$, which means $\frac{d}{dx}\left(\frac{dy}{dx}\right)$, the derivative of $\frac{dy}{dx}$.

What Does the Second Derivative Tell Us?

Recall that the derivative of a function tells you whether a function is increasing or decreasing:

> If $f' > 0$ on an interval, then f is *increasing* over that interval.
> If $f' < 0$ on an interval, then f is *decreasing* over that interval.

Since f'' is the derivative of f',

> If $f'' > 0$ on an interval, then f' is *increasing* over that interval.
> If $f'' < 0$ on an interval, then f' is *decreasing* over that interval.

So the question becomes: what does it mean for f' to be increasing or decreasing? The case in which f' is increasing is shown in Figure 2.30, where the curve is bending upward, or is *concave up*. In the case when f' is decreasing, shown in Figure 2.31, the graph is bending downward, or is *concave down*.

> $f'' > 0$ means f' is increasing, so the graph of f is concave up on that interval,
> $f'' < 0$ means f' is decreasing, so the graph of f is concave down on that interval.

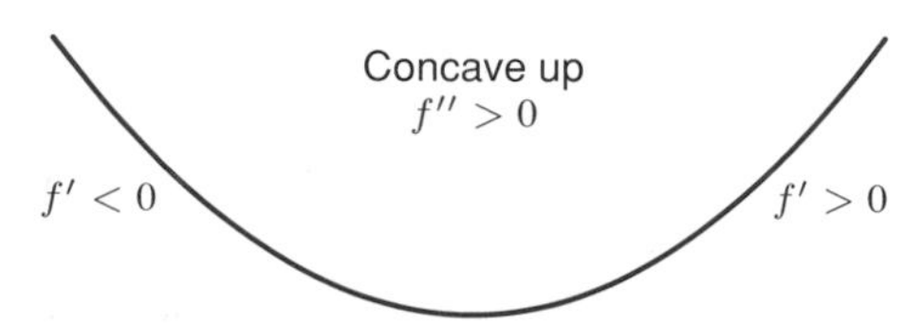

Figure 2.30: Meaning of f'': The slope increases from left to right, so f'' is positive and f is concave up

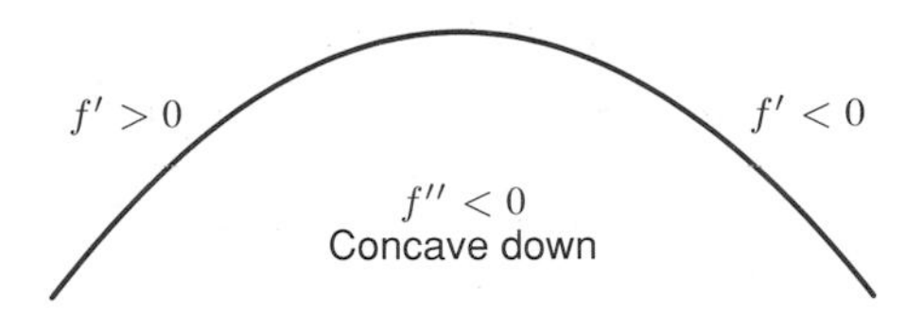

Figure 2.31: Meaning of f'': The slope decreases from left to right, so f'' is negative and f is concave down

Example 1 For the functions whose graphs are given in Figure 2.32, decide where their second derivatives are positive and where they are negative.

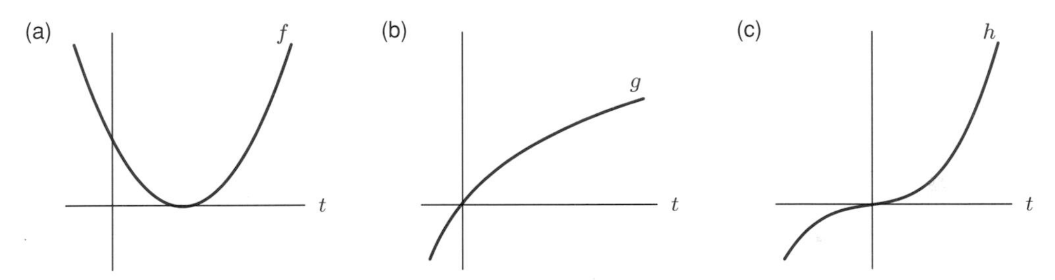

Figure 2.32: What signs do the second derivatives have?

Solution From the graphs it appears that:

(a) $f'' > 0$ everywhere, because the graph of f is concave up everywhere.
(b) $g'' < 0$ everywhere, because the graph is concave down everywhere.
(c) $h'' > 0$ for $t > 0$, because the graph of h is concave up there; $h'' < 0$ for $t < 0$, because the graph of h is concave down there.

Interpretation of the Second Derivative as a Rate of Change

If we think of the derivative as a rate of change, then the second derivative is a rate of change of a rate of change. If the second derivative is positive, the rate of change is increasing; if the second derivative is negative, the rate of change is decreasing.

The second derivative is often a matter of practical concern. In 1985 a newspaper headline reported the Secretary of Defense as saying that Congress and the Senate had cut the defense budget. As his opponents pointed out, however, Congress had merely cut the rate at which the defense budget was increasing.[2] In other words, the derivative of the defense budget was still positive (the budget was increasing), but the second derivative was negative (the budget's rate of increase had slowed).

Example 2 A population, P, growing in a confined environment often follows a logistic growth curve, like the graph shown in Figure 2.33. Describe how the rate at which the population is increasing changes over time. What is the practical interpretation of t_0 and L?

[2]In the *Boston Globe*, March 13, 1985, Representative William Gray (D–Pa.) was reported as saying: "It's confusing to the American people to imply that Congress threatens national security with reductions when you're really talking about a reduction in the increase."

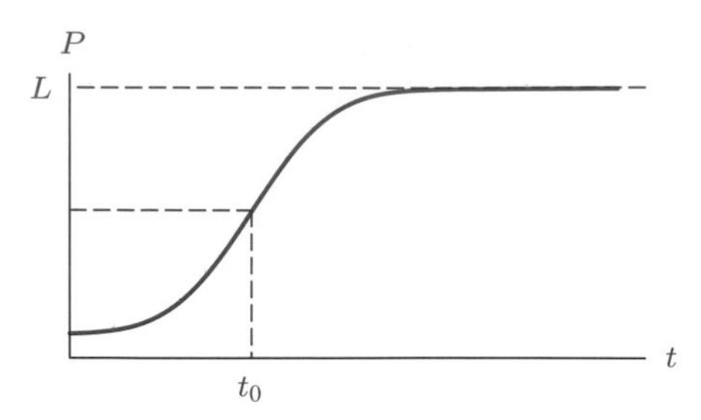

Figure 2.33: Logistic growth curve

Solution Initially, the population is increasing, and at an increasing rate. Thus, initially dP/dt is increasing and so $d^2P/dt^2 > 0$. At t_0, the rate at which the population is increasing is a maximum. Thus at time t_0 the population is growing fastest. Beyond t_0, the rate at which the population is growing is decreasing, and so $d^2P/dt^2 < 0$. At t_0, the concavity changes from positive to negative, and $d^2P/dt^2 = 0$.

The quantity L represents the limiting value of the population which is approached as t tends to infinity. L is called the *carrying capacity* of the environment and represents the maximum population that the environment can support.

Example 3 Table 2.13 shows the number of abortions per year, A, performed in the US in year t (as reported to the Center for Disease Control and Prevention). Suppose these data points lie on a smooth curve $A = f(t)$.

TABLE 2.13 *Abortions reported in the US (1972–1985)*

Year, t	1972	1976	1980	1985
Number of abortions reported, A	586,760	988,267	1,297,606	1,328,570

(a) Estimate dA/dt for the time intervals shown between 1972 and 1985.

(b) What can you say about the sign of d^2A/dt^2 during the period 1972–1985?

Solution (a) For each time interval we can calculate the average rate of change of the number of abortions per year over this interval. For example, between 1972 and 1976

$$\frac{dA}{dt} \approx \begin{matrix}\text{Average rate}\\ \text{of change}\end{matrix} = \frac{988{,}267 - 586{,}760}{1976 - 1972} \approx 100{,}377.$$

Values of dA/dt are listed in Table 2.14:

TABLE 2.14 *Rate of change of number of abortions reported*

Time	1972–1976	1976–1980	1980–1985
Average rate of change, dA/dt	100,377	77,335	6,193

(b) Since the values of dA/dt are decreasing for 1976–1985, d^2A/dt^2 is negative for this period. For 1972–1976, the sign of d^2A/dt^2 is less clear; abortion data from 1968 would help. The graph of A against t in Figure 2.34 confirms this; the graph is concave down for 1976–1985, but straight for 1972–1976. The fact that dA/dt is positive tells us that the number of abortions reported has increased during the period 1972–1985. The fact that d^2A/dt^2 is negative for 1976–1985 tells us that the rate of increase has slowed over this period.

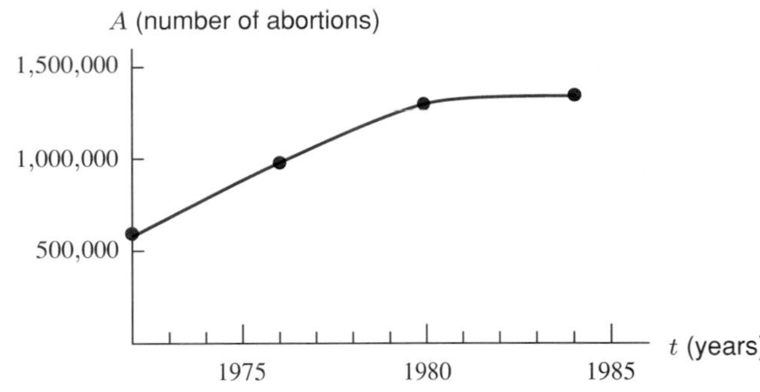

Figure 2.34: How the number of reported abortions in the US is changing with time

Velocity and Acceleration

The velocity, v, of a moving body is the rate at which the position, s, of the body is changing with respect to time:

$$v = \frac{ds}{dt}.$$

The acceleration of a moving body describes how fast the velocity is changing with time, so

$$a = \frac{dv}{dt}$$

or, since the velocity is itself a derivative,

$$a = \frac{dv}{dt} = \frac{d^2s}{dt^2}.$$

Example 4 A particle is moving along a straight line. If its distance, s, to the right of a fixed point is given by Figure 2.35, estimate:

(a) When the particle is moving to the right and when it is moving to the left.

(b) When the particle has positive acceleration and when it has negative acceleration.

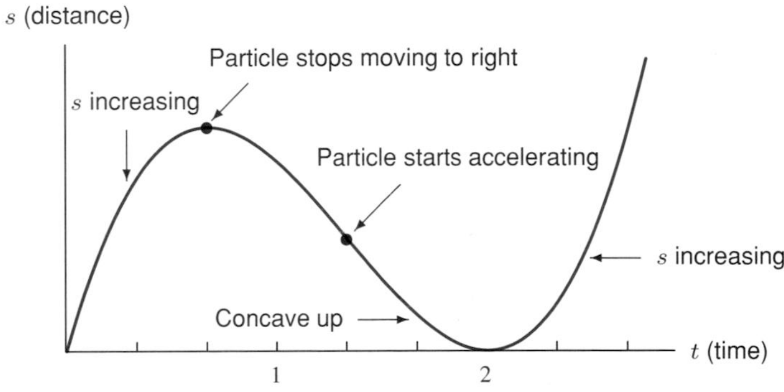

Figure 2.35: Distance of particle to right of fixed point

Solution (a) The particle is moving to the right whenever s is increasing. From the graph, this appears to be for $0 < t < \frac{2}{3}$ and for $t > 2$. For $\frac{2}{3} < t < 2$, the value of s is decreasing, so the particle is moving to the left.

(b) The particle has positive acceleration whenever the curve is concave up, which appears to be for $t > \frac{4}{3}$. The particle has negative acceleration when the curve is concave down, for $t < \frac{4}{3}$.

Problems for Section 2.5

1. (a) If f'' is positive on an interval, then f' is ______________ on that interval, and f is ______________ on that interval.
   (b) If f'' is negative on an interval, then f' is ______________ on that interval, and f is ______________ on that interval.
2. (a) Sketch a curve whose first and second derivatives are everywhere positive.
   (b) Sketch a curve whose second derivative is everywhere negative but whose first derivative is everywhere positive.
   (c) Sketch a curve whose second derivative is everywhere positive but whose first derivative is everywhere negative.
   (d) Sketch a curve whose first and second derivatives are everywhere negative.
3. (a) Sketch a smooth curve whose slope is both positive and increasing at first but later is positive and decreasing.
   (b) Sketch the graph of the first derivative of the curve in part (a).
   (c) Sketch the graph of the second derivative of the curve in part (a).
4. "Winning the war on poverty" has been described cynically as slowing the rate at which people are slipping below the poverty line. Assuming that this is happening:
   (a) Sketch a graph of the total number of people in poverty against time.
   (b) If N is the number of people below the poverty line at time t, what are the signs of dN/dt and d^2N/dt^2?
5. In economics "total utility" refers to the total satisfaction from consuming some commodity. According to the economist Samuelson[3]:

   > As you consume more of the same good, the total (psychological) utility increases. However, . . . with successive new units of the good, your total utility will grow at a slower and slower rate because of a fundamental tendency for your psychological ability to appreciate more of the good to become less keen.

   (a) Sketch the total utility as a function of the number of units consumed.
   (b) In terms of derivatives, what is Samuelson telling us?
6. Let $P(t)$ represent the price of a share of stock of a corporation at time t. What does each of the following statements tell us about the signs of the first and second derivatives of $P(t)$?
   (a) "The price of the stock is rising faster and faster."
   (b) "The price of the stock is close to bottoming out."
7. IBM-Peru uses second derivatives to assess the relative success of various advertising campaigns. They assume that all campaigns produce some increase in sales. If a graph of sales against time shows a positive second derivative during a new advertising campaign, what does this suggest to IBM management? Why? What does a negative second derivative during a new campaign suggest?
8. Given the following data:

x	0	0.2	0.4	0.6	0.8	1.0
$f(x)$	3.7	3.5	3.5	3.9	4.0	3.9

   (a) Estimate $f'(0.6)$ and $f'(0.5)$. (b) Estimate $f''(0.6)$.
   (c) Where do you think the maximum and minimum values of f occur in the interval $0 \leq x \leq 1$?

[3]From Paul A. Samuelson, *Economics*, 11th edition (New York: McGraw-Hill, 1981).

9. An industry is being charged by the Environmental Protection Agency (EPA) with dumping unacceptable levels of toxic pollutants in a lake. Over a several month period, an engineering firm makes daily measurements of the rate at which pollutants are being discharged into the lake.

   Suppose the engineers produce a graph similar to either Figure 2.36(a) or Figure 2.36(b). For each case, give an idea of what argument the EPA might make in court against the industry and of the industry's defense.

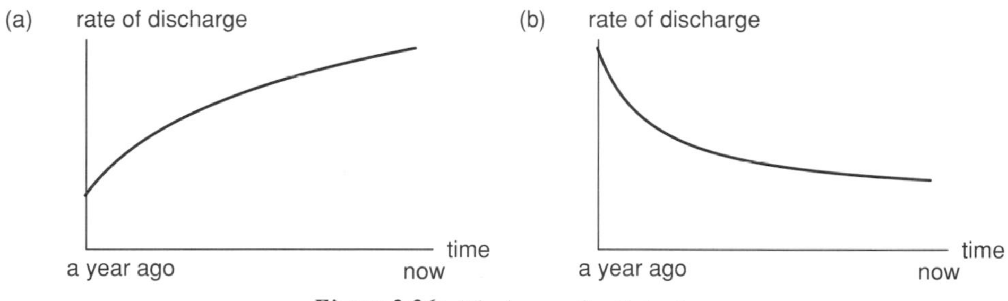

Figure 2.36: Discharge of pollutants

10. The graph of f' (not f) is given in Figure 2.37. At which of the marked values of x is
    (a) $f(x)$ greatest? (b) $f(x)$ least? (c) $f'(x)$ greatest? (d) $f'(x)$ least?
    (e) $f''(x)$ greatest? (f) $f''(x)$ least?

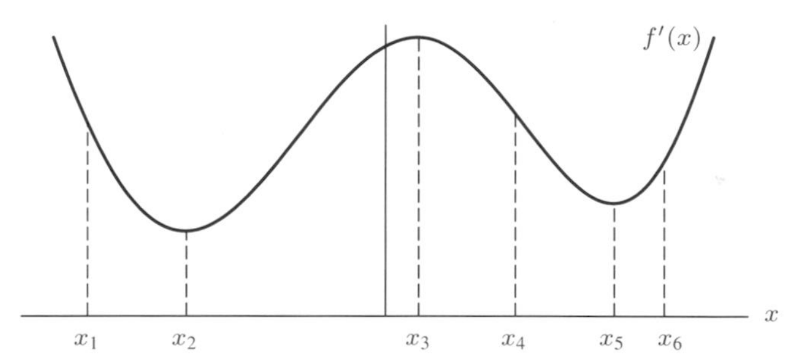

Figure 2.37: Note that this is a graph of f' against x

11. Which of the points labeled by letters in the graph of f in Figure 2.38 have
    (a) f' and f'' nonzero and of the same sign?
    (b) At least two of f, f', and f'' equal to zero?

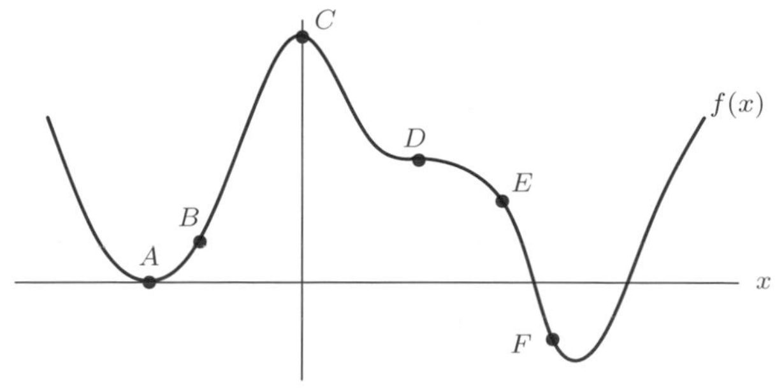

Figure 2.38

2.6 APPROXIMATIONS AND LOCAL LINEARITY

As you know, near any point the graphs of most functions look almost like a line whose slope is the derivative of the function at the point. If we are studying a complicated function over a small interval, we can approximate the graph of the function by this line. This property of most functions is called *local linearity*—local because it holds only in a small region of the domain of the function.

The purpose of this section is to work backwards and use local linearity to calculate approximate values for the function. First, as a reminder, look at $f(x) = x^2$. Although $f(x) = x^2$ is definitely *not* a linear function, the values in Table 2.15 look *almost* linear. For a constant change in x of 0.001, there is a nearly constant change in $f(x)$ of 0.002. Thus, near $x = 1$, the x^2 function appears nearly linear with a slope of 2. In the next example we use local linearity to estimate values of x^2 quickly.

TABLE 2.15 *Table of values for* $f(x) = x^2$

x	x^2
1	1
1.001	1.002001
1.002	1.004004
1.003	1.006009
1.004	1.008016
1.005	1.010025

Example 1 Use the fact that $f(x) = x^2$ is locally linear near $x = 1$ to give a rule of thumb for estimating the square of a number near 1. Use this rule to estimate $(1.013)^2$, $(1.00007)^2$, $(0.989)^2$, and $(0.994)^2$.

Solution We have observed that $f(x) = x^2$ is approximately linear near $x = 1$ with slope 2. (See Figure 2.39.) Thus, differences from 1 are doubled when we square a number near 1. Therefore,

$$(1.013)^2 \approx 1 + 2(0.013) = 1.026$$
$$(1.00007)^2 \approx 1 + 2(0.00007) = 1.00014$$
$$(0.989)^2 = (1 - 0.011)^2 \approx 1 - 2(0.011) = 0.978$$
$$(0.9994)^2 = (1 - 0.0006)^2 \approx 1 - 2(0.0006) = 0.9988$$

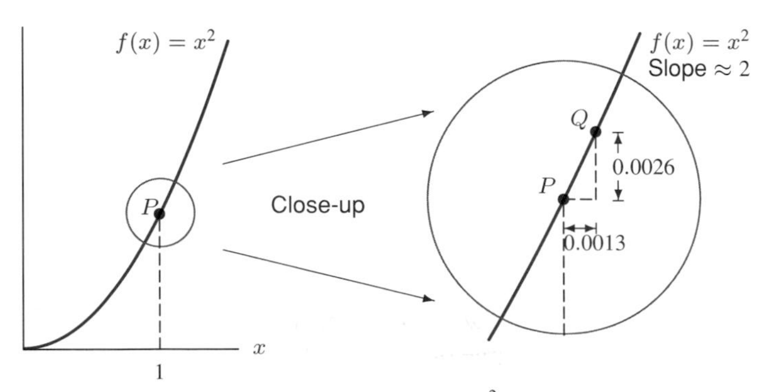

Figure 2.39: Zooming in on a portion of $f(x) = x^2$ until the graph is almost straight

Example 2 Check the approximation $(1.0013)^2 \approx 1.0026$ with a calculator. How large is the error? Is the approximation an over- or underestimate?

Solution $(1.0013)^2 = 1.00260169$, so the error is 0.00000169. The approximation is smaller than the true value.

As another example, look at the square root function. It, too, looks locally linear near any point. In Table 2.16 are values near $x = 4$. Observe that every change in x of 0.001 gives rise to a change in $f(x)$ of about 0.00025. Thus we have locally linear behavior with a slope of about $0.00025/0.001 = \frac{1}{4}$. We can use this locally linear behavior to estimate a value in the table between two given values (this is called *interpolation*) or to extend the table to a value beyond the given values (this is called *extrapolation*).

TABLE 2.16 *Table of values for* $f(x) = \sqrt{x}$

x	$f(x) = \sqrt{x}$
4	2.0000000
4.001	2.0002500
4.002	2.0005000
4.003	2.0007500
4.004	2.0010000
4.005	2.0012496

Example 3 Using the fact that $\sqrt{x}$ is approximately linear near $x = 4$ with slope $\frac{1}{4}$, estimate $\sqrt{4.0036}$, and $\sqrt{4.007}$.

Solution Since the slope is $1/4$, the change between 4.003 and 4.0036 should be multiplied by $1/4$ to get the change between the square roots. Since $\sqrt{4.003} = 2.0007500$, we interpolate to get

$$\sqrt{4.0036} \approx 2.0007500 + (1/4)(0.0006) = 2.0009000.$$

To extrapolate out to $x = 4.007$, we can either start at 4 and multiply the change between 4 and 4.007 by $1/4$, or we could start at 4.005 and multiply the change between 4.005 and 4.007 by $1/4$. If we do the latter, using $\sqrt{4.005} = 2.0012496$, we get

$$\sqrt{4.007} \approx 2.0012496 + (1/4)(0.002) = 2.0017496$$

Example 4 Use a calculator to see how good the estimate $\sqrt{4.007} \approx 2.0017496$ is. How large is the error? Is the approximation too large or too small?

Solution $\sqrt{4.007} \approx 2.0017492$, so the error is 0.0000004. The approximation is too large.

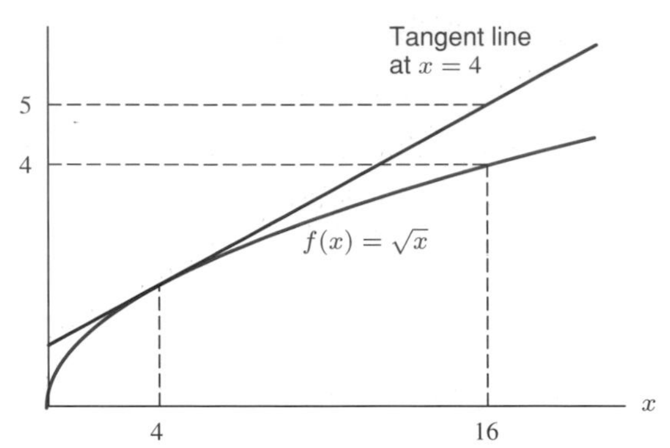

Figure 2.40: Local linearization: Approximating $f(x) = \sqrt{x}$ by its tangent line at $x = 4$

> This method is called *local linearization* or the *tangent line approximation* of a function. We are using the fact that near the point of contact, a curve and its tangent line are not very far apart. The values we have been calculating are coordinates of points on the tangent line.

Of course, there's no reason to expect that the curve will look like the tangent line if we go too far away, and usually it doesn't. For example, using the same linearization to find $\sqrt{16}$, we get $\sqrt{16} \approx 2 + 12(\frac{1}{4}) = 5$. As you can see, this is not so good! (See Figure 2.40.) The problem is that we have traveled too far from the place where the curve looks like a line with slope $\frac{1}{4}$. In general, extrapolating too far from known data is dangerous.

Example 5 Local linearization will give values too small for the function x^2 and too large for the function $\sqrt{x}$. Draw pictures to explain why. Can you formulate a conjecture that tells when the tangent line approximation will be too large or too small?

Solution The graph of x^2 is concave up and lies above its tangent line; therefore, the linearization will always be too small. The graph of $\sqrt{x}$ is concave down and lies below its tangent line, and therefore the linearization will be too large.

Of course, the main use of this method of approximation is not for quantities like $(1.0013)^2$ and $\sqrt{4.007}$, which are much better found using a calculator. The real purpose of such approximations is to make estimates which cannot easily be made in other ways, as in the following example.

Example 6 Climbing health care costs have been a source of concern for some time. The data[4] in Table 2.17 shows the average yearly per-capita (i.e., per person) health care expenditures for various years since 1970. Use this data to estimate average per-capita expenditures in 1988 and 1995.

[4] Data from the Department of Health and Human Services, reported by John Wright, ed., in the *Universal Almanac 1990* (Kansas City, Mo.: Andrews & McMeel, 1990)

TABLE 2.17 *Health care costs*

Year	Per capita expenditure ($)
1970	349
1975	591
1980	1055
1985	1596
1987	1987[5]

Solution Health care costs have clearly been increasing throughout the last 20 years. Between 1985 and 1987 they increased $(1987 - 1596)/2 = \$195.50$ per year. To make estimates beyond 1987 we will assume that costs continue to climb at the same rate. Therefore,

$$\begin{aligned}\text{Costs in 1988} &= \text{Costs in 1987} + \text{Change in Costs}\\ &\approx \$1987 + \$195.50 = \$2182.50.\end{aligned}$$

Since 1995 is 8 years beyond 1987,

$$\text{Costs in 1995} \approx \$1987 + \$195.50(8) = \$3551.$$

You should realize that the estimate for 1988 in the preceding example is much more likely to be close to the true value than the estimate for 1995. The further we extrapolate from the given data, the more errors we are likely to introduce. It is unlikely the rate of change of health care costs will stay at $195.50/year all the way until 1995.

Graphically, what we have done is to extend the line joining the points for 1985 and 1987 to make projections for the future. (See Figure 2.41.) You might be concerned that we used only the last two pieces of data to make the estimates. Isn't there valuable information to be gained from the rest of the data? Yes, indeed—though there's no fixed way of taking this information into account. You might look at the rate of change for the years before 1985 and take an average. Alternatively, you might draw a line through the data and use that. However, in each of these methods you are estimating a rate of change of health care costs with time. In other words, you are estimating the derivative of the cost function.

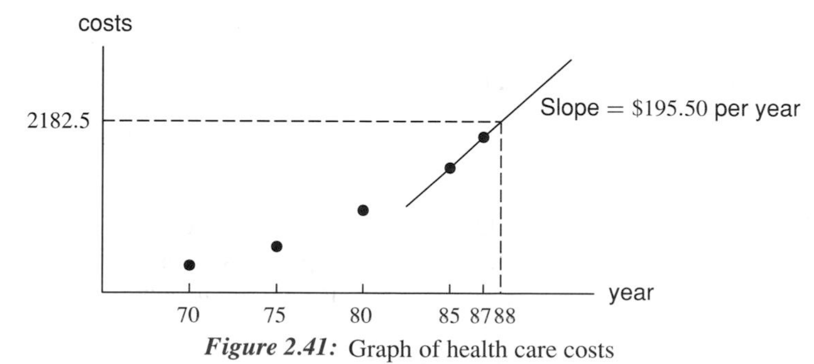

Figure 2.41: Graph of health care costs

[5] Yes, the last number is correct. The expenditures in 1987 were $1987 per person.

Problems for Section 2.6

1. Find the equation of the tangent line to $f(x) = \sqrt{x}$ at $x = 4$, using the fact that $f'(4) = 1/4$. Confirm that the point used in the approximation of $\sqrt{4.007}$ lies on this line. (See Example 3, page 133.)

2. Use the fact that the derivative of $f(x) = x^3$ is 12 at $x = 2$ to fill in approximate values of x^3 near 2 in the table below. Check your answers by finding x^3 exactly with a calculator.

x	2.000	2.001	2.002	2.003	2.004	2.005
x^3	8.000	8.012	8.024			

3. Use a calculator to construct a table of values for $\tan x$, rounded to three decimals, for $x = 0.80$, $0.81, 0.82, \ldots, 0.90$. Is the table approximately linear?

4. Given the following data:

x	0	0.2	0.4	0.6	0.8	1.0
$f(x)$	3.7	3.5	3.5	3.9	4.0	3.9

   (a) Estimate an equation of the tangent line to $y = f(x)$ at $x = 0.6$.
   (b) Using this equation, estimate $f(0.7)$, $f(1.2)$, and $f(1.4)$. Which of these estimates do you feel most confident about? Why?

5. Given the following data about a function, f:

x	6.5	7.0	7.5	8.0	8.5	9.0
$f(x)$	10.3	8.2	6.5	5.2	4.1	3.2

   (a) Estimate $f'(7.0)$, $f'(8.5)$ and $f'(6.75)$.
   (b) Estimate the rate of change of f' at $x = 7$.
   (c) Find, approximately, an equation of the tangent line to the graph of f at $x = 7$.
   (d) Estimate $f(6.8)$.

6. If a child were to step off from the edge of a spinning merry-go-round, she would move in the direction of the line tangent to the path of the merry-go-round at the point where she steps off. Suppose a merry-go-round of radius 10 is centered at the origin and the child steps off from the point $(6, 8)$. Figure 2.42 shows the child's path from above. Estimate the derivative of $y = \sqrt{100 - x^2}$ at $x = 6$. Use that value to find the slope of the line along which she will move (considering her horizontal motion only, and not her height above the ground). Your answer should be around $-6/8$. Why?

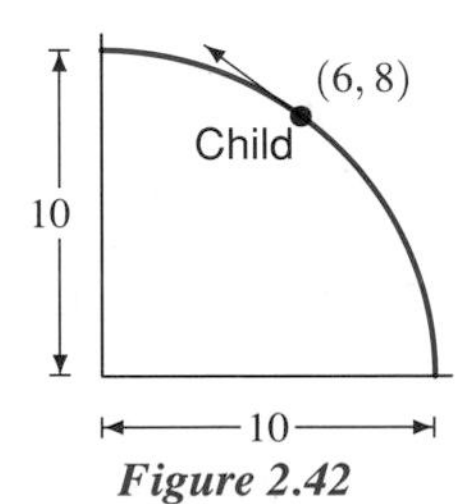

Figure 2.42

7. The child in Problem 6 moved along the path of the tangent line to the circle she had been following. This phenomenon, a result of *inertia*, would occur regardless of the original path of an object—i.e., the momentum of an object is always along the tangent line to the path it is following.

   Suppose a man was riding "The Quadratic Express," a train from $(-2, 4)$ to $(2, 4)$ along a track which followed the curve $y = x^2$. If he happened to be riding on top of the train and stepped off just as he was passing the origin, along what line would he move? What if he stepped off at $(1, 1)$? (Again, we are considering horizontal motion only.)

8. In Example 6, page 134, we assumed that health care costs were increasing linearly. We will now assume that health care costs are increasing exponentially.

   (a) Fit an exponential function to the data for 1985 and 1987 in Table 2.17, and use this function to predict health care costs in 1988 and 1995.

   (b) How close are the linear and exponential predictions for 1988? For 1995? How will the two predictions compare further into the future?

9. The number of hours, H, of daylight in Madrid as a function of date is approximated by the formula

$$H = 12 + 2.4\sin[0.0172(t - 80)]$$

   where t is the number of days since the start of the year. Figure 2.43 shows a one-month portion of the graph of H.

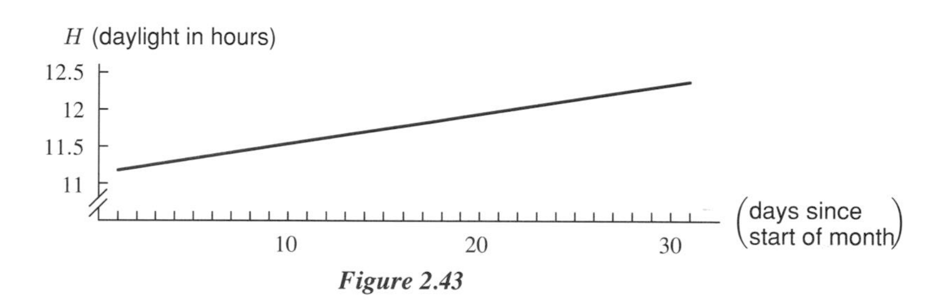

Figure 2.43

   (a) Comment on the shape of the graph. Why does it look like a straight line?

   (b) What month does this graph show? How do you know?

   (c) What is the approximate slope of this line? What does the slope represent in practical terms?

10. Suppose you put a yam in a hot oven, maintained at a constant temperature of 200°C. As the yam picks up heat from the oven, its temperature rises.[6]

   (a) Draw a possible graph of the temperature T of the yam against time t (minutes) since it is put into the oven. Explain any interesting features of the graph, and in particular explain its concavity.

   (b) Suppose that, at $t = 30$, the temperature T of the yam is 120° and increasing at the (instantaneous) rate of 2°/min. Using this information, plus what you know about the shape of the T graph, estimate the temperature at time $t = 40$.

   (c) Suppose in addition you are told that at $t = 60$, the temperature of the yam is 165°. Can you improve your estimate of the temperature at $t = 40$?

   (d) Assuming all the data given so far, estimate the time at which the temperature of the yam is 150°.

[6]From Peter D. Taylor, *Calculus: The Analysis of Functions* (Toronto: Wall & Emerson, Inc., 1992)

2.7 NOTES ON THE LIMIT

Our investigation of the problem of speed has taken us a long way: we ended up with a notion of instantaneous rate of change for any function. As mentioned previously, we have not really simplified the measurement of such a quantity; we have just hidden the difficulties in the notion of limit. Let's look a bit harder at the limit concept.

The expression

$$\frac{f(a+h)-f(a)}{h}$$

is called the difference quotient of f. It measures the average rate of change of f over the interval $a \le x \le a+h$. As the interval shrinks, that is, when h gets smaller and smaller, these average rates may begin to approach a particular number. If there is a number L, say, such that we can make the average rate as close to L as we please simply by choosing a small enough interval, then L is said to be the *limit* of the difference quotients as h approaches 0. In symbols,

$$L = \lim_{h\to 0} \frac{f(a+h)-f(a)}{h}.$$

The thing to note here is that we are asserting that as h gets small, the difference quotient approaches the number L. We are not saying *anything* about what happens when h *equals* 0. Indeed, when $h = 0$, the difference quotient becomes $0/0$, which is meaningless. The limit was introduced just to avoid such trouble.

> We define
>
> $$\lim_{h\to c} g(h)$$
>
> to be a number L (if one exists) such that $g(h)$ is as close to L as we please whenever h is sufficiently close to c (but $h \neq c$).

Limits have many uses in calculus—finding the instantaneous rate of change is just one of them. Actually, we can investigate the limit of any function at any point we choose. For example, the value of

$$\lim_{h\to 2} h^2$$

is 4. It's 4 because we can make h^2 as close to 4 as we like by taking h sufficiently close to 2. (Look at the values of 1.9^2, 1.99^2, 1.999^2, and 2.1^2, 2.01^2, 2.001^2.) Since the limit doesn't "ask" what happens *at* $h = 2$, it is not sufficient to simply put in 2 to find the answer. The limit describes behavior of a function *near* a point, not *at* the point.

Not every limit statement is as transparent as $\lim_{h\to 2} h^2 = 4$. For example, how would you find

$$\lim_{h\to 0}(1+h)^{1/h}?$$

You might let $h = 0.01, 0.001, 0.0001$, and so on. Try it—the result is remarkable! Notice that substituting $h = 0$ makes the expression undefined, so the only way to investigate this is to use values near but not equal to zero. (See Problem 10 on page 140.)

Example 1 Use a graph to find $\lim_{\theta \to 0} \frac{\sin\theta}{\theta}$ for θ in radians.

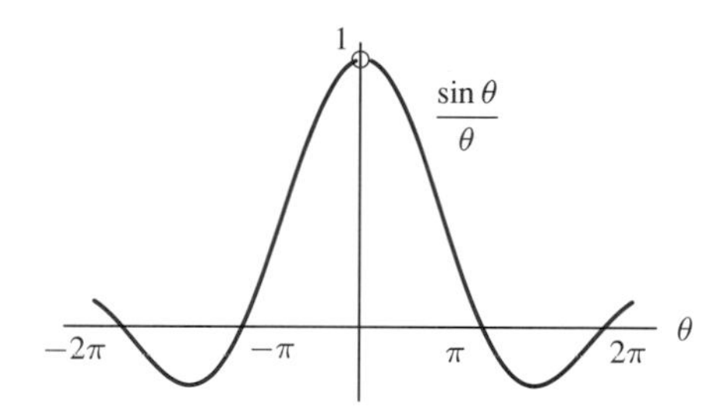

Figure 2.44: Find the limit as $\theta \to 0$

Solution Figure 2.44 shows that as θ gets closer to 0, $\frac{\sin\theta}{\theta}$ gets close to 1, so we say $\lim_{\theta \to 0} \frac{\sin\theta}{\theta} = 1$.

One- and Two-Sided Limits

One other technical comment about limits is needed. When we write

$$\lim_{h \to 2} f(h),$$

we want this to represent the number that $f(h)$ approaches as h gets close to 2 *from both sides*. We examine values of $f(h)$ as h approaches to 2 through values greater than 2 (e.g. 2.1, 2.01, 2.001) and values less than 2 (e.g. 1.9, 1.99, 1.999). If $f(h)$ approaches the same number as we approach 2 from both sides, then that number is called the limit of f as h approaches 2. It is almost always the case that these "left" and "right" limits are the same. If they are not, we say the limit does not exist.

If, for some reason, we want h to approach 2 only through values greater than 2, we write

$$\lim_{h \to 2^+} f(h)$$

for the number that $f(h)$ approaches (assuming such a number exists). Similarly,

$$\lim_{h \to 2^-} f(h)$$

denotes the number (if it exists) obtained by letting $h \to 2$ through values less than 2. We call $\lim_{h \to 2^+}$ a *right-hand limit* and $\lim_{h \to 2^-}$ a *left-hand limit.*

Limits and Continuity

Appendix B describes informally what it means for a function to be continuous. The appendix explains how to recognize whether a function is continuous from its graph: the graph should have no jumps, breaks, or holes. Numerically, a function is continuous if nearby values of the independent variable give values of the function that are as close together as we want.

We now give a more precise definition of continuity. Suppose $f(x)$ is a continuous function of x on some interval, and $x = a$ is in that interval. Then if x is close to a, we know that $f(x)$ is close to $f(a)$. In fact, the closer x gets to a, the closer $f(x)$ gets to $f(a)$. Thus, as $x \to a$, the limit of $f(x)$ must be $f(a)$. Mathematicians use this idea and the limit notation to give a formal definition of continuity.

A function f is **continuous** at $x = a$ if

$$\lim_{x \to a} f(x) = f(a).$$

Note that to calculate $\lim_{h \to 2} h^2$ you were told not simply to substitute 2 (although, as you can see, substituting 2 does give the right answer). The reason is that *if* you know the function $f(h) = h^2$ is continuous (which it is), then you *can* calculate the limit by substitution. In fact, the criterion for continuity is that the limit and the value of the function at the point are the same.

Problems for Section 2.7

1. Consider the limit

$$\lim_{x \to 0+} x^x.$$

Calculate this limit either by evaluating x^x for smaller and smaller positive values of x ($x = 0.1, 0.01, 0.001, \ldots$) or by zooming in on the graph of $y = x^x$ near $x = 0$.

Use a graph to evaluate each of the limits in Problems 2–5.

2. $\lim_{\theta \to 0} \dfrac{\sin 2\theta}{\theta}$, ($\theta$ in radians)

3. $\lim_{\theta \to 0} \dfrac{\cos \theta - 1}{\theta}$, ($\theta$ in radians)

4. $\lim_{\theta \to 0} \dfrac{\sin \theta}{\theta}$, ($\theta$ in degrees)

5. $\lim_{\theta \to 0} \dfrac{\theta}{\tan 3\theta}$, ($\theta$ in radians)

We define the absolute value of x, $|x|$, by $|x| = x$ (if $x \geq 0$) and $|x| = -x$ (if $x < 0$). Evaluate the limits in Problems 6–7 graphically.

6. $\lim_{x \to 0} \dfrac{|x|}{x}$

7. $\lim_{x \to 0} x \ln |x|$

8. Is $f(x) = |x| / x$ continuous?

9. If θ is in radians, is

$$g(\theta) = \begin{cases} \dfrac{\sin \theta}{\theta} & \theta \neq 0 \\ 0 & \theta = 0 \end{cases}$$

continuous? Why or why not?

10. Calculate

$$\lim_{x \to 0} (1 + x)^{1/x}$$

using the graph of $y = (1 + x)^{1/x}$. You should recognize the answer you get. What is it?

11. Using a graph, calculate

$$\lim_{x \to 0} (1 + 2x)^{1/x}.$$

12. Does $\lim_{x \to 0+} x e^{1/x}$ exist? Why or why not?

2.8 NOTES ON DIFFERENTIABILITY

A function is said to be *differentiable* at any point in its domain at which it has a derivative. Most of the functions we will be dealing with will have a derivative at every point; they are said to be *differentiable everywhere*.

How Do You Know Whether a Function Is Differentiable?

If a function has a derivative at a point, its graph must have a tangent line there. The slope of the tangent line is the derivative. Thus, we can recognize a differentiable function from its graph by observing that a function is differentiable at a point if when you "zoom in," or magnify, the graph at that point, you see a nonvertical straight line.

Occasionally—but not often—we will meet a function which fails to have a derivative at a few points; for example, a discontinuous function whose graph has a break at some point cannot have a derivative at that point. Most of the functions we will deal with are continuous and have graphs with no breaks, jumps, or holes.

Are There Continuous Functions Which Are Not Differentiable at Some Point?

The answer is yes; the points of nondifferentiability occur where the graph of the function does not look like a nonvertical straight line when you zoom in. Two of the ways in which a function can fail to be differentiable at a point are if it has a graph with:

- A sharp "corner" at that point.
- A vertical tangent line.

Look at the function graphed in Figure 2.45. There is no tangent at A because the graph has a "corner" there. As x approaches a from the left, the slope of the line joining P to A converges to some positive number. As x approaches a from the right, the slope of the line joining P to A converges to some negative number. Thus the slopes approach different numbers as you approach $x = a$ from different sides. Therefore the function is not differentiable at $x = a$.

At B, there is no sharp corner, but as x approaches b from the left or right, the slope of the lines joining B to Q do not converge: they just keep growing larger and larger. This reflects the fact that the graph has a vertical tangent at B. However, since the slope of a vertical line is not defined, the function is not differentiable at $x = b$ either. The function appears to be differentiable, however, at all points except $x = a$ and $x = b$.

If the graph of f looks more and more straight and nonvertical as the region near x is magnified, then f possesses a derivative there. On the other hand, if no amount of magnification will cause a "corner" to straighten out at some point, or if it has a vertical tangent line at that point, then f does not have a derivative there.

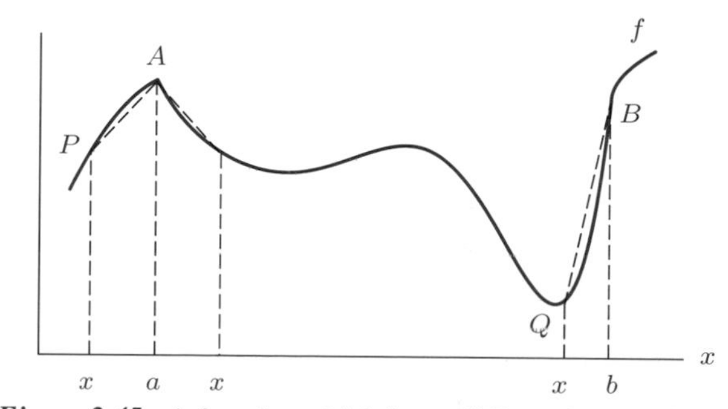

Figure 2.45: A function which is not differentiable at A or B

The Absolute Value Function

The best-known function with a "corner" is the absolute value function

$$f(x) = |x| = \begin{cases} x & \text{if } x \geq 0, \\ -x & \text{if } x < 0. \end{cases}$$

The graph of this function is in Figure 2.46. Even closeup views of the graph of $f(x)$ look the same, so this is a corner which can't be straightened out by zooming in.

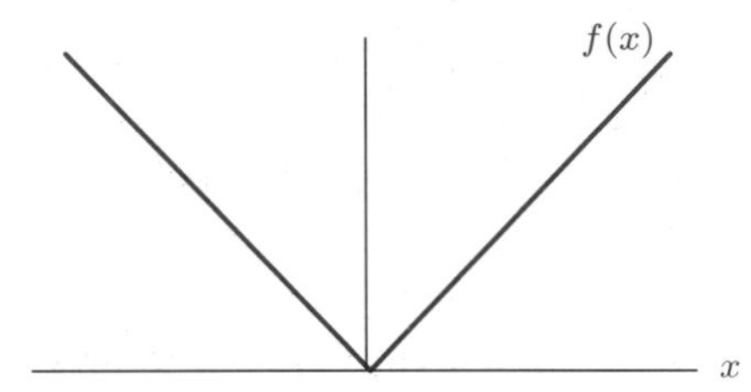

Figure 2.46: Absolute value function: $f(x) = |x|$

Example 1 Try to compute the derivative of the function $f(x) = |x|$ at $x = 0$. Is f differentiable there?

Solution To find the slope at $x = 0$, we want to look at

$$\lim_{h \to 0} \frac{|h| - 0}{h} = \lim_{h \to 0} \frac{|h|}{h}.$$

As h approaches 0 from the right, since h is always positive, $|h| = h$, and the ratio is always 1. On the other side, as h approaches 0 from the left, then $|h| = -h$, and the ratio is -1. These limits correspond to the fact that the slope of the right-hand part of the graph is 1, and the slope of the left-hand part is -1. Since the limits are different on each side, the limit of the difference quotient fails to exist. Thus, the absolute value function is not differentiable at $x = 0$. If you imagine a high magnification of the curve near $x = 0$, you will always see the corner there. Try it on a graphing calculator, or see Figure 2.46. By the way, this function has a derivative everywhere *except* at $x = 0$.

Example 2 Look at the graph of $f(x) = (x^2 + 0.0001)^{1/2}$ shown in Figure 2.47. The graph of f appears to have a sharp corner at $x = 0$. Does f have a derivative at $x = 0$?

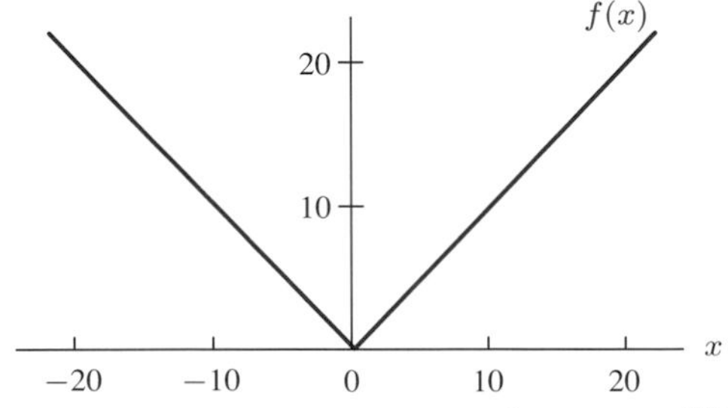

Figure 2.47: Graph of $f(x) = (x^2 + 0.0001)^{1/2}$

Solution We want to look at

$$\lim_{h \to 0} \frac{(h^2 + 0.0001)^{1/2} - (0.0001)^{1/2}}{h}.$$

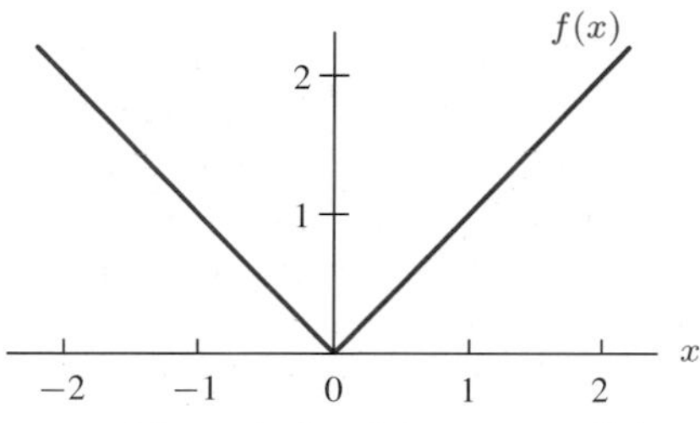

Figure 2.48: Close-ups of $f(x) = (x^2 + 0.0001)^{1/2}$ showing differentiability at $x = 0$

As $h \to 0$ from positive or negative numbers, the difference quotient approaches 0. (Try evaluating it for $h = 0.001, 0.0001$, etc.) So there is a derivative at $x = 0$ (namely 0). How can this be if f has a corner at $x = 0$?

The answer lies in the fact that what appears to be a corner is in fact smooth—when you zoom in, the graph of f looks like a straight line with slope 0! (See Figure 2.48.)

Example 3 Investigate the differentiability of $f(x) = x^{1/3}$ at $x = 0$.

Solution This function is perfectly smooth at $x = 0$ (no lumps or bumps) but appears to have a vertical tangent there. Zoom in using a graphing calculator, or see Figure 2.49. Looking at the difference quotient at $x = 0$, we see

$$\lim_{h \to 0} \frac{(0+h)^{1/3} - 0^{1/3}}{h} = \lim_{h \to 0} \frac{h^{1/3}}{h} = \lim_{h \to 0} \frac{1}{h^{2/3}}.$$

As $h \to 0$ the denominator becomes small, so the fraction grows without bound. Hence, the function fails to have a derivative at $x = 0$.

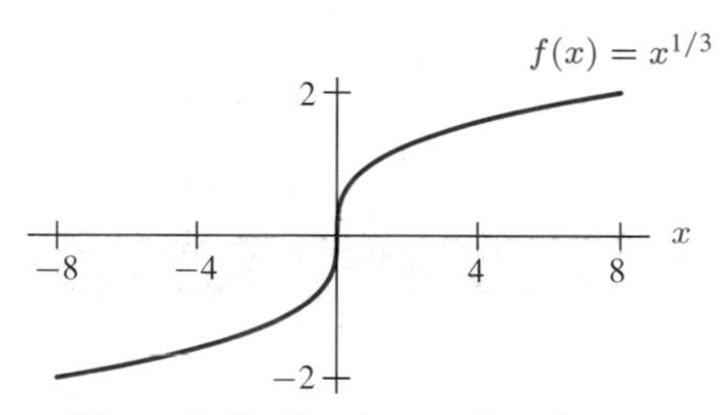

Figure 2.49: Continuous function not differentiable at $x = 0$: Vertical tangent

---

Example 4 Consider the function given by the formulas

$$f(x) = \begin{cases} x + 1 & \text{when} \quad x \leq 1 \\ 3x - 1 & \text{when} \quad x > 1 \end{cases}$$

Draw the graph of f. Is f continuous? Is f differentiable at $x = 1$? Explain.

Solution The graph in Figure 2.50 has no breaks in it, and therefore the function is continuous. However, the graph has a corner at $x = 1$ which no amount of magnification will remove. To the left of $x = 1$, the slope is 1; to the right of $x = 1$, the slope is 3. Thus the function is not differentiable at $x = 1$. This function is called *piecewise linear* because each part of it is linear. Such functions are often used to make approximations, because although they are not smooth, they are easy to work with.

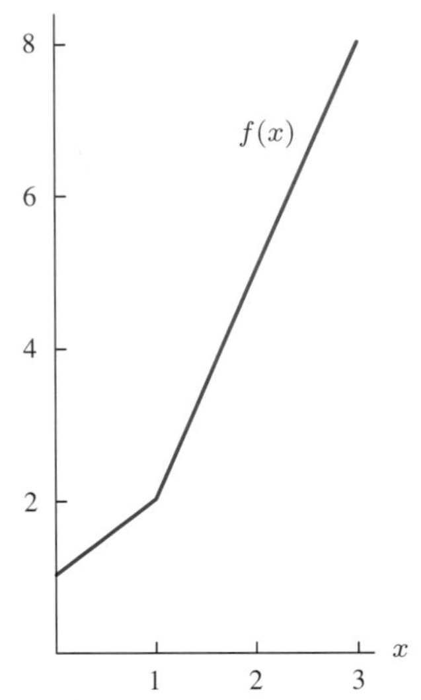

Figure 2.50: Continuous function not differentiable at $x = 1$

A great deal of interest has been sparked in the last few years in the study of curves which do not possess derivatives *anywhere*. These curves arise in the modeling of natural processes that are random and chaotic, such as the path of a water molecule in a glass of water. As the molecule bounces off of its neighbors haphazardly, it traces out a path with many jagged, nondifferentiable corners. Although the path may be smooth between collisions, it can be modeled effectively by a curve that is not smooth anywhere. The coastlines of Maine and Washington are also examples. They never straighten out, no matter how close we look.

Problems for Section 2.8

Decide if the functions in Problems 1–3 are differentiable at $x = 0$. Try zooming in on a graphing calculator, or calculating the derivative $f'(0)$ from the definition.

1. $f(x) = (x + |x|)^2 + 1$

2. $f(x) = \begin{cases} x\sin(1/x) + x & \text{if } x \neq 0 \\ 0 & \text{for } x = 0 \end{cases}$

3. $f(x) = \begin{cases} x^2\sin(1/x) & \text{if } x \neq 0 \\ 0 & \text{for } x = 0 \end{cases}$

4. The acceleration due to gravity, g, varies with height above the surface of the earth. If you go down below the surface of the earth, g varies in a different way. It can be shown that, as a function of r, the distance from the center of the earth, g is given by

$$g = \begin{cases} \dfrac{GMr}{R^3} & \text{for } r < R \\[2ex] \dfrac{GM}{r^2} & \text{for } r \geq R \end{cases}$$

where R is the radius of the earth, M is the mass of the earth, and G is the gravitational constant.

(a) Is g a continuous function of r? Explain your answer.
(b) Sketch a graph of g against r.
(c) Is g a differentiable function of r? Explain your answer.

5. Sometimes, odd behavior can be hidden beneath the surface of a rather normal-looking function. Consider the following function:

$$f(x) = \begin{cases} 0 & \text{if } x < 0 \\ x^2 & \text{if } x \geq 0. \end{cases}$$

(a) Sketch a graph of this function. Does it have any vertical segments or corners? Is it differentiable everywhere? If so, sketch the derivative f' of this function. [Hint: You may want to use the result of Example 4 on page 115.]
(b) Is the derivative differentiable everywhere? If not, at what point(s) is it not differentiable? Draw the second derivative wherever it exists. Is the second derivative differentiable? Continuous?

6. Graph the function defined by

$$g(r) = \begin{cases} 1 + \cos(\pi r/2) & \text{for} \quad -2 \leq r \leq 2 \\ 0 & \text{for} \quad r < -2 \quad \text{or} \quad r > 2. \end{cases}$$

(a) Is g continuous at $r = 2$? Explain your answer.
(b) Do you think g is differentiable at $r = 2$? Explain your answer.

7. An electric charge, Q, in a circuit is given as a function of time by

$$Q = \begin{cases} C & t \leq 0 \\ Ce^{-t/RC} & t > 0 \end{cases}$$

where C and R are positive constants. The electric current, I, is the rate of change of charge, so

$$I = \frac{dQ}{dt}.$$

(a) Is the charge, Q, a continuous function of time?
(b) Do you think the current, I, is defined for all times, t? [Hint: To graph this function, take, for example, $C = 1$ and $R = 1$.]

8. The potential, ϕ, of a charge distribution at a point on the y-axis is given by

$$\phi = \begin{cases} 2\pi\sigma\left(\sqrt{y^2 + a^2} - y\right) & \text{for } y \geq 0 \\ 2\pi\sigma\left(\sqrt{y^2 + a^2} + y\right) & \text{for } y < 0 \end{cases}$$

where σ and a are positive constants. [Hint: To graph this function, take, for example, $2\pi\sigma = 1$ and $a = 1$.]

(a) Is ϕ continuous at $y = 0$?
(b) Do you think ϕ is differentiable at $y = 0$?

REVIEW PROBLEMS FOR CHAPTER TWO

1. Using a calculator or computer, sketch a graph of $f(x) = x \sin x$ on the interval $-10 \le x \le 10$.
   (a) How many zeros does $f(x)$ have on this interval?
   (b) Is f increasing or decreasing at $x = 1$? At $x = 4$?
   (c) On which interval is the average rate of change greater: $0 \le x \le 2$ or $6 \le x \le 8$?
   (d) Is the instantaneous rate of change greater at $x = -9$ or $x = 1$?

2. For the function $f(x) = \log x$, estimate $f'(1)$. From the graph of $f(x)$, would you expect your estimate to be greater than or less than $f'(1)$?

3. Let $f(x) = x^2$. In Section 2.2 we calculated an estimate of $f'(1)$ by tabulating values of x^2 near $x = 1$ (see Table 2.3 on page 105). Do the same for $x = -1$ and for $x = 5$. In each case estimate the derivative at that point. Recalling from Example 4 of Section 2.3 that $f'(2) \approx 4$ and $f'(3) \approx 6$, guess a general formula for the derivative of $f(x) = x^2$.

4. Construct tables of values, rounded to three decimals, for $f(x) = x^3$ near $x = 1$, near $x = 3$ and near $x = 5$, as in Problem 3. In each case estimate the derivatives $f'(1)$, $f'(3)$, and $f'(5)$. Then guess a general formula for $f'(x)$.

5. For $f(x) = \ln x$, construct tables, rounded to four decimals, near $x = 1$, $x = 2$, $x = 5$, and $x = 10$ as in Problem 3. Use the tables to estimate $f'(1)$, $f'(2)$, $f'(5)$, and $f'(10)$. Then guess a general formula for $f'(x)$.

6. (a) If f is even and $f'(10) = 6$, what must $f'(-10)$ equal?
   (b) If f is any even function and $f'(0)$ exists, what must $f'(0)$ equal?

7. If g is an odd function and $g'(4) = 5$, what must $g'(-4)$ equal?

For Problems 8–9, sketch a graph of the derivative function of the given functions.

8.

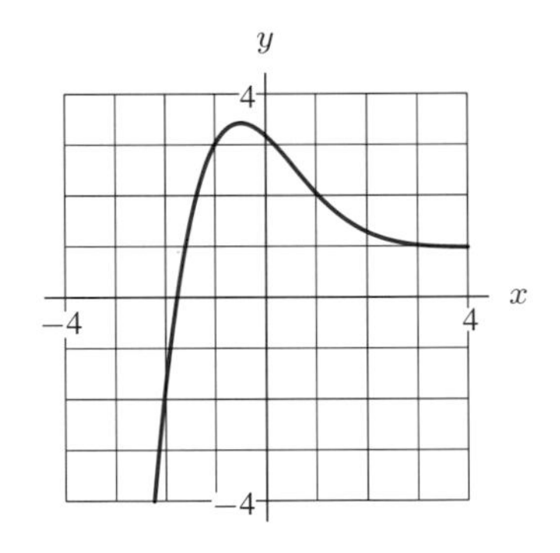

9.

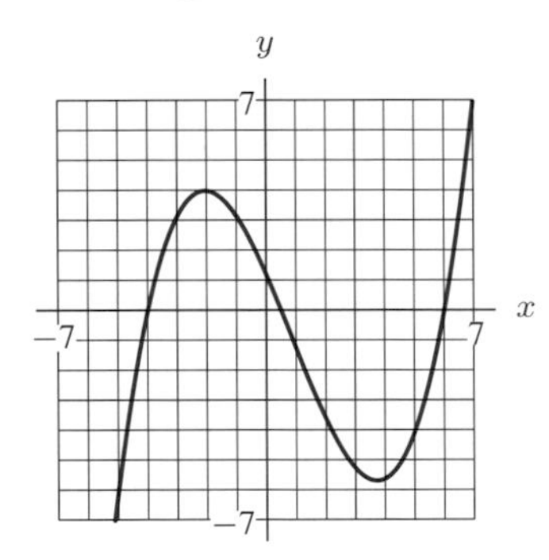

Sketch the graphs of the derivatives of the functions shown in Problems 10–14. Be sure your sketches are consistent with the important features of the original functions.

10.

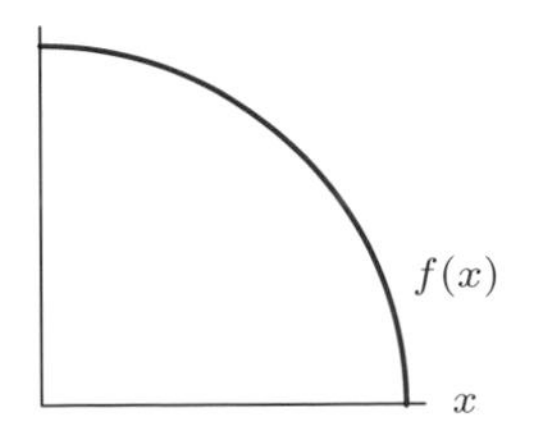

11.

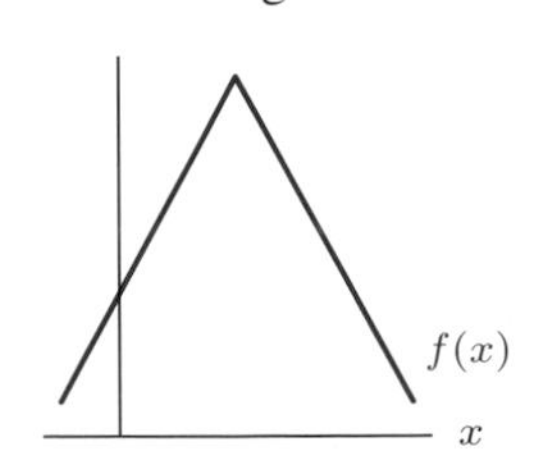

12.

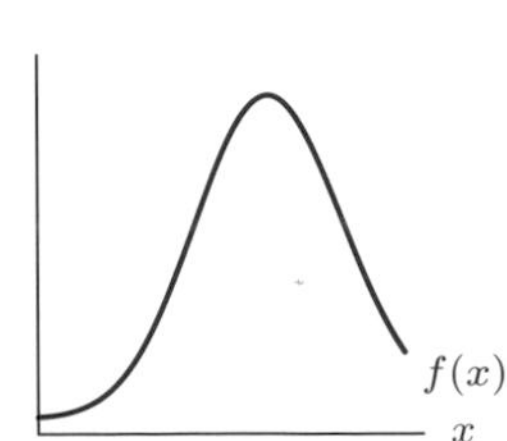

13.

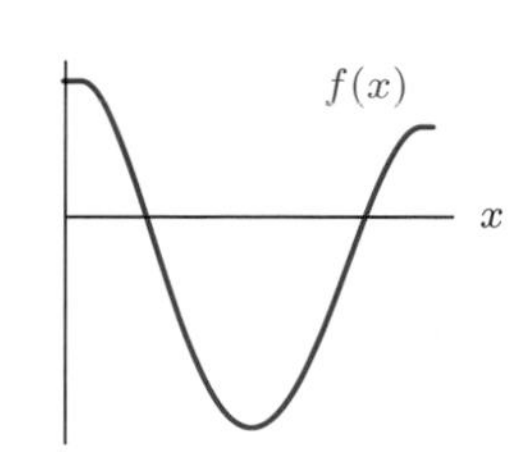

14.

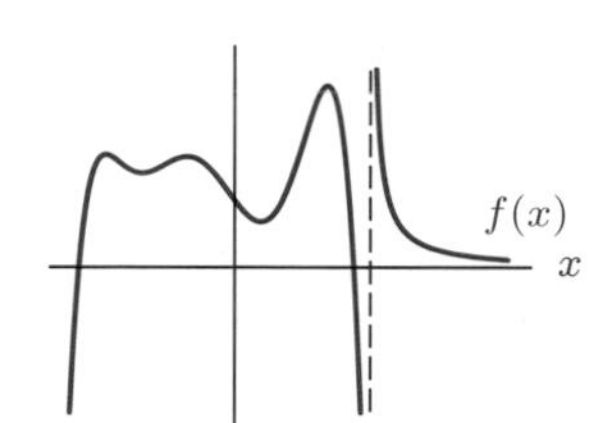

15. Draw a possible graph of $y = f(x)$ given the following information about its derivative:

- For $x < -2$, $f'(x) > 0$ and the derivative is increasing.
- For $-2 < x < 1$, $f'(x) > 0$ and the derivative is decreasing.
- At $x = 1$, $f'(x) = 0$.
- For $x > 1$, $f'(x) < 0$ and the derivative is decreasing (getting more and more negative).

16. Students were asked to evaluate $f'(4)$ from the following table which shows the values of the function f:

x	1	2	3	4	5	6
$f(x)$	4.2	4.1	4.2	4.5	5.0	5.7

- Student A estimated the derivative as $f'(4) \approx \dfrac{f(5) - f(4)}{5 - 4} = 0.5$.
- Student B estimated the derivative as $f'(4) \approx \dfrac{f(4) - f(3)}{4 - 3} = 0.3$.
- Student C suggested that they should split the difference and estimate the average of these two results, that is, $f'(4) \approx \frac{1}{2}(0.5 + 0.3) = 0.4$.

(a) Sketch the graph of f, and indicate how the three estimates are represented on the graph.
(b) Explain which answer is likely to be best.
(c) Use Student C's method to find an algebraic formula which approximates $f'(x)$ using increments of size h.

17. Each of the graphs in Figure 2.51 shows the position of a particle moving along the x-axis as a function of time, $0 \le t \le 5$. The vertical scales of the graphs are the same. During this time interval, which particle has

(a) Constant velocity?
(b) The greatest initial velocity?
(c) The greatest average velocity?
(d) Zero average velocity?
(e) Zero acceleration?
(f) Positive acceleration throughout?

(I) x t 5

(II)

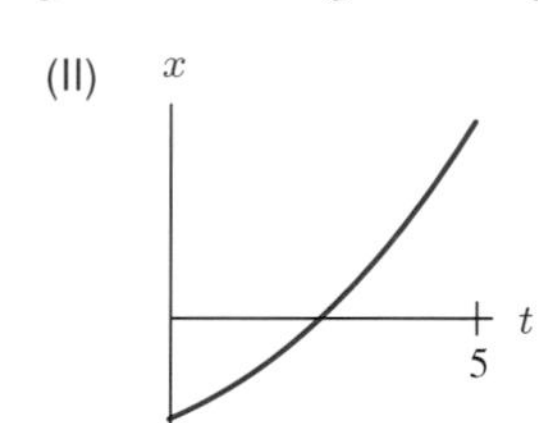

(III)

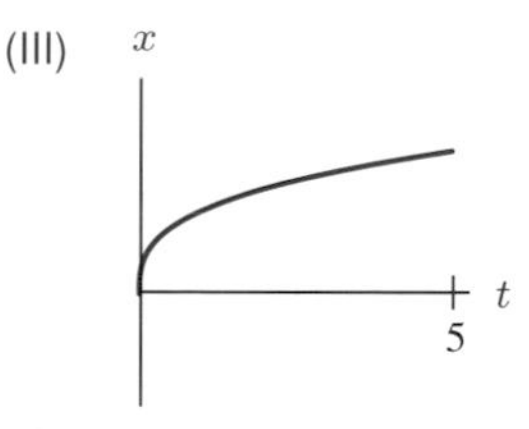

(IV)

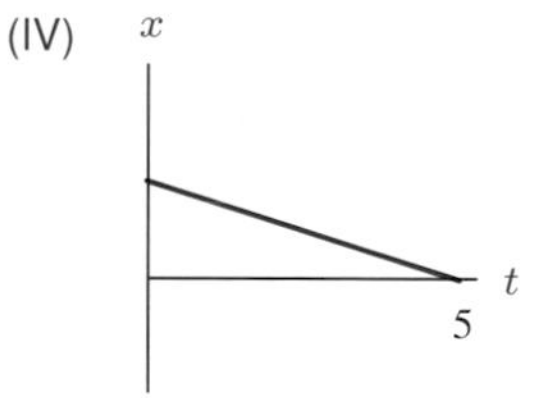

Figure 2.51

18. Let $g(x) = \sqrt{x}$ and $f(x) = kx^2$, where k is a constant.

(a) Find the slope of the tangent line to the graph of g at the point $(4, 2)$.
(b) Find the equation of this tangent line.
(c) If the graph of f contains the point $(4, 2)$, find k.
(d) Where does the graph of f intersect the tangent line?

19. A circle with center at the origin and radius of length $\sqrt{19}$ has equation $x^2 + y^2 = 19$. Graph the circle.

(a) Just from looking at the graph, what can you say about the slope of the line tangent to the circle at the point $(0, \sqrt{19})$? What about the slope of the tangent at $(\sqrt{19}, 0)$?

(b) Estimate the slope of the tangent to the circle at the point $(2, -\sqrt{15})$ by graphing the tangent carefully at that point.

(c) Use the result of part (b) and the symmetry of the circle to find slopes of the tangents drawn to the circle at $(-2, \sqrt{15})$, $(-2, -\sqrt{15})$, and $(2, \sqrt{15})$.

20. A person with a certain liver disease first exhibits larger and larger concentrations of certain enzymes (called SGOT and SGPT) in the blood. As the disease progresses, the concentration of these enzymes drops, first to the predisease level and eventually to zero (when almost all of the liver cells have died). Monitoring the levels of these enzymes allows doctors to track the progress of a patient with this disease. If $C = f(t)$ is the concentration of the enzymes in the blood as a function of time:

(a) Sketch a possible graph of $C = f(t)$.
(b) Mark on the graph the intervals where $f' > 0$ and where $f' < 0$.
(c) What does $f'(t)$ represent, in practical terms?

21. The population of a herd of deer is modeled by

$$P(t) = 4000 + 400 \sin\left(\frac{\pi}{6}t\right) + 180 \sin\left(\frac{\pi}{3}t\right)$$

where t is measured in months from the first of April.

(a) Use a calculator or computer to sketch a graph showing how this population varies with time.

Use the graph to answer the following questions.

(b) When is the herd largest? How many deer are in it at that time?
(c) When is the herd smallest? How many deer are in it then?
(d) When is the herd growing the fastest? When is it shrinking the fastest?
(e) How fast is the herd growing on April 1?

22. A continuous function defined for all x has the following properties:

- f is increasing.
- f is concave down.
- $f(5) = 2$.
- $f'(5) = \frac{1}{2}$.

(a) Sketch a possible graph for f.
(b) How many zeros does f have?
(c) What can you say about the location of the zeros?
(d) What is $\lim_{x \to -\infty} f(x)$?
(e) Is it possible that $f'(1) = 1$?
(f) Is it possible that $f'(1) = \frac{1}{4}$?

23. Roughly sketch the shape of the graph of a quadratic polynomial, f, if it is known that:

- $(1, 3)$ is on the graph of f.
- $f'(0) = 3$, $f'(2) = 1$, $f'(3) = 0$.

24. Roughly sketch the shape of the graph of a cubic polynomial, f, if it is known that:

- $(0, 3)$ is on the graph of f.
- $f'(0) = 4$, $f'(1) = 0$, $f'(2) = -\frac{4}{3}$, $f'(4) = 4$.

CHAPTER THREE

KEY CONCEPT: THE DEFINITE INTEGRAL

We started Chapter 2 by calculating the velocity from the distance traveled. This led us to the notion of the derivative, or rate of change, of a function. Now we will consider the reverse problem: given the velocity, how can we calculate the distance traveled? This will lead us to our second key concept, the *definite integral,* which computes the total change in a function from its rate of change. We will then discover that the definite integral can be used to calculate not only distance but also other quantities, such as the area under a curve and the average value of a function.

We end this chapter with the Fundamental Theorem of Calculus, which tells us that we can use a definite integral to get information about a function from its derivative. Calculating derivatives and calculating definite integrals are, in a sense, reverse processes.

3.1 HOW DO WE MEASURE DISTANCE TRAVELED?

If the velocity is a constant, we can find the distance using the formula

$$\text{Distance} = \text{Velocity} \times \text{Time}.$$

In this section we will see how to estimate the distance when the velocity is not a constant.

A Thought Experiment: How Far Did the Car Go?

One Second Velocity Data

Suppose a car is moving with increasing velocity. First, let's suppose we are given the velocity every second, as in Table 3.1:

TABLE 3.1 *Velocity of car every second*

Time (sec)	0	1	2	3	4	5
Velocity (ft/sec)	20	30	38	44	48	50

How far has the car moved? Since we don't know how fast the car is moving at every moment, we can't figure the distance out exactly, but we can make an estimate. Since the velocity is increasing, the car goes at least 20 feet during the first second. Likewise, it goes at least 30 feet during the next second, at least 38 feet during the third second, at least 44 feet during the fourth second, and at least 48 feet during the last second. During the five-second period it goes at least

$$20 + 30 + 38 + 44 + 48 = 180 \text{ feet}.$$

Thus, 180 feet is an underestimate of the total distance moved during the five seconds.

To get an overestimate, we can reason this way: In the first second the car moved at most 30 feet, in the next second it moved at most 38 feet, in the third second it moved at most 44 feet, and so on. Therefore, altogether it moved at most

$$30 + 38 + 44 + 48 + 50 = 210 \text{ feet}.$$

Thus, the total distance moved is between 180 and 210 feet:

$$180 \text{ feet} \leq \text{Total distance traveled} \leq 210 \text{ feet}.$$

There is a difference of 30 feet between our upper and lower estimates.

Half Second Velocity Data

What if we wanted a more accurate estimate? Then we should ask for more frequent velocity measurements, say every 0.5 seconds. See Table 3.2.

TABLE 3.2 *Velocity of car every half second*

Time (sec)	0	0.5	1.0	1.5	2.0	2.5	3.0	3.5	4.0	4.5	5.0
Velocity (ft/sec)	20	26	30	35	38	42	44	46	48	49	50

As before, we get a lower estimate for each half second by using the velocity at the beginning of that half second. During the first half second the velocity is at least 20 ft/sec, and so the car travels at least $(20)(0.5) = 10$ feet. During the next half second the car moves at least $(26)(0.5) = 13$ feet, and so on. So now we can say

$$\begin{aligned}\text{Lower estimate} &= (20)(0.5) + (26)(0.5) + (30)(0.5) + (35)(0.5) + (38)(0.5)\\ &\quad + (42)(0.5) + (44)(0.5) + (46)(0.5) + (48)(0.5) + (49)(0.5)\\ &= 189 \text{ feet}.\end{aligned}$$

Notice that this is higher than our old lower estimate of 180 feet.

We get a new upper estimate by considering the velocity at the end of each half second. During the first half second the velocity is at most 26 ft/sec, and so the car moves at most $(26)(0.5) = 13$ feet; in the next half second it moves at most $(30)(0.5) = 15$ feet, and so on. Thus,

$$\begin{aligned}\text{Upper estimate} &= (26)(0.5) + (30)(0.5) + (35)(0.5) + (38)(0.5) + (42)(0.5)\\ &\quad + (44)(0.5) + (46)(0.5) + (48)(0.5) + (49)(0.5) + (50)(0.5)\\ &= 204 \text{ feet}.\end{aligned}$$

This is lower than our old upper estimate of 210 feet. Now we know that

$$189 \text{ feet} \leq \text{Total distance traveled} \leq 204 \text{ feet}.$$

Notice that the difference between our new upper and lower estimates is now 15 feet, half of what it was before. By halving our interval of measurement, we have halved the difference between the upper and lower estimates.

Visualizing Distance on the Velocity Graph: One Second Data

We can represent both upper and lower estimates on a graph of the velocity, as well as see how changing the interval of measurement changes the accuracy of our estimates.

The velocity can be graphed by plotting the data in Table 3.1 and drawing a smooth curve through them. (See Figure 3.1.) The area of the first dark rectangle is $(20)(1) = 20$, the lower

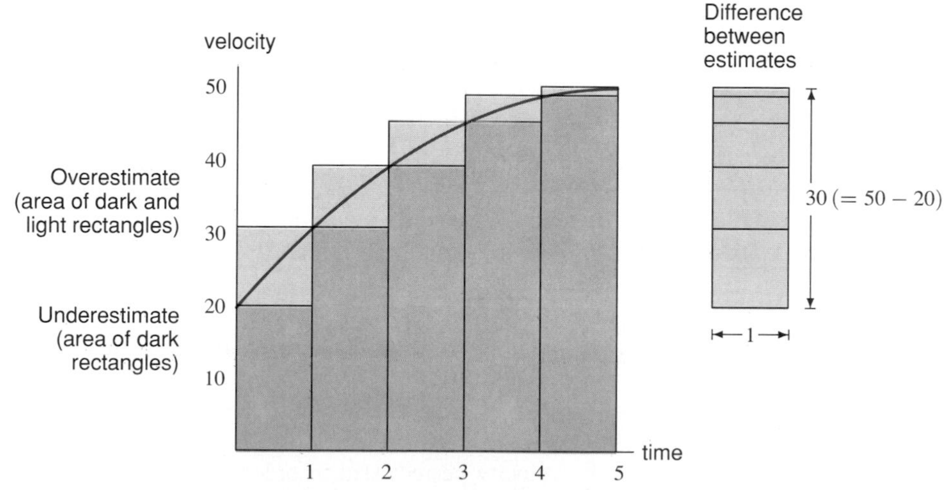

Figure 3.1: Velocity measured each second

estimate of the distance moved in the first second. The area of the second dark rectangle is $(30)(1) = 30$, the lower estimate for the distance moved in the next second. Therefore, the total area of the dark rectangles represents the lower estimate for the total distance moved during the five seconds.

If the dark and light rectangles are considered together, the first area is $(30)(1) = 30$, the upper estimate for the distance moved in the first second. Continuing this calculation shows that the upper estimate for the total distance is represented by the area of the dark and light rectangles together. Therefore, the area of the light rectangles alone represents the difference between the two estimates.

To calculate the difference between the two estimates, look at Figure 3.1 and imagine the light rectangles all pushed to the right and stacked on top of each other; this gives a rectangle of width 1 and height 30. Notice that the height, 30, is just the difference between the initial and final values of the velocity, $30 = 50 - 20$; and the width, 1, is the time interval between velocity measurements.

Visualizing Distance on the Velocity Graph: Half Second Data

The data for the velocities measured each half second is in Figure 3.2. The area of the dark rectangles again represents the lower estimate, and the dark and light rectangles together represent the upper estimate. As before, the difference between the two estimates is represented by the area of the light rectangles. This difference can be calculated by stacking the light rectangles vertically, giving a rectangle of the same height as before but of half the width. Its area is therefore half what it was before. Again, the height of this stack is $50 - 20 = 30$, and its width is the time interval 0.5.

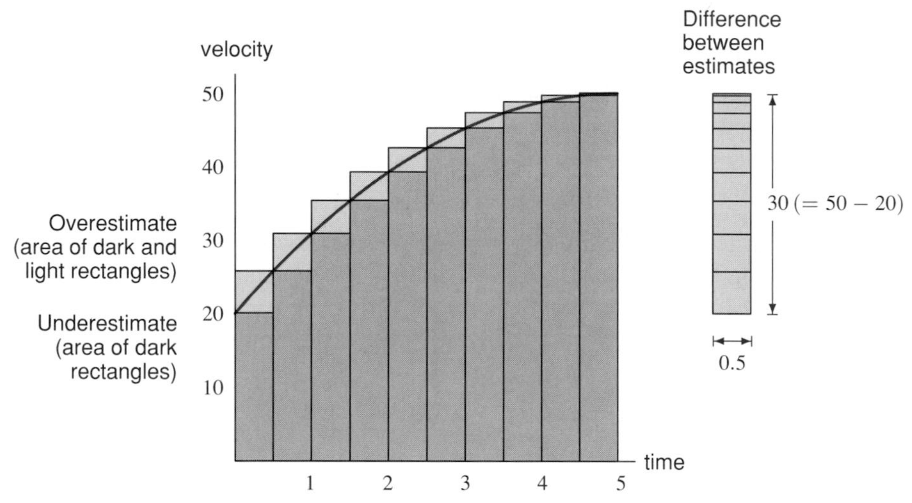

Figure 3.2: Velocity measured each half second

Example 1 What would be the difference between the upper and lower estimates if the velocity were given every tenth of a second? Every thousandth of a second?

Solution Every tenth of a second: difference $= (50 - 20)(1/10) = 3$ feet.
Every thousandth of a second: difference $= (50 - 20)(1/1000) = 0.03$ feet.

Example 2 How frequently must the velocity be recorded in order to estimate the total distance traveled to within 0.1 foot?

Solution The difference between the velocity at the beginning and end of the observation period is $50 - 20 = 30$. If the time between the measurements is h, then the difference between the upper and lower estimates is $(30)h$. We want

$$(30)h < 0.1$$

giving

$$h < \frac{0.1}{30} \approx 0.0033.$$

So if the measurements are made less than 0.0033 seconds apart, the distance estimate will be accurate to within 0.1 feet.

Determining Distance Precisely

In this section we will obtain a precise expression for the distance traveled. We express the exact distance traveled as a limit of estimates in much the same way as we expressed velocity as a limit of average velocities.

Suppose we want to know the distance traveled by a moving object over the time interval $a \leq t \leq b$. Let the velocity at time t be given by the function $v = f(t)$. Suppose we take measurements of $f(t)$ at equally spaced times $t_0, t_1, t_2, \ldots, t_n$ with time $t_0 = a$ and time $t_n = b$. The time interval between any two consecutive measurements is

$$\Delta t = \frac{b-a}{n}$$

where Δt (read "delta t") means the change, or increment, in t.

During the first time interval, the velocity can be approximated by $f(t_0)$, so the distance traveled is approximately

$$f(t_0)\Delta t.$$

During the second time interval, the velocity is about $f(t_1)$, so the distance traveled is about

$$f(t_1)\Delta t.$$

Continuing in this way and adding up all the estimates, we get an estimate for the total distance. In the last interval, the velocity is approximately $f(t_{n-1})$, so the last term is $f(t_{n-1})\Delta t$:

$$\text{Total distance traveled between } a \text{ and } b \approx f(t_0)\Delta t + f(t_1)\Delta t + f(t_2)\Delta t + \cdots + f(t_{n-1})\Delta t.$$

This is called a *left-hand sum* because we used the value of velocity from the left end of each time interval. It is represented by the sum of the areas of the rectangles in Figure 3.3. We can also calculate a *right-hand sum* by using the value of the velocity at the right end of each time interval. In that case the estimate for the first interval is $f(t_1)\Delta t$, for the second interval it is $f(t_2)\Delta t$, and so on. The estimate for the last interval is now $f(t_n)\Delta t$, so

$$\text{Total distance traveled between } a \text{ and } b \approx f(t_1)\Delta t + f(t_2)\Delta t + f(t_3)\Delta t + \cdots + f(t_n)\Delta t.$$

The right-hand sum is represented by the area of the rectangles in Figure 3.4.

If f is an increasing function, the left-hand sum will be an underestimate of the total distance traveled, since for each time interval we use the velocity at the start of that interval to compute the

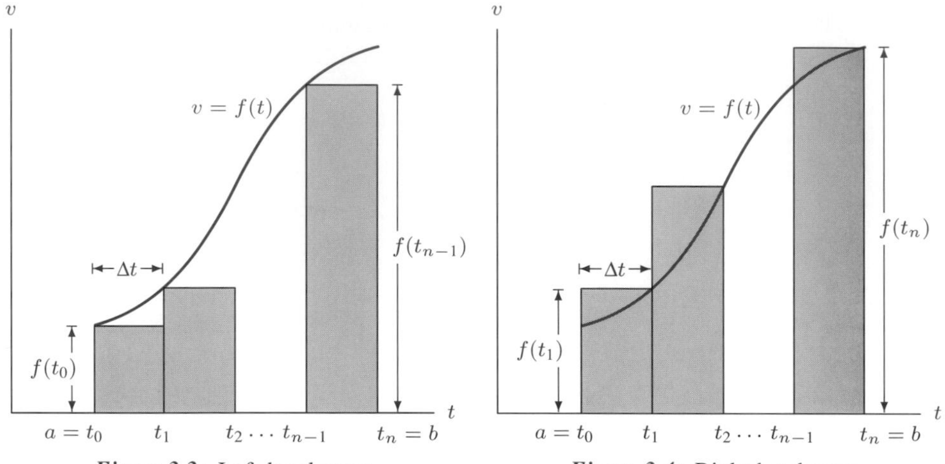

Figure 3.3: Left-hand sums

Figure 3.4: Right-hand sums

distance traveled, whereas in fact the velocity continues to increase after that measurement. We get an overestimate by using the velocity at the right end of each time interval. If f is decreasing, as in Figure 3.5, then the roles of the two sums are reversed: the left-hand sum is an overestimate, and the right-hand sum is an underestimate.

For either increasing or decreasing functions, the exact value of the distance traveled lies somewhere between the two estimates. Thus the accuracy of our estimate depends on how close these two sums are. For a function which is increasing throughout or decreasing throughout the interval $[a, b]$:

$$\left|\begin{array}{c}\text{Difference between}\\ \text{upper and lower estimates}\end{array}\right| = \left|\begin{array}{c}\text{Difference between}\\ f(a) \text{ and } f(b)\end{array}\right| \times \Delta t = |f(b) - f(a)| \cdot \Delta t.$$

(Absolute values are used to make the difference non-negative.) By making the measurements close enough together, we can make Δt as small as we like; therefore, we can make the difference between our lower and upper estimates as small as we like.

To find exactly the total distance traveled between a and b, we take the limit of the sums, as n, the number of subdivisions of the interval $[a, b]$, goes to infinity. The sum of the areas of the rectangles approaches the area under the curve between $t = a$ and $t = b$. So we conclude that

$$\begin{aligned}\begin{array}{l}\text{Total distance traveled}\\ \text{between } t = a \text{ and } t = b\end{array} &= \lim_{n\to\infty} (\text{Left-hand sum})\\ &= \lim_{n\to\infty} \left[f(t_0)\Delta t + f(t_1)\Delta t + \cdots + f(t_{n-1})\Delta t\right]\\ &= \text{Area under curve } f(t) \text{ from } t = a \text{ to } t = b\end{aligned}$$

and

$$\begin{aligned}\begin{array}{l}\text{Total distance traveled}\\ \text{between } t = a \text{ and } t = b\end{array} &= \lim_{n\to\infty} (\text{Right-hand sum})\\ &= \lim_{n\to\infty} \left[f(t_1)\Delta t + f(t_2)\Delta t + \cdots + f(t_n)\Delta t\right]\\ &= \text{Area under curve } f(t) \text{ from } t = a \text{ to } t = b.\end{aligned}$$

Thus, if n is large enough, both the left-hand and the right-hand sums are accurate estimates of the distance traveled. This method of calculating the distance by taking the limit of a sum works even if the velocity is not increasing throughout, or decreasing throughout, the interval.

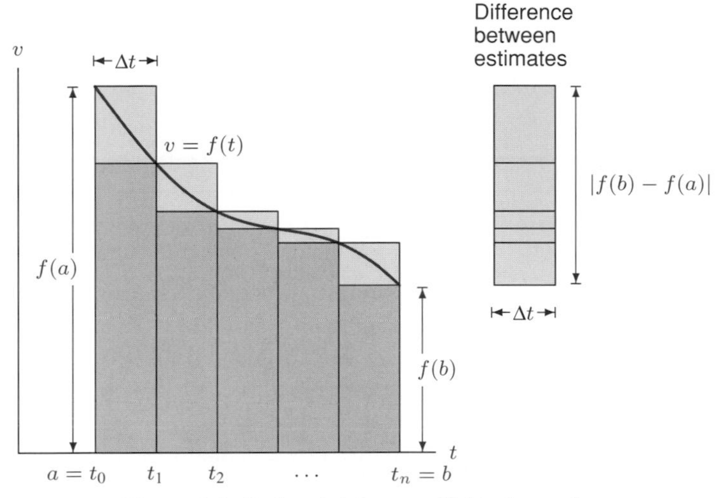

Figure 3.5: Left and right sums if f is decreasing

Problems for Section 3.1

1. A car comes to a stop five seconds after the driver slams on the brakes. While the brakes are on, the following velocities are recorded:

Time since brakes applied (sec)	0	1	2	3	4	5
Velocity (ft/sec)	88	60	40	25	10	0

   (a) Give lower and upper estimates of the distance the car traveled after the brakes were applied.
   (b) On a sketch of velocity against time, show the lower and upper estimates and the difference between them.

2. Roger decides to run a marathon. Roger's friend Jeff rides behind him on a bicycle and clocks his pace every 15 minutes. Roger starts out strong, but after an hour and a half he is so exhausted that he has to stop. The data Jeff collected is summarized below:

Time spent running (min)	0	15	30	45	60	75	90
Speed (mph)	12	11	10	10	8	7	0

   (a) Assuming that Roger's speed is always decreasing, give upper and lower estimates for the distance Roger ran during the first half hour.
   (b) Give upper and lower estimates for the distance Roger ran in total.
   (c) How often would Jeff have needed to measure Roger's pace in order to find lower and upper estimates within 0.1 mile of the actual distance that he ran?

3. Coal gas is produced at a gasworks. Pollutants in the gas are removed by scrubbers, which become less and less efficient as time goes on. Measurements made at the start of each month showing the rate at which pollutants are escaping in the gas are as follows:

Time (months)	0	1	2	3	4	5	6
Rate pollutants are escaping (tons/month)	5	7	8	10	13	16	20

(a) Make an overestimate and an underestimate of the total quantity of pollutants that escaped during the first month.
(b) Make an overestimate and an underestimate of the total quantity of pollutants that escaped during the first six months.
(c) How often would measurements have to be made in order to find overestimates and underestimates which differ by less than 1 ton from the exact quantity of pollutants that escaped during the first six months?

4. In Figure 3.6, use the grid to estimate the area of the region bounded by the curve, the horizontal axis and the lines $x = \pm 3$. Get an upper and a lower estimate that are within 4 square units of one another. Explain what you are doing.

5. (a) In Figure 3.7 estimate the shaded area with an error of at most 0.1.
(b) How can you approximate this shaded area to any desired degree of accuracy?

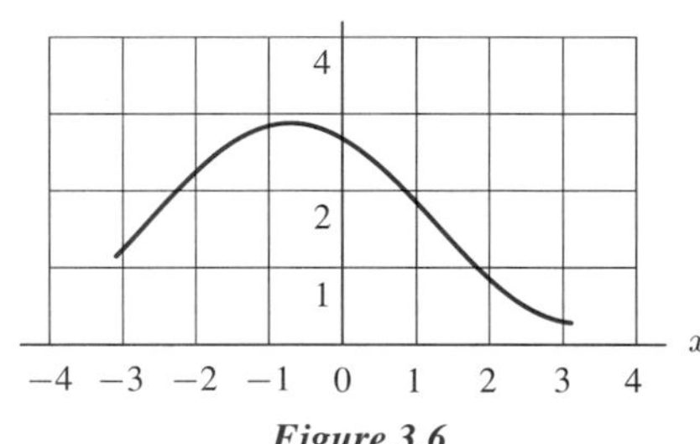

Figure 3.6

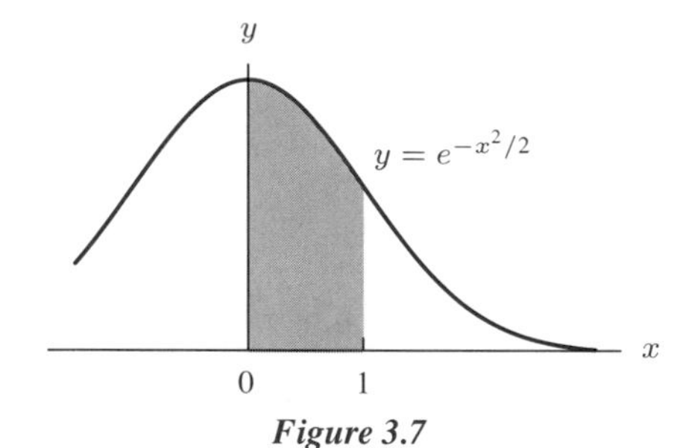

Figure 3.7

6. Figure 3.8 shows the graph of the velocity, v, of an object (in m/sec). Estimate the total distance the object traveled between $t = 0$ and $t = 6$.

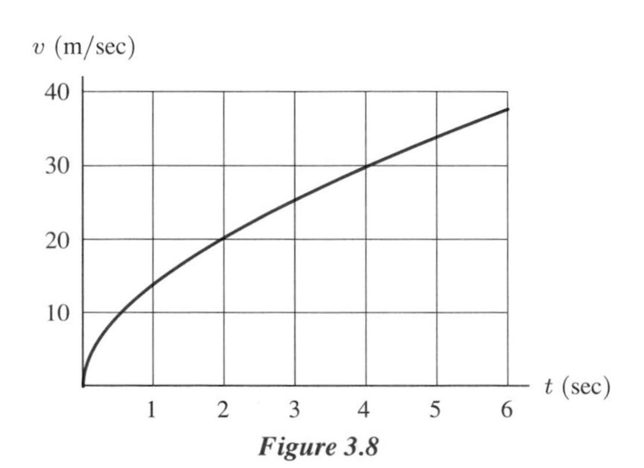

Figure 3.8

7. Your velocity is given by

$$v(t) = \sin(t^2) \quad \text{for } 0 \le t \le 1.1.$$

Estimate the distance traveled during this time (to one decimal place).

8. For $0 \le t \le 1$, a bug is crawling at a velocity, v, determined by the formula

$$v = \frac{1}{1+t},$$

where t is in hours and v is in meters/hour. Estimate the distance that the bug crawls during this hour. Your estimate should be within 0.1 meter of the true answer.

9. You jump out of an airplane. Before your parachute opens you fall faster and faster, but your acceleration decreases as you fall because of air resistance. The table below gives your acceleration, a (in m/sec^2), after t seconds.

t	0	1	2	3	4	5
a	9.81	8.03	6.53	5.38	4.41	3.61

(a) Give upper and lower estimates of your speed at $t = 5$.
(b) Get a new estimate by taking the average of your upper and lower estimates. What does the concavity of the graph of acceleration tell you about your new estimate?

10. An object has zero initial velocity and a constant acceleration of 32 ft/sec^2. Find a formula for its velocity as a function of time. Then use left and right sums to find upper and lower bounds on the distance that the object travels in four seconds. How can the precise distance be found? [Hint: Find the area under a curve.]

3.2 THE DEFINITE INTEGRAL

In Section 3.1 we saw how distance traveled can be approximated by sums and expressed exactly as the limit of a sum. This process of forming sums and taking their limits has many interpretations. This section will be devoted to looking at this process in more detail. In particular, we will show how these sums can be defined for any function f, whether or not it represents a velocity. This is similar to what we did for the derivative: first we introduced the limit of difference quotients as the solution to the velocity problem, and then we studied the limit in its own right.

Suppose we have a function $f(t)$ which is continuous for $a \le t \le b$, except perhaps at a few points, and bounded everywhere. We divide the interval from $a \le t \le b$ into n equal subdivisions, and we call the width of an individual subdivision Δt. Thus,

$$\Delta t = \frac{b-a}{n}.$$

We will let $t_0, t_1, t_2, \ldots, t_n$ be endpoints of the subdivisions, as in Figures 3.9 and 3.10. As before, we construct two sums:

$$\text{Left-hand sum} = f(t_0)\Delta t + f(t_1)\Delta t + \cdots + f(t_{n-1})\Delta t$$

and

$$\text{Right-hand sum} = f(t_1)\Delta t + f(t_2)\Delta t + \cdots + f(t_n)\Delta t.$$

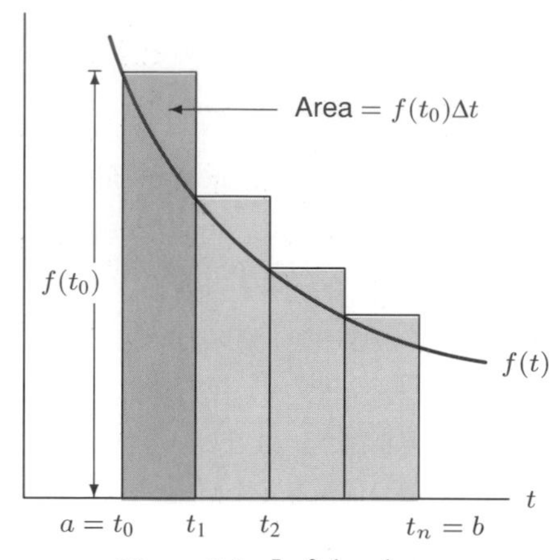

Figure 3.9: Left-hand sum

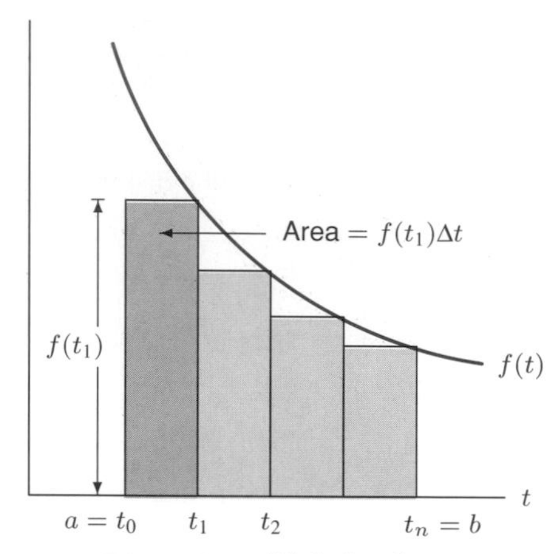

Figure 3.10: Right-hand sum

These sums may be represented graphically as the areas in Figures 3.9 and 3.10, provided $f(t) \geq 0$ and $a < b$. In Figure 3.9, the first rectangle has width Δt and height $f(t_0)$, since the top of its left edge just touches the curve, and hence it has area $f(t_0)\Delta t$; the second rectangle has width Δt and height $f(t_1)$ and hence has area $f(t_1)\Delta t$; and so on. The sum of all these areas is the left-hand sum. The right-hand sum, shown in Figure 3.10, is constructed in the same way, except that each rectangle touches the curve on its right edge instead of its left.

Writing Left and Right Sums Using Sigma Notation

Both the left-hand and right-hand sums can be written more compactly using *sigma*, or summation, notation. The symbol $\sum$ is a capital sigma, or Greek letter "S." We write

$$\text{Right-hand sum} = \sum_{i=1}^{n} f(t_i)\Delta t = f(t_1)\Delta t + f(t_2)\Delta t + \cdots + f(t_n)\Delta t$$

where the $\sum$ tells us to add terms of the form $f(t_i)\Delta t$. The "$i = 1$" at the base of the sigma sign tells us to start at $i = 1$, and the "n" at the top tells us to stop at $i = n$.

In the left-hand sum we start at $i = 0$ and stop at $i = n - 1$, so we write

$$\text{Left-hand sum} = \sum_{i=0}^{n-1} f(t_i)\Delta t = f(t_0)\Delta t + f(t_1)\Delta t + \cdots + f(t_{n-1})\Delta t.$$

Taking the Limit to Define the Definite Integral

In the previous section we took the limit of these sums as n went to infinity, and we do the same here. Whenever the integrand is continuous for $a \leq x \leq b$, and in fact for any function you are ever likely to meet, the limits of the left- and right-hand sums will exist and be equal. In such cases, we define the *definite integral* to be the limit of these sums. When $f(t) \geq 0$ and $a < b$, the definite integral represents the area between the graph of $f(t)$ and the t-axis, from $t = a$ to $t = b$. The definite integral is well approximated by a left- or right-hand sum if n is large enough.

The **definite integral** of f from a to b, written

$$\int_a^b f(t)\,dt,$$

is the limit of the left-hand or right-hand sums with n subdivisions as n gets arbitrarily large. In other words,

$$\int_a^b f(t)\,dt = \lim_{n\to\infty} (\text{Left-hand sum}) = \lim_{n\to\infty} \left(\sum_{i=0}^{n-1} f(t_i)\Delta t\right)$$

and

$$\int_a^b f(t)\,dt = \lim_{n\to\infty} (\text{Right-hand sum}) = \lim_{n\to\infty} \left(\sum_{i=1}^{n} f(t_i)\Delta t\right).$$

Each of these sums is called a *Riemann sum*, f is called the *integrand*, and a and b are called the *limits of integration*.

The "$\int$" notation comes from an old-fashioned "S," which stands for "sum" in the same way that $\sum$ does. The "dt" in the integral comes from the factor Δt. Notice that the limits on the $\sum$ sign are 0 and $n-1$ for the left-hand sum, or 1 and n for the right-hand sum, whereas the limits on the $\int$ sign are a and b.

Example 1 Find the left-hand and right-hand sums with $n = 2$ and $n = 10$ for $\int_1^2 \frac{1}{t}\,dt$. How do the values of these sums compare with the true value of the integral?

Solution Here $a = 1$ and $b = 2$, so for $n = 2$, $\Delta t = (2-1)/2 = 0.5$. Therefore, $t_0 = 1$, $t_1 = 1.5$ and $t_2 = 2$. (See Figure 3.11.) Thus,

$$\begin{aligned}\text{Left-hand sum} &= f(1)\Delta t + f(1.5)\Delta t\\ &= 1(0.5) + \frac{1}{1.5}(0.5)\\ &\approx 0.8333,\end{aligned}$$

and

$$\begin{aligned}\text{Right-hand sum} &= f(1.5)\Delta t + f(2)\Delta t\\ &= \frac{1}{1.5}(0.5) + \frac{1}{2}(0.5)\\ &\approx 0.5833.\end{aligned}$$

From Figure 3.11 you can see that the left-hand sum is bigger than the area under the curve and the right-hand sum is smaller, so the area under the curve $f(t) = 1/t$ from $t = 1$ to $t = 2$ is between 0.5833 and 0.8333. Thus,

$$0.5833 < \int_1^2 \frac{1}{t}\,dt < 0.8333.$$

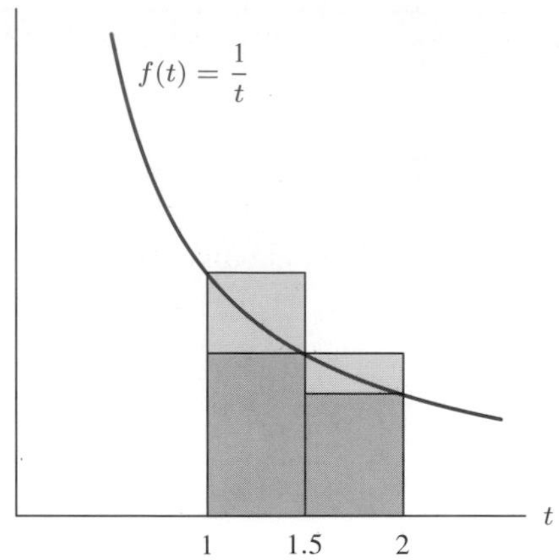

Figure 3.11: Approximating $\int_1^2 \frac{1}{t}\,dt$ with $n = 2$

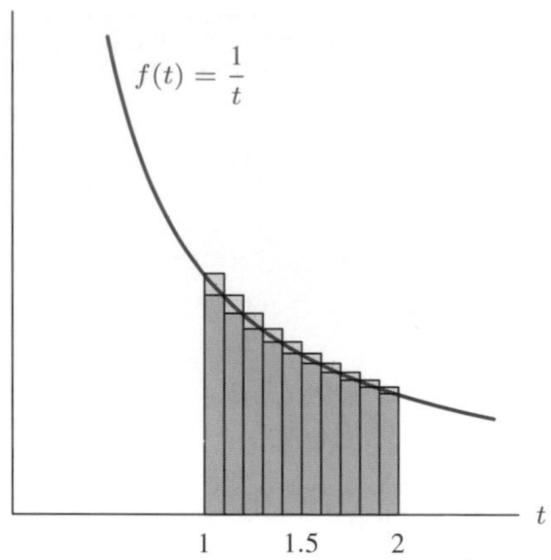

Figure 3.12: Approximation of $\int_1^2 \frac{1}{t}\,dt$ with $n = 10$

When $n = 10$, $\Delta t = 0.1$ (see Figure 3.12), so

$$\begin{aligned}
\text{Left-hand sum} &= f(1)\Delta t + f(1.1)\Delta t + \cdots + f(1.9)\Delta t \\
&= \left(1 + \frac{1}{1.1} + \cdots + \frac{1}{1.9}\right) 0.1 \\
&\approx 0.7188, \\
\text{Right-hand sum} &= f(1.1)\Delta t + f(1.2)\Delta t + \cdots + f(2)\Delta t \\
&= \left(\frac{1}{1.1} + \frac{1}{1.2} + \cdots + \frac{1}{2}\right) 0.1 \\
&\approx 0.6688.
\end{aligned}$$

From Figure 3.12 you can see that the left-hand sum is larger than the area under the curve, and the right-hand sum smaller, so

$$0.6688 < \int_1^2 \frac{1}{t}\,dt < 0.7188.$$

Notice that the left- and right-hand sums trap the true value of the integral between them. As the subdivisions become finer, the left- and right-hand sums get closer together.

Convergence of Left and Right Sums

A function which is either increasing throughout an interval or decreasing throughout that interval is said to be *monotonic* there. It is usually easy to tell whether or not a function is monotonic just by looking at its graph.

When f Is Monotonic

If f is monotonic, the left- and right-hand sums trap the exact value of the integral between them. Let us continue the example

$$\int_1^2 \frac{1}{t}\,dt.$$

The left- and right-hand sums for $n = 2$, 10, 50, and 250 are listed in Table 3.3.

TABLE 3.3 *Left- and right-hand sums for* $\int_1^2 \frac{1}{t}\,dt$

n	Left-hand sum	Right-hand sum
2	0.8333	0.5833
10	0.7188	0.6688
50	0.6982	0.6882
250	0.6941	0.6921

Because the function $f(t) = 1/t$ is decreasing, the left-hand sums converge down on the integral from above, and the right-hand sums converge up from below. From the last row of the table we can deduce that

$$0.6921 < \int_1^2 \frac{1}{t}\,dt < 0.6941$$

so $\int_1^2 \frac{1}{t}\,dt \approx 0.69$ to two decimal places. To approximate definite integrals accurately, we evaluate left- and right-hand sums with a large number of subdivisions, using a computer or calculator.

When f Is Not Monotonic

If f is not monotonic, the definite integral is not always bracketed between the left- and right-hand sums, but in practice you can still see the convergence quite easily. For example, Table 3.4 gives sums for the integral $\int_0^{2.5} \sin(t^2)\,dt$.

TABLE 3.4 *Left- and right-hand sums for* $\int_0^{2.5} \sin(t^2)\,dt$

n	Left-hand sum	Right-hand sum
2	1.2500	1.2085
10	0.4614	0.4531
50	0.4324	0.4307
250	0.4307	0.4304
1000	0.4306	0.4305

Although $\sin(t^2)$ is certainly not monotonic, by the time we get to $n = 250$, it is pretty clear that $\int_0^{2.5} \sin(t^2)\,dt \approx 0.43$ to two decimal places. As you can see, the sums wander around a bit at the beginning, and it's not certain that they won't wander later on. Nonetheless, it is a good rule of thumb that when the digits begin to stabilize, or remain the same as n increases, you are getting close to the true value of the integral. This rule of thumb is not foolproof, but it works well in practice. For $\int_0^{2.5} \sin(t^2)\,dt$, we do not have a guarantee of the accuracy, as we do for monotonic functions. Notice that 0.43 does not lie between 1.2500 and 1.2085, the left- and right-hand sums for $n = 2$, or even between 0.4614 and 0.4531, the two sums for $n = 10$. If the integrand is not monotonic, the left- and right-hand sums may both be larger (or smaller) than the integral. (See Problems 19 and 20, page 163, to see how this can happen.)

If f is not monotonic, it is often possible to construct upper and lower estimates for an integral by first splitting the interval $[a, b]$ into subintervals on which f is monotonic.

Error in Approximating a Definite Integral by Riemann Sums

For any approximation, it is useful to have an idea of how good it is. We define the *error* in an approximation to be the magnitude of the difference between the approximate and the true values. (Notice that error doesn't mean mistake here.) Since the true value of the integral is between the upper estimate and the lower estimate, the error must be less than the difference between these estimates. So, for a monotonic function f on the interval $[a, b]$:

$$\text{Error} \leq \left| \begin{array}{c} \text{Difference between} \\ \text{upper and lower estimates} \end{array} \right| = |f(b) - f(a)| \cdot \Delta t,$$

where $\Delta t = (b - a)/n$. Thus, we can estimate the error if the function is monotonic.

If our approximation is taken to be the average of the lower and upper estimates (as in Example 2 below) then our bound for the error is halved.

Example 2 Compute $\int_1^3 t^2\, dt$ to one decimal place.

Solution Since $f(t) = t^2$ is an increasing function on the interval from 1 to 3, the left-hand sum will underestimate the integral and the right-hand sum will overestimate it, as shown in Figure 3.13.

How large should n be? The problem asks us to find a value for the integral to one decimal place, that is, a value which we are certain is within 0.05 of the true value. [1] (Note: This is *not* telling us that Δt is 0.05.) One way is to use a computer or calculator to calculate left- and right-hand sums for various values of n.

For $n = 100$, the left-hand sum is 8.5868 and the right-hand sum is 8.7468, so

$$8.5868 < \int_1^3 t^2\, dt < 8.7468.$$

The left- and right-hand sums differ by 0.16, and averaging them gives

$$\int_1^3 t^2\, dt \approx 8.6668,$$

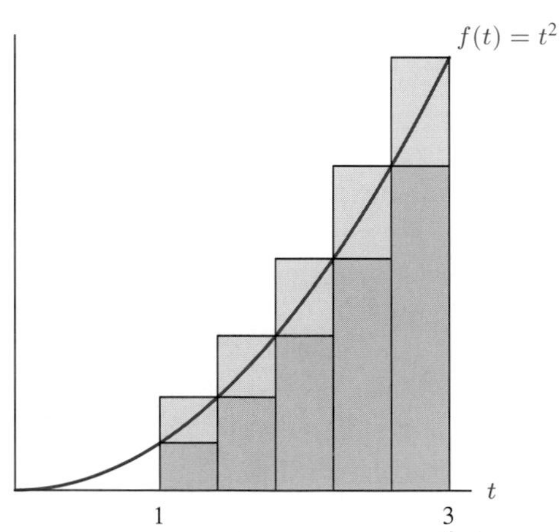

Figure 3.13: Approximation of $\int_1^3 t^2\, dt$

---

[1] See Appendix A for a discussion of decimal place accuracy.

which differs from the true value by at most $0.16/2 = 0.08$. However, we want more accuracy, so we need a larger value of n. Taking $n = 5000$ makes the left-hand sum 8.6650 (rounding down) and the right-hand sum 8.6683 (rounding up), so

$$8.6650 < \int_1^3 t^2\,dt < 8.6683.$$

Averaging the left and right sums gives

$$\int_1^3 t^2\,dt \approx 8.6667.$$

Since the left and right sums differ by 0.0033, the average must be within 0.00165 of the true value. (In fact, much smaller values of n will do. See Problem 18 below.)

Problems for Section 3.2

For Problems 1–6 draw up a table of left- and right-hand sums with 2, 10, 50, and 250 subdivisions. Observe the limit to which your sums are tending as the number of subdivisions gets larger, and estimate the value of the definite integral.

1. $\int_0^1 x^3\,dx$
2. $\int_0^{\pi/2} \cos x\,dx$
3. $\int_0^1 e^{t^2}\,dt$
4. $\int_1^2 x^x\,dx$
5. $\int_2^3 \sin(t^2)\,dt$
6. $\int_{0.2}^3 \sin\left(1/x\right)\,dx$

In Problems 7–17, compute the definite integrals to one decimal place. In each case, say how many subdivisions you are using. [Note that, except for Problem 17, each function is monotonic over the given interval.]

7. $\int_0^5 x^2\,dx$
8. $\int_1^4 \frac{1}{\sqrt{1+x^2}}\,dx$
9. $\int_1^{1.5} \sin x\,dx$
10. $\int_0^{\pi/4} \frac{d\theta}{\cos\theta}$
11. $\int_1^5 (\ln x)^2\,dx$
12. $\int_{1.1}^{1.7} e^t \ln t\,dt$
13. $\int_1^2 2^x\,dx$
14. $\int_1^2 (1.03)^t\,dt$
15. $\int_{-2}^{-1} \cos^3 y\,dy$
16. $\int_0^1 \tan(z^2)\,dz$
17. $\int_{-3}^3 e^{-t^2}\,dt$

18. Consider the integral $\int_1^3 t^2\,dt$. Use the error estimate at the end of this section to calculate the smallest number of subdivisions that will ensure that the average of the upper and lower estimates for this integral is accurate to one decimal place (in other words, the lower and upper estimates should differ by less than 0.1).
19. Using the graph of $2 + \cos x$, for $0 \le x \le 4\pi$, list the following quantities in increasing order: the value of the integral $\int_0^{4\pi} (2 + \cos x)\,dx$, the left-hand and right-hand sum with $n = 2$ subdivisions.

20. Sketch the graph of a function f (you do not need to give a formula for f) on an interval $[a, b]$ with the property that with $n = 2$ subdivisions,

$$\int_a^b f(x)\,dx < \text{Left-hand sum } < \text{Right-hand sum}.$$

3.3 THE DEFINITE INTEGRAL AS AREA AND AVERAGE

The Definite Integral as an Area

If $f(x)$ is positive we can interpret each term $f(x_0)\Delta x, f(x_1)\Delta x, \ldots$ in a left- or right-hand Riemann sum as the area of a rectangle. As the width Δx of the rectangles approaches zero, the rectangles fit the curve of the graph more exactly, and the sum of their areas gets closer and closer to the area under the curve, shaded in Figure 3.14. Thus, we know that:

> **When $f(x)$ is positive** and $a < b$:
>
> $$\begin{matrix}\text{Area under graph of } f \\ \text{between } a \text{ and } b\end{matrix} = \int_a^b f(x)\,dx.$$

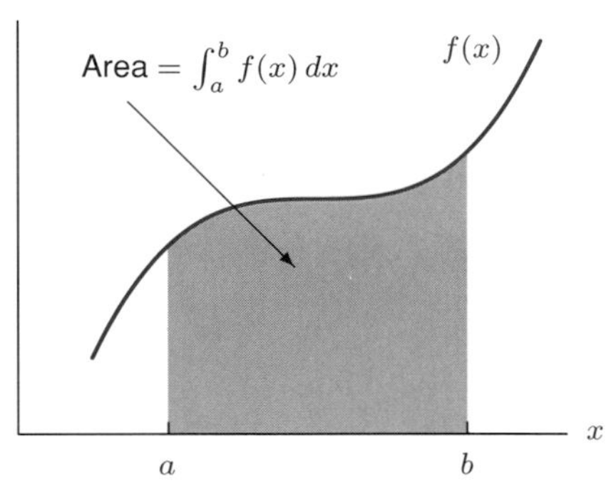

Figure 3.14: The definite integral $\int_a^b f(x)\,dx$

Example 1 Consider the integral $\displaystyle\int_{-1}^{1} \sqrt{1 - x^2}\,dx$.

(a) Interpret the integral as an area, and find its exact value.

(b) Estimate the integral using a left-hand or right-hand sum with 100 subdivisions. Compare your answer to the exact value.

Solution (a) The integral is the area under the graph of $y = \sqrt{1 - x^2}$ between -1 and 1, which is a semicircle of radius 1 and area $\pi/2$ (see Figure 3.15).

(b) With $n = 100$, the left-hand and right-hand sums are 1.569. For comparison, $\pi/2 = 1.5707\ldots$

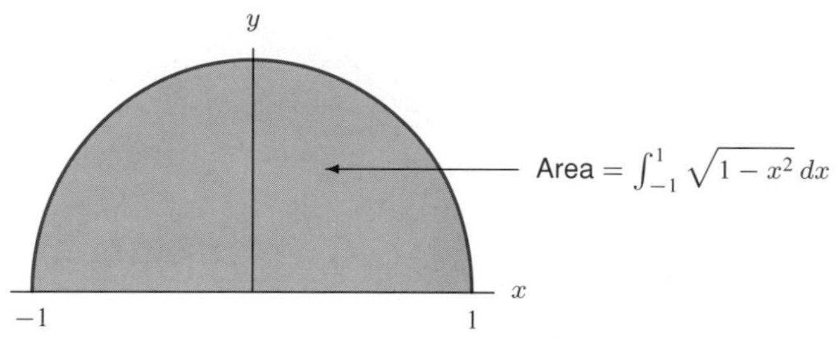

Figure 3.15: Area interpretation of $\int_{-1}^{1} \sqrt{1-x^2}\,dx$

When $f(x)$ is not positive

We have assumed in drawing Figure 3.14 that the graph of $f(x)$ lies above the x-axis. If the graph lies below the x-axis, then each value of $f(x)$ is negative, so each $f(x)\Delta x$ is negative, and the area gets counted negatively. In that case, the definite integral is the negative of the area.

Example 2 What is the relation between the definite integral $\displaystyle\int_{-1}^{1} (x^2 - 1)\,dx$, and the area between the parabola $y = x^2 - 1$ and the x-axis?

Solution The parabola lies below the axis between $x = -1$ and $x = 1$. (See Figure 3.16.) So,

$$\int_{-1}^{1} (x^2 - 1)\,dx = -\text{Area} \approx -1.33.$$

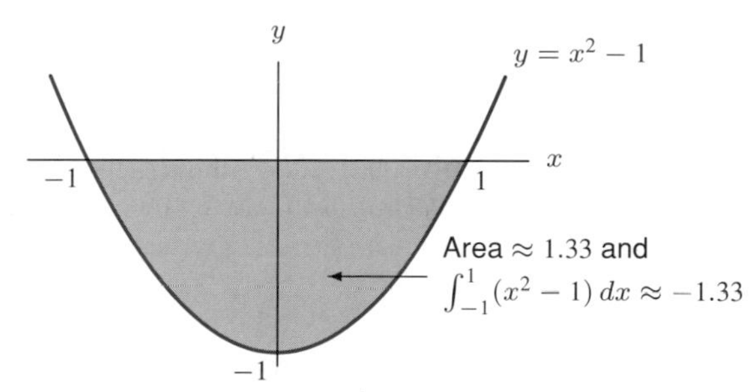

Figure 3.16: Integral $\int_{-1}^{1} (x^2 - 1)\,dx$ is negative of shaded area

> **When $f(x)$ is positive for some x values and negative for others**, and $a < b$:
> $\displaystyle\int_a^b f(x)\,dx$ is the sum of the areas above the x-axis, counted positively, and areas below the x-axis, counted negatively.

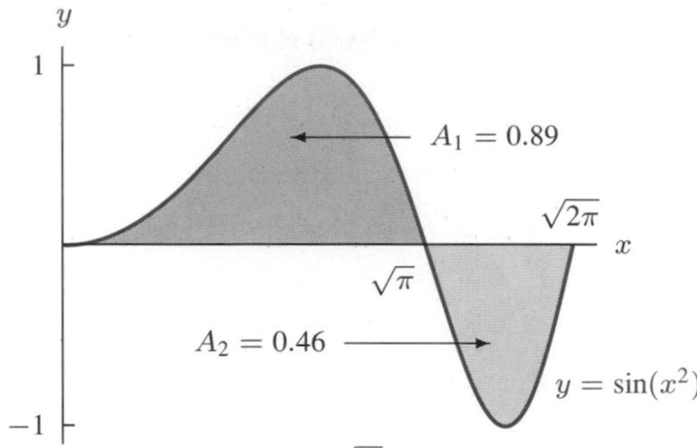

Figure 3.17: Integral $\int_0^{\sqrt{2\pi}} \sin(x^2)\,dx = A_1 - A_2$

Example 3 Interpret the definite integral $\displaystyle\int_0^{\sqrt{2\pi}} \sin(x^2)\,dx$ in terms of areas.

Solution The integral is the area above the x-axis, A_1, minus the area below the x-axis, A_2, in Figure 3.17. Approximating the integral with $n = 1000$ (see Table 3.4, page 161) shows

$$\int_0^{\sqrt{2\pi}} \sin(x^2)\,dx \approx 0.43.$$

The graph of $y = \sin(x^2)$ crosses the x-axis where $x^2 = \pi$, that is, at $x = \sqrt{\pi}$. The next crossing is at $x = \sqrt{2\pi}$. Breaking the integral into two parts and calculating each one separately gives, after rounding,

$$\int_0^{\sqrt{\pi}} \sin(x^2)\,dx = 0.89 \quad \text{and} \quad \int_{\sqrt{\pi}}^{\sqrt{2\pi}} \sin(x^2)\,dx = -0.46$$

so $A_1 = 0.89$ and $A_2 = 0.46$. Then, as we would expect,

$$\int_0^{\sqrt{2\pi}} \sin(x^2)\,dx = A_1 - A_2 = 0.89 - 0.46 = 0.43.$$

The Definite Integral as an Average

We know how to find the average of n numbers: add them and divide by n. But how do we find the average value of a continuously varying function? Let us consider an example. Suppose $C = f(t)$ is the temperature at time t, measured in hours since midnight, and that we want to calculate the average temperature over a 24-hour period. One way to start would be to average the temperatures at n times, $t_1, t_2, \ldots, t_n$, during the day.

$$\text{Average temperature} \approx \frac{f(t_1) + f(t_2) + \cdots + f(t_n)}{n}.$$

The larger we make n, the better the approximation. We can rewrite this expression as a Riemann sum over the interval $0 \le t \le 24$ if we use the fact that $\Delta t = 24/n$, so $n = 24/\Delta t$:

$$\begin{aligned}\text{Average temperature} &\approx \frac{f(t_1) + f(t_2) + \cdots + f(t_n)}{24/\Delta t} \\ &= \frac{f(t_1)\Delta t + f(t_2)\Delta t + \cdots + f(t_n)\Delta t}{24} \\ &= \frac{1}{24}\sum_{i=1}^{n} f(t_i)\Delta t.\end{aligned}$$

As $n \to \infty$, the Riemann sum tends towards an integral and also approximates the average temperature better. Thus, in the limit

$$\begin{aligned}\text{Average temperature} &= \lim_{n\to\infty} \frac{1}{24}\sum_{i=1}^{n} f(t_i)\Delta t \\ &= \frac{1}{24}\int_0^{24} f(t)\,dt.\end{aligned}$$

Thus we have found a way of expressing the average temperature in terms of an integral. Generalizing for any function f, we define

$$\text{Average value of } f \text{ from } a \text{ to } b = \frac{1}{b-a}\int_a^b f(x)\,dx$$

How to Visualize the Average on a Graph

The definition of average value tells us that

$$(\text{Average value of } f)\cdot(b-a) = \int_a^b f(x)\,dx.$$

Thus, if we interpret the integral as the area under the graph of f, then we can think of the average value of f as the height of the rectangle with the same area that is on the same base, $(b-a)$. (See Figure 3.18.)

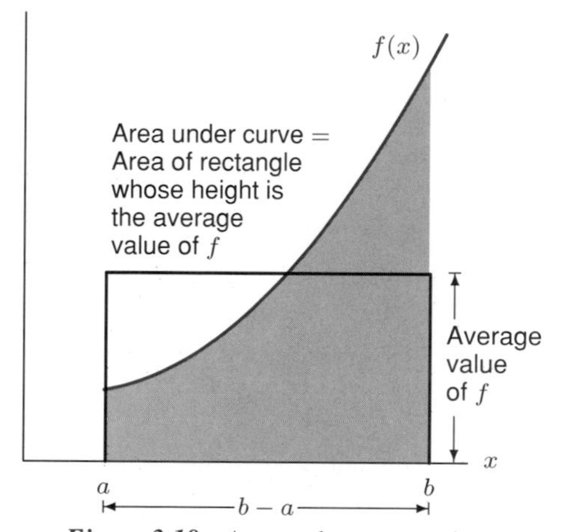

Figure 3.18: Area and average value

The Notation and Units for the Definite Integral

Just as the Leibniz notation dy/dx for the derivative reminds us that the derivative is the limit of a ratio of differences, the notation for the definite integral helps us recall the meaning of the integral. The symbol

$$\int_a^b f(x)\,dx$$

reminds us that an integral is a limit of sums (the integral sign is a misshapen S) of terms of the form "$f(x)$ times a small difference of x." Officially, dx is not a separate entity, but a part of the whole integral symbol. Thus, just as one thinks of d/dx as a single symbol meaning "the derivative with respect to x of...," one can think of $\int_a^b \ldots dx$ as a single symbol meaning "the integral of ... with respect to x."

However, most scientists and mathematicians informally think of dx as an "infinitesimally" small bit of x which in this context is multiplied by a function value $f(x)$. This viewpoint is often the key to interpreting the meaning of a definite integral. For example, if $f(t)$ is the velocity of a moving particle at time t, then $f(t)dt$ may by thought of informally as velocity $\times$ time, giving the distance traveled by the particle during a small bit of time dt. The integral $\int_a^b f(t)\,dt$ may then be thought of as the sum of all these small distances, giving us the net change in position of the particle between $t = a$ and $t = b$.

The notation for the integral also helps us determine what units should be used for the numerical value of the integral. Since the terms being added up are products of the form "$f(x)$ times a difference in x," the unit of measurement for $\int_a^b f(x)\,dx$ is the product of the units for x and the units for $f(x)$. Thus if $f(t)$ is velocity measured in meters/second and t is time measured in seconds, then

$$\int_a^b f(t)\,dt$$

has for units (meters/sec)$\times$(sec)=meters, which is what we expect since the value of the integral represents change in position. Similarly, if we graph $y = f(x)$ with the same units of measurement of length along the x and y axes, then $f(x)$ and x are measured in the same units so

$$\int_a^b f(x)\,dx$$

is measured in square units, say cm$\times$cm=cm^2. Again this is what we would expect since in this context the integral represents an area. Finally for the average value,

$$\frac{1}{b-a}\int_a^b f(x)\,dx,$$

the units for dx inside the integral are canceled by the units for $1/(b-a)$ outside the integral, leaving only the units for $f(x)$. This is as it should be since the average value of f should be measured in the same units as $f(x)$.

Example 4 Suppose that $C(t)$ represents the cost per day to heat your house measured in dollars per day, where t is time measured in days and $t = 0$ corresponds to January 1, 1993. Interpret $\displaystyle\int_0^{90} C(t)\,dt$ and $\displaystyle\frac{1}{90-0}\int_0^{90} C(t)\,dt.$

Solution The units for the integral $\displaystyle\int_0^{90} C(t)\,dt$ are (dollars/day)$\times$(days)=dollars. The integral represents the total cost in dollars to heat your house for the first 90 days of 1993, namely the months of January, February, and March. The second expression is measured in (1/days)(dollars) or dollars per day, the same units as $C(t)$. It represents the average cost per day to heat your house during the first 90 days of 1993.

Problems for Section 3.3

1. If $f(x)$ is measured in pounds and x is measured in feet, what are the units of measurement for $\int_a^b f(x)\,dx$?
2. If $f(t)$ is measured in meters/second^2 and t is measured in seconds, what are the units of measurement for $\int_a^b f(t)\,dt$?
3. If $f(t)$ is measured in dollars per year and t is measured in years, what are the units of measurement for $\int_a^b f(t)\,dt$?
4. Oil is leaking out of a ruptured tanker at a rate of $r = f(t)$ gallons per minute. Write a definite integral expressing the total quantity of oil which leaks out of the tanker in the first hour.
5. Find the area under one arch of the curve $y = \sin x$.
6. Find the area between the line $y = 1$ and one arch of the curve $y = \sin\theta$.
7. Find the area between the parabola $y = 4 - x^2$ and the x-axis.
8. Find the area between $y = x^2 - 9$ and the x-axis.
9. (a) Sketch a graph of $f(x) = x(x+2)(x-1)$.
   (b) Find the total area between the graph and the x-axis.
   (c) Find $\int_{-2}^{1} f(x)\,dx$ and interpret it in terms of areas.
10. Find the area under the curve $y = \cos\sqrt{x}$ for $0 \le x \le 2$.
11. Compute the definite integral $\int_0^4 \cos\sqrt{x}\,dx$ and interpret the result in terms of areas.
12. (a) Sketch a graph of $f(x) = \sin(x^2)$ and mark on it the points where $x = \sqrt{\pi}, \sqrt{2\pi}, \sqrt{3\pi}, \sqrt{4\pi}$.
    (b) Use your graph and the area interpretation of the definite integral to decide which of the four numbers

$$\int_0^{\sqrt{n\pi}} \sin(x^2)\,dx \qquad n = 1, 2, 3, 4$$

    is largest. Which is smallest? How many of the numbers are positive?
13. Find the average value of $g(t) = 1 + t$ over the interval $[0, 2]$.
14. Find the average value of $g(t) = e^t$ over the interval $[0, 10]$.
15. If $f(x) = 2$, show that the average value of $f(x)$ over the interval $[a, b]$ is 2.
16. (a) What is the average value of $f(x) = \sqrt{1 - x^2}$ over the interval $0 \le x \le 1$?
    (b) How can you tell whether this average value is more or less than 0.5 without doing any calculations?
17. (a) Without computing any integrals, explain why the average value of $f(x) = \sin x$ on $[0, \pi]$ must be between 0.5 and 1.
    (b) Compute this average to 2 decimal places of accuracy.

18. The value, V, of a Tiffany lamp, worth \$225 in 1965, increases at 15% per year. Its value t years after 1965 is given by

$$V = 225(1.15)^t.$$

Find the average value of the lamp over the period 1965–2000.

19. The number of hours, H, of daylight in Madrid as a function of date is approximated by the formula

$$H = 12 + 2.4\sin[0.0172(t - 80)],$$

where t is the number of days since the start of the year. Find the average number of hours of daylight in Madrid:
(a) in January (b) in June (c) over a whole year
(d) Comment on the relative magnitudes of your answers to parts (a), (b), and (c). Why are they reasonable?

20. A bar of metal is cooling from 1000°C to room temperature, 20°C. The temperature, H, of the bar t minutes after it starts cooling is given, in °C, by

$$H = 20 + 980e^{-0.1t}.$$

(a) Find the temperature of the bar at the end of one hour.
(b) Find the average value of the temperature over the first hour.
(c) Is your answer to part (b) greater or smaller than the average of the temperatures at the beginning and the end of the hour? Explain this in terms of the concavity of the graph of H.

21. For the function f graphed in Figure 3.19:
(a) Suppose you know $\int_0^2 f(x)\,dx$. What is $\int_{-2}^2 f(x)\,dx$?
(b) Suppose you know $\int_0^5 f(x)\,dx$ and $\int_2^5 f(x)\,dx$. What is $\int_0^2 f(x)\,dx$?
(c) Suppose you know $\int_{-2}^5 f(x)\,dx$ and $\int_{-2}^2 f(x)\,dx$. What is $\int_0^5 f(x)\,dx$?

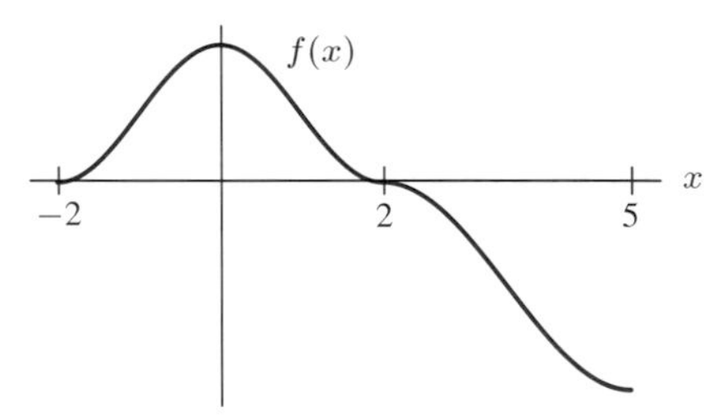

Figure 3.19

22. For the function f graphed in Figure 3.19:
(a) Suppose you know $\int_{-2}^2 f(x)\,dx$ and $\int_0^5 f(x)\,dx$. What is $\int_2^5 f(x)\,dx$?
(b) Suppose you know $\int_{-2}^5 f(x)\,dx$ and $\int_{-2}^0 f(x)\,dx$. What is $\int_2^5 f(x)\,dx$?
(c) Suppose you know $\int_2^5 f(x)\,dx$ and $\int_{-2}^5 f(x)\,dx$. What is $\int_0^2 f(x)\,dx$.

23. For the function f in Figure 3.19, write an expression involving one or more definite integrals that denote(s):
(a) The average value of f for $0 \le x \le 5$.
(b) The average value of $|f|$ for $0 \le x \le 5$.

24. For the function f in Figure 3.19, consider the average value of f over the following intervals:
(I) $0 \le x \le 1$ (II) $0 \le x \le 2$ (III) $0 \le x \le 5$ (IV) $-2 \le x \le 2$
(a) For which interval is the average value of f least?
(b) For which interval is the average value of f greatest?
(c) For which pair of intervals are the average values equal?

3.4 THE FUNDAMENTAL THEOREM OF CALCULUS

The Definite Integral of a Rate Gives Total Change

We have seen how the definite integral of a velocity function can be interpreted as total distance traveled. If $v(t)$ is velocity and $s(t)$ is position, then $v(t) = s'(t)$ and we know that

$$\text{Change in position} = s(b) - s(a) = \int_a^b s'(t)\,dt.$$

In this section we will generalize this result to show that the integral of the rate of change of any quantity will give the total change in that quantity. Suppose $F'(t)$ is the rate of change of some quantity $F(t)$ with respect to time, and that we are interested in the total change in $F(t)$ between $t = a$ and $t = b$. We divide the interval $a \le t \le b$ into n subintervals, each of length Δt. For each small interval, we will calculate the change in $F(t)$, written ΔF, and then add all these up. For each subinterval we assume the rate of change of $F(t)$ is approximately constant, so that we can say

$$\Delta F \approx \text{Rate of change of } F \times \text{Time}.$$

For the first subinterval, from t_0 to t_1, the rate of change of $F(t)$ is approximately $F'(t_0)$, so

$$\Delta F \approx F'(t_0)\,\Delta t.$$

Similarly, for the second interval

$$\Delta F \approx F'(t_1)\,\Delta t.$$

Summing over all the subintervals,

$$\text{Total change in } F = \sum \Delta F \approx \sum_{i=0}^{n-1} F'(t_i)\,\Delta t.$$

Thus, we have approximated the change in $F(t)$ as a left-hand sum. By a similar argument, we can approximate the change in $F(t)$ by a right-hand sum:

$$\text{Total change in } F = \sum \Delta F \approx \sum_{i=1}^{n} F'(t_i)\,\Delta t.$$

The total change in $F(t)$ between the times $t = a$ and $t = b$ can be written as $F(b) - F(a)$. Thus, taking the limit as n goes to infinity, we get the following result:

$$F(b) - F(a) = \begin{array}{c}\text{Total change in } F(t)\\ \text{between } t = a \text{ and } t = b\end{array} = \int_a^b F'(t)\,dt.$$

This result is one of the most important in calculus because it makes the connection between the derivative and the definite integral. It is called the Fundamental Theorem of Calculus and is often stated as follows:

The Fundamental Theorem of Calculus

If f is continuous and $f(t) = \frac{dF(t)}{dt}$, then

$$\int_a^b f(t)\,dt = F(b) - F(a).$$

In words:

The definite integral of a rate gives total change.

The Fundamental Theorem provides a precise way of computing certain definite integrals.

Example 1 Compute $\int_1^3 2x\,dx$ by two different methods.

Solution Using left- and right-hand sums, we can approximate this integral as accurately as we want. With $n = 100$, for example, the left-sum is 7.96 and the right sum is 8.04. Using $n = 500$ we learn

$$7.992 < \int_1^3 2x\,dx < 8.008.$$

The Fundamental Theorem, on the other hand, allows us to compute the integral exactly. We take $f(x) = 2x$. By Example 4, on page 115, we know that if $F(x) = x^2$, then $F'(x) = 2x$. So we take $f(x) = 2x$ and $F(x) = x^2$ and obtain

$$\int_1^3 2x\,dx = F(3) - F(1) = 3^2 - 1^2 = 8.$$

The Fundamental Theorem can also be used when the rate, $F'(t)$, is known and we want to find the total change $F(b) - F(a)$. If we know $F(a)$, the theorem enables us to reconstruct the function F from knowledge about its derivative $F' = f$.

Example 2 After t hours, a population of bacteria is growing at a rate of 2^t million bacteria per hour. Estimate the total increase in the bacteria population during the first hour.

Solution Since the rate at which the population is growing is $F'(t) = 2^t$, we have

$$\text{Change in population} = F(1) - F(0) = \int_0^1 2^t\,dt.$$

Since the rate is a monotonic function, the integral will be trapped between the left- and right-hand sums. For $n = 100$,

$$\text{Left sum} = 1.4377 < \int_0^1 2^t\,dt < 1.4477 = \text{Right sum}$$

so

$$\text{Change in population} = \int_0^1 2^t\,dt \approx 1.4 \text{ million bacteria.}$$

Example 3 Suppose you are given that $F'(t) = t\cos t$ and $F(0) = 2$. Find the values of $F(b)$ at the points $b = 0, 0.1, 0.2, \ldots, 1.0$.

Solution We apply the Fundamental Theorem with $f(t) = t\cos t$ and $a = 0$ to get values for $F(b)$. Since

$$F(b) - F(0) = \int_0^b F'(t)\,dt = \int_0^b t\cos t\,dt$$

and $F(0) = 2$, we have

$$F(b) = 2 + \int_0^b t\cos t\,dt.$$

Using Riemann sums to estimate the definite integral $\int_0^b t\cos t\,dt$ for each of the values $b = 0, 0.1, 0.2, \ldots, 1.0$ gives the approximate values for F in Table 3.5:

TABLE 3.5 *Approximate values for F*

b	0	0.1	0.2	0.3	0.4	0.5	0.6	0.7	0.8	0.9	1.0
$F(b)$	2.000	2.005	2.020	2.044	2.077	2.117	2.164	2.216	2.271	2.327	2.382

Notice from the table that the function $F(b)$ is increasing between $b = 0$ and $b = 1$. This could have been predicted without the use of the Fundamental Theorem from the fact that $t\cos t$, the derivative of $F(t)$, is positive for t between 0 and 1.

Example 4 The graph of the derivative F' of some function F is given in Figure 3.20. If you are told that $F(20) = 150$, estimate the maximum value attained by F.

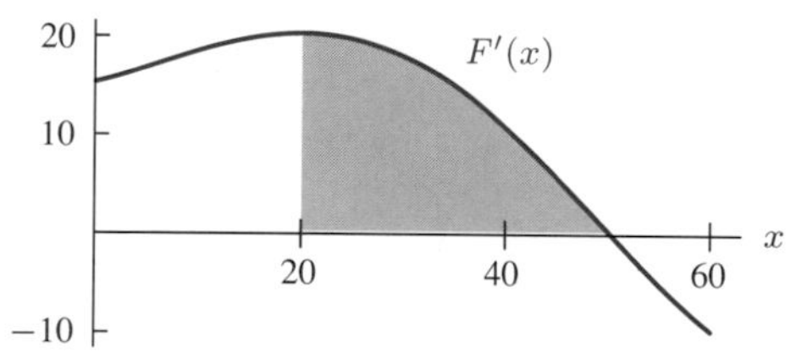

Figure 3.20: Graph of the derivative F' of some function F

Solution Let's begin by getting a rough idea of how F behaves. We know that $F(x)$ increases for $x < 50$ because the derivative of F is positive for $x < 50$. Similarly, $F(x)$ decreases for $x > 50$ because $F'(x)$ is negative for $x > 50$. Therefore, the graph of F rises until the point at which $x = 50$, and then it begins to fall. Evidently, the highest point on the graph of F is at $x = 50$, and so the maximum value attained by F is $F(50)$. To evaluate $F(50)$, we use the Fundamental Theorem:

$$F(50) - F(20) = \int_{20}^{50} F'(x)\,dx,$$

which gives

$$F(50) = F(20) + \int_{20}^{50} F'(x)\,dx = 150 + \int_{20}^{50} F'(x)\,dx.$$

The definite integral equals the area of the shaded region under the graph of F', which we can estimate is roughly 300. Therefore, the greatest value attained by F is $F(50) \approx 150 + 300 = 450$.

Example 5 Biological activity in a pond is reflected in the rate at which carbon dioxide, CO_2, is added to or withdrawn from the water. Plants take CO_2 out of the water during the day for photosynthesis and put CO_2 into the water at night. Animals put CO_2 into the water all the time as they breathe. Biologists are interested in how the net rate at which CO_2 enters or leaves a pond varies during the day. Figure 3.21 shows this rate as a function of time of day.[2] The rate is measured in millimoles (mmol) of CO_2 per liter of water per hour; time is measured in hours past dawn. At dawn, there were 2.600 mmol of CO_2 per liter of water.

(a) What can be concluded from the fact that the rate is negative during the day and positive at night?

(b) Some scientists have suggested that plants and animals respire (breathe) at a constant rate at night, and that plants photosynthesize at a constant rate during the day. Does Figure 3.21 support this view?

(c) When was the CO_2 content of the water at its lowest? How low did it go?

(d) How much CO_2 was released into the water during the 12 hours of darkness? Compare this quantity with the amount of CO_2 withdrawn from the water during the 12 hours of daylight. How can you tell by looking at the graph whether the CO_2 in the pond is in equilibrium?

(e) Estimate the CO_2 content of the water at three hour intervals throughout the day. Use your estimates to plot a graph of CO_2 content throughout the day.

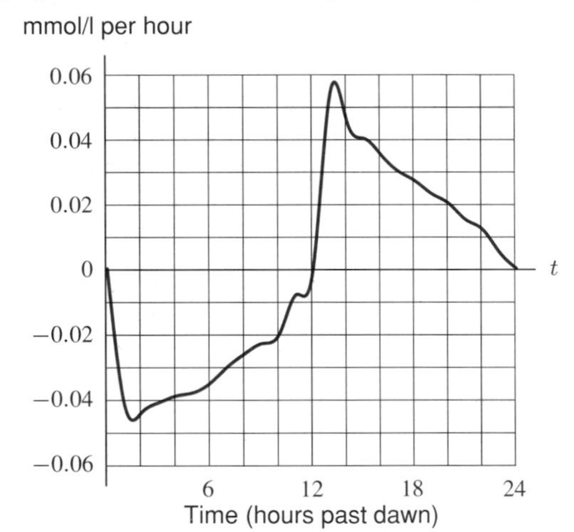

Figure 3.21: Rate at which CO_2 is entering or leaving water, $f'(t)$

Solution (a) CO_2 is being taken out of the water during the day and returned at night. The pond must therefore contain some plants. (The data is in fact from pond water containing both plants and animals.)

(b) Suppose t is the number of hours past dawn. The graph in Figure 3.21 shows that the CO_2 content changes at a greater rate for the first 6 hours of daylight, $0 < t < 6$, than it does for the final 6 hours of daylight, $6 < t < 12$. It turns out that plants photosynthesize more vigorously in the morning than in the afternoon. Similarly, CO_2 content changes more rapidly in the first half of the night, $12 < t < 18$, than in the 6 hours just before dawn, $18 < t < 24$. The reason seems to be that at night plants quickly use up most of the sugar that they synthesized during the day, and then their respiration rate is inhibited.

[2]Data from R. J. Beyers, *The Pattern of Photosynthesis and Respiration in Laboratory Microsystems* (Mem. 1st. Ital. Idrobiol., 1965).

(c) We are now being asked about the total quantity of CO_2 in the pond, rather than the rate at which it is changing. We will let $f(t)$ denote the CO_2 content of the pond water (in mmol/l) at t hours past dawn. Then Figure 3.21 is a graph of the derivative $f'(t)$. There are 2.600 mmol/l of CO_2 in the water at dawn, so $f(0) = 2.600$.

The CO_2 content $f(t)$ decreases during the 12 hours of daylight, $0 < t < 12$, when $f'(t) < 0$, and then $f(t)$ increases for the next 12 hours. Thus, $f(t)$ is at a minimum when $t = 12$, at dusk. By the Fundamental Theorem,

$$f(12) = f(0) + \int_0^{12} f'(t)\,dt = 2.600 + \int_0^{12} f'(t)\,dt.$$

We must approximate the definite integral by a Riemann sum. From the graph in Figure 3.21, we estimate the values of the function $f'(t)$ in Table 3.6.

TABLE 3.6 *Rate, $f'(t)$, at which CO_2 is entering or leaving water*

t	$f'(t)$	t	$f'(t)$	t	$f'(t)$	t	$f'(t)$	t	$f'(t)$	t	$f'(t)$
0	0.000	4	−0.039	8	−0.026	12	0.000	16	0.035	20	0.020
1	−0.042	5	−0.038	9	−0.023	13	0.054	17	0.030	21	0.015
2	−0.044	6	−0.035	10	−0.020	14	0.045	18	0.027	22	0.012
3	−0.041	7	−0.030	11	−0.008	15	0.040	19	0.023	23	0.005

The left Riemann sum with $n = 12$ terms, corresponding to $\Delta t = 1$, gives

$$\int_0^{12} f'(t)\,dt \approx (0.000)(1) + (-0.042)(1) + (-0.044)(1) + \cdots + (-0.008)(1) = -0.346.$$

At 12 hours past dawn, the CO_2 content of the pond water reaches its lowest level, which is approximately

$$2.600 - 0.346 = 2.254 \text{ mmol/l}.$$

(d) The increase in CO_2 during the 12 hours of darkness equals

$$f(24) - f(12) = \int_{12}^{24} f'(t)\,dt.$$

Using Riemann sums to estimate this integral, we find that about 0.306 mmol/l of CO_2 was released into the pond during the night. In part (c) we calculated that about 0.346 mmol/l of CO_2 was absorbed from the pond during the day. If the pond is in equilibrium, we would expect the daytime absorption to equal the nighttime release. These quantities are sufficiently close (0.346 and 0.306) that the difference could be due to measurement error.

If the pond is in equilibrium, the area between the rate curve in Figure 3.21 and the axis for $0 \le t \le 12$ will equal the area between the rate curve and the axis for $12 \le t \le 24$. (The axis is the horizontal line at 0.00.) In this experiment the areas do look approximately equal.

(e) We must evaluate

$$f(b) = f(0) + \int_0^{b} f'(t)\,dt = 2.600 + \int_0^{b} f'(t)\,dt$$

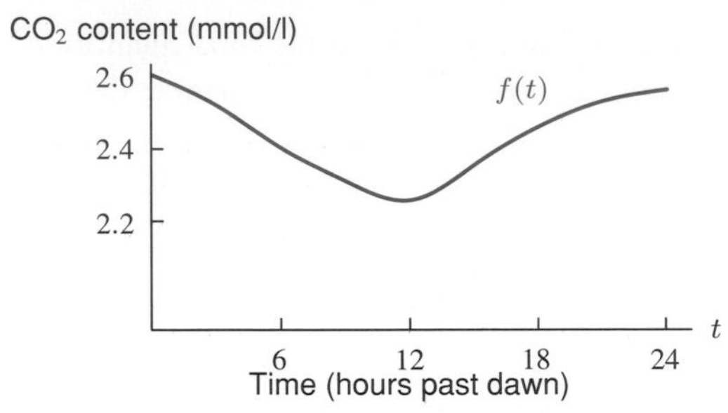

Figure 3.22: CO_2 content in pond water throughout the day

for the values $b = 0, 3, 6, 9, 12, 15, 18, 21, 24$. Left Riemann sums with $\Delta t = 1$ give the values for the CO_2 content in Table 3.7. The graph is shown in Figure 3.22.

TABLE 3.7 CO_2 *content throughout the day*

b (hours after dawn)	0	3	6	9	12	15	18	21	24
$f(b)$ (CO_2 content)	2.600	2.514	2.396	2.305	2.254	2.353	2.458	2.528	2.560

Problems for Section 3.4

1. Figure 3.23 shows the graph of f. If $F' = f$ and $F(0) = 0$, find $F(b)$ for $b = 1, 2, 3, 4, 5, 6$.

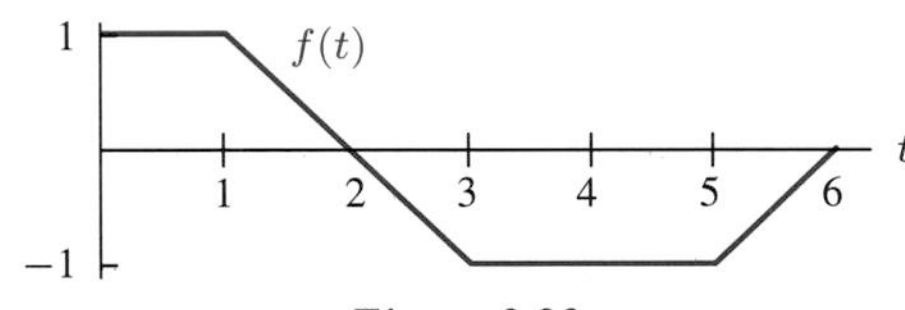

Figure 3.23

2. The graph of $y = f(x)$ is given in Figure 3.24.

   (a) What is $\int_{-3}^{0} f(x)\,dx$?

   (b) If the area of the shaded region is A, what is $\int_{-3}^{4} f(x)\,dx$?

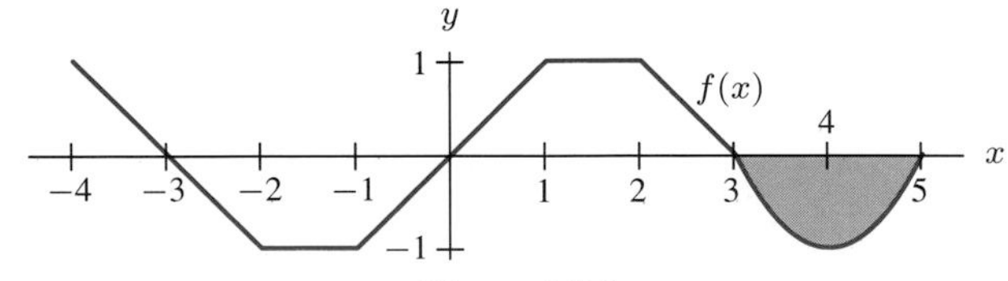

Figure 3.24

For Problems 3–5, suppose $F(0) = 0$ and $F'(\theta) = f(\theta) = \sin(\theta^2)$ for $0 \le \theta \le 2.5$.

3. Approximate $F(b)$ for $b = 0,\ 0.5,\ 1,\ 1.5,\ 2,\ 2.5$.
4. Using a graph of F', decide where F is increasing and where F is decreasing.
5. Does F have a maximum value for $0 \le \theta \le 2.5$? If so, at what value of θ does it occur, and approximately what is that maximum value?

6. If $F(t) = t(\ln t) - t$, it can be shown that $f(t) = F'(t) = \ln t$. Find $\int_{10}^{12} \ln t \, dt$ two ways:
   (a) Using left and right sums. (Approximate to one decimal place.)
   (b) Using the Fundamental Theorem of Calculus.

7. (a) Let $F(x) = \sin x$. Use graphs of $\sin x$ and $\cos x$ to explain why it is reasonable that $F'(x) = \cos x$.
   (b) Now find $\int_0^{\pi/2} \cos x \, dx$ two ways:
      (i) Using left and right sums. (Approximate to one decimal place.)
      (ii) Using the Fundamental Theorem of Calculus.

Problems 8–9 concern the graph of f' in Figure 3.25.

8. Which is greater, $f(0)$ or $f(1)$?

9. List the following quantities in increasing order:

$$\frac{f(4) - f(2)}{2}, \quad f(3) - f(2), \quad f(4) - f(3).$$

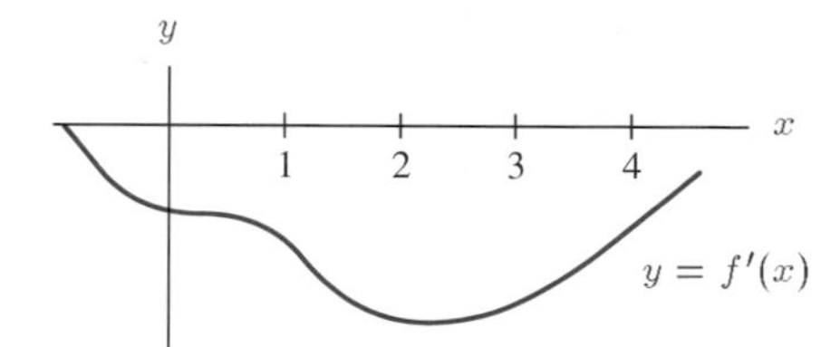

Figure 3.25: Note: This is the graph of f', not f.

10. A news broadcast in early 1993 said the average American's annual income is changing at a rate given in dollars per month by $r(t) = 40(1.002)^t$ where t is in months from January 1, 1993. If this trend continues, what change in income can the average American expect during 1993?

11. A cup of coffee at 90°C is put into a 20°C room when $t = 0$. If the coffee's temperature is changing at a rate given in °C per minute by

$$r(t) = -7e^{-0.1t}, \quad t \text{ in minutes,}$$

estimate, to one decimal place, the coffee's temperature when $t = 10$.

12. Water is leaking out of a tank at a rate of $R(t)$ gallons/hour, where t is measured in hours.
   (a) Write a definite integral that expresses the total amount of water that leaks out in the first two hours.
   (b) Figure 3.26 is a graph of $R(t)$. On a sketch, shade in the region whose area represents the total amount of water that leaks out in the first two hours.
   (c) Give an upper and lower estimate of the total amount of water that leaks out in the first two hours.

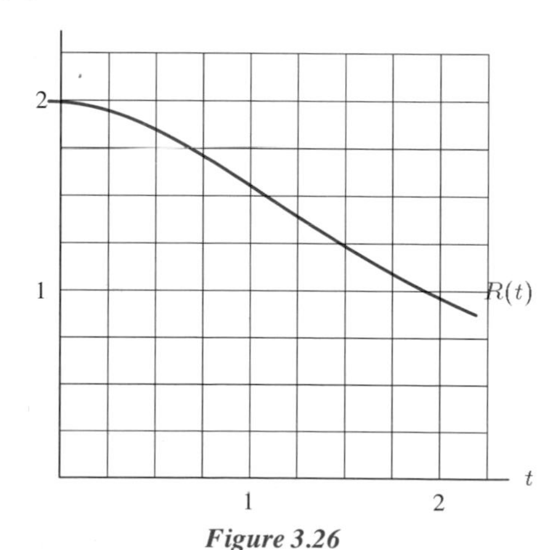

Figure 3.26

13. The rate at which the world's oil is being consumed is continuously increasing. Suppose the rate (in billions of barrels per year) is given by the function $r = f(t)$, where t is measured in years and $t = 0$ is the start of 1990.
    (a) Write a definite integral which represents the total quantity of oil used between the start of 1990 and the start of 1995.
    (b) Suppose $r = 32e^{0.05t}$. Using a left-hand sum with five subdivisions, find an approximate value for the total quantity of oil used between the start of 1990 and the start of 1995.
    (c) Interpret each of the five terms in the sum from part (b) in terms of oil consumption.

14. A bicyclist is pedaling along a straight road with velocity, v, given in Figure 3.27. Suppose the cyclist starts 5 miles from a lake, and that positive velocities take her away from the lake and negative velocities towards the lake. When is the cyclist farthest from the lake, and how far away is she then?

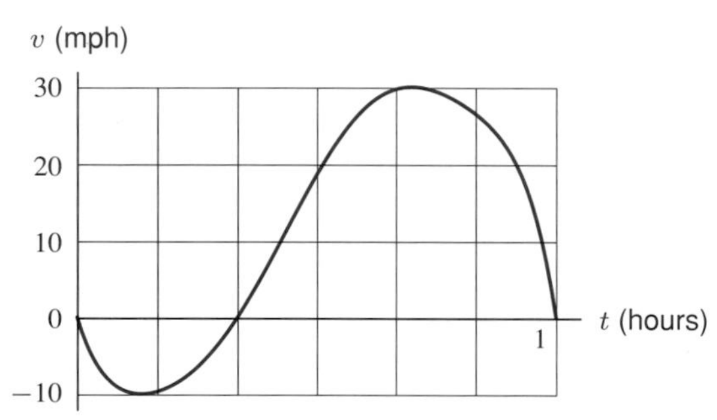

Figure 3.27

For Problems 15–18, mark the following quantities on a copy of the graph of f in Figure 3.28.

15. A length representing $f(b) - f(a)$.
16. A slope representing $\dfrac{f(b) - f(a)}{b - a}$.
17. An area representing $F(b) - F(a)$, where $F' = f$.
18. A length roughly approximating $\dfrac{F(b) - F(a)}{b - a}$, where $F' = f$.

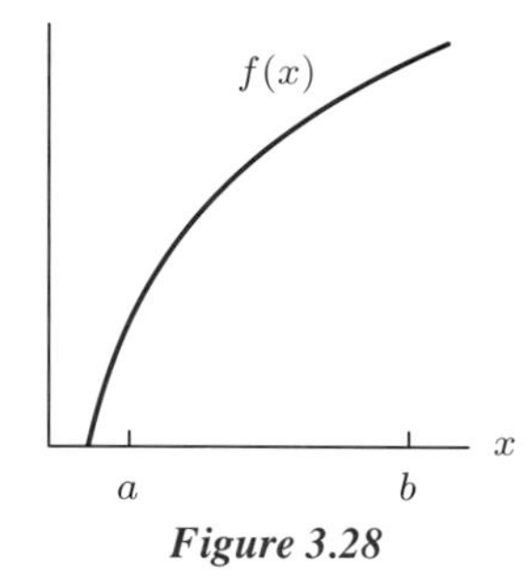

Figure 3.28

For Problems 19–21, assuming $F' = f$, mark the following quantities on a copy of the graph of F in Figure 3.29.

19. A slope representing $f(a)$.
20. A length representing $\int_a^b f(x)\,dx$.
21. A slope representing $\dfrac{1}{b - a}\int_a^b f(x)\,dx$.

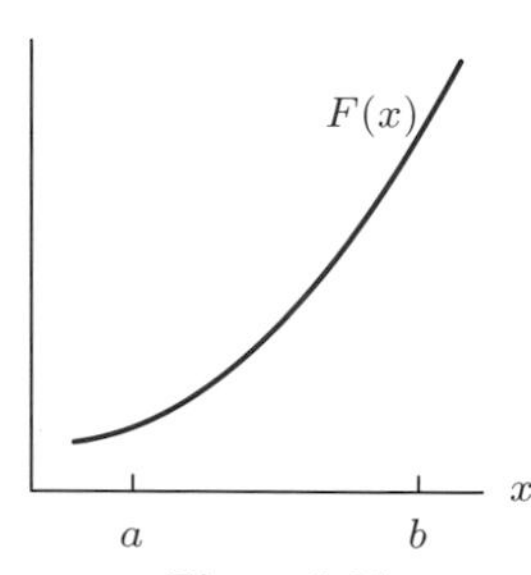

Figure 3.29

22. In Section 2.1, page 98, we defined the average velocity of a particle over the interval $a \le t \le b$ as the change in position during this interval divided by $(b - a)$. If the velocity of the particle is $v = f(t)$, the average velocity is also given by $\frac{1}{b-a}\int_a^b f(t)dt$. Show why these two ways of calculating the average velocity always lead to the same answer.

3.5 FURTHER NOTES ON THE LIMIT

When we defined the definite integral, we introduced a limit of the type

$$\lim_{n\to\infty}.$$

This limit is different from the limit

$$\lim_{h\to 0}$$

that we used in the definition of the derivative. What is the difference? If we take the limit of a function as $h \to 0$, we are looking for a value approached by the function as $h = 0.1, 0.01, 0.001, 0.0001, \ldots$, as well as $h = -0.1, -0.01, -0.001, -0.0001, \ldots$. In the case of the limit as $n \to \infty$, we're interested in the behavior of the function for values of n such as $n = 10^6, 10^{10}, 10^{100}$, or $3 \times 10^{1000}, \ldots$ In other words, we want to know what happens to the function as n becomes arbitrarily large. One difference between taking the limit as $h \to 0$ and as $n \to \infty$ is that we cannot consider integers "larger than ∞," so we don't examine the limit at ∞ from both sides, whereas we do look at numbers approaching 0 (or any finite number) from either side. In more precise language,

> We define
>
> $$\lim_{n\to\infty} f(n)$$
>
> to be a number L (if one exists) such that $f(n)$ is as close to L as we please whenever n is sufficiently large.

Here's an example:

$$\lim_{n\to\infty}\frac{1}{n}.$$

It's not hard to see that the value of the limit is $L = 0$: you can make $1/n$ as close to 0 as you like by taking a sufficiently large n. For example, to make $1/n$ smaller than $1/1000$, take $n > 1000$; to make $1/n$ smaller than 0.0000000001, take $n > 10^{10}$; and so on.

Here's another example:

$$\lim_{n\to\infty} n^2.$$

As n increases in size, the function n^2 doesn't approach any particular number—it too increases in size. In this case, we say that n^2 has no limit as $n \to \infty$, or that n^2 increases without bound.

Examples such as $\lim_{n\to\infty} n^2$ rarely arise when the limit is used to calculate a definite integral. Usually you can compute the value of the definite integral as accurately as you wish by taking sufficiently large values for n. Roughly, the reason that the sum

$$\sum_{i=1}^{n} f(x_i)\,\Delta x$$

doesn't increase without bound as n gets larger, even though you're adding together more and more terms, is that the terms are getting smaller fast enough.

The limit is important because it allows us to avoid the philosophical problems of infinite sums. You might, at first, want to define the definite integral as an infinite sum of infinitesimally small terms. But what would it mean to add up infinitely many numbers? The Riemann sum definition of the definite integral as the limit of a finite sum, with the number of terms tending to infinity, avoids this problem.

Problems for Section 3.5

For Problems 1–3, use a graph to find the limit.

1. $\lim_{t\to\infty} te^{-t}$
2. $\lim_{t\to\infty} t^3e^{-t}$
3. $\lim_{x\to\infty} \dfrac{\sin x}{x}$.

4. (a) Using a calculator or computer, plot the graph of
$$y = \left(1+\frac{1}{x}\right)^x \quad \text{for } x > 0.$$
(b) By extending the domain of your graph, investigate the value of y as x becomes very large; in other words, find
$$\lim_{x\to\infty}\left(1+\frac{1}{x}\right)^x.$$
(c) You should recognize your answer to part (b). What is it?

5. (a) What is wrong with the following "proof"?
$$\lim_{n\to\infty}\left(1+\frac{1}{n}\right) = 1$$
so
$$\lim_{n\to\infty}\left(1+\frac{1}{n}\right)^n = \lim_{n\to\infty} 1^n = 1.$$
(b) Explain why, if you take n large enough (for example, $n = 10^{20}$ or more), your calculator thinks that
$$\lim_{n\to\infty}\left(1+\frac{1}{n}\right)^n = 1.$$

6. The Fibonacci sequence has been famous for many years. It is defined as follows: $F_0 = 0$, $F_1 = 1$, $F_2 = F_1 + F_0$, $F_3 = F_2 + F_1, \ldots, F_{n+1} = F_n + F_{n-1}, \ldots$. Work out and write in a column the Fibonacci numbers through F_{20}. Then, to the right of F_n, write the value of F_n/F_{n+1} evaluated to seven decimal places. As you work your way down the column, try to guess the value of each entry before you find it with your calculator. Work out the values of $(F_{n+1}/F_{n+2}) - (F_n/F_{n+1})$ as fractions for the first few values of n. What do you see? Is that consistent with what you noticed as you were writing the decimal values? Use algebra to see if you can prove that what you have noticed will indeed go on as n goes to infinity. (Fibonacci was the pseudonym of Leonardo of Pisa, a mathematician who died in about 1240.)

REVIEW PROBLEMS FOR CHAPTER THREE

For Problems 1–6, find the integrals to one decimal place. In each case say how many subdivisions you used.

1. $\int_0^{10} 2^{-x}\, dx$

2. $\int_0^{\pi/2} \sin(2\theta^2 + 1)\, d\theta$

3. $\int_0^1 \sqrt{1 + \cos^2 t}\, dt$

4. $\int_{-1}^1 \frac{x^2 + 1}{x^2 - 4}\, dx$

5. $\int_2^3 \frac{-1}{(r+1)^2}\, dr$

6. $\int_0^1 \arcsin z\, dz$

7. Statisticians sometimes use values of the function

$$F(b) = \int_0^b e^{-x^2}\, dx.$$

(a) What is $F(0)$?
(b) Does the value of F increase or decrease as b increases? (Assume $b \geq 0$.)
(c) Estimate $F(1)$, $F(2)$, and $F(3)$.

8. If $F(x) = e^{x^2}$, it can be shown that $f(x) = F'(x) = 2xe^{x^2}$. Find $\int_0^1 2xe^{x^2}\, dx$ two ways:

(a) Using left and right sums. (Approximate to one decimal place.)
(b) Using the Fundamental Theorem of Calculus.

9. If $F(t) = \frac{1}{2}\sin^2 t$, it can be shown that $f(t) = F'(t) = \sin t \cos t$. Find $\int_{0.2}^{0.4} \sin t \cos t\, dt$ two ways:

(a) Using left and right sums. (Approximate to one decimal place.)
(b) Using the Fundamental Theorem of Calculus.

10. A car accelerates smoothly from 0 to 60 mph in 10 seconds.

(a) How often would you have to measure its velocity to be able to determine within 1 foot the total distance the car travels in the 10-second period?
(b) Suppose the car's velocity as a function of time is given in Figure 3.30. Estimate how far the car travels during the 10-second period.

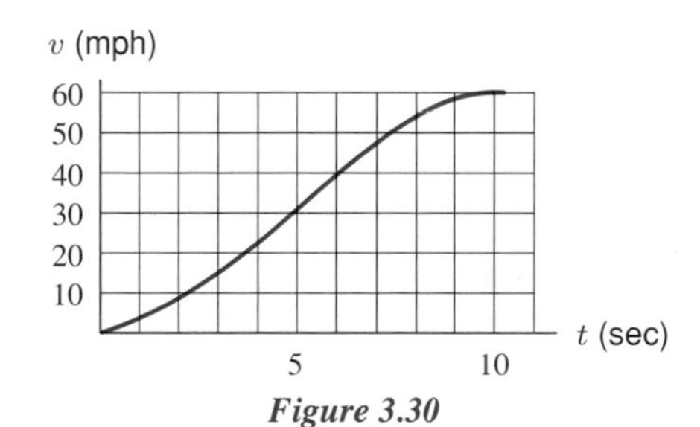

Figure 3.30

11. The Environmental Protection Agency was recently asked to investigate a spill of radioactive iodine. Measurements showed the ambient radiation levels at the site to be four times the maximum acceptable limit, so the EPA ordered an evacuation of the surrounding neighborhood.

It is known that the level of radiation from an iodine source decreases according to the formula

$$R(t) = R_0 e^{-0.004t}$$

where R is the radiation level (in millirems/hour) at time t, R_0 is the initial radiation level (at $t = 0$), and t is the time measured in hours.

(a) How long will it take for the site to reach an acceptable level of radiation?
(b) How much total radiation (in millirems) will have been emitted by that time, assuming the maximum acceptable limit is 0.6 millirems/hour?

12. Each of the graphs in Figure 3.31 represents the velocity, v, of a particle moving along the x-axis for time $0 \leq t \leq 5$. The vertical scales of all graphs are the same. Identify the graph(s) showing which particle(s)

(a) has a constant acceleration.
(b) ends up farthest to the left of where it started.
(c) ends up the farthest from its starting point.
(d) experiences the greatest initial acceleration.
(e) has the greatest average velocity.
(f) has the greatest average acceleration.

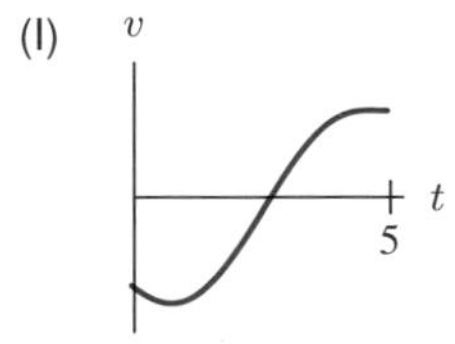

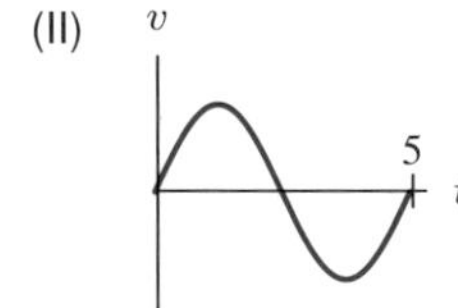

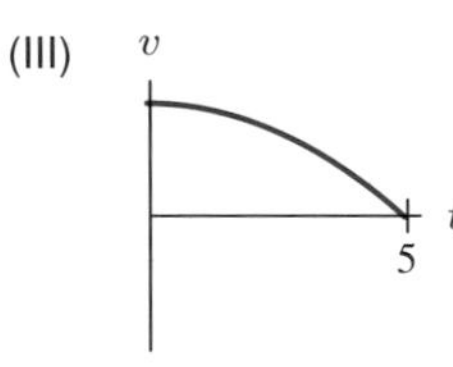

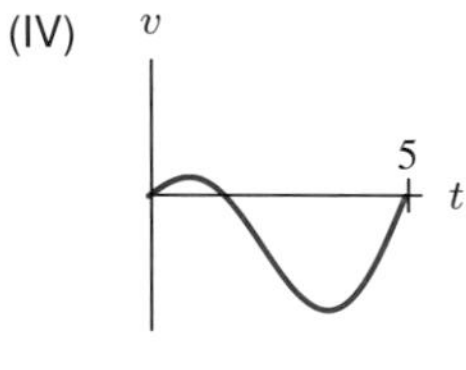

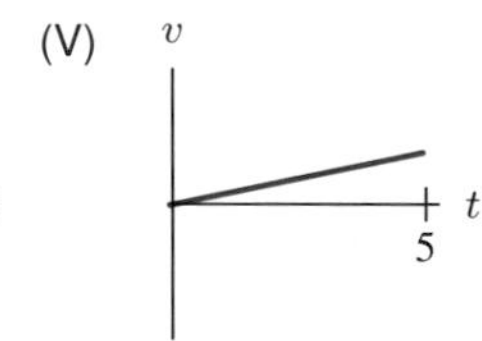

Figure 3.31

13. Assume the population, P, of Mexico (in millions), is given by

$$P = 67.38(1.026)^t,$$

where t is the number of years since 1980.

(a) What was the average population of Mexico between 1980 and 1990?
(b) What is the average of the population of Mexico in 1980 and the population in 1990?
(c) Explain, in terms of the concavity of the graph of P (see Figure 1.18 on page 17), why your answer to part (b) is larger or smaller than your answer to part (a).

14. As this country's relatively rich coal deposits are depleted, a larger fraction of the country's coal will come from strip-mining, and it will become necessary to strip-mine larger and larger areas for each ton of coal. The graph in Figure 3.32 shows an estimate of the number of acres/million tons that will be defaced during strip-mining as a function of the number of million tons removed, starting from the present day.

(a) Estimate the total number of acres defaced in extracting the next 4 million tons of coal (measured from the present day). Draw four rectangles under the curve, and compute their area.

(b) Reestimate the number of acres defaced using rectangles above the curve.

(c) Use your answers to parts (a) and (b) to get a better estimate of the actual number of acres defaced.

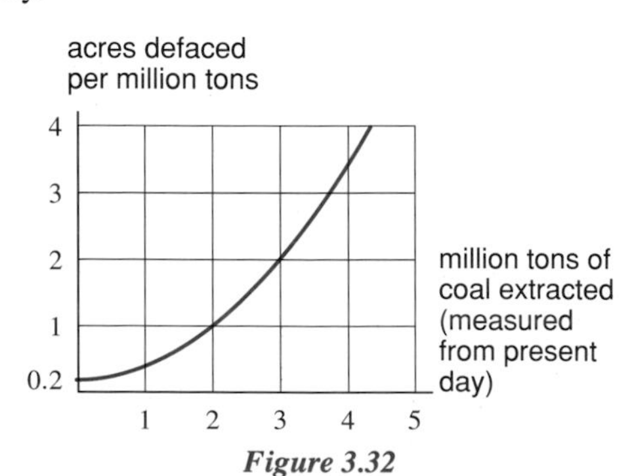

Figure 3.32

15. In an imaginary galaxy, a star is growing at a rate given by

$$r(t) = (\tan t)^{t+1} \quad \text{unit mass per century.}$$

A second star grows at a rate given by

$$p(t) = 2^t \quad \text{unit mass per century.}$$

(a) Which star gained more mass during the first 1.25 centuries?
(b) Which is greater, $r'(0.4)$ or $p'(0.4)$? What is the physical meaning of $r'(0.4)$?
(c) Give a physical interpretation for the definite integral

$$\int_0^1 (2^t - (\tan t)^{t+1})\, dt.$$

16. Find the shaded area in Figure 3.33. You are given that $g(x) = e^{-x/4}$, and that $f(x)$ is a trigonometric function.

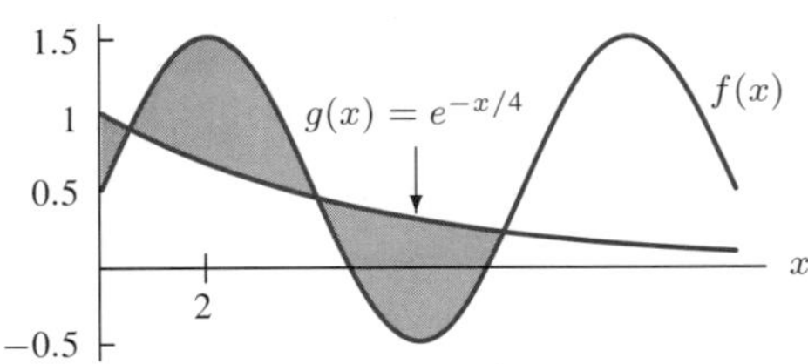

Figure 3.33

17. How long is one arch of the sine curve (i.e., that portion of the graph of $y = \sin x$ from $x = 0$ to $x = \pi$)? Use a calculator and common sense to get an approximate answer. (You are not expected to use a formula for arc length.)

18. The Montgolfier brothers (Joseph and Etienne) were eighteenth-century pioneers in the field of hot-air ballooning. Had they had the appropriate instruments, they might have left us a record of one of their early experiments, like that shown in Figure 3.34. The graph shows their vertical velocity, v, with upward as positive.

(a) Over what intervals was the acceleration positive? Negative? Zero?
(b) What was the greatest altitude achieved, and at what time?
(c) At what time was the upward acceleration greatest?
(d) At what time was the deceleration greatest?
(e) What might have happened during this flight to explain the answer to part (d)?
(f) This particular flight ended on top of a hill. How do you know that it did, and what was the height of the hill above the starting point?

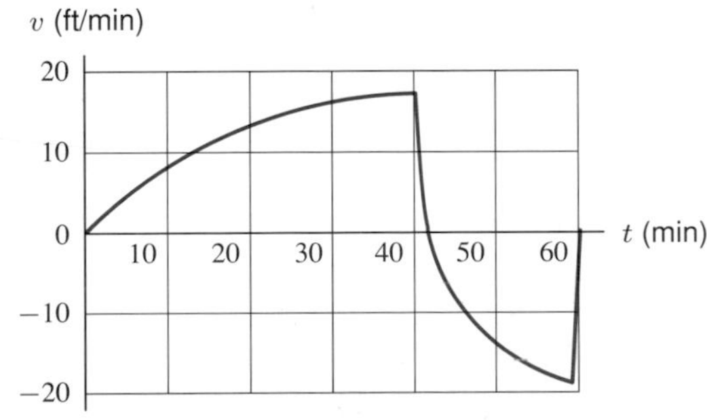

Figure 3.34

19. A mouse moves back and forth in a tunnel, attracted to bits of cheddar cheese alternately introduced to and removed from the ends (right and left) of the tunnel. The graph of the mouse's velocity, v, is given in Figure 3.35, with velocity being positive moving towards the right end of the tunnel, and negative towards the left end. Assuming that the mouse starts ($t = 0$) at the center of the tunnel, use the graph to estimate the time(s) at which:

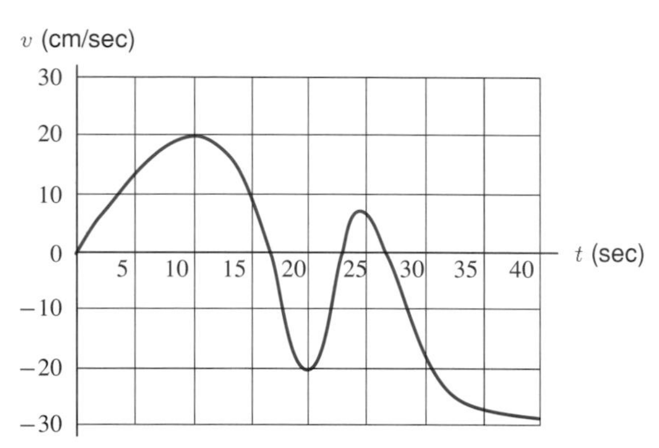

Figure 3.35

(a) The mouse changes direction.
(b) The mouse is moving most rapidly to the right; to the left.
(c) The mouse is farthest to the right of center; farthest to the left.
(d) The mouse's speed (i.e., the magnitude of its velocity) is decreasing.
(e) The mouse is at the center of the tunnel.

20. The graph of some function f is given in Figure 3.36. List, from *least* to *greatest*:

(a) $f'(1)$.
(b) The average value of $f(x)$, $0 \leq x \leq a$.
(c) The average value of the rate of change in $f(x)$, for $0 \leq x \leq a$.
(d) $\int_0^a f(x)\,dx$.

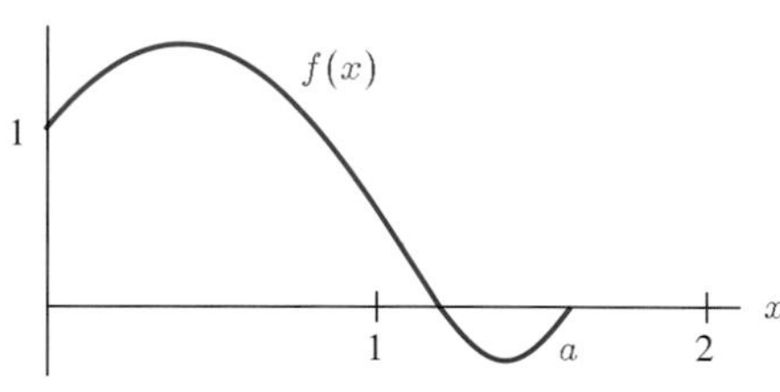

Figure 3.36

21. If you jump out of an airplane and your parachute fails to open, your downward velocity t seconds after the jump is approximated by

$$v(t) = \frac{g}{k}(1 - e^{-kt})$$

where $g = 9.8$ m/sec^2 and $k = 0.2$ sec^{-1}.

(a) Write an expression for the distance you fall in T seconds.
(b) If you jump from 5000 meters above the ground, estimate, using left- and right-hand sums, how many seconds you fall before hitting the ground.

CHAPTER FOUR

SHORT-CUTS TO DIFFERENTIATION

In this chapter we make a systematic study of the derivatives of functions given by formulas. These functions include powers, polynomials, exponential, logarithmic, and trigonometric functions. The chapter also contains general rules, such as the product, quotient, and chain rules, which allow us to differentiate combinations of functions.

4.1 FORMULAS FOR DERIVATIVE FUNCTIONS

In Chapter 2, we defined the derivative function

$$f'(x) = \lim_{h \to 0} \frac{f(x+h) - f(x)}{h}$$

and saw how the derivative represented a slope and a rate of change. We also learned how to find the derivative of a function given by a graph (by estimating the slope of the tangent at each point) and how to estimate the derivative of a function given by a table (by finding the average rate of change of the function between data values).

Useful Notation: We will write $\frac{d}{dx}(x^3)$, for example, to mean the derivative of x^3 with respect to x. Similarly, $\frac{d}{d\theta}\left(\sin(\theta^2)\right)$ denotes the derivative of $\sin(\theta^2)$ with θ as the variable.

Derivative of a Constant Function

The graph of a constant function $f(x) = c$ is a horizontal line, with a slope of 0 everywhere. Therefore, its derivative is 0 everywhere. (See Figure 4.1.)

If $f(x) = c$, then $f'(x) = 0$.

For example, $\frac{d}{dx}(5) = 0$, $\frac{d}{dx}(\pi) = 0$.

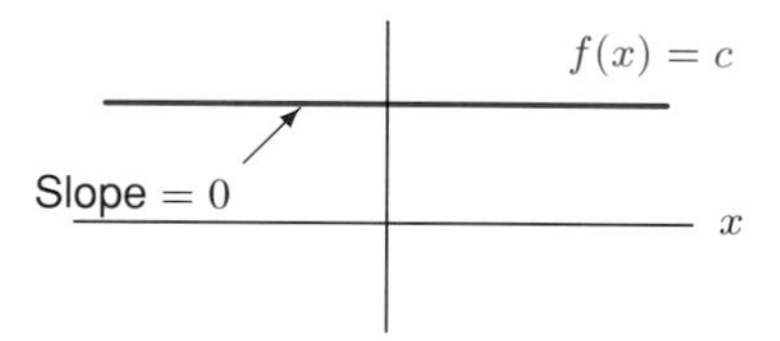

Figure 4.1: A constant function

Derivative of a Linear Function

We already know that the slope of a straight line is constant. This tells us that the derivative of a linear function is constant.

If $f(x) = b + mx$, then $f'(x) = \text{Slope} = m$.

For example, $\frac{d}{dx}(5 - \frac{3}{2}x) = -\frac{3}{2}$.

Alternatively, we can use the definition of the derivative:

$$
\begin{aligned}
f'(x) &= \lim_{h\to 0} \frac{f(x+h)-f(x)}{h} \\
&= \lim_{h\to 0} \frac{[b+m(x+h)]-[b+mx]}{h} = \lim_{h\to 0} \frac{mh}{h} = \lim_{h\to 0} m \qquad \text{(since } h \neq 0\text{)} \\
&= m.
\end{aligned}
$$

To find the second derivative of a linear function, we must differentiate m, which is a constant. Thus, if f is linear,

$$f''(x) = 0.$$

So, as we would expect for a straight line, f is neither concave up nor concave down.

Derivative of a Constant Times a Function

In Figure 4.2, you see the graph of $y = f(x)$ and of three multiples: $y = 3f(x)$, $y = \frac{1}{2}f(x)$, and $y = -2f(x)$. What is the relationship between the derivatives of these functions? In other words, for a particular x value, how are the slopes of these graphs related?

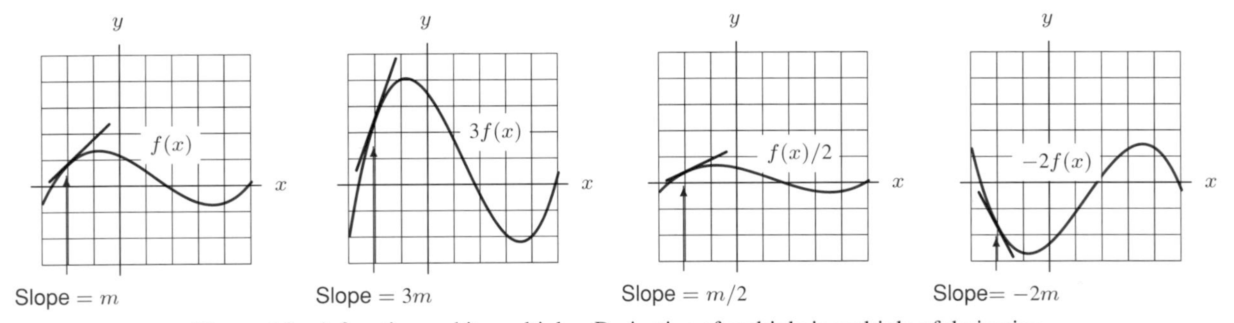

Figure 4.2: A function and its multiples: Derivative of multiple is multiple of derivative

Multiplying by a constant stretches or shrinks the graph (and flips it over the x-axis if the constant is negative). The zeros of the function remain the same and the peaks and valleys occur at the same x values. What does change is the slope of the curve at each point. If the graph has been stretched, the "rises" have all been increased by the same factor, whereas the "runs" remain the same. Thus, the slopes are all steeper by the same factor. If the graph has been shrunk, the slopes are all smaller by the same factor. If the graph has been flipped over the x-axis, the slopes will all have their signs reversed. In other words, if a function is multiplied by a constant, c, so is its derivative:

Derivative of a Constant Multiple

$$\frac{d}{dx}[cf(x)] = cf'(x).$$

Algebraic Justification: This result can be obtained algebraically, too:

$$\frac{d}{dx}\left[cf(x)\right] = \lim_{h\to 0}\frac{cf(x+h)-cf(x)}{h} = \lim_{h\to 0} c\,\frac{f(x+h)-f(x)}{h}$$
$$= c\lim_{h\to 0}\frac{f(x+h)-f(x)}{h} = cf'(x).$$

You may wonder why c can be taken across the limit sign: the reason is that c is a constant. If a function's value gets close to a certain number, c times that function gets close to c times that number. The function in this case is $\left[f(x+h)-f(x)\right]/h$.

Derivatives of Sums and Differences

Suppose we have two functions, $f(x)$ and $g(x)$, with the values listed in Table 4.1. Values of the sum $f(x)+g(x)$ are in the same table.

TABLE 4.1 *Sum of Functions*

x	$f(x)$	$g(x)$	$f(x)+g(x)$
0	100	0	100
1	110	0.2	110.2
2	130	0.4	130.4
3	160	0.6	160.6
4	200	0.8	200.8
5	250	1.0	251.0
6	310	1.2	311.2
7	380	1.4	381.4

You can see that when the increments of $f(x)$ and the increments of $g(x)$ are added together, they give the increments of $f(x)+g(x)$. For example, as x increases from 0 to 1, $f(x)$ increases by 10 and $g(x)$ increases by 0.2, while $f(x)+g(x)$ increases by $110.2 - 100 = 10.2$. Similarly, as x increases from 5 to 6, $f(x)$ increases by 60 and $g(x)$ by 0.2, while $f(x)+g(x)$ increases by $311.2 - 251.0 = 60.2$.

From this example, you can see that the rate at which $f(x)+g(x)$ is increasing is the sum of the rates at which $f(x)$ and $g(x)$ are increasing. Stating this with derivatives:

Derivative of Sum

$$\frac{d}{dx}\left[f(x)+g(x)\right] = f'(x)+g'(x).$$

Similarly for the difference:

Derivative of Difference

$$\frac{d}{dx}\left[f(x) - g(x)\right] = f'(x) - g'(x).$$

Algebraic Justification: Let's justify the sum rule using the definition of the derivative:

$$\begin{aligned}\frac{d}{dx}\left[f(x)+g(x)\right] &= \lim_{h\to 0}\frac{\left[f(x+h)+g(x+h)\right]-\left[f(x)+g(x)\right]}{h}\\ &= \lim_{h\to 0}\left[\underbrace{\frac{f(x+h)-f(x)}{h}}_{\text{Limit of this is } f'(x)} + \underbrace{\frac{g(x+h)-g(x)}{h}}_{\text{Limit of this is } g'(x)}\right]\\ &= f'(x)+g'(x).\end{aligned}$$

You may have noticed that we have used the fact that the limit of a sum is the sum of the limits—a fact which is true, although we have not proved it. (Can you see why it is true?)

Problems for Section 4.1

1. Let $f(x) = -3x + 2$ and $g(x) = 2x + 1$.
   (a) If $k(x) = f(x) + g(x)$, find a formula for $k(x)$ and verify the sum rule by comparing $k'(x)$ with $f'(x) + g'(x)$.
   (b) If $j(x) = f(x) - g(x)$, determine a formula for $j(x)$ and compare $j'(x)$ with $f'(x) - g'(x)$.
2. Let $r(t) = 2t - 4$. If $s(t) = 3r(t)$, verify the constant multiple rule by showing that $s'(t) = 3r'(t)$.
3. (a) If $f(x) = 5x - 3$, $g(x) = -2x + 1$, find the derivative of $f(g(x))$.
   (b) Use your answer to part (a) to make a conjecture about the derivative of the composition of two linear functions. Explain why what you say is true for any two linear functions.
4. If $r(x) = f(x) + 2g(x) + 3$, and $f'(x) = g(x)$ and $g'(x) = r(x)$, express
   (a) $r'(x)$ in terms of $f(x)$ and $g(x)$. Your answer should not involve r, r', f', or g'.
   (b) $f'(x)$ in terms of $f(x)$ and $r(x)$. Your answer should not involve r', f', g, or g'.
5. State in your own words why the limit of a sum is the sum of the limits.

4.2 POWERS AND POLYNOMIALS

Positive Integral Powers of x

First, let's look at what graphs can tell us about the derivative of $f(x) = x^n$, with n a positive integer. We'll start with $n = 2$ and $n = 3$. The graphs of $f(x) = x^2$ and $g(x) = x^3$ are in Figure 4.3.

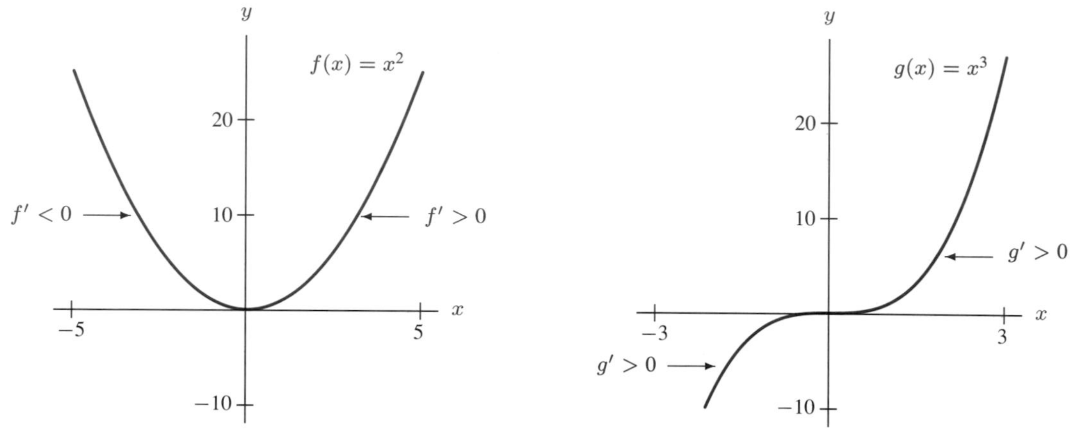

Figure 4.3: Graphs of $f(x) = x^2$ and $g(x) = x^3$

Derivative of x^2 and x^3

The shape of the graph of $f(x) = x^2$ in Figure 4.3 shows that its derivative is negative when $x < 0$, zero when $x = 0$, and positive for $x > 0$. The graph of $g(x) = x^3$, on the other hand, suggests that its derivative is positive when $x < 0$, zero when $x = 0$, and becomes positive again when $x > 0$.

We'll calculate the derivative of $f(x) = x^2$ using the definition:

$$\begin{aligned} f'(x) &= \lim_{h\to 0} \frac{f(x+h) - f(x)}{h} = \lim_{h\to 0} \frac{(x+h)^2 - x^2}{h} \\ &= \lim_{h\to 0} \frac{x^2 + 2xh + h^2 - x^2}{h} = \lim_{h\to 0} \frac{2xh + h^2}{h} \\ &= \lim_{h\to 0} \frac{h(2x+h)}{h}. \end{aligned}$$

To take the limit, we look at what happens when h is close to 0, but we do not let $h = 0$. Therefore, we can divide by h and say

$$f'(x) = \lim_{h\to 0} \frac{h(2x+h)}{h} = \lim_{h\to 0} (2x+h) = 2x,$$

because as h gets close to zero, $2x + h$ gets close to $2x$. So

$$f'(x) = \frac{d}{dx}(x^2) = 2x.$$

Now $f'(x) = 2x$ has the behavior we expected: it is negative for $x < 0$, zero when $x = 0$, and positive for $x > 0$. The graph of f' is in Figure 4.4, along with the graph of f.

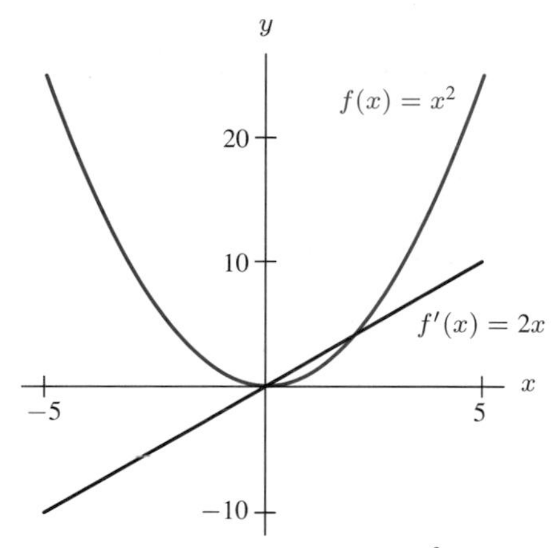

Figure 4.4: Graphs of $f(x) = x^2$ and its derivative

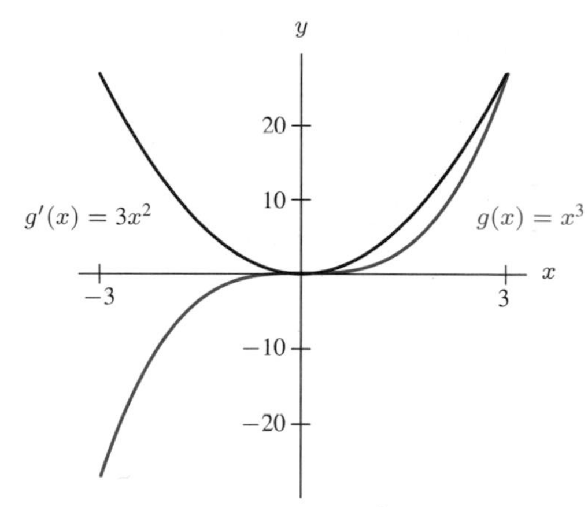

Figure 4.5: Graphs of $g(x) = x^3$ and its derivative

Example 1 Find the derivative of $g(x) = x^3$.

Solution

$$
\begin{aligned}
g'(x) &= \lim_{h \to 0} \frac{g(x+h) - g(x)}{h} = \lim_{h \to 0} \frac{(x+h)^3 - x^3}{h} \\
\text{Multiplying out} \longrightarrow &= \lim_{h \to 0} \frac{x^3 + 3x^2h + 3xh^2 + h^3 - x^3}{h} \\
&= \lim_{h \to 0} \frac{3x^2h + 3xh^2 + h^3}{h} \\
\text{Dividing by } h \longrightarrow &= \lim_{h \to 0} (3x^2 + 3xh + h^2) = 3x^2,
\end{aligned}
$$

Looking at what happens as $h \to 0$

so

$$g'(x) = \frac{d}{dx}(x^3) = 3x^2.$$

Again, $g'(x) = 3x^2$ is zero when $x = 0$, but positive everywhere else. (See Figure 4.5.)

Further Examples

Similar calculations will show you that

$$\frac{d}{dx}(x^4) = 4x^3, \quad \frac{d}{dx}(x^5) = 5x^4,$$

and so on. These results might well lead you to conjecture that, for a positive integer n, the following general relation holds:

$$\boxed{\frac{d}{dx}(x^n) = nx^{n-1},}$$

and indeed this turns out to be true.

Justification of $\frac{d}{dx}(x^n) = nx^{n-1}$, for n a Positive Integer

To calculate the derivatives of x^2 and x^3, we had to multiply out $(x+h)^2$ and $(x+h)^3$. Here we must expand $(x+h)^n$. Let's look back at the previous expansions:

$$(x+h)^2 = x^2 + 2xh + h^2, \qquad (x+h)^3 = x^3 + 3x^2h + 3xh^2 + h^3,$$

and multiply out a few more examples:

$$(x+h)^4 = x^4 + 4x^3h + 6x^2h^2 + 4xh^3 + h^4,$$
$$(x+h)^5 = x^5 + 5x^4h + \underbrace{10x^3h^2 + 10x^2h^3 + 5xh^4 + h^5}_{\text{Terms involving } h^2 \text{ and higher powers of } h}.$$

In general, we can say

$$(x+h)^n = x^n + nx^{n-1}h + \underbrace{\cdots\cdots + h^n}_{\text{Terms involving } h^2 \text{ and higher powers of } h}.$$

Now to find the derivative,

$$\begin{aligned}
\frac{d}{dx}(x^n) &= \lim_{h\to 0}\frac{(x+h)^n - x^n}{h} \\
&= \lim_{h\to 0}\frac{(x^n + nx^{n-1}h + \cdots + h^n) - x^n}{h} \\
&= \lim_{h\to 0}\frac{nx^{n-1}h + \overbrace{\cdots + h^n}^{\text{Terms involving } h^2 \text{ and higher powers of } h}}{h} \\
&= \lim_{h\to 0}\frac{h(nx^{n-1} + \cdots + h^{n-1})}{h}.
\end{aligned}$$

When you factor out h from terms involving h^2 and higher powers of h, each term will still have an h in it. Performing the division, we get:

$$\frac{d}{dx}(x^n) = \lim_{h\to 0}(nx^{n-1} + \overbrace{\cdots + h^{n-1}}^{\text{Terms involving } h \text{ and higher powers of } h}).$$

But, as $h \to 0$, all terms involving an h will go to 0, so

$$\frac{d}{dx}(x^n) = \lim_{h\to 0}(nx^{n-1} + \underbrace{\cdots + h^{n-1}}_{\text{These terms go to 0}}) = nx^{n-1}.$$

Example 2 Differentiate $5x^2 - 7x^3$.

Solution

$$\begin{aligned}
\frac{d}{dx}(5x^2 - 7x^3) &= \frac{d}{dx}(5x^2) - \frac{d}{dx}(7x^3) && \text{Derivative of difference} \\
&= 5\frac{d}{dx}(x^2) - 7\frac{d}{dx}(x^3) && \text{Derivative of multiple} \\
&= 5(2x) - 7(3x^2) = 10x - 21x^2.
\end{aligned}$$

Negative and Fractional Powers

We can calculate the derivatives of functions such as $g(x) = x^{-1}$ and $k(x) = x^{-2}$ using the definition of the derivative.

Example 3 Find $\frac{d}{dx}(x^{-2})$.

Solution Let $k(x) = x^{-2}$. Then, provided $x \neq 0$,

$$\begin{aligned}
\frac{d}{dx}\left(x^{-2}\right) = \frac{d}{dx}\left(\frac{1}{x^2}\right) &= \lim_{h\to 0}\frac{k(x+h)-k(x)}{h} = \lim_{h\to 0}\left(\frac{\frac{1}{(x+h)^2}-\frac{1}{x^2}}{h}\right)\\
&= \lim_{h\to 0}\frac{1}{h}\left[\frac{x^2-(x+h)^2}{(x+h)^2x^2}\right] \quad \text{Combining fractions over common denominator}\\
&= \lim_{h\to 0}\frac{1}{h}\left[\frac{x^2-(x^2+2xh+h^2)}{(x+h)^2x^2}\right] \quad \text{Multiplying out}\\
&= \lim_{h\to 0}\frac{-2xh-h^2}{h(x+h)^2x^2} \quad \text{Simplifying numerator}\\
&= \lim_{h\to 0}\frac{-2x-h}{(x+h)^2x^2} \quad \text{Dividing numerator and denominator by } h\\
&= \frac{-2x}{x^4} \quad \text{Letting } h\to 0\\
&= -\frac{2}{x^3} = -2x^{-3}.
\end{aligned}$$

So

$$\frac{d}{dx}(x^{-2}) = -2x^{-3}.$$

The graphs of $k(x) = x^{-2}$ and $k'(x) = -2x^{-3}$ are shown in Figure 4.6. Does the graph of k' have the features you expect?

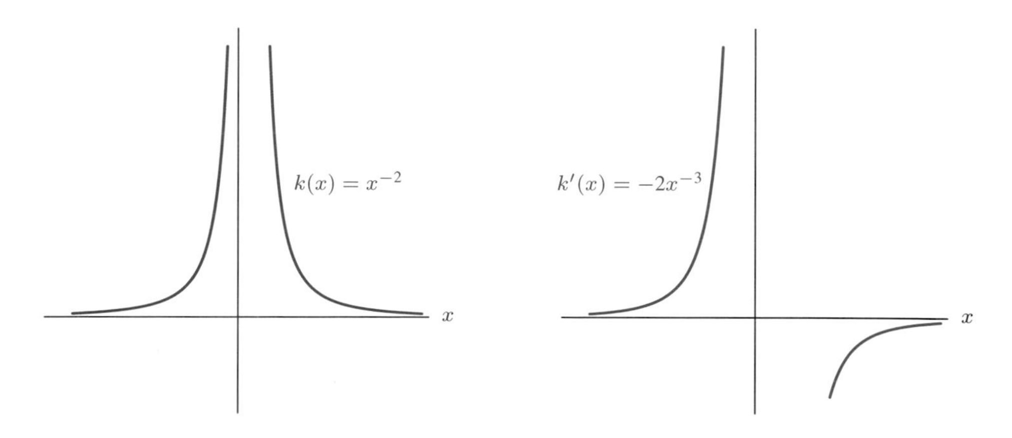

Figure 4.6: Graphs of $k(x) = x^{-2}$ and its derivative

This example suggests that the rule

$$\frac{d}{dx}(x^n) = nx^{n-1}$$

may hold for n a negative integer also, and indeed it does. It can be proved using a method similar to that for positive integral powers. In fact, this rule holds for any constant real number n, though we cannot yet easily justify it.

Example 4 Differentiate (a) $\frac{1}{x^3}$, (b) $x^{1/2}$, (c) $\frac{1}{\sqrt[3]{x}}$.

Solution (a) For $n=-3$: $\frac{d}{dx}\left(\frac{1}{x^3}\right) = \frac{d}{dx}(x^{-3}) = -3x^{-3-1} = -3x^{-4} = -\frac{3}{x^4}$, provided $x \neq 0$.

(b) For $n=1/2$: $\frac{d}{dx}\left(x^{1/2}\right) = \frac{1}{2}x^{(1/2)-1} = \frac{1}{2}x^{-1/2} = \frac{1}{2\sqrt{x}}$.

(c) For $n=-1/3$: $\frac{d}{dx}\left(\frac{1}{\sqrt[3]{x}}\right) = \frac{d}{dx}\left(x^{-1/3}\right) = -\frac{1}{3}x^{(-1/3)-1} = -\frac{1}{3}x^{-4/3} = -\frac{1}{3x^{4/3}}$.

Derivatives of Polynomials

We already know how to differentiate powers, constant multiples, and sums. For example,

$$\frac{d}{dx}(3x^5) = 3\frac{d}{dx}(x^5) = 3 \cdot 5x^4 = 15x^4$$

and

$$\frac{d}{dx}(x^5 + x^3) = \frac{d}{dx}(x^5) + \frac{d}{dx}(x^3) = 5x^4 + 3x^2.$$

Using these rules together, we can differentiate any polynomial.

Example 5 Find the derivatives of (a) $5x^2 + 3x + 2$, (b) $\sqrt{3}x^7 - \frac{x^5}{5} + \pi$.

Solution (a)

$$\begin{aligned}\frac{d}{dx}(5x^2 + 3x + 2) &= 5\frac{d}{dx}(x^2) + 3\frac{d}{dx}(x) + \frac{d}{dx}(2) \\ &= 5 \cdot 2x + 3 \cdot 1 + 0 \qquad \text{Since the derivative of a constant, } d(2)/dx\text{, is zero} \\ &= 10x + 3.\end{aligned}$$

(b)

$$\begin{aligned}\frac{d}{dx}\left(\sqrt{3}x^7 - \frac{x^5}{5} + \pi\right) &= \sqrt{3}\frac{d}{dx}(x^7) - \frac{1}{5}\frac{d}{dx}(x^5) + \frac{d(\pi)}{dx} \\ &= \sqrt{3} \cdot 7x^6 - \frac{1}{5} \cdot 5x^4 + 0 \qquad \text{Since } \pi \text{ is a constant, } d\pi/dx = 0 \\ &= 7\sqrt{3}x^6 - x^4.\end{aligned}$$

We can also use these rules to differentiate expressions which are not polynomials.

Example 6 Differentiate (a) $5\sqrt{x} - \dfrac{10}{x^2} + \dfrac{1}{2\sqrt{x}}$, (b) $0.1x^3 + 2x^{\sqrt{2}}$.

Solution (a)

$$\frac{d}{dx}\left(5\sqrt{x} - \frac{10}{x^2} + \frac{1}{2\sqrt{x}}\right) = \frac{d}{dx}\left(5x^{1/2} - 10x^{-2} + \frac{1}{2}x^{-1/2}\right)$$
$$= 5 \cdot \frac{1}{2}x^{-1/2} - 10(-2)x^{-3} + \frac{1}{2}\left(-\frac{1}{2}\right)x^{-3/2}$$
$$= \frac{5}{2\sqrt{x}} + \frac{20}{x^3} - \frac{1}{4x^{3/2}}.$$

(b)

$$\frac{d}{dx}(0.1x^3 + 2x^{\sqrt{2}}) = 0.1\frac{d}{dx}(x^3) + 2\frac{d}{dx}(x^{\sqrt{2}}) = 0.3x^2 + 2\sqrt{2}x^{\sqrt{2}-1}.$$

Example 7 Find and interpret the second derivatives of
(a) $f(x) = x^2$, (b) $g(x) = x^3$, (c) $k(x) = x^{1/2}$.

Solution (a) For $f(x) = x^2$, $f'(x) = 2x$, so $f''(x) = \dfrac{d}{dx}(2x) = 2$. Since f'' is always positive, f is concave up, as expected for a parabola opening upwards. (See Figure 4.7.)

(b) For $g(x) = x^3$, $g'(x) = 3x^2$, so $g''(x) = \dfrac{d}{dx}(3x^2) = 3\dfrac{d}{dx}(x^2) = 3 \cdot 2x = 6x$. This is positive for $x > 0$ and negative for $x < 0$, which means x^3 is concave up for $x > 0$ and concave down for $x < 0$. (See Figure 4.8.)

(c) For $k(x) = x^{1/2}$, $k'(x) = \frac{1}{2}x^{(1/2)-1} = \frac{1}{2}x^{-1/2}$, so

$$k''(x) = \frac{d}{dx}\left(\frac{1}{2}x^{-1/2}\right) = \frac{1}{2} \cdot (-\frac{1}{2})x^{-(1/2)-1} = -\frac{1}{4}x^{-3/2}.$$

Now k' and k'' only make sense on the domain of k, i.e. $x \geq 0$. In fact, k' and k'' are undefined when $x = 0$, since $x^{-1/2}$ and $x^{-3/2}$ are not defined for $x = 0$. We see that $k''(x)$ is negative for $x > 0$, so k is concave down. (See Figure 4.9.)

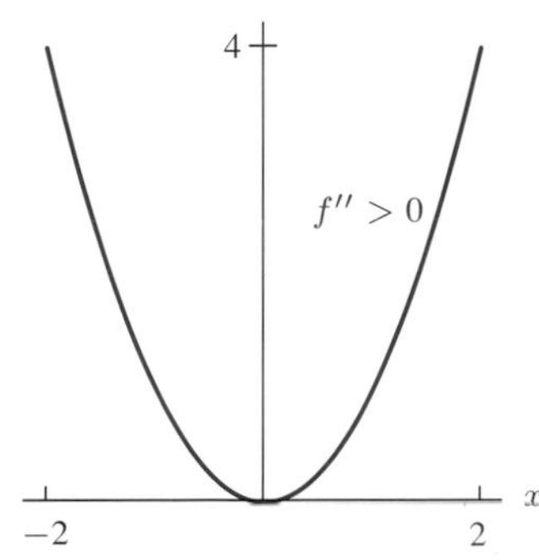

Figure 4.7: $f(x) = x^2$ and $f''(x) = 2$

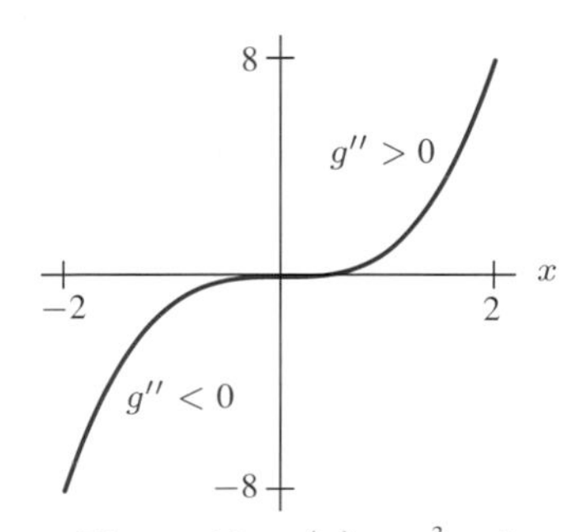

Figure 4.8: $g(x) = x^3$ and $g''(x) = 6x$

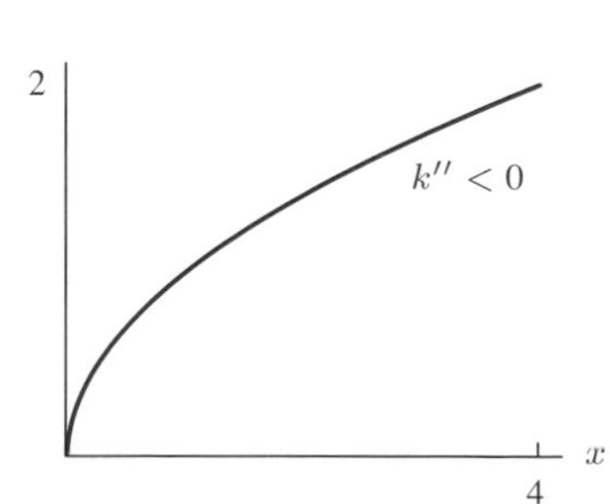

Figure 4.9: $k(x) = x^{1/2}$ and $k''(x) = -\frac{1}{4}x^{-3/2}$

Example 8 If the position of a body, in meters, is given as a function of time t, in seconds, by

$$s = -4.9t^2 + 5t + 6,$$

find the velocity and acceleration of the body at time t.

Solution The velocity, v, is the derivative of the position:

$$v = \frac{ds}{dt} = \frac{d}{dt}(-4.9t^2 + 5t + 6) = -9.8t + 5,$$

and the acceleration, a, is the derivative of the velocity:

$$a = \frac{dv}{dt} = \frac{d}{dt}(-9.8t + 5) = -9.8.$$

Example 9 Figure 4.10 shows the graph of a cubic polynomial. Both graphically and algebraically, describe the behavior of the derivative of this cubic.

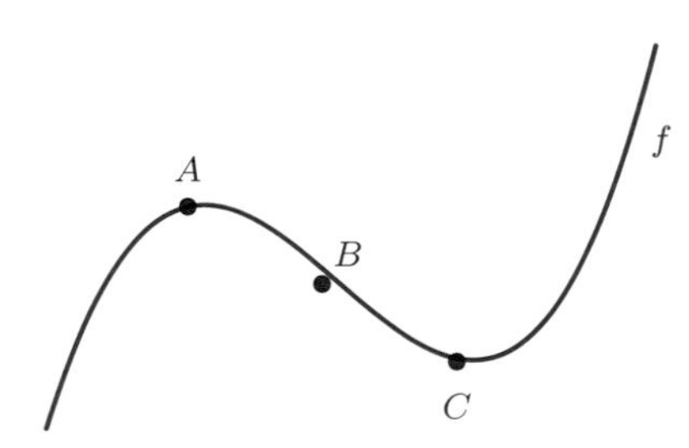

Figure 4.10: The cubic of Example 9

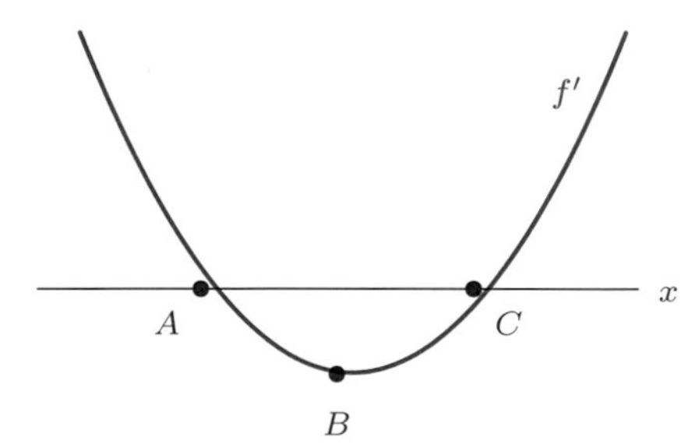

Figure 4.11: Derivative of the cubic of Example 9

Solution Graphical approach: Suppose you move along the curve from left to right. To the left of A, the slope is positive; it starts very positive and decreases until the curve reaches A, where the slope is 0. Between A and C the slope is negative. Between A and B the slope is decreasing (getting more negative); it is most negative at B. Between B and C the slope is negative but increasing; at C the slope is zero. From C to the right, the slope is positive and increasing. The graph of the derivative function is shown in Figure 4.11.

Algebraic approach: f is a cubic that goes to $+\infty$ as $x \to +\infty$ so

$$f(x) = ax^3 + bx^2 + cx + d$$

with $a > 0$. Hence,

$$f'(x) = 3ax^2 + 2bx + c,$$

whose graph is a parabola opening upward, as in Figure 4.11.

Problems for Section 4.2

For Problems 1–18, find the derivative of the function given.

1. $y = x^{12}$
2. $y = x^{-12}$
3. $y = x^{4/3}$
4. $y = x^{3/4}$
5. $y = x^{-3/4}$
6. $f(x) = \dfrac{1}{x^4}$
7. $f(x) = \sqrt[4]{x}$
8. $f(x) = x^e$
9. $y = 4x^{3/2} - 5x^{1/2}$
10. $y = 6x^3 + 4x^2 - 2x$
11. $y = -3x^4 - 4x^3 - 6x + 2$
12. $y = 3t^5 - 5\sqrt{t} + \dfrac{7}{t}$
13. $y = 3t^2 + \dfrac{12}{\sqrt{t}} - \dfrac{1}{t^2}$
14. $y = z^2 + \dfrac{1}{2z}$
15. $y = \dfrac{x^2 + 1}{x}$
16. $g(z) = \dfrac{z^7 + 5z^6 - z^3}{z^2}$
17. $f(t) = \dfrac{t^2 + t^3 - 1}{t^4}$
18. $y = \dfrac{\theta - 1}{\sqrt{\theta}}$
19. Find the functions in Problems 1–9 whose derivatives do not exist at $x = 0$.

For Problems 20–29, determine if the derivative rules from this section apply. If they do, find the derivative. If they don't apply, indicate why.

20. $y = \sqrt{x}$
21. $y = (x + 3)^{1/2}$
22. $y = 3x^2 + 4$
23. $y = \dfrac{1}{3z^2} + \dfrac{1}{4}$
24. $y = \dfrac{1}{3\sqrt{x}} + \dfrac{1}{4}$
25. $y = \dfrac{1}{3x^2 + 4}$
26. $y = 3^x$
27. $y = x^3$
28. $y = x^{\pi}$
29. $y = \pi^x$
30. If $f(t) = 2t^3 - 4t^2 + 3t - 1$, find $f'(t)$ and $f''(t)$.
31. (a) Find the *eighth* derivative of
$$f(x) = x^7 + 5x^5 - 4x^3 + 6x - 7.$$
Think ahead! (The nth derivative is the result of differentiating n times.)
(b) Find the seventh derivative of $f(x)$.
32. Find the equation of the line tangent to the graph of f at $(1, 1)$, where f is given by $f(x) = 2x^3 - 2x^2 + 1$.
33. The graph of the equation $y = x^3 - 9x^2 - 16x + 1$ has a slope equal to 5 at exactly two points. Find the coordinates of the points.
34. If $f(x) = 13 - 8x + \sqrt{2}x^2$ and $f'(r) = 4$, find r.
35. On what intervals is the function $f(x) = x^4 - 4x^3$ both decreasing and concave up?
36. If $f(x) = 4x^3 + 6x^2 - 23x + 7$, find the intervals on which $f'(x) \geq 1$.
37. Given $p(x) = x^n - x$, find the intervals over which p is a decreasing function when: (a) $n = 2$ (b) $n = \frac{1}{2}$ (c) $n = -1$
38. Using a graph to help you, find the equations of all lines through the origin tangent to the parabola
$$y = x^2 - 2x + 4.$$
Sketch your solutions.

39. Is there any value of n which makes $y = x^n$ a solution to the equation $13x\frac{dy}{dx} = y$? If so, what value?

40. Using the definition of derivative, justify the formula

$$\frac{d}{dx}(x^n) = nx^{n-1}.$$

(a) For $n = -1$; for $n = -3$.
(b) For any negative integer n.

41. The gravitational attraction, F, between the earth and a satellite of mass m at a distance r from the center of the earth is given by

$$F = \frac{GMm}{r^2},$$

where M is the mass of the earth, and G is a constant. Find the rate of change of force with respect to distance.

42. The period, T, of a pendulum is given in terms of its length, l, by

$$T = 2\pi\sqrt{\frac{l}{g}},$$

where g is the acceleration due to gravity (a constant).

(a) Find $\frac{dT}{dl}$.
(b) What is the sign of $\frac{dT}{dl}$? What does this tell you about the period of pendulums?

43. A ball is dropped from the top of the Empire State building to the ground below. The height, y, of the ball above the ground (in feet) is given as a function of time, t, (in seconds) by

$$y = 1250 - 16t^2.$$

(a) Find the velocity of the ball at time t. What is the sign of the velocity? Why is this to be expected?
(b) Show that the acceleration of the ball is a constant. What are the value and sign of this constant?
(c) When does the ball hit the ground, and how fast is it going at that time? Give your answer in feet per second and in miles per hour (1 ft/sec = 15/22 mph).

44. (a) Use the formula for the area of a circle of radius r, $A = \pi r^2$, to find $\frac{dA}{dr}$.
(b) The result from part (a) should look familiar. What does $\frac{dA}{dr}$ represent geometrically? Draw a picture.
(c) Use the difference quotient to explain the observation you made in part (b).

45. What is the formula for $V(r)$, the volume of a sphere of radius r? Find $\frac{dV}{dr}$. What is the geometrical meaning of $\frac{dV}{dr}$?

46. We know that the graph of a tangent line lies close to the graph of the function near the point of tangency. However, usually, the further we go from the point of tangency, the greater the distance between the graph of the function and the tangent line. Let's investigate this for $f(x) = 1/x$. Find the value of f at $x = 2$. Find the tangent line to the curve at $x = 1$, and use it to approximate the value of f at $x = 2$. Now find the tangent to the curve at $x = 100$ and use it to approximate the value of f at $x = 2$. Which tangent line lies closer to the curve at $x = 2$? Is there something wrong with our basic idea? Explain.

4.3 THE EXPONENTIAL FUNCTION

What would you expect the graph of the derivative of the exponential function $f(x) = a^x$ to look like? The graph of the exponential function is shown in Figure 4.12. The function increases slowly when $x < 0$ and more rapidly for $x > 0$, so the values of f' are small for $x < 0$ and larger for $x > 0$. Since the function is increasing for all real values of x, the graph of the derivative must lie above the x-axis. After some reflection, you may decide that the graph of f' must resemble the graph of f itself. We will see how this observation holds for $f(x) = 2^x$ and $g(x) = 3^x$.

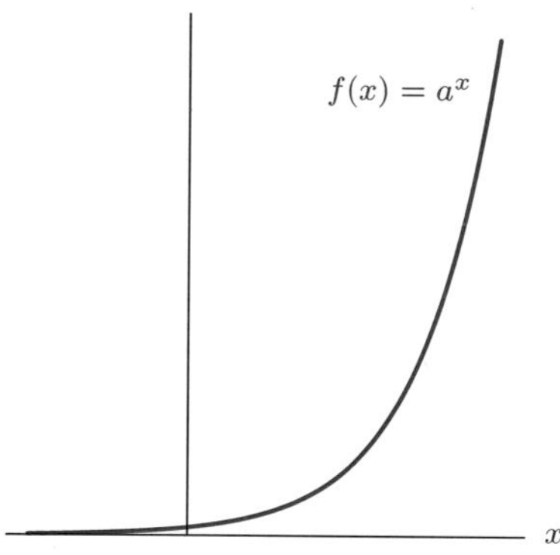

Figure 4.12: $f(x) = a^x$, with $a > 1$

The Derivatives of 2^x and 3^x

In Chapter 2, we saw that the derivative of $f(x) = 2^x$ at $x = 0$ is given by

$$f'(0) = \lim_{h \to 0} \frac{2^h - 2^0}{h} = \lim_{h \to 0} \frac{2^h - 1}{h} \approx 0.6931,$$

where we found the limit by evaluating $(2^h - 1)/h$ for small values of h. Similarly, we can calculate the derivative at $x = 1$ and $x = 2$:

$$f'(1) = \lim_{h \to 0} \frac{2^{1+h} - 2^1}{h} \approx 1.3863,$$

$$f'(2) = \lim_{h \to 0} \frac{2^{2+h} - 2^2}{h} \approx 2.7726.$$

Can you see the relationship between these values of the derivative? If you notice that $1.3863 \approx 2(0.6931)$ and $2.7726 \approx 4(0.6931)$, you see

$$f'(0) \approx 0.6931 = 0.6931 \cdot 2^0,$$
$$f'(1) \approx 1.3863 \approx 0.6931 \cdot 2^1,$$
$$f'(2) \approx 2.7726 \approx 0.6931 \cdot 2^2.$$

Can you see a pattern? It looks as though $f'(x) \approx 0.6931 \cdot 2^x$. Let's look at the derivative function to see why this happens:

$$f'(x) = \lim_{h \to 0} \left(\frac{2^{x+h} - 2^x}{h} \right) = \lim_{h \to 0} \left(\frac{2^x 2^h - 2^x}{h} \right) = \lim_{h \to 0} 2^x \left(\frac{2^h - 1}{h} \right)$$

$$= 2^x \lim_{h \to 0} \left(\frac{2^h - 1}{h} \right). \qquad \text{Since } x \text{ and } 2^x \text{ are fixed during this calculation}$$

We have already shown $\lim_{h \to 0} \frac{2^h - 1}{h} \approx 0.6931$, so we have shown that

$$\frac{d}{dx}(2^x) = f'(x) \approx (0.6931)2^x.$$

The graphs of $f(x) = 2^x$ and $f'(x) \approx (0.6931)2^x$ are shown in Figure 4.13. Notice that they do indeed resemble one another.

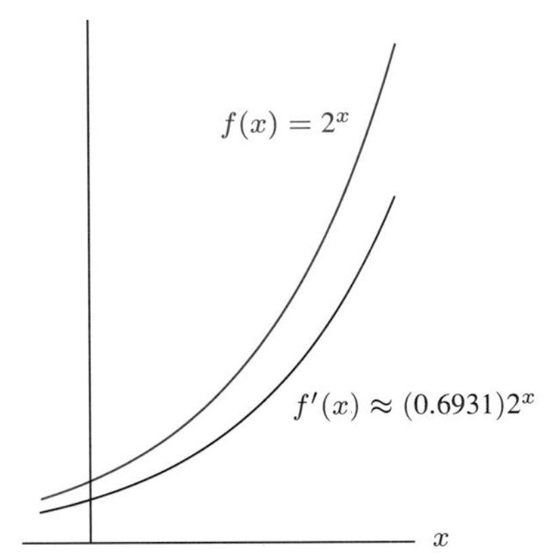

Figure 4.13: Graph of $f(x) = 2^x$ and its derivative

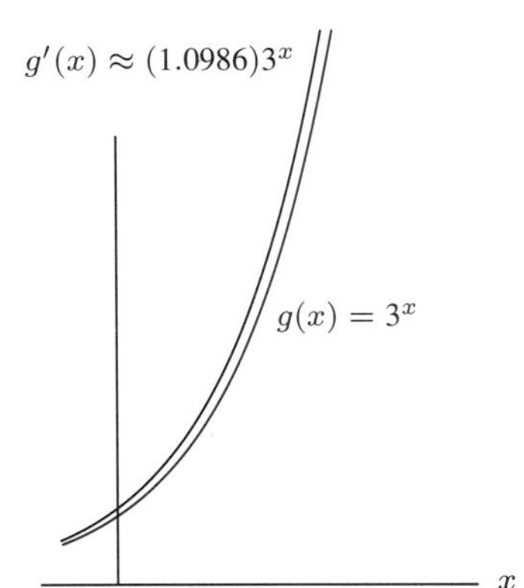

Figure 4.14: Graph of $g(x) = 3^x$ and its derivative

Example 1 Find the derivative of $g(x) = 3^x$ and plot g and g' on the same axes.

Solution As before

$$g'(x) = \lim_{h \to 0} \frac{3^{x+h} - 3^x}{h} = \lim_{h \to 0} \frac{3^x 3^h - 3^x}{h} = \lim_{h \to 0} 3^x \left(\frac{3^h - 1}{h} \right).$$

Using a calculator gives

$$\lim_{h \to 0} \left(\frac{3^h - 1}{h} \right) \approx 1.0986,$$

so

$$g'(x) \approx (1.0986)3^x.$$

The graphs are shown in Figure 4.14.

Notice that for both $f(x) = 2^x$ and $g(x) = 3^x$, *the derivative is proportional to the original function.* For $f(x) = 2^x$, $f'(x) \approx (0.6931)2^x$, so the constant of proportionality is less than 1, and the graph of the derivative is below the graph of the original function. For $g(x) = 3^x$, $g'(x) \approx (1.0986)3^x$ and the constant is greater than 1, so the graph of the derivative is above that of the original function.

Note on Round-Off Error and Limits

If you try to evaluate $(2^h - 1)/h$ on a calculator by taking smaller and smaller values of h, the values of $(2^h - 1)/h$ will at first get closer to 0.6931. However, they will eventually move away from 0.6931 because of the *round-off error* (i.e., errors introduced by the fact that the calculator can only hold a certain number of digits).

As you try smaller and smaller values of h, how do you know when to stop? Unfortunately, there is no fixed rule. A calculator can only *suggest* the value of a limit, but can never confirm that this value is correct. In this case, you can be pretty sure that the limit is about 0.6931 because the values of $(2^h - 1)/h$ hover around 0.6931 for a while. But to be absolutely sure this is correct, you would have to find the limit by theoretical means.

The Derivative of a^x and the Definition of e

The calculation of the derivative of $f(x) = a^x$, for $a > 0$, works just the same as before and leads to

$$f'(x) = \lim_{h\to 0} \frac{a^{x+h} - a^x}{h} = a^x \lim_{h\to 0} \frac{a^h - 1}{h}.$$

The quantity $\lim\limits_{h\to 0} (a^h - 1)/h$ doesn't depend on x, and so, for any particular a, is a constant. Therefore the derivative is again proportional to the original function, with constant of proportionality

$$\lim_{h\to 0} \frac{a^h - 1}{h}.$$

We can't use a calculator to evaluate this limit without knowing the value of a. However, when $a = 2$, we know that the limit (0.6931) is less than 1, and the derivative is smaller than the original function. When $a = 3$, the limit (1.0986) is more than 1, and the derivative is greater than the original function. Is there an in-between case, when derivative and function are exactly equal? In other words:

$$\text{Is there a value of } a \text{ that makes } \frac{d}{dx}(a^x) = a^x?$$

If so, we have found a function with the remarkable property that it is equal to its own derivative.

So let's look for such an a. This means we want to find a such that

$$\lim_{h\to 0} \frac{a^h - 1}{h} = 1$$

or, for small h,

$$\frac{a^h - 1}{h} \approx 1.$$

Solving for a suggests that we can calculate a as follows:

$$a^h - 1 \approx h \qquad \text{or} \qquad a^h \approx 1 + h \qquad \text{so} \qquad a \approx (1 + h)^{1/h}.$$

TABLE 4.2

h	$(1+h)^{1/h}$
0.001	2.7169239
0.0001	2.7181459
0.00001	2.7182682

Taking small values of h, as in Table 4.2, you can see $a \approx e = 2.718\ldots$, the number introduced in Chapter 1. Thus, we see that for small h:

$$e \approx (1+h)^{1/h} \approx 2.718.$$

In fact, it can be shown that

$$e = \lim_{h\to 0}(1+h)^{1/h} = 2.718\ldots \quad \text{and} \quad \lim_{h\to 0}\frac{e^h - 1}{h} = 1,$$

which means that e^x is its own derivative:

$$\boxed{\frac{d}{dx}(e^x) = e^x.}$$

It turns out that the constants involved in the derivatives of 2^x and 3^x are natural logarithms. In fact, since $0.6931 \approx \ln 2$ and $1.0986 \approx \ln 3$, this suggests that:

$$\frac{d}{dx}(2^x) = (\ln 2)2^x \qquad \text{and} \qquad \frac{d}{dx}(3^x) = (\ln 3)3^x.$$

In Section 4.7, we will show that, in general,

$$\boxed{\frac{d}{dx}(a^x) = (\ln a)a^x.}$$

Figure 4.15 shows the graph of the derivative of 2^x below the graph of the function, and the graph of the derivative of 3^x above the graph of the function. With $e \approx 2.718$, the function e^x and its derivative are identical.

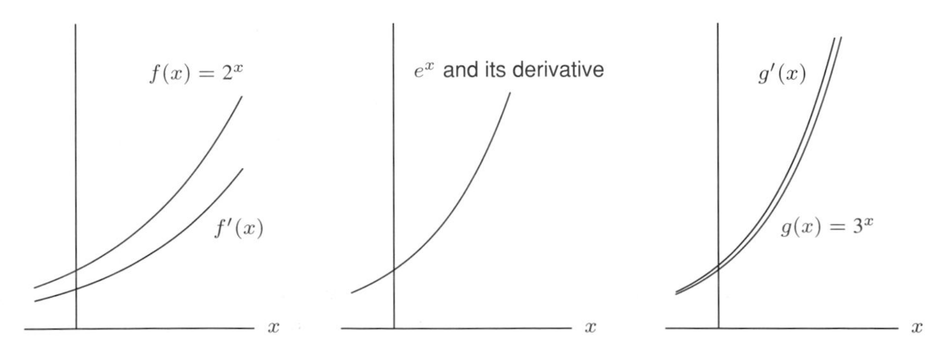

Figure 4.15: Graphs of the functions 2^x, e^x, and 3^x and their derivatives

Example 2 Differentiate $2 \cdot 3^x + 5e^x$.

Solution

$$\begin{aligned}\frac{d}{dx}(2 \cdot 3^x + 5e^x) &= \frac{d}{dx}(2 \cdot 3^x) + \frac{d}{dx}(5e^x) \\ &= 2\frac{d}{dx}(3^x) + 5\frac{d}{dx}(e^x) \\ &\approx 2(1.0986)3^x + 5e^x \\ &= (2.1972)3^x + 5e^x.\end{aligned}$$

Numerical Demonstration of $\frac{d}{dx}(2^x) = (0.693)\,2^x$

Example 3 Construct a table of values of 2^x for $x = 1.0, 1.1, \ldots, 1.5$. Use difference quotients with $h = 0.001$ to estimate its derivative at these points. Compare the values of 2^x and its derivative. What do you notice?

Solution While constructing the table, keep all the digits your calculator holds (usually 8 to 12) throughout the calculation. The results are displayed to three decimal places in Table 4.3.

TABLE 4.3 *Comparing values of* 2^x *and difference quotients (with* $h = 0.001$*)*

x	2^x	Difference quotient $= \frac{2^{x+h} - 2^x}{h}$	$\frac{\text{Difference quotient}}{2^x}$
1.0	2.000	1.387	0.693
1.1	2.144	1.486	0.693
1.2	2.297	1.593	0.693
1.3	2.462	1.707	0.693
1.4	2.639	1.830	0.693
1.5	2.828	1.961	0.693

The difference quotients in the third column approximate the derivative of 2^x. The fourth column contains the ratio (Difference quotient)/2^x. The remarkable thing about the exponential function is that *all* the entries in the last column are 0.693 (to three decimal places). This suggests that the derivative of 2^x is proportional to the values of 2^x with constant of proportionality 0.693.

We have seen that the derivative of a^x is proportional to a^x. Many quantities have rates of change which are proportional to themselves; for example, the pressure in the atmosphere as a function of altitude and the growth of a population both have this property. The fact that the constant of proportionality is 1 when $a = e$ makes e a particularly useful base for exponential functions.

Problems for Section 4.3

Find the derivatives of the functions in Problems 1–18.

1. $y = 5t^2 + 4e^t$
2. $f(x) = 2e^x + x^2$
3. $f(x) = 2^x + 2 \cdot 3^x$
4. $y = 4 \cdot 10^x - x^3$
5. $y = 3x - 2 \cdot 4^x$
6. $y = \dfrac{3^x}{3} + \dfrac{33}{\sqrt{x}}$
7. $f(x) = e^2 + x^e$
8. $f(x) = e^{1+x}$
9. $f(t) = e^{t+2}$
10. $y = e^{\theta - 1}$
11. $z = (\ln 4)e^x$
12. $z = (\ln 4)4^x$
13. $f(z) = (\ln 3)z^2 + (\ln 4)e^z$
14. $f(t) = (\ln 3)^t$
15. $f(x) = x^3 + 3^x$
16. $y = 5 \cdot 5^t + 6 \cdot 6^t$
17. $y = \pi^2 + \pi^x$
18. $f(x) = x^{\pi^2} + (\pi^2)^x$

Which of the functions in Problems 19–29 can be differentiated using the rules we have developed so far? Perform the differentiation if you can; otherwise, indicate why the rules discussed so far do not apply.

19. $y = x^2 + 2^x$
20. $y = \sqrt{x} - (\frac{1}{2})^x$
21. $y = x^2 \cdot 2^x$
22. $y = \dfrac{2^x}{x}$
23. $y = e^{x+5}$
24. $y = e^{5x}$
25. $f(s) = 5^s e^s$
26. $y = 4^{(x^2)}$
27. $f(z) = (\sqrt{4})^z$
28. $f(\theta) = 4^{\sqrt{\theta}}$
29. $f(x) = 4^{(3^x)}$
30. Construct a table of values of 3^x for $x = 0, 0.1, 0.2, \ldots, 0.5$, similar to Table 4.3 on page 203. Use difference quotients with $h = 0.001$ to estimate the derivative. Compute the ratio of the derivative and the function. What do you notice?
31. (a) Find the slope of the graph of $f(x) = 1 - e^x$ at the point where it crosses the x-axis.
    (b) Find the equation of the tangent line to the curve at this point.
    (c) Find the equation of the line which is perpendicular to the tangent line at this point. (This line is known as the *normal* line.)
32. Find the value of c in Figure 4.16, where the line l tangent to the graph of $y = 2^x$ at $(0, 1)$ intersects the x-axis.

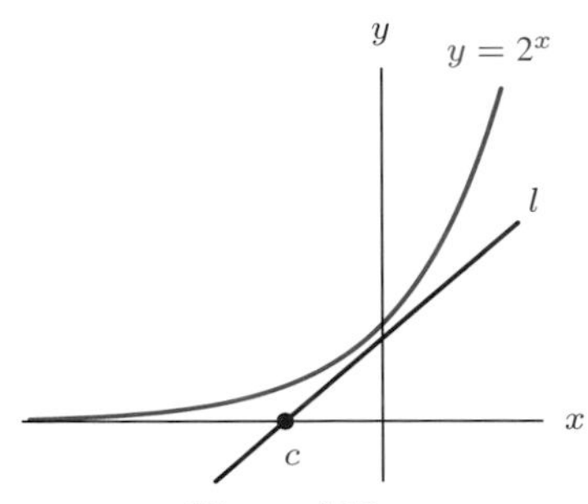

Figure 4.16

33. With an inflation rate of 5%, prices are described by

$$P = P_0(1.05)^t$$

where P_0 is the price in dollars when $t = 0$ and t is time in years. Suppose $P_0 = 1$. How fast (in cents/year) is the price of the good rising when $t = 10$?

34. Since January 1, 1960, the population of Slim Chance has been described by the formula $P = 35{,}000(0.98)^t$, where P is the population of the city t years after 1960. At what rate was the population of the city changing on January 1, 1983?

35. Certain pieces of antique furniture increased very rapidly in price in the 1970s and 1980s. For example, the value of a particular rocking chair is well approximated by

$$V = 75(1.35)^t,$$

where V is in dollars and t is the number of years since 1975. Find the rate, in dollars per year, that the price is increasing as a function of time.

36. The *Global 2000 Report* gave the world's population, P, as 4.1 billion in 1975 and growing at 2% annually.
    (a) Give a formula for P in terms of time, t, measured in years since 1975.
    (b) Find $\frac{dP}{dt}$, $\left.\frac{dP}{dt}\right|_{t=0}$, and $\left.\frac{dP}{dt}\right|_{t=15}$. What do each of these represent in practical terms?

37. In this section, we stated that, for $a > 0$

$$\frac{d}{dx}(a^x) = (\ln a)a^x.$$

Use this expression for the derivative to explain for what values of a the function a^x is increasing and for what values it is decreasing.

38. Find the quadratic function $g(x) = ax^2 + bx + c$ which best fits the function $f(x) = e^x$ at $x = 0$, in the sense that

$$g(0) = f(0), \quad \text{and} \quad g'(0) = f'(0), \quad \text{and} \quad g''(0) = f''(0).$$

Using a computer or calculator, sketch graphs of f and g on the same axes. What do you notice?

39. Using the tangent line to e^x at $x = 0$, show that

$$e^x \geq 1 + x$$

for all values of x. A sketch will be helpful.

40. Find all solutions of the equation

$$2^x = 2x.$$

How do you know that you found all solutions?

4.4 THE PRODUCT AND QUOTIENT RULES

You now know how to find derivatives of powers and exponentials, and of sums and constant multiples of functions. This section will show you how to find the derivatives of products and quotients.

Using Δ Notation

To express the difference quotients of general functions, some additional notation is helpful. We write Δf for a small change in the value of f,

$$\Delta f = f(x+h) - f(x).$$

In this notation, the derivative is the limit of the ratio $\Delta f/h$:

$$f'(x) = \lim_{h\to 0} \frac{\Delta f}{h}.$$

The Product Rule

Suppose we know the derivatives of $f(x)$ and $g(x)$ and want to calculate the derivative of the product, $f(x)g(x)$. The derivative of the product is calculated by taking the limit, namely,

$$\frac{d[f(x)g(x)]}{dx} = \lim_{h\to 0} \frac{f(x+h)g(x+h) - f(x)g(x)}{h}.$$

To picture the quantity $f(x+h)g(x+h) - f(x)g(x)$, imagine the rectangle with sides $f(x+h)$ and $g(x+h)$ in Figure 4.17, where $\Delta f = f(x+h) - f(x)$ and $\Delta g = g(x+h) - g(x)$.

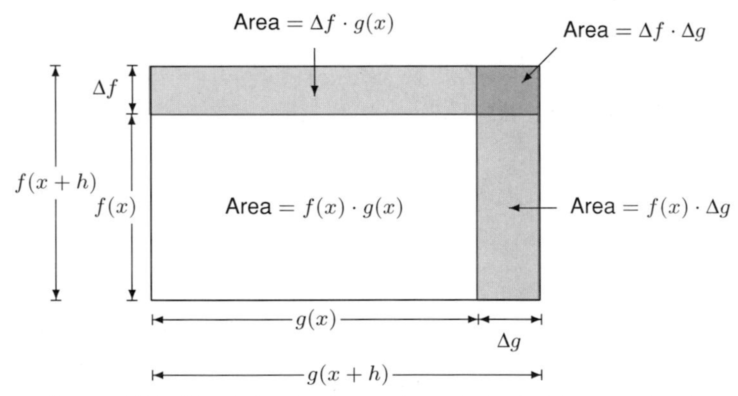

Figure 4.17: Illustration for the product rule (with Δf, Δg positive)

Then

$$\begin{aligned} f(x+h)g(x+h) - f(x)g(x) &= \text{(Area of whole rectangle)} - \text{(Unshaded area)} \\ &= \text{Area of the three shaded rectangles} \\ &= \Delta f \cdot g(x) + f(x) \cdot \Delta g + \Delta f \cdot \Delta g \end{aligned}$$

Now divide by h:

$$\frac{f(x+h)g(x+h) - f(x)g(x)}{h} = \frac{\Delta f}{h} \cdot g(x) + f(x) \cdot \frac{\Delta g}{h} + \frac{\Delta f \cdot \Delta g}{h}.$$

To evaluate the limit as $h \to 0$, let's examine the three terms on the right separately. Notice that

$$\lim_{h\to 0} \frac{\Delta f}{h} \cdot g(x) = f'(x)g(x) \quad \text{and} \quad \lim_{h\to 0} f(x) \cdot \frac{\Delta g}{h} = f(x)g'(x).$$

In the third term we multiply the top and bottom by h to get $\frac{\Delta f}{h} \cdot \frac{\Delta g}{h} \cdot h$. Then,

$$\lim_{h\to 0} \frac{\Delta f \cdot \Delta g}{h} = \lim_{h\to 0} \frac{\Delta f}{h} \cdot \frac{\Delta g}{h} \cdot h = \lim_{h\to 0} \frac{\Delta f}{h} \cdot \lim_{h\to 0} \frac{\Delta g}{h} \cdot \lim_{h\to 0} h = f'(x) \cdot g'(x) \cdot 0 = 0.$$

Therefore, we conclude that

$$\begin{aligned}
\lim_{h\to 0} \frac{f(x+h)g(x+h) - f(x)g(x)}{h} &= \lim_{h\to 0} \left(\frac{\Delta f}{h} \cdot g(x) + f(x) \cdot \frac{\Delta g}{h} + \frac{\Delta f \cdot \Delta g}{h} \right) \\
&= \lim_{h\to 0} \frac{\Delta f}{h} \cdot g(x) + \lim_{h\to 0} f(x) \cdot \frac{\Delta g}{h} + \lim_{h\to 0} \frac{\Delta f \cdot \Delta g}{h} \\
&= f'(x)g(x) + f(x)g'(x).
\end{aligned}$$

Thus we have the following rule:

The Product Rule

$$(fg)' = f'g + fg'.$$

In words:

The derivative of a product is the derivative of the first factor multiplied by the second, plus the first factor multiplied by the derivative of the second.

Another justification of the product rule is given in Problem 11 on page 237. In the alternative notation, if $u = f(x)$ and $v = g(x)$, we have the following:

The Product Rule

$$\frac{d(uv)}{dx} = \frac{du}{dx} \cdot v + u \cdot \frac{dv}{dx}$$

Example 1 Differentiate (a) x^2e^x, (b) $(3x^2+5x)e^x$, (c) $\frac{e^x}{x^2}$.

Solution (a)

$$\frac{d(x^2e^x)}{dx} = \left(\frac{d(x^2)}{dx}\right)e^x + x^2\frac{d(e^x)}{dx} = 2xe^x + x^2e^x = (2x+x^2)e^x.$$

(b)

$$\frac{d((3x^2+5x)e^x)}{dx} = \left(\frac{d(3x^2+5x)}{dx}\right)e^x + (3x^2+5x)\frac{d(e^x)}{dx}$$
$$= (6x+5)e^x + (3x^2+5x)e^x = (3x^2+11x+5)e^x.$$

(c) First we must write $\frac{e^x}{x^2}$ as the product $x^{-2}e^x$:

$$\frac{d}{dx}\left(\frac{e^x}{x^2}\right) = \frac{d(x^{-2}e^x)}{dx} = \left(\frac{d(x^{-2})}{dx}\right)e^x + x^{-2}\frac{d(e^x)}{dx}$$
$$= -2x^{-3}e^x + x^{-2}e^x = (-2x^{-3}+x^{-2})e^x.$$

The Quotient Rule

Suppose we want to differentiate a function of the form $Q(x) = \frac{f(x)}{g(x)}$. (Of course, we have to avoid points where $g(x) = 0$.) We'll find a formula for Q' in terms of f' and g'.

Since $f(x) = Q(x)g(x)$, and assuming that $Q(x)$ is differentiable,[1] we can use the product rule:

$$f'(x) = Q'(x)g(x) + Q(x)g'(x)$$
$$= Q'(x)g(x) + \frac{f(x)}{g(x)}g'(x).$$

Solving for $Q'(x)$:

$$Q'(x) = \frac{f'(x) - \frac{f(x)}{g(x)}g'(x)}{g(x)}.$$

Multiplying the top and bottom by $g(x)$ to simplify gives

$$\left[\frac{f(x)}{g(x)}\right]' = \frac{f'(x)g(x) - f(x)g'(x)}{(g(x))^2}$$

So we have the following rule:

The Quotient Rule

$$\left(\frac{f}{g}\right)' = \frac{f'g - fg'}{g^2}.$$

In words:

The derivative of a quotient is the derivative of the numerator times the denominator minus the numerator times the derivative of the denominator all over the denominator squared.

[1]The method in Example 5 on page 215 can be used to show that $Q(x)$ must be differentiable.

Alternatively, if $u = f(x)$ and $v = g(x)$, this can be written as follows:

The Quotient Rule

$$\frac{d}{dx}\left(\frac{u}{v}\right) = \frac{\dfrac{du}{dx} \cdot v - u \cdot \dfrac{dv}{dx}}{v^2}.$$

Example 2 Differentiate (a) $\dfrac{5x^2}{x^3+1}$, (b) $\dfrac{1}{1+e^x}$, (c) $\dfrac{e^x}{x^2}$.

Solution (a)

$$\begin{aligned}\frac{d}{dx}\left(\frac{5x^2}{x^3+1}\right) &= \frac{\left(\dfrac{d}{dx}(5x^2)\right)(x^3+1) - 5x^2\dfrac{d}{dx}(x^3+1)}{(x^3+1)^2}\\ &= \frac{10x(x^3+1) - 5x^2(3x^2)}{(x^3+1)^2}\\ &= \frac{-5x^4+10x}{(x^3+1)^2}.\end{aligned}$$

(b)

$$\begin{aligned}\frac{d}{dx}\left(\frac{1}{1+e^x}\right) &= \frac{\left(\dfrac{d}{dx}(1)\right)(1+e^x) - 1\dfrac{d}{dx}(1+e^x)}{(1+e^x)^2}\\ &= \frac{0(1+e^x) - 1(0+e^x)}{(1+e^x)^2}\\ &= \frac{-e^x}{(1+e^x)^2}.\end{aligned}$$

(c) This is the same as part (c) of Example 1, but this time we will do it by the quotient rule.

$$\begin{aligned}\frac{d}{dx}\left(\frac{e^x}{x^2}\right) &= \frac{\left(\dfrac{d(e^x)}{dx}\right)x^2 - e^x\left(\dfrac{d(x^2)}{dx}\right)}{(x^2)^2} = \frac{e^x x^2 - e^x(2x)}{x^4}\\ &= e^x\left(\frac{x^2-2x}{x^4}\right) = e^x\left(\frac{x-2}{x^3}\right).\end{aligned}$$

This is, in fact, the same answer as before, although it looks different. Can you show that it is the same?

Problems for Section 4.4

1. If $f(x) = x^2(x^3 + 5)$, find $f'(x)$ two ways: by using the product rule and by multiplying out. Do you get the same result? Should you?
2. If $f(x) = 2^x \cdot 3^x$, find $f'(x)$ two ways: by using the product rule and by multiplying out. Do you get the same result?

For Problems 3–18, find the derivative. In some cases, it may be to your advantage to simplify first.

3. $f(x) = xe^x$
4. $f(x) = \dfrac{x}{e^x}$
5. $y = x \cdot 2^x$
6. $y = \sqrt{x} \cdot 2^x$
7. $f(x) = (x^2 - \sqrt{x})3^x$
8. $w = (t^3+5t)(t^2-7t+2)$
9. $z = (s^2 - \sqrt{s})(s^2 + \sqrt{s})$
10. $y = (t^2 + 3)e^t$
11. $z = \dfrac{3t+1}{5t+2}$
12. $y = (t^3 - 7t^2 + 1)e^t$
13. $z = \dfrac{t^2 + 5t + 2}{t+3}$
14. $w = \dfrac{y^3 - 6y^2 + 7y}{y}$
15. $f(x) = \dfrac{x^2+3}{x}$
16. $y = \dfrac{\sqrt{t}}{t^2+1}$
17. $f(x) = \dfrac{1+x}{2+3x+4x^2}$
18. $f(z) = \dfrac{3z^2}{5z^2+7z}$

19. If $f(x) = (3x + 8)(2x - 5)$, find $f'(x)$ and $f''(x)$.
20. Differentiate $f(t) = e^{-t}$ by writing it as $f(t) = \dfrac{1}{e^t}$.
21. Differentiate $f(x) = e^{2x}$ by writing it as $f(x) = e^x \cdot e^x$.
22. Differentiate $f(x) = e^{3x}$ by writing it as $f(x) = e^x \cdot e^{2x}$ and using the result of Problem 21.
23. Based on your answers to Problems 21 and 22, guess the derivative of e^{4x}.
24. (a) Differentiate $y = \dfrac{e^x}{x}$, $y = \dfrac{e^x}{x^2}$, and $y = \dfrac{e^x}{x^3}$.

    (b) What do you anticipate the derivative of $y = \dfrac{e^x}{x^n}$ will be? Confirm your guess.
25. Find the equation of the tangent line at $x = 1$ to $y = f(x)$ where $f(x) = \dfrac{3x^2}{5x^2+7x}$.
26. Given: $\left\{ \begin{array}{ll} H(3) = 1 & F(3) = 5 \\ H'(3) = 3 & F'(3) = 4 \end{array} \right\}$ find: $\left\{ \begin{array}{ll} \text{(a) } G'(3) & \text{if } G(z) = F(z) \cdot H(z) \\ \text{(b) } G'(3) & \text{if } G(w) = F(w)/H(w) \end{array} \right.$
27. (a) Suppose

$$f(x) = \frac{x}{x^2-1}.$$

    For what values of x does this formula define f? Find a formula for $f'(x)$. Explain why $f(1.01)$ is so large, and why $f'(1.01)$ is even larger in magnitude.

    (b) Now suppose

$$g(x) = \frac{x^2+3x-4}{x^2-1}.$$

For what values of x does this formula define g? Find a formula for $g'(x)$. Explain why $g(1.01)$ and $g'(1.01)$ don't seem to be very large.

28. Using the product rule and the fact that $\dfrac{d(x)}{dx} = 1$, show that $\dfrac{d(x^2)}{dx} = 2x$ and $\dfrac{d(x^3)}{dx} = 3x^2$.

29. Find $f'(x)$ for the following functions by using the product rule rather than by multiplying out.
   (a) $f(x) = (x-1)(x-2)$.
   (b) $f(x) = (x-1)(x-2)(x-3)$.
   (c) $f(x) = (x-1)(x-2)(x-3)(x-4)$.

30. Use the answer from Problem 29 to find $f'(x)$ for the following function.
$$f(x) = (x-r_1)(x-r_2)(x-r_3)\cdots(x-r_n)$$
where $r_1, r_2, \ldots, r_n$ are any real numbers.

31. The function
$$f(x) = e^x$$
has the properties
$$f'(x) = f(x) \text{ and } f(0) = 1.$$
Show that $f(x)$ is the only function with these properties.
[Hint: Assume $g'(x) = g(x)$, and $g(0) = 1$, for some function $g(x)$. Define $h(x) = g(x)/e^x$, and compute $h'(x)$. Use the fact that if $k'(x) = 0$ for all x, then $k(x)$ is a constant function.]

32. When an electric current passes through two resistors with resistance r_1 and r_2, connected together at both ends, the combined resistance, R, can be calculated from the equation
$$\frac{1}{R} = \frac{1}{r_1} + \frac{1}{r_2}.$$
Find the rate at which the combined resistance changes with respect to changes in r_1. Assume that r_2 is constant.

33. The quantity, q, of a certain skateboard sold depends on the selling price, p, so we write $q = f(p)$. You are given that $f(140) = 15{,}000$ and $f'(140) = -100$.
   (a) What does $f(140) = 15{,}000$ and $f'(140) = -100$ tell you about the sale of the skateboards?
   (b) The total revenue, R, earned by the sale of skateboards is given by $R = pq$. Find $\left.\dfrac{dR}{dp}\right|_{p=140}$.
   (c) What is the sign of $\left.\dfrac{dR}{dp}\right|_{p=140}$? If the skateboards are currently selling for \$140, how should the price be changed to increase revenues?

34. Let $f(v)$ be the gas consumption (in liters/km) of a car going at velocity v (in km/hr). In other words, $f(v)$ tells you how many liters of gas the car uses to go one kilometer, if it is going at velocity v. You are told that
$$f(80) = 0.05 \text{ and } f'(80) = 0.0005.$$
   (a) Let $g(v)$ be the distance the same car goes on one liter of gas at velocity v. What is the relationship between $f(v)$ and $g(v)$? Find $g(80)$ and $g'(80)$.

(b) Let $h(v)$ be the gas consumption in liters per hour. In other words, $h(v)$ tells you how many liters of gas the car uses in one hour if it is going at velocity v. What is the relationship between $h(v)$ and $f(v)$? Find $h(80)$ and $h'(80)$.
(c) How would you explain the practical meaning of the values of these functions and their derivatives to a driver who knows no calculus?

35. A museum has decided to sell one of its paintings and to invest the proceeds. The price the painting will fetch changes with time, and is denoted by $P(t)$, where t is the number of years since 1990. If the picture is sold between 1990 and 2010 (so $0 \leq t \leq 20$), and the money from the sale is invested in a bank account earning 5% annual interest compounded once a year, the balance, $B(t)$, in the account in the year 2010 is given by

$$B(t) = P(t)(1.05)^{20-t}.$$

(a) Explain why $B(t)$ is given by this formula.
(b) Show that the formula for $B(t)$ is equivalent to

$$B(t) = (1.05)^{20}\frac{P(t)}{(1.05)^t}.$$

(c) Find $B'(10)$, given that $P(10) = 150{,}000$ and $P'(10) = 5000$.

4.5 THE CHAIN RULE

Composite functions such as $\sin(3t)$ or e^{-x^2} occur frequently in practice. In this section we will see how to differentiate such functions.

The Derivative of a Composition of Functions

Suppose $f(g(x))$ is a composite function, with f being the outside function and g being the inside. Let us write

$$z = g(x) \quad \text{and} \quad y = f(z), \quad \text{so} \quad y = f(g(x)).$$

Then a small change in x, called Δx, generates a small change in z, called Δz. In turn, Δz generates a small change in y called Δy. Provided Δx and Δz are not zero, we can say:

$$\frac{\Delta y}{\Delta x} = \frac{\Delta y}{\Delta z} \cdot \frac{\Delta z}{\Delta x}.$$

Since $\dfrac{dy}{dx} = \lim_{\Delta x \to 0} \dfrac{\Delta y}{\Delta x}$, this suggests that in the limit as Δx, Δy, and Δz get smaller and smaller, we have:

The Chain Rule

$$\frac{dy}{dx} = \frac{dy}{dz} \cdot \frac{dz}{dx}.$$

Since $\dfrac{dy}{dz} = f'(z)$ and $\dfrac{dz}{dx} = g'(x)$, this suggests that

$$\frac{d}{dx}f(g(x)) = f'(z) \cdot g'(x)$$

or, substituting $z = g(x)$, we have the following version:

The Chain Rule

$$\frac{d}{dx}f(g(x)) = f'(g(x)) \cdot g'(x).$$

In words:

The derivative of a composite function is the product of the derivatives of the outside and inside functions. The derivative of the outside function must be evaluated at the inside function.

A justification of the chain rule is given in Problem 12 on page 237.

Example 1 Suppose the length, L cm, of a steel bar depends on the air temperature, H°C, which itself depends on time t, measured in hours. If the length increases by 2 cm for every degree increase in temperature and the temperature is increasing at 3°C per hour, how fast is the length of the bar increasing?

Solution We are told that

$$\begin{array}{l}\text{Rate length increasing} \\ \text{with respect to temperature}\end{array} = \frac{dL}{dH} = 2\text{ cm}/°\text{C}$$

$$\begin{array}{l}\text{Rate temperature increasing} \\ \text{with respect to time}\end{array} = \frac{dH}{dt} = 3°\text{C}/\text{hr}.$$

We want to calculate the rate at which the length is increasing with respect to time, or dL/dt. We think of L as a function of H and H as a function of t. By the chain rule we know that

$$\frac{dL}{dt} = \frac{dL}{dH} \cdot \frac{dH}{dt} = \left(2\frac{\text{cm}}{°\text{C}}\right) \cdot \left(3\frac{°\text{C}}{\text{hr}}\right) = 6\text{ cm/hr}.$$

Thus, the length is increasing at 6 cm/hr.

Example 1 above shows us how to interpret the chain rule in practical terms. The next examples show how it is used to compute derivatives.

Example 2 Find the derivative of the following functions: (a) $(4x^2+1)^7$, (b) e^{3x}.

Solution (a) Here $z = g(x) = 4x^2 + 1$ is the inside function; $f(z) = z^7$ is the outside function. Since $g'(x) = 8x$ and $f'(z) = 7z^6$, we have

$$\frac{d}{dx}\left[(4x^2+1)^7\right] = 7z^6 \cdot 8x = 7(4x^2+1)^6 \cdot 8x = 56x(4x^2+1)^6.$$

(b) Let $z = g(x) = 3x$ and $f(z) = e^z$. Then $g'(x) = 3$ and $f'(z) = e^z$, so

$$\frac{d}{dx}\left(e^{3x}\right) = e^z \cdot 3 = 3e^{3x}.$$

Example 3 Differentiate (a) $(x^2+1)^{100}$, (b) $\sqrt{3x^2+5x-2}$, (c) $\frac{1}{x^2+x^4}$, (d) $\sqrt{e^x+1}$, (e) e^{x^2}.

Solution (a) Here $z = g(x) = x^2 + 1$ is the inside function; $f(z) = z^{100}$ is the outside function. Now $g'(x) = 2x$ and $f'(z) = 100z^{99}$, so

$$\frac{d[(x^2+1)^{100}]}{dx} = 100z^{99} \cdot 2x = 100(x^2+1)^{99} \cdot 2x = 200x(x^2+1)^{99}.$$

(b) Here $z = g(x) = 3x^2 + 5x - 2$ and $f(z) = \sqrt{z}$, so $g'(x) = 6x + 5$ and $f'(z) = \frac{1}{2\sqrt{z}}$. Hence

$$\frac{d(\sqrt{3x^2+5x-2})}{dx} = \frac{1}{2\sqrt{z}} \cdot (6x+5) = \frac{1}{2\sqrt{3x^2+5x-2}} \cdot (6x+5).$$

(c) Let $z = g(x) = x^2 + x^4$ and $f(z) = 1/z$, so $g'(x) = 2x + 4x^3$ and $f'(z) = -z^{-2} = -\frac{1}{z^2}$. Then

$$\frac{d}{dx}\left(\frac{1}{x^2+x^4}\right) = -\frac{1}{z^2}(2x+4x^3) = -\frac{2x+4x^3}{(x^2+x^4)^2}.$$

We could have done this problem using the quotient rule. Try it and see that you get the same answer!

(d) Let $z = g(x) = e^x + 1$ and $f(z) = \sqrt{z}$. Therefore $g'(x) = e^x$ and $f'(z) = \frac{1}{2\sqrt{z}}$. We get

$$\frac{d(\sqrt{e^x+1})}{dx} = \frac{1}{2\sqrt{z}}e^x = \frac{e^x}{2\sqrt{e^x+1}}.$$

(e) To evaluate e^{x^2} you must first evaluate x^2 and then take e to that power. So $z = g(x) = x^2$ and $f(z) = e^z$. Therefore, $g'(x) = 2x$, and $f'(z) = e^z$, giving

$$\frac{d(e^{x^2})}{dx} = e^z \cdot 2x = e^{x^2} \cdot 2x = 2xe^{x^2}.$$

Example 4 Find the derivative of e^{2x} by the chain rule and by the product rule.

Solution Chain rule: Let the inside function be $z = g(x) = 2x$ and the outside function be $f(z) = e^z$, so

$$\frac{d(e^{2x})}{dx} = f'(g(x)) \cdot g'(x) = e^{2x} \cdot 2 = 2e^{2x}.$$

Product rule: Write $e^{2x} = e^x \cdot e^x$ so

$$\frac{d(e^{2x})}{dx} = \frac{d(e^x e^x)}{dx} = \left(\frac{d(e^x)}{dx}\right)e^x + e^x\left(\frac{d(e^x)}{dx}\right) = e^x \cdot e^x + e^x \cdot e^x = 2e^{2x}.$$

Using the Product and Chain Rules to Differentiate a Quotient

If you prefer, you can differentiate a quotient by the product and chain rules, instead of by the quotient rule. The resulting formulas may look different, but they will be equivalent.

Example 5 Find $k'(x)$ if

$$k(x) = \frac{x}{(x^2+1)^2}.$$

Solution One way is to use the quotient rule:

$$\begin{aligned} k'(x) &= \frac{1\cdot(x^2+1)^2 - x\cdot 2(x^2+1)\cdot 2x}{\left[(x^2+1)^2\right]^2} \\ &= \frac{(x^2+1)[x^2+1-4x^2]}{(x^2+1)^4} && \text{Factoring } (x^2+1) \text{ out from top} \\ &= \frac{1-3x^2}{(x^2+1)^3}. && \text{Canceling } (x^2+1) \end{aligned}$$

Alternatively, you can write the original function as a product,

$$k(x) = x\frac{1}{(x^2+1)^2} = x\cdot(x^2+1)^{-2},$$

and use the product rule:

$$k'(x) = 1\cdot(x^2+1)^{-2} + x\cdot\frac{d}{dx}\left[(x^2+1)^{-2}\right].$$

Now use the chain rule to differentiate $(x^2+1)^{-2}$. Let $z = x^2+1$, and $f(z) = z^{-2}$, so

$$\frac{d}{dx}\left[(x^2+1)^{-2}\right] = -2z^{-3}\cdot 2x = -2(x^2+1)^{-3}\cdot 2x = \frac{-4x}{(x^2+1)^3}.$$

Therefore,

$$k'(x) = \frac{1}{(x^2+1)^2} + x\cdot\frac{-4x}{(x^2+1)^3} = \frac{1}{(x^2+1)^2} - \frac{4x^2}{(x^2+1)^3}.$$

Notice that if the product and chain rules are used, no canceling is necessary. If you put these two fractions over a common denominator, you will get the same answer as from the quotient rule.

Problems for Section 4.5

Find the derivative of the functions in Problems 1–24.

1. $f(x) = (x+1)^{99}$
2. $f(x) = \sqrt{1-x^2}$
3. $w = (t^2+1)^{100}$
4. $w = (t^3+1)^{100}$
5. $w = (\sqrt{t}+1)^{100}$
6. $f(t) = e^{3t}$

7. $y = e^{3w/2}$

8. $y = e^{-4t}$

9. $y = \sqrt{s^3 + 1}$

10. $w = e^{\sqrt{s}}$

11. $y = te^{-t^2}$

12. $f(z) = \sqrt{z}e^{-z}$

13. $f(z) = \dfrac{\sqrt{z}}{e^z}$

14. $z = 2^{5t-3}$

15. $f(t) = te^{5-2t}$

16. $f(z) = \dfrac{1}{(e^z + 1)^2}$

17. $f(\theta) = \dfrac{1}{1 + e^{-\theta}}$

18. $f(x) = 6e^{5x} + e^{-x^2}$

19. $f(w) = (5w^2 + 3)e^{w^2}$

20. $w = (t^2 + 3t)(1 - e^{-2t})$

21. $f(y) = \sqrt{10^{(5-y)}}$

22. $f(x) = e^{-(x-1)^2}$

23. $f(y) = e^{e^{(y^2)}}$

24. $f(t) = 2 \cdot e^{-2e^{2t}}$

25. Find the equation of the tangent line at $x = 1$ to $y = f(x)$ where $f(x)$ is the function in Problem 18.

26. For what values of x is the graph of $y = e^{-x^2}$ concave down?

27. Given: $\left\{\begin{array}{ll} F(2) = 1 & G(4) = 2 \\ F(4) = 3 & G(3) = 4 \\ F'(2) = 5 & G'(4) = 6 \\ F'(4) = 7 & G'(3) = 8 \end{array}\right\}$ find: $\left\{\begin{array}{lll} \text{(a)} & H(4) & \text{if } H(x) = F(G(x)) \\ \text{(b)} & H'(4) & \text{if } H(x) = F(G(x)) \\ \text{(c)} & H(4) & \text{if } H(x) = G(F(x)) \\ \text{(d)} & H'(4) & \text{if } H(x) = G(F(x)) \\ \text{(e)} & H'(4) & \text{if } H(x) = F(x)/G(x) \end{array}\right.$

28. Suppose $f(x) = (2x + 1)^{10}(3x - 1)^7$. Find a formula for $f'(x)$. Then decide on a reasonable way to simplify your result, and find a formula for $f''(x)$.

29. Is $x = \sqrt[3]{2t + 5}$ a solution to the equation $3x^2 dx/dt = 2$? Why or why not?

30. One gram of radioactive carbon-14 decays according to the formula

$$Q = e^{-0.000121t},$$

where Q is the number of grams of carbon-14 remaining after t years.

(a) Find the rate at which carbon-14 is decaying (in grams/year).

(b) Sketch the rate you found in part (a) against time.

31. The temperature, H, in degrees Fahrenheit (°F), of a can of soda that is put into a refrigerator to cool is given as a function of time, t, in hours, by

$$H = 40 + 30e^{-2t}.$$

(a) Find the rate at which the temperature of the soda is changing (in °F/hour).

(b) What is the sign of $\dfrac{dH}{dt}$? Why?

(c) When, for $t \geq 0$, is the magnitude of $\dfrac{dH}{dt}$ largest? In terms of the can of soda, why is this?

32. If you invest P dollars in a bank account at an annual interest rate of $r\%$, after t years you will have B dollars, where

$$B = P\left(1 + \frac{r}{100}\right)^t.$$

(a) Find $\dfrac{dB}{dt}$, assuming P and r are constant. In terms of money, what does $\dfrac{dB}{dt}$ represent?

(b) Find $\dfrac{dB}{dr}$, assuming P and t are constant. In terms of money, what does $\dfrac{dB}{dr}$ represent?

33. The theory of relativity predicts that an object whose mass is m_0 when it is at rest will appear heavier when moving at speeds near the speed of light. When the object is moving at speed v, its mass m is given by

$$m = \frac{m_0}{\sqrt{1-(v^2/c^2)}},$$

where c is the speed of light.

(a) Find $\dfrac{dm}{dv}$.

(b) In terms of physics, what does $\dfrac{dm}{dv}$ tell you?

34. Under certain conditions, when air expands, the pressure p and the volume v satisfy the relationship $pv^{1.4}$ = constant. Suppose that at some instant the pressure is 50 lbs/in^2 and the volume is 32 in^3 and decreasing at the rate of 4 in^3/sec. What is the rate of change of pressure at that instant?

35. The charge, Q, on a capacitor which starts discharging at time $t = 0$ is given by

$$Q = \begin{cases} C & \text{for } t \le 0 \\ Ce^{-t/RC} & \text{for } t > 0, \end{cases}$$

where R and C are positive constants depending on the circuit. The current, I, flowing in the circuit is given by $I = dQ/dt$.

(a) Find the current I for $t < 0$ and for $t > 0$.

(b) Is it possible to define I at $t = 0$?

(c) Is the function Q differentiable at $t = 0$?

4.6 THE TRIGONOMETRIC FUNCTIONS

Derivatives of the Sine and Cosine

Since the sine and cosine functions are periodic, their derivatives must be periodic also. (Why?) Let's look at the graph of $f(x) = \sin x$ in Figure 4.18 and estimate the derivative function graphically.

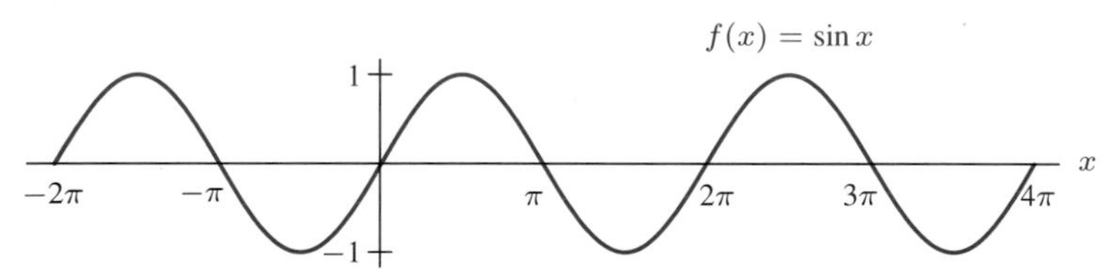

Figure 4.18: The sine function

First you might ask yourself where the derivative is zero. (At $x = \pm\pi/2, \pm 3\pi/2, \pm 5\pi/2$, etc.) Then ask yourself where the derivative is positive and where it is negative. (Positive for $-\pi/2 < x < \pi/2$; negative for $\pi/2 < x < 3\pi/2$, etc.) Since the largest positive slopes are at $x = 0, 2\pi$, and so on, and the largest negative slopes are at $x = \pi, 3\pi$, and so on, you get something like the graph in Figure 4.19.

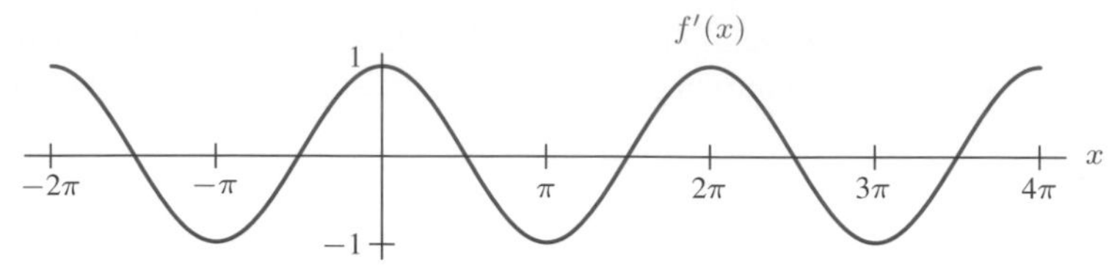

Figure 4.19: Derivative of $f(x) = \sin x$

The graph of the derivative looks suspiciously like the graph of the cosine function and this might lead you to conjecture, quite correctly, that the derivative of the sine is the cosine.

Of course, we cannot be sure, just from the graphs, that the derivative of the sine really is the cosine. Even if the graph of the derivative of the sine looks exactly like the cosine, we can't be sure it isn't the graph of something else. However, we'll assume for now that the derivative of the sine *is* the cosine and confirm the result at the end of the section.

One thing we can do now is to check that the derivative function in Figure 4.19 has amplitude 1 (as it ought to if it is the cosine). That means we have to show that the derivative of $f(x) = \sin x$ is 1 when $x = 0$. The next example shows this is true when x is in radians.

Example 1 Using a calculator, estimate the derivative of $f(x) = \sin x$ at $x = 0$. Make sure your calculator is set in radians.

Solution Since $f(x) = \sin x$,

$$f'(0) = \lim_{h\to 0} \frac{\sin(0+h) - \sin 0}{h} = \lim_{h\to 0} \frac{\sin h}{h}.$$

Table 4.4 contains values of $(\sin h)/h$ which suggest that this limit is 1, so

$$f'(0) = \lim_{h\to 0} \frac{\sin h}{h} = 1.$$

TABLE 4.4

h (radians)	$\dfrac{\sin h}{h}$
±0.1	0.99833
±0.01	0.99998
±0.001	1.00000
±0.0001	1.00000

Warning: It is important to notice that in the previous example h was in *radians*; any conclusions we have drawn about the derivative of $\sin x$ are valid *only* when x is in radians.

Example 2 Starting with the graph of the cosine function, sketch a graph of its derivative.

Solution The graph of $g(x) = \cos x$ is in Figure 4.20(a). Its derivative is 0 at $x = 0, \pm\pi, \pm 2\pi$, and so on; it is positive for $-\pi < x < 0$, $\pi < x < 2\pi$, and so on, and it is negative for $0 < x < \pi$, $2\pi < x < 3\pi$, and so on. The derivative is in Figure 4.20(b).

(a)

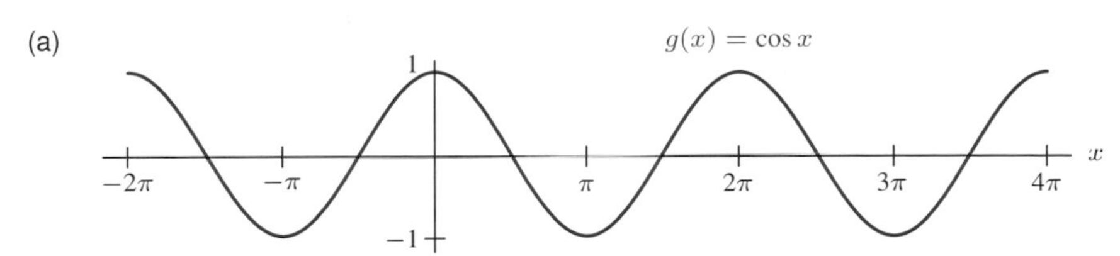

(b)

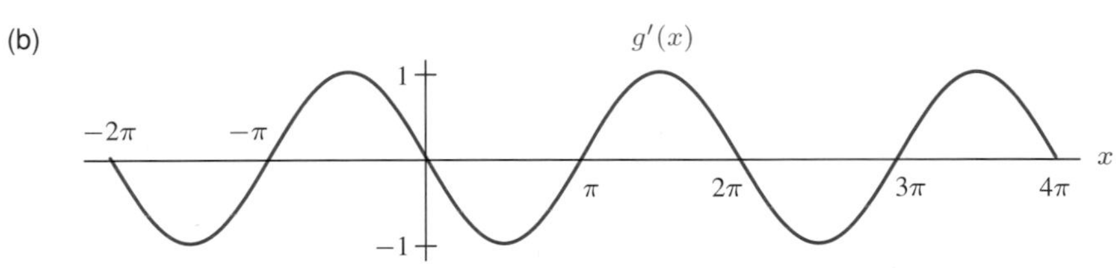

Figure 4.20: $g(x) = \cos x$ and its derivative, $g'(x)$

As we did with the sine, we'll use the graphs to make a conjecture. The derivative of the cosine in Figure 4.20(b) looks exactly like the graph of sine, except reflected about the x-axis. But how can we be sure that the derivative is $-\sin x$?

Example 3 Use the relation $\frac{d}{dx}(\sin x) = \cos x$ to show that $\frac{d}{dx}(\cos x) = -\sin x$.

Solution Since the cosine function is the sine function shifted to the left by $\pi/2$, we would expect the derivative of the cosine to be the derivative of the sine, shifted to the left by $\pi/2$. Since

$$\cos x = \sin\left(x + \frac{\pi}{2}\right),$$

using the chain rule gives

$$\frac{d}{dx}(\cos x) = \frac{d}{dx}\left(\sin\left(x + \frac{\pi}{2}\right)\right) = \cos\left(x + \frac{\pi}{2}\right).$$

But $\cos(x + \pi/2)$ is the cosine shifted to the left by $\pi/2$, which gives a sine curve reflected about the x-axis. Thus,

$$\frac{d}{dx}(\cos x) = \cos\left(x + \frac{\pi}{2}\right) = -\sin x.$$

For x in radians,

$$\frac{d}{dx}(\sin x) = \cos x \quad \text{and} \quad \frac{d}{dx}(\cos x) = -\sin x.$$

Now let's look at these derivative formulas numerically.

Example 4 Construct a table of values for $\sin x$ for $x = 0, 0.1, 0.2, \ldots, 0.6$ (in radians). Use difference quotients with $h = 0.001$ to estimate the derivative at these points and compare them with the values of $\cos x$.

Solution The point of this example is to notice that the entries in the second column of Table 4.5 below agree almost exactly with those in the third column. The fact that the difference quotients approximate $\cos x$ so closely suggests that $\frac{d}{dx}(\sin x) = \cos x$.

TABLE 4.5

x	Difference quotient $= \dfrac{\sin(x+h) - \sin x}{h}$	$\cos x$
0	1.000	1.000
0.1	0.995	0.995
0.2	0.980	0.980
0.3	0.955	0.955
0.4	0.921	0.921
0.5	0.877	0.878
0.6	0.825	0.825

Example 5 Differentiate (a) $2\sin 3\theta$, (b) $\cos^2 x$, (c) $\cos(x^2)$, (d) $e^{-\sin t}$.

Solution Use the chain rule:

(a) $$\frac{d(2\sin 3\theta)}{d\theta} = 2\frac{d(\sin 3\theta)}{d\theta} = 2(\cos 3\theta)\frac{d(3\theta)}{d\theta} = 2(\cos 3\theta)3 = 6\cos 3\theta.$$

(b) $$\frac{d(\cos^2 x)}{dx} = \frac{d\left[(\cos x)^2\right]}{dx} = 2(\cos x) \cdot \frac{d(\cos x)}{dx} = 2(\cos x)(-\sin x) = -2\cos x \sin x.$$

(c) $$\frac{d}{dx}\left[\cos(x^2)\right] = -\sin(x^2) \cdot \frac{d}{dx}(x^2) = -2x\sin(x^2).$$

(d) $$\frac{d(e^{-\sin t})}{dt} = e^{-\sin t}\frac{d(-\sin t)}{dt} = -(\cos t)e^{-\sin t}.$$

Derivative of the Tangent Function

Since $\tan x = \dfrac{\sin x}{\cos x}$, we can differentiate $\tan x$ using the quotient rule. Writing $(\sin x)'$ for $d(\sin x)/dx$, we have:

$$\frac{d(\tan x)}{dx} = \frac{(\sin x)'(\cos x) - (\sin x)(\cos x)'}{\cos^2 x} = \frac{\cos^2 x + \sin^2 x}{\cos^2 x} = \frac{1}{\cos^2 x}.$$

For x in radians,

$$\frac{d}{dx}(\tan x) = \frac{1}{\cos^2 x}.$$

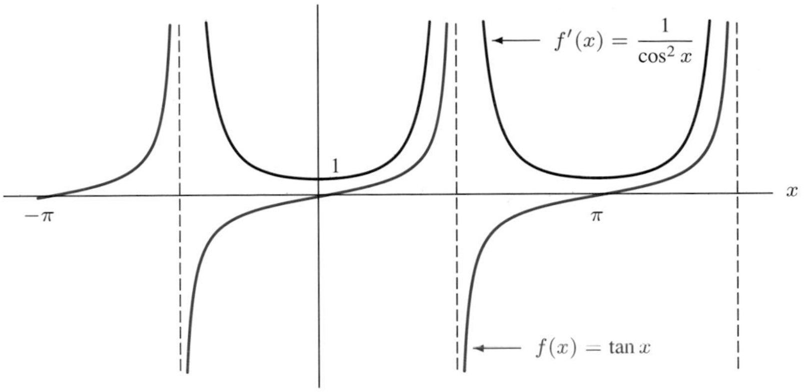

Figure 4.21: Tan x and its derivative

The graphs of $f(x) = \tan x$ and $f'(x) = 1/\cos^2 x$ are in Figure 4.21. Does it seem reasonable to you that f' is always positive? Are the asymptotes of f' where you expect?

Example 6 Differentiate (a) $2\tan 3t$, (b) $\tan(1-\theta)$, (c) $\dfrac{1+\tan t}{1-\tan t}$.

Solution (a) Use the chain rule:

$$\frac{d}{dt}(2\tan 3t) = 2\frac{1}{\cos^2 3t}\frac{d(3t)}{dt} = \frac{6}{\cos^2 3t}.$$

(b) Use the chain rule:

$$\frac{d(\tan(1-\theta))}{d\theta} = \frac{1}{\cos^2(1-\theta)} \cdot \frac{d(1-\theta)}{d\theta} = \frac{-1}{\cos^2(1-\theta)}.$$

(c) Use the quotient rule:

$$\begin{aligned}
\frac{d}{dt}\left(\frac{1+\tan t}{1-\tan t}\right) &= \frac{\left(\dfrac{d(1+\tan t)}{dt}\right)(1-\tan t) - (1+\tan t)\dfrac{d(1-\tan t)}{dt}}{(1-\tan t)^2} \\
&= \frac{\dfrac{1}{\cos^2 t}(1-\tan t) - (1+\tan t)\left(-\dfrac{1}{\cos^2 t}\right)}{(1-\tan t)^2} \\
&= \frac{2}{\cos^2 t \cdot (1-\tan t)^2}.
\end{aligned}$$

Informal Justification of $\frac{d}{dx}(\sin x) = \cos x$

To find the derivative of $\sin x$, we need to estimate

$$\frac{\sin(x+h) - \sin x}{h}.$$

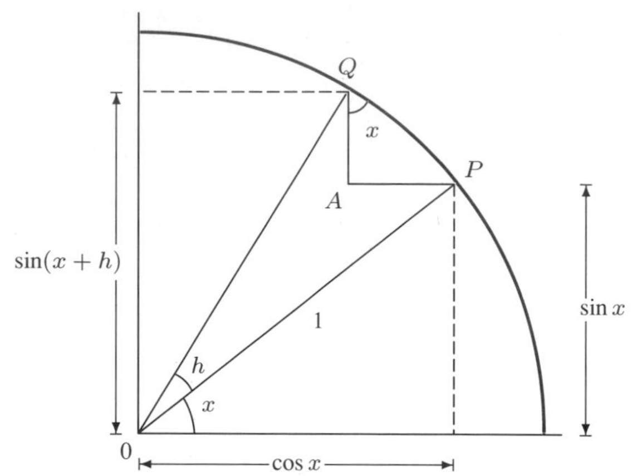

Figure 4.22: Unit circle showing $\sin(x+h)$ and $\sin x$

In Figure 4.22, the quantity $\sin(x+h) - \sin x$ is represented by the length QA. The arc QP is of length h, so

$$\frac{\sin(x+h) - \sin x}{h} = \frac{QA}{\text{Arc } QP}.$$

Now, if h is small, QAP is approximately a right triangle because the arc QP is almost a straight line. Furthermore, using geometry, you can show that angle $AQP \approx x$. Thus, for small h:

$$\frac{\sin(x+h) - \sin x}{h} = \frac{QA}{\text{Arc } QP} \approx \cos x.$$

As $h \to 0$, the approximation gets better, so

$$\frac{d}{dx}(\sin x) = \lim_{h \to 0} \frac{\sin(x+h) - \sin x}{h} = \cos x.$$

Other derivations of this result are given in Problems 33 and 34 on page 224.

Problems for Section 4.6

Find the derivatives of the functions in Problems 1–20.

1. $f(x) = \sqrt{1 - \cos x}$
2. $f(x) = \cos(\sin x)$
3. $f(x) = \sin(3x)$
4. $z = \cos(4\theta)$
5. $w = \sin(e^t)$
6. $f(x) = x^2 \cos x$
7. $f(x) = e^{\cos x}$
8. $f(y) = e^{\sin y}$
9. $z = \theta e^{\cos \theta}$
10. $f(x) = 2x \sin(3x)$
11. $f(x) = \sin(2x) \cdot \sin(3x)$
12. $y = e^{\theta} \sin(2\theta)$
13. $f(x) = e^{-2x} \cdot \sin x$
14. $z = \sqrt{\sin t}$
15. $y = \sin^5 \theta$
16. $g(z) = \tan(e^z)$
17. $z = \tan(e^{-3\theta})$
18. $w = e^{-\sin \theta}$
19. $h(t) = t \cos t + \tan t$
20. $f(\theta) = \theta^2 \sin \theta + 2\theta \cos \theta - 2 \sin \theta$

21. Construct a table of values for $\cos x$, $x = 0, 0.1, 0.2, \ldots, 0.6$ as in Example 4, page 220. Estimate the derivative at these points, using $h = 0.001$, and compare it with $-\sin x$.

22. A boat at anchor is bobbing up and down in the sea. The vertical distance, y, in feet, between the sea floor and the boat is given as a function of time, t, in minutes, by

$$y = 15 + \sin 2\pi t.$$

 (a) Find the vertical velocity, v, of the boat at time t.
 (b) Make a rough sketch of y and v against t.

23. On page 70 the depth, y, of water in Boston harbor was given by

$$y = 5 + 4.9 \cos\left(\frac{\pi}{6}t\right),$$

where t is the number of hours since midnight.

 (a) Find $\frac{dy}{dt}$. What does $\frac{dy}{dt}$ represent, in terms of water level?
 (b) For $0 \le t \le 24$, when is $\frac{dy}{dt}$ zero? (Figure 1.74 on page 70 may help.) Explain what it means (in terms of water level) for $\frac{dy}{dt}$ to be zero.

24. The function $y = A \sin\left(\sqrt{\frac{k}{m}}t\right)$ represents the oscillations of a mass m at the end of a spring. The constant k measures the stiffness of the spring.

 (a) Find a time at which the mass is farthest from its equilibrium position. Find a time at which the mass is moving fastest. Find a time at which the mass is accelerating fastest.
 (b) What is the period, T, of the oscillation?
 (c) Find dT/dm. What does the sign of dT/dm tell you?

25. Consider the function $f(x) = e^{\sin x}$. Without plotting points, sketch the graph of this function by figuring out some of its properties. Is it periodic? Does it grow unboundedly like an exponential function? Does it ever achieve a negative value? Does it have any zeros? Does it have a maximum value? A minimum value? Check your answer using a graphing calculator or computer.

26. Repeat Problem 25 with $f(x) = \sin(e^x)$.

27. Find $\lim_{h\to 0} (\sin h)/h$ when h is in degrees. Explain why this should be larger or smaller than the limit obtained with h in radians [Hint: Consider the numerical values (ignoring for the moment the units of measure) of the numerator and denominator of $\sin h/h$ for h in degrees and h in radians.]

28. Suppose x is measured in degrees. Calculate the derivative of $y = \sin x$ when $x = 35°$.

29. If $w(t) = u(v(t))$ and $v(t) = \cos t$, and $u'(-1) = 2$, find $w'(\pi)$.

30. If $w(t)$ is given as in Problem 29, but $u'(-1)$ is not given, can you still find $w'(\pi)$? Explain why, both algebraically and intuitively.

31. Consider the functions $k(p) = \tan(2p)$ and $r(p) = \tan(\frac{1}{2}p)$ for $-\pi < p < \pi$.

 (a) Solve the equations $k'(p) = 1$ and $r'(p) = 1$.
 (b) Interpret the results in part (a) graphically, in terms of slopes.

32. For $f(x) = \sin x$, find the equations of the tangent lines at $x = 0$ and at $x = \pi/3$. Use each tangent line to approximate $\sin \pi/6$. Would you expect these results to be equally accurate,

since they are taken equally far away from $x = \pi/6$ but on opposite sides? If the accuracy is different, can you account for the difference?

33. We will use the identities

$$\sin(a+b) = \sin a \cos b + \sin b \cos a,$$
$$\cos(a+b) = \cos a \cos b - \sin a \sin b,$$

to calculate the derivatives of $\sin x$ and $\cos x$.

(a) Use the definition of the derivative to show that if $f(x) = \sin x$,

$$f'(x) = \sin x \lim_{h\to 0} \frac{\cos h - 1}{h} + \cos x \lim_{h \to 0} \frac{\sin h}{h}.$$

(b) Evaluate the limits in part (a) with your calculator to conclude that $f'(x) = \cos x$.

(c) If $g(x) = \cos x$, use the definition of the derivative to show that $g'(x) = -\sin x$.

34. In this problem you will calculate the derivative of $\tan\theta$ rigorously (and without using the derivatives of $\sin\theta$ or $\cos\theta$). You will then use your result for $\tan\theta$ to calculate the derivatives of $\sin\theta$ and $\cos\theta$. Figure 4.23 shows $\tan\theta$ and $\Delta(\tan\theta)$, which is the change in $\tan\theta$, namely $\tan(\theta + \Delta\theta) - \tan\theta$.

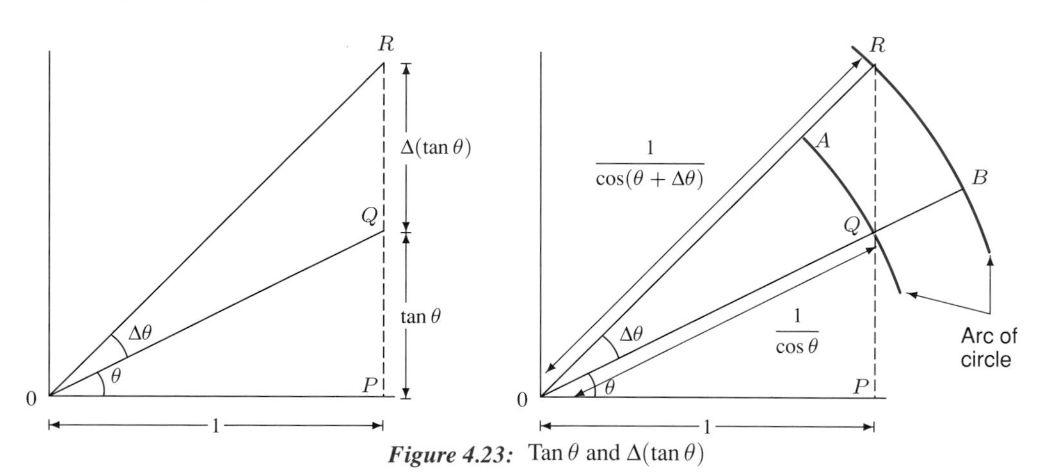

Figure 4.23: Tan θ and $\Delta(\tan\theta)$

(a) By paying particular attention to how the two figures relate and using the fact that

Area of Sector OAQ $\leq$ Area of Triangle OQR $\leq$ Area of Sector OBR

explain why

$$\frac{\Delta\theta}{2\pi}\cdot\pi\left(\frac{1}{\cos\theta}\right)^2 \leq \frac{1}{2}\cdot 1\cdot\Delta(\tan\theta) \leq \frac{\Delta\theta}{2\pi}\cdot\pi\left(\frac{1}{\cos(\theta+\Delta\theta)}\right)^2.$$

[Hint: A sector of circle with angle α at the center has area $\alpha/(2\pi)$ times the area of the whole circle.]

(b) Use part (a) to show as $\Delta\theta \to 0$ that

$$\frac{\Delta\tan\theta}{\Delta\theta} \to \left(\frac{1}{\cos\theta}\right)^2,$$

and hence that $\dfrac{d(\tan\theta)}{d\theta} = \left(\dfrac{1}{\cos\theta}\right)^2.$

(c) Derive the identity $(\tan\theta)^2 + 1 = \left(\dfrac{1}{\cos\theta}\right)^2$. Then differentiate both sides of this identity with respect to θ, using the chain rule and the result of part (b) to show that $\dfrac{d}{d\theta}(\cos\theta) = -\sin\theta$.

(d) Differentiate both sides of the identity $(\sin\theta)^2 + (\cos\theta)^2 = 1$ and use the result of part (c) to show that $\dfrac{d}{d\theta}(\sin\theta) = \cos\theta$.

4.7 APPLICATIONS OF THE CHAIN RULE

In this section we will use the chain rule to verify the derivatives of fractional powers and to find the derivatives of logarithms, exponentials, and the inverse trigonometric functions.

Derivative of $x^{1/2}$

Earlier we calculated the derivative of x^n, with n an integer, but we have been using the result for non-integral values of n as well. We will now calculate the derivative of one of the powers we've so far taken on faith, $f(x) = x^{1/2}$. Since

$$[f(x)]^2 = x,$$

the derivative of $[f(x)]^2$ and the derivative of x must be equal, so

$$\frac{d}{dx}\left[f(x)\right]^2 = \frac{d}{dx}(x).$$

We can use the chain rule with $f(x)$ as the inside function to obtain:

$$2f(x) \cdot f'(x) = 1.$$

Solving for $f'(x)$ gives

$$f'(x) = \frac{1}{2f(x)} = \frac{1}{2x^{1/2}}.$$

$$\boxed{\frac{d(x^{1/2})}{dx} = \frac{1}{2x^{1/2}}.}$$

Example 1 Differentiate (a) $x^{1/3}$, (b) $x^{1/n}$.

Solution (a) Since

$$(x^{1/3})^3 = x,$$

differentiating both sides gives

$$3(x^{1/3})^2 \cdot \frac{d}{dx}(x^{1/3}) = 1.$$

Thus,

$$\frac{d}{dx}(x^{1/3}) = \frac{1}{3x^{2/3}} = \frac{1}{3}x^{-2/3}.$$

(b) Since

$$(x^{1/n})^n = x,$$

differentiating both sides gives

$$n(x^{1/n})^{n-1} \cdot \frac{d}{dx}(x^{1/n}) = 1.$$

Thus,

$$\frac{d}{dx}(x^{1/n}) = \frac{1}{nx^{(n-1)/n}} = \frac{1}{nx^{1-(1/n)}} = \frac{1}{n}x^{(1/n)-1}.$$

Derivative of ln x

We'll again use the chain rule, and differentiate an identity which involves $\ln x$. Since $e^{\ln x} = x$, using the chain rule gives:

$$\frac{d}{dx}(e^{\ln x}) = \frac{d}{dx}(x),$$

$$e^{\ln x} \cdot \frac{d}{dx}(\ln x) = 1. \qquad \text{Since } e^x \text{ is outside function and } \ln x \text{ is inside function.}$$

Thus,

$$\frac{d}{dx}(\ln x) = \frac{1}{e^{\ln x}} = \frac{1}{x},$$

so

$$\boxed{\frac{d(\ln x)}{dx} = \frac{1}{x}.}$$

Example 2 Differentiate (a) $\ln(x^2+1)$, (b) $t^2 \ln t$, (c) $\sqrt{1+\ln(1-y)}$.

Solution (a) Using the chain rule:

$$\frac{d(\ln(x^2+1))}{dx} = \frac{1}{x^2+1}\frac{d(x^2+1)}{dx} = \frac{2x}{x^2+1}.$$

(b) Using the product rule:

$$\frac{d(t^2\ln t)}{dt} = \left(\frac{d(t^2)}{dt}\right)\ln t + t^2\frac{d(\ln t)}{dt} = 2t\ln t + t^2 \cdot \frac{1}{t} = 2t\ln t + t.$$

(c)

$$\frac{d\sqrt{1+\ln(1-y)}}{dy} = \frac{1}{2\sqrt{1+\ln(1-y)}} \cdot \frac{d[1+\ln(1-y)]}{dy} \qquad \text{Using the chain rule}$$

$$= \frac{1}{2\sqrt{1+\ln(1-y)}} \cdot \frac{1}{1-y} \cdot \frac{d(1-y)}{dy} \qquad \text{Using the chain rule again}$$

$$= \frac{-1}{2(1-y)\sqrt{1+\ln(1-y)}}.$$

Derivative of a^x

Earlier we showed that the derivative of a^x is proportional to a^x. Now we show that the constant of proportionality is $\ln a$. We will use the identity

$$\ln(a^x) = x \ln a.$$

Differentiating, using $\frac{d}{dx}(\ln x) = \frac{1}{x}$ and the chain rule on the left and remembering that $\ln a$ is a constant, we obtain:

$$\frac{d}{dx}(\ln a^x) = \frac{1}{a^x} \cdot \frac{d}{dx}(a^x) = \ln a.$$

Solving gives the result we expected in Section 4.3:

$$\boxed{\frac{d(a^x)}{dx} = (\ln a)a^x.}$$

Derivative of Inverse Trigonometric Functions

In Section 1.10 we defined the inverse sine and tangent as follows:

$\arcsin x$ is the number between $-\pi/2$ and $\pi/2$ (inclusive) whose sine is x.
$\arctan x$ is the number between $-\pi/2$ and $\pi/2$ whose tangent is x.

To find $\frac{d}{dx}(\arctan x)$ we use the identity $\tan(\arctan x) = x$. Differentiating using the chain rule gives

$$\frac{1}{\cos^2(\arctan x)} \cdot \frac{d}{dx}(\arctan x) = 1,$$

so

$$\frac{d}{dx}(\arctan x) = \cos^2(\arctan x).$$

Using the identity $1 + \tan^2\theta = \frac{1}{\cos^2\theta}$, we have

$$\cos^2(\arctan x) = \frac{1}{1 + \tan^2(\arctan x)} = \frac{1}{1 + x^2}.$$

Thus we have

$$\boxed{\frac{d}{dx}(\arctan x) = \frac{1}{1 + x^2}.}$$

By a similar argument, one can obtain the result:

$$\boxed{\frac{d}{dx}(\arcsin x) = \frac{1}{\sqrt{1 - x^2}}.}$$

Example 3 Differentiate (a) $\arctan(t^2)$, (b) $\arcsin(\tan\theta)$.

Solution Use the chain rule:

(a) $\dfrac{d}{dt}[\arctan(t^2)] = \dfrac{1}{1+(t^2)^2}\cdot\dfrac{d}{dt}(t^2) = \dfrac{2t}{1+t^4}.$

(b) $\dfrac{d}{dt}[\arcsin(\tan\theta)] = \dfrac{1}{\sqrt{1-(\tan\theta)^2}}\cdot\dfrac{d}{d\theta}(\tan\theta) = \dfrac{1}{\sqrt{1-\tan^2\theta}}\cdot\dfrac{1}{\cos^2\theta}.$

Problems for Section 4.7

For Problems 1–16, find the derivative of the given function. In some cases, it may be to your advantage to simplify before differentiating.

1. $f(x) = \ln(1-x)$
2. $f(t) = \ln(t^2+1)$
3. $f(z) = \dfrac{1}{\ln z}$
4. $f(\theta) = \ln(\cos\theta)$
5. $f(x) = \ln(1-e^{-x})$
6. $f(\alpha) = \ln(\sin\alpha)$
7. $f(x) = \ln(e^x+1)$
8. $f(t) = \ln(\ln t) + \ln(\ln 2)$
9. $f(x) = \ln(e^{7x})$
10. $f(x) = e^{(\ln x)+1}$
11. $f(w) = \ln(\cos(w-1))$
12. $f(t) = \ln(e^{\ln t})$
13. $f(y) = \arcsin(y^2)$
14. $g(t) = \arctan(3t-4)$
15. $h(w) = w\arcsin w$
16. $g(\alpha) = \sin(\arcsin\alpha)$
17. Using the chain rule, find $\dfrac{d}{dx}(\arcsin x)$.
18. Using the chain rule, find $\dfrac{d}{dx}(\log x)$. (Recall that $\log x = \log_{10} x$.)
19. To compare the acidity of different solutions, chemists use the pH. The pH (which is a single number, not the product of p and H) is defined in terms of the concentration, x, of hydrogen ions in the solution as
$$\text{pH} = -\log x.$$
Find the rate of change of pH with hydrogen ion concentration when the pH is 2. [Hint: Use the result of Problem 18.]
20. Hungary is one of the few countries in the world where the population is decreasing, currently at about 0.2% a year. Thus, if t is time in years since 1990, the population P, in millions, of Hungary can be approximated by
$$P = 10.8(0.998)^t.$$
   (a) What does this model predict the population of Hungary will be in the year 2000?
   (b) How fast (in people/year) does this model predict Hungary's population will be decreasing in the year 2000?

21. Imagine you are zooming in on the graph of each of the following functions near the origin:

$y = x$	$y = \sqrt{x}$	$y = x^2$	$y = \sin x$
$y = x \sin x$	$y = \tan x$	$y = \sqrt{x/(x+1)}$	$y = x^3$
$y = \ln(x+1)$	$y = \frac{1}{2}\ln(x^2+1)$	$y = 1 - \cos x$	$y = \sqrt{2x - x^2}$

Which of them look the same? Group together those functions which become indistinguishable, and give the equations of the lines they look like.

22. (a) Find the equation of the tangent line to $y = \ln x$ at $x = 1$.
(b) Use it to calculate approximate values for $\ln(1.1)$ and $\ln(2)$.
(c) Using a graph, explain whether the approximate values you have calculated are smaller or larger than the true values. Would the same result have held if you had used the tangent line to estimate $\ln(0.9)$ and $\ln(0.5)$? Why?

23. (a) Find the equation of the best quadratic approximation to $y = \ln x$ at $x = 1$. The best quadratic approximation will have the same first and second derivatives as $y = \ln x$ at $x = 1$.
(b) Use a computer or calculator to graph the approximation and $y = \ln x$ on the same set of axes. What do you notice?
(c) Use your quadratic approximation to calculate approximate values for $\ln(1.1)$ and $\ln(2)$.

24. (a) For $x > 0$, find the derivative of $f(x) = \arctan x + \arctan(1/x)$ and simplify.
(b) What does your result tell you about f?

25. For the amusement of the guests, some hotels have elevators on the outside of the building. Suppose such a hotel is 300 feet high. You are standing by a window 100 feet above the ground and 150 feet away from the hotel, and the elevator descends at a constant speed of 30 ft/sec, starting at time $t = 0$, where t is time in seconds. Let θ be the angle between the line of your horizon and your line of sight to the elevator. (See Figure 4.24.)
(a) Find a formula for $h(t)$, the elevator's height above the ground as it descends from the top of the hotel.
(b) Using your answer to part (a), express θ as a function of time t and find the rate of change of θ with respect to t.
(c) If the rate of change of θ is a measure of how fast the elevator appears to you to be moving, at what height will the elevator be when it appears to be moving fastest?

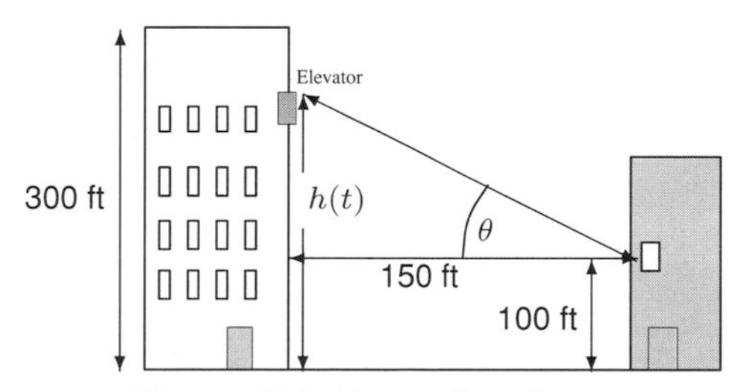

Figure 4.24: Descending elevator

4.8 IMPLICIT FUNCTIONS

In earlier chapters, most functions were written in the form $y = f(x)$; here y is said to be an *explicit* function of x. An equation such as

$$x^2 + y^2 = 4$$

is said to give y as an *implicit* function of x. Its graph is the circle in Figure 4.25. Since there are x values which correspond to two y values, y is not a function of x on the whole circle. Solving gives

$$y = \pm\sqrt{4 - x^2},$$

where $y = \sqrt{4 - x^2}$ represents the top half of the circle and $y = -\sqrt{4 - x^2}$ represents the bottom half. On the top half and the bottom half of the circle, considered separately, y is a function of x.

But let's consider the circle as a whole. The equation does represent a curve which has a tangent line at each point. The slope of this tangent can be found by differentiating the equation of the circle with respect to x:

$$\frac{d(x^2)}{dx} + \frac{d(y^2)}{dx} = \frac{d(4)}{dx}.$$

If you think of y as a function of x and use the chain rule, you get

$$2x + 2y\frac{dy}{dx} = 0.$$

Solving gives

$$\frac{dy}{dx} = -\frac{x}{y}.$$

The derivative here depends on both x and y (instead of just on x). This is because for each x value (except for $x = \pm 2$) there are two y values, and the curve has a different slope at each one. Figure 4.25 shows that for x and y both positive, we are on the top right half of the curve, and the slope should be negative (as the formula predicts). For x positive and y negative, we are on the bottom right half of the curve and the slope should be positive (as the formula predicts).

Differentiating the equation of the circle has given us the slope of the curve at all points except two, namely $(2, 0)$ and $(-2, 0)$, where the tangent is vertical. In general, an implicitly defined function leads to a derivative whenever the expression for the derivative does not have a zero in the denominator.

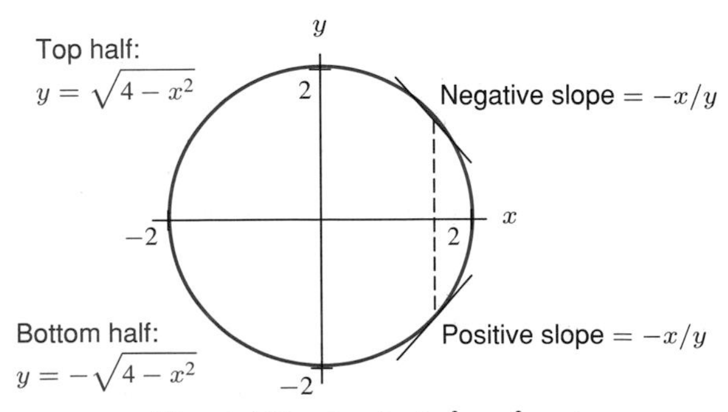

Figure 4.25: Graph of $x^2 + y^2 = 4$

Example 1 Make a table of x and approximate y values for the equation $y^3 - xy = -6$ near $x = 7, y = 2$. Your table should include the x values $6.8, 6.9, 7.0, 7.1$, and 7.2.

Solution We would like to solve for y in terms of x, but we cannot isolate y by factoring. There is a formula for solving cubics, somewhat like the quadratic formula, but it is too complicated to be useful here. Instead, first observe that $x = 7$, $y = 2$ does satisfy the equation. (Check this!)

Then find dy/dx by implicit differentiation:

$$\begin{aligned}
\frac{d}{dx}(y^3) - \frac{d}{dx}(xy) &= \frac{d}{dx}(-6) \\
3y^2\frac{dy}{dx} - y - x\frac{dy}{dx} &= 0 \\
3y^2\frac{dy}{dx} - x\frac{dy}{dx} &= y \\
(3y^2 - x)\frac{dy}{dx} &= y \\
\frac{dy}{dx} &= \frac{y}{3y^2 - x}.
\end{aligned}$$

When $x = 7$ and $y = 2$, we have

$$\frac{dy}{dx} = \frac{2}{12 - 7} = \frac{2}{5}.$$

(See Figure 4.26.) The equation of the tangent line at $(7, 2)$ is

$$y - 2 = \frac{2}{5}(x - 7)$$

or

$$y = 0.4x - 0.8.$$

Since the tangent lies very close to the curve near the point $(7, 2)$, we use the equation of the tangent line to calculate the approximate y values below:

x	6.8	6.9	7.0	7.1	7.2
Approximate y	1.92	1.96	2.00	2.04	2.08

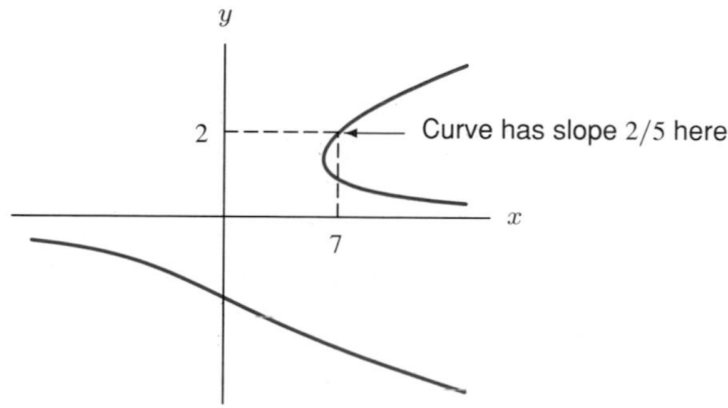

Figure 4.26: Graph of $y^3 - xy = -6$

The point of the preceding example is that although an equation such as $y^3 - xy = -6$ may lead to a complicated curve which is difficult to deal with algebraically, it still looks like a straight line locally, just like the graph of any differentiable function. This means that the result we get for dy/dx from implicit differentiation is very useful in analyzing the relationship between x and y, at least locally. In particular, we can use implicit differentiation to locate special points, such as where the tangent is horizontal or vertical.

Example 2 Find all points where the tangent line to $y^3 - xy = -6$ is either horizontal or vertical.

Solution From the previous example, $\dfrac{dy}{dx} = \dfrac{y}{3y^2 - x}$. The tangent is horizontal when the numerator of dy/dx equals 0, so $y = 0$. Since we also must satisfy $y^3 - xy = -6$, we get $0^3 - x \cdot 0 = -6$, which is impossible. We conclude that there are no points on the curve where the tangent line is horizontal.

The tangent is vertical when the denominator of dy/dx is 0, giving $3y^2 - x = 0$. Thus, $x = 3y^2$ at any point with a vertical tangent line. Again, we must also satisfy $y^3 - xy = -6$, so

$$\begin{aligned} y^3 - (3y^2)y &= -6, \\ -2y^3 &= -6, \\ y &= \sqrt[3]{3} \approx 1.442. \end{aligned}$$

We can then find x by substituting $y = \sqrt[3]{3}$ in $y^3 - xy = -6$. We get $3 - x(\sqrt[3]{3}) = -6$, so $x = 9/(\sqrt[3]{3}) \approx 6.240$.

Using our expression for dy/dx to find the points where the tangent is vertical or horizontal, as in the previous example, is a first step in obtaining an overall (global, not just local) picture of the curve $y^3 - xy = -6$. However, filling in the rest of the graph, even roughly, by using the sign of dy/dx to tell us where the curve is increasing and where it is decreasing, can be difficult. Another method of getting the graph is described in Problem 16 on page 233.

Problems for Section 4.8

For Problems 1–8, find dy/dx.

1. $x^2 + xy - y^3 = xy^2$
2. $x^2 + y^2 = \sqrt{7}$
3. $\sqrt{x} + \sqrt{y} = 25$
4. $x^{2/3} + y^{2/3} = a^{2/3}$ (a is a constant.)
5. $\sin(xy) = 2x + 5$
6. $x \ln y + y^3 = \ln x$
7. $e^{\cos y} = x^3 \arctan y$
8. $\cos^2 y + \sin^2 y = y + 2$

For Problems 9–11, find the equations of the tangent lines to the following curves at the indicated points.

9. $xy^2 = 1$ at $(1, -1)$
10. $y^2 = \dfrac{x^2}{xy - 4}$ at $(4, 2)$
11. $x^{2/3} + y^{2/3} = a^{2/3}$ at $(a, 0)$

12. Assume y is a differentiable function of x and that $y + \sin y + x^2 = 9$. Find dy/dx at the point $x = 3, y = 0$.

13. Show that the power rule for derivatives applies to rational powers of the form $y = x^{m/n}$ by raising both sides to the nth power and using implicit differentiation.

14. (a) Find the equations of the tangent lines to the circle $x^2 + y^2 = 25$ at the points where $x = 4$.
    (b) Find the equations of the normal lines to this circle at the same points. (The normal line to a curve at a point is perpendicular to the tangent there.)
    (c) At what point do the two normal lines intersect?

15. (a) Find the slope of the tangent line to the ellipse $\frac{x^2}{25} + \frac{y^2}{9} = 1$ at the point (x, y).
    (b) Are there any points where the slope is not defined?

16. Solve the equation $y^3 - xy = -6$ for x in terms of y, and show how this enables you to sketch a graph of the equation by viewing x as a function of y. (Put the x-axis horizontally.) For which values of x are there three y values? Two y values? One y value? Identify the points where the tangent line is vertical. Compare your answer to the result of Example 2, page 232.

17. Consider the equation $x^3 + y^3 - xy^2 = 5$.
    (a) Find dy/dx by implicit differentiation.
    (b) Give a table of approximate y values near $x = 1, y = 2$ for $x = 0.96, 0.98, 1, 1.02, 1.04$.
    (c) Find the y value for $x = 0.96$ by substituting $x = 0.96$ in the equation and solving for y using a computer or calculator. Compare with your answer in part (b).
    (d) Find all points where the tangent line is horizontal or vertical.

18. Consider the curve $xe^{5y} = 3y$.
    (a) Find dy/dx.
    (b) Find the equation of the tangent line to the curve at $(0, 0)$.
    (c) If $x = 0.1$, estimate y using the tangent line.

4.9 NOTES ON THE TANGENT LINE APPROXIMATION

Since the graph of a function and its tangent line have the same slope at the point of tangency (namely, the derivative at that point), the tangent line stays close to the graph of the function near that point. As we have seen, this means that we can approximate values of the function by values on the tangent line. This can be helpful because values on the tangent line may be easier to calculate.

The slope of the tangent to the graph of $y = f(x)$ at $x = a$ is $f'(a)$, so the equation of the tangent line is

$$y - f(a) = f'(a)(x - a)$$

or

$$y = f(a) + f'(a)(x - a).$$

(Consider Figure 4.27.) Now we approximate the values of f by the y values from the tangent line, giving the result in the following box.

The Tangent Line Approximation

For values of x near a,

$$f(x) \approx f(a) + f'(a)(x - a)$$

We are thinking of a as fixed, so that $f(a)$ and $f'(a)$ are constant. This "almost equality" is called the *local linearization* of f near $x = a$.

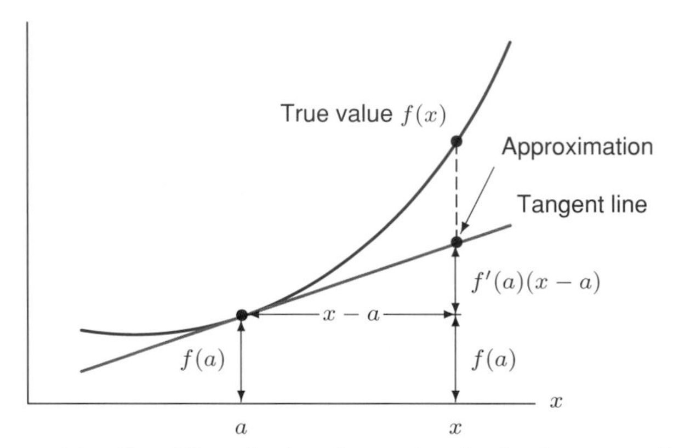

Figure 4.27: Local linearization: Approximation by the tangent line

---

Example 1 What is the tangent line approximation for $\sin x$ near $x = 0$?

Solution The local linearization of f near $x = 0$ is

$$f(x) \approx f(0) + f'(0)(x - 0).$$

When $f(x) = \sin x$, $f'(x) = \cos x$, so $f(0) = \sin 0 = 0$ and $f'(0) = \cos 0 = 1$. Our approximation is thus

$$\sin x \approx x.$$

This means that, near $x = 0$, the function $f(x) = \sin x$ can be approximated by the function $y = x$. Another way of saying this is that if you zoom in on the graphs of the functions $\sin x$ and x near the origin, you won't be able to tell them apart. (See Figure 4.28.) As before, we are assuming x is in radians unless specifically stated otherwise.

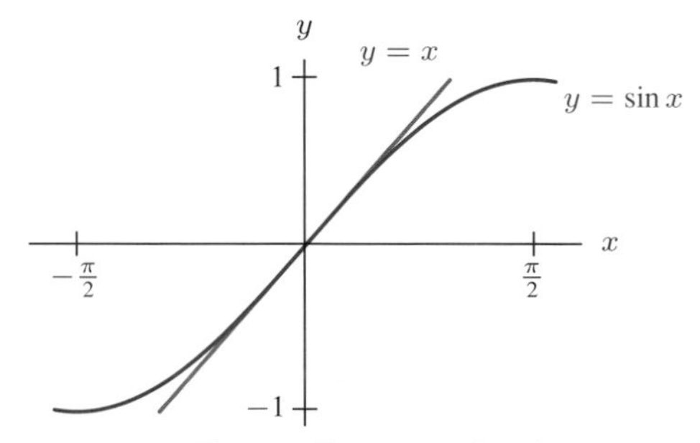

Figure 4.28: Tangent line approximation to $y = \sin x$

Example 2 Use local linearity to find $\lim_{h \to 0} \frac{\sin 3h}{h}$.

Solution When h is small, $3h$ is small too. Thus Example 1 tells us that $\sin 3h \approx 3h$, so $\frac{\sin 3h}{h} \approx \frac{3h}{h} = 3$. The approximation gets better and better as h gets smaller, so

$$\lim_{h \to 0} \frac{\sin 3h}{h} = 3.$$

Example 3 What is the local linearization of e^{kx} near $x = 0$?

Solution With $f(x) = e^{kx}$, we see that the local linearization of f near $x = 0$,

$$f(x) \approx f(0) + f'(0)(x - 0),$$

becomes

$$e^{kx} \approx e^{k \cdot 0} + f'(0)x,$$

where $f'(0)$ is the derivative of e^{kx} at $x = 0$. By the chain rule, if $f(x) = e^{kx}$, then $f'(x) = ke^{kx}$, so $f'(0) = ke^{k \cdot 0} = k$. Thus our formula reduces to

$$e^{kx} \approx 1 + kx.$$

This is the local linearization of e^{kx} near $x = 0$. In other words, if you zoom in on the functions $f(x) = e^{kx}$ and $y = 1 + kx$ near the origin, you won't be able to tell them apart.

Example 4 Linearize $\sqrt{1+x}$ near $x = 0$.

Solution With $f(x) = \sqrt{1+x}$, the chain rule gives $f'(x) = 1/(2\sqrt{1+x})$, so $f(0) = 1$ and $f'(0) = 1/2$. Therefore the local linearization of f near $x = 0$,

$$f(x) \approx f(0) + f'(0)(x - 0),$$

becomes

$$\sqrt{1+x} \approx 1 + \frac{x}{2}.$$

This means that, near $x = 0$, the function $\sqrt{1+x}$ can be approximated by its tangent line $y = 1 + x/2$. (See Figure 4.29.)

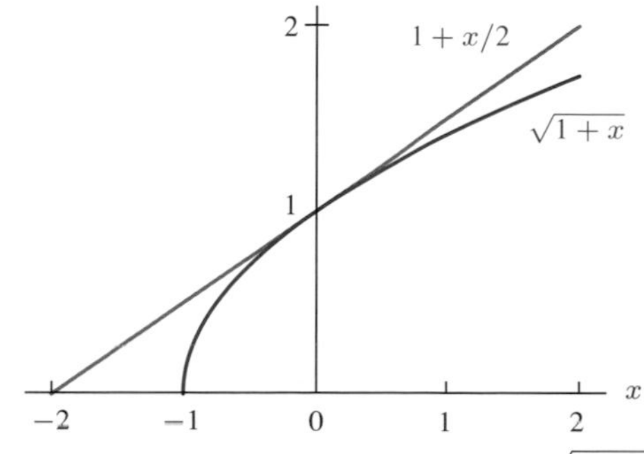

Figure 4.29: Tangent approximation to $\sqrt{1+x}$

Example 5 The "Rule of 70" is a rule of thumb to estimate how long it takes money in a bank to double. The Rule of 70 says that the time it takes the amount of money in an account bearing $i\%$ annual interest compounded yearly to double is approximately $70/i$ years for small values of i. Find the local linearization of $\ln(1+x)$, and use it to verify this rule.

Solution Let $r = i/100 = i\%$. (For example if $i = 5\%$, $r = 0.05$.) Then the balance $\$B$ after t years is given by

$$B = P(1+r)^t,$$

where $\$P$ is the original deposit. If we are doubling our money, then $B = 2P$, so we wish to solve for t in the equation $2P = P(1+r)^t$, which is equivalent to

$$2 = (1+r)^t.$$

Taking natural logarithms of both sides and solving for t yields

$$\ln 2 = t\ln(1+r),$$
$$t = \frac{\ln 2}{\ln(1+r)}.$$

We now approximate $\ln(1+r)$ near $r = 0$. Let $f(r) = \ln(1+r)$. Then $f'(r) = 1/(1+r)$. Thus, $f(0) = 0$ and $f'(0) = 1$, so

$$f(r) \approx f(0) + f'(0)r$$

becomes

$$\ln(1+r) \approx r.$$

Therefore,

$$t = \frac{\ln 2}{\ln(1+r)} \approx \frac{\ln 2}{r} = \frac{100\ln 2}{i} \approx \frac{69.3}{i} \approx \frac{70}{i},$$

as claimed. We expect this approximation to hold for small values of r and i; it turns out that values of i up to 10 give good enough answers for most everyday purposes.

This example illustrates a useful feature of local linearization; such estimates as the Rule of 70, or $\sin x \approx x$ for small x are fairly accurate and very useful for quick calculations.

Problems for Section 4.9

1. What is the local linearization of $1/x$ near $x = 1$?
2. Show that $1 - x/2$ is the local linearization of $1/\sqrt{1+x}$ near $x = 0$.
3. Show that $e^{-x} \approx 1 - x$ near $x = 0$.
4. Using the local linearization of e^x, evaluate $\lim_{h\to 0} \frac{e^{kh}-1}{h}$.
5. Interpret the local linearization $e^{rt} \approx 1 + rt$ in terms of compound interest if r is the nominal annual interest rate and t is the time in years.
6. (a) Show that $1 + kx$ is the local linearization of $(1+x)^k$ near $x = 0$.
   (b) Someone claims that the square root of 1.1 is about 1.05. Without using a calculator, do you think that this estimate is about right?
   (c) Is the actual number above or below 1.05?

7. (a) Use the identity $\cos x = 1 - 2\sin^2(x/2)$ and the tangent line approximation to $\sin x$ near $x = 0$ to obtain

$$\cos x \approx 1 - \frac{x^2}{2}.$$

(b) Using this approximation, evaluate

(i) $\lim_{h\to 0} \frac{\cos h - 1}{h}$ (ii) $\lim_{h\to 0} \frac{\cos h - 1}{h^2}$

8. Multiply the local linearization of e^x near $x = 0$ by itself to obtain an approximation for e^{2x}. Compare this with the actual local linearization of e^{2x}. Explain why these two approximations are consistent, and discuss which one is more accurate.

9. (a) Show that $1 - x$ is the local linearization of $\frac{1}{1+x}$ near $x = 0$.

(b) From your answer to part (a), show that near $x = 0$,

$$\frac{1}{1+x^2} \approx 1 - x^2.$$

(c) Without differentiating, what do you think the derivative of $\frac{1}{1+x^2}$ is at $x = 0$?

10. From the local linearizations of e^x and $\sin x$ near $x = 0$, write down the local linearization of the function $e^x \sin x$. From this result, write down the derivative of $e^x \sin x$ at $x = 0$. Using this technique, write down the derivative of $e^x \sin x/(1 + x)$ at $x = 0$.

11. Use local linearization to derive the product rule,

$$[f(x)g(x)]' = f'(x)g(x) + f(x)g'(x).$$

[Hint: Use the definition of the derivative, and use local linearization for $f(x + h)$ and $g(x + h)$.]

12. Derive the chain rule using local linearization. [Hint: In other words, differentiate $f(g(x))$, using $g(x + h) \approx g(x) + g'(x)h$ and $f(z + k) \approx f(z) + f'(z)k$.]

REVIEW PROBLEMS FOR CHAPTER FOUR

Find the derivatives for the functions in Problems 1–21. Assume a and k are constants.

1. $f(x) = (3x^2 + \pi)(e^x - 4)$

2. $g(x) = 2x - \frac{1}{\sqrt[3]{x}} + 3^x - e$

3. $f(z) = \frac{z^2 + 1}{\sqrt{z}}$

4. $h(r) = \frac{r^2}{2r + 1}$

5. $g(t) = e^{(1+3t)^2}$

6. $f(t) = 2te^t - \frac{1}{\sqrt{t}}$

7. $h(x) = xe^{\tan x}$

8. $g(w) = \frac{1}{2^w + e^w}$

9. $f(y) = \ln\ln(2y^3)$

10. $f(x) = (2 - 4x - 3x^2)(6x^e - 3\pi)$

11. $r(\theta) = \sin[(3\theta - \pi)^2]$

12. $s(\theta) = \sin^2(3\theta - \pi)$

13. $g(\theta) = \sqrt{a^2 - \sin^2\theta}$

14. $g(x) = \dfrac{x^2 + \sqrt{x} + 1}{x^{3/2}}$

15. $w(\theta) = \dfrac{\theta}{\sin^2\theta}$

16. $p(\theta) = \dfrac{\sin(5 - \theta)}{\theta^2}$

17. $h(t) = \ln\left(e^{-t} - t\right)$

18. $g(x) = x^k + k^x$

19. $s(x) = \arctan(2 - x)$

20. $r(\theta) = e^{\left(e^{\theta} + e^{-\theta}\right)}$

21. $r(y) = \dfrac{y}{(\cos y) + a}$

22. Given: $r(2) = 4$, $s(2) = 1$, $s(4) = 2$, $r'(2) = -1$, $s'(2) = 3$, $s'(4) = 3$. Compute the following derivatives, or state what additional information you would need to be able to compute the derivative.

(a) $H'(2)$ if $H(x) = r(x) \cdot s(x)$

(b) $H'(2)$ if $H(x) = \sqrt{r(x)}$

(c) $H'(2)$ if $H(x) = r(s(x))$

(d) $H'(2)$ if $H(x) = s(r(x))$

23. Imagine you are zooming in on the graphs of the following functions near the origin:

$y = \arcsin x \qquad y = \sin x - \tan x \qquad y = x - \sin x \qquad y = \arctan x$

$y = \dfrac{\sin x}{1 + \sin x} \qquad y = \dfrac{x^2}{x^2 + 1} \qquad y = \dfrac{1 - \cos x}{\cos x} \qquad y = \dfrac{x}{x^2 + 1}$

$y = \dfrac{\sin x}{x} - 1 \qquad y = -x\ln x \qquad y = e^x - 1 \qquad y = x^{10} + \sqrt[10]{x}$

$y = \dfrac{x}{x + 1}$

Which of them look the same? Group together those functions which become indistinguishable, and give the equation of the line they look like. [Note: $(\sin x)/x - 1$ and $-x\ln x$ never quite make it to the origin.]

24. The graphs of $\sin x$ and $\cos x$ intersect once between 0 and $\pi/2$. What is the angle between the two curves at the point where they intersect? (You need to think about how the angle between two curves should be defined.)

25. The acceleration due to gravity, g, at a distance r from the center of the earth is given by

$$g = \frac{GM}{r^2},$$

where M is the mass of the earth and G is a constant.

(a) Find $\dfrac{dg}{dr}$.

(b) What is the practical interpretation (in terms of acceleration) of $\dfrac{dg}{dr}$? Why would you expect it to be negative?

(c) You are told that $M = 6 \times 10^{24}$ and $G = 6.67 \times 10^{-20}$ where M is in kilograms and r in kilometers. What is the value of $\dfrac{dg}{dr}$ at the surface of the earth ($r = 6400$ km) ?

(d) What does this tell you about whether or not it is reasonable to assume g is constant near the surface of the earth?

26. In 1975, the population of Mexico was about 84 million and growing at 2.6% annually, while the population of the US was about 250 million and growing at 0.7% annually. Which population was growing faster, if we measure growth rates in people/year? Explain your answer.

27. Suppose the distance, s, of a moving body from a fixed point is given as a function of time by $s = 20e^{t/2}$.

(a) Find the velocity, v, of the body as a function of t.
(b) Find v as a function of s, and hence show that s satisfies the differential equation $s' = \frac{1}{2}s$.

28. Air pressure at sea level is 30 inches of mercury. At an altitude of h feet above sea level, the air pressure, P, in inches of mercury, is given by

$$P = 30e^{-3.23\times 10^{-5}h}$$

(a) Sketch a rough graph of P against h.
(b) Find the equation of the tangent line to the graph at $h = 0$.
(c) A rough rule of thumb used by travelers is that air pressure drops about 1 inch for every 1000-foot increase in height above sea level. Write a formula for the air pressure given by this rule of thumb.
(d) What is the relation between your answers to parts (b) and (c)? Explain why the rule of thumb works.
(e) Are the predictions made by the rule of thumb too large or too small? Why?

29. Suppose the depth of the water, y, in meters, in the Bay of Fundy, Canada, is given as a function of time, t, in hours after midnight, by the function

$$y = 10 + 7.5\cos(0.507t).$$

Is the tide rising or falling, and how fast (in meters/hour), at each of the following times?
(a) 6:00 am (b) 9:00 am (c) Noon (d) 6:00 pm

30. (a) Using the chain rule, find g' for $g(x) = e^{n\ln x}$, where n is any real number.
(b) Using the fact that $x^n = e^{n\ln x}$, show you can use g' to verify the power rule,

$$\frac{d(x^n)}{dx} = nx^{n-1},$$

for all real numbers, n and for $x > 0$.

31. (a) Assuming that $\frac{d}{dx}(\ln x) = \frac{1}{x}$, show that

$$\lim_{n\to\infty}\left(1+\frac{1}{n}\right)^n = e$$

by writing $(1 + 1/n)^n = e^{n\ln(1+1/n)}$ and showing that $n\ln(1 + 1/n)$ is the slope of the secant to $f(x) = \ln x$ between $x = 1$ and $x = (1 + 1/n)$.
(b) What does the formula

$$\lim_{n\to\infty}\left(1+\frac{1}{n}\right)^n = e$$

tell you about compound interest?

(c) Explain in terms of compound interest why it is reasonable to expect that

$$\left(1+\frac{1}{n+1}\right)^{n+1} > \left(1+\frac{1}{n}\right)^{n}$$

for all positive integers n.

(d) Demonstrate the formula

$$\left(1+\frac{1}{n+1}\right)^{n+1} > \left(1+\frac{1}{n}\right)^{n}$$

using the method of part (a). Explain your reasoning with a sketch.

32. Given a number $a > 1$, the equation

$$a^x = 1 + x$$

has the solution $x = 0$ for all a. Are there any other solutions? How does your answer depend on the value of a? [Hint: Graph the functions on both sides of the equation.]

33. Suppose you put a yam in a hot oven, maintained at a constant temperature 200°C. Suppose that at time $t = 30$ minutes, the temperature T of the yam is 120° and is increasing at a (instantaneous) rate of 2°/min. Newton's law of cooling (or, in our case, warming) implies that the temperature at time t will be given by a formula of the form

$$T(t) = 200 - ae^{-bt}.$$

Find a and b.

34. A spherical cell is growing at a constant rate of 400 μm^3/day (1 μm$= 10^{-6}$ m). At what rate is its radius increasing when the radius is 10 μm?

35. When the growth of a spherical cell depends on the flow of nutrients through the surface, it is reasonable to assume that the growth rate dV/dt is proportional to the surface area, S. Assume that for a particular cell $dV/dt = \frac{1}{3} \cdot S$. At what rate is its radius r increasing?

36. Suppose the total number of people, N, who have contracted a disease by a time t days after its outbreak is given by

$$N = \frac{1{,}000{,}000}{1 + 5{,}000e^{-0.1t}}.$$

(a) In the long run, how many people will have had the disease?

(b) Is there any day on which more than a million people fall sick? Half a million? Quarter of a million? (Note: You do not have to try to find out on what days these things happen.)

37. (a) Provide a three dimensional analogue for the geometrical demonstration of the formula for the derivative of a product, given in Figure 4.17 on page 206. In other words, find a formula for the derivative of $F(x) \cdot G(x) \cdot H(x)$ using Figure 4.30.

(b) Verify your results by writing $F(x) \cdot G(x) \cdot H(x)$ as $[F(x) \cdot G(x)] \cdot H(x)$ and using the product rule twice.

(c) Generalize your result to n functions: what is the derivative of

$$f_1(x) \cdot f_2(x) \cdot f_3(x) \cdots f_n(x)?$$

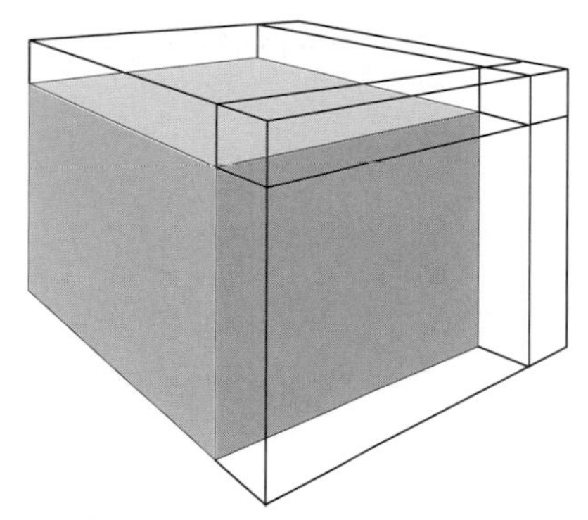

Figure 4.30: A graphical representation of the 3-dimensional product rule.

Problems 38–39 involve Boyle's Law which states that for a fixed quantity of gas at a constant temperature, the pressure, P, and the volume, V, are inversely related. Thus, for some constant k

$$PV = k.$$

38. Suppose a fixed quantity of gas is allowed to expand at constant temperature. Find the rate of change of pressure with volume.

39. Suppose a certain quantity of gas occupies 10 cc at a pressure of 2 atmospheres. Suppose the pressure is increased, while keeping the temperature constant.

(a) Does the volume increase or decrease?
(b) If the pressure is increasing at a rate of 0.05 atmospheres/minute at a time when the pressure is 2 atmospheres, find the rate at which the volume is changing at that moment. What are the units of your answer?

40. The analysis of blood flow through the heart leads to a function of the form

$$f(r) = -2|r| + \sqrt{1 - 4r^2 + 4|r|}.$$

Since the absolute value function $|r|$ is not differentiable at $r = 0$, we might expect that $f(r)$ would not be differentiable at $r = 0$.

(a) By zooming in, investigate the differentiability of $f(r)$ at $r = 0$.
(b) Using the fact that $|r| = r$ for $r \geq 0$, find a formula, without absolute values for $f(r)$, for $r \geq 0$. Use this to find the slope of $f(r)$ to the right of $r = 0$.
(c) Use the fact that $|r| = -r$ for $r < 0$ to find the slope of $f(r)$ to the left of $r = 0$.
(d) What do your answers to parts (b) and (c) tell you about the differentiability of $f(r)$ at $r = 0$?

CHAPTER FIVE

USING THE DERIVATIVE

In Chapter 2 we introduced the derivative and some of its interpretations. In Chapter 4 we saw how to differentiate all of the standard functions, including powers, exponentials, logarithms, and the trigonometric functions. Now we will use first and second derivatives to analyze the behavior of families of functions, and to compute solutions to equations using Newton's Method.

As we saw in Chapter 2, the connection between the derivative and the original function is given by the following:

- If $f' > 0$ on an interval, then f is increasing on that interval.
- If $f' < 0$ on an interval, then f is decreasing on that interval.
- If $f'' > 0$ on an interval, then the graph of f is concave up on that interval.
- If $f'' < 0$ on an interval, then the graph of f is concave down on that interval.

We can do more with these principles now than we could in Chapter 2 because we now have formulas for the derivatives of the standard functions.

5.1 USING THE FIRST DERIVATIVE

Why Is it Useful to Know Where a Function is Increasing and Decreasing?

The following example shows how to use the derivative of a function to understand its graph. When we graph a function on a computer or graphing calculator, we see only part of the picture. The derivative can often direct our attention to important features on the graph.

Example 1 Sketch a helpful graph of the function $f(x) = x^3 - 9x^2 - 48x + 52$.

Solution Since f is a cubic, we expect a graph that is roughly S-shaped. Plotting this function with $-10 \leq x \leq 10$, $-10 \leq y \leq 10$, gives the two nearly vertical lines in Figure 5.1. We know that there is more going on than this, but how do we tell where to look?

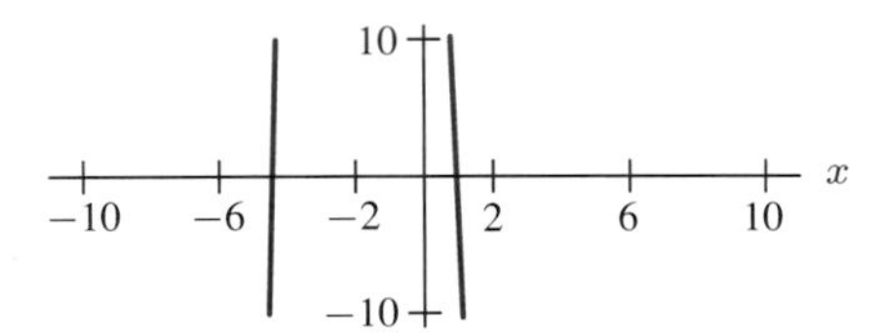

Figure 5.1: Unhelpful graph of $f(x) = x^3 - 9x^2 - 48x + 52$

We'll use the derivative to decide where the function is increasing and where it is decreasing. The derivative of f is

$$f'(x) = 3x^2 - 18x - 48.$$

To find where $f' > 0$ or $f' < 0$, we first find where $f' = 0$, that is, where $3x^2 - 18x - 48 = 0$. Solving the quadratic, we get $x = -2$ and $x = 8$. Since $f' = 0$ *only* at $x = -2$ and $x = 8$, and since f' is continuous, f' cannot change sign between these points. How can we tell the sign of f' between these points? The easiest way is to pick a point and substitute into f'. For example, since $f'(-3) = 33 > 0$, we know f' is positive for $x < -2$, so f is increasing for $x < -2$. Similarly, since $f'(0) = -48$ and $f'(10) = 72$, we know that f decreases between $x = -2$ and $x = 8$ and increases for $x > 8$. Summarizing the behavior of f on each interval:

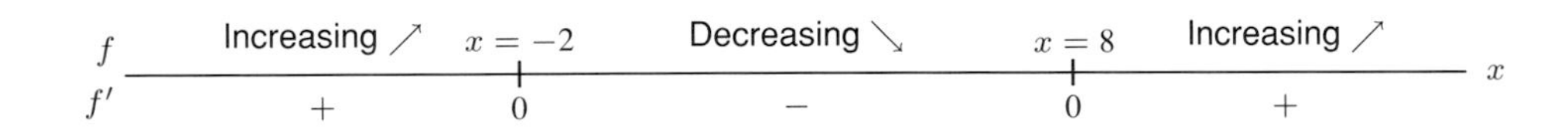

We find that $f(-2) = 104$ and $f(8) = -396$. Hence on the interval $-2 < x < 8$ the function decreases from a high of 104 to a low of -396. (Now we see why not much showed up in our first calculator graph.) One more point on the graph is easy to get: the y intercept, $f(0) = 52$. With just these three points we can get a much more helpful graph. By setting the domain and range in our calculator to $-10 \leq x \leq 20$ and $-400 \leq y \leq 400$, we get Figure 5.2. Where does the graph in Figure 5.1 fit into this one?

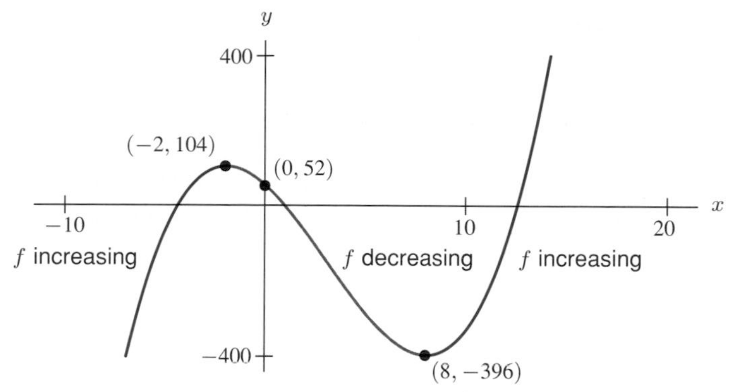

Figure 5.2: Helpful graph of $f(x) = x^3 - 9x^2 - 48x + 52$

Critical Points

In the preceding example, the points $x = -2$ and $x = 8$, where $f'(x) = 0$, played a key role. Now we will give a name to such points.

> For any function f, a point p in the domain of f where $f'(p) = 0$ or $f'(p)$ is undefined is called a **critical point** of the function. In addition, the point $(p, f(p))$ on the graph of f is also called a critical point. A **critical value** of f is the value, $f(p)$, of the function at a critical point, p.

Notice that "critical point of f" can refer either to special points in the domain of f or to special points on the graph of f. You will know which meaning is intended from the context.

What Do the Critical Points Tell Us?

Geometrically, at a critical point where $f'(p) = 0$, the line tangent to the graph of f at p is horizontal. At a critical point where $f'(p)$ is undefined, there is no horizontal tangent to the graph—there's either a vertical tangent or no tangent at all. (For example, $x = 0$ is a critical point for the absolute value function $f(x) = |x|$.) However, most of the functions we will work with will be differentiable everywhere, and therefore most of our critical points will be of the $f'(p) = 0$ variety.

The points where $f'(p) = 0$ (or $f'(p)$ is undefined) divide the domain of f into intervals on which the sign of the derivative stays the same, either positive or negative. Therefore, *between two successive critical points the graph of a function cannot change direction; it either goes up or down.*

A function may have any number of critical points or none at all. (See Figures 5.3–5.5.)

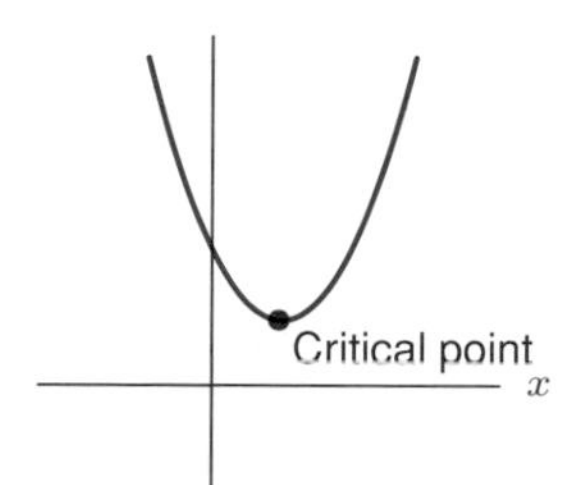

Figure 5.3: A quadratic: One critical point

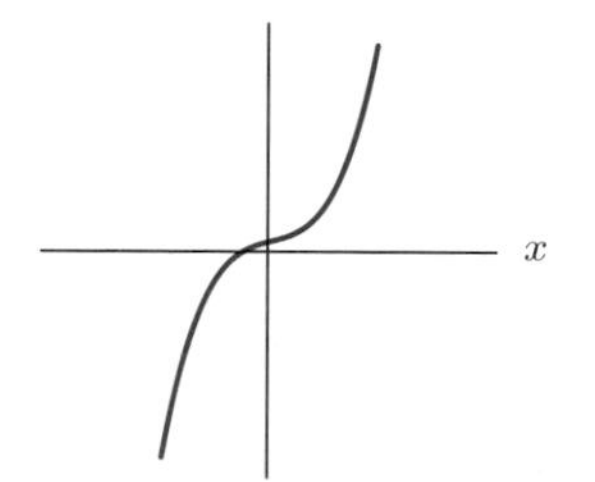

Figure 5.4: $f(x) = x^3 + x + 1$: No critical points

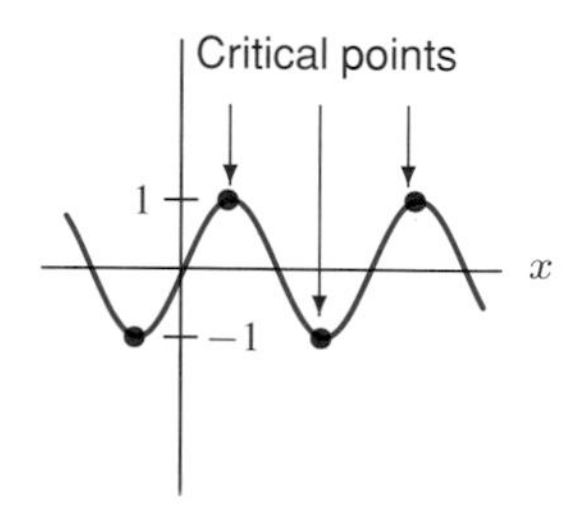

Figure 5.5: $f(x) = \sin x$: Infinitely many critical points

Local Maxima and Minima

What happens to a function at a critical point? Suppose that $f'(p) = 0$. We know that *at* p the graph has a horizontal tangent, but what happens *near* p? If f' has different signs on either side of p, then the graph changes direction at p, so the graph must look like one of those in Figure 5.6.

In the left-hand graph in Figure 5.6 we say that f has a local minimum at p, and in the right-hand graph we say that f has a local maximum at p. We use the adjective "local" because we are describing only what happens near p.

> Suppose p is a critical point of a function f. Then
> - f has a **local minimum** at p if, near p, the values of f get no smaller than $f(p)$.
> - f has a **local maximum** at p if, near p, the values of f get no larger than $f(p)$.

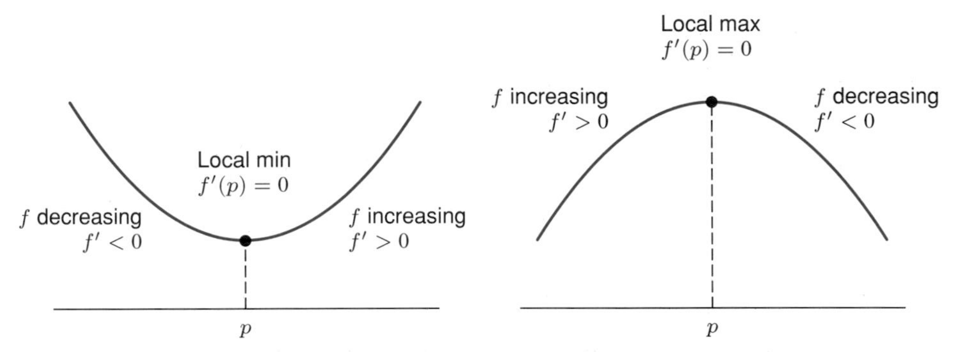

Figure 5.6: Changes in direction: Local maxima and minima

How Do We Decide Which Critical Points Are Local Maxima and Which Are Local Minima?

As Figure 5.6 illustrates, we have the following criterion.

> **The First-Derivative Test for Local Maxima and Minima**
>
> Suppose p is a critical point in the domain of f and f is continuous at p. If f' changes sign at p, then f has either a local minimum or a local maximum at p.
> - If f' is negative to the left of p and positive to the right of p, then f has a local minimum at p.
> - If f' is positive to the left of p and negative to the right of p, then f has a local maximum at p.

Example 2 Identify the local maxima and minima of $f(x) = x^3 - 9x^2 - 48x + 52$. (This is a continuation of Example 1.)

Solution The function has derivative $f'(x) = 3x^2 - 18x - 48 = 3(x^2 - 6x - 16) = 3(x - 8)(x + 2)$. To the left of $x = -2$, f' is positive. Between $x = -2$ and $x = 8$, f' is negative. To the right of $x = 8$, f' is positive. By the first-derivative test, f has a local maximum at $x = -2$ and a local minimum at $x = 8$. (See Figure 5.7.)

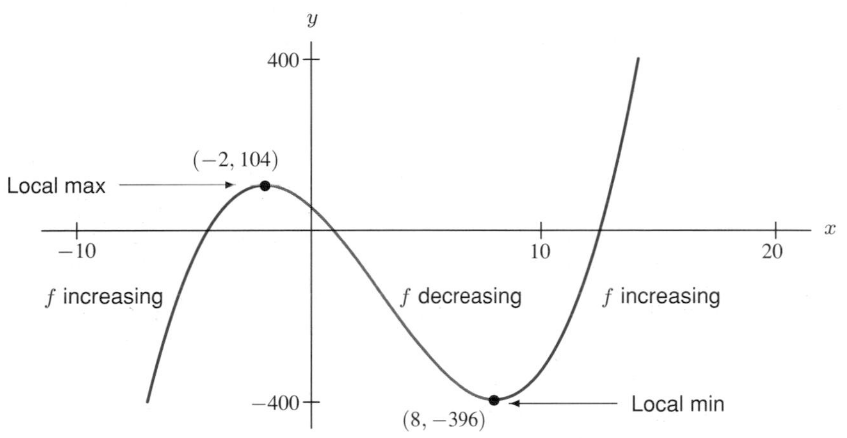

Figure 5.7: Local maxima and minima of $f(x) = x^3 - 9x^2 - 48x + 52$

Warning!

The sign of f' doesn't *have* to change at a critical point. Consider $f(x) = x^3$, whose graph is in Figure 5.8. The derivative , $f'(x) = 3x^2$, is positive on both sides of $x = 0$, so f increases on both sides of $x = 0$, and there is neither a local maximum nor a local minimum at $x = 0$. In other words, *a function doesn't have to have a local maximum or local minimum at every critical point.*

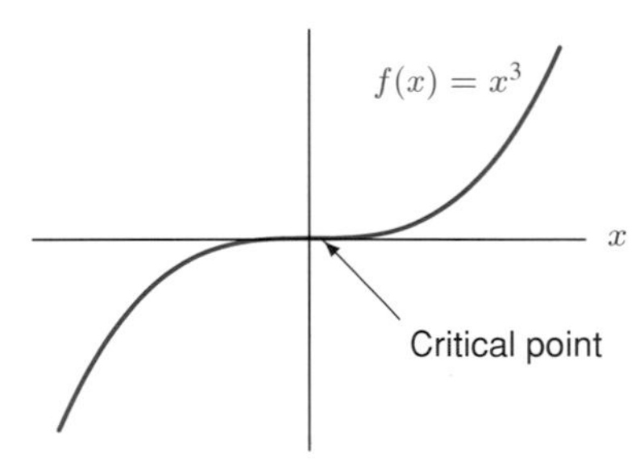

Figure 5.8: Critical point which is not a local maximum or minimum.

Global Maxima and Minima

The local maxima and minima tell us where a function is locally largest or smallest. However we are often more interested in where the function is absolutely largest or smallest in a given domain. We say

- f has a **global minimum** at p if all values of f are greater than or equal to $f(p)$.
- f has a **global maximum** at p if all values of f are less than or equal to $f(p)$.

How Do We Find Global Maxima and Minima?

If f is a continuous function defined on a closed interval $a \leq x \leq b$ (i.e., an interval containing its endpoints), Figure 5.9 shows that the global maximum or minimum of f occurs either at a local maximum or a local minimum, or at one of the endpoints, $x = a$ or $x = b$.

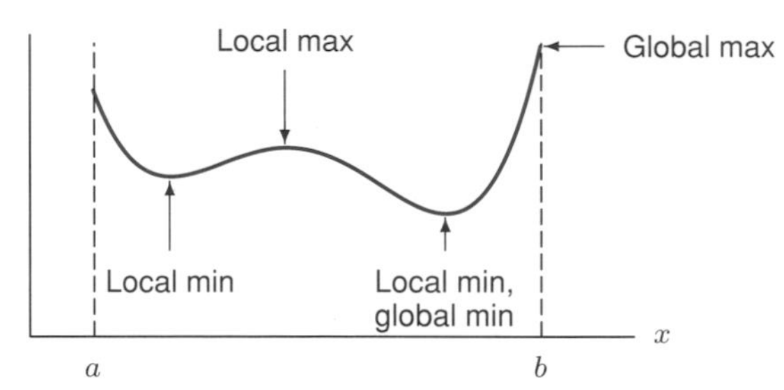

Figure 5.9: Global maximum and minimum on a closed interval $a \leq x \leq b$

To find the global maximum and minimum of a continuous function on a closed interval: Compare values of the function at all the critical points in the interval and at the endpoints.

What if the function is defined on an open interval $a < x < b$ (i.e., an interval not including its endpoints), or on the entire real line? We say there is no global maximum in Figure 5.10 because the function has no actual largest value. The global minimum in Figure 5.10 coincides with the local minimum and is marked. Figure 5.10 and Figure 5.11 show that there may not be a global maximum or minimum on an open interval.

To find the global maximum and minimum of a continuous function on an open interval or on the entire real line: Find the value of the function at all the critical points and sketch a graph.

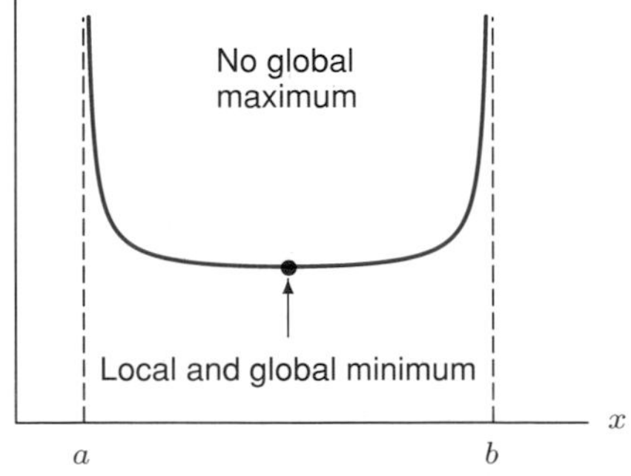

Figure 5.10: Global maximum and minimum on $a < x < b$

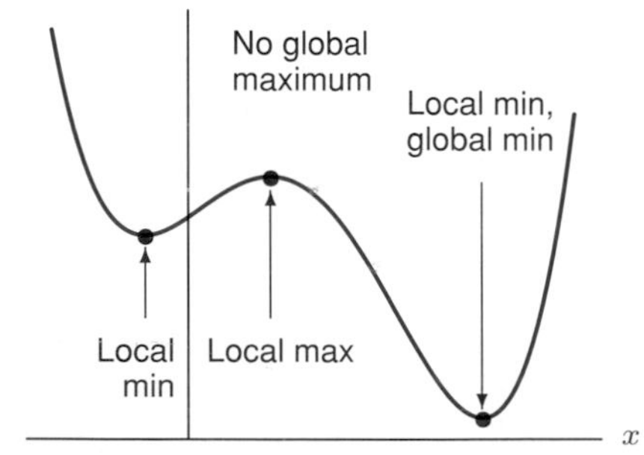

Figure 5.11: Global maximum and minimum when the domain is the real line.

Warning!

Notice that local maxima or minima of a function f occur at critical points, where $f'(p) = 0$ (or $f'(p)$ is undefined). Since global maxima or minima can occur at endpoints (where f' is not necessarily 0 or undefined), *not every global maximum or minimum is a local maximum or minimum.*

Example 3 Find the global maxima and minima of $f(x) = x^3 - 9x^2 - 48x + 52$ on the following intervals:

(a) $-5 \le x \le 12$ (b) $-5 \le x \le 14$ (c) $-5 \le x < \infty$.

Solution (a) We have calculated the critical points of this function previously using

$$f'(x) = 3x^2 - 18x - 48 = 3(x+2)(x-8)$$

so $x = -2$ and $x = 8$ are critical points. Evaluating f at the critical points and the endpoints, we discover

$$\begin{aligned} f(-5) &= (-5)^3 - 9(-5)^2 - 48(-5) + 52 = -58 \\ f(-2) &= 104 \\ f(8) &= -396 \\ f(12) &= -92. \end{aligned}$$

So the global maximum on $[-5, 12]$ is 104 and occurs at $x = -2$, and the global minimum on $[-5, 12]$ is -396 and occurs at $x = 8$.

(b) For the interval $[-5, 14]$, we compare

$$f(-5) = -58, \quad f(-2) = 104, \quad f(8) = -396$$

with the value of the function at the new endpoint:

$$f(14) = 360.$$

The global maximum is now 360 and occurs at $x = 14$, and the global minimum is still -396 and occurs at $x = 8$. Notice that since the function is increasing for $x > 8$, changing the right-hand end of the interval from $x = 12$ to $x = 14$ alters the global maximum but not the global minimum. See Figure 5.12.

(c) Figure 5.12 shows that for $-5 \le x < \infty$ there is no global maximum, because we can make $f(x)$ as large as we please by choosing x large enough. The global minimum remains -396 at $x = 8$.

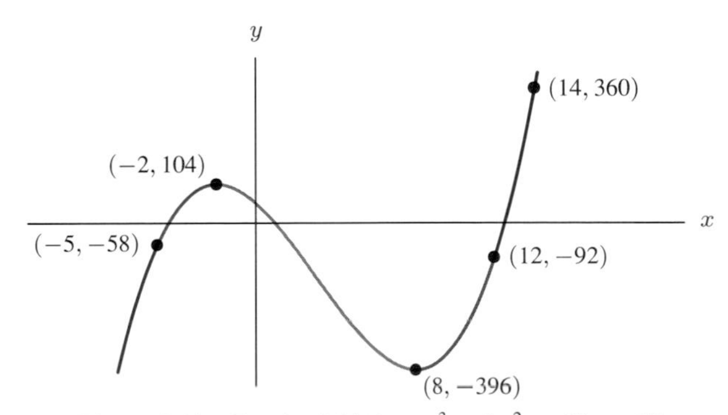

Figure 5.12: Graph of $f(x) = x^3 - 9x^2 - 48x + 52$

Example 4 Use a graph of the function $f(x) = \dfrac{1}{x(x-1)}$ to find its local maxima and minima, and confirm your result analytically. What are its global maxima and minima on $0 < x < 1$, on $-\infty < x < 0$, and on $1 < x < \infty$?

Solution The graph in Figure 5.13 shows that this function has no local minima but that there is a local maximum at about $x = \frac{1}{2}$. Explaining this result analytically means showing that it is what we would expect using the formula for the derivative. Since

$$f'(x) = -\frac{2x-1}{\big(x(x-1)\big)^2},$$

we have $f'(x) = 0$ where $2x - 1 = 0$. Thus the only critical point in the domain is $x = \frac{1}{2}$.

Furthermore, $f'(x) > 0$ where $0 < x < \frac{1}{2}$ and $f'(x) < 0$ where $\frac{1}{2} < x < 1$; thus, by the first-derivative test, the critical point $x = \frac{1}{2}$ is a local maximum. It is also a global maximum for f on $0 < x < 1$. The function does not have a global minimum on $0 < x < 1$ since $f(x)$ tends to $-\infty$ both as x tends to 0 from the right and as x tends to 1 from the left.

For $-\infty < x < 0$ or $1 < x < \infty$, there are no critical points and no local or global maxima or minima. Although $1/(x(x-1)) \to 0$ both as $x \to \infty$ and as $x \to -\infty$, we don't say 0 is a minimum (either local or global) because $1/\big(x(x-1)\big)$ never actually *equals* 0.

Notice that although $f' > 0$ everywhere that it is defined for $x < \frac{1}{2}$, f is not increasing throughout this interval. The problem is that f and f' are not defined at $x = 0$, so we cannot conclude from $f' > 0$ that f is increasing throughout this interval.

Example 5 The graph of $f(x) = \sin x + 2e^x$ is in Figure 5.14. Using the derivative, explain why there are no local maxima or minima for $x \geq 0$.

Solution Local maxima and minima can occur only where

$$f'(x) = \cos x + 2e^x = 0.$$

But $\cos x$ is always between -1 and 1, and $2e^x \geq 2$ for $x \geq 0$, so $f'(x)$ cannot be 0 for any $x \geq 0$. Therefore there are no local maxima or minima there. However, $x = 0$ gives the global minimum for $x \geq 0$. The point $x = 0$ is not a local minimum because it is not a critical point.

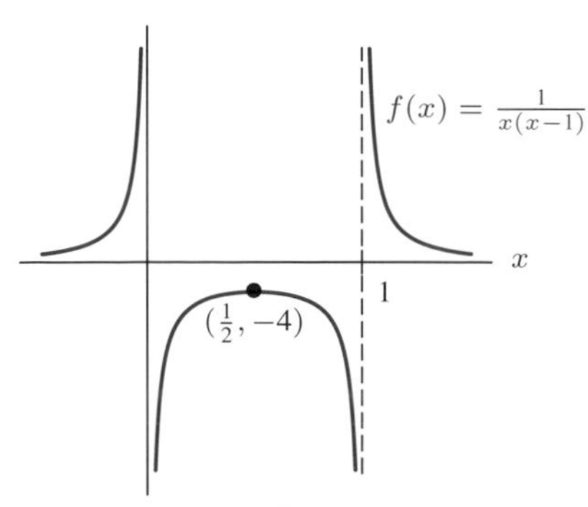

Figure 5.13: Find maxima and minima

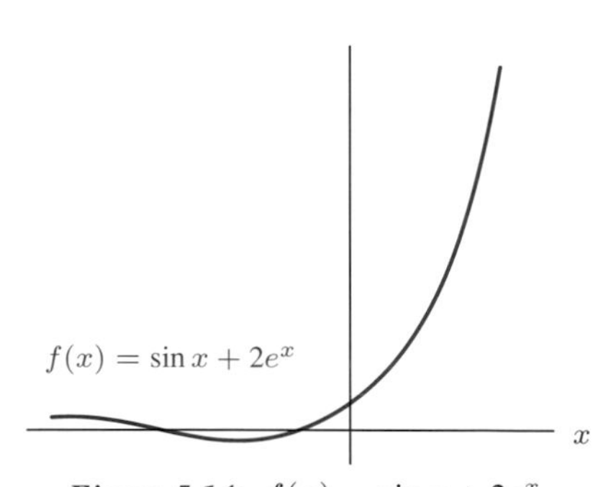

Figure 5.14: $f(x) = \sin x + 2e^x$

Problems for Section 5.1

1. Indicate all critical points on the graph of f in Figure 5.15 and determine which correspond to local maxima of f, which to local minima, and which to neither.

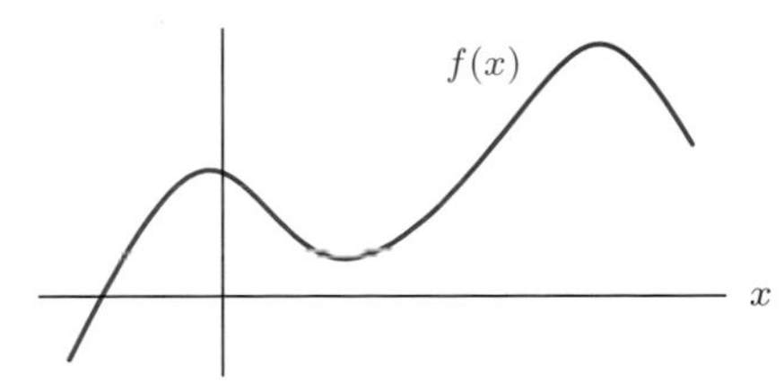

Figure 5.15

2. Sketch graphs of two continuous functions f and g, each of which has exactly five critical points, the points A–E in Figure 5.16, and which satisfy the following conditions:

   (a) $\lim_{x \to -\infty} f(x) = \infty$ and $\lim_{x \to \infty} f(x) = \infty$

   (b) $\lim_{x \to -\infty} g(x) = -\infty$ and $\lim_{x \to \infty} g(x) = 0$

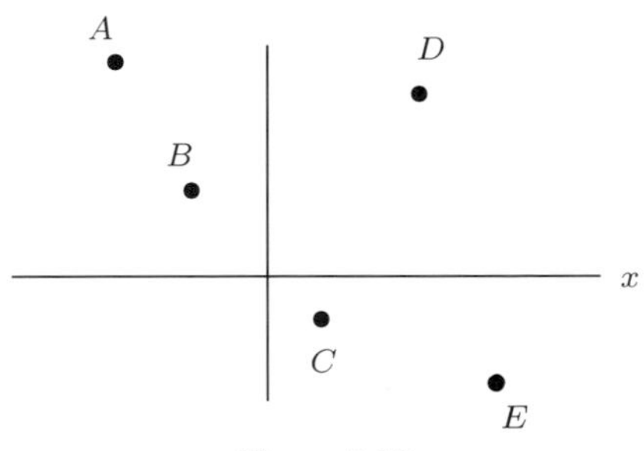

Figure 5.16

3. Indicate on the graph of the derivative function f' in Figure 5.17 the x-values that are critical points of the function f itself. At which critical points does f have local maxima, local minima, or neither?

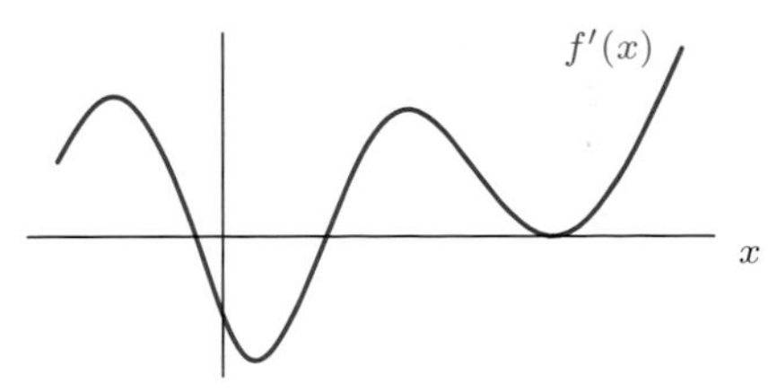

Figure 5.17

4. How many real roots does the equation $x^5 + x + 7 = 0$ have? How do you know?

In each of Problems 5–10, find all critical points and then use the first-derivative test to determine local maxima and minima. When you are finished, use a calculator or computer to sketch a graph of each function to check your work.

5. $f(x) = 2x^3 + 3x^2 - 36x + 5$
6. $f(x) = 3x^4 - 4x^3 + 6$
7. $f(x) = (x^2 - 4)^7$
8. $f(x) = (x^3 - 8)^4$
9. $f(x) = \dfrac{x}{x^2 + 1}$
10. $f(x) = 2x^2e^{5x} + 1$

11. Plot the graph of $f(x) = x^3 - e^x$ using a graphing calculator or computer to find all local and global maxima and minima for:
    (a) $-1 \le x \le 4$ (b) $-3 \le x \le 2$
12. For $y = f(x) = x^{10} - 10x$, and $0 \le x \le 2$, find the value(s) of x for which:
    (a) $f(x)$ has a local maximum or local minimum. Indicate which ones are maxima and which are minima.
    (b) $f(x)$ has a global maximum or global minimum. Indicate which ones are maxima and which are minima.
13. For $f(x) = x - \ln x$, and $0.1 \le x \le 2$, find the value(s) of x for which:
    (a) $f(x)$ has a local maximum or local minimum. Indicate which ones are maxima and which are minima.
    (b) $f(x)$ has a global maximum or global minimum. Indicate which ones are maxima and which are minima.
14. For $f(x) = \sin^2 x - \cos x$, and $0 \le x \le \pi$, find, to two decimal places, the value(s) of x for which:
    (a) $f(x)$ has a local maximum or local minimum. Indicate which ones are maxima and which are minima.
    (b) $f(x)$ has a global maximum or global minimum. Indicate which ones are maxima and which are minima.
15. (a) Water is flowing at a constant rate into a cylindrical container standing vertically. Sketch a graph showing the depth of water against time.
    (b) Water is flowing at a constant rate into a cone-shaped container standing on its point. Sketch a graph showing the depth of the water against time.
16. Choose the constants a and b in the function

$$f(x) = axe^{bx}$$

    such that $f(\frac{1}{3}) = 1$ and the function has a maximum at $x = \frac{1}{3}$.
17. Use the derivative formulas and algebra to find the intervals over which the function $f(x) = (x + 50)/(x^2 + 525)$ is increasing and is decreasing. It is possible to solve this problem on a graphing calculator or computer, but it is somewhat difficult; describe the difficulty.
18. Sketch the graph of

$$f(x) = 2x^3 - 9x^2 + 12x + 1.$$

    Use the graph to decide how many solutions the following equations have:
    (a) $f(x) = 10$ (b) $f(x) = 5$ (c) $f(x) = 0$ (d) $f(x) = 2e$
    You need not find these solutions.

19. Use derivatives and the fact that $\sin 0 = 0$ to show that $\sin x \leq x$ for all $x \geq 0$.

20. Suppose f has a continuous derivative everywhere. From the values of $f'(\theta)$ in the table below, estimate the θ values with $1 < \theta < 2.1$ at which $f(\theta)$ has a local maximum or minimum, and identify which is which.

θ	$f'(\theta)$
1.0	2.37
1.1	0.31
1.2	−2.00
1.3	−3.45
1.4	−3.34
1.5	−1.70

θ	$f'(\theta)$
1.6	0.76
1.7	2.80
1.8	3.61
1.9	2.76
2.0	0.69
2.1	−1.62

21. (a) On a computer or calculator, graph $f(\theta) = \theta - \sin\theta$. Can you tell whether the function has any zeros in the interval $0 \leq \theta \leq 1$?
   (b) Find f'. What does the sign of f' tell you about the zeros of f in the interval $0 \leq \theta \leq 1$?

22. Assume the function f is differentiable everywhere and has just one critical point, at $x = 3$. In parts (a) through (d), you are given additional conditions. In each case decide whether $x = 3$ is a local maximum, a local minimum, or neither. Explain your reasoning. Also sketch possible graphs for all four cases.
   (a) $f'(1) = 3$ and $f'(5) = -1$
   (b) $\lim_{x\to\infty} f(x) = \infty$ and $\lim_{x\to-\infty} f(x) = \infty$
   (c) $f(1) = 1$, $f(2) = 2$, $f(4) = 4$, $f(5) = 5$
   (d) $f'(2) = -1$, $f(3) = 1$, $\lim_{x\to\infty} f(x) = 3$

23. (a) Show that $x > 2\ln x$ for all $x > 0$.
   [Hint: Find the minimum of $f(x) = x - 2\ln x$.]
   (b) Use the above result to show that $e^x > x^2$ for all positive x.
   (c) Is $x > 3\ln x$ for all positive x?

24. Show that for all values of x

$$x^4 - 4x > -4.$$

25. The rabbit population on a small Pacific island is approximated by

$$P(t) = \frac{2000}{1 + e^{(5.3-0.4t)}}$$

with t measured in years since 1774, when Captain James Cook left 10 rabbits on the island. Using a calculator or computer:
   (a) Graph P. Does the population level off?
   (b) Estimate when the rabbit population grew most rapidly. How large was the population at that time?
   (c) What natural causes could lead to the shape of the graph of P?

5.2 USING THE SECOND DERIVATIVE

Recall that the sign of the second derivative tells us whether the graph of f is concave up or concave down. If $f'' > 0$ on an interval, then the graph of f is concave up on the interval; and if $f'' < 0$ then the graph is concave down on the interval.

Why Is It Useful to Know the Concavity of f?

Geometrically, if a curve is concave up near some point then it lies above its tangent line at that point and if it is concave down it lies below its tangent line at that point. See Figure 5.18. The next example shows how we can use this to get estimates for functions.

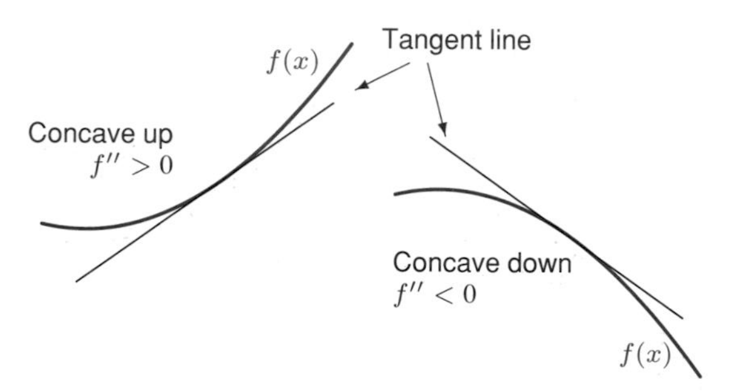

Figure 5.18: Concavity and the tangent line

Example 1 Explain why we know that $e^x \geq 1 + x$ for all values of x.

Solution Figure 5.19 shows that the graph of the function $f(x) = e^x$ is concave up everywhere, and the equation of its tangent line at the point $(0, 1)$ is $y = x + 1$. Since the graph always lies above its tangent, we have the inequality

$$e^x \geq 1 + x.$$

Problem 30 on page 263 shows how this inequality can be used to compute e.

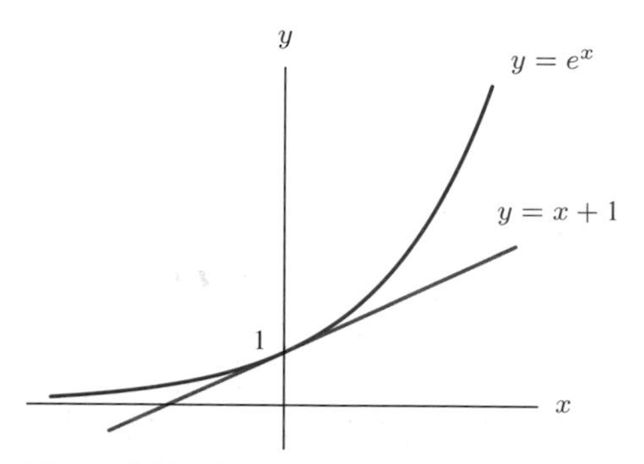

Figure 5.19: Graph showing that $e^x \geq 1 + x$

In the next example, we will see that knowing the concavity of f can be enough to determine the end behavior of a function.

Example 2 If f is increasing and its graph is concave up, what can you conclude about $\lim_{x\to\infty} f(x)$? What can you say about $\lim_{x\to\infty} g(x)$ if g is decreasing and its graph is concave down?

Solution We must have $\lim_{x\to\infty} f(x) = \infty$ and $\lim_{x\to\infty} g(x) = -\infty$. Can you draw a sketch to show why this must be true?

The Second-Derivative Test for Local Maxima and Minima

Knowing the concavity can also be useful in testing if a critical point is a local maximum or a local minimum. Suppose p is a critical point of f, so that $f'(p) = 0$ and the graph of f has a horizontal tangent line at p. If the graph is concave up at p, then f has a local minimum at p. Likewise, if the graph is concave down, f has a local maximum. (See Figure 5.20.) Thus we have the following result:

The Second-Derivative Test for Local Maxima and Minima

- If $f'(p) = 0$ and $f''(p) > 0$ then f has a local minimum at p.
- If $f'(p) = 0$ and $f''(p) < 0$ then f has a local maximum at p.

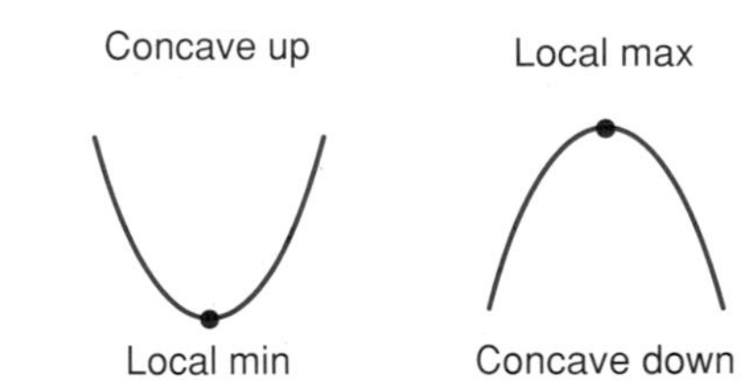

Figure 5.20: Local maxima and minima and concavity

Here are some examples of the use of the second-derivative test.

Example 3 Classify as local maxima or local minima the critical points of

(a) $f(x) = x^3 - 9x^2 - 48x + 52$ (b) $g(x) = xe^{-x}$ (c) $h(x) = x + \dfrac{1}{x}$

Solution (a) As we saw in Example 2 on page 246,

$$f'(x) = 3x^2 - 18x - 48$$

and the critical points of f are $x = -2$ and $x = 8$. We have

$$f''(x) = 6x - 18.$$

Thus

$$f''(8) = 6(8) - 18 = 30 > 0\,.$$

So f has a local minimum at $x = 8$. Since

$$f''(-2) = 6(-2) - 18 = -30 < 0,$$

f has a local maximum at $x = -2$. (A graph of this function is in Figure 5.22.)

(b) We have

$$g'(x) = e^{-x} - xe^{-x} = (1 - x)e^{-x}\,.$$

Hence $x = 1$ is the only critical point. We see that g' changes from positive to negative at $x = 1$ since e^{-x} is always positive, so by the first-derivative test g has a local maximum at $x = 1$. If we wish to use the second-derivative test, we compute

$$g''(x) = (x - 2)e^{-x}$$

and thus $g''(1) = (-1)e^{-1} < 0$, so again $x = 1$ gives a local maximum. (See Figure 5.23.)

(c) For $h(x) = x + \dfrac{1}{x}$ we calculate

$$h'(x) = 1 - \frac{1}{x^2}$$

and so the critical points of h are at $x = \pm 1$. Now

$$h''(x) = \frac{2}{x^3}$$

so $h''(1) = 2 > 0$ and $x = 1$ gives a local minimum. On the other hand $h''(-1) = -2 < 0$ so $x = -1$ gives a local maximum.

Warning!

The second-derivative test does not tell us anything if both $f'(p) = 0$ and $f''(p) = 0$. For example, if $f(x) = x^3$ and $g(x) = x^4$, both $f'(0) = g'(0) = 0$ and $f''(0) = g''(0) = 0$. However, the point $x = 0$ is a minimum for g but is neither a maximum nor a minimum for f. When the second derivative test fails to give information about a critical point p because $f''(p) = 0$, the first-derivative test can still be useful. For example, g' changes sign from negative to positive at 0, so we know g has a local minimum there.

Inflection Points

In Section 5.1 we studied points where the slope changes sign, which led us to critical points. Now we will look at points where the concavity changes.

> A point at which the graph of a function changes concavity is called an **inflection point** of f.

The words "inflection point of f" can refer either to a point in the domain of f or to a point on the graph of f. The context of the problem will tell you which is meant.

How Do You Locate an Inflection Point?

Since the concavity changes at an inflection point, the sign of f'' changes there. It is positive on one side of the inflection point, and negative on the other; so at the inflection point, f'' is zero or undefined. (See Figure 5.21.)

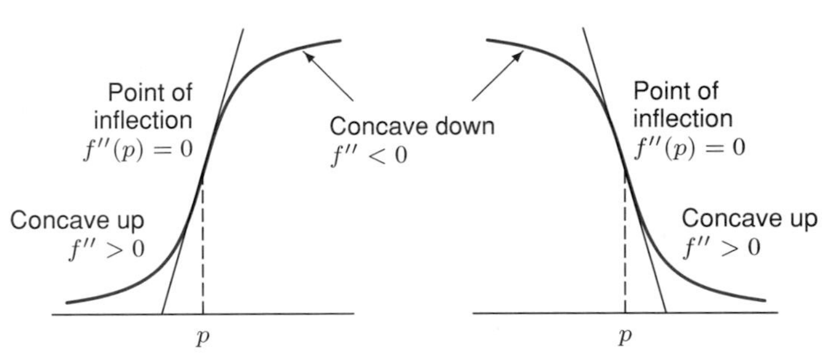

Figure 5.21: Change in concavity

Example 4 Find the inflection points of $f(x) = x^3 - 9x^2 - 48x + 52$.

Solution From the graph of $f(x)$ in Figure 5.22 we see that the function must have an inflection point. We calculate $f'(x) = 3x^2 - 18x - 48$, and $f''(x) = 6x - 18$, so $f''(x) = 0$ when $x = 3$. Further, $f''(x) < 0$ when $x < 3$ and $f''(x) > 0$ when $x > 3$, so the graph changes concavity at $x = 3$. Hence $x = 3$ is an inflection point.

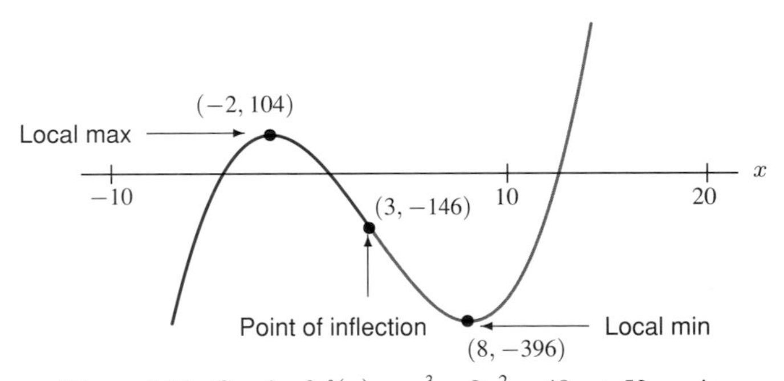

Figure 5.22: Graph of $f(x) = x^3 - 9x^2 - 48x + 52$, again

Example 5 Find the inflection points for $g(x) = xe^{-x}$ and sketch the graph.

Solution We have $g'(x) = (1 - x)e^{-x}$ and $g''(x) = (x - 2)e^{-x}$. Hence $g'' < 0$ when $x < 2$ and $g'' > 0$ when $x > 2$, so $x = 2$ is an inflection point. The graph is sketched in Figure 5.23.

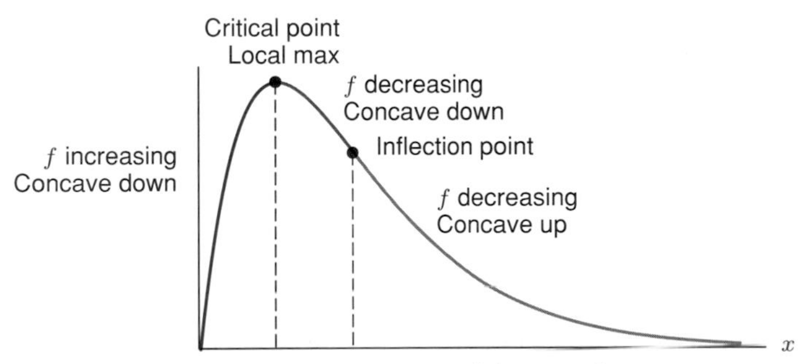

Figure 5.23: Graph of $g(x) = xe^{-x}$

Warning!

Not every point x where $f''(x) = 0$ (or f'' is undefined) is an inflection point (just as not every point where $f' = 0$ is a local maximum or minimum). For instance $f(x) = x^4$ has $f''(x) = 12x^2$ so $f''(0) = 0$, but $f'' > 0$ when $x > 0$ and when $x < 0$, so there is *no* change in concavity at $x = 0$. See Figure 5.24.

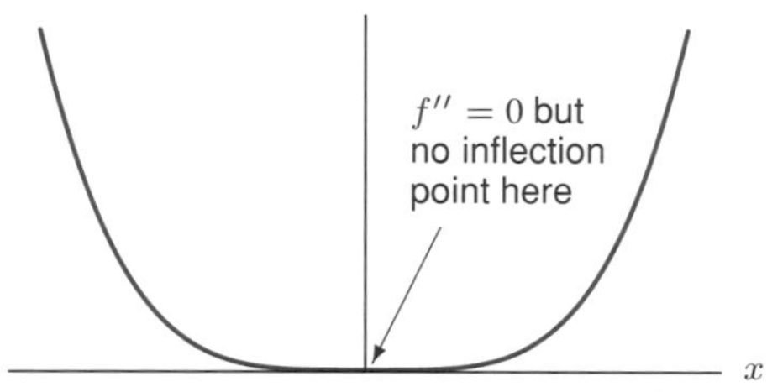

Figure 5.24: Graph of $f(x) = x^4$

The Relation Between Inflection Points and Local Maxima and Minima of the Derivative

Inflection points can also be interpreted in terms of first derivatives. Recall that the graph of f is concave up where f' is increasing and concave down where f' is decreasing. So the concavity changes where f' changes from increasing to decreasing or from decreasing to increasing; that is, where f' has a local maximum or a local minimum. Another way to see this is to notice that if p is an inflection point, then $f''(p) = 0$ (or $f''(p)$ is undefined) and hence p is a critical point of the derivative function f'. Further, since f'' changes sign at p, this critical point is a local maximum or minimum of f'.

> A function f has an inflection point at p
> - If f' has a local minimum or a local maximum at p
> - If f'' changes sign at p

Example 6 Sketch the graph of $f(x) = x + \sin x$, and find the points where it is increasing most rapidly and where it is increasing least rapidly.

Solution To see where $f(x)$ is increasing most and least rapidly, look at the graph of the derivative, $f'(x) = 1 + \cos x$ shown in Figure 5.25. Notice that the graph touches the x-axis at $x = \ldots, -3\pi, -\pi, \pi, 3\pi, 5\pi, \ldots$; these are the critical points of f. Everywhere else the graph of f' is above the x-axis, and so $f'(x) > 0$; thus f is increasing between its critical points. See Figure 5.26.

Where is f increasing most rapidly? At the points $x = \ldots, -2\pi, 0, 2\pi, 4\pi, 6\pi, \ldots$, because these points are local maxima for f' and f' has the same value at each of them. Likewise f is increasing least rapidly at the points $x = \ldots, -3\pi, -\pi, \pi, 3\pi, 5\pi, \ldots$, since these points are local minima for f'.

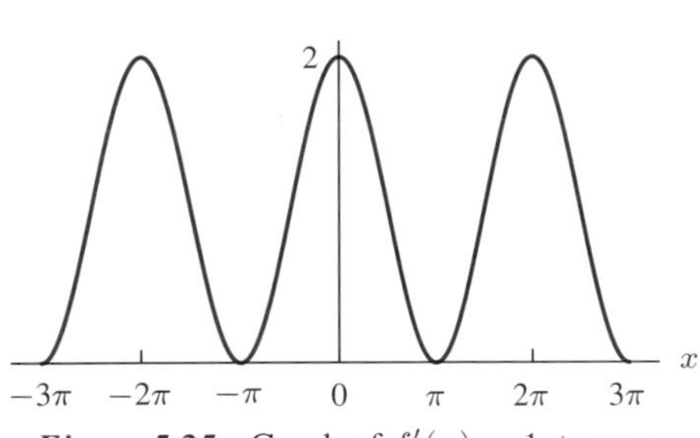

Figure 5.25: Graph of $f'(x) = 1 + \cos x$

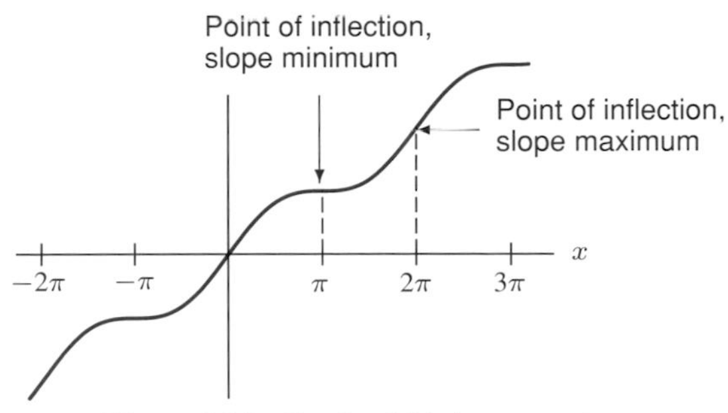

Figure 5.26: Graph of $f(x) = x + \sin x$

Notice that both the points where f is increasing most rapidly, and the points where it is increasing least rapidly, are inflection points of f. See Figure 5.26. This is because they are local maxima or minima of f', so f' changes from increasing to decreasing or from decreasing to increasing, and hence f changes concavity.

Example 7 Suppose that water is being poured into the vase in Figure 5.27 at a constant rate measured in volume per unit time. Graph $y = f(t)$, the depth of the water against time, t. Explain the concavity, and indicate the inflection points.

Solution At first the water level, y, rises quite slowly because the base of the vase is wide, and so it takes a lot of water to make the depth increase. However, as the vase narrows, the rate at which the water is rising increases. This means that initially y is increasing at an increasing rate, and the graph is concave up. The rate of increase in the water level is at a maximum when the water reaches the middle of the vase, where the diameter is smallest; this is an inflection point. After that, the rate at which y increases starts to decrease again, and so the graph is concave down. (See Figure 5.28.)

Figure 5.27: A vase

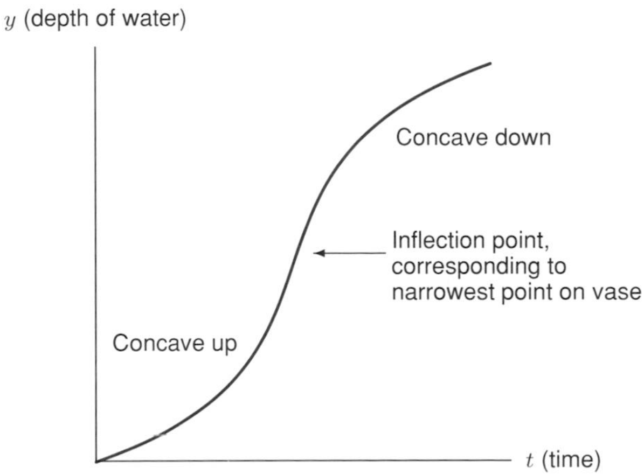

Figure 5.28: Graph of depth of water in the vase, y, against time, t

---

Problems for Section 5.2

1. Indicate on the graph of the derivative f' in Figure 5.29 the x values that are inflection points of the function f itself.

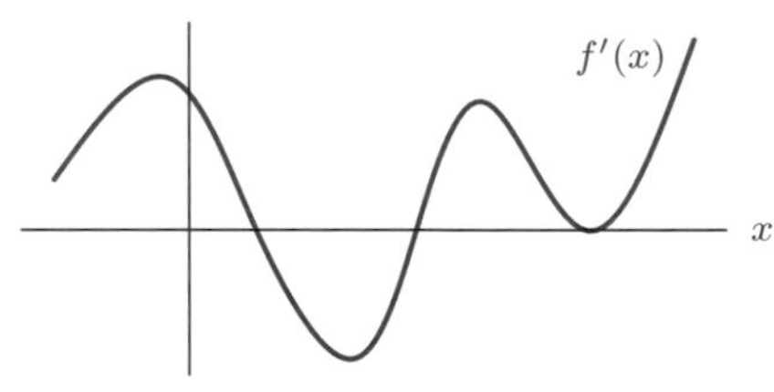

Figure 5.29

2. Indicate on the graph of the second derivative f'' in Figure 5.30 the x values that are inflection points of the function f itself.

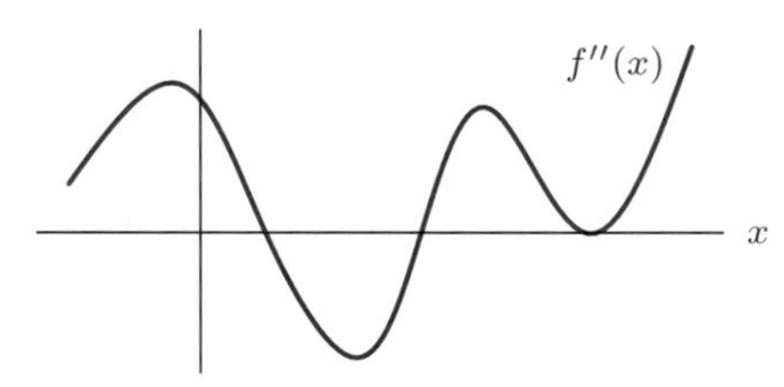

Figure 5.30

3. A function f has derivative $f'(x) = \cos(x^2) + 2x - 1$. Does f have a local maximum, a local minimum, or neither at its critical point $x = 0$?

Sketch the functions in Problems 4–7. Indicate clearly the coordinates of all maxima and minima and points of inflection.

4. $y = 0.5xe^{-10x}$
5. $y = 2 + 3\cos x, \quad \text{for} \quad 0 \leq x \leq 6\pi$
6. $f(x) = 3x^5 - 5x^3, \quad \text{for} \quad |x| \leq 1.5$
7. $f(x) = \sin x + \cos x, \quad \text{for} \quad 0 \leq x \leq 6\pi$

8. For $f(x) = e^{x/2} - \ln(x^2 + 1)$, find the coordinates of all intercepts, maxima, minima, and inflection points to two decimal places.
9. For $f(x) = x^{10} - 10x, \quad 0 \leq x \leq 2$, find the value(s) of x for which $f(x)$ is increasing most rapidly; decreasing most rapidly.
10. For $f(x) = x - \ln x, \quad 0.1 \leq x \leq 2$, find the value(s) of x for which $f(x)$ is increasing most rapidly; decreasing most rapidly.
11. For $f(x) = x - \cos x, \quad 0 \leq x \leq \pi$, find the value(s) of x for which
    (a) $f(x)$ is greatest; $f(x)$ is least.
    (b) $f(x)$ is increasing most rapidly; decreasing most rapidly.
    (c) The slopes of the lines tangent to the graph of f are increasing most rapidly.

12. For the function, f, given in the graph in Figure 5.31:
    (a) Sketch $f'(x)$.
    (b) Where does $f'(x)$ change its sign?
    (c) Where does $f'(x)$ have local maxima or minima?

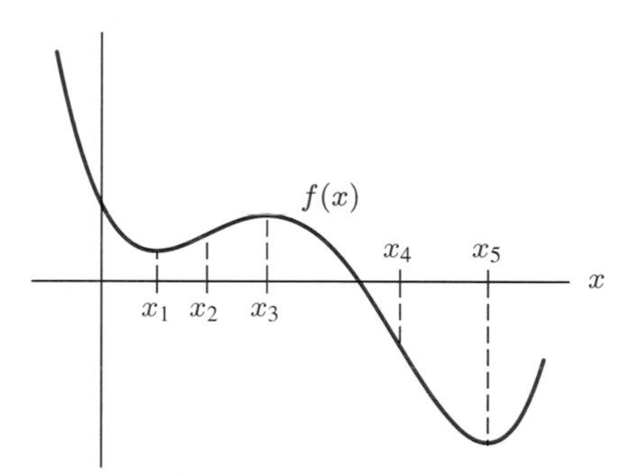

Figure 5.31

13. Using your answer to Problem 12 as a guide, write a short paragraph (using complete sentences) which describes the relationship between the following features of a function f:
    (a) The local maxima and minima of f.
    (b) The points at which f changes concavity.
    (c) The sign changes of f'.
    (d) The local maxima and minima of f'.

14. Suppose you know that a certain function f has $f(0) = 0$ and derivative given by

$$f'(x) = (\ln x)^2 - 2(\sin x)^4 \qquad \text{on the interval} \qquad 0 < x \leq 7.5$$

Use a calculator or computer to graph f' and its derivative. Clearly state where $f(x)$ is increasing and where it is decreasing, and where $f(x)$ is concave up and where it is concave down. Use this information to sketch a graph of $f(x)$.

15. Use concavity to show that $\ln x \leq x - 1$.

16. Use the fact that $\ln x$ and e^x are inverse functions to show that the inequalities $e^x \geq 1 + x$ and $\ln x \leq x - 1$ are equivalent.

17. For $f(x) = e^{1/x}$, with $x \neq 0$:
    (a) Describe the behavior of the function as $x \to \infty$, as $x \to -\infty$, as $x \to 0^+$, and as $x \to 0^-$ (i.e., as $x \to 0$ from the right and from the left, respectively).
    (b) For what values of x is the function increasing? Decreasing? Neither?
    (c) For what values of x is the graph of the function concave up? Concave down? Neither?
    (d) Use the preceding results to sketch a graph of f.

18. For $f(x) = \ln(x^3 + 1)$, with $x > -1$:
    (a) Describe the behavior of the function as $x \to \infty$ and as $x \to (-1)^+$ (i.e., as $x \to -1$ from the right).
    (b) For what values of x is the function increasing? Decreasing? Neither?
    (c) For what values of x is the the graph of the function concave up? Concave down? Neither?
    (d) Use the preceding results to sketch a graph of f.

19. Assume the polynomial f has exactly two local maxima and one local minimum.
    (a) Sketch a possible graph of f.
    (b) What is the largest number of zeros f could have?
    (c) What is the least number of zeros f could have?
    (d) What is the least number of inflection points f could have?
    (e) Is the degree of f even or odd? How can you tell?
    (f) What is the smallest degree f could have?
    (g) Find a possible formula for $f(x)$.

20. Sketch a smooth curve $f(x)$ for $0 \le x \le 10$ satisfying the following:
    - $|f''(x)| \le 0.5$ for all x.
    - $f'(x)$ takes on the values -2 and $+2$ somewhere.
    - f'' is not constant.

21. Let f be a function with positive values. Set $g = 1/f$.
    (a) If f is increasing in an interval around x_0, what about g?
    (b) If f has a local maximum at x_1, what about g?
    (c) If f is concave down at x_2, what about g?

For Problems 22–27, sketch a possible graph of $y = f(x)$, using the given information about the derivatives $y' = f'(x)$ and $y'' = f''(x)$. Assume that the function is defined and continuous for all real x.

22.

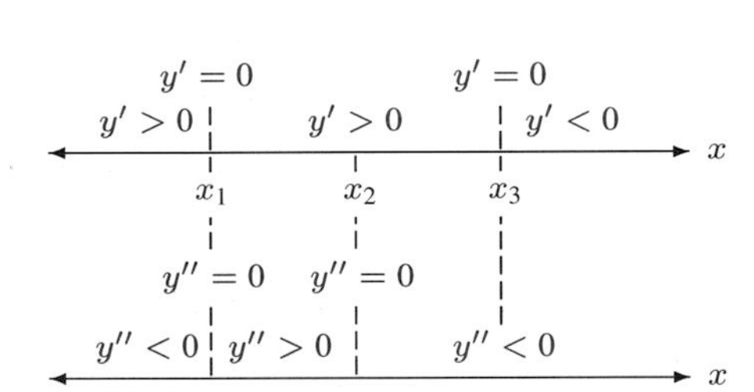

23.

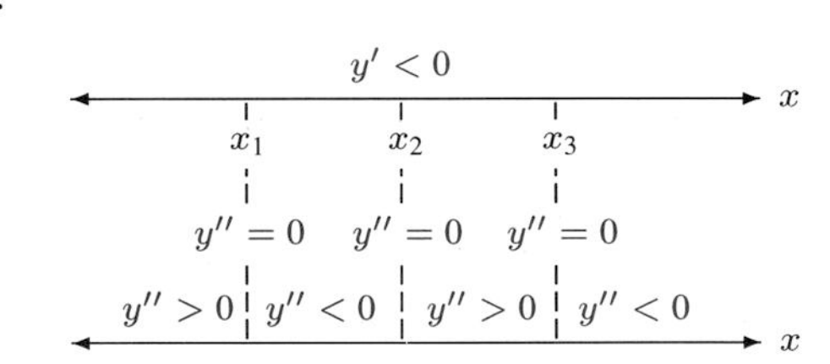

24.

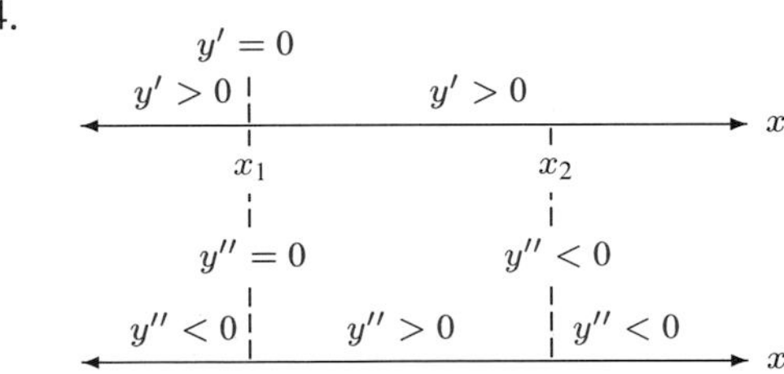

25.

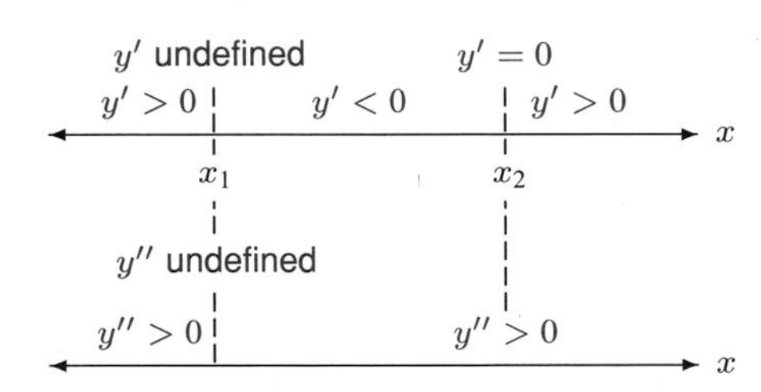

26.

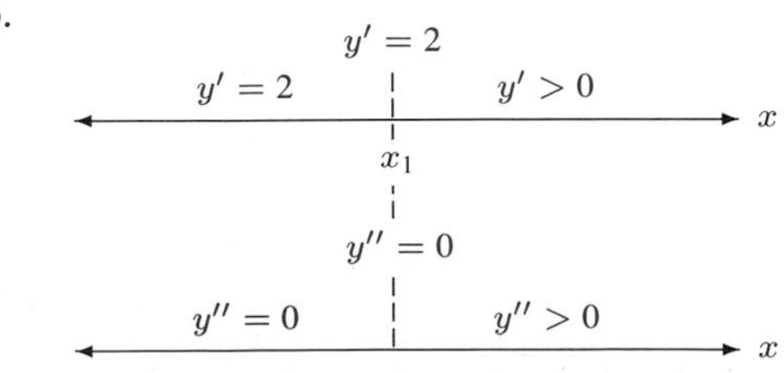

27.

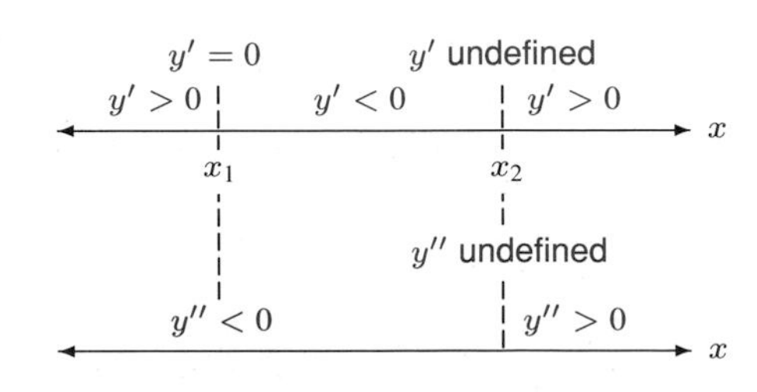

28. If water is flowing at a constant rate (i.e., constant volume per unit time) into the Grecian urn in Figure 5.32, sketch a graph of the depth of the water against time. Mark on the graph the time at which the water reaches the widest point of the vase.

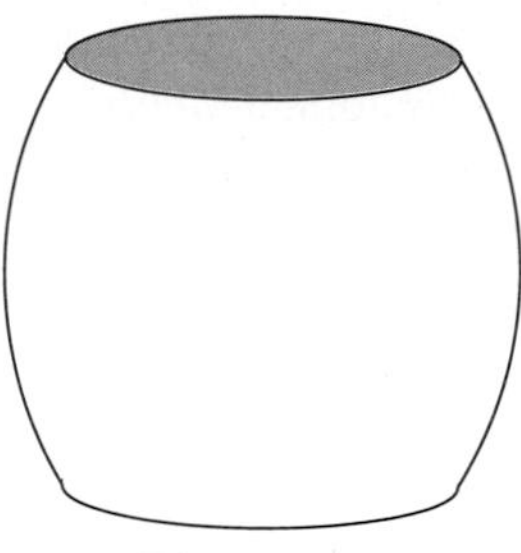

Figure 5.32

29. If water is flowing at a constant rate (i.e., constant volume per unit time) into the vase in Figure 5.33, sketch a graph of the depth of the water against time. Mark on the graph the time at which the water reaches the corner of the vase.

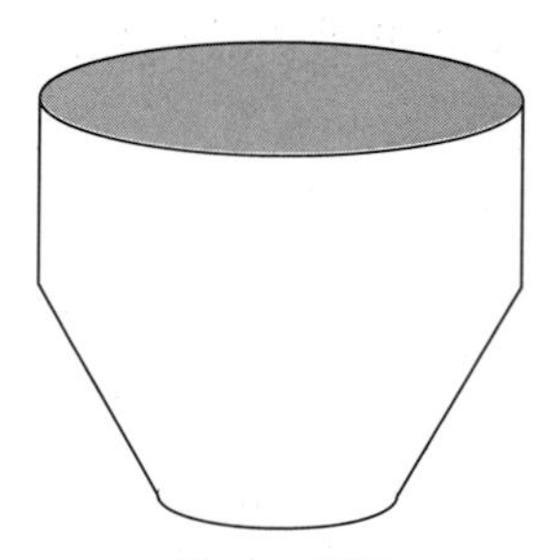

Figure 5.33

30. Given that $e^x \geq 1 + x$ for all x, let $x = 1/n$ and show that for every positive integer n,

$$e > \left(1 + \frac{1}{n}\right)^n.$$

Consider $x = -1/(n+1)$, and show that

$$e < \left(1 + \frac{1}{n}\right)^{n+1}$$

for all positive integers n.

Use your calculator with some specific choices of n to prove that $2.7 < e < 2.8$. Be clear about what choice of n you are making. Suppose you want to calculate e to 10 decimal places this way. Figure out a specific value of n that will enable you to pinch e into an interval of length less than 10^{-10}. (Try this on your calculator. Did you get a reasonable answer? If not, why not?)

5.3 FAMILIES OF CURVES: A QUALITATIVE STUDY

We saw in Chapter 1 that knowledge of one function can provide knowledge of the graphs of many others. The shape of the graph of $y = x^2$ also tells us, indirectly, about the shape of the graphs of $y = x^2 + 2$, $y = (x+2)^2$, $y = 2x^2$, and countless other functions. We say that all functions of the form $y = a(x+b)^2 + c$ form a *family of functions*; their graphs are identical to that of $y = x^2$ but for shifts and stretches determined by the values of a, b, and c. The constants a, b, c are called *parameters*. Different values of the parameters give different members of the family.

A reason for studying families of functions is their use in mathematical modeling. Confronted with the problem of modeling some phenomenon, a crucial first step involves recognizing families of functions which might fit the available data.

Curves of the Form $y = ax^3 + bx^2 + cx + d$

This is the family of cubics. From the specific examples we encountered in earlier chapters, we expect the graphs in this family to be S-shaped.

The effect of the parameter d is easy to see. It is the y-intercept of the curve, and changing d shifts the entire curve up or down without changing its shape. Since this is easy enough to account for, we set $d = 0$ and study the family $y = ax^3 + bx^2 + cx$; each of these curves passes through the origin.

One effect of the parameter a, called the leading coefficient, is also easy to observe. No matter what the size of a, provided $a \neq 0$, for large positive and negative values of x the term ax^3 dominates $bx^2 + cx$, and so for such values the curve looks like $y = ax^3$. If $a = 0$, we have a quadratic polynomial whose graph will be a parabola. For simplicity, we will assume that $a > 0$.

For the family $y = ax^3 + bx^2 + cx$, we have

$$\frac{dy}{dx} = 3ax^2 + 2bx + c, \qquad \frac{d^2y}{dx^2} = 6ax + 2b.$$

The formula for dy/dx is a quadratic polynomial, so it may equal 0 at exactly zero, one, or two points. Hence a given cubic may have *no* critical points (e.g., $y = x^3 + x$), *one* critical point (e.g., $y = x^3$), or *two* critical points (e.g., $y = x^3 - x$).

However, every cubic has exactly one inflection point. It occurs where $d^2y/dx^2 = 0$, which is at

$$x = \frac{-b}{3a}.$$

The second derivative changes from negative to positive at this point, so the graph changes from concave down to concave up there.

Since the graph of a cubic is S-shaped, it must look like one of the three possible cases in Figure 5.34. (Remember, we're assuming that $a > 0$.) The reasoning behind these sketches is as follows: in case (a), there is no critical point, so the curve must be always increasing. This is so because dy/dx can never change sign and hence must be always positive, or always negative. Since $a > 0$ and therefore $y \to +\infty$ as $x \to +\infty$, dy/dx must be always positive. A similar argument goes for case (b), but we allow the curve to flatten out at one point. It is not hard to show that in this case the critical point and the inflection point coincide. For case (c) the curve must get from $y = -\infty$ to $y = +\infty$ as x goes from $-\infty$ to $+\infty$, changing concavity once and having two horizontal tangents. If we disregard the size of the critical values (how high the peak, how low the valley) the shape must be something like case (c).

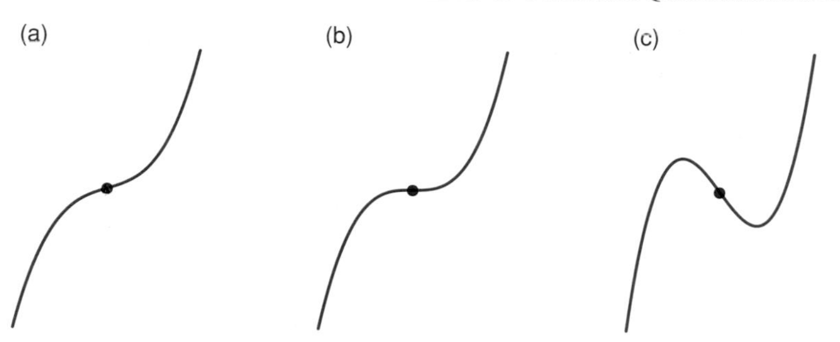

Figure 5.34: A cubic with (a) no critical points, (b) one critical point, (c) two critical points (The dots mark the inflection points.)

Curves of the Form $y = xe^{-bx}$

We will consider only $x \geq 0$. The graph of one member of the family, with $b = 1$, is shown in Figure 5.35. Such a graph represents a quantity that increases rapidly at first, and then decreases toward zero. For example, the number of bacteria in a person during the course of an infection, from onset to cure, might be described in this way. The question is, how exactly does the shape of the curve change when b varies? For simplicity we will consider only curves with $b > 0$. Note that all curves of the family pass through the origin. The behavior of the curves as $x \to \infty$ is also pretty clear. Since $b > 0$ (by assumption) and e^{bx} grows much faster than x as $x \to \infty$, we must have

$$xe^{-bx} = \frac{x}{e^{bx}} \to 0 \quad \text{as} \quad x \to \infty.$$

Geometrically, this means that the x-axis is a horizontal asymptote.

Next we find and classify critical points, and then we check concavity. We compute

$$\frac{dy}{dx} = x\left(-be^{-bx}\right) + e^{-bx} = (1 - bx)e^{-bx},$$

$$\frac{d^2y}{dx^2} = (1 - bx)\left(-be^{-bx}\right) + (-b)e^{-bx} = -b(2 - bx)e^{-bx}.$$

There is a critical point where $dy/dx = 0$, that is , where $(1 - bx)e^{-bx} = 0$. Since e^{-bx} is never zero, the only critical point is $x = 1/b$. Looking at the expression for dy/dx, we can see that if $x < 1/b$ then $dy/dx > 0$, and if $x > 1/b$, then $dy/dx < 0$. Hence there is a local maximum at $x = 1/b$. The maximum value is

$$y = \frac{1}{b}e^{-b(1/b)} = \frac{1}{be}.$$

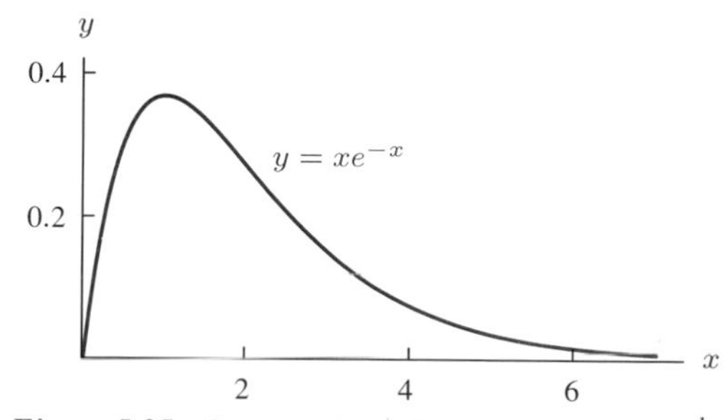

Figure 5.35: One member of the family $y = xe^{-bx}$

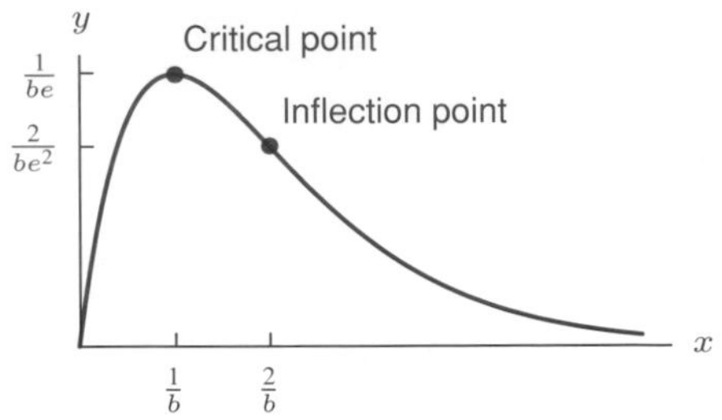

Figure 5.36: Graph of $y = xe^{-bx}$

The second derivative shows that $d^2y/dx^2 = 0$ only where $2 - bx = 0$. Solving for x gives $x = 2/b$. It is easy to see from the expression for d^2y/dx^2 that the second derivative changes from negative to positive at $x = 2/b$, so the graph changes from concave down to concave up at $x = 2/b$, which is therefore an inflection point. (See Figure 5.36.) At that point

$$y = \frac{2}{b}e^{-b(\frac{2}{b})} = \frac{2}{be^2}.$$

The most obvious effect of varying b is to move the critical point (local maximum) of the curve. If b is large, then $1/b$ is small and so xe^{-bx} reaches a low maximum very quickly. If b is small, then $1/b$ is very large, so the curve rises very high before the decay towards zero begins. The inflection point moves as b varies, too. It will always be twice as far from the y-axis as the critical point. See Figure 5.37.

Figure 5.38 shows several curves in the family on the same set of axes. Note also that all the curves have the same tangent line at the origin. This is because the derivative dy/dx has the same value at $x = 0$ for every b, namely $dy/dx = 1$. (Note that the scales on the x and y axes are different.)

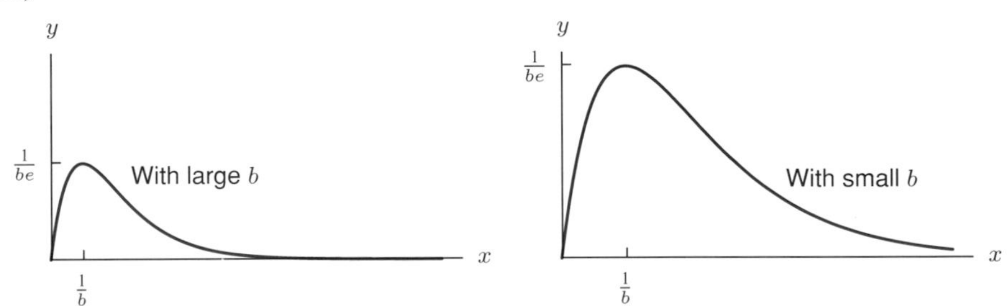

Figure 5.37: Graph of $y = xe^{-bx}$ with large b and small b

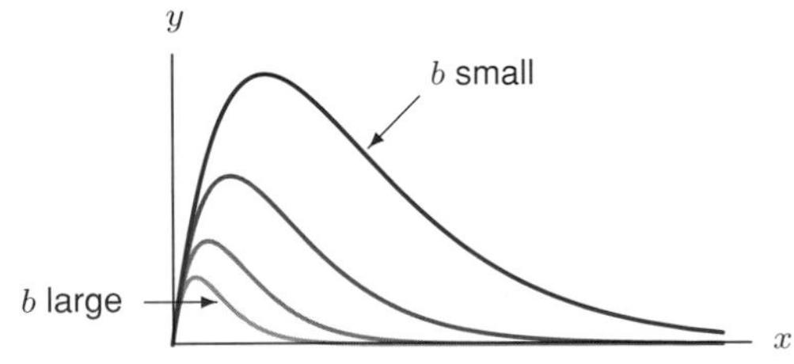

Figure 5.38: Graph of $y = xe^{-bx}$, with b varying.

Curves of the Form $y = a\left(1 - e^{-bx}\right)$

The graph of one member of the family, with $a = 2$ and $b = 1$, is shown in Figure 5.39. Such a graph represents a quantity which is increasing but leveling off. For example, a body falling against air

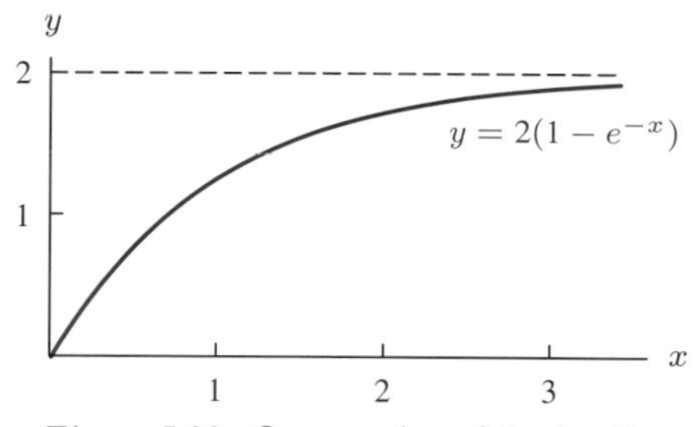

Figure 5.39: One member of the family $y = a(1 - e^{-bx})$

resistance speeds up and its velocity levels off as it approaches the terminal velocity. Similarly, if a pollutant pouring into a lake builds up toward a saturation level, its concentration may be described in this way. The graph might also represent the temperature of a cooking yam.

How does the shape change if a and b vary? For simplicity, we will assume that a and b are positive. We are interested in how the shape of the graph changes when one parameter is fixed and the other is allowed to vary.

First, fix $b > 0$. As $x \to \infty$, $e^{-bx} \to 0$ and so $y = a(1 - e^{-bx})$ approaches $y = a$ from below; thus $y = a$ is a horizontal asymptote. Physically, a may represent the terminal velocity of a falling object or the saturation level of a pollutant in a pond.

What happens with a fixed and b varying? How does the graph look when b is small and when b is large? As before, we try to locate and classify critical points and check concavity. We compute

$$\frac{dy}{dx} = abe^{-bx}, \qquad \frac{d^2y}{dx^2} = -ab^2e^{-bx}.$$

Since e^{-bx} is never zero there are no critical points. We see that dy/dx is always positive and d^2y/dx^2 is always negative; hence each curve is always increasing and always concave down.

How sharply does the curve rise and how soon does it get close to a? The answer to these questions depends on the size of b. Since $dy/dx = abe^{-bx}$, the slope of the tangent to the curve at $x = 0$ is ab. How long does it take the curve to climb halfway up from $y = 0$ to $y = a$? Where $y = a/2$, we have

$$a(1 - e^{-bx}) = \frac{a}{2}, \qquad \text{which leads to} \qquad x = \frac{\ln 2}{b}.$$

If b is large then $(\ln 2)/b$ is small, so in a very short distance the curve is already half way up to a. If b is small, then $(\ln 2)/b$ is large and you have to go a long way out to get up to $a/2$. See Figure 5.40. Fixing a and letting b vary from small to large gives the family in Figure 5.41.

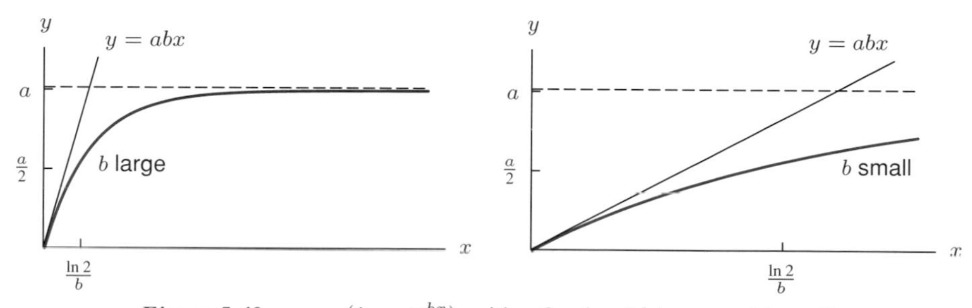

Figure 5.40: $y = a(1 - e^{-bx})$, with a fixed and b large, and b small

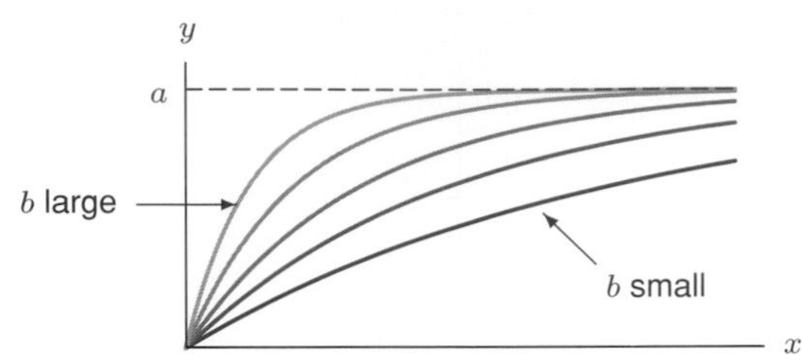

Figure 5.41: $y = a(1 - e^{-bx})$, with a fixed, b varying

Curves of the Form $y = e^{-(x-a)^2/b}$

This is a multiple of the *normal density* function, used in probability and statistics. In applications, it is common to scale the function by multiplying it by a positive constant, but we will just consider the two parameter family depending on a and b, and we will assume that $b > 0$. The graph of any member of the family is a bell-shaped curve as shown in Figure 5.42 (where $a = 1$ and $b = 2$).

As we saw in Chapter 1, the role of a is to move the curve to the right or left. Notice that for any curve in the family, y is always positive. Since $b > 0$, $y \to 0$ as $x \to \pm\infty$ for any choice of a. Thus, the x-axis is always a horizontal asymptote. To investigate the critical points and points of inflection, we calculate

$$\frac{dy}{dx} = -\frac{2(x-a)}{b}e^{-(x-a)^2/b}$$

and, using the product rule,

$$\begin{aligned}
\frac{d^2y}{dx^2} &= -\frac{2}{b}\left(\left(\frac{d}{dx}(x-a)\right)e^{-(x-a)^2/b} + (x-a)\frac{d}{dx}\left(e^{-(x-a)^2/b}\right)\right)\\
&= -\frac{2}{b}\left(e^{-(x-a)^2/b} + (x-a)\left(-\frac{2(x-a)}{b}e^{-(x-a)^2/b}\right)\right)\\
&= -\frac{2}{b}e^{-(x-a)^2/b}\left(1 + (x-a)\left(-\frac{2(x-a)}{b}\right)\right)\\
&= \frac{2}{b}e^{-(x-a)^2/b}\left(\frac{2(x-a)^2}{b} - 1\right).
\end{aligned}$$

There is a critical point where $dy/dx = 0$, that is, where

$$\frac{dy}{dx} = -\frac{2(x-a)}{b}e^{-(x-a)^2/b} = 0.$$

Since $e^{-(x-a)^2/b}$ is never zero, the only critical point is where $x = a$. At that point, $y = 1$. (Why?) Looking at the expression for dy/dx, we can see that if $x < a$, then $dy/dx > 0$, and if

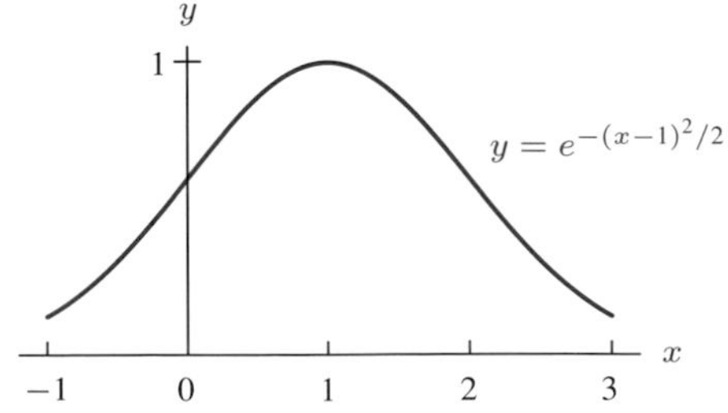

Figure 5.42: Graph of one member of the family $y = e^{-(x-a)^2/b}$

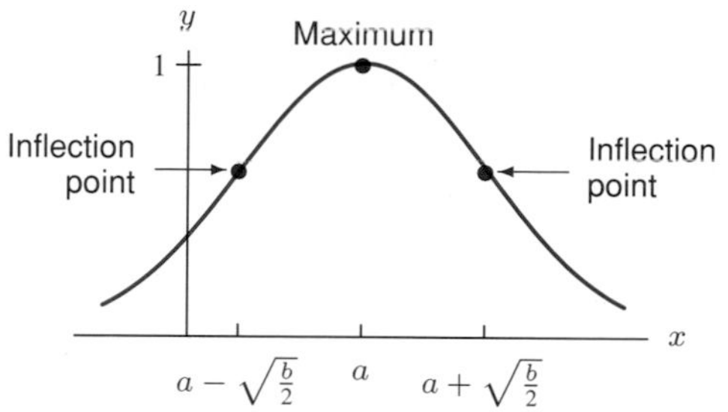

Figure 5.43: Graph of $y = e^{-(x-a)^2/b}$: bell-shaped curve with peak at $x = a$

$x > a$, then $dy/dx < 0$. Hence there is a local maximum at $x = a$. Inflection points occur where the second derivative changes sign; thus, we wish to find values of x for which $d^2y/dx^2 = 0$. Since $e^{-(x-a)^2/b}$ is never zero, the expression for d^2y/dx^2 tells us that this only occurs where

$$\frac{2(x-a)^2}{b} - 1 = 0.$$

Solving for x gives

$$x = a \pm \sqrt{\frac{b}{2}}.$$

Looking at the expression for d^2y/dx^2, we see that d^2y/dx^2 is negative for $x = a$ and positive for very large x, both positive and negative. (Why?) Therefore the concavity changes at $x = a + \sqrt{b/2}$ and at $x = a - \sqrt{b/2}$, so we have inflection points there. These points are symmetrically located at a distance of $\sqrt{b/2}$ on either side of $x = a$. (See Figure 5.43.) At these points $y = e^{-1/2} \approx 0.6$.

With this information we can see the effect of the parameters. The parameter a determines the location of the center of the bell and the parameter b determines how narrow or wide the bell is. (See Figure 5.44.) If b is small, then the inflection points are close to a and the bell is sharply peaked near a; if b is large, the inflection points are further away from a and the bell is more spread out.

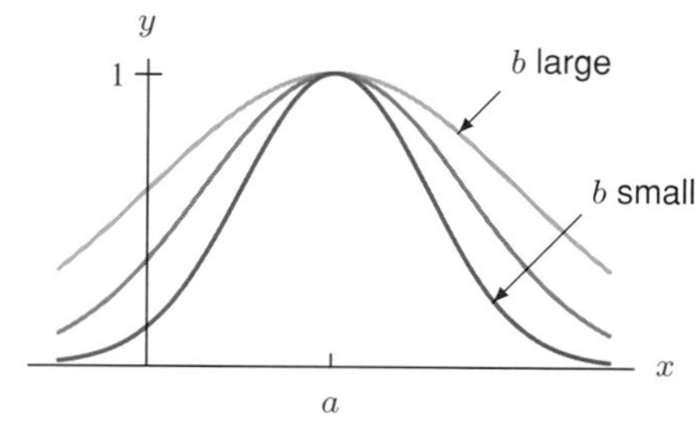

Figure 5.44: Graph of $y = e^{-(x-a)^2/b}$ with various values of b (a fixed)

Curves of the Form $y = e^{-ax} \sin bx$

As usual, we will assume that the parameters a and b are positive. The shape of $y = \sin bx$ is a sine curve with period $2\pi/b$. So if b is small, the period will be large and the graph of the function will be a stretched-out sine curve. If b is large, the period is small and the graph oscillates rapidly. What is the effect of multiplying by e^{-ax}? The e^{-ax} represents the amplitude of the wave. Since e^{-ax} decreases as x increases, the amplitude of the wave continuously shrinks, or is *damped*. These curves represent *damped harmonic motion*, such as we see with a pendulum oscillating with friction.

To see the shape of $y = e^{-ax}\sin bx$, we draw $y = e^{-ax}$ and $y = -e^{-ax}$ (its reflection in the x-axis) and fit an ever-shrinking version of $y = \sin bx$ between these two curves. (See Figure 5.45.)

As you can see in Figure 5.46, if a is large relative to b then the damping takes place very fast. If a is small relative to b, then the damping takes longer, so there may be many oscillations before we notice much damping. It is important to note however, that no matter the size of a or b, the limiting value is zero (i.e., $\lim_{x\to\infty} e^{-ax}\sin bx = 0$), so the oscillations die out as $x \to \infty$.

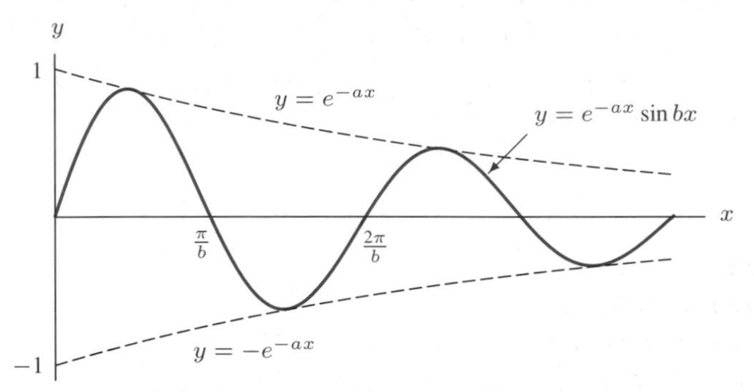

Figure 5.45: Graph of $y = e^{-ax}\sin bx$

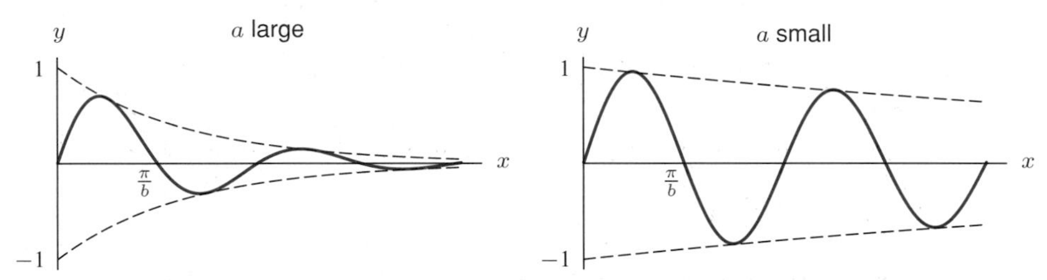

Figure 5.46: $y = e^{-ax}\sin bx$, with b fixed, and a large and a small.

Problems for Section 5.3

1. Under what conditions on a, b, and c is the cubic

$$f(x) = x^3 + ax^2 + bx + c$$

increasing everywhere?

2. Consider the function $p(x) = x^3 - ax$, where a is constant.
   (a) If $a < 0$, show that $p(x)$ is always increasing.
   (b) If $a > 0$, show that $p(x)$ has a local maximum and a local minimum.
   (c) Sketch and label typical graphs for the cases when $a < 0$ and when $a > 0$.

3. Consider the function $p(x) = x^3 - ax$, where a is constant and $a > 0$.
   (a) Find the local maxima and minima of p.
   (b) What effect does increasing the value of a have on the positions of the maxima and minima?
   (c) On the same axes, sketch and label the graphs of p for three positive values of a.

4. What effect does increasing the value of a have on the graph of $f(x) = x^2 + 2ax$? Consider roots, maxima and minima, and both positive and negative values of a.

5. (a) Find all critical points of $f(x) = x^4 + ax^2 + b$.
   (b) Under what conditions on a and b will this function have exactly one critical point? What is the one critical point, and is it a local maximum, a local minimum, or neither?
   (c) Under what conditions on a and b will this function have exactly three critical points? What are they and which are local maxima and which are local minima?
   (d) Is it ever possible for this function to have two critical points? No critical points? More than three critical points? Give an explanation in each case.

6. The number, N, of people who have heard a rumor spread by mass media is modeled by the following function of time, t:

$$N = a(1 - e^{-kt}).$$

Suppose that there are 200,000 people in the population who hear the rumor eventually. If 10% of them heard in the first day, find a and k, assuming t is measured in days.

7. The velocity, v, of a skydiver is given by

$$v = \frac{mg}{k}\left(1 - e^{-kt/m}\right)$$

where t is time, m is the mass of the parachutist, g is the acceleration due to gravity (a constant), and k is a positive constant. If t is measured in seconds and v is in feet per second, $g = 32$ ft/sec^2.

   (a) What is the skydiver's terminal velocity? That is, in terms of the constants in the problem, what is $\lim_{t\to\infty} v$?

Suppose you are a skydiver weighing 150 lb and that in the first 10 seconds you reach 80% of your terminal velocity. (Notice that mass and weight are different. Your mass m is your weight divided by g, the acceleration due to gravity.)

   (b) Find m and k.
   (c) Find your terminal velocity in ft/sec and in mph. (Note: 1 mph $= 22/15$ ft/sec.)

8. Suppose the temperature, T, of a yam put into a hot oven maintained at 200°C is given as a function of time by

$$T = a(1 - e^{-kt}) + b$$

where T is in degrees Celsius and t is in minutes.

   (a) If the yam starts at 20°C, find a and b.
   (b) If the temperature of the yam is initially increasing at 2°C per minute, find k.

9. Consider the family of curves

$$y = A(x + B)^2$$

   (a) Assuming $B = 0$, what is the effect of varying A on the graph? Consider:
      (i) $A > 0$ and $A < 0$.
      (ii) $A > 1$ and $0 < A < 1$.
   (b) Assuming $A = 1$, what is the effect of varying B on the graph?
   (c) Write a couple of sentences describing the role played by A and B in determining the shape of the graph. Illustrate your answer with sketches.

10. Consider the family
$$y = \frac{A}{x+B}.$$
(a) If $B = 0$, what is the effect of varying A on the graph?
(b) If $A = 1$, what is the effect of varying B?
(c) On one set of axes, graph the function for several values of A and B.

11. Consider the family of curves
$$y = ae^{-bx^2}, \qquad \text{for positive } a, b.$$
Analyze the effect of varying a and b on the shape of the curve. Illustrate your answer with sketches.

12. Consider the *surge function*
$$y = axe^{-bx}, \qquad \text{for positive } a, b.$$
(a) Find the maxima, minima, and points of inflection.
(b) How does varying a and b affect the shape of the graph?
(c) On one set of axes, graph this function for several values of a and b.

13. Consider the family of curves
$$y = A\sin(Bx + C).$$
What is the effect of varying A, B, and C on the shape of the graph? For each parameter, sketch some graphs for various values (holding the others constant).

14. (a) Assuming $x \geq 0$ and $a > 0$, find the maximum and minimum values of
$$y = e^{-ax}\sin x$$
in terms of a.
(b) Describe in words how the position and values of these maxima and minima change if a increases. Give an intuitive justification for this.

15. The study of resonance leads to the family
$$y = \frac{1}{(1-x^2)^2 + 2ax^2}, \qquad x \geq 0, \quad a > 0.$$
(a) Find and classify the critical points, and then explain why the family is most interesting for $0 < a < 1$.
(b) Show that for a very near 0, there is a critical point at approximately $(1, 1/2a)$.
(c) Show that $y < 1/(1-x^2)^2$ for $x > 0$.
(d) Sketch the curves for $a = 0.05, 0.10, 1.0$, and 3.0 on one set of axes, together with $y = 1/(1-x^2)^2$, the curve with $a = 0$.
(e) What is the geometric significance of the parameter a?

5.4 ECONOMIC APPLICATIONS: MARGINALITY

Management decisions within a particular firm or industry usually depend on the costs and revenues involved. In this section we will look at applications of the derivative to the cost and revenue functions.

> The **cost function**, $C(q)$, gives the total cost of producing a quantity q of some good.

What sort of function do you expect C to be? The more goods that are made, the higher the cost, so C is an increasing function. For most goods, such as cars or cases of soda, q can only be an integer, so the graph of C might look like that in Figure 5.47.

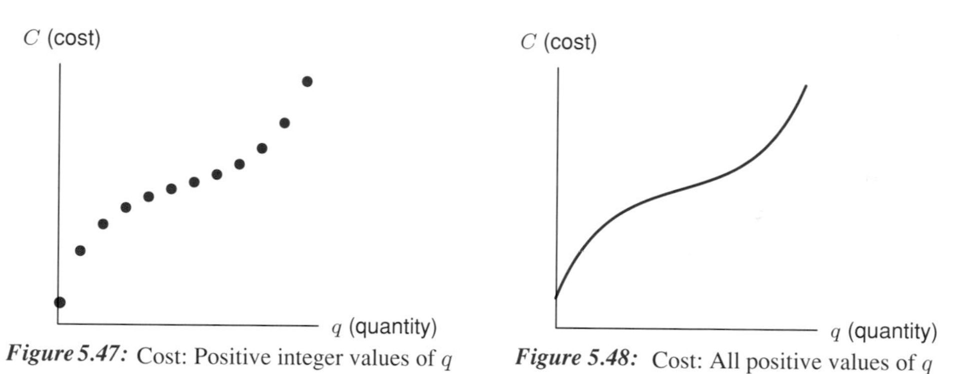

Figure 5.47: Cost: Positive integer values of q

Figure 5.48: Cost: All positive values of q

However, economists usually imagine the graph of C as the smooth curve drawn through these points, as in Figure 5.48. This is equivalent to assuming that $C(q)$ is defined for all nonnegative values of q, not just for integers.

Cost functions usually have the general shape shown in Figure 5.48. The intercept on the C-axis represents the *fixed costs*, which are incurred even if nothing is produced. (This includes, for instance, the machinery needed to begin production.) The cost function increases quickly at first and then more slowly because producing larger quantities of a good is usually more efficient than producing smaller quantities—this is called economy of scale. At still higher production levels, the cost function starts to increase faster again as resources become scarce, and sharp increases may occur when new factories have to be built. Thus, $C(q)$ starts out concave down and becomes concave up later on.

> The **revenue function**, $R(q)$, represents the total revenue received by the firm from selling a quantity q of some good.

If the price, p, is a constant, then

$$\text{Revenue} = \text{Price} \times \text{Quantity}, \quad \text{so} \quad R = pq$$

and the graph of R is a straight line through the origin. See Figure 5.49. In practice, for large values of q, the market may become glutted, causing the price to drop, giving R the shape in Figure 5.50.

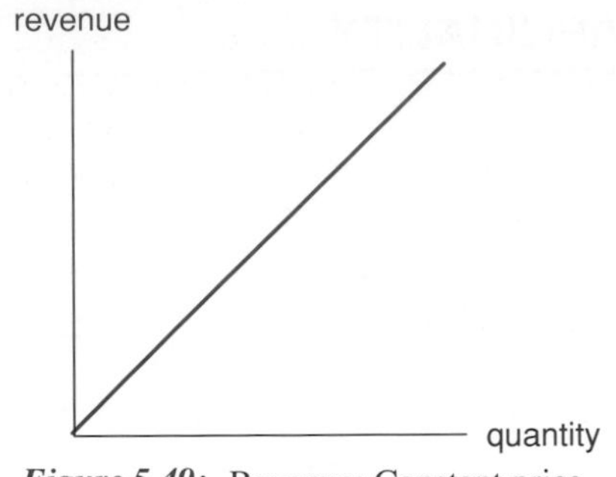

Figure 5.49: Revenue: Constant price

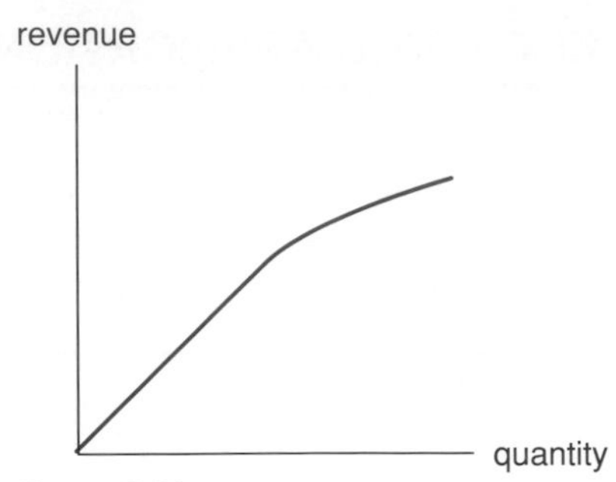

Figure 5.50: Revenue: Decreasing price

Decisions are often made by considering the profit, usually written as π (to distinguish it from the price, p; this π has nothing to do with the area of a circle, and merely stands for the Greek equivalent of the letter "p") and defined by

$$\text{Profit} = \text{Revenue} - \text{Cost} \quad \text{so} \quad \pi = R - C.$$

Example 1 If $C(q)$ and $R(q)$ are given by the graphs in Figure 5.51, for what values of q does the firm make a profit?

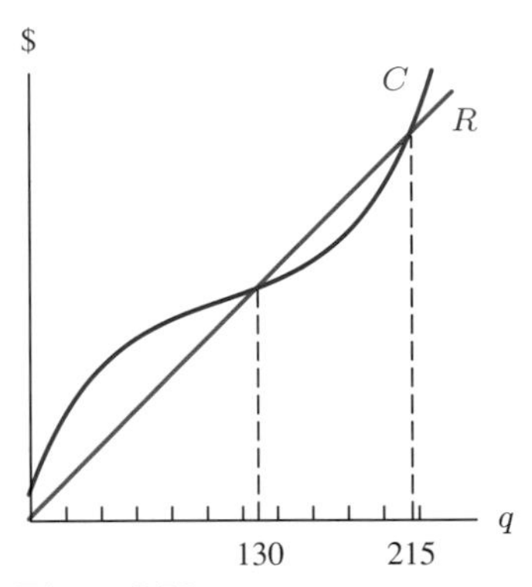

Figure 5.51: Costs and revenues for Example 1

Solution The firm makes a profit whenever revenues are greater than costs, that is, when $R > C$. The graph of R is above the graph of C approximately when $130 < q < 215$, so production between $q = 130$ and $q = 215$ will generate a profit.

Marginal Analysis

Many economic decisions are based on an analysis of the costs and revenues "at the margin." Let's look at this idea through an example.

Suppose you are running an airline and you are trying to decide whether to offer an additional flight. How should you decide? We'll assume that the decision is to be made purely on financial

Does this represent a local maximum or minimum of π? We can tell by looking at what is going on at production levels of 649 units and 651 units. When $q = 649$ we have $MR = \$1.106$, which is greater than the (constant) marginal cost of \$1.1. This means that producing one more unit will bring in more revenue than its cost, so profit will increase. When $q = 651$, $MR = \$1.094$, which is now *less* than MC, so it is not profitable to produce the 651st unit. We conclude that $q = 650$ is a local maximum for the profit function π. The profit earned by producing and selling this quantity is $\pi(650) = R(650) - C(650) = \$1982.50 - \$1015 = \$967.50$.

To check for global maxima we need to look at the endpoints. If $q = 0$, the only cost is \$300 (the fixed costs) and there is no revenue, so $\pi(0) = -300$. At the upper limit of $q = 1000$, $R(1000) = \$2000$, $C(1000) = \$1400$ and so $\pi(1000) = \$600$. Therefore, the maximum profit is where $MR = MC$, and the minimum profit occurs when $q = 0$.

Problems for Section 5.4

1. An agricultural worker in Uganda is interested in planting clover to increase the number of bees making their home in the region. There are 100 bees in the region naturally, and for every acre put under clover, 20 more bees are found in the region.
   (a) Draw a graph of the total number, $N(x)$, of bees as a function of x, the number of acres devoted to clover.
   (b) Explain, both geometrically and algebraically, the shape of the graph of:
      (i) The marginal rate of increase of the number of bees with acres of clover, $N'(x)$.
      (ii) The average number of bees per acre of clover, $N(x)/x$.
2. A manufacturer reports the total cost and total revenue functions shown in Figure 5.57. Sketch graphs, as a function of quantity, of:
   (a) Total profit, (b) Marginal cost, (c) Marginal revenue.
   Label the points q_1 and q_2 on your graphs.
3. Let $C(q)$ be the total cost of producing a quantity q of a certain good. (See Figure 5.58.)
   (a) What is the meaning of $C(0)$?
   (b) Describe in words how the marginal cost changes as the quantity produced increases.
   (c) Explain the concavity of the graph (in terms of economics).
   (d) Explain the economic significance (in terms of marginal cost) of the point at which the concavity changes.
   (e) Do you think the graph of $C(q)$ looks like this for all types of goods?

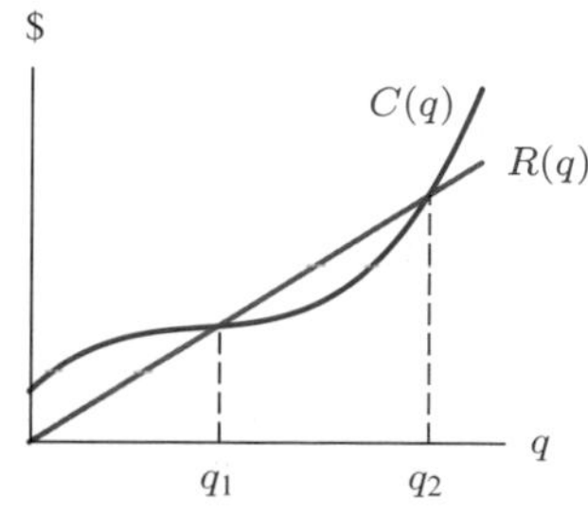

Figure 5.57: Cost and revenue for a good

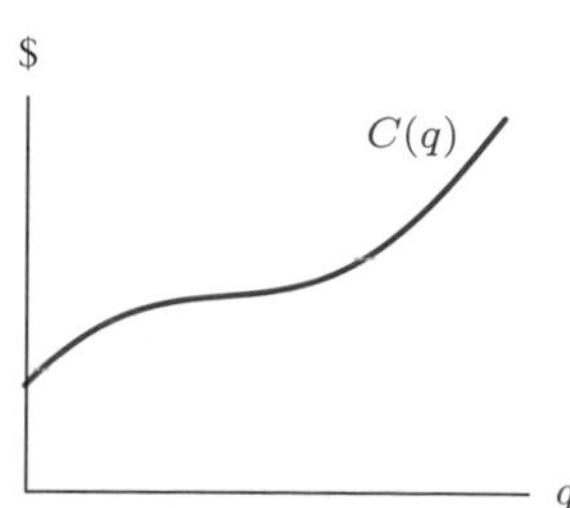

Figure 5.58: Cost of producing a good

4. Assume a manufacturer's cost of producing a good is given by the graph of $C(q)$ in Figure 5.59. Assume also that the manufacturer can sell the product for a price p each (regardless of the quantity sold), so that the total revenue from selling a quantity q is $R(q) = pq$. (See Figure 5.59.)
   (a) The difference $\pi(q) = R(q) - C(q)$ is the total profit. For which quantity q_0 is the profit a maximum? Mark your answer on a sketch of the graph.
   (b) What is the relationship between p and $C'(q_0)$? (Note that p is the slope of the line $R(q) = pq$.) Explain your result both graphically and analytically. ($\pi(q)$ has a maximum at $q = q_0$, so $\pi'(q_0) = 0$.) What does this mean in terms of economics?
   (c) Graph $C'(q)$ and p (as a horizontal line) on the same axes. Mark q_0 on the q-axis.

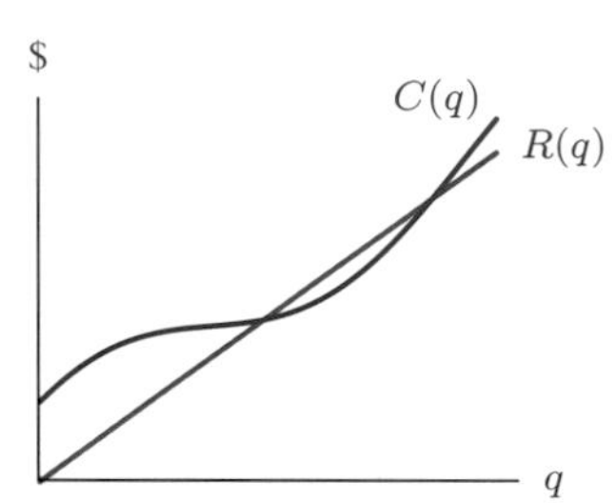

Figure 5.59: Cost and revenue for a good

5. Suppose the total cost $C(q)$ of producing q goods is given by:

$$C(q) = 0.01q^3 - 0.6q^2 + 13q.$$

   (a) What is the fixed cost?
   (b) What is the maximum profit if each item is sold at a price of \$7? (Assume you can sell everything you produce.)
   (c) At a fixed production level of 34 goods, for each \$1 increase in price, 2 less goods are sold. Should you raise the price, and if so by how much?

6. Suppose a company manufactures only one product. The quantity produced, q, of this product depends on the amount of capital, K, invested (i.e., the number of machines the company owns, the size of its building, and so on) and the amount of labor, L, available. It is often assumed that q can be expressed as a function of K and L by a *Cobb-Douglas* production function:

$$q = cK^{\alpha}L^{\beta}$$

where c, α, β are positive constants, with $0 < \alpha < 1$ and $0 < \beta < 1$.

In this problem we will see how the Russian government could use a Cobb-Douglas function to estimate how many people a newly privatized industry might employ. A company in such an industry will have only a small amount of capital available to it, and will need to use all of it; K is therefore fixed. Suppose L is measured in man-hours, and that each man-hour costs the company w rubles (a ruble is the unit of Russian currency). Suppose that the company has no other costs besides labor, and that each unit of the good can be sold for a fixed price of p rubles. How many man-hours of labor should the company use in order to maximize its profit?

Problems 7 – 10 involve the *average cost* of manufacturing a quantity q of a good, which is defined to be

$$a(q) = \frac{C(q)}{q}.$$

7. A reasonably realistic model of a firm's costs is given by the *short-run Cobb-Douglas cost curve*

$$C(q) = Kq^{1/a} + F,$$

where a is a positive constant, F is the fixed costs, and K measures the technology available to the firm.

   (a) Show that C is concave down if $a > 1$.
   (b) Assuming that average cost is minimized when average cost equals marginal cost, find what value of q minimizes the average cost.

8. Suppose a firm produces a quantity q of some good and that the average cost per item is given by:

$$a(q) = 0.01q^2 - 0.6q + 13, \quad \text{for} \quad q > 0.$$

   (a) What is the total cost, $C(q)$, of producing q goods?
   (b) What is the minimum marginal cost? What is the practical interpretation of this result?
   (c) At what production level is the average cost a minimum? What is the lowest average cost?
   (d) Compute the marginal cost at $q = 30$. How does this relate to your answer to part (c)? Explain this relationship both analytically and qualitatively.

9. Suppose you are given the graph of the average cost $a(q)$ in Figure 5.60.

   (a) Sketch a graph of the marginal cost $C'(q)$.
   (b) Show that if

$$a(q) = b + mq$$

   then

$$C'(q) = b + 2mq.$$

10. $C(q)$ is the total cost of producing a quantity q. The average cost $a(q)$ is given in Figure 5.61. The following rule is used by economists to determine the marginal cost $C'(q_0)$, for any q_0:

   - Construct the tangent t_1 to $a(q)$ at q_0.
   - Let t_2 be the line with the same y-intercept as t_1 but with twice the slope of t_1.

   Then $C'(q_0)$ is as shown in Figure 5.61. Explain why this rule works.

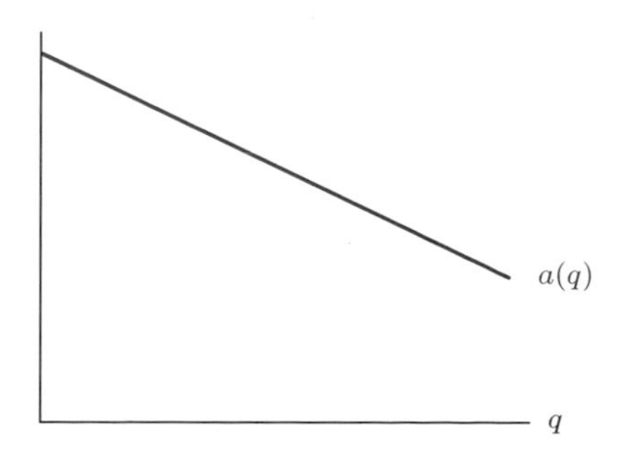

Figure 5.60: Average cost of a good

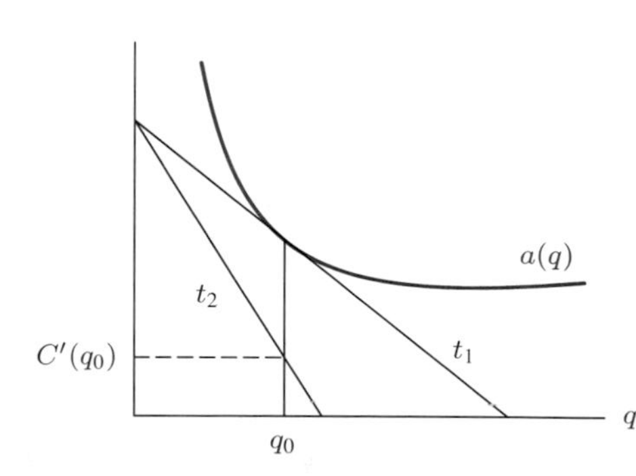

Figure 5.61

5.5 OPTIMIZATION

The world is full of problems where it is important to find the maximum or minimum value of some quantity. For example, engineers want to cut the strongest beam from a log of wood, scientists want to calculate which wavelength carries the maximum radiation at a given temperature, and urban planners want to design traffic patterns to minimize delays. All the techniques for finding such values make up the field called *optimization*. In this section and the next, you'll see how the derivative provides an efficient way of solving many optimization problems.

Example 1 A grapefruit is tossed straight up with an initial velocity of 100 ft/sec. Its height at time t is given by

$$y = -16t^2 + 100t + 6.$$

How high does it go before returning to the ground?

Solution We want to maximize the height, y, of the grapefruit above the ground. (See Figure 5.62.) Using the derivative we can find exactly when the grapefruit is at the highest point. We can think of this in two ways. By common sense, at the peak of the grapefruit's flight, the velocity, dy/dt, must be zero. Alternately, we are looking for a global maximum of y, so we look for critical points where $dy/dt = 0$. We have

$$\frac{dy}{dt} = 100 - 32t = 0 \quad \text{and so} \quad t = \frac{100}{32} = 3.125 \text{ sec.}$$

Thus, we have the time at which the height is a maximum; the maximum value of y is then

$$y = -16(3.125)^2 + 100(3.125) + 6 = 162.25 \text{ feet.}$$

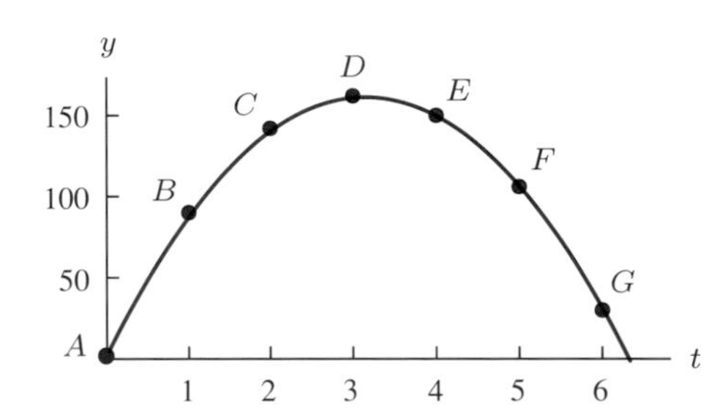

Figure 5.62: Height of the grapefruit

Example 2 When an arrow is shot into the air, its range R is defined as the horizontal distance from the archer to the point at which the arrow hits the ground. (See Figure 5.63.) Clearly R depends on how fast the arrow is going and the initial angle that the arrow makes with the ground. If we fix the initial speed, some angles will yield larger R values than others: shooting the arrow nearly vertically will give a (dangerously) small R. A nearly horizontal angle will also give a small R as the arrow hits the ground before it has had a chance to travel far horizontally.

What initial angle will give a maximum for R? To answer this question we need to know how R depends on the angle, θ. If the ground is horizontal and we neglect air resistance, it can be shown that

$$R = \frac{{v_0}^2 \sin(2\theta)}{g}$$

where v_0 is the initial velocity of the arrow, g is the (constant) acceleration due to gravity, and θ is measured from the horizontal, so $0 \le \theta \le \pi/2$.

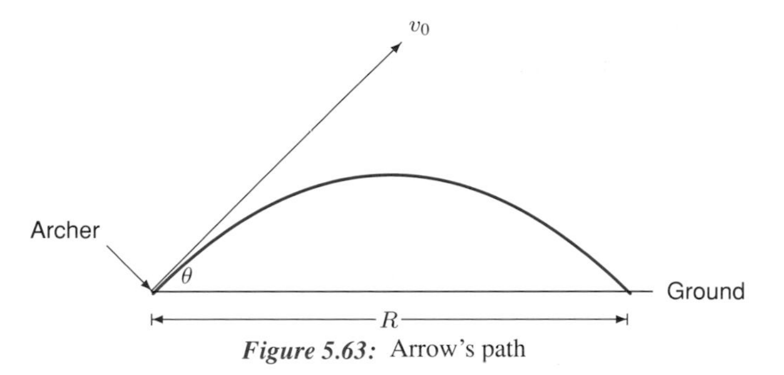

Figure 5.63: Arrow's path

Solution We can find the maximum of this function without using calculus. The maximum value of R occurs when $\sin(2\theta) = 1$, so $\theta = \pi/4$, giving $R = v_0^2/g$.

Let's see how we can do the same problem with calculus. We want to find the global maximum of R for $0 \le \theta \le \pi/2$. First we look for critical points of R in the interval $0 \le \theta \le \pi/2$:

$$\frac{dR}{d\theta} = 2\frac{v_0^2 \cos(2\theta)}{g}.$$

Setting $dR/d\theta$ equal to 0 we find

$$0 = \cos(2\theta) \quad \text{giving} \quad 2\theta = \pm\frac{\pi}{2}, \pm\frac{3\pi}{2}, \pm\frac{5\pi}{2}, \ldots$$

so $\pi/4$ is the only critical point in the interval $0 \le \theta \le \pi/2$. The critical value is $R\left(\pi/4\right) = {v_0}^2/g$. Now we must check the value of R at the endpoints $\theta = 0$ and $\theta = \pi/2$ (when the arrow is shot exactly horizontally and exactly vertically). Since $R = 0$ at each endpoint, the critical point $\theta = \pi/4$ gives both a local and a global maximum on $0 \le \theta \le \pi/2$. Therefore, if you want the arrow to go furthest, you should shoot it at an angle of $\pi/4$ or 45°.

Minimizing Gas Consumption

The next example looks at a way of setting driving speeds to maximize fuel efficiency. We will assume some information about gas consumption, g (in gallons/hour), as a function of velocity, v (in mph). Notice that we are measuring gas consumption in gallons per *hour*, so g represents the rate at which gas is being consumed by the engine.

For low speeds, gas consumption drops as the speed increases because an engine runs inefficiently at low speeds. At higher speeds, gas consumption increases again. If gas consumption at different speeds were measured experimentally, we might get a graph like that in Figure 5.64.[1]

Looking at the graph, you can see the minimum gas per *hour* is used by going about 30 mph. But that is not necessarily what we want for fuel efficiency. If we go as slow as 30 mph, the journey might take so long that we'd use up less gas by going faster. In fact, if we assume that the distance we drive is fixed, then what we want to minimize is the gas consumption per *mile*, not the gas consumption per hour.

[1] Adapted from Peter D. Taylor, *Calculus: The Analysis of Functions* (Toronto: Wall & Emerson, 1992).

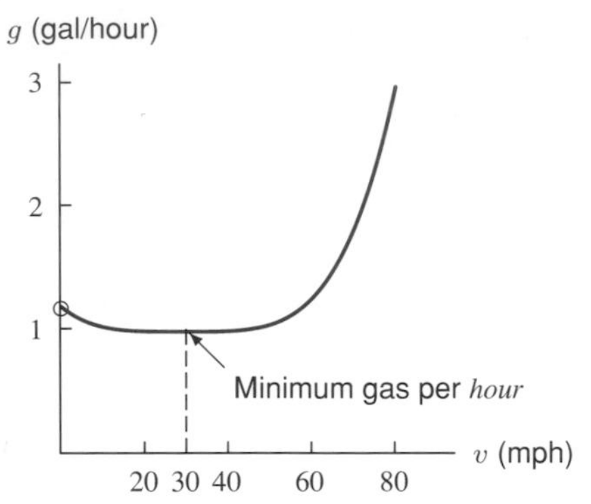

Figure 5.64: Gas consumption versus velocity

Example 3 Using Figure 5.64, estimate the velocity which gives the minimum gas consumption, G, in gallons per mile.

Solution First we must find G, the average gas consumption per mile. We know that g is measured in gallons/hour and that

$$\frac{\text{gallons/hour}}{\text{miles/hour}} = \frac{\text{gallons}}{\text{mile}}.$$

Since g is in gallons/hour and v is in mph, we can say that

$$G = \frac{g}{v} \text{ gallons/mile.}$$

Thus we want to find the minimum value of g/v when g and v are related by the graph in Figure 5.64. What does g/v represent graphically? Figure 5.65 shows that g/v is the slope of the line from the origin to the point P. Where on the curve should P be to make the slope a minimum? From the possible positions of the line shown in Figure 5.65, you can see that the slope of the line is a minimum when the line is tangent to the curve. From Figure 5.66, you can see that the velocity at this point is about 55 mph. As we expected, this is larger than 30 mph. Thus to minimize gas consumption per mile, you should drive about 55 mph.

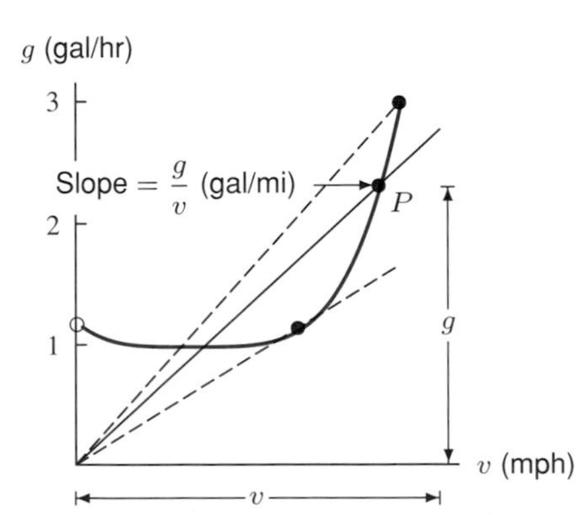

Figure 5.65: Graphical representation of gas consumption per mile, $G = \frac{g}{v}$

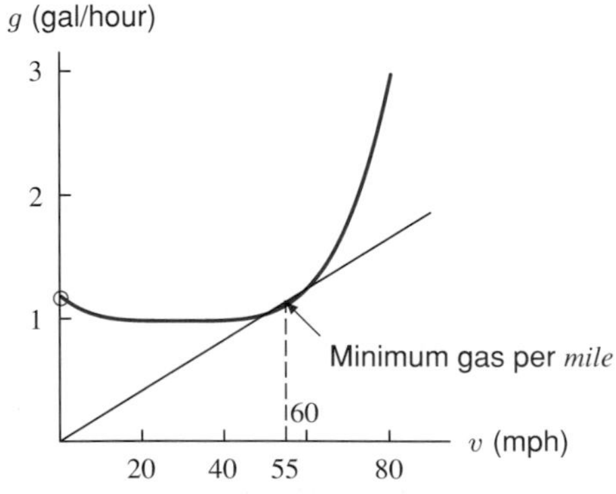

Figure 5.66: Velocity for maximum fuel efficiency

Finding Upper and Lower Bounds

A problem which is closely related to finding maxima and minima is finding the *bounds* of a function. In Example 1, the height of the grapefruit ranges from 0 (when it hits the ground) to nearly 162.25 feet. Thus

$$0 \le y \le 162.25$$

and we say that the function y is *bounded below* by 0 and *bounded above* by 162.25. Of course, we could also say that

$$-10 \le y \le 170$$

so that y is also bounded below by -10 and above by 170. However, we consider the 0 and 162.25 to be the *best possible bounds* because they describe more accurately how the function y behaves. (The bounds of a function are discussed in more detail in Appendix B, page 666.)

Example 4 Suppose the distance of an object above its equilibrium position at $y = 0$ is given as a function of time, t, by

$$y = e^{-t}\cos t.$$

Find the greatest distance the object goes above and below the equilibrium for $t \ge 0$.

Solution We are looking for the bounds of the function. What does the graph of the function look like? As in Section 5.3, you can think of it as a cosine curve with a decreasing amplitude of e^{-t}; in other words, it is a cosine curve squashed between the graphs of $y = e^{-t}$ and $y = -e^{-t}$, forming a wave with lower and lower crests and shallower and shallower troughs. (See Figure 5.67.)

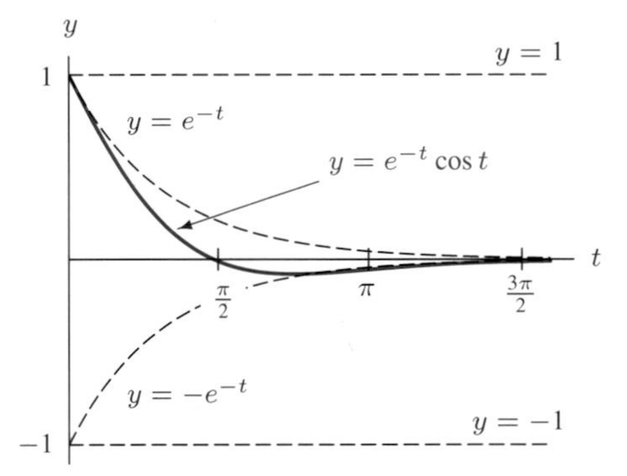

Figure 5.67: $f(t) = e^{-t}\cos t$ for $t \ge 0$

From the graph you can see that for $t \ge 0$, the graph lies between the horizontal lines $y = -1$ and $y = 1$. This means that -1 and 1 are bounds:

$$-1 \le e^{-t}\cos t \le 1$$

The line $y = 1$ is the best possible upper bound because the graph does come up that high (at $t = 0$). However, we can find a better lower bound if we find the global minimum value of f for $t \ge 0$; this minimum occurs in the first trough between $t = \pi/2$ and $t = 3\pi/2$ because later troughs are squashed closer to the t-axis. At that point $dy/dt = 0$. Using the product rule

$$\frac{dy}{dt} = (-e^{-t})\cos t + e^{-t}(-\sin t) = -e^{-t}(\cos t + \sin t) = 0$$

Since e^{-t} is never 0, we must have

$$\cos t + \sin t = 0$$

so

$$\tan t = -1 \quad \text{giving} \quad t = \frac{3\pi}{4}$$

Thus, the minimum you can see on the graph occurs at $t = 3\pi/4$. The value of y at that minimum is

$$y = e^{-\frac{3\pi}{4}} \cos\left(\frac{3\pi}{4}\right) \approx -0.067.$$

Rounding down so that the inequalities still hold for all $t \geq 0$ gives

$$-0.07 \leq e^{-t} \cos t \leq 1.$$

Notice how much smaller in magnitude the lower bound is than the upper: 0.07 (less than $\frac{1}{10}$) versus 1. This is a reflection of how quickly the factor e^{-t} causes the oscillation to die out.

Problems for Section 5.5

1. When you cough, your windpipe contracts. The speed, v, with which air comes out depends on the radius, r, of your windpipe. If R is the normal (rest) radius of your windpipe, then for $r \leq R$, the speed is given by:

$$v = a(R - r)r^2$$

where a is a positive constant. What value of r maximizes the speed?

2. The temperature change, T, in a patient generated by a dose, D, of a drug is given by

$$T = \left(\frac{C}{2} - \frac{D}{3}\right) D^2$$

where C is a positive constant.

(a) What dosage maximizes the temperature change?

(b) The sensitivity of the body, at dosage D, to the drug is defined as dT/dD. What dosage maximizes sensitivity?

3. A smokestack deposits soot on the ground with a concentration inversely proportional to the square of the distance from the stack. With two smokestacks 20 miles apart, the concentration of the combined deposits on the line joining them, at a distance x from one stack, is given by

$$S = \frac{k_1}{x^2} + \frac{k_2}{(20 - x)^2}$$

where k_1 and k_2 are positive constants which depend on the quantity of smoke each stack is emitting. If $k_1 = 7k_2$, find the point on the line joining the stacks where the concentration of the deposit is a minimum.

4. Some numbers are smaller than their squares (2 is smaller than 4), some are larger than their squares (1/3 is larger than 1/9), and some are equal to their squares (0 and 1). Any number x with $0 \leq x \leq 1$ will be greater than or equal to its square. Among all such numbers, find the ones that differ from their squares the most, that is, for which the function $s(x) = x - x^2$ is a maximum.

5. Given numbers a_1, a_2, a_3, define

$$f(x) = (x - a_1)^2 + (x - a_2)^2 + (x - a_3)^2.$$

Where is $f(x)$ a minimum? Why is this result reasonable?

6. (a) For any numbers a_1, a_2, a_3, a_4, find the value of x that minimizes

$$f(x) = (x - a_1)^2 + (x - a_2)^2 + (x - a_3)^2 + (x - a_4)^2.$$

(b) Use your answer to part (a) and to Problem 5 to guess the value of x that minimizes

$$f(x) = (x - a_1)^2 + (x - a_2)^2 + (x - a_3)^2 + \cdots + (x - a_n)^2.$$

7. Let $f(v)$ be the amount of energy consumed by a flying bird, measured in joules per second (a joule is a unit of energy), as a function of its speed v (in meters/sec). See Figure 5.68.

(a) Suggest a reason for the shape of this graph (in terms of the way birds fly).

Now let $a(v)$ be the amount of energy consumed by the same bird, measured in joules *per meter*.

(b) What is the relationship between $f(v)$ and $a(v)$?
(c) Where is $a(v)$ a minimum?
(d) Should the bird try to minimize $f(v)$ or $a(v)$ when it is flying? Why?

8. When birds lay eggs, they do so in clutches of several at a time. When the eggs hatch, each clutch gives rise to a brood of baby birds. We want to determine the clutch size which maximizes the number of birds surviving to adulthood per brood. If the clutch is small, there are few baby birds in the brood; if the clutch is large, there are so many baby birds to feed that most die of starvation. The number of surviving birds per brood as a function of clutch size is shown by the benefit curve in Figure 5.69.[2]

(a) Estimate the clutch size which maximizes the number of survivors per brood.
(b) Suppose also that there is a biological cost to having a larger clutch: the female survival rate is reduced by large clutches. This cost is represented by the dotted line in Figure 5.69. If we take cost into account by assuming that the optimal clutch size in fact maximizes the vertical distance between the curves, what is the new optimal clutch size?

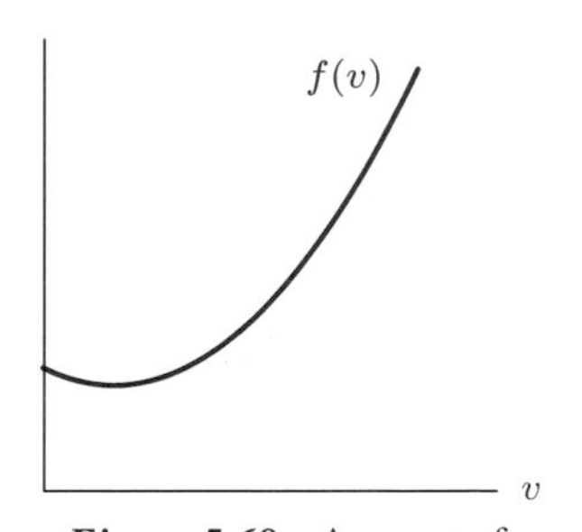

Figure 5.68: Amount of energy consumed by a flying bird

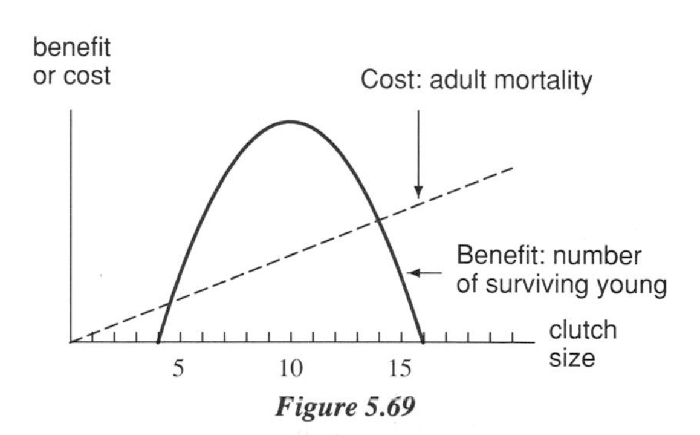

Figure 5.69

[2]Data from C. M. Perrins and D. Lack reported by J. R. Krebs and N. B. Davies in *An Introduction to Behavioural Ecology* (Oxford: Blackwell, 1987).

9. If you invest x dollars in a certain project, your return is $R(x)$. Suppose you want to choose x to maximize your return per dollar invested,[3] which is

$$r(x) = \frac{R(x)}{x}.$$

(a) Suppose the graph of $R(x)$ has the form in Figure 5.70, with $R(0) = 0$. Illustrate on a copy of this graph that the maximum value of $r(x)$ is obtained at a point on the graph of $R(x)$ at which the line from the origin to the point is tangent to the graph.
(b) It is also true that the maximum of $r(x)$ should occur at a point at which the slope of the $r(x)$ graph is zero. On the same set of axes as part (a), draw a rough version of the graph of $r(x)$, corresponding to your R graph, and illustrate that the maximum occurs where the slope is 0.
(c) Show, by taking the derivative of the preceding formula for $r(x)$, that the conditions in part (a) and (b) are equivalent: the point where the line from the origin is tangent to the graph of R is the same point where the graph of r has zero slope.

10. A bird such as a starling feeds worms to its young. To collect worms, the bird flies to a site where worms are to be found, picks up several in its beak, and flies back to its nest. The *loading curve* in Figure 5.71 shows how the number of worms (the load) a starling collects depends on the time it has been searching for them.[4] The curve is concave down because the bird can pick up worms more efficiently when its beak is empty; when its beak is partly full, the bird becomes much less efficient. The traveling time (from nest to site and back) is represented by the distance PO in Figure 5.71. Suppose the bird wants to maximize the rate at which it brings worms to the nest, where

$$\text{Rate worms arrive at nest} = \frac{\text{Load}}{\text{Traveling time} + \text{Searching time}}$$

(a) Draw a line in Figure 5.71 whose slope is this rate.
(b) Using the graph, estimate the load which maximizes this rate.
(c) If the traveling time is increased, does the optimal load increase or decrease? Why?

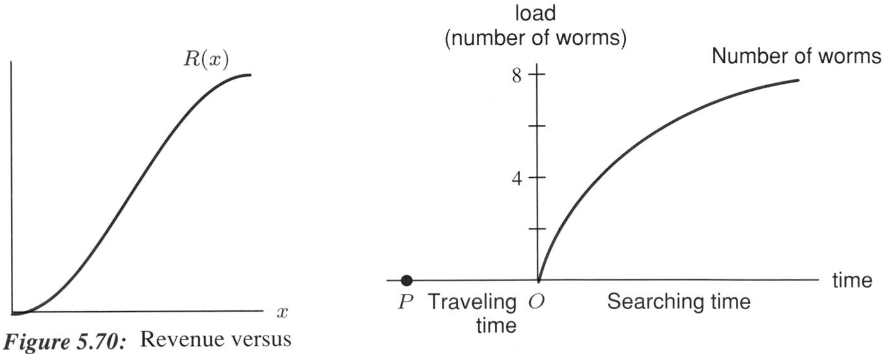

Figure 5.70: Revenue versus investment

Figure 5.71: Bird's loading curve

[3]From Peter D. Taylor, *Calculus: The Analysis of Functions* (Toronto: Wall & Emerson, 1992).
[4]Alex Kacelnick(1984). Reported by J. R. Krebs and N. B. Davis, *An Introduction to Behavioural Ecology* (Oxford: Blackwell, 1987).

Find the best possible bounds for each of the functions in Problems 11–15.

11. e^{-x^2}, for $|x| \le 0.3$

12. $\ln(1+x)$, for $x \ge 0$

13. $\ln(1+x^2)$, for $-1 \le x \le 2$

14. $x^3 - 4x^2 + 4x$, for $0 \le x \le 4$

15. $x + \sin x$, for $0 \le x \le 2\pi$

16. (a) For which positive number x is $x^{1/x}$ largest? Justify your answer. [Hint: You may want to write $x^{1/x} = e^{\ln(x^{1/x})}$.]
    (b) For which positive integer n is $n^{1/n}$ largest? Justify your answer.
    (c) Use your answer to parts (a) and (b) to decide which is larger: $3^{1/3}$ or $\pi^{1/\pi}$.

17. The *arithmetic mean* of two numbers a and b is defined as $(a+b)/2$; the *geometric mean* of two positive numbers a and b is defined as $\sqrt{ab}$.
    (a) For two positive numbers, which of the two means is larger? Justify your answer. [Hint: Define $f(x) = (a+x)/2 - \sqrt{ax}$ for fixed a.]
    (b) For three positive numbers a, b, c, the arithmetic and geometric mean are $(a+b+c)/3$ and $\sqrt[3]{abc}$, respectively. Which of the two means of three numbers is larger? [Hint: Redefine $f(x)$ for fixed a *and* b.]

5.6 MORE OPTIMIZATION: INTRODUCTION TO MODELING

Finding global maxima and minima is made much easier by having a formula for the function to be maximized or minimized. The process of translating a problem into a function whose formula we know and a domain over which it is to be optimized is called *modeling*. By looking at the examples that follow, you will get the flavor of some kinds of modeling.

Example 1 What are the dimensions of an aluminum can that can hold 40 in^3 of juice and that uses the least material (i.e., aluminum)? Assume that the can is cylindrical, and is capped on both ends.

Solution It is *always* a good idea to think about a problem in general terms before trying to solve it. Since we're trying to use as little material as possible, why not make the can very small, say, the size of a peanut? We can't, since the can must hold 40 in^3. If we make the can short, to try to use less material in the sides, where the label is, we'll have to make it fat as well, so that it can hold 40 in^3. You can see that saving aluminum in the sides might actually use more aluminum in the top and bottom in a short, fat can. See Figure 5.72(a).

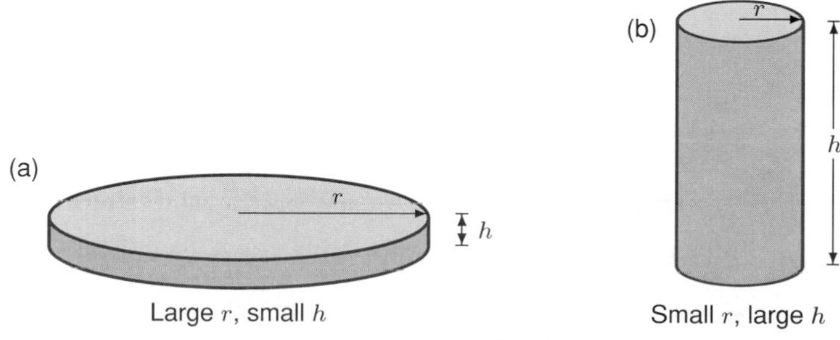

Figure 5.72: Various cylindrical-shaped cans

TABLE 5.1 *Amount of material used in the can for various choices of r and h*

r	h	Material used, M (in^2)
0.2	318.31	400.03
1.0	12.73	86.28
2.0	3.18	65.13
3.0	1.41	83.22
4.0	0.80	120.53
10.0	0.13	636.32

If we try to save material by making the top and bottom small, the can has to be tall to accommodate the 40 in^3 of juice. So any savings we get by using a small top and bottom might be outweighed by the height of the sides. See Figure 5.72(b). This is, in fact, true, as Table 5.1 shows. The table gives the amount of material used in the can for some choices of r and h. You can see that r and h change in opposite directions, and that more material is used at the extremes (very large or very small r and h) than in the middle. From the table you can see that the optimal radius for our can lies somewhere in $1.0 \le r \le 3.0$. If we consider the material used, M, as a function of the radius, r, a graph of this function looks like Figure 5.73. The graph shows that the global minimum we want is at a critical point.

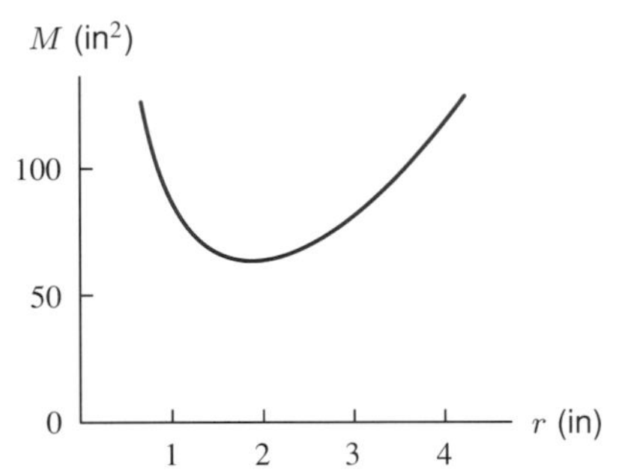

Figure 5.73: Total material used in can, M, as a function of radius of ends, r

Both the table and the graph were obtained from a model, which in this case is a formula for the material used in making the can. Finding such a formula depends on a knowledge of the geometry of a cylinder, in particular of its area and volume. We'll label the height of the can h and the radius of the circular ends r. Now we construct the model:

$$\begin{aligned} M &= \text{Material used in the can} \\ &= \text{Material in ends} + \text{Material in the side} \end{aligned}$$

where

$$\begin{aligned} \text{Material in ends} &= 2 \times \text{Area of a circle with radius } r = 2 \cdot \pi r^2 \\ \text{Material in the side} &= \text{Area of cylinder with height } h \text{ and radius } r = 2\pi r h. \end{aligned}$$

As we mentioned, h is not independent of r, but changes in the opposite direction: if r grows, h shrinks and vice-versa. To find a precise relationship, we use the fact that the volume of the cylindrical can, $\pi r^2 h$, is equal to the constant 40 in^3, so

$$\pi r^2 h = 40 \quad \text{giving} \quad h = \frac{40}{\pi r^2}.$$

So

$$\text{Material used in the side} = 2\pi r h = 2\pi r \frac{40}{\pi r^2} = \frac{80}{r}.$$

Thus we obtain the formula for the total material used in a can of radius r:

$$M = 2\pi r^2 + \frac{80}{r}.$$

This function could be used to create Table 5.1 and Figure 5.73. Note that the domain of this function is all $r > 0$, so that there are no endpoints to consider.

Now we use calculus to find the minimum of M. We look for critical points:

$$\frac{dM}{dr} = 4\pi r - \frac{80}{r^2} = 0 \quad \text{at a critical point,} \quad \text{so} \quad 4\pi r = \frac{80}{r^2}.$$

Therefore,

$$\pi r^3 = 20 \quad \text{so} \quad r = \left(\frac{20}{\pi}\right)^{1/3} \approx 1.85 \text{ inches,}$$

giving

$$h = \frac{40}{\pi r^2} \approx \frac{40}{\pi (1.85)^2} \approx 3.70 \text{ inches.}$$

Thus, the material used, $M(1.85)$, is about 64.74 in^2. Figure 5.73 shows that M has one critical point which is the global minimum, so 64.74 in^2 must be that global minimum.

Practical Tips for Modeling Optimization Problems

1. Think about the problem. Make sure that you know what quantity or function is to be optimized.
2. If possible, make several sketches showing how the elements that vary are related. Label your sketches clearly by assigning variables to quantities which change.
3. Try to obtain a formula for the function to be optimized in terms of the variables that you identified in the previous step. If necessary, eliminate from this formula all but one variable. Identify the domain over which this variable varies.
4. Apply the techniques of Section 5.1 to obtain candidates for local maxima and minima, and evaluate the function at these points and at the endpoints, if any, to find the global maxima and minima.

The next example, another problem in geometry, illustrates this approach.

Example 2 Alaina wants to get to the bus stop as quickly as possible. The bus stop is across a grassy park, 2000 feet west and 600 feet north of her starting position. Alaina can walk west along the edge of the park on the sidewalk at a speed of 6 ft/sec. She can also travel through the grass in the park, but only at a rate of 4 ft/sec (the park is a favorite place to walk dogs, so she must move with care). What path will get her to the bus stop the fastest?

Solution

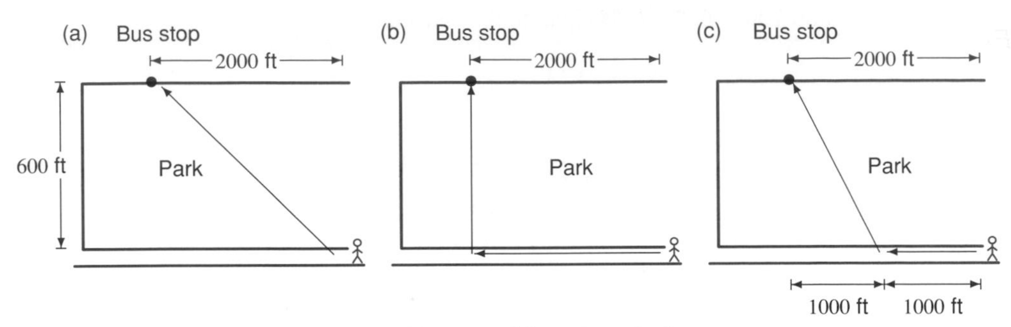

Figure 5.74: Three possible paths to the bus stop

Thinking about the problem, you might assume that she should take a path that is the shortest distance. Unfortunately, the path that follows the shortest distance to the bus stop is entirely in the park, where her speed is slow. (See Figure 5.74(a).) That distance is $\sqrt{2000^2 + 600^2} \approx 2100$ feet, which will take her about 525 seconds to traverse. She could instead walk quickly the entire 2000 feet along the sidewalk, which would leave her just the 600-foot northward journey through the park. (See Figure 5.74(b).) This method would take $2000/6 + 600/4 \approx 483$ seconds total walking time.

But can she do even better? Perhaps another combination of sidewalk and park will give a shorter travel time. For example, what is the travel time if she walks 1000 feet west along the sidewalk and the rest of the way through the park? (See Figure 5.74(c).) (Answer: about 454 seconds.)

Let's make a model for this problem. We can label the distance that she walks west along the sidewalk x and the distance she walks through the park y, as in Figure 5.75.

Then the total time, t, will be

$$t = t_{\text{sidewalk}} + t_{\text{park}}.$$

Since

$$\text{Time} = \text{Distance}/\text{Speed},$$

and she can walk 6 ft/sec on the sidewalk and 4 ft/sec in the park, we have that:

$$t = \frac{x}{6} + \frac{y}{4}.$$

Now, by the Pythagorean Theorem, $y = \sqrt{(2000 - x)^2 + 600^2}$ feet through the park. The total time for both pieces is therefore

$$t = \frac{x}{6} + \frac{\sqrt{(2000 - x)^2 + 600^2}}{4}.$$

If you try searching for critical points of this function, you run into quite a bit of algebra. You may find it easier to graph the function on a calculator or computer and estimate the critical point, which turns out to be $x \approx 1463$ feet, giving the minimum total time of about 445 seconds.

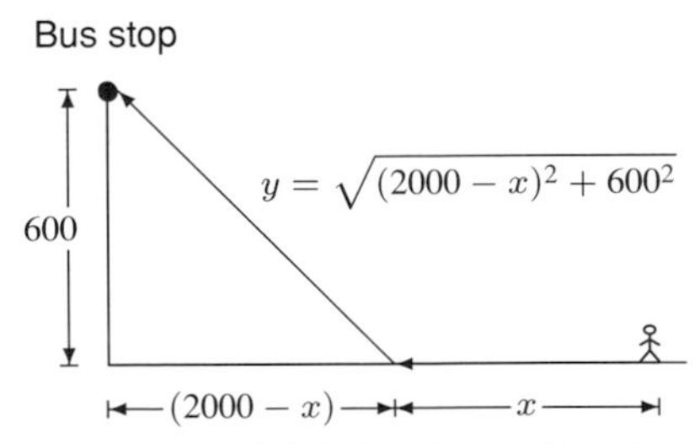

Figure 5.75: Modeling time to bus stop

Example 3 One hallway which is 4 feet wide meets another hallway which is 8 feet wide in a right angle. (See Figure 5.76.) What is the length of the longest ladder which can be carried horizontally around the corner?

Solution We imagine the ladder being carried on its side and ignore its width. To allow the longest ladder possible we carry the ladder around the corner so that it just touches both walls (at A and C) and just touches the corner at B. Let's draw some lines that do this. (See Figure 5.76.) As you can see, the length of the line $\overline{ABC}$ decreases as the corner is turned, then increases again. The minimum length would be the *largest* length of a ladder that could make it around the corner. Certainly a smaller ladder would work (it wouldn't touch A, B, and C simultaneously), but a larger one would not fit.

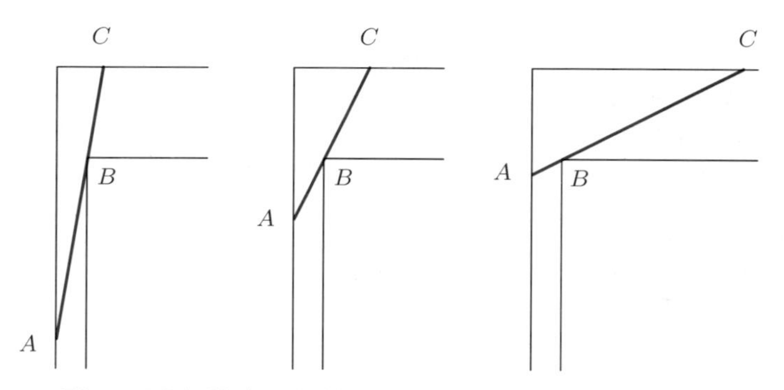

Figure 5.76: Various ladders that touch both walls and corner

Thus, the way to approach this problem of finding the largest ladder that can make it around the corner in the hallway is to find the *smallest* length of the line $\overline{ABC}$. That smallest length will be the length of the longest possible ladder. We will express the length, l, as a function of θ, the angle between the line and the wall of the narrow hall. (See Figure 5.77.)

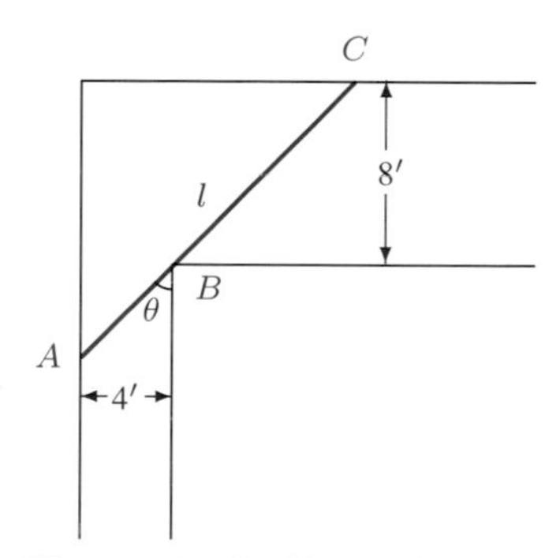

Figure 5.77: Ladder and hallway

We have

$$l = \overline{AB} + \overline{BC}.$$

Now $\overline{AB} = 4/\sin\theta$ and $\overline{BC} = 8/\cos\theta$, so

$$l = \frac{4}{\sin\theta} + \frac{8}{\cos\theta}.$$

Here the domain of θ is $0 < \theta < \pi/2$. The next step is to look for a minimum of l by searching for critical points.

$$\frac{dl}{d\theta} = -\frac{4}{(\sin\theta)^2}(\cos\theta) - \frac{8}{(\cos\theta)^2}(-\sin\theta).$$

When $dl/d\theta = 0$,

$$-4\frac{\cos\theta}{(\sin\theta)^2} + 8\frac{\sin\theta}{(\cos\theta)^2} = 0$$

so

$$2(\sin\theta)^3 = (\cos\theta)^3$$

giving

$$\frac{(\sin\theta)^3}{(\cos\theta)^3} = \frac{1}{2}$$

so

$$\tan\theta = \sqrt[3]{0.5} \approx 0.79$$

and

$$\theta \approx 0.67 \text{ radians.}$$

Thus $\theta \approx 0.67$ is a critical point, and an investigation of what happens for θ near 0 and θ near $\pi/2$ shows that l has a global minimum there. The minimum value of l is therefore

$$l = \frac{4}{\sin(0.67)} + \frac{8}{\cos(0.67)} \approx 16.65 \text{ feet,}$$

so the ladder can be at most 16.65 feet long and still make it around the corner.

Problems for Section 5.6

1. If you have 100 feet of fencing and want to enclose a rectangular area up against a long, straight wall, what is the largest area you can enclose?
2. A square-bottomed box with a top is to have a fixed volume V. What dimensions minimize the surface area?
3. A square-bottomed box without a top is to have a fixed volume V. What dimensions minimize the surface area?
4. The illumination at a point near a source of light is proportional to the intensity of the source and inversely proportional to the square of the distance of the point from the source. Suppose two sources of light, one twice the intensity of the other, are a distance d apart. Where on the line between them is the illumination least?
5. Suppose you run a small independent furniture business. Your assistant signs a deal with a customer to deliver however many chairs the customer orders. The price will be \$90 per chair up to 300 chairs, and above 300, the price will be reduced by \$0.25 per chair (on the whole order) for every additional chair over 300 ordered, up to 100. What are the largest and smallest revenues your company can make under this deal?
6. The cost of fuel to propel a boat through the water (in dollars per hour) is proportional to the cube of the speed. A certain ferry boat uses \$100 worth of fuel per hour when cruising at 10 miles per hour. Apart from fuel, the cost of running this ferry (labor, maintenance, and so on) is \$675 per hour. At what speed should it travel so as to minimize the cost *per mile* traveled?

7. On a record (or CD or diskette) the amount of information (for example, the number of minutes of music) stored on a track that goes once around the record is the same no matter whether the track is near the center or at the edge.
   (a) Why is this reasonable?
   (b) Let b be the maximum possible density of information/cm which can be stored in a track, let R cm be the fixed outer radius, and let a be the number of tracks per radial cm. What should the inner radius be to maximize the total amount of information stored?
8. The graphs of $y = \sqrt{x}$, $x = 9$, and $y = 0$ bound a region in the first quadrant. Find the dimensions of the rectangle of maximum *area* and the dimensions of the rectangle of maximum *perimeter* that can be inscribed in this region. (The sides of the rectangles should be parallel to the axes.)
9. Which point on the parabola $y = x^2$ is nearest to $(1, 0)$? Find the coordinates to two decimals. [Hint: Minimize the square of the distance—this avoids square roots.]
10. Find the point(s) on the ellipse
$$\frac{x^2}{9} + y^2 = 1$$
   (a) Closest to the point $(2, 0)$. (b) Closest to the focus $(\sqrt{8}, 0)$.
   [Hint: Minimize the square of the distance—this avoids square roots.]
11. Sketch the graph of $f(x) = 2 - 2\sin x$. Express the distance from the origin to a point on the graph of f as a function of x. Find the point on the graph of f that is closest to the origin. Label this point on the graph, draw the line segment from the origin to this point, and state how long this minimum distance is. [Hint: Use a graphing calculator]
12. On the same side of a straight river are two towns, and the townspeople want to build a pumping station, S, that supplies water to them. The pumping station is to be at the river's edge with pipes extending straight to the two towns. The distances are shown in Figure 5.78. Where should the pumping station be located to minimize the total length of pipe?

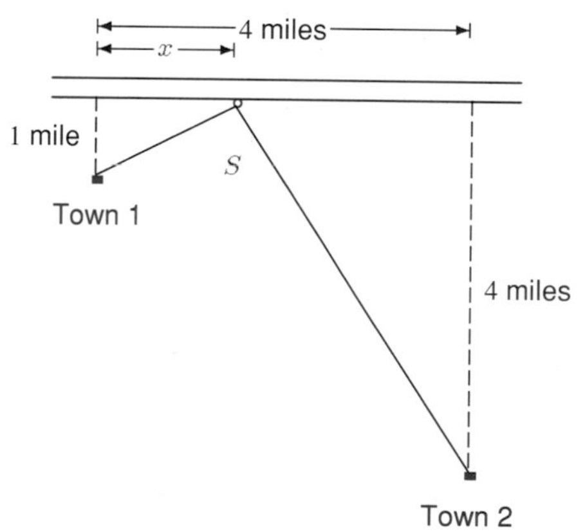

Figure 5.78

13. (a) Bueya spends her days as follows: She is at Washington University in the morning, at her job in East St. Louis in the afternoon, and late in the evening she has a beer in her favorite bar. She has lunch and dinner at home. Where on the road should she look for an apartment to minimize her daily traveling distance? (See Figure 5.79)

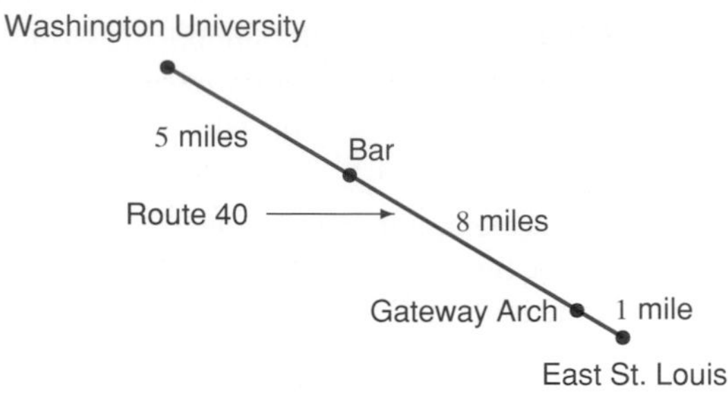

Figure 5.79

(b) Her colleague Marie-Josée goes to a gym near the Gateway Arch before breakfast (which she eats at home), and she spends the rest of the day the same way as Bueya. Where on the road should she look for an apartment?

14. Assume that a pigeon is released from a boat (point B in Figure 5.80) floating on a lake. Because of falling air over the cool water, the energy required to fly one meter over the lake is twice the corresponding energy e required for flying over the bank ($e = 3$ joule/meter). To minimize the energy required to fly from B to the loft, L, the pigeon will head to a certain point P on the bank and then fly along the bank to L.
    (a) Express the energy required to fly from B to L as a function of the angle θ (the angle BPA).
    (b) What is the optimal angle θ?
    (c) Does your answer change if $\overline{AL}$, $\overline{AB}$, and e have different numerical values?

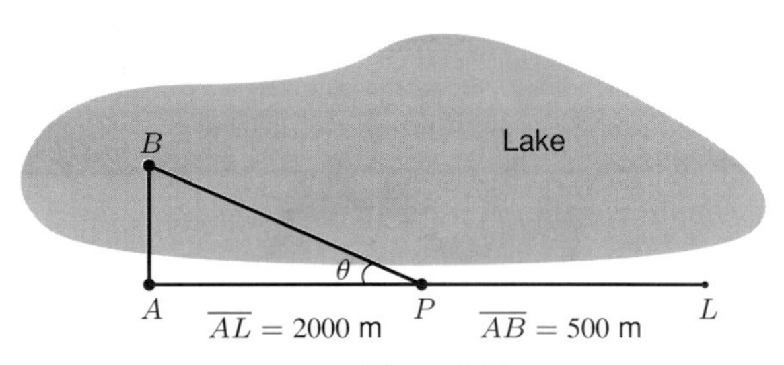

Figure 5.80

15. To get the best view of the Statue of Liberty in Figure 5.81, you should be at the position where θ is a maximum. If the statue stands 92 meters high, including the pedestal, which is 46 meters high, how far from the base should you be? [Hint: Find a formula for θ in terms of your distance from the base. Use this function to maximize θ, noting that $0 \le \theta \le \pi/2$.]

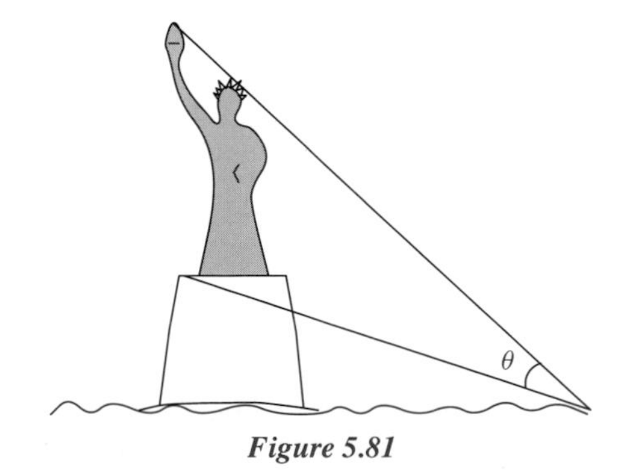

Figure 5.81

5.7 NEWTON'S METHOD

Many problems in mathematics involve finding the root of an equation. For example, we might have to locate the zeros of a polynomial, or determine the point of intersection of two curves. Here we will see a numerical method for approximating solutions which cannot be calculated any other way.

One such method is described in Appendix A: bisection. Although it is very simple, the bisection method has two major drawbacks. First, it cannot locate a root where the curve is tangent to, but does not cross, the x-axis. Second, it is relatively slow in the sense that it requires a considerable number of iterations to achieve a desired level of accuracy. Although speed may not be important in solving a single equation, a practical problem may involve solving thousands of equations as a parameter changes. In such a case, any reduction in the number of steps can be important.

Using Newton's Method

We now consider a powerful root-finding method developed by Newton. Suppose we have a function $y = f(x)$. The equation $f(x) = 0$ has a root at $x = r$, as shown in Figure 5.82. We begin with an initial estimate, x_0, for this root. (This can be a guess.) We will now obtain a better estimate x_1. To do this, construct the tangent line to the curve at the point $x = x_0$, and extend it until it crosses the x-axis, as shown in Figure 5.82. The point where it crosses the axis is usually much closer to r, and we use that point as the next estimate, x_1. Having found x_1, we now repeat the process using x_1 instead of x_0. We construct another tangent line to the curve at $x = x_1$, extend it until it crosses the x-axis, use that x-intercept as the next approximation, x_2, and so on. The resulting sequence of x-intercepts usually converges rapidly to the root r.

Let's see how this looks algebraically. We know that the slope of the tangent line at the initial estimate x_0 is $f'(x_0)$, and so the equation of the tangent line is

$$y - f(x_0) = f'(x_0)(x - x_0).$$

At the point where this tangent line crosses the x-axis, we have $y = 0$ and $x = x_1$, so that

$$0 - f(x_0) = f'(x_0)(x_1 - x_0).$$

Solving for x_1, we obtain

$$x_1 = x_0 - \frac{f(x_0)}{f'(x_0)}$$

provided that $f'(x_0)$ is not zero. We now repeat this argument and find that the next approximation is

$$x_2 = x_1 - \frac{f(x_1)}{f'(x_1)}.$$

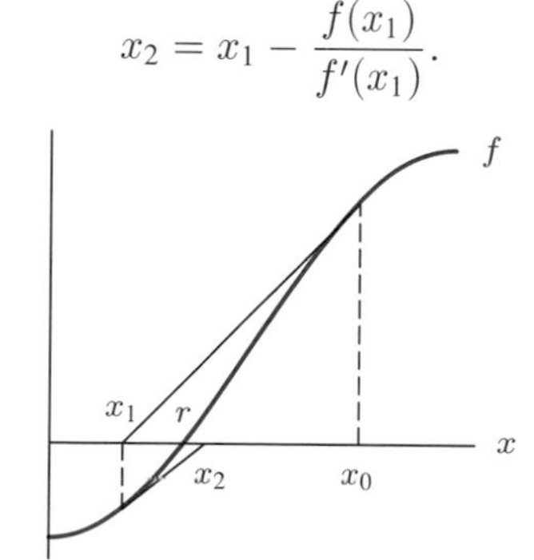

Figure 5.82: Newton's method: Successive approximations $x_0, x_1, x_2, \ldots$ to the root, r.

Summarizing, for any $n = 0, 1, 2, \ldots$, we obtain the following result.

Newton's Method

$$x_{n+1} = x_n - \frac{f(x_n)}{f'(x_n)}$$

provided that $f'(x_n)$ is not zero.

Example 1 Use Newton's method to find the fifth root of 23. (By calculator, this is 1.872171231, correct to nine decimal places.)

Solution To use Newton's method, we need an equation of the form $f(x) = 0$ having $23^{1/5}$ as a root. Since $23^{1/5}$ is a root of $x^5 = 23$ or $x^5 - 23 = 0$, we take $f(x) = x^5 - 23$. The root of this equation is less than 2 (since $2^5 = 32$), so we will choose $x_0 = 2$ as our initial estimate. Now $f'(x) = 5x^4$, so we can set up Newton's method as

$$x_{n+1} = x_n - \frac{x_n^5 - 23}{5x_n^4}.$$

In this case, we can simplify using a common denominator, to obtain

$$x_{n+1} = \frac{4x_n^5 + 23}{5x_n^4}.$$

Therefore, starting with $x_0 = 2$, we find that $x_1 = 1.8875$. This leads to $x_2 = 1.872418193$ and $x_3 = 1.872171296$. These values are in Table 5.2. Since $f(1.872171231) > 0$ and $f(1.872171230) < 0$, the root lies between 1.872171230 and 1.872171231. Therefore, in just four iterations of Newton's method, we have achieved nine-decimal accuracy.

As a general guideline for Newton's method, once the first correct decimal place is found, each successive iteration approximately doubles the number of correct digits.

What happens if we select a very poor initial estimate? In the preceding example, suppose x_0 were 10 instead of 2. The results are in Table 5.3. Notice that even with $x_0 = 10$, the sequence of values moves reasonably quickly toward the solution: we achieve seven-decimal place accuracy by the eleventh iteration.

TABLE 5.2 *Newton's method:* $x_0 = 2$

n	x_n	$f(x_n)$
0	2	9
1	1.8875	0.957130661
2	1.872418193	0.015173919
3	1.872171296	0.000004020
4	1.872171231	0.000000027

TABLE 5.3 *Newton's method:* $x_0 = 10$

n	x_n	n	x_n
0	10	6	2.679422313
1	8.000460000	7	2.232784753
2	6.401419079	8	1.971312452
3	5.123931891	9	1.881654220
4	4.105818871	10	1.872266333
5	3.300841811	11	1.872171240

Example 2 Find the first point of intersection of the curves given by $f(x) = \sin x$ and $g(x) = e^{-x}$.

Solution The graphs in Figure 5.83 make it clear that there are an infinite number of points of intersection, all with $x > 0$. In order to find the first one numerically, we consider the function

$$F(x) = f(x) - g(x) = \sin x - e^{-x}$$

whose derivative is $F'(x) = \cos x + e^{-x}$. The point we want must be fairly close to $x = 0$, so we start with $x_0 = 0$. The values in Table 5.4 are approximations to the root. Since $F(0.588532744) > 0$ and $F(0.588532743) < 0$, the root lies between 0.588532743 and 0.588532744. (Remember, your calculator must be set in radians.)

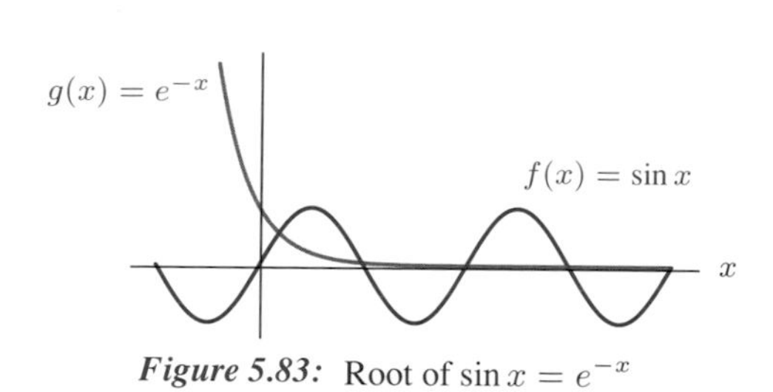

Figure 5.83: Root of $\sin x = e^{-x}$

TABLE 5.4
Successive approximations to root of $\sin x = e^{-x}$

n	x_n
0	0
1	0.5
2	0.585643817
3	0.588529413
4	0.588532744
5	0.588532744

When Does Newton's Method Fail?

In most practical situations, Newton's method works well. Occasionally, however, the sequence x_0, x_1, x_2, ... fails to converge or fails to converge to the root you want. Sometimes, for example, the sequence can jump from one root to another. This is particularly likely to happen if the magnitude of the derivative $f'(x_n)$ is small for some x_n. In this case, the tangent line is nearly horizontal and so x_{n+1} will be far from x_n. (See Figure 5.84.)

If the equation $f(x) = 0$ has *no* root, then the sequence will not converge. In fact, the sequence obtained by applying Newton's method to $f(x) = 1 + x^2$ is one of the best known examples of *chaotic behavior* and has attracted considerable research interest recently. (See Figure 5.85.)

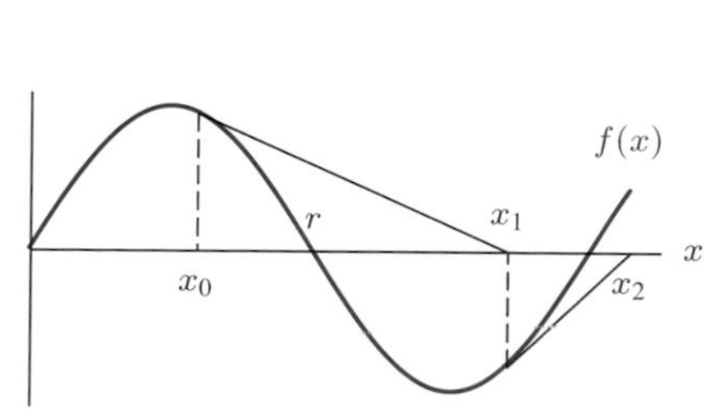

Figure 5.84: Problems with Newton's method: Converges to wrong root

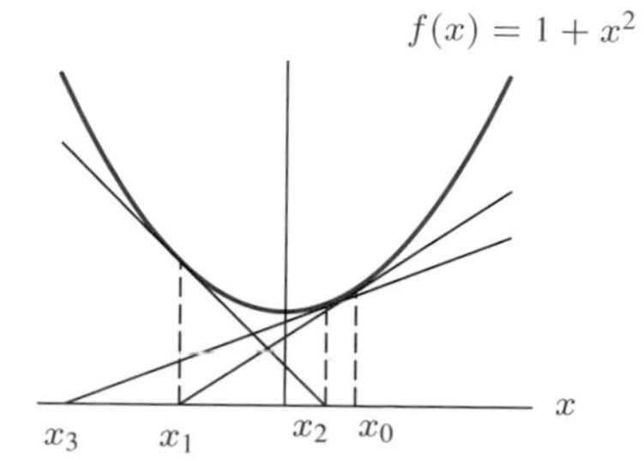

Figure 5.85: Problems with Newton's method: Chaotic behavior

Problems for Section 5.7

1. Suppose you want to find a solution of the equation
$$x^3 + 3x^2 + 3x - 6 = 0.$$
Consider $f(x) = x^3 + 3x^2 + 3x - 6$.
   (a) Find $f'(x)$, and use it to show that $f(x)$ increases everywhere.
   (b) How many roots does the original equation have?
   (c) For each root, find an interval which contains it.
   (d) Find each root to two decimal places, using Newton's method.

For Problems 2–4, use Newton's method to find the given quantities to two decimal places:

2. $\sqrt[3]{50}$
3. $\sqrt[4]{100}$
4. $10^{-1/3}$

For Problems 5–8, solve each equation and give each answer to two decimal places:

5. $\sin x = 1 - x$
6. $\cos x = x$
7. $e^{-x} = \ln x$
8. $e^x \cos x = 1$, for $0 < x < \pi$

9. Find, to two decimal places, all solutions of $\ln x = 1/x$.
10. How many zeros do the following functions have? For each zero, find an upper and a lower bound which differ by no more than 0.1.
   (a) $f(x) = x^3 + x - 1$ (b) $f(x) = \sin x - \frac{2}{3}x$ (c) $f(x) = 10xe^{-x} - 1$
11. Find the largest zero of
$$f(x) = x^3 + x - 1$$
to six decimal places, using Newton's method. How do you know your approximation is as good as you claim?
12. For any positive number, a, the problem of calculating the square root, $\sqrt{a}$, is often done by applying Newton's method to the function $f(x) = x^2 - a$. Apply the method to obtain an expression for x_{n+1} in terms of x_n. Use this to approximate $\sqrt{a}$ for $a = 2, 10, 1000$, and π, correct to four decimal places, starting at $x_0 = a/2$ in each case.
13. What is the smallest positive zero of the function $f(x) = \sin x$? Apply Newton's method, with initial guess $x_0 = 3$, to see how fast it converges to π.
   (a) Compute the first two approximations, x_1 and x_2; compare x_2 with π.
   (b) Newton's method works very well here. Explain why. To do this, you will have to outline the basic idea behind Newton's method.
   (c) If you used the bisection method with 3 and 4 as original estimates, how many bisections would you need to get an estimate as good as x_2?
14. Newton's method can be very sensitive to your initial estimate, x_0. Using a Newton's method program on your calculator or computer, try the following experiment with $f(x) = \sin x - \frac{2}{3}x$.
   (a) Use Newton's method with the following initial estimates:
$$x_0 = 0.904, \quad x_0 = 0.905, \quad x_0 = 0.906.$$
   (b) What happens?

REVIEW PROBLEMS FOR CHAPTER FIVE

For each of the functions in Problems 1–4, do the following:

(a) Find f' and f''.

(b) Find the critical points of f.

(c) Find any inflection points.

(d) Evaluate f at the critical points and at the endpoints. Identify the local and global maxima and minima of f.

(e) Sketch f. Indicate clearly where f is increasing or decreasing, and its concavity.

1. $f(x) = x^3 - 3x^2 \quad (-1 \le x \le 3)$

2. $f(x) = x + \sin x \quad (0 \le x \le 2\pi)$

3. $f(x) = e^{-x}\sin x \quad (0 \le x \le 2\pi)$

4. $f(x) = x^{-2/3} + x^{1/3} \quad (1.2 \le x \le 3.5)$

For each of the functions in Problems 5–7, find the limits as x tends to $+\infty$ and $-\infty$, and then proceed as in Problems 1–4 (i.e., find f', etc.).

5. $f(x) = 2x^3 - 9x^2 + 12x + 1$

6. $f(x) = \dfrac{4x^2}{x^2+1}$

7. $f(x) = xe^{-x}$

8. Sketch a graph of $e^{-x^2/2}$, marking local maxima and minima and points of inflection.

For each of the functions in Problems 9–14, use derivatives to find and identify local maxima and minima and points of inflection and sketch its graph. Confirm your answers using a calculator or computer.

9. $f(x) = x^3 + 3x^2 - 9x - 15$

10. $f(x) = x^5 - 15x^3 + 10$

11. $f(x) = x - 2\ln x \quad$ for $x > 0$

12. $f(x) = x^2 e^{5x}$

13. $f(x) = e^{-x^2}$

14. $f(x) = \dfrac{x^2}{x^2+1}$

Find the best possible bounds for the functions in Problems 15–16:

15. $e^{-x}\sin x, \quad$ for $x \ge 0$

16. $x\sin x, \quad$ for $0 \le x \le 2\pi$

For the graphs of f' in Problems 17–20 decide:

(a) Over what intervals is f increasing? Decreasing?

(b) Does f have maxima or minima? If so, which, and where?

17.

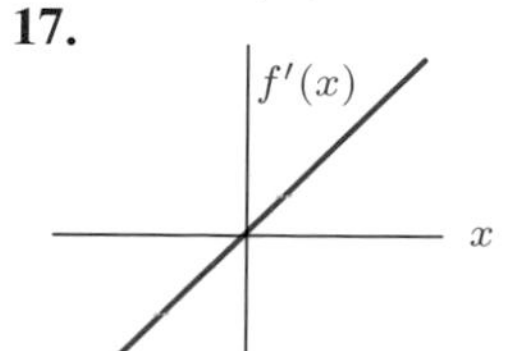

18.

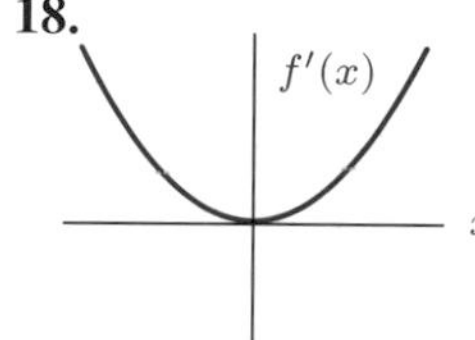

19.

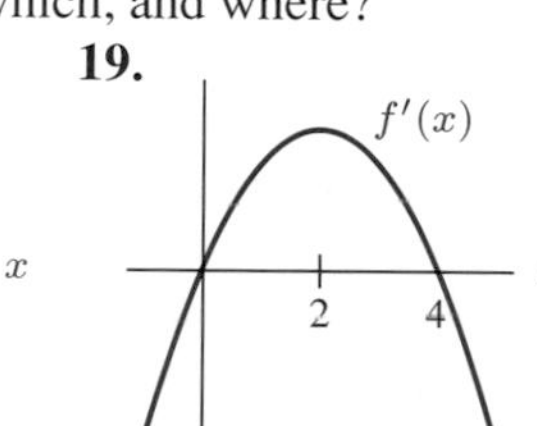

20.

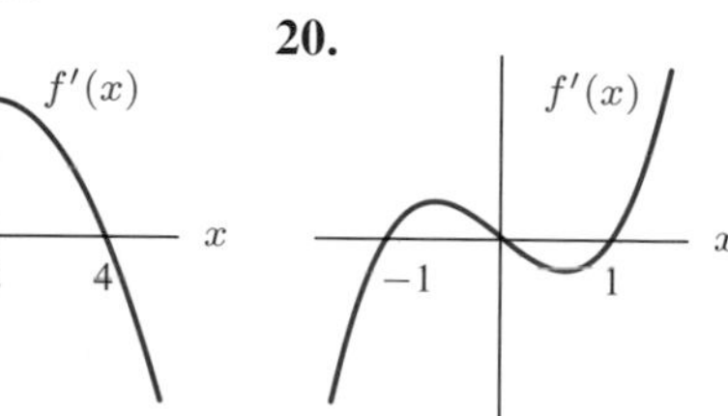

21. Consider the vase in Figure 5.86. Assume the vase is filled with water at a constant rate (i.e., constant volume per unit time).

(a) Graph $y = f(t)$, the depth of the water, against time, t. Show on your graph the points at which the concavity changes.

(b) Where does $y = f(t)$ grow fastest? Slowest? Estimate the ratio between these two growth rates.

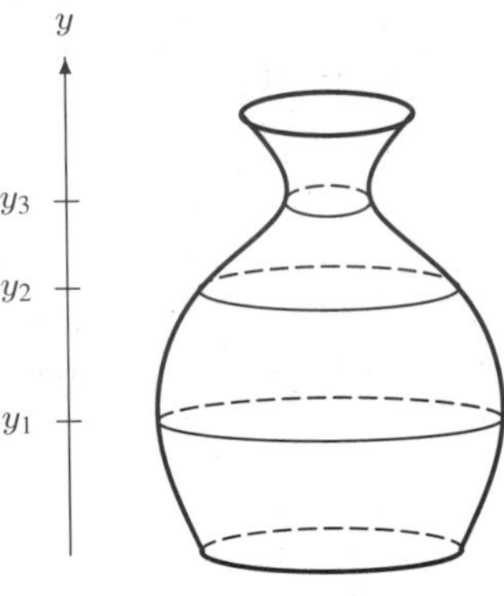

Figure 5.86

22. Sketch several members of the family

$$y = x^3 - ax^2$$

on the same coordinate plane. Show that the critical points lie on the curve $y = -\frac{1}{2}x^3$.

23. Suppose $g(t) = (\ln t)/t$ for $t > 0$.

(a) Does g have either a global maximum or a global minimum on $0 < t < \infty$? If so, where, and what are their values?

(b) What does your answer to part (a) tell you about the number of solutions to the equation

$$\frac{\ln x}{x} = \frac{\ln 5}{5}?$$

(Note: There are many ways to investigate the number of solutions to this equation. We are asking you to draw a conclusion from your answer to part (a).)

(c) Find the solution(s).

24. For $a > 0$, the line

$$a(a^2 + 1)y = a - x$$

forms a triangle in the first quadrant with the x- and y-axes.

(a) Find, in terms of a, the x- and y-intercepts of the line.

(b) Find the area of the triangle, as a function of a.

(c) Find the value of a making the area a maximum.

(d) What is this greatest area?

(e) If you want the triangle to have area $1/5$, what choices do you have for a?

25. Let $C(q)$ be the total cost of producing a quantity q of a certain good. The *average cost* is $a(q) = C(q)/q$.

(a) Interpret $a(q)$ graphically, as the slope of a line in Figure 5.87.
(b) Find on the graph the quantity q_0 where $a(q)$ is minimal.
(c) What is the relationship between $a(q_0)$ and the marginal cost, $C'(q_0)$? Explain your result both graphically and analytically (using the fact that $a'(q_0) = 0$, since $a(q)$ has a minimum at q_0). What does this result mean, in terms of economics?
(d) Graph $C'(q)$ and $a(q)$ on the same axes. Mark q_0 on the q-axis.

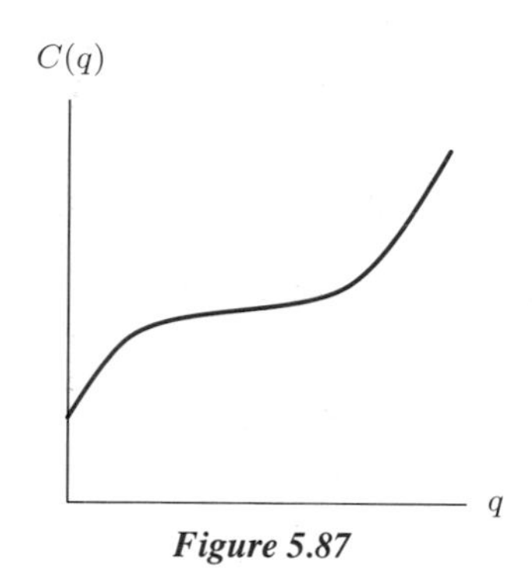

Figure 5.87

26. Consider a large tank of water, with temperature $W(t)$. The ambient temperature $A(t)$ (i.e., the temperature of the surrounding air), is given by the graph in Figure 5.88.

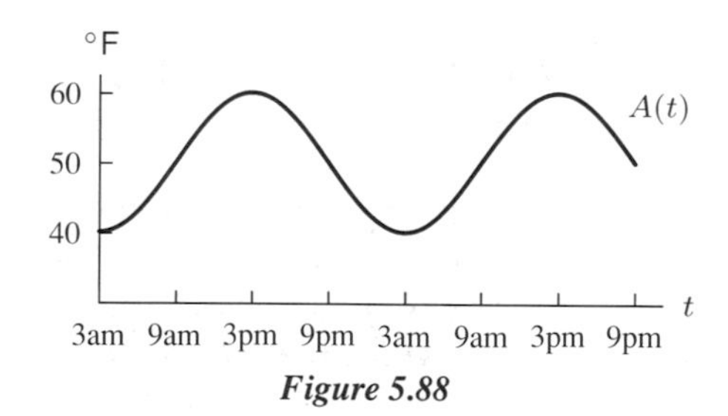

Figure 5.88

The temperature of the water is affected by the temperature of the surrounding air.

(a) How does the temperature of the water change if the water is colder than the surrounding air? What if the water is warmer?
(b) Using your answer in part (a), sketch a possible graph for $W(t)$ on the same axes as $A(t)$.
(c) Explain the relationship between the maxima and minima of $W(t)$ and the points where the two graphs intersect.
(d) What is the relationship between the rate at which the temperature of the water changes and the difference $A(t) - W(t)$?
(e) What is the relationship between the inflection points of $W(t)$ and the points where $A(t) - W(t)$ has a maxima or minima?
(f) Assume the tank is refilled with cold water (35°F) at 3 am. Sketch a possible graph for $W(t)$. Pay attention to the concavity.

27. Any body radiates energy at various wavelengths. The power of the radiation (per meter^2 of surface) and the distribution of the radiation among the wavelengths varies with temperature.

The function in Figure 5.89 represents the intensity of the radiation of a black body at a temperature $T = 3000°$ Kelvin, as a function of the wavelength. The intensity of the radiation is highest in the infrared range, that is, at wavelengths longer than that of visible light (0.4–0.7μm). Max Planck's radiation law, announced to the Berlin Physical Society on October 19, 1900, states that

$$r(\lambda) = \frac{a}{\lambda^5(e^{b/\lambda} - 1)},$$

where a and b are empirical constants chosen to best fit the experimental data.

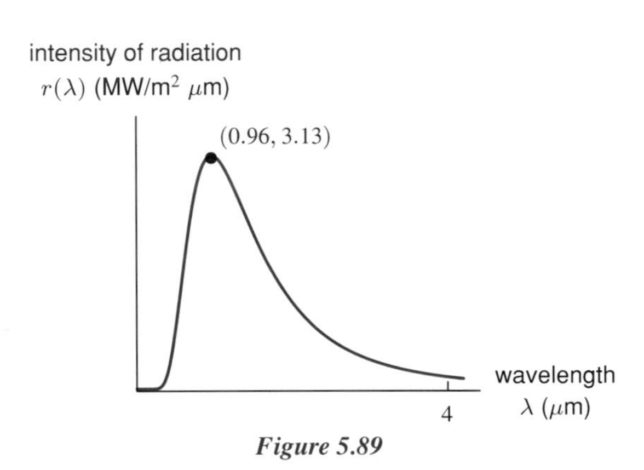

Figure 5.89

Find a and b so that the formula fits the graph.

(Later in 1900 Planck was able to derive his radiation law entirely from theory. He found that $a = 2\pi c^2 h$ and $b = \frac{hc}{Tk}$ where c = speed of light, h = Planck's constant, and k = Boltzmann's constant.)

28. Populations with a growth limit have been modeled with the logistic family

$$y = \frac{A}{1 + Be^{-Cx}} \qquad -\infty < x < \infty \quad \text{and} \quad A, B, C > 0.$$

(a) Sketch a graph of $g(x) = A/\left(1 + e^{-Cx}\right)$. What is the significance of the parameter A?
(b) Evaluate $g(-x) + g(x)$ and say what the sum means geometrically.
(c) What happens to the graph of g if A is kept constant and C is increased?
(d) Show that the curve $y = A/\left(1 + Be^{-Cx}\right)$ is a horizontal shift of the graph of g.

CHAPTER SIX

RECONSTRUCTING A FUNCTION FROM ITS DERIVATIVE

Chapter 2 showed us how to calculate the rate of change, or derivative, of a function. In Chapter 3 we defined the definite integral to reconstruct changes in the original function from its derivative. For example, in Chapter 2 we saw how to calculate velocity given position, and in Chapter 3 we saw how to reconstruct distance from velocity. In this chapter we will look in more detail at how to reconstruct a function from its derivative.

We already know how to get a rough idea of the graph of f from the graph of f'. Given a starting value for f, say $f(0) = 3$, we sketch a graph of f whose slope at any point is the value of f'. We also know how to estimate values of f numerically: changes in f are obtained from f' using the Fundamental Theorem of Calculus.

In this chapter we start to see how to go from f' to f algebraically: given a formula for f', what is the formula for f?

6.1 THE DEFINITE INTEGRAL REVISITED

In Chapter 3 we defined the definite integral to allow us to reconstruct from the velocity the distance traveled by an object. If f is bounded and continuous (except perhaps at a few points) on the interval $a \leq x \leq b$, we divide the interval into n equal subdivisions of length $\Delta x = (b-a)/n$ using the points $a = x_0, x_1, \ldots, x_n = b$, and then we form the Riemann sums:

$$\text{Left-hand sum} = f(x_0)\Delta x + f(x_1)\Delta x + \cdots + f(x_{n-1})\Delta x = \sum_{i=0}^{n-1} f(x_i)\Delta x.$$

$$\text{Right-hand sum} = f(x_1)\Delta x + f(x_2)\Delta x + \cdots + f(x_n)\Delta x = \sum_{i=1}^{n} f(x_i)\Delta x.$$

We expect that as n goes to infinity and the subdivisions become finer, both sums will approach the same number. This common limit is defined to be the *definite integral*:

$$\int_a^b f(x)\,dx = \lim_{n\to\infty}(\text{Left-hand sum}) = \lim_{n\to\infty}\sum_{i=0}^{n-1} f(x_i)\Delta x$$
$$= \lim_{n\to\infty}(\text{Right-hand sum}) = \lim_{n\to\infty}\sum_{i=1}^{n} f(x_i)\Delta x.$$

The function f is called the *integrand*, and the numbers a and b are the *limits of integration*. In Chapter 3 we only considered the case $a < b$, but we now allow $a \geq b$. We still set $x_0 = a$, $x_n = b$, and $\Delta x = (b-a)/n$. If $a > b$, the quantity Δx is negative and the names left- and right-sum no longer apply. (The points $a = x_0, x_1, \ldots, x_n = b$ still divide the interval into equal subdivisions, but they go from right to left, not left to right.) However, the definition in the preceding box still applies.

Interpretations of the Definite Integral

Two interpretations of the integral $\int_a^b f(x)\,dx$ were discussed in Chapter 3:

- The integral represents the **area** between the graph of f and the x-axis from a to b, assuming $f \geq 0$ and $a < b$. (See Figure 6.1.)
- The **average value** of f over the interval $[a, b]$ is defined in terms of the integral:

$$\begin{matrix}\text{Average value of } f \\ \text{from } a \text{ to } b\end{matrix} = \frac{1}{b-a}\int_a^b f(x)\,dx.$$

The average value of f is the height of the rectangle with base $(b-a)$ and whose area equals the area under the graph of $f(x)$ between $x = a$ and $x - b$. (See Figure 6.2.)

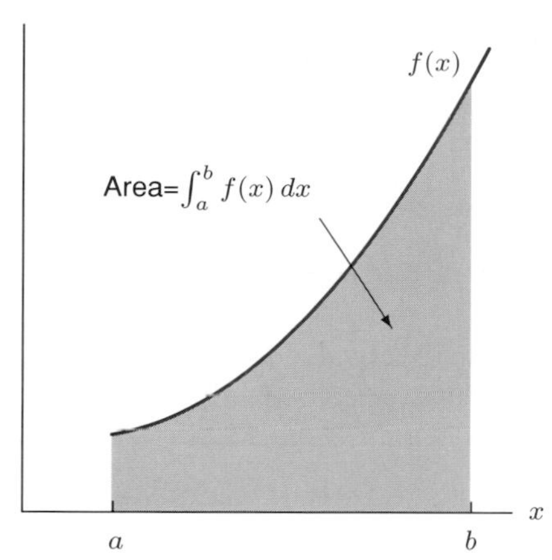

Figure 6.1: Area as an integral

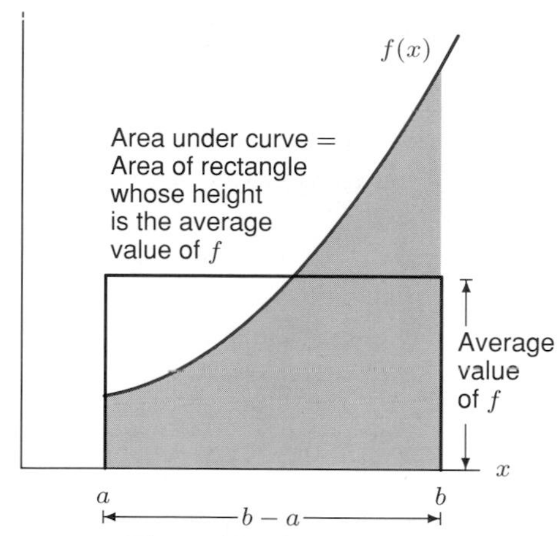

Figure 6.2: Average value

The Definite Integral of a Rate of Change: When $f = F'$

If f is the derivative of a function F, then we can interpret $f(x)$ as the instantaneous rate of change of the quantity $F(x)$. The integral $\int_a^b f(x)\,dx$ then represents the total change in $F(x)$ as x changes from a to b:

$$\begin{matrix}\text{Total change in } F(x) \\ \text{between } a \text{ and } b\end{matrix} = \int_a^b F'(x)\,dx = \int_a^b f(x)\,dx.$$

Since the change in $F(x)$ can also be expressed as $F(b) - F(a)$, we have the result presented at the end of Chapter 3:

> **The Fundamental Theorem of Calculus**
>
> If f is the derivative of F, that is, $f = F'$, and if f is continuous, then
>
> $$\int_a^b f(x)\,dx = \int_a^b F'(x)\,dx = F(b) - F(a).$$
>
> In words:
>
> The definite integral of a rate of change gives total change.

If $f = F'$, we call F an **antiderivative** of f. This theorem has tremendous practical importance: it says that *if* we have a formula for F, we don't need to calculate the left and right sums to find a definite integral, but instead we can simply subtract $F(a)$ from $F(b)$. Equally important is the fact that if we estimate the integral $\int_a^b f(x)\,dx$, this theorem gives us an estimate for $F(b) - F(a)$ which can be used to extract information about F from its derivative f.

Example 1 Use the Fundamental Theorem to find $\displaystyle\int_0^{\pi} \sin x\,dx$.

Solution We take $f(x) = \sin x$ and find an antiderivative, $F(x)$. Since

$$\frac{d}{dx}(-\cos x) = -(-\sin x) = \sin x,$$

we take $F(x) = -\cos x$. Then

$$\int_0^\pi \sin x\, dx = F(\pi) - F(0) = -\cos\pi - (-\cos 0) = -(-1) - (-1) = 2.$$

Notice that the Fundamental Theorem gives us an exact value for the definite integral easily. The drawback to using this method is that you need to find an antiderivative, and that is not always easy.

Applications of the Fundamental Theorem

The Fundamental Theorem of Calculus tells us that the total distance traveled by a moving object in a given time interval may be represented by a definite integral. The examples below show how representing a quantity as a definite integral, and thereby as an area, can be helpful even if you don't evaluate the integral.

Example 2 Two cars start from rest at a traffic light and accelerate for several minutes. Figure 6.3 shows their velocities as a function of time. (a) Which car is ahead after one minute? (b) Which car is ahead after two minutes?

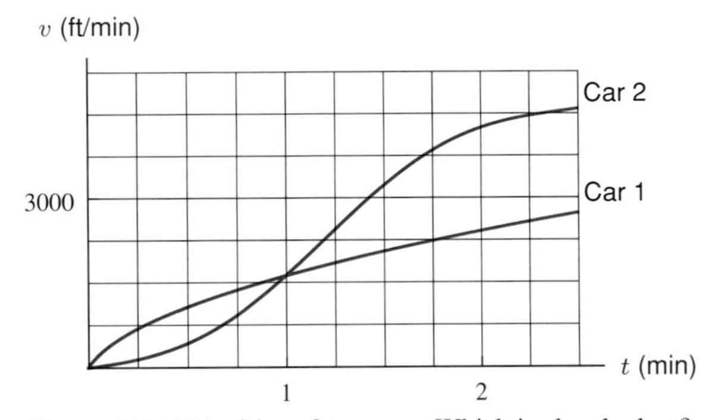

Figure 6.3: Velocities of two cars. Which is ahead when?

Solution (a) For the first minute car 1 goes faster than car 2, and therefore car 1 must be ahead at the end of one minute.

(b) At the end of two minutes the situation is less clear, since car 1 was going faster for the first minute and car 2 for the second. However, if $v = f(t)$ is the velocity of a car after t minutes, then we know that

$$\text{Distance traveled in two minutes} = \int_0^2 f(t)\, dt,$$

since the integral of velocity is distance traveled. This definite integral may also be interpreted as the area under the graph of f between 0 and 2. Since the area representing the distance traveled by car 2 is clearly larger than the area for car 1 (see Figure 6.3), we know that car 2 has traveled farther than car 1.

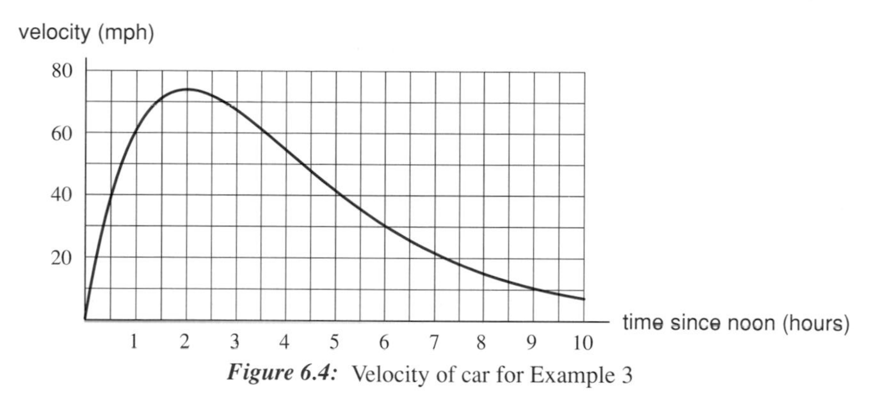

Figure 6.4: Velocity of car for Example 3

Example 3 A car starts at noon and travels with the velocity shown in Figure 6.4. A truck starts at 1 pm from the same place and travels at a constant velocity of 50 mph.

(a) How far away is the car when the truck starts?

(b) During the period when the car is ahead of the truck, when is the distance between them greatest, and what is that greatest distance?

(c) When does the truck overtake the car, and how far have both traveled then?

Solution To find distances from the velocity graph, we use the fact that if t is the time measured from noon, and v is the velocity,

$$\text{Distance traveled by car at time } T = \int_0^T v\,dt = \text{Area under velocity graph between 0 and } T.$$

The truck's motion can be represented on the same graph by the horizontal line $v = 50$, starting at $t = 1$. The distance traveled by the truck is then the rectangular area under this line, and the distance between the two vehicles is the difference between these areas. One small rectangle on the graph corresponds to moving at 10 mph for a half hour (i.e., to a distance of 5 miles).

(a) The distance traveled by the car when the truck starts is represented by the shaded area in Figure 6.5, which totals about seven rectangles or about 35 miles.

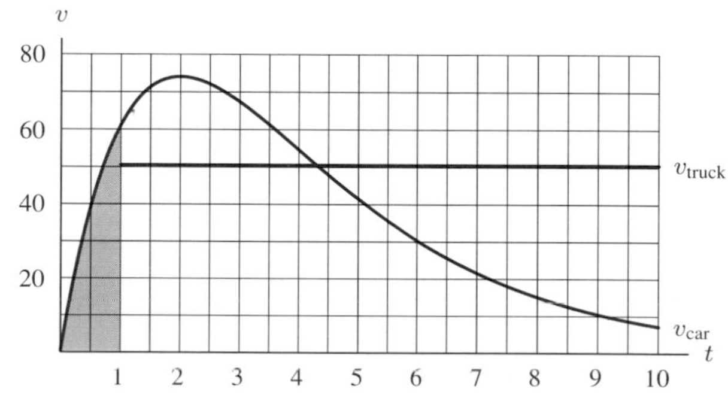

Figure 6.5: Shaded area represents distance traveled by car (noon to 1pm)

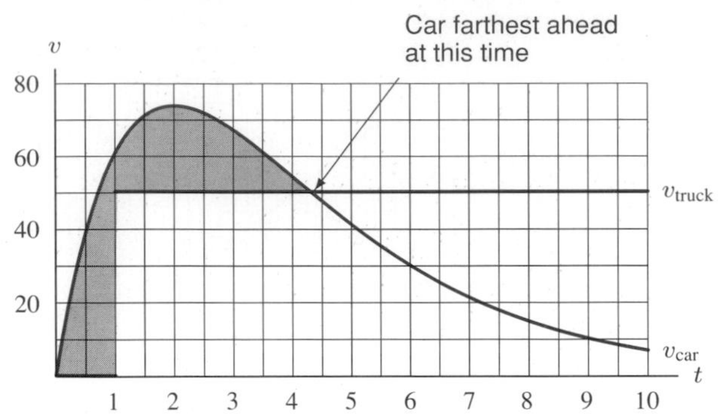

Figure 6.6: Shaded area = distance by which car is ahead at 4:18 pm

(b) The car starts ahead of the truck, and the distance between them increases as long as the velocity of the car is greater than the velocity of the truck. Later, when the truck's velocity exceeds the car's, the truck starts to gain on the car. Thus the distance between the car and the truck will increase as long as $v_{\text{car}} > v_{\text{truck}}$, and it will decrease when $v_{\text{car}} < v_{\text{truck}}$. Therefore, the maximum distance occurs when $v_{\text{car}} = v_{\text{truck}}$, that is, when $t \approx 4.3$ hours (at about 4:18 pm). (See Figure 6.6.)

The distance traveled by the car is the area under the v_{car} graph between $t = 0$ and $t = 4.3$; the distance traveled by the truck is the area under the v_{truck} line between $t = 1$ (when it started) and $t = 4.3$. The distance between the car and truck is represented by the shaded area in Figure 6.6, which is approximately

$$35 \text{ miles } + 50 \text{ miles} = 85 \text{ miles}.$$

(c) The truck overtakes the car when both have traveled the same distance. This occurs when the area under the curve up to that time equals the area under the line up to that time. Since the areas under the curve and the line overlap (see Figure 6.7), they are equal when the lightly shaded area equals the heavily shaded area (which we know is about 85 miles). This happens when $t \approx 8.3$ hours, or about 8:18 pm.

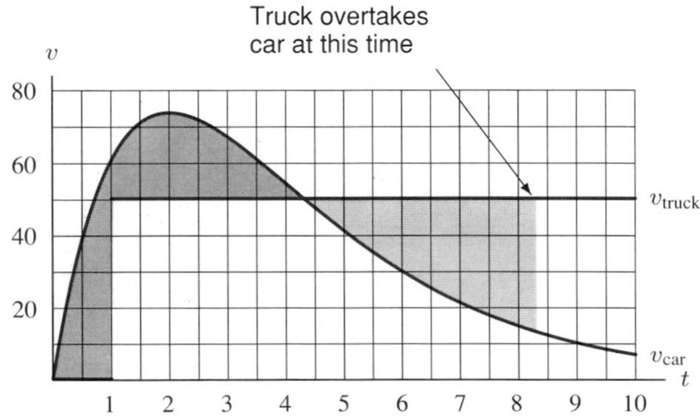

Figure 6.7: Truck overtakes car when dark and light shaded areas are equal

Problems for Section 6.1

1. Estimate $\int_1^2 x^2\,dx$ using left- and right-hand sums with four subdivisions. How far from the true value of the integral could your estimate be?
2. If the graph of f is in Figure 6.8:

   (a) What is $\int_1^6 f(x)\,dx$?

   (b) What is the average value of f on $[1,6]$?

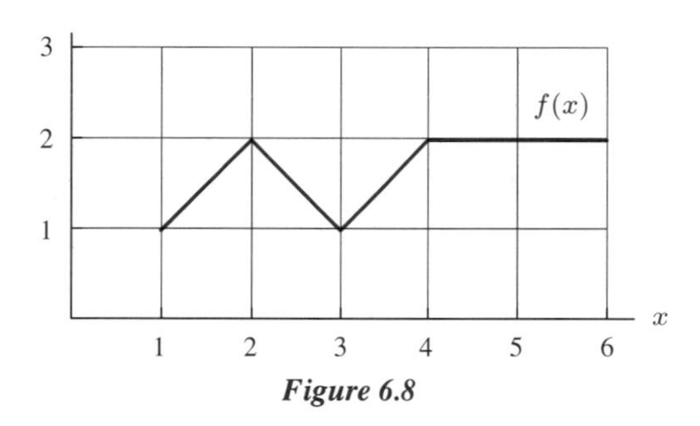

Figure 6.8

3. What is the average value of $f(x) = 4x + 7$ on $[1,3]$?
4. Given $f(x) = x$, find intervals $[a, b]$ satisfying:

   (a) $\int_a^b f(x)\,dx$ is less than the average value of f on the interval $[a, b]$.

   (b) $\int_a^b f(x)\,dx$ is equal to the average value of f on the interval $[a, b]$.

   (c) $\int_a^b f(x)\,dx$ is greater than the average value of f on the interval $[a, b]$.

5. Use the Fundamental Theorem of Calculus to find $\int_0^2 (3x^2 + 1)\,dx$.
6. Without computing the integral, decide if

$$\int_0^{2\pi} e^{-x} \sin x\,dx$$

   is positive or negative, and explain your decision. [Hint: Sketch $e^{-x} \sin x$.]

7. For the two cars in Example 2, page 308, estimate

   (a) the distances moved by car 1 and car 2 during the first minute.

   (b) the time at which the two cars have gone the same distance.

8. Consider the car and the truck in Example 3, page 309.

   (a) How fast is the distance between the car and the truck increasing or decreasing at 3 pm?

   (b) What is the practical significance (in terms of the distance between the car and the truck) of the fact that the car's velocity is maximized at about 2 pm?

9. Consider the car and the truck in Example 3, page 309, but suppose the truck starts at noon. (Everything else remains the same.)

   (a) Sketch a new graph showing the velocities of both car and truck against time.

   (b) How many times do the two graphs intersect? What does each intersection mean in terms of the distance between the two?

10. If the velocity of a particle at time t is given by $v(t) = \cos t$, use the Fundamental Theorem of Calculus to find the total distance traveled by the particle between $t = 0$ and $t = \pi/2$.

11. A car moves along a straight line with velocity, in feet/second, given by

$$v(t) = 6 - 2t \quad \text{for } t \geq 0.$$

   (a) Describe the car's motion in words. (When is it moving forward, backward, and so on?)
   (b) Suppose the car's position is measured from its starting point. When is it farthest forward? Backward?
   (c) Find s, the car's position measured from its starting point, as a function of time.

12. A car moves along a straight line with velocity given by

$$v(t) = 2 + 10t \text{ ft/sec}$$

for $0 \leq t \leq 10$ seconds.

   (a) Graph $v(t)$ for $0 \leq t \leq 10$, and explain how you can find the total distance the car has traveled between $t = 0$ and $t = 10$ seconds geometrically (i.e., from the graph). Find this distance using the formula for the area of a trapezoid.
   (b) Find the function $s(t)$ that gives the position of the car as a function of time. Explain the meaning of any new constants.
   (c) Use your function $s(t)$ to find the total distance traveled by the car between $t = 0$ and $t = 10$ seconds. Compare with your answer in part (a).
   (d) Explain how this verifies the Fundamental Theorem of Calculus.

13. A new sales agent finds that as she gains experience, she increases the number of large appliances she sells each month. In the first month she sells only seven, but each month she sells two more than the month before, so that the number she sells in month t is $2t + 5$.

   (a) Find the average number of large appliances she sells per month over the first year arithmetically, by calculating the number of appliances sold each month and then taking the average over 12 months.
   (b) Now find the average by integration as though the sales function applied for all values of t (instead of just for integers).
   (c) How well do the two results compare?
   (d) If the answer you found in part (a) is viewed as the true answer, and the integral answer as an approximation, why would anyone want to use the integral answer instead of the true answer?
   (e) Draw a picture showing both answers as the area of some region. Mark on your picture a region representing the error in the integral answer.

14. Water is run into a large tank through a hose at a constant rate. After 5 minutes a hole is opened in the bottom of the tank, and water starts to flow out. Initially the flow rate through the hole is twice as great as the rate through the hose, but as the water level in the tank goes down, the flow rate through the hole decreases; after another 10 minutes the water level in the tank appears to be constant. Plot graphs of the flow rates through the hose and through the hole against time on the same pair of axes. Show how the volume of water in the tank at any time can be interpreted as an area (or the difference between two areas) on the graph. In particular, interpret the steady-state volume of water in the tank.[1]

15. The graph of $f(x) = 1/(x + 1)$ is in Figure 6.9. For $x \geq 0$, define a new function $F(x)$ as follows: $F(x)$ is the area of the region bounded by the x-axis, the y-axis, the graph of f, and the vertical line at x.

[1] From *Calculus: The Analysis of Functions*, by Peter D. Taylor (Toronto: Wall & Emerson, Inc., 1992)

(a) For positive x and h, sketch the area represented by $F(x+h) - F(x)$.
(b) Using your sketch, show that

$$h \cdot f(x) > F(x+h) - F(x) > h \cdot f(x+h).$$

(c) Using part (b), find $F'(x)$ as the limit of the difference quotient.

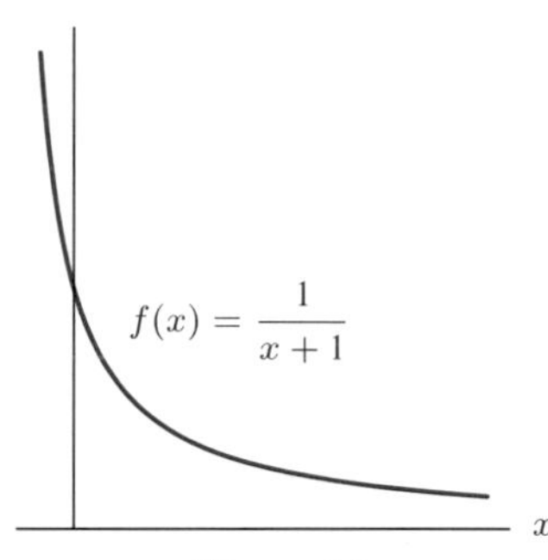

Figure 6.9

16. Figure 6.10 is a graph of the annual yield, $y(t)$ (in bushels per year), from an orchard t years after planting. The trees take about 10 years to get established, but for the next 20 years they give a substantial yield. After about 30 years, however, age and disease start to take their toll, and the annual yield falls off.[2]

    (a) Represent on a sketch of Figure 6.10 the total yield $Y(T)$ until time T, and write an expression for $Y(T)$ in terms of $y(t)$.

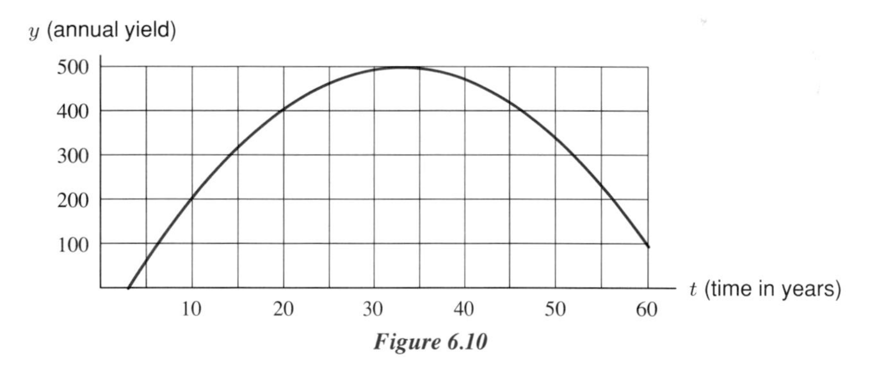

Figure 6.10

    (b) Sketch a graph of $Y(T)$ against T.
    (c) Write an expression for the average annual yield, $a(T)$, up until time T.
    (d) The important question is: When should the orchard be cut down and replanted? Assume that you want to maximize your average revenue per year, and that fruit prices remain constant, so that this is achieved by maximizing your average annual yield.
        (i) What condition on $Y(T)$ maximizes the average annual yield?
        (ii) What condition on $y(T)$ maximizes the average annual yield? Estimate the T value which yields the maximum average annual yield.

[2]From *Calculus: The Analysis of Functions*, by Peter D. Taylor (Toronto: Wall & Emerson, Inc., 1992)

6.2 PROPERTIES OF THE DEFINITE INTEGRAL

This section contains properties of the definite integral that will enable us to say something about the value of an integral without having to actually evaluate it. This is analogous to being able to say that $\sqrt{24}$ is less than $\sqrt{37}$ without knowing the exact value of either $\sqrt{24}$ or $\sqrt{37}$.

Facts About Limits of Integration

If a, b, and c are any numbers, then

1. $\displaystyle\int_b^a f(x)\,dx = -\int_a^b f(x)\,dx.$
2. $\displaystyle\int_a^c f(x)\,dx + \int_c^b f(x)\,dx = \int_a^b f(x)\,dx.$

In words:

1. The integral from b to a is the negative of the integral from a to b.
2. The integral from a to c plus the integral from c to b is the integral from a to b.

By interpreting the integrals as areas, we can justify these results for $f \geq 0$. In fact, they are true for all functions for which the integrals make sense.

Why is $\displaystyle\int_b^a f(x)\,dx = -\int_a^b f(x)\,dx$?

By definition, both integrals are approximated by sums of the form $\sum f(x_i)\Delta x$. The x_i's are the same in each case: the only difference in the sums for $\int_b^a f(x)\,dx$ and $\int_a^b f(x)\,dx$ is that in the first, $\Delta x = (a-b)/n = -(b-a)/n$ and in the second $\Delta x = (b-a)/n$. Since everything else about the sums is the same, we must have $\int_b^a f(x)\,dx = -\int_a^b f(x)\,dx$.

Why is $\displaystyle\int_a^c f(x)\,dx + \int_c^b f(x)\,dx = \int_a^b f(x)\,dx$?

Suppose $a < c < b$. Figure 6.11 suggests that $\int_a^c f(x)\,dx + \int_c^b f(x)\,dx = \int_a^b f(x)\,dx$ since the area under f from a to c plus the area under f from c to b together make up the whole area under f from a to b.

Actually, this property holds for all numbers a, b, and c, not just those satisfying $a < c < b$. (See Figure 6.12.) For example, the area under f from 3 to 6 is equal to the area from 3 to 8 *minus* the area from 6 to 8, so

$$\int_3^6 f(x)\,dx = \int_3^8 f(x)\,dx - \int_6^8 f(x)\,dx = \int_3^8 f(x)\,dx + \int_8^6 f(x)\,dx.$$

Figure 6.11: Additivity of the definite integral ($a < c < b$)

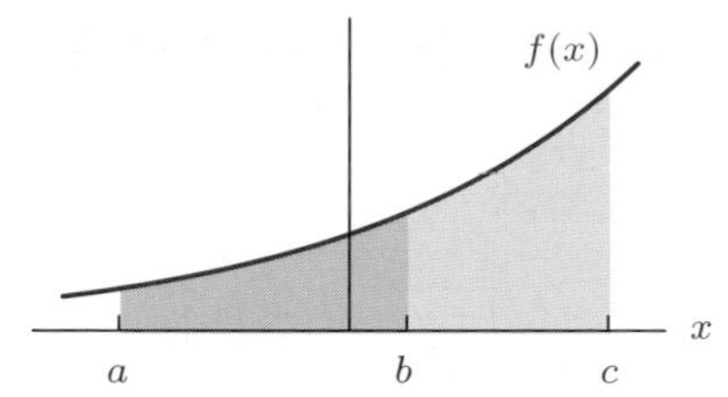

Figure 6.12: Additivity of the definite integral ($a < b < c$)

Example 1 Suppose you know that $\int_0^{1.25} \cos(x^2)\,dx = 0.98$ and $\int_0^1 \cos(x^2)\,dx = 0.90$. (See Figure 6.13.) What are the values of the following integrals?

(a) $\int_1^{1.25} \cos(x^2)\,dx$ (b) $\int_{-1}^{1} \cos(x^2)\,dx$ (c) $\int_{1.25}^{-1} \cos(x^2)\,dx$

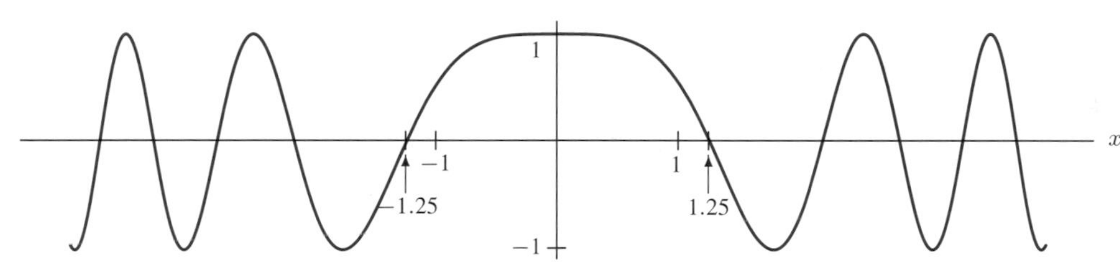

Figure 6.13: Graph of $f(x) = \cos x^2$

Solution (a) Since $\int_0^{1.25} \cos(x^2)\,dx = \int_0^1 \cos(x^2)\,dx + \int_1^{1.25} \cos(x^2)\,dx$ by the additivity property, we get $0.98 = 0.90 + \int_1^{1.25} \cos(x^2)\,dx$, so $\int_1^{1.25} \cos(x^2)\,dx = 0.08$.

(b) $\int_{-1}^{1} \cos(x^2)\,dx = \int_{-1}^{0} \cos(x^2)\,dx + \int_0^1 \cos(x^2)\,dx$.
By the symmetry of $\cos(x^2)$ about the y-axis, $\int_{-1}^{0} \cos(x^2)\,dx = \int_0^1 \cos(x^2)\,dx = 0.90$.
So $\int_{-1}^{1} \cos(x^2)\,dx = 0.90 + 0.90 = 1.80$.

(c) $\int_{1.25}^{-1} \cos(x^2)\,dx = -\int_{-1}^{1.25} \cos(x^2)\,dx = -\int_{-1}^{0} \cos(x^2)\,dx - \int_0^{1.25} \cos(x^2)\,dx = -0.90 - 0.98 = -1.88$.

Facts about Sums and Constant Multiples of the Integrand

Let f and g be functions and let c be a constant.

1. $\displaystyle\int_a^b (f(x) \pm g(x))\,dx = \int_a^b f(x)\,dx \pm \int_a^b g(x)\,dx.$
2. $\displaystyle\int_a^b cf(x)\,dx = c\int_a^b f(x)\,dx.$

In words:

1. The integral of the sum (or difference) of two functions is the sum (or difference) of their integrals.
2. The integral of a constant times a function is that constant times the integral of the function.

Why Do these Properties Hold?

Both can be understood by thinking of the definition of the definite integral as the limit of a sum of areas of rectangles.

For property 1, let's suppose that f and g are positive on the interval $[a, b]$ so that the area under $f(x)+g(x)$ is approximated by the sum of the areas of rectangles like the one shaded in Figure 6.14. The area of this rectangle is

$$[f(x_i) + g(x_i)]\Delta x = f(x_i)\Delta x + g(x_i)\Delta x.$$

Since $f(x_i)\Delta x$ is the area of a rectangle under the graph of f, and $g(x_i)\Delta x$ is the area of a rectangle under the graph of g, the area under $f(x) + g(x)$ is the sum of the areas under $f(x)$ and $g(x)$.

For property 2, notice that multiplying a function by c stretches or flattens the graph in the vertical direction by a factor of c. Thus, it stretches or flattens the height of each approximating rectangle by c, and hence multiplies the area by c.

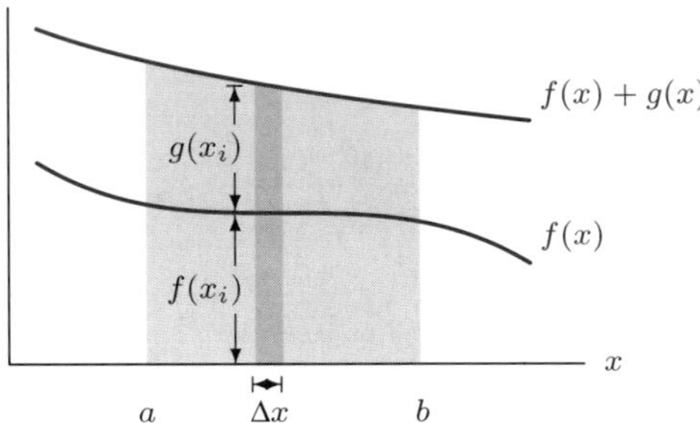

Figure 6.14: Area $= \int_a^b [f(x) + g(x)]\, dx = \int_a^b f(x)\, dx + \int_a^b g(x)\, dx$

Example 2 Evaluate the definite integral $\displaystyle\int_0^2 (1 + 3x)\, dx$.

Solution We can break this integral up as follows:

$$\int_0^2 (1 + 3x)\, dx = \int_0^2 1\, dx + \int_0^2 3x\, dx = \int_0^2 1\, dx + 3\int_0^2 x\, dx.$$

This expresses our original integral in terms of two simpler integrals. From the area interpretation of the integral, we see that

$$\int_0^2 1\, dx = 2,$$

since it represents the area under the horizontal line $y = 1$ between $x = 0$ and $x = 2$ (see Figure 6.15). Similarly,

$$\int_0^2 x\, dx = \frac{1}{2} \cdot 2 \cdot 2 = 2$$

because it is the area of the triangle in Figure 6.16. Therefore,

$$\int_0^2 (1 + 3x)\, dx = \int_0^2 1\, dx + 3\int_0^2 x\, dx = 2 + 3(2) = 8.$$

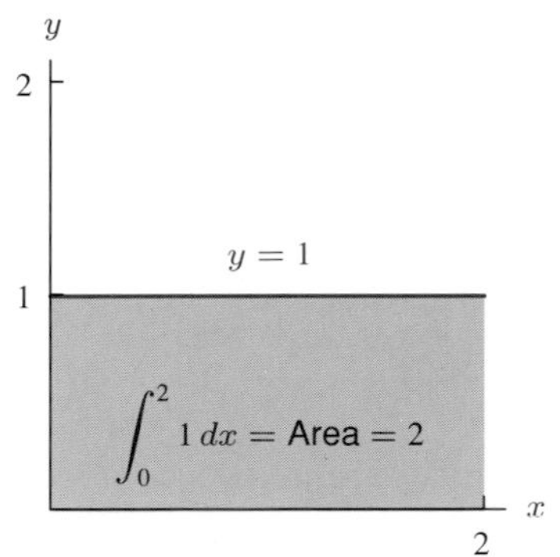

Figure 6.15: Area representing $\int_0^2 1\,dx$

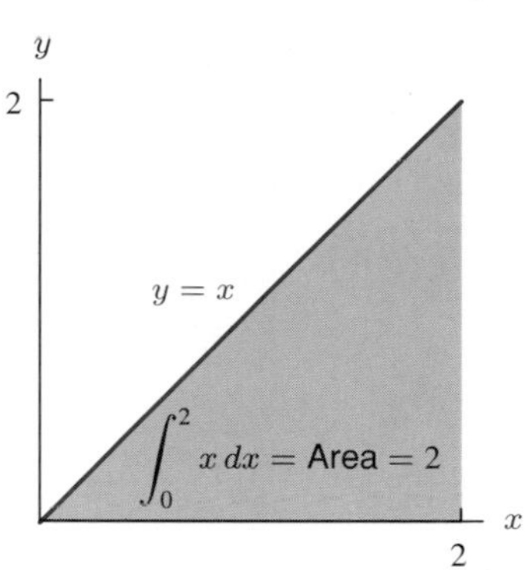

Figure 6.16: Area representing $\int_0^2 x\,dx$

How to Compare Integrals

Although we need a computer or calculator to evaluate $\int_0^{\sqrt{\pi}} \sin(x^2)\,dx$ accurately, we can say immediately that it is less than $\sqrt{\pi}$, since, for all x,

$$\sin(x^2) \le 1.$$

This means that the area under $y = \sin(x^2)$ is less than the rectangular area under the line $y = 1$ between $x = 0$ and $x = \sqrt{\pi}$ (see Figure 6.17). The rectangle has arca $\sqrt{\pi} \cdot 1 = \sqrt{\pi}$, so

$$\int_0^{\sqrt{\pi}} \sin(x^2)\,dx \le \sqrt{\pi}.$$

More generally, we have the following fact:

Comparing Definite Integrals

If $f(x) \le g(x)$ for $a \le x \le b$, then

$$\int_a^b f(x)\,dx \le \int_a^b g(x)\,dx.$$

If $0 \le f(x) \le g(x)$, this result says that the area under the graph of f is smaller than the area under the graph of g. See Figure 6.18.

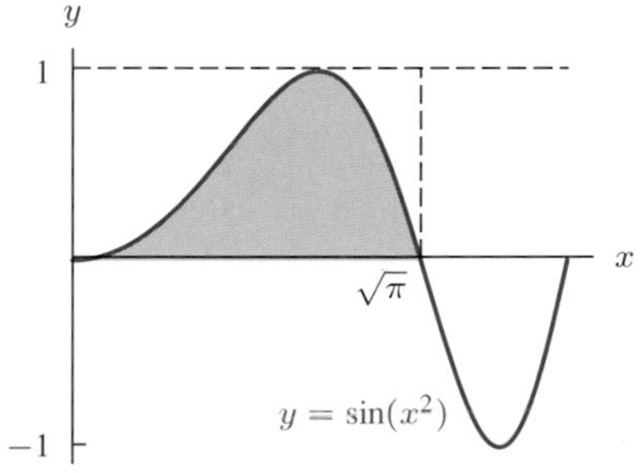

Figure 6.17: Graph showing that $\int_0^{\sqrt{\pi}} \sin(x^2)\,dx < \sqrt{\pi}$

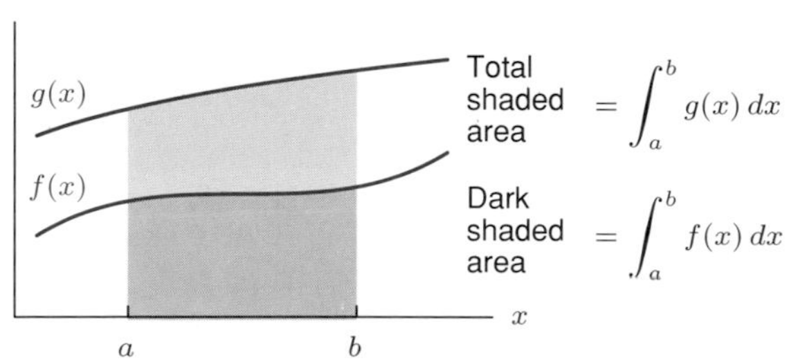

Figure 6.18: If $f(x) \le g(x)$ then $\int_a^b f(x)\,dx \le \int_a^b g(x)\,dx$

Problems for Section 6.2

1. Suppose you know $\int_1^3 3x^2\,dx = 26$ and $\int_1^3 2x\,dx = 8$. What is $\int_1^3 (x^2 - x)\,dx$?

2. (a) Construct a formula, using geometry, for
$$\int_a^b 1\,dx$$
in terms of a and b. Assume $a \le b$.

   (b) Use the result of part (a) to find: (i) $\int_2^5 1\,dx$ (ii) $\int_{-3}^8 1\,dx$ (iii) $\int_1^3 23\,dx$.

3. (a) Construct a formula, using geometry, for
$$\int_a^b x\,dx$$
in terms of a and b. Assume that $0 \le a \le b$.

   (b) Use the result of part (a) to find: (i) $\int_2^5 x\,dx$ (ii) $\int_{-3}^8 x\,dx$ (iii) $\int_1^3 5x\,dx$.

4. Find $\displaystyle\int_{-1}^1 |x|\,dx$ geometrically.

5. The function for the *standard normal distribution*, which is often used in statistics, has the formula
$$\frac{1}{\sqrt{2\pi}}e^{-x^2/2}$$
and the graph in Figure 6.19. Statistics books usually contain tables such as the one below, showing only the area under the curve from 0 to b, for different values of b.

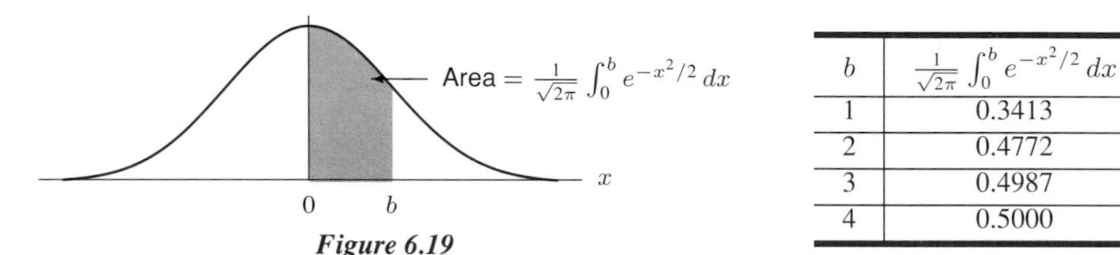

Figure 6.19

b	$\frac{1}{\sqrt{2\pi}}\int_0^b e^{-x^2/2}\,dx$
1	0.3413
2	0.4772
3	0.4987
4	0.5000

Use the information given in the table and the symmetry of the standard normal curve about the y-axis to find:

(a) $\displaystyle\frac{1}{\sqrt{2\pi}}\int_1^3 e^{-x^2/2}\,dx.$ (b) $\displaystyle\frac{1}{\sqrt{2\pi}}\int_{-2}^3 e^{-x^2/2}\,dx.$

6. Use the property $\displaystyle\int_b^a f(x)\,dx = -\int_a^b f(x)\,dx$ to show that $\displaystyle\int_a^a f(x)\,dx = 0$.

7. Is $\displaystyle\int_{-1}^1 e^{x^2}\,dx$ positive, negative, or zero? Explain.

8. Explain why $\displaystyle 0 < \int_0^1 e^{x^2}\,dx < 3$.

9. Let $F(x) = \int_0^x 2t\,dt$. Then $F(x)$ is the area under the line $y = 2t$ from the origin to x.

   (a) Construct a table showing the values of F for $x = 0, 1, 2, 3, 4, 5$.

(b) Is F increasing or decreasing when $x > 0$? Concave up or down? Explain.
(c) When $t < 0$, the line $y = 2t$ is below the t-axis (the horizontal axis). Explain why $F(-1)$ is positive.

10. Suppose the graph of f is in Figure 6.20 and its formula is:

$$f(x) = \begin{cases} -x+1, & \text{for } 0 \le x \le 1; \\ x-1, & \text{for } 1 < x \le 2. \end{cases}$$

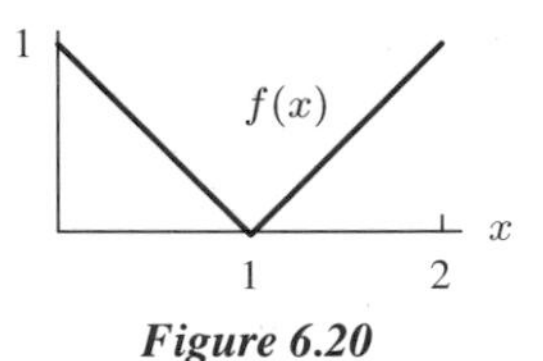

Figure 6.20

(a) Find a function F such that $F' = f$ and $F(1) = 1$.
(b) Use geometry to show the area under the graph of f above the x-axis between $x = 0$ and $x = 2$ is equal to $F(2) - F(0)$.
(c) Use parts (a) and (b) to verify the Fundamental Theorem of Calculus.

6.3 CONSTRUCTING ANTIDERIVATIVES GRAPHICALLY AND NUMERICALLY

The Family of Antiderivatives

Recall that f is called an *antiderivative* of f'. For example, if v is the velocity of a car and s is its position, then $v = ds/dt$ and s is an antiderivative of v. Notice that a function has many antiderivatives. If s is an antiderivative of v, so are $s+1$, $s+2$, $s+3$, since, for example,

$$\frac{d}{dt}(s+1) = \frac{ds}{dt} + 0 = v.$$

Similarly, $s + C$ is an antiderivative of v for any constant C. Hence a function has a *family* of antiderivatives. In terms of the car, adding C to s is equivalent to adding C to the odometer reading. Adding a fixed distance to the odometer reading simply means measuring distance from a different point, which doesn't alter the car's velocity.

Visualizing Antiderivatives Using Slopes

Suppose we have the graph of f', and we want to sketch an approximate graph of f. We are looking for the graph of f whose slope at any point is equal to the value of f' there. Where f' is above the x-axis, f is climbing; where f' is below the x-axis, f is falling. If f' is increasing, f is concave up; if f' is decreasing, f is concave down.

Example 1 The graph of f' is given in Figure 6.21. Sketch a graph of f in the cases when $f(0) = 0$ and $f(0) = 1$.

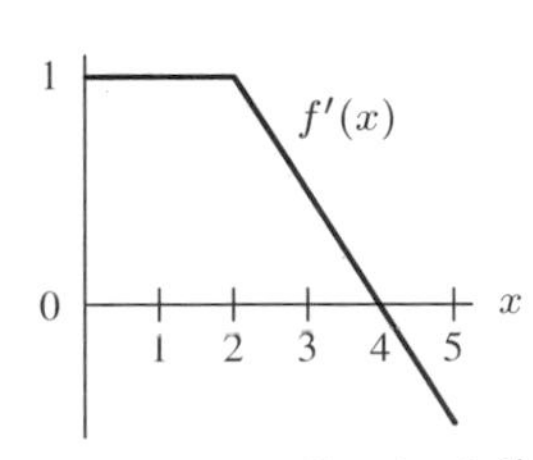

Figure 6.21: Graph of f'

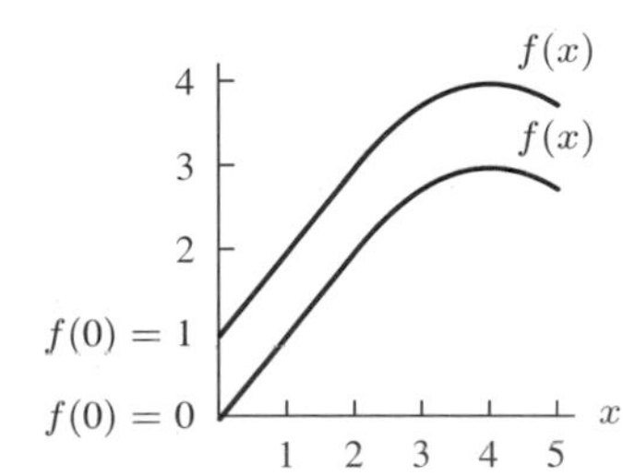

Figure 6.22: Two different f's which have the same derivative f'

Solution For $0 \leq x \leq 2$, f has a constant slope of 1 and so the graph of f is a straight line. For $2 \leq x \leq 4$, f is increasing but more slowly; it has a maximum at $x = 4$ and decreases thereafter. (See Figure 6.22.)

Notice that the solutions with $f(0) = 0$ and $f(0) = 1$ start at different points on the vertical axis but have the same shape.

Example 2 For the function f' given in Figure 6.23, sketch a graph of three antiderivative functions f, where (a) $f(0) = 0$, (b) $f(0) = 1$, (c) $f(0) = 2$.

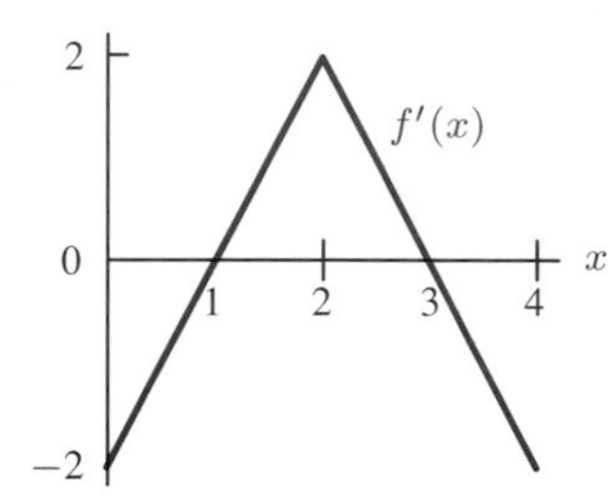

Figure 6.23: Slope function, f'

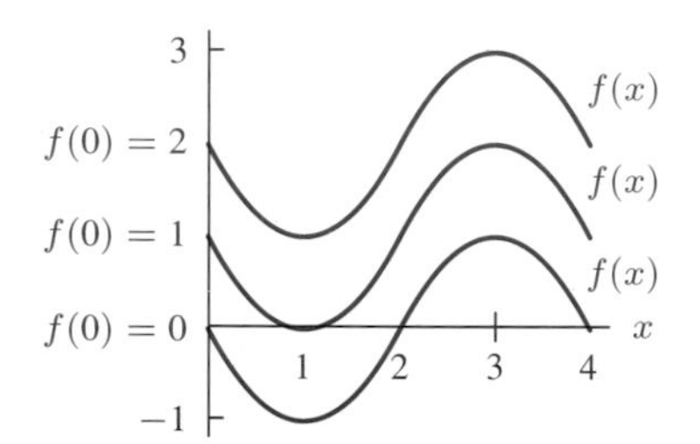

Figure 6.24: Antiderivatives f

Solution To draw a graph of f, start at the point on the vertical axis specified by the initial condition ($f(0) = 0$, $f(0) = 1$, or $f(0) = 2$) and move across the plane with a slope given by the value of f'. (See Figure 6.24.) Different initial conditions lead to different graphs for f, but for a given x value they all have the same slope (because the value of $f'(x)$ is the same for each). Thus, the different f curves are obtained from one another by a vertical shift. Notice that where f' is positive ($1 < x < 3$), f is increasing; where f' is negative ($0 < x < 1$ or $3 < x < 4$), f is decreasing. Where f' is increasing ($0 < x < 2$), f is concave up; where f' is decreasing ($2 < x < 4$), f is concave down. In addition, at the point where f' has a maximum ($x = 2$), f has a point of inflection.

In Example 4, we will see how to calculate exactly how much f increases or decreases using the Fundamental Theorem of Calculus.

Example 3 Figure 6.25 shows a graph of a function f'. Sketch the family of antiderivatives, f.

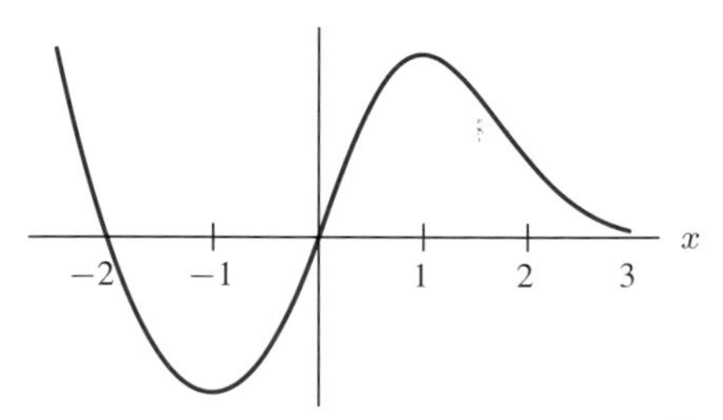

Figure 6.25: Graph of slope function, f'

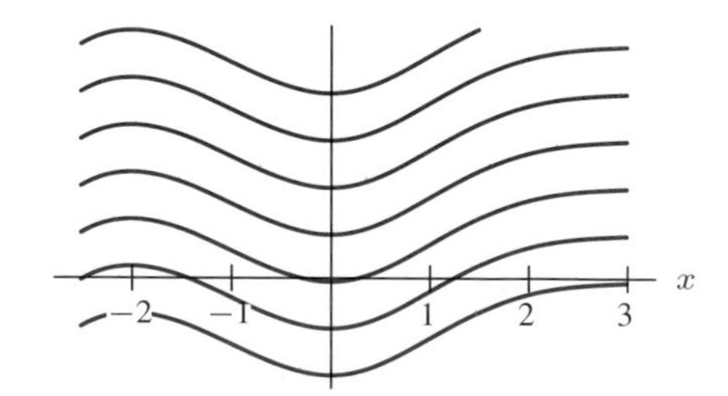

Figure 6.26: Graph of antiderivatives, f

Solution Let's first sketch any one antiderivative f. We get all the others by shifting this graph up or down by a constant. The function f' gives us the slope of f. Looking at the graph of f' we see that f has a horizontal tangent at $x = -2$ and $x = 0$ because $f' = 0$ there. Since $f'(-2) = 0$ and f'

f' is decreasing at $x = -2$, f has a local maximum there; f' is increasing at $x = 0$, so f has a local minimum there. Furthermore, f has inflection points at $x = -1$ and $x = 1$ because f' has local maxima and minima at these points. (See Figure 6.26.) The x-axis appears to be a horizontal asymptote for f', because $f' \to 0$ as $x \to \infty$; this means that the graph of f levels off as $x \to \infty$. The antiderivatives of f' will all have these properties, since they are all of the form $f(x) + C$ for some particular antiderivative f and some constant C.

Computing Values of an Antiderivative Using the Fundamental Theorem

A graph of f' shows where f is increasing and where f is decreasing, as well as where f has a local maximum or minimum (where $f' = 0$). We can find the value of the function at the maximum or minimum, by using the Fundamental Theorem of Calculus to calculate changes in the value of the function.

Example 4 Figure 6.27 is the graph of the derivative $f'(x)$ of a function $f(x)$. It is given that $f(0) = 100$. Sketch the graph of $f(x)$, showing all critical points and inflection points of f and giving their coordinates.

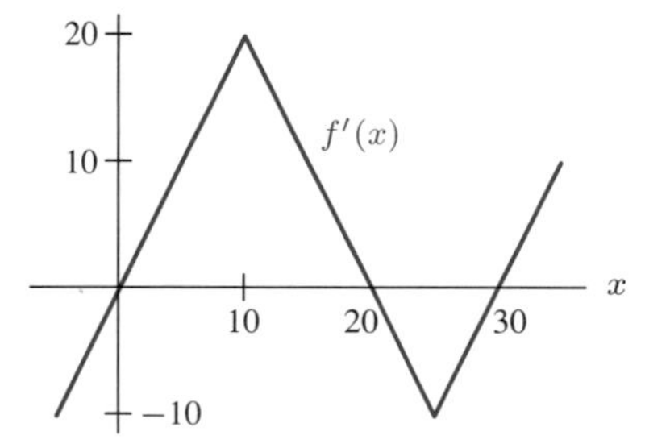

Figure 6.27: Graph of derivative

Solution The critical points of f occur at $x = 0$, $x = 20$, and $x = 30$, where $f'(x) = 0$. The inflection points of f occur at $x = 10$ and $x = 25$, where $f'(x)$ has extremes. To find the coordinates of the critical points and inflection points of f we must evaluate $f(x)$ for $x = 0, 10, 20, 25, 30$. First, we use the Fundamental Theorem to express the values of $f(x)$ in terms of definite integrals. Second, we evaluate the definite integrals. We will evaluate the definite integrals by calculating the areas of triangular regions under the graph of $f'(x)$. (See Figure 6.28.)

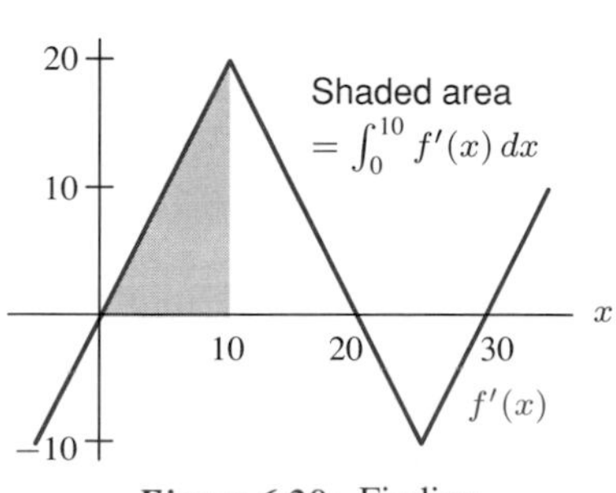

Figure 6.28: Finding $f(10) = f(0) + \int_0^{10} f'(x)\,dx$

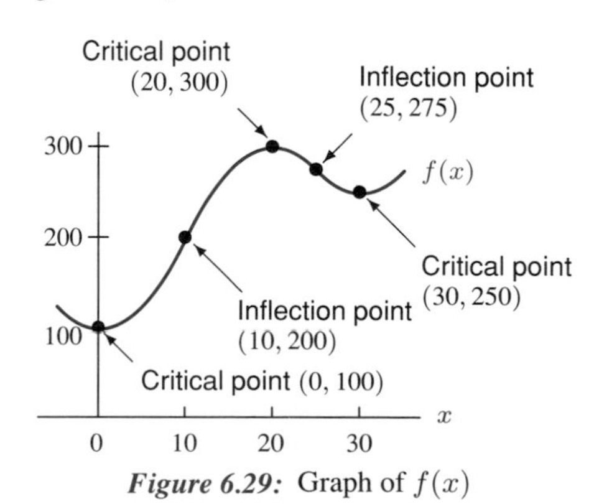

Figure 6.29: Graph of $f(x)$

$$
\begin{aligned}
f(0) &= 100 \text{ (given)} \\
f(10) &= f(0) + \int_0^{10} f'(x)\,dx = 100 + \text{(Shaded area in Figure 6.28)} \\
&= 100 + \frac{1}{2}(10)(20) = 200 \\
f(20) &= f(10) + \int_{10}^{20} f'(x)\,dx = 200 + \frac{1}{2}(10)(20) = 300 \\
f(25) &= f(20) + \int_{20}^{25} f'(x)\,dx = 300 - \frac{1}{2}(5)(10) = 275 \\
f(30) &= f(25) + \int_{25}^{30} f'(x)\,dx = 275 - \frac{1}{2}(5)(10) = 250
\end{aligned}
$$

The graph of f is sketched in Figure 6.29.

Problems for Section 6.3

For each function in Problems 1–4, sketch two functions F such that $F' = f$. In one case let $F(0) = 0$ and in the other, let $F(0) = 1$.

1.

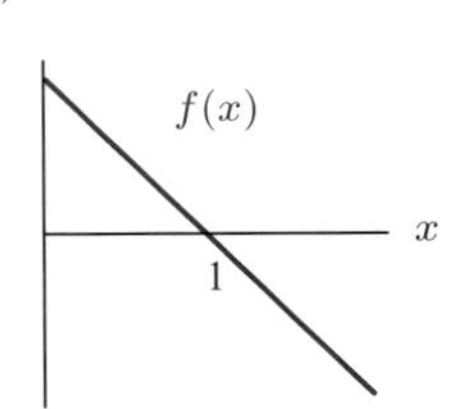

2.

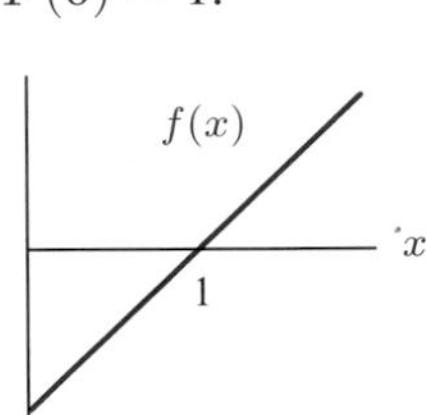

3.

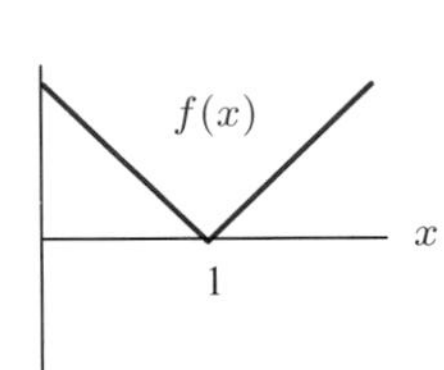

4.

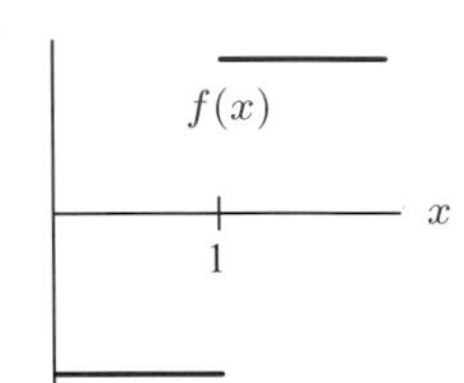

For the functions in Problems 5–6, graph a function $F(x)$ such that $F'(x) = f(x)$ and $F(0) = 0$.

5.

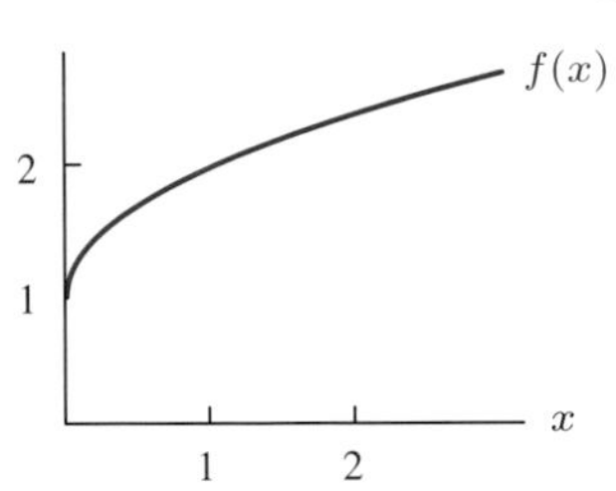

6.

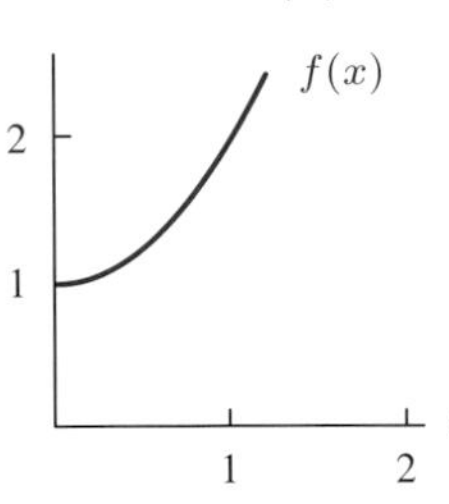

For each function in Problems 7–9, sketch two functions F where $F'(x) = f(x)$. In one case, let $F(0) = 0$, and in the other, let $F(0) = 1$. In each case, mark x_1, x_2, and x_3 on the x-axis of your graph.

7.

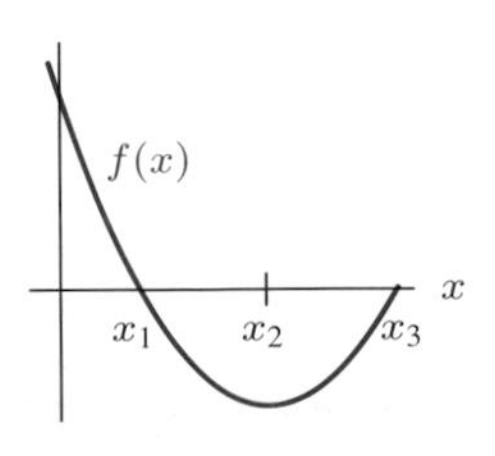

8.

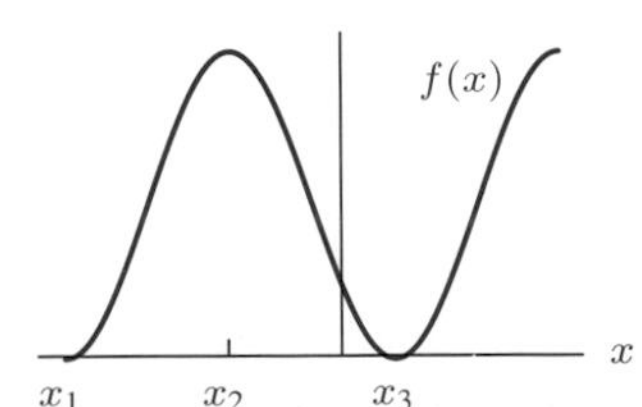

9.

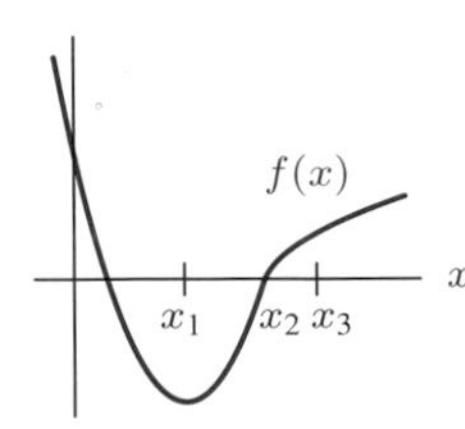

10. Sketch the derivative and antiderivative of the function in Figure 6.30. Make the antiderivative go through the origin. Mark the points x_1, $x_2, \ldots, x_6$ on the x-axis of each graph.

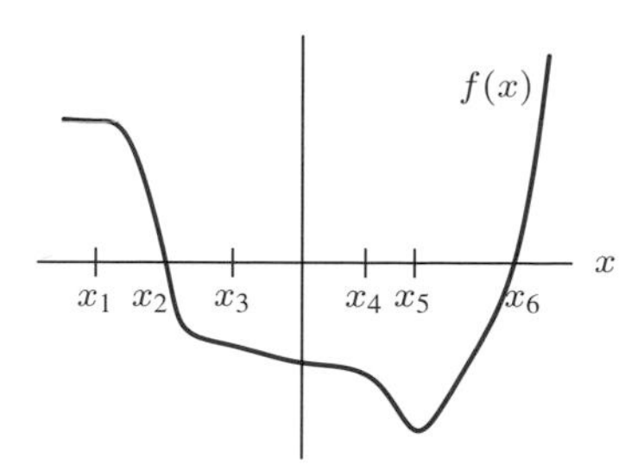

Figure 6.30

11. The vertical velocity of a cork bobbing up and down on the waves in the sea is given by Figure 6.31. Upward is considered positive. Describe the motion of the cork at each of the labeled points. At which point(s), if any, is the acceleration zero? Sketch a graph of the height of the cork above the sea floor as a function of time.

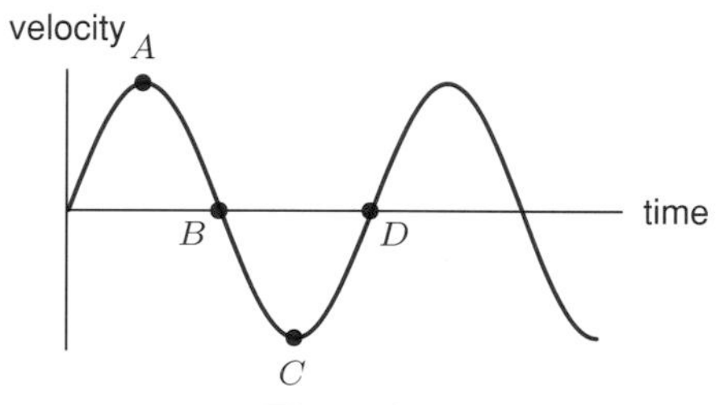

Figure 6.31

12. Figure 6.32 shows the graph of the derivative $g'(x)$ of a function $g(x)$. It is given that $g(0) = 50$. Sketch the graph of $g(x)$, showing all critical points and inflection points of g and giving their coordinates.

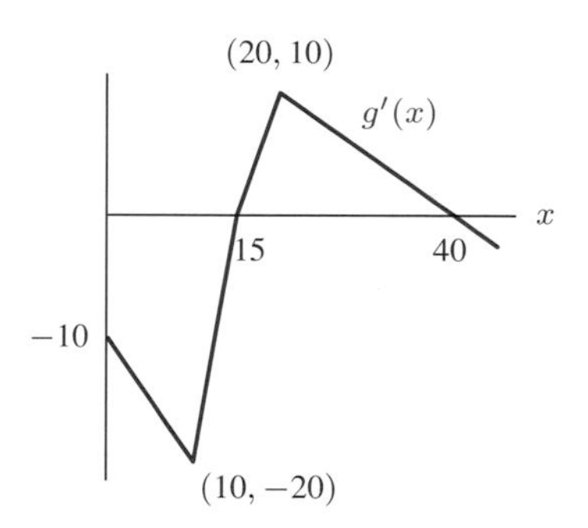

Figure 6.32

13. The graph of dy/dt against t is in Figure 6.33. Suppose the three shaded regions each have area 2. Given that $y = 0$ when $t = 0$, draw the graph of y against t, indicating all special features the graph might have (known heights, maxima and minima, inflection points, etc.). Pay particular attention to the relationship between the graphs. Mark $t_1, t_2, \ldots, t_5$ on the t axis.[3]

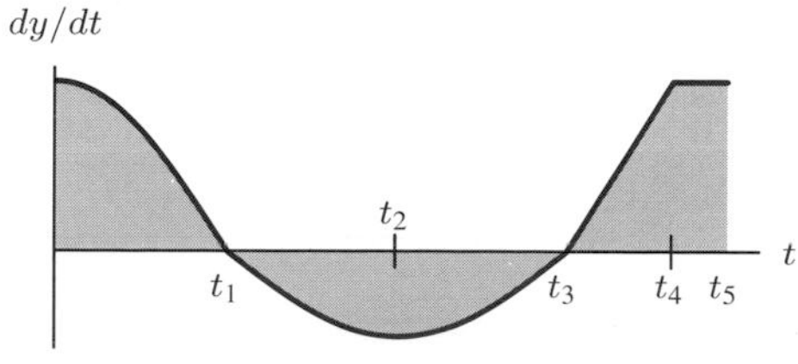

Figure 6.33

[3]From *Calculus: The Analysis of Functions*, by Peter D. Taylor (Toronto: Wall & Emerson, Inc., 1992)

14. Repeat Problem 13 for the graph of dy/dt given in Figure 6.34. (Again the three shaded regions each have area 2.)

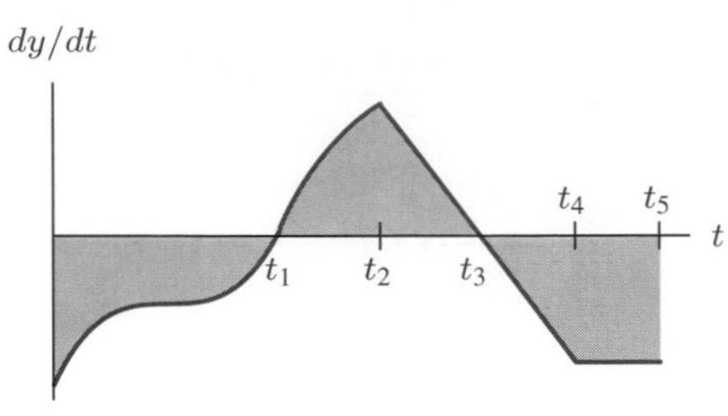

Figure 6.34

15. If the graph of f is as shown in Figure 6.35, sketch a graph of the function $F(x) = \int_0^x f(t)\,dt$.

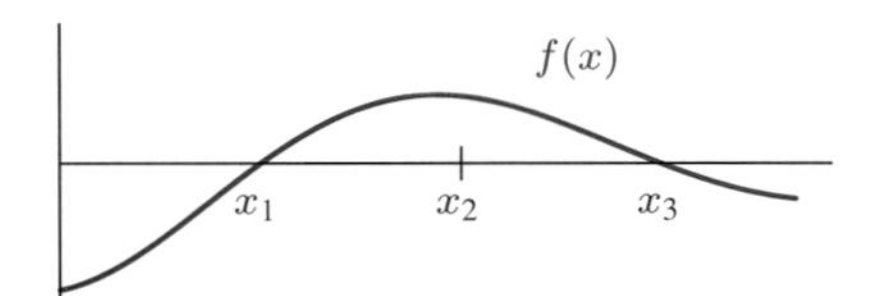

Figure 6.35

16. The graph of f'' is given in Figure 6.36. Draw graphs of f and f', assuming both go through the origin, and use them to decide at which of the labeled x values:

(a) $f(x)$ is greatest.
(b) $f(x)$ is least.
(c) $f'(x)$ is greatest.
(d) $f'(x)$ is least.
(e) $f''(x)$ is greatest.
(f) $f''(x)$ is least.

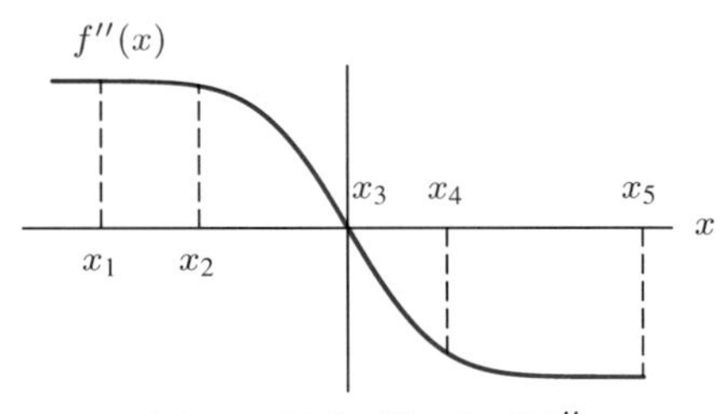

Figure 6.36: Graph of f''

17. Suppose the velocity, v, of a car traveling along a straight road is approximated by the graph in Figure 6.37. Find a formula for the distance of the car from its starting point as a function of time in hours. Draw a graph of this function.

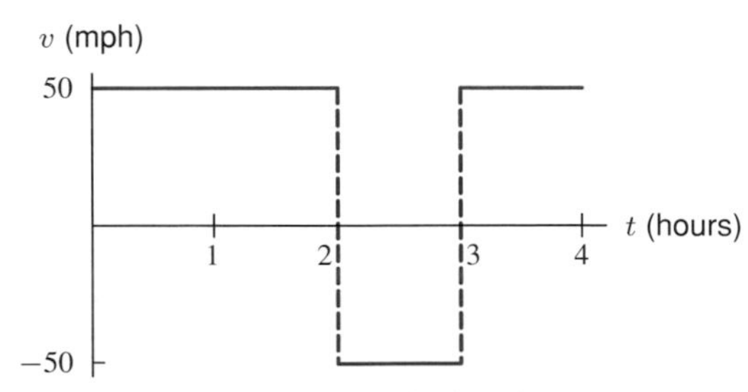

Figure 6.37: Velocity of car

18. The Quabbin Reservoir in the western part of Massachusetts provides most of Boston's water. The graph in Figure 6.38 represents the flow of water in and out of the Quabbin Reservoir throughout 1993.

    (a) Sketch a possible graph for the quantity of water in the reservoir, as a function of time.

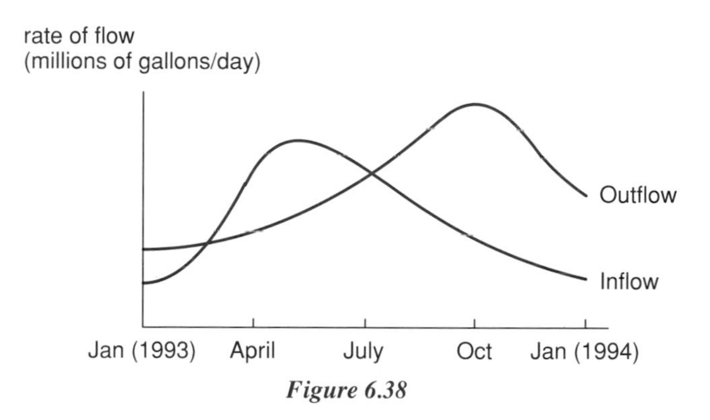

Figure 6.38

    (b) When, in the course of 1993, was the quantity of water in the reservoir largest? Smallest? Mark and label these points on the graph you drew in part (a).
    (c) When was the quantity of water decreasing most rapidly? Again, mark and label this time on both graphs.
    (d) By July 1994 the quantity of water in the reservoir was about the same as in January 1993. Draw plausible graphs for the flow into and the flow out of the reservoir for the first half of 1994. Explain your graph.

6.4 CONSTRUCTING ANTIDERIVATIVES ALGEBRAICALLY

Chapter 2 began with the idea that velocity is the derivative of distance and that acceleration is the derivative of velocity. In this section we will analyze the motion of an object falling freely under the influence of gravity. This involves going "backward" from acceleration to velocity to position.

Uniform Motion

To set the stage, let's briefly consider a familiar problem. An object moving in a straight line with constant velocity is said to be in *uniform motion*. Imagine a car moving at 50 mph. How far does it go in a given time? The answer is given by

$$\text{Distance} = \text{Rate} \times \text{Time}$$

or

$$s = 50t$$

where s is the distance of the car (in miles) from a fixed reference point and t is the time in hours. Alternatively, we can describe the motion by writing the equation

$$\frac{ds}{dt} = 50.$$

We can view this as an equation, a *differential equation*, for the *function* s, and we say that the solution to this equation results from our answering the question: What function $s = f(t)$ do we differentiate to get 50? We know from our experience in taking derivatives that one answer is $s = 50t$ because:

$$\frac{d}{dt}(50t) = 50\frac{d}{dt}(t) = 50.$$

Solving the differential equation means working backwards from ds/dt to s itself. Notice that the solution $s = 50t$ is the same as the one we obtained before by using Distance = Rate $\times$ Time.

In addition, if C is a constant, then

$$s = 50t + C$$

is also a possible expression for s because the derivative of C is zero.

The equation $s = 50t + C$ tells us that $s = C$ when $t = 0$. Thus, the constant C represents the initial distance of the car from the reference point. (After all, the car does not necessarily start at the reference point.) Call this initial distance s_0. Then a solution to the *initial value problem*

$$\frac{ds}{dt} = 50, \quad \text{and} \quad s = s_0 \quad \text{when} \quad t = 0,$$

is

$$s = 50t + s_0.$$

We will soon see that this is the only solution to this initial value problem.

Uniformly Accelerated Motion

Now we will consider an object moving with *constant acceleration* along a straight line, or *uniformly accelerated motion*. It has been known since Galileo's time that this is the kind of motion experienced by an object moving under the influence of gravity. Thus, if v is the upward velocity and t is the time,

$$\frac{dv}{dt} = -g$$

where g is the constant *acceleration due to gravity*, whose value is determined by experiment. In today's most frequently used units, its value is approximately

$$g = 32 \text{ ft/sec}^2, \quad g = 9.81 \text{ m/sec}^2, \quad \text{or} \quad g = 981 \text{ cm/sec}^2.$$

The negative sign comes from the fact that velocity is measured upward, whereas gravity acts downward.

Example 1 A stone is dropped from a 100-foot-high building. Find, as a function of time, its position and velocity. When does it hit the ground, and how fast is it going at that time?

Solution If we measure distance, s, in feet above the ground, and the velocity, v, in ft/sec upward, then the acceleration due to gravity is 32 ft/sec^2 downward, so

$$\frac{dv}{dt} = -32,$$

where t is measured in seconds from when the stone was dropped. From what we know about taking derivatives, we must have

$$v = -32t + C$$

where C is some constant. Since $v = C$ when $t = 0$, the constant C represents the initial velocity, v_0. The fact that the stone is dropped rather than thrown off the top of the building tells us that the initial velocity is zero, so $C = v_0 = 0$. Substituting gives

$$0 = -32(0) + C \quad \text{so} \quad C = 0.$$

Thus,

$$v = -32t.$$

But now we can write

$$v = \frac{ds}{dt} = -32t.$$

Finding s means finding a function whose derivative is $-32t$. One possible function is $-16t^2$ because

$$\frac{d}{dt}(-16t^2) = -16\frac{d}{dt}(t^2) = -16(2t) = -32t.$$

As before, we could also have

$$s = -16t^2 + K$$

where K is another constant.

Since the stone starts at the top of the building, $s = 100$ when $t = 0$. Substituting into $s = -16t^2 + K$ gives

$$100 = -16(0^2) + K,$$

so

$$K = 100,$$

and therefore

$$s = -16t^2 + 100.$$

Thus, we have found both v and s as functions of time.

The stone hits the ground when $s = 0$, so we must solve

$$0 = -16t^2 + 100$$

giving $t^2 = 100/16$ or $t = \pm 10/4 = \pm 2.5$ sec. Since t must be positive, $t = 2.5$ sec. At that time, $v = -32(2.5) = -80$ ft/sec. (Note that the velocity is negative because we are considering up as the positive direction and down as the negative direction.)

Example 2 An object is thrown vertically upward with a speed of 10 m/sec from a height of 2 meters. Find the highest point it reaches and when it hits the ground.

Solution We must find the position as a function of time. In this example, the velocity is in m/sec, so we use $g = 9.8$ m/sec^2. Measuring distance in meters upward from the ground, we have

$$\frac{dv}{dt} = -9.8.$$

As before, v is a function whose derivative is constant, so

$$v = -9.8t + C.$$

Since the initial velocity is 10 m/sec upward, we know that $v = 10$ when $t = 0$. Substituting gives

$$10 = -9.8(0) + C \quad \text{so} \quad C = 10.$$

Thus,

$$v = -9.8t + 10.$$

To find s, we use

$$v = \frac{ds}{dt} = -9.8t + 10$$

and look for a function that has $-9.8t + 10$ as its derivative. Let's think about each term separately. Since

$$\frac{d}{dt}\left(-9.8\frac{t^2}{2}\right) = -9.8\frac{2t}{2} = -9.8t \quad \text{and} \quad \frac{d}{dt}(10t) = 10,$$

a possible formula is $s = -9.8\frac{t^2}{2} + 10t = -4.9t^2 + 10t$. As before, we can add any constant, so

$$s = -4.9t^2 + 10t + K.$$

To find K, we use the fact that the object starts at a height of 2 meters, so $s = 2$ when $t = 0$. Substituting gives

$$2 = -4.9(0)^2 + 10(0) + K, \quad \text{so} \quad K = 2,$$

and therefore

$$s = -4.9t^2 + 10t + 2.$$

The object reaches its highest point when the velocity is 0, or

$$v = -9.8t + 10 = 0$$

which occurs when

$$t = \frac{10}{9.8} \approx 1.02 \text{ sec.}$$

When $t = 1.02$ seconds,

$$s = -4.9(1.02)^2 + 10(1.02) + 2 \approx 7.10 \text{ m.}$$

So the maximum height reached is 7.10 meters. The object reaches the ground when $s = 0$:

$$0 = -4.9t^2 + 10t + 2.$$

Solving this using the quadratic formula gives

$$t \approx -0.18 \text{ and } t \approx 2.22 \text{ sec.}$$

Since the time at which the object hits the ground must be positive, $t \approx 2.22$ seconds.

Antiderivatives and Differential Equations

We solved the problem of uniformly accelerated motion by working backward from the derivative of a function to the function itself. In the process, we solved a *differential equation* of the form

$$\frac{dy}{dx} = f(x).$$

Here f is a known function and we found the unknown function $y = F(x)$ with $F'(x) = \frac{dy}{dx} = f(x)$. In other words, we found an antiderivative of f. For example:

An antiderivative of $f(x) = x^2$ is $F(x) = \frac{1}{3}x^3$ because $\frac{d}{dx}\left(\frac{1}{3}x^3\right) = \frac{1}{3} \cdot 3x^2 = x^2$.

An antiderivative of $g(x) = x^3$ is $G(x) = \frac{1}{4}x^4$ because $\frac{d}{dx}\left(\frac{1}{4}x^4\right) = \frac{1}{4} \cdot 4x^3 = x^3$.

An antiderivative of $h(x) = e^x$ is $H(x) = e^x$. (Why?)

We obtained these antiderivatives "by inspection"; that is, our experience with differentiating gives us a chance of answering the reverse question: "Which function do I differentiate to get . . .?" We can easily make a table of functions and their derivatives and then use the table to solve

$$\frac{dy}{dx} = f(x)$$

whenever we happen to find $f(x)$ in the list of derivatives. What if we can't find a given $f(x)$ in our table? In other words, what do we do if $f(x)$ is not the derivative of a function we know? This will be discussed in Chapter 7. In this section we will look at some other important questions.

What Is the Solution to the Differential Equation $\frac{dy}{dx} = 0$?

What about a function whose derivative is *always* zero? A constant function has this property; are there any other functions whose derivative is always zero? Common sense says no. A function whose derivative is zero everywhere on an interval must have a horizontal tangent line at every point of its graph, and the only way this can happen is if the graph itself is a horizontal line, that is, if the function is constant. Put another way, if we think of the derivative as a velocity, and if the velocity is always zero, then the object is standing still. Whatever its initial position is, that's where it stays; the position function is constant. A rigorous proof of this result using the definition of the derivative is surprisingly subtle.

> If the derivative of a function is zero everywhere on an interval, then the function is constant in the interval:
>
> If $F'(x) = 0$ on an interval, then $F(x) = C$ on this interval.

What Is the Most General Antiderivative of f?

We know that if a function f has an antiderivative F, then it has a family of antiderivatives of the form $F(x) + C$, where C is any constant. You might wonder if there are any others. Let's see. Suppose that we have two functions F and G with $F' = f$ and $G' = f$: that is, F and G are both antiderivatives of the same function f. Since $F' = G'$ we have $F' - G' = 0$ or $(F - G)' = 0$. But this means that we must have $F - G =$ constant, so $F - G = C$, or $F(x) = G(x) + C$, where C is a constant. Thus, any two antiderivatives of the same function do differ by a constant.

> If F and G are both antiderivatives of a function f, then $F(x) = G(x) + C$ for some constant C. Equivalently, if $y = F(x)$ is a solution to the differential equation
>
> $$\frac{dy}{dx} = f(x),$$
>
> then the most general solution is $y = F(x) + C$, where C is a constant.

We used this fact implicitly in our study of motion when we wrote that the general solution of $ds/dt = 50$ was $s = 50t + C$. We can also make sense of this from a geometric point of view. When we are asked to solve

$$\frac{dy}{dx} = f(x)$$

for the unknown function $y = F(x)$, what we are told about F is that the slope of its graph is given by f. We now know that the graph of *any* solution to this differential equation is obtained by shifting the graph of F vertically. Hence, the solution of a differential equation is actually not one function but a *family* of functions, not one curve but a *family* of parallel curves of the form $F(x) + C$.

Example 3 Find and graph the general solution of the differential equation

$$\frac{dy}{dx} = \sin x + 2.$$

Solution We are asking for a function whose derivative is $\sin x + 2$. Since the derivative of a sum is the sum of the derivatives, we can try to find antiderivatives of $\sin x$ and 2 separately and then add the results. An antiderivative of $\sin x$ is $-\cos x$, since $\frac{d}{dx}(-\cos x) = \sin x$, and an antiderivative of 2 is $2x$. Thus, a solution of the equation is

$$y = -\cos x + 2x.$$

The general solution is therefore

$$y = -\cos x + 2x + C,$$

where C is any constant. Figure 6.39 shows several curves in this family.

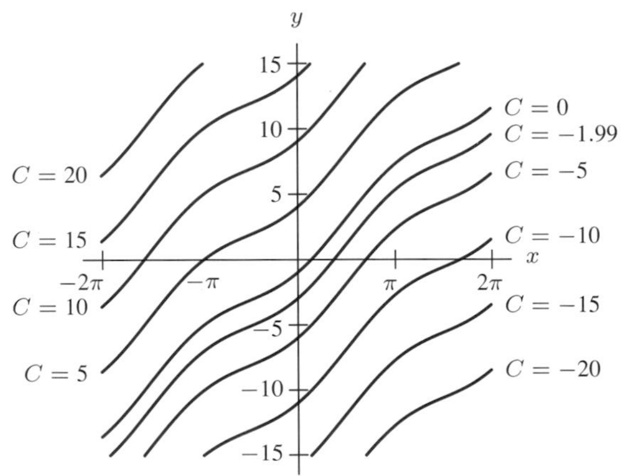

Figure 6.39: Solution curves of $\frac{dy}{dx} = \sin x + 2$

How Can We Pick Out One Antiderivative of a Function *f*?

Picking out one antiderivative is equivalent to selecting a value of C, which, as we saw in the motion examples, requires knowing an extra piece of information, such as the initial velocity or initial position. In general, we can pick out one particular curve in the family of curves corresponding to solutions of $dy/dx = f(x)$ by specifying, *as an extra condition*, that we want the curve that passes through a given point (x_0, y_0). Provided that F is defined at x_0, there will be exactly one solution curve of $dy/dx = f(x)$ that passes through a given point (x_0, y_0). Equivalently, there will be exactly one antiderivative F of f for which $F(x_0) = y_0$; imposing an extra condition allows us to determine the constant C.

We say that the differential equation *plus* an extra condition

$$\frac{dy}{dx} = f(x), \qquad y(x_0) = y_0$$

constitutes an *initial-value problem*. An initial-value problem usually has a unique solution.

Notation: $y(x_0) = y_0$ is shorthand for $y = y_0$ when $x = x_0$. Although it involves using the same letter for the variable y and the function f, this notation is frequently used for differential equations.

Example 4 Find the solution of the initial value problem

$$\frac{dy}{dx} = \sin x + 2, \quad y(3) = 5.$$

Solution We have already seen that the general solution to the differential equation is $y = -\cos x + 2x + C$. The initial condition will allow us to determine the constant C. Substituting $y(3) = 5$ gives

$$5 = y(3) = -\cos 3 + 2 \cdot 3 + C$$

so C is given by

$$C = 5 + \cos 3 - 6 \approx -1.99$$

Thus, the (unique) solution is

$$y = -\cos x + 2x - 1.99.$$

Figure 6.39 shows this particular solution (marked with $C = -1.99$).

Using Antiderivatives to Compute Definite Integrals: The Indefinite Integral

If we want to compute $\int_1^3 2x\,dx$, we can either use Riemann sums or the Fundamental Theorem. Since

$$\frac{d}{dx}(x^2) = 2x,$$

taking $F(x) = x^2$ tells us that

$$\int_1^3 2x\,dx = F(3) - F(1) = 3^2 - 1^2 = 8.$$

Which antiderivative should you use to compute a definite integral? It makes no difference because the constant cancels out when you subtract $F(a)$ from $F(b)$. For example, if we had used $x^2 + C$ in computing $\int_1^3 2x\,dx$, we would have arrived at the same answer as before:

$$\int_1^3 2x\,dx = F(3) - F(1) = (3^2 + C) - (1^2 + C) = 8.$$

The general antiderivative is written $F(x)+C$. Because of the connection with the definite integral, we introduce a notation for this general antiderivative that looks like the definite integral without the limits and is called the *indefinite integral*:

$$\int f(x)\,dx = F(x) + C.$$

It is important to understand the difference between

$$\int_a^b f(x)\,dx \qquad \text{and} \qquad \int f(x)\,dx.$$

The first is a number, and the second is actually a *family* of functions. They have such similar notations because the second is helpful in computing the first. Because the notation is similar, the word "integration" is frequently used for the process of finding the antiderivative as well as of finding the definite integral. The context usually makes clear which is intended.

We will also introduce a shorthand notation for $F(b)-F(a)$: we will write it as

$$F(x)\Big|_a^b$$

For example:

$$\int_1^3 2x\,dx = x^2\Big|_1^3 = 3^2 - 1^2 = 8.$$

Example 5 Find: (a) $\int x^2\,dx$, (b) $\int e^t\,dt$.

Solution (a) $\int x^2\,dx$ means the general antiderivative of x^2, so

$$\int x^2\,dx = \frac{x^3}{3} + C.$$

(b) $\int e^t\,dt$ means the general antiderivative of e^t, so

$$\int e^t\,dt = e^t + C.$$

Problems for Section 6.4

For each of the functions in Problems 1–28, find an antiderivative.

1. $f(x) = 5$
2. $f(x) = 5x$
3. $f(x) = x^2$
4. $g(t) = t^2 + t$
5. $h(t) = \cos t$
6. $g(z) = \sqrt{z}$, for $z > 0$
7. $h(z) = \dfrac{1}{z}$, for $z > 0$
8. $r(t) = \dfrac{1}{t^2}$
9. $g(z) = \dfrac{1}{z^3}$

10. $f(z) = e^z$

11. $g(t) = \sin t$

12. $f(t) = 2t^2 + 3t^3 + 4t^4$

13. $p(t) = t^3 - \frac{t^2}{2} - t$

14. $q(y) = y^4 + \frac{1}{y}$, for $y > 0$

15. $f(x) = 5x - \sqrt{x}$, for $x > 0$

16. $f(t) = \frac{t^2 + 1}{t}$

17. $p(\theta) = 2\sin(2\theta)$

18. $r(t) = e^t + 5e^{5t}$

19. $q(t) = (t+1)^2$

20. $f(x) = 5^x$

21. $p(t) = \cos t + \frac{1}{\cos^2 t}$

22. $h(t) = 2t\cos(t^2)$

23. $f(t) = t\cos(t^2)$

24. $r(x) = 3x^2\cos(x^3 + 7)$

25. $g(y) = e^2 + 2^y$

26. $g(\theta) = \sin\theta + \frac{1}{1+\theta^2}$

27. $f(x) = xe^{x^2}$

28. $f(x) = 6\sqrt{x} - \frac{1}{x^2} + \frac{10}{x}$, for $x > 0$

For each of the functions in Problems 29–34, find a function F such that $F' = f$ and $F(0) = 2$.

29. $f(x) = 3$

30. $f(x) = e^x$

31. $f(x) = x^2$

32. $f(x) = \cos x$

33. $f(x) = \sin x$

34. $f(x) = (x-1)^2$

Find the indefinite integrals in Problems 35–46.

35. $\int 3x\,dx$

36. $\int (4t + 7)\,dt$

37. $\int \cos\theta\,d\theta$

38. $\int 5e^z\,dz$

39. $\int \left(x + \frac{1}{\sqrt{x}}\right) dx$

40. $\int \sin t\,dt$

41. $\int (\pi + x^{11})\,dx$

42. $\int \left(t\sqrt{t} + \frac{1}{t\sqrt{t}}\right) dt$

43. $\int \cos(x+1)\,dx$

44. $\int e^{2r}\,dr$

45. $\int \frac{1}{e^z}\,dz$

46. $\int \left(\frac{y^2 - 1}{y}\right)^2 dy$

47. Ice is forming on a pond at a rate given by
$$\frac{dy}{dt} = k\sqrt{t},$$
where y is the thickness of the ice in inches at time t measured in hours since the ice started forming, and k is a positive constant. Find y as a function of t.

48. If a car goes from 0 to 80 mph in six seconds with constant acceleration, what is that acceleration?

49. A car going 80 ft/sec (about 55 mph) brakes to a stop in five seconds. Assume the deceleration is constant.
    (a) Graph the velocity against time, t, for $0 \le t \le 5$ seconds.
    (b) Represent, as an area on the graph, the total distance traveled from the time the brakes are applied until the car comes to a stop.
    (c) Find this area and hence the distance traveled.
    (d) Now find the total distance traveled using antidifferentiation.

50. A 727 jet needs to be flying 200 mph to take off. If it can accelerate from 0 to 200 mph in 30 seconds, how long must the runway be? (Assume constant acceleration.)

51. Suppose a car going at 30 ft/sec decelerates at a constant rate of 5 ft/sec^2.
    (a) Draw up a table showing the velocity of the car every half second. When does the car come to rest?
    (b) Using your table, find left and right sums which estimate the total distance traveled before the car comes to rest. Which is an overestimate, and which is an underestimate?
    (c) Sketch a graph of velocity against time. Show on the graph an area representing the distance traveled before the car comes to rest, and hence calculate its distance.
    (d) Now find a formula for the velocity of the car as a function of time, and hence find the total distance traveled by antidifferentiation. What is the relationship between your answer to parts (c) and (d) and your estimates in part (b)?

52. A stone thrown upwards from the top of a 320-foot cliff at 128 ft/sec eventually falls to the beach below.
    (a) How long does the stone take to reach its highest point?
    (b) What is its maximum height?
    (c) How long before the stone hits the beach?
    (d) What is the velocity of the stone on impact?

53. On the moon, the acceleration due to gravity is about 1.6 m/sec^2 (compared to $g \approx 9.8$ m/sec^2 on earth). If you drop a rock on the moon (with initial velocity 0), find formulas for:
    (a) Its velocity, $v(t)$, at time t.
    (b) The distance, $s(t)$, it falls in time t.

54. (a) Imagine throwing a rock straight up in the air. What should its initial velocity be if the rock reaches a maximum height of 100 feet above its starting point?
    (b) Now imagine being transplanted to the moon and throwing a moon rock vertically upward with the same velocity as in part (a). How high will it go? (On the moon, $g = 5$ ft/sec^2.)

6.5 NOTES ON THE EQUATIONS OF MOTION: WHY ACCELERATION?

The problem of a body moving freely under the influence of gravity near the surface of the earth intrigued mathematicians and philosophers from Greek times onward and was finally solved by Galileo and Newton. The question to be answered was: How do the velocity and the position of the body vary with time? We define s to be the position, or height, of the body above a fixed point (often ground level); v then is the velocity of the body measured upward. We assume that the acceleration of the body is a constant, $-g$ (the negative sign means that the acceleration is downward), so

$$\text{Acceleration} = \frac{dv}{dt} = -g.$$

Thus, velocity is the antiderivative of $-g$:

$$v = -gt + C.$$

If the initial velocity is v_0, then $C = v_0$, so

$$v = -gt + v_0.$$

How about the position? We know that

$$\frac{ds}{dt} = v = -gt + v_0.$$

Therefore, we can find s by antidifferentiating again, giving:

$$s = -\frac{gt^2}{2} + v_0 t + C.$$

If the initial position is s_0, then we must have

$$s = -\frac{gt^2}{2} + v_0 t + s_0.$$

Our derivation of the formulas for the velocity and the position of the body took little effort. It hides an almost 2000-year struggle to understand the mechanics of falling bodies, from Aristotle's *Physics* to Galileo's *Dialogues Concerning Two New Sciences*. Before sketching Galileo's solution of the problem, we make a few more observations on uniform and uniformly accelerated motion.

Though it is an oversimplification of his ideas, we can say that Aristotle's conception of motion was primarily in terms of *change of position*. This seems entirely reasonable; it is what we commonly observe, and this view dominated discussions of motion for centuries. But it misses a subtlety that was brought to light by Descartes, Galileo, and, with a different emphasis, by Newton. That subtlety is now usually referred to as the *principle of inertia*.

This principle holds that a body traveling undisturbed at constant velocity in a straight line will continue in this motion indefinitely. Stated another way, it says that one cannot distinguish in any absolute sense (that is, by performing an experiment), between being at rest and moving with constant velocity in a straight line. If you are reading this book in a closed room and have no external reference points, there is no experiment that will tell you, one way or the other, whether you are at rest or whether you, the room, and everything in it are moving with constant velocity in a straight line. Therefore, as Newton saw, an understanding of motion should be based on *change of velocity* rather than change of position. Since acceleration is the rate of change of velocity, it is acceleration that must play a central role in the description of motion.

How does acceleration come about? How does the velocity change? Through the action of *forces*. Newton placed a new emphasis on the importance of forces. Newton's laws of motion do not say what a force *is*, they say how it *acts*. His first law is the principle of inertia, which says what happens in the *absence* of a force — there is no change in velocity. His second law says that a force acts to produce a change in velocity, that is, an acceleration. It states that $F = ma$, where m is the mass of the object, F is the net force, and a is the acceleration produced by this force.

Let's return to Galileo for a moment. He demonstrated that a body falling under the influence of gravity does so with constant acceleration. Furthermore, assuming we can neglect air resistance, this constant acceleration is independent of the mass of the body. This last fact was the outcome of Galileo's famous observation that a heavy ball and a light ball dropped off the Leaning Tower of Pisa hit the ground at the same time. Whether or not he actually performed this experiment, Galileo presented a very clear thought experiment in the *Dialogues* to prove the same point. (This point was counter to Aristotle's more commonsense notion that the heavier ball would reach the ground first.) Galileo showed that the mass of the object need not appear as a variable in the equation of motion. Thus, the same constant acceleration equation applies to all bodies falling under the influence of gravity.

Nearly a hundred years after Galileo's experiment, Newton formulated his laws of motion and gravity, which led to a differential equation describing the motion of a falling body. According to Newton, acceleration is caused by force, and in the case of falling bodies, that force is the force of gravity. Newton's law of gravity says that the gravitational force between two bodies is attractive and given by

$$F = \frac{GMm}{r^2},$$

where G is the gravitational constant, m and M are the masses of the two bodies, and r is the distance between them. This is the famous *inverse square law*. For a falling body, we take M to be the mass of the earth and r to be the distance from the body to the center of the earth. So, actually, r changes as the body falls, but for anything we can easily observe (say, a ball dropped from the Tower of Pisa), it won't change significantly over the course of the motion. Hence, as an approximation, it is reasonable to assume that the force is constant. According to Newton's second law,

$$\text{Force} = \text{Mass} \times \text{Acceleration}.$$

Since the gravitational force is acting downwards

$$-\frac{GMm}{r^2} = m\frac{d^2s}{dt^2}.$$

Hence,

$$\frac{d^2s}{dt^2} = -\frac{GM}{r^2} = \text{constant}.$$

If we define $g = GM/r^2$, then

$$\frac{d^2s}{dt^2} = -g.$$

The fact that the mass cancels out of Newton's equations of motion reflects Galileo's experimental observation that the acceleration due to gravity is independent of the mass of the body.

Problems for Section 6.5

1. An object is dropped from a 400-foot tower. When does it hit the ground and how fast is it going at the time of the impact?
2. Consider the object in Problem 1, and suppose it falls off the same 400-foot tower. What would the acceleration due to gravity have to be to make it reach the ground in half the time?
3. A ball that is dropped from a window hits the ground in five seconds. How high is the window? (Give your answer in feet.)
4. On the moon the acceleration due to gravity is 5 ft/sec^2. An astronaut jumps into the air with an initial upward velocity of 10 ft/sec. How high does he go? How long is the astronaut off the ground?
5. Galileo was the first person to show that the distance traveled by a body falling from rest is proportional to the square of the time it has traveled, and independent of the mass of the body. Derive this result from the fact that the acceleration due to gravity is a constant.
6. While attempting to understand the motion of bodies under gravity, Galileo stated that:

   > The time in which any space is traversed by a body starting at rest and uniformly accelerated is equal to the time in which that same space would be traversed by the same body moving at a uniform speed whose value is the mean of the highest velocity and the velocity just before acceleration began.

   (a) Write Galileo's statement in symbols, defining all the symbols you use.
   (b) Verify Galileo's statement for a body dropped off a 100-foot building accelerating from rest under gravity until it hits the ground.
   (c) Show why Galileo's statement is true in general.

7. In his *Dialogues Concerning Two New Sciences*, Galileo wrote:

   > The distances traversed during equal intervals of time by a body falling from rest stand to one another in the same ratio as the odd numbers beginning with unity.

   Assume, as is now believed, that $s = -(gt^2)/2$, where s is the total distance traveled in time t, and g is the acceleration due to gravity.

   (a) How far does a falling body travel in the first second (between $t = 0$ and $t = 1$)? During the second second (between $t = 1$ and $t = 2$)? The third second? The fourth second?

   (b) What do your answers tell you about the truth of Galileo's statement?

8. The acceleration due to gravity 2 meters from the ground is 9.8 m/sec^2. What is the acceleration due to gravity 100 meters from the ground? At 100,000 meters? (The radius of the earth is 6.4×10^6 meters.)

REVIEW PROBLEMS FOR CHAPTER SIX

Are the statements in Problems 1–3 true or false? Why?

1. If F and G are two different antiderivatives of $f(x) = 3x^2$, then their graphs never intersect.

2. $\int_0^2 x\,dx = 2\int_0^1 x\,dx$.

3. $\int_0^x f(t)\,dt$ and $\int_1^x f(t)\,dt$ differ by a constant.

4. Write between one and two pages on the two different methods (Riemann sums and antiderivatives) of integration you have seen. Explain the idea behind each one, using an example to illustrate your discussion. Lastly, explain how they are related.

One way to approach this is to imagine that you're trying to explain to a friend who has just missed a week of class what the "big idea" of integration was.

For Problems 5–7, find $f(x)$.

5. $f'(x) = 2x^3 - \dfrac{1}{x} - \dfrac{1}{x^2}$ **6.** $f'(x) = e^x + x^e$ **7.** $f'(x) = e^\pi + \dfrac{1}{\sqrt{x^3}}$

For Problems 8–9 the graph of $f'(x)$ is given. Sketch a possible graph for $f(x)$. Mark the points $x_1 \ldots x_4$ on your graph and label local maxima, local minima and points of inflection.

8.

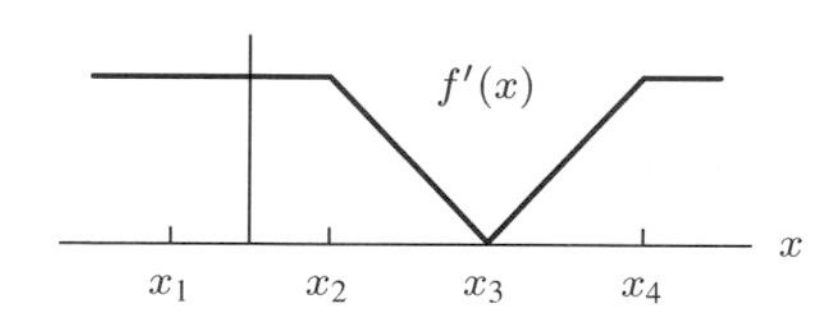

9.

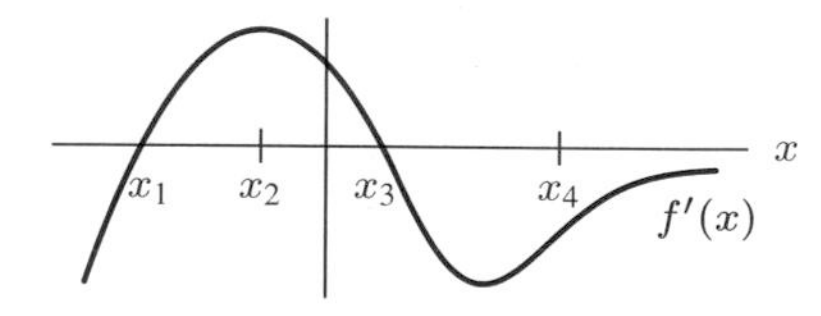

Find antiderivatives for the functions in Problems 10–25 by guessing an answer. Check by differentiation. You may use the fact that an antiderivative of x^r is $x^{r+1}/(r+1)$ for $r \neq -1$ and that an antiderivative of $1/x$ is $\ln|x|$.

10. $f(x) = x + x^5 + x^{-5}$ **11.** $g(x) = \dfrac{1}{x} + \dfrac{1}{x^2} + \dfrac{1}{x^3}$ **12.** $f(x) = x^6 - \dfrac{1}{7x^6}$

13. $g(t) = 5 + \cos t$ **14.** $g(\alpha) = 3\sin(3\alpha)$ **15.** $h(r) = 2\sqrt{r} + \dfrac{1}{2\sqrt{r}}$

16. $g(x) = (x+1)^3$

17. $h(t) = \dfrac{(t-1)^2}{t^2}$

18. $p(y) = \dfrac{1}{y} + y + 1$

19. $f(z) = e^z + 3$

20. $f(x) = 2xe^{x^2}$

21. $g(x) = e^x + e^{1+x}$

22. $g(\theta) = \sin\theta - 2\cos\theta$

23. $p(r) = 2\pi r$

24. $f(x) = 2^x + \dfrac{1}{2^x}$

25. $h(t) = \cos t(1+\sin t)^{29}$

Find the indefinite integrals in Problems 26–33 by guessing an answer. Check by differentiation. You may use the fact that an antiderivative of x^r is $x^{r+1}/(r+1)$ for $r \neq -1$ and that an antiderivative of $1/x$ is $\ln|x|$.

26. $\int (\sqrt{t} + t^9)\, dt$

27. $\int 2\cos(2x)\, dx$

28. $\int \dfrac{1}{\cos^2\theta}\, d\theta$

29. $\int (3e^x + 2^x)\, dx$

30. $\int (x-2)^2\, dx$

31. $\int \dfrac{1}{x+5}\, dx$

32. $\int \cos\theta e^{\sin\theta}\, d\theta$

33. $\int \sqrt{x}\left(1 - \dfrac{1}{x^{3/2}}\right) dx$

34. A cat, walking along the window ledge of a New York apartment, knocks off a flower pot, which falls to the street 200 feet below. How fast is the flower pot traveling when it hits the street? (Give your answer in ft/sec and in mph, given that 1 ft/sec = 15/22 mph.)

35. Of all the cars surveyed in the April 1991 issue of *Car and Driver*, the Acura NSX, going at 70 mph, stops in the shortest distance: 157 feet. Find the acceleration, assuming it is constant.

36. An object is thrown vertically upward from the ground with a velocity of 160 ft/sec.

(a) Sketch a graph of the velocity of the object (with upward as positive) against time.
(b) Mark on the graph the points at which the object reaches its highest point and when it hits the ground.
(c) Express the maximum height reached by the object as an area on the graph, and hence find this height.
(d) Now express the velocity of the object as a function of time, and find the greatest height by antidifferentiation.

37. An object is thrown vertically upward with a velocity of 80 ft/sec.

(a) Make a table showing its velocity every second.
(b) When does it reach its highest point? When does it hit the ground?
(c) Using your table, write left and right sums which under- and overestimate the height the object attains.
(d) Use antidifferentiation to find the greatest height it reaches.

38. A car, initially moving at 60 mph, decelerates at a constant rate and stops in a distance of 200 feet. What is the rate at which it decelerates? (Give your answer in ft/sec^2. Note that 1 mph $= 22/15$ ft/sec.)

39. A car moving at a velocity of v_0 decelerates at a constant rate, a, and stops in a distance D. Express D in terms of v_0 and a.

40. Quantum mechanics predicts that the force, F, between two gas molecules separated by a distance r is given by

$$F = -\frac{A}{r^7} + \frac{B}{r^{13}}$$

for some positive constants A and B. A sketch of F against r is in Figure 6.40. The potential energy, V, of the gas molecules satisfies $F = -dV/dr$, and $V \to 0$ as $r \to \infty$.

(a) Find a formula for V.

(b) Sketch a graph of V against r. Mark r_0 on it.

(c) What does r_0 represent, both in terms of the force and in terms of potential energy of the gas molecules?

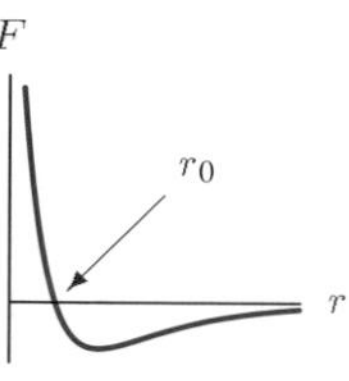

Figure 6.40

41. Consider a bacteria population whose birth rate, B, is given by the curve in Figure 6.41 as a function of time in hours. The birth rate is in births per hour. The curve marked D gives the death rate (in deaths per hour) of the same population.

(a) Explain what the shape of each of these graphs tells you about the population.

(b) Use the graphs to find the time at which the net rate of increase of the population is at a maximum.

(c) Suppose at time $t = 0$ the population has size N. Sketch the graph of the total number born by time t. Also sketch the graph of the number alive at time t. Use the given graphs to find the time at which the population size is a maximum.

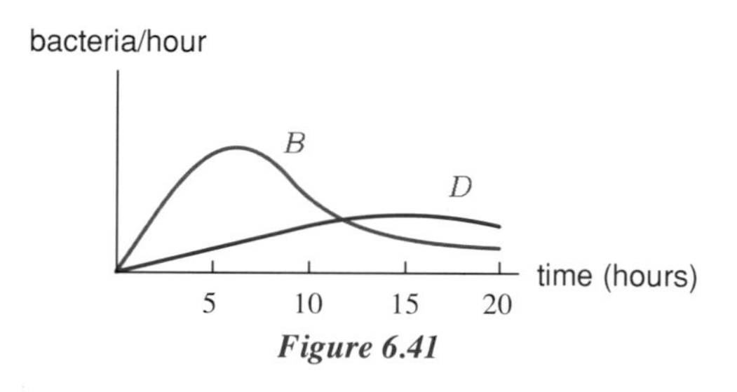

Figure 6.41

42. Water flows at a constant rate into the left side of the W-shaped container shown in Figure 6.42. Sketch a graph of the height, H, of the water in the left side of the container as a function of time t.

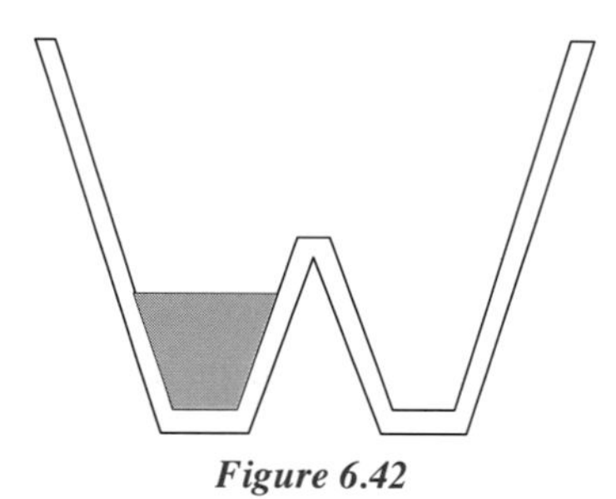

Figure 6.42

43. A student once gave the following nice rule for falling bodies: If at a certain time its velocity is u (measured downwards) and t seconds later its velocity is w, then the distance it has fallen during that interval is

$$s = \frac{u + w}{2} t.$$

That is, the distance can be calculated by assuming the velocity is constant over the time interval and equal to the average of the velocities at the start and finish of the interval.[4]

(a) Using the fact that the acceleration of a falling body is constant (equal to g), draw a graph showing the velocity, v, against time, t. Assume that it started moving at time $t = 0$ with an initial velocity of zero.
(b) How can the distance moved by time t be represented by an area on the graph?
(c) Use area to derive the student's rule.
(d) Suppose the body starts with velocity v_0 (where $v_0 > 0$). How does this change the picture? Does this change the student's rule?
(e) So far we have neglected air resistance. If we take air resistance into account, the equation $s = (u + w)t/2$ becomes an inequality. Explain.

44. (a) Find the average $a(T)$ of $f(t) = \sin(t)$ between 0 and T.
(b) Graph $a(t)$ and $f(t)$ on the same axis. Where is $a(t)$ increasing? Where is $a(t)$ smaller than $f(t)$? Where is $a(t)$ maximum? Where is $a(t) = f(t)$? Explain.

45. Repeat Problem 44 with $f(t) = t\sin(t)$.

46. Consider the average $a(T)$ of an arbitrary function $f(t)$ between 0 and T. Explain why local minima and maxima of $a(T)$ occur at values of T where the graphs of $a(T)$ and $f(T)$ intersect.

47. Whether a resource is distributed evenly among members of a population is often an important political or economic question. How can we measure this? How can we decide if the distribution of wealth in this country is becoming more or less equitable over time? How can we measure which country has the most equitable income distribution? This problem describes a way of doing this.

Suppose the resource is distributed evenly. Then any 20% of the population will have 20% of the resource. Similarly, any 30% will have 30% of the resource and so on. If, however, the resource is not distributed evenly, the poorest P% of the population (in terms of this resource) will not have P% of the goods. Suppose $F(x)$ represents the fraction of the resources owned by the poorest fraction x of the population. Thus $F(0.4) = 0.1$ means that the poorest 40% of the population owns 10% of the resource.

(a) What would F be if the resource was distributed evenly?
(b) What must be true of any such F? What must $F(0)$ and $F(1)$ equal? Is F increasing or decreasing? Concave up or concave down?
(c) Gini's index of inequality, G, is one way to measure how evenly the resource is distributed. It is defined by

$$G = 2\int_0^1 [x - F(x)]\,dx.$$

Show graphically what G represents.

48. A car is coming down US Route 95. It moves so that at time t its acceleration is $30\sin t$ ft/sec^2, where t is in seconds. At time $t = 0$, its velocity is 30 ft/sec.

(a) Find the car's velocity as a function of time.
(b) How far did the car travel between $t = 0$ and $t = 3\pi$?
(c) At $t = 10\pi$, a policeman stops the car and arrests the driver for reckless driving. Explain in ordinary English why this is appropriate.
(d) Should the charge also include speeding? (The speed limit is 55 mph $= 80\frac{2}{3}$ ft/sec.)

[4]From *Calculus: The Analysis of Functions*, by Peter D. Taylor (Toronto: Wall & Emerson, Inc., 1992)

CHAPTER SEVEN

THE INTEGRAL

We have seen that the derivative has many useful interpretations, and the same is true of the definite integral. Until now, we have approximated definite integrals by left- and right-hand sums. In this chapter we will learn to find definite integrals exactly, using the Fundamental Theorem of Calculus, and we will develop more efficient numerical methods.

7.1 ANTIDERIVATIVES AND THE FUNDAMENTAL THEOREM

Why Is the Fundamental Theorem Useful?

Knowing the Fundamental Theorem of Calculus, which tells us that when $F' = f$,

$$\int_a^b f(x)\,dx = F(b) - F(a),$$

gives us a whole new way of calculating definite integrals. To find $\int_a^b f(x)\,dx$, we first try to find F, and then calculate $F(b) - F(a)$. This method of computing definite integrals has an important advantage over using left- and right-hand sums: it gives an exact answer quickly. However, the method only works in situations where you can find $F(x)$. This is not always easy; for example, none of the functions we have encountered so far is an antiderivative of $\sin(x^2)$. Nonetheless, because the method is so helpful when it does work, we will spend the next few sections trying to make it work as often as we can.

Example 1 Compute $\displaystyle\int_1^3 2x\,dx$ using the Fundamental Theorem.

Solution We need a function with derivative $2x$. Since

$$\frac{d}{dx}x^2 = 2x,$$

$F(x) = x^2$ is an antiderivative of $f(x) = 2x$. Thus,

$$\int_1^3 2x\,dx = F(3) - F(1) = 3^2 - 1^2 = 8.$$

Finding Antiderivatives

Finding antiderivatives of functions is like taking square roots of numbers: if you pick a number at random, such as 7 or 493, you will have trouble saying what its square root is without a calculator. But if you happen to pick a number such as 25 or 64, which you know is a perfect square, then you can find its square root exactly. Similarly, if you happen to pick a function which you recognize as a derivative, then you can find its antiderivative easily.

We saw this when we noticed that $2x$ was the derivative of x^2; this told us that x^2 was an antiderivative of $2x$. If we divide by 2, then we find that

$$\text{An antiderivative of } x \text{ is } \frac{x^2}{2}.$$

To check this statement, take the derivative of $x^2/2$:

$$\frac{d}{dx}\left(\frac{x^2}{2}\right) = \frac{1}{2}\cdot\frac{d}{dx}x^2 = \frac{1}{2}\cdot 2x = x.$$

What about an antiderivative of x^2? The derivative of x^3 is $3x^2$, so the derivative of $x^3/3$ is $3x^2/3 = x^2$. Thus,

$$\text{An antiderivative of } x^2 \text{ is } \frac{x^3}{3}.$$

Can you see the pattern? It looks like

$$\text{An antiderivative of } x^n \text{ is } \frac{x^{n+1}}{n+1}.$$

(We assume $n \neq -1$, or we would have $x^0/0$, which doesn't make sense.) It is easy to check this formula by differentiation:

$$\frac{d}{dx}\left(\frac{x^{n+1}}{n+1}\right) = \frac{(n+1)x^n}{n+1} = x^n.$$

Recall that the indefinite integral of a function is the family of all its antiderivatives. Thus, in indefinite integral notation, we have shown that

$$\int x^n \, dx = \frac{x^{n+1}}{n+1} + C, \quad n \neq -1.$$

So what about when $n = -1$? In other words, what is the antiderivative of $1/x$? Fortunately, we know a function whose derivative is $1/x$, namely, the natural logarithm. Thus, since

$$\frac{d}{dx}(\ln x) = \frac{1}{x},$$

we know that

$$\int \frac{1}{x} \, dx = \ln x + C, \quad \text{for } x > 0.$$

If $x < 0$, then $\ln x$ is not defined, so it can't be an antiderivative of $1/x$. In this case, we can use $\ln(-x)$:

$$\frac{d}{dx} \ln(-x) = (-1)\frac{1}{-x} = \frac{1}{x}$$

so

$$\int \frac{1}{x} \, dx = \ln(-x) + C, \quad \text{for } x < 0.$$

So $\ln x$ is an antiderivative of $1/x$ if $x > 0$, and $\ln(-x)$ is an antiderivative of $1/x$ if $x < 0$. Since $|x| = x$ when $x > 0$ and $|x| = -x$ when $x < 0$:

$$\text{An antiderivative of } \frac{1}{x} \text{ is } \ln|x|.$$

Therefore

$$\int \frac{1}{x} \, dx = \ln|x| + C.$$

Since the exponential function is its own derivative, it is also its own antiderivative; thus

$$\int e^x \, dx = e^x + C.$$

Also, antiderivatives of the sine and cosine are easy to guess. Since

$$\frac{d}{dx}\sin x = \cos x \qquad \text{and} \qquad \frac{d}{dx}\cos x = -\sin x,$$

we get

$$\boxed{\int \cos x \, dx = \sin x + C} \qquad \text{and} \qquad \boxed{\int \sin x \, dx = -\cos x + C.}$$

Example 2 Find $\int (3x + x^2)\, dx$.

Solution We know that $x^2/2$ is an antiderivative of x and that $x^3/3$ is an antiderivative of x^2. Putting these together, we get

$$\int (3x + x^2)\, dx = 3\left(\frac{x^2}{2}\right) + \frac{x^3}{3} + C.$$

You should always check your antiderivatives by differentiation—it's easy to do. Here

$$\frac{d}{dx}\left(\frac{3}{2}x^2 + \frac{x^3}{3} + C\right) = \frac{3}{2}\cdot 2x + \frac{3x^2}{3} = 3x + x^2.$$

The preceding example illustrates that the sum and constant multiplication rules of differentiation work in reverse:

Facts about Antiderivatives: Sums and Constant Multiples

In indefinite integral notation,

1. $\int [f(x) \pm g(x)]\, dx = \int f(x)\, dx \pm \int g(x)\, dx$
2. $\int cf(x)\, dx = c\int f(x)\, dx.$

In words,

1. An antiderivative of the sum (or difference) of two functions is the sum (or difference) of their antiderivatives.
2. An antiderivative of a constant times a function is the constant times an antiderivative of the function.

Example 3 Find $\int (\sin x + 3\cos x)\, dx$.

Solution

$$\int (\sin x + 3\cos x)\, dx = \int \sin x\, dx + 3\int \cos x\, dx = -\cos x + 3\sin x + C.$$

Check by differentiating:

$$\frac{d}{dx}(-\cos x + 3\sin x + C) = \sin x + 3\cos x.$$

Example 4 Find $\int_0^1 \sin x\, dx$.

Solution The indefinite integral is

$$\int \sin x\, dx = -\cos x + C,$$

so

$$\int_0^1 \sin x\, dx = -\cos x\Big|_0^1 = (-\cos 1) - (-\cos 0) \approx -0.5403 + 1 = 0.4597.$$

Problems for Section 7.1

For Problems 1–9, find an antiderivative $F(x)$ with $F'(x) = f(x)$ and $F(0) = 0$. Is there only one possible solution in each case?

1. $f(x) = 3$
2. $f(x) = 2x$
3. $f(x) = -7x$
4. $f(x) = \frac{1}{4}x$
5. $f(x) = x^2$
6. $f(x) = \sqrt{x}$
7. $f(x) = 2 + 4x + 5x^2$
8. $f(x) = 5 - \frac{x^6}{6} - x^7$
9. $f(x) = \sin x$

Find an antiderivative of each function in Problems 10–15.

10. $f(x) = \frac{1}{\sqrt{x}}$
11. $f(x) = x^3 + 5\sqrt{x} - \frac{2}{x^2}$
12. $g(x) = x^{3/2} + x^{-3/2}$
13. $k(t) = 7 - \frac{t^8}{8} + \frac{1}{t}$
14. $h(y) = \frac{y^2+1}{y}$
15. $F(\theta) = \cos\theta + \sin\theta$
16. Find an antiderivative $F(x)$ of $f(x) = e^x$ that satisfies $F(1) = 0$.
17. Show that $\frac{x^2 \sin x}{2}$ is not an antiderivative of $x \cos x$.
18. (a) Show that $\frac{x^2/2 + 6x}{x^3/3}$ is not an antiderivative of $\frac{x+6}{x^2}$.
    (b) Find an antiderivative of $\frac{x+6}{x^2}$.
19. Show that $\arctan 2y$ is an antiderivative of $\frac{2}{1+4y^2}$.

Find the indefinite integrals in Problems 20–38.

20. $\int (5x + 7)\, dx$
21. $\int \left(\frac{3}{t} - \frac{2}{t^2}\right) dt$
22. $\int \left(3\cos\psi + 3\sqrt{\psi}\right) d\psi$
23. $\int \left(x^{3/2} + \frac{\sqrt{x}}{5} - \frac{2}{x}\right) dx$
24. $\int \left(\frac{x+1}{x}\right) dx$
25. $\int \left(\frac{x^2+x+1}{x}\right) dx$
26. $\int (e^x + 5)\, dx$
27. $\int (3\cos x - 7\sin x)\, dx$

28. $\int \left(x + \frac{2}{x} + \pi \sin x \right) dx$

29. $\int (2e^x - 8\cos x)\, dx$

30. $\int \frac{1}{\cos^2 x}\, dx$

31. $\int 2^x\, dx$ [Hint: What is $\frac{d}{dx}(2^x)$?]

32. $\int (x+1)^2\, dx$

33. $\int (x+1)^3\, dx$

34. $\int (x+1)^9\, dx$ [Hint: What is $\frac{d}{dx}(x+1)^{10}$?]

35. $\int \frac{1}{x+1}\, dx$ [Hint: What is $\frac{d}{dx}(\ln|x+1|)$?]

36. $\int \frac{1}{2x-1}\, dx$ [Hint: What is $\frac{d}{dx} \ln|2x-1|$?]

37. $\int (e^{5+x} + e^{5x})\, dx$ [Hint: What is $\frac{d}{dx}(e^{5x})$?]

38. $\int (\cos 2x - 2\sin x)\, dx$ [Hint: What is $\frac{d}{dx}(\sin 2x)$?]

39. Find antiderivatives of: (a) e^{2t} [Hint: What is $\frac{d}{dt}e^{2t}$?] (b) $\frac{1}{e^{3\theta}}$

40. Consider the function $f(x) = x^3$. Sketch the graph of f and three of its antiderivatives F, where $F(0) = 0, F(0) = 1, F(0) = -1$.

Using the Fundamental Theorem, evaluate the definite integrals in Problems 41–50 exactly [as in $\ln(3\pi)$] and numerically [$\ln(3\pi) \approx 2.243$]:

41. $\int_2^5 (x^3 - \pi x^2)\, dx$

42. $\int_0^1 \sin\theta\, d\theta$

43. $\int_1^2 \frac{1+y^2}{y}\, dy$

44. $\int_0^2 \left(\frac{x^3}{3} + 2x \right) dx$

45. $\int_0^{\pi/4} (\sin t + \cos t)\, dt$

46. $\int_{-3}^{-1} \frac{2}{r^3}\, dr$

47. $\int_0^1 2e^x\, dx$

48. $\int_0^{\pi/4} \frac{1}{\cos^2 x}\, dx$

49. $\int_{-1}^1 2^x\, dx$
[Hint: What is $\frac{d}{dx}2^x$?]

50. $\int_{\pi/6}^0 (\sin x + \cos 2x)\, dx$
[Hint: What is $\frac{d}{dx}(\sin 2x)$?]

51. (a) Find the following derivatives:
(i) $\frac{d}{dx}\sin(5x)$ (ii) $\frac{d}{dx}\cos(x^2)$ (iii) $\frac{d}{dx}e^{\sin x}$
(iv) $\frac{d}{dx}\sin(\cos x)$ (v) $\frac{d}{dx}[\ln(\cos x)]$ (vi) $\frac{d}{dx}\ln[\cos(x^4)]$
(b) Use the answers to part (a) to find the following indefinite integrals:
(i) $\int \cos(5x)\, dx$ (ii) $\int x\sin(x^2)\, dx$ (iii) $\int e^{\sin x}\cos x\, dx$
(iv) $\int \sin x \cos(\cos x)\, dx$ (v) $\int \tan x\, dx$ (vi) $\int x^3 \tan(x^4)\, dx$
[Hint: Express tan in terms of sin and cos.]
(c) We shall formalize the technique used in these problems in Section 7.2. Can you make any generalizations now about what is happening in these integrals?

52. (a) Find derivatives for the following functions:

    (i) $\ln(1+x) - \ln(2+x)$ (ii) $\ln(\cos x)$ (iii) e^{x^2} (iv) $(1+x^2)^{15}$

    (b) Using your answers to part (a), find antiderivatives for the following functions:

    (i) $\tan x$ [Hint: Express tan in terms of sin and cos.] (ii) $xe^{x^2/2}$

    (iii) $x(1+x^2)^{14}$ (iv) $\dfrac{1}{(1+x)(2+x)}$ (v) $x(1+x^2)^{12}$ (vi) $\dfrac{1}{(1+x)(3+x)}$

53. The average value of the function $v(x) = 6/x^2$ on the interval $[1, c]$ is equal to 1. Find the value of c.

54. (a) What is the average value of $f(t) = \sin t$ over $0 \le t \le 2\pi$? Why is this a reasonable answer?

    (b) Find the average value of $f(t) = \sin t$ over $0 \le t \le \pi$.

55. Sketch the parabola $y = x(x-\pi)$ and the curve $y = \sin x$, showing their points of intersection. Find the area between the two graphs.

56. Consider the costs of drilling an oil well. There are two types of costs: *fixed costs* (independent of the depth of the well) and *marginal costs* (the increment in costs that comes from drilling one more meter). By using both pieces of information, you can determine the total cost, C. The marginal costs depend on the depth at which you are drilling; drilling becomes more expensive, per meter, as you dig deeper into the earth. Suppose the fixed costs are 1,000,000 riyals (the riyal is the unit of currency of Saudi Arabia), and the marginal costs are

$$C'(x) = 4000 + 10x$$

in riyals/meter, where x is the depth in meters. Find the total cost of drilling a well x meters deep.

57. One of the earliest pollution problems brought to the attention of the Environmental Protection Agency (EPA) was the case of the Sioux Lake in eastern South Dakota. For years a small paper plant located nearby had been discharging waste containing carbon tetrachloride (CCl_4) into the waters of the lake. At the time the EPA learned of the situation, the chemical was entering at a rate of 16 cubic yards/year.

    The agency immediately ordered the installation of filters designed to slow (and eventually stop) the flow of CCl_4 from the mill. Implementation of this program took exactly three years, during which the flow of pollutant was steady at 16 cubic yards/year. Once the filters were installed, the flow declined. From the time the filters were installed until the time the flow stopped, the rate of flow was well approximated by

$$\text{Rate (in cubic yards/year)} = t^2 - 14t + 49$$

where t is time measured in years since the EPA learned of the situation.

    (a) Draw a graph showing the rate of CCl_4 flow into the lake as a function of time, beginning at the time the EPA first learned of the situation.

    (b) How many years elapsed between the time the EPA learned of the situation and the time the pollution flow stopped entirely?

    (c) How much CCl_4 entered the waters during the time shown in the graph in part (a)?

7.2 INTEGRATION BY SUBSTITUTION: PART I

In Chapter 4, we learned rules which enable you to differentiate any function obtained by combining constants, powers of x, $\sin x$, $\cos x$, a^x, and $\ln x$, using addition, multiplication, division, or composition of functions. These are called *elementary* functions. Thus, to differentiate

$$f(x) = (\cos x)\sqrt{x^3+1},$$

you recognize $f(x)$ as a product of $\cos x$ and $\sqrt{x^3+1}$ and apply the product rule; to differentiate $\sqrt{x^3+1}$, you think of it as a composition and use the chain rule.

Are there similar rules for antidifferentiation? For example, to antidifferentiate a sum you antidifferentiate each term separately, so there is a sum rule. But what about a product rule? Or a chain rule? In the next few sections, we'll introduce two methods of antidifferentiation: the substitution method and integration by parts, which reverse the chain and product rules, respectively.

You'll find that there's a great difference between looking for derivatives and looking for antiderivatives. Every elementary function has an elementary derivative found by using the differentiation rules. However, most elementary functions do not have an elementary antiderivative. Some examples are $\sqrt{x^3+1}$, $(\sin x)/x$, and e^{-x^2}. These are not exotic functions, but rather simple functions that arise naturally. On the other hand, there are many functions that do have elementary antiderivatives.

The Guess-and-Check Method

A good strategy for finding simple antiderivatives is to *guess* an answer (using your knowledge of differentiation rules) and then *check* your answer by differentiating it. If you get the expected result, then you're done; otherwise you may have to revise your guess and check again.

The method of guess-and-check is most useful in looking for what could be the result of applying the chain rule. According to the chain rule,

$$\frac{d}{dx}(f(g(x))) = \underbrace{f'}_{\text{Derivative of outside}}\overbrace{(g(x))}^{\text{Inside}} \cdot \underbrace{g'(x)}_{\text{Derivative of inside}}.$$

Thus any function which is the result of applying the chain rule will be the product of two factors: the "derivative of the outside" and the "derivative of the inside." If you recognize a function of this form, its antiderivative is $f(g(x))$.

Example 1 Find $\int 3x^2\cos(x^3)\,dx$.

Solution The function $3x^2\cos(x^3)$ looks like the result of applying the chain rule: there is an "inside" function x^3 and its derivative $3x^2$ appearing as a factor. So let's guess $\sin(x^3)$ for the antiderivative. Differentiating to check gives

$$\frac{d}{dx}(\sin(x^3)) = \cos(x^3)\cdot(3x^2).$$

Since this is what we began with, we have

$$\int 3x^2\cos(x^3)\,dx = \sin(x^3) + C.$$

The basic idea behind this method, trying to find an inside function whose derivative appears as a factor, works even when the derivative is missing a constant factor, as shown in the next example.

Example 2 Find $\int te^{(t^2+1)}\,dt$.

Solution For our guess, we'll pick $e^{(t^2+1)}$, since taking the derivative of an exponential results in the reappearance of the exponential, together with other terms from the chain rule. Now we check:

$$\frac{d}{dt}\left(e^{(t^2+1)}\right) = \left(e^{(t^2+1)}\right)\cdot 2t.$$

Our original guess was off by a factor of 2. No problem: just guess $\frac{1}{2}e^{(t^2+1)}$ and check again:

$$\frac{d}{dt}\left(\frac{1}{2}e^{(t^2+1)}\right) = \frac{1}{2}e^{(t^2+1)}\cdot 2t = e^{(t^2+1)}\cdot t.$$

Thus,

$$\int te^{(t^2+1)}\,dt = \frac{1}{2}e^{(t^2+1)} + C.$$

Example 3 Find $\int x^3\sqrt{x^4+5}\,dx$.

Solution Here the inside function is x^4+5, and its derivative appears as a factor (with the exception of the missing 4). So the integrand we have is more or less of the form

$$g'(x)\sqrt{g(x)}$$

with $g(x) = x^4+5$. Thus, we might guess that an antiderivative is

$$\frac{(g(x))^{3/2}}{3/2} = \frac{(x^4+5)^{3/2}}{3/2}.$$

Let's check and see:

$$\frac{d}{dx}\left(\frac{(x^4+5)^{3/2}}{3/2}\right) = \frac{3}{2}\frac{(x^4+5)^{1/2}}{3/2}\cdot 4x^3 = 4x^3(x^4+5)^{1/2},$$

so $\frac{(x^4+5)^{3/2}}{3/2}$ is too big by a factor of 4. The correct antiderivative is

$$\frac{1}{4}\frac{(x^4+5)^{3/2}}{3/2} = \frac{1}{6}(x^4+5)^{3/2}.$$

Thus

$$\int x^3\sqrt{x^4+5}\,dx = \frac{1}{6}(x^4+5)^{3/2} + C.$$

As a final check:

$$\frac{d}{dx}\left(\frac{1}{6}(x^4+5)^{3/2}\right) = \frac{1}{6}\cdot\frac{3}{2}(x^4+5)^{1/2}\cdot 4x^3 = x^3(x^4+5)^{1/2}.$$

As we have seen in the preceding examples, antidifferentiating a function often involves "correcting for" constant factors: if differentiation produces an extra factor of 4, antidifferentiation will require a factor of $\frac{1}{4}$.

The Method of Substitution

There is a formalization of guess-and-check—called *substitution*— that helps to find antiderivatives when the integrand is complicated. The method works as follows:

To Make a Substitution

Let w be the "inside function" and $dw = w'(x)\,dx = \dfrac{dw}{dx}dx$.

Let's redo the three examples that we just did by guess-and-check.

Example 4 Find $\displaystyle\int 3x^2\cos(x^3)\,dx$.

Solution As before, we look for an inside function whose derivative appears—in this case x^3. We let $w = x^3$. Then $dw = w'(x)\,dx = 3x^2\,dx$. The original integrand can now be completely rewritten in terms of the new variable w:

$$\int 3x^2\cos(x^3)\,dx = \int \cos\underbrace{(x^3)}_{w}\cdot\underbrace{3x^2\,dx}_{dw} = \int \cos w\,dw = \sin w + C = \sin(x^3) + C.$$

By changing the variable to w, the integrand becomes simpler, $\cos w$, and can be antidifferentiated more easily. The final step, after antidifferentiating, is to convert back to the original variable, x.

Why Does Substitution Work?

The substitution method makes it look as if you can treat dw and dx as separate entities, and even cancel them in the equation $dw = (dw/dx)dx$. Let's see why this works. Suppose we have an integral of the form $\int f(g(x))g'(x)\,dx$, where $g(x)$ is the inside function and $f(x)$ is the outside function. If F is an antiderivative of f, so $F' = f$, then by the chain rule $\frac{d}{dx}(F(g(x))) = f(g(x))g'(x)$, so

$$\int f(g(x))g'(x)\,dx = F(g(x)) + C.$$

Now write $w = g(x)$ and $dw/dx = g'(x)$ on both sides of this equation:

$$\int f(w)\frac{dw}{dx}\,dx = F(w) + C.$$

On the other hand, knowing that $F' = f$ tells us that

$$\int f(w)\,dw = F(w) + C.$$

So the following two integrals are equal:

$$\int f(w)\frac{dw}{dx}\,dx = \int f(w)\,dw.$$

Thus, substituting w for the inside function and writing $dw = w'(x)dx$ leaves the indefinite integral unchanged.

Let's revisit the second example that we did by guess-and-check.

Example 5 Find $\int te^{(t^2+1)}\, dt$.

Solution Here the inside function is $t^2 + 1$, so we'll try

$$w = t^2 + 1.$$

Then

$$dw = w'(t)\, dt = 2t\, dt.$$

Notice that the original integrand has only $t\, dt$, not $2t\, dt$. But extra multiplicative constants are not a problem. We simply write

$$\frac{1}{2}\, dw = t\, dt$$

and then substitute:

$$\int te^{(t^2+1)}\, dt = \int e^{\overbrace{(t^2+1)}^{w}} \cdot \underbrace{t\, dt}_{\frac{dw}{2}} = \int e^w \frac{1}{2}\, dw = \frac{1}{2}\int e^w\, dw = \frac{1}{2}e^w + C = \frac{1}{2}e^{(t^2+1)} + C.$$

This gives the same answer as we found using guess-and-check.

Now we'll redo the third of the examples that we solved previously by guess-and-check.

Example 6 Find $\int x^3\sqrt{x^4+5}\, dx$.

Solution The inside function is $x^4 + 5$, with derivative $4x^3$. But the original problem has an x^3, not a $4x^3$ in it. However, when the only thing missing is a constant factor, we can still use substitution. As before, we let w be the inside function:

$$w = x^4 + 5,$$

so

$$dw = w'(x)\, dx = 4x^3\, dx,$$

giving

$$\frac{1}{4}dw = x^3\, dx.$$

Thus,

$$\int x^3\sqrt{x^4+5}\, dx = \int \sqrt{w}\, \frac{1}{4}dw = \frac{1}{4}\int w^{1/2}\, dw = \frac{1}{4}\cdot\frac{w^{3/2}}{\frac{3}{2}} + C = \frac{1}{6}(x^4+5)^{3/2} + C.$$

Once again, we get the same result as with guess-and-check.

Warning

As you saw in the preceding examples, you can apply the substitution method when a *constant* factor is missing from the derivative of the inside function. However, you may not be able to use substitution if anything other than a constant factor is missing. For example, to find

$$\int x^2\sqrt{x^4+5}\,dx$$

setting $w = x^4 + 5$ does you no good because you can't easily get $x^2\,dx$ in terms of w. The moral of the story is that to use substitution, you want to have the derivative of the inside function, *to within a constant factor*, in the integrand.

Example 7 Find $\int e^{\cos\theta}\sin\theta\,d\theta$.

Solution This time we let $w = \cos\theta$, since its derivative is $-\sin\theta$, and there is a factor of $\sin\theta$ in the integrand. Then

$$dw = w'(\theta)\,d\theta = -\sin\theta\,d\theta$$

and so

$$-dw = \sin\theta\,d\theta.$$

Thus

$$\int e^{\cos\theta}\sin\theta\,d\theta = \int e^w\,(-dw) = (-1)\int e^w\,dw = -e^w + C = -e^{\cos\theta} + C.$$

You might be wondering why we didn't put $(-1)\int e^w\,dw = -e^w - C$ in the preceding example. Since the constant C is arbitrary, it doesn't really matter whether we add or subtract it, but the convention is always to add it onto whatever antiderivative you have calculated. If we think of $-e^{\cos\theta}$ as one of the antiderivatives of $e^{\cos\theta}\sin\theta$, then we write

$$-e^{\cos\theta} + C$$

to remind ourselves that there are lots of other antiderivatives.

Example 8 Find $\int \frac{e^t}{1+e^t}\,dt$.

Solution If you observe that the derivative of $1+e^t$ is e^t, then $w = 1+e^t$ is a good choice. Then $dw = e^t\,dt$, so that

$$\begin{aligned}\int \frac{e^t}{1+e^t}\,dt = \int \frac{1}{1+e^t}\,e^t\,dt = \int \frac{1}{w}\,dw &= \ln|w| + C\\ &= \ln|1+e^t| + C\\ &= \ln(1+e^t) + C \qquad \text{Since } (1+e^t) \text{ is always positive}\end{aligned}$$

Since the numerator is $e^t\,dt$, you might have wanted to try $w = e^t$. This substitution leads to the integral $\int[1/(1+w)]dw$, which is better than the original integral, but requires another substitution, $u = 1 + w$, to finish. There are often several different ways of doing an integral by substitution.

Notice the pattern in the previous example: having a function in the denominator and its derivative in the numerator leads to a natural logarithm. The next example follows the same pattern.

Example 9 Find $\int \tan\theta\, d\theta$.

Solution Recall that $\tan\theta = (\sin\theta)/(\cos\theta)$. If $w = \cos\theta$, then $dw = -\sin\theta\, d\theta$, so

$$\int \tan\theta\, d\theta = \int \frac{\sin\theta}{\cos\theta}\, d\theta = \int -\frac{dw}{w} = -\ln|w| + C = -\ln|\cos\theta| + C.$$

Some people prefer the substitution method over guess-and-check since it is more systematic, but both methods achieve the same result. For simple problems, guess-and-check can be faster.

Problems for Section 7.2

1. (a) Find the derivatives of $\sin(x^2+1)$ and $\sin(x^3+1)$.
   (b) Use your answer to part (a) to find antiderivatives of:
   (i) $x\cos(x^2+1)$ (ii) $x^2\cos(x^3+1)$
   (c) Find the antiderivatives of:
   (i) $x\sin(x^2+1)$ (ii) $x^2\sin(x^3+1)$

For each of the functions in Problems 2–8, find $\int f(x)\, dx$.

2. $f(x) = 2x\cos(x^2)$
3. $f(x) = \sin 3x$
4. $f(x) = \sin(2-5x)$
5. $f(x) = e^{\sin x}\cos x$
6. $f(x) = \dfrac{x}{x^2+1}$
7. $f(x) = \dfrac{1}{3\cos^2 2x}$
8. $f(x) = \sin x\left(\sqrt{2+3\cos x}\right)$

9. For each of the integrals in Problems 12–23, state the substitution you would use, but do not perform the integration.
10. For each of the integrals in Problems 24–33, state the substitution you would use, but do not perform the integration.
11. For each of the integrals in Problems 34–40, state the substitution you would use, but do not perform the integration.

Find the integrals in Problems 12–42. Remember, you can check your answers.

12. $\displaystyle\int xe^{-x^2}\, dx$
13. $\displaystyle\int y(y^2+5)^8\, dy$
14. $\displaystyle\int t^2(t^3-3)^{10}\, dt$
15. $\displaystyle\int x(x^2-4)^{7/2}\, dx$
16. $\displaystyle\int \frac{dy}{y+5}$
17. $\displaystyle\int (2t-7)^{73}\, dt$
18. $\displaystyle\int x(x^2+3)^2\, dx$
19. $\displaystyle\int (x^2+3)^2\, dx$
20. $\displaystyle\int \frac{1}{\sqrt{4-x}}\, dx$
21. $\displaystyle\int \sin\theta(\cos\theta+5)^7\, d\theta$
22. $\displaystyle\int x^2 e^{x^3+1}\, dx$
23. $\displaystyle\int \sin^3\alpha\cos\alpha\, d\alpha$

24. $\int \sqrt{\cos 3t} \sin 3t \, dt$

25. $\int \frac{(\ln z)^2}{z} \, dz$

26. $\int \sin^6 \theta \cos \theta \, d\theta$

27. $\int \sin^6(5\theta) \cos(5\theta) \, d\theta$

28. $\int \frac{\cos \sqrt{x}}{\sqrt{x}} \, dx$

29. $\int \frac{e^t + 1}{e^t + t} \, dt$

30. $\int \frac{1 + e^x}{\sqrt{x + e^x}} \, dx$

31. $\int \frac{y}{y^2 + 4} \, dy$

32. $\int \frac{e^x}{2 + e^x} \, dx$

33. $\int \frac{e^{\sqrt{y}}}{\sqrt{y}} \, dy$

34. $\int \tan 2x \, dx$

35. $\int \frac{x + 1}{x^2 + 2x + 19} \, dx$

36. $\int \frac{x \cos(x^2)}{\sqrt{\sin(x^2)}} \, dx$

37. $\int x^2(1 + 2x^3)^2 \, dx$

38. $\int y^2(1 + y)^2 \, dy$

39. $\int \frac{t}{1 + 3t^2} \, dt$

40. $\int \frac{(t + 1)^2}{t^2} \, dt$

41. $\int \frac{e^x - e^{-x}}{e^x + e^{-x}} \, dx$

42. $\int (2x+1)e^{x^2} e^x \, dx$ [Hint: Rewrite $e^{x^2} e^x = e^?$.]

43. Find $\int 4x(x^2 + 1) \, dx$ using two methods:
   (a) Do the multiplication first, and then antidifferentiate.
   (b) Now do the integral using the substitution $w = x^2 + 1$.
   (c) Explain why the two expressions formed in parts (a) and (b) are different. Are they both correct?

44. Throughout much of this century, the yearly consumption of electricity in the US has been increasing exponentially at a continuous rate of 7% per year. Assuming this trend continues, and that the electrical energy consumed in 1900 was 1.4 million megawatt-hours (a megawatt-hour is a measure of electrical energy),
   (a) Write an expression for electricity consumption as a function of time, t, measured in years since 1900.
   (b) Find the average yearly electrical consumption throughout this century.
   (c) During what year was electrical consumption closest to the average for the century?
   (d) Without doing the calculation for part (c), how could you have predicted which half of the century the answer would be in?

45. If we assume that wind resistance is proportional to velocity, then the velocity, v, of a body of mass m falling vertically is given by

$$v = \frac{mg}{k}\left(1 - e^{-kt/m}\right),$$

where g is the acceleration due to gravity and k is a constant. Find the height, h, above the surface of the earth as a function of time. Assume the body starts at height h_0.

7.3 INTEGRATION BY SUBSTITUTION: PART II

In this section we'll see how to use substitution to evaluate *definite* integrals, using the Fundamental Theorem of Calculus. Then we'll look at some examples where it is harder to see what substitution to use, but where the method works nonetheless.

Definite Integrals by Substitution

Example 1 Compute $\int_0^2 xe^{x^2}\,dx$.

Solution To evaluate a definite integral using the Fundamental Theorem of Calculus, we first need to find an antiderivative of $f(x) = xe^{x^2}$. The inside function is x^2, so let $w = x^2$. Then $dw = 2x\,dx$, so $\frac{1}{2}\,dw = x\,dx$. Thus,

$$\int xe^{x^2}\,dx = \int e^w \frac{1}{2}\,dw = \frac{1}{2}e^w + C = \frac{1}{2}e^{x^2} + C.$$

Now we find the definite integral

$$\int_0^2 xe^{x^2}\,dx = \frac{1}{2}e^{x^2}\Big|_0^2 = \frac{1}{2}(e^4 - e^0) = \frac{1}{2}(e^4 - 1).$$

There is another way to look at the same problem. After we established that

$$\int xe^{x^2}\,dx = \frac{1}{2}e^w + C,$$

our next two steps were to replace w by x^2, and then x by 2 and 0. We could have directly replaced the original limits of integration, $x = 0$ and $x = 2$, by the corresponding w limits. Since $w = x^2$, the w limits are $w = 0^2 = 0$ and $w = 2^2 = 4$, so we get

$$\int_{x=0}^{x=2} xe^{x^2}\,dx = \frac{1}{2}\int_{w=0}^{w=4} e^w\,dw = \frac{1}{2}e^w\Big|_0^4 = \frac{1}{2}\left(e^4 - e^0\right) = \frac{1}{2}(e^4 - 1).$$

As you would expect, both methods give the same answer.

To Use Substitution to Find Definite Integrals,

either

- Compute the indefinite integral, expressing an antiderivative in terms of the original variable, and then evaluate the result at the original limits,

or

- Convert the original limits to new limits in terms of the new variable and do not convert the antiderivative back to the original variable.

Example 2 Evaluate $\int_0^{\pi/4} \frac{\tan^3\theta}{\cos^2\theta}\,d\theta$.

Solution To use substitution, we must decide what w should be. There are two possible inside functions, $\tan\theta$ and $\cos\theta$. Now

$$\frac{d}{d\theta}(\tan\theta) = \frac{1}{\cos^2\theta} \quad \text{and} \quad \frac{d}{d\theta}(\cos\theta) = -\sin\theta,$$

and since the integral contains a factor of $1/\cos^2\theta$ but not of $\sin\theta$, we let $w = \tan\theta$ and $dw = (1/\cos^2\theta)d\theta$.

When $\theta = 0$, $w = \tan 0 = 0$, and when $\theta = \pi/4$, $w = \tan\pi/4 = 1$, so

$$\int_0^{\pi/4} \frac{\tan^3\theta}{\cos^2\theta}\,d\theta = \int_0^{\pi/4} (\tan\theta)^3 \cdot \frac{1}{\cos^2\theta}\,d\theta = \int_0^1 w^3\,dw = \frac{1}{4}w^4\bigg|_0^1 = \frac{1}{4}.$$

Example 3 Evaluate $\displaystyle\int_1^3 \frac{dx}{5-x}$.

Solution Let $w = 5 - x$, so $dw = -dx$. When $x = 1$, $w = 4$, and when $x = 3$, $w = 2$, so

$$\int_1^3 \frac{dx}{5-x} = \int_4^2 \frac{-dw}{w} = -\ln|w|\bigg|_4^2 = -(\ln 2 - \ln 4) = \ln\left(\frac{4}{2}\right) = \ln 2 \approx 0.69.$$

Notice that we write the limit $w = 4$ at the bottom, even though it is larger than $w = 2$, because $w = 4$ corresponds to the lower limit $x = 1$.

More Complex Substitutions

In the examples of substitution presented so far, we had to guess an expression for w and then we had to be lucky enough to have dw (or some constant multiple of it) just lying around elsewhere in the problem, waiting for us. What if we are not so lucky? It turns out that we can often get somewhere by letting w be some messy expression contained inside, say, a cosine or under a root, even if we cannot see immediately how such a substitution helps. So go ahead and try!

Example 4 Find $\displaystyle\int (x+7)\sqrt[3]{3-2x}\,dx$.

Solution Here, instead of the derivative of the inside function (which is -2), we have the factor $(x+7)$. However, substituting $w = 3 - 2x$ turns out to help. Then $dw = -2\,dx$, so $(-1/2)\,dw = dx$. Now we must convert everything to w, including $x + 7$. Well, if $w = 3 - 2x$, then $2x = 3 - w$, so $x = 3/2 - w/2$, and therefore we can write $x + 7$ in terms of w. Thus

$$\begin{aligned}
\int (x+7)\sqrt[3]{3-2x}\,dx &= \int \left(\frac{3}{2} - \frac{w}{2} + 7\right)\sqrt[3]{w}\left(-\frac{1}{2}\right)dw \\
&= -\frac{1}{2}\int \left(\frac{17}{2} - \frac{w}{2}\right) w^{1/3}\,dw \\
&= -\frac{1}{4}\int (17 - w)\,w^{1/3}\,dw \\
&= -\frac{1}{4}\int \left(17w^{1/3} - w^{4/3}\right)dw \\
&= -\frac{1}{4}\left(17\frac{w^{4/3}}{4/3} - \frac{w^{7/3}}{7/3}\right) + C \\
&= -\frac{1}{4}\left(\frac{51}{4}(3-2x)^{4/3} - \frac{3}{7}(3-2x)^{7/3}\right) + C.
\end{aligned}$$

Looking back over the solution, you can see the reason this substitution works is that it converts $\sqrt[3]{3-2x}$, the messiest part of the integrand, to $\sqrt[3]{w}$, which can be integrated.

Example 5 Find $\int \sqrt{1+\sqrt{x}}\,dx$.

Solution This time, the derivative of the inside function is nowhere to be seen. However, try $w = 1+\sqrt{x}$. Then $w-1=\sqrt{x}$, so $(w-1)^2 = x$. Therefore $2(w-1)\,dw = dx$. We then have

$$\begin{aligned}
\int \sqrt{1+\sqrt{x}}\,dx &= \int \sqrt{w}\,2(w-1)\,dw \\
&= 2\int w^{1/2}(w-1)\,dw \\
&= 2\int (w^{3/2} - w^{1/2})\,dw \\
&= 2\left(\frac{2}{5}w^{5/2} - \frac{2}{3}w^{3/2}\right) + C \\
&= 2\left(\frac{2}{5}(1+\sqrt{x})^{5/2} - \frac{2}{3}(1+\sqrt{x})^{3/2}\right) + C.
\end{aligned}$$

Notice that the substitution in the preceding example again converts the inside of the messiest function into something simple. In addition, since the derivative of the inside function is not waiting for us, we have to solve for x so that we can get dx entirely in terms of w.

Substitution can be used to express one definite integral in terms of another, even in situations where we don't know what the indefinite integral is.

Example 6 Derive a formula for the area of the ellipse

$$\frac{x^2}{a^2} + \frac{y^2}{b^2} = 1$$

assuming $a, b > 0$.

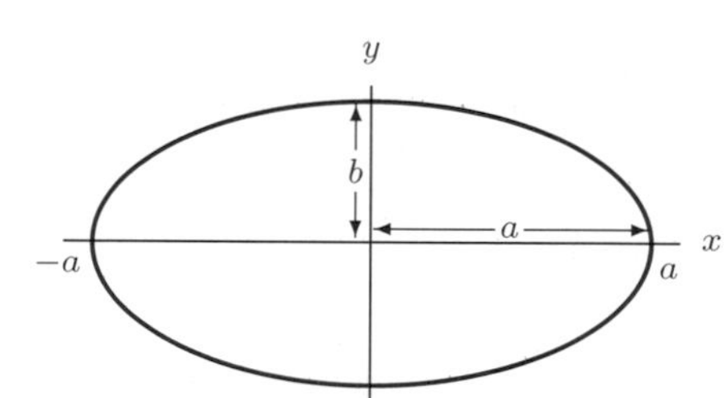

Figure 7.1: The ellipse: $\frac{x^2}{a^2} + \frac{y^2}{b^2} = 1$

Solution Solving for y gives

$$y = \pm b\sqrt{1 - \frac{x^2}{a^2}},$$

where the positive square root gives the upper half of the ellipse, and the negative square root gives the lower. (See Figure 7.1.) Thus, the area of the ellipse is given by

$$\text{Area} = 2\int_{-a}^{a} b\sqrt{1-\frac{x^2}{a^2}}\,dx$$

Substitution allows us to convert this integral to one we can compute. Let $w = x/a$, so $dx = a\,dw$. When $x=-a$, $w=-1$, and when $x=a$, $w=1$. Thus

$$2\int_{-a}^{a} b\sqrt{1-\frac{x^2}{a^2}}\,dx = 2\int_{-1}^{1} b\sqrt{1-w^2}\,a\,dw = 2ab\int_{-1}^{1}\sqrt{1-w^2}\,dw.$$

Now the integral $\int_{-1}^{1} \sqrt{1-w^2}\,dw$ is the area of half a circle of radius 1 (because $y = \sqrt{1-w^2}$ is the top half of the circle $w^2 + y^2 = 1$). Thus,

$$\int_{-1}^{1} \sqrt{1-w^2}\,dw = \frac{\pi}{2}$$

so

$$\text{Area of ellipse} = 2ab \int_{-1}^{1} \sqrt{1-w^2}\,dw = \pi ab.$$

(Alternatively, calculating $\int_{-1}^{1} \sqrt{1-w^2}\,dw$ using left-hand sums with 100 subdivisions gives the value of this integral as 1.57..., which agrees with $\pi/2$ to two decimal places.)

Problems for Section 7.3

1. Use substitution to express each of the following integrals as a multiple of $\int_a^b \frac{1}{w}\,dw$ for some a and b. Then evaluate the integrals.

   (a) $\displaystyle\int_0^1 \frac{x}{1+x^2}\,dx$. [Hint: Try $w = 1 + x^2$.]

   (b) $\displaystyle\int_0^{\pi/4} \frac{\sin x}{\cos x}\,dx$.

For Problems 2–7, use substitution to calculate the definite integrals.

2. $\displaystyle\int_0^{1/2} \cos \pi x\,dx$

3. $\displaystyle\int_1^8 \frac{e^{\sqrt[3]{x}}}{\sqrt[3]{x^2}}dx$

4. $\displaystyle\int_0^2 \frac{x}{(1+x^2)^2}\,dx$

5. $\displaystyle\int_{-1}^{e-2} \frac{1}{t+2}dt$

6. $\displaystyle\int_1^4 \frac{\cos\sqrt{x}}{\sqrt{x}}\,dx$

7. $\displaystyle\int_{-\pi/4}^{\pi/4} \cos^2\theta \sin^5\theta\,d\theta$
   [Hint: $\cos^2\theta + \sin^2\theta = 1$.]

For Problems 8–21, evaluate the definite integrals. Whenever possible, use the Fundamental Theorem of Calculus, perhaps after a substitution. Otherwise use left- and right-hand sums.

8. $\displaystyle\int_1^3 \frac{1}{x}\,dx$

9. $\displaystyle\int_{-1}^3 (x^3 + 5x)\,dx$

10. $\displaystyle\int_1^2 \frac{\sin t}{t}\,dt$

11. $\displaystyle\int_0^{\pi} \sin\theta(\cos\theta + 5)^7\,d\theta$

12. $\displaystyle\int_{-1}^1 \frac{1}{1+y^2}\,dy$

13. $\displaystyle\int_0^1 x(1+x^2)^{20}\,dx$

14. $\displaystyle\int_0^1 \frac{x}{1+5x^2}\,dx$

15. $\displaystyle\int_0^{\pi/12} \sin 3\alpha\,d\alpha$

16. $\displaystyle\int_1^2 \frac{x^2+1}{x}\,dx$

17. $\displaystyle\int_4^1 x\sqrt{x^2+4}\,dx$

18. $\displaystyle\int_0^1 \frac{1}{x^2+2x+1}\,dx$

19. $\displaystyle\int_0^{1/\sqrt{2}} \frac{x}{\sqrt{1-x^4}}\,dx$

20. $\displaystyle\int_{-2}^0 \frac{2x+4}{x^2+4x+5}\,dx$

21. $\displaystyle\int_{1/4}^1 \sin\frac{1}{t}\,dt$

In Problems 22–26, explain why the two antiderivatives are really, despite their apparent dissimilarity, different expressions of the same problem. You do not need to evaluate the integrals.

22. $\int \frac{e^x\,dx}{1+e^{2x}}$ and $\int \frac{\cos x\,dx}{1+\sin^2 x}$

23. $\int \frac{\ln x}{x}\,dx$ and $\int x\,dx$

24. $\int \sqrt{x+1}\,dx$ and $\int \frac{\sqrt{1+\sqrt{x}}}{\sqrt{x}}\,dx$

25. $\int e^{\sin x}\cos x\,dx$ and $\int \frac{e^{\arcsin x}}{\sqrt{1-x^2}}\,dx$

26. $\int (\sin x)^3 \cos x\,dx$ and $\int (x^3+1)^3 x^2\,dx$

27. Find $\int \frac{dx}{x^2+4x+5}$ by completing the square and using the substitution $x+2=\tan\theta$.

28. Integrate:
 (a) $\int \frac{1}{\sqrt{x}}\,dx$ (b) $\int \frac{1}{\sqrt{x+1}}\,dx$ (c) $\int \frac{1}{\sqrt{x}+1}\,dx$

29. Find the area under the graph of $f(x)=xe^{x^2}$ between $x=0$ and $x=2$.

30. Find the average value of $f(x)=\frac{1}{x+1}$ on the interval $x=0$ to $x=2$. Sketch a graph showing the function and the average value.

31. (a) Find $\int \sin\theta\cos\theta\,d\theta$.
 (b) You probably solved part (a) by making the substitution $w=\sin\theta$ or $w=\cos\theta$. (If not, go back and do it that way.) Now find $\int \sin\theta\cos\theta\,d\theta$ by making the *other* substitution.
 (c) There is yet another way of finding this integral which involves the trigonometric identities

$$\sin 2\theta = 2\sin\theta\cos\theta \text{ and } \cos 2\theta = \cos^2\theta - \sin^2\theta.$$

 Find $\int \sin\theta\cos\theta\,d\theta$ using one of these identities and then the substitution $w=2\theta$.
 (d) You should now have three different expressions for the indefinite integral $\int \sin\theta\cos\theta\,d\theta$. Are they really different? Are they all correct? Explain the reasons for your answer.

32. If we assume that wind resistance is proportional to the square of velocity, then the downward velocity, v, of a falling body is given by

$$v=\sqrt{\frac{g}{k}}\left(\frac{e^{t\sqrt{gk}}-e^{-t\sqrt{gk}}}{e^{t\sqrt{gk}}+e^{-t\sqrt{gk}}}\right).$$

Use the substitution $w=e^{t\sqrt{gk}}+e^{-t\sqrt{gk}}$ to find the height, h, of the body above the surface of the earth as a function of time. Assume the body starts at a height h_0.

33. The rate at which water is flowing into a tank is $r(t)$ gallons/minute, with t in minutes.
 (a) Write an expression approximating the amount of water entering the tank during the interval from time t to time $t+\Delta t$, where Δt is small.
 (b) Write a sum approximating the total amount of water entering the tank between $t=0$ and $t=5$. Write an exact expression for this amount.
 (c) By how much has the amount of water in the tank changed between $t=0$ and $t=5$ if $r(t)=20e^{0.02t}$?
 (d) If $r(t)$ is as given in part (c), and if the tank contains 3000 gallons initially, find a formula for $Q(t)$, the amount of water in the tank at time t.

7.4 INTEGRATION BY PARTS

We have seen how the method of substitution reverses the chain rule. Now we will introduce a method called *integration by parts* that is based on the product rule.

Example 1 Find $\int xe^x \, dx$.

Solution We are looking for a function whose derivative is xe^x. If we think about how the product rule works, we might be led to guess xe^x, because we know that when we take the derivative we will get two terms, one of which will be xe^x:

$$\frac{d}{dx}(xe^x) = \frac{d}{dx}(x)e^x + x\frac{d}{dx}(e^x) = e^x + xe^x.$$

Of course, our guess is wrong because of the extra e^x. Maybe we can adjust our guess by subtracting from xe^x something to cancel out the extra e^x. This leads us to try $xe^x - e^x$. Let's check it:

$$\frac{d}{dx}(xe^x - e^x) = \frac{d}{dx}(xe^x) - \frac{d}{dx}(e^x) = e^x + xe^x - e^x = xe^x.$$

It works, so $\int xe^x \, dx = xe^x - e^x + C$.

Example 2 Find $\int \theta \cos\theta \, d\theta$.

Solution We guess the antiderivative is $\theta \sin\theta$ and use the product rule to check:

$$\frac{d}{d\theta}(\theta \sin\theta) = \frac{d(\theta)}{d\theta}\sin\theta + \theta\frac{d}{d\theta}(\sin\theta) = \sin\theta + \theta\cos\theta.$$

To correct for the extra $\sin\theta$ term, we must subtract from our original guess something whose derivative is $\sin\theta$. Since $\frac{d}{d\theta}(\cos\theta) = -\sin\theta$, we'll try:

$$\frac{d}{d\theta}(\theta\sin\theta + \cos\theta) = \frac{d}{d\theta}(\theta\sin\theta) + \frac{d}{d\theta}(\cos\theta) = \sin\theta + \theta\cos\theta - \sin\theta = \theta\cos\theta.$$

Thus, $\int \theta\cos\theta \, d\theta = \theta\sin\theta + \cos\theta + C$.

The General Formula for Integration by Parts

We can formalize the process illustrated in the last two examples in the following way. We begin with the product rule:

$$\frac{d}{dx}(uv) = u'v + uv'$$

where u and v are functions of x with derivatives u' and v', respectively. We rewrite this as:

$$uv' = \frac{d}{dx}(uv) - u'v$$

and then integrate both sides:

$$\int uv' \, dx = \int \frac{d}{dx}(uv) \, dx - \int u'v \, dx.$$

Since an antiderivative of $\frac{d}{dx}(uv)$ is just uv, we get the following formula:

Integration by Parts

$$\int uv'\,dx = uv - \int u'v\,dx.$$

This formula is useful when the integrand can be viewed as a product and when the integral on the right-hand side is simpler than that on the left.

In effect, we were using integration by parts in the previous two examples. In Example 1, we let $xe^x = (x)\cdot(e^x) = uv'$, then $u = x$ and $v' = e^x$. Thus, $u' = 1$ and $v = e^x$, so

$$\int \underbrace{(x)}_{u}\underbrace{(e^x)}_{v'}\,dx = \underbrace{(x)}_{u}\underbrace{(e^x)}_{v} - \int \underbrace{(1)}_{u'}\underbrace{(e^x)}_{v}\,dx = xe^x - e^x + C.$$

So uv represents our first guess, and $\int u'v\,dx$ represents the correction to our guess.

Notice what would have happened in Example 1 if we took $u = x$ and $v = e^x + C_1$. Then

$$\begin{aligned}\int xe^x\,dx &= x(e^x + C_1) - \int (e^x + C_1)\,dx\\ &= xe^x + C_1x - e^x - C_1x + C\\ &= xe^x - e^x + C,\end{aligned}$$

as before. Therefore you see that it is not necessary to take a general antiderivative for v; any antiderivative will do.

You may wonder what would have happened in Example 1 if we had picked u and v' the other way around. If $u = e^x$ and $v' = x$, then $u' = e^x$ and $v = x^2/2$. The formula for integration by parts then gives

$$\int xe^x\,dx = \frac{x^2}{2}e^x - \int \frac{x^2}{2}\cdot e^x\,dx,$$

which is true but hardly helpful since the integral on the right is worse than the one on the left. The moral of the story is that you want to choose u and v' to make the integral on the right easier than that on the left.

Advice on How to Choose *u* and *v*′

- Whatever you let v' be, you need to be able to find v.
- It helps if u' is simpler than u (or at least no more complicated than u).
- It helps if v is simpler than v' (or at least no more complicated than v').

If we pick $v' = x$ in Example 1, then $v = x^2/2$, which is certainly "worse" than v'.

There are some examples which don't look like good candidates for integration by parts, because they don't appear to involve products, but for which the process works well. Such examples often involve $\ln x$ or the inverse trig functions. Here is one:

Example 3 Find $\displaystyle\int_2^3 \ln x\, dx$.

Solution This doesn't look like a product unless you write $\ln x = (1)(\ln x)$. Then we might say $u = 1$ so $u' = 0$, which certainly makes things simpler. But if $v' = \ln x$, what is v? If we knew, we wouldn't need integration by parts. Let's try the other way: if $u = \ln x$, $u' = 1/x$ and if $v' = 1$, $v = x$, so

$$\int_2^3 \underbrace{(\ln x)}_{u}\,\underbrace{(1)}_{v'}\,dx = \underbrace{(\ln x)}_{u}\,\underbrace{(x)}_{v}\Big|_2^3 - \int_2^3 \underbrace{\left(\frac{1}{x}\right)}_{u'}\cdot\underbrace{(x)}_{v}\,dx$$

$$= x\ln x\Big|_2^3 - \int_2^3 1\,dx = (x\ln x - x)\Big|_2^3$$

$$= 3\ln 3 - 3 - 2\ln 2 + 2 = 3\ln 3 - 2\ln 2 - 1.$$

Notice that when doing a definite integral by parts, you must remember to put the limits of integration (here 2 and 3) on the uv term (in this case $x \ln x$) as well as on the integral $\int u'v\,dx$.

Example 4 Find $\displaystyle\int x^6 \ln x\, dx$.

Solution View $x^6 \ln x$ as uv' where $u = \ln x$ and $v' = x^6$. (Notice we did not let $u = x^6$, $v' = \ln x$.) Then $v = \frac{1}{7}x^7$ and $u' = 1/x$, so integration by parts gives us:

$$\int x^6 \ln x\,dx = \int (\ln x)x^6\,dx = (\ln x)\left(\frac{1}{7}x^7\right) - \int \frac{1}{7}x^7\cdot\frac{1}{x}\,dx$$

$$= \frac{1}{7}x^7\ln x - \frac{1}{7}\int x^6\,dx$$

$$= \frac{1}{7}x^7\ln x - \frac{1}{49}x^7 + C.$$

In Example 4 we did not choose $v' = \ln x$, because it is not immediately clear what v is. In fact, we had to use integration by parts in Example 3 to find out what the antiderivative of $\ln x$ was. Also, using $u = \ln x$, as we have done, gives $u' = 1/x$, which most people would consider simpler than $u = \ln x$. This shows that u does not have to be the factor written on the left (here x^6). However, if you do choose $u = x^6$ and $v' = \ln x$, the integration by parts still works—it's just less pleasant.

Example 5 Find $\displaystyle\int x^2 \sin 4x\, dx$.

Solution If we let $v' = \sin 4x$, then $v = -\frac{1}{4}\cos 4x$, which is no worse than v'. Also letting $u = x^2$, we get $u' = 2x$, which is simpler than $u = x^2$. Using integration by parts:

$$\int x^2 \sin 4x\, dx = x^2\left(-\frac{1}{4}\cos 4x\right) - \int 2x\left(-\frac{1}{4}\cos 4x\right)\, dx$$
$$= -\frac{1}{4}x^2\cos 4x + \frac{1}{2}\int x\cos 4x\, dx.$$

The trouble is we still have to grapple with $\int x\cos 4x\, dx$. This can be done by using integration by parts again with a new u and v, namely $u = x$ and $v' = \cos 4x$:

$$\int x\cos 4x\, dx = x\left(\frac{1}{4}\sin 4x\right) - \int 1\cdot\frac{1}{4}\sin 4x\, dx$$
$$= \frac{1}{4}x\sin 4x - \frac{1}{4}\cdot\left(-\frac{1}{4}\cos 4x\right) + C$$
$$= \frac{1}{4}x\sin 4x + \frac{1}{16}\cos 4x + C.$$

Thus,

$$\int x^2\sin 4x\, dx = -\frac{1}{4}x^2\cos 4x + \frac{1}{2}\int x\cos 4x\, dx$$
$$= -\frac{1}{4}x^2\cos 4x + \frac{1}{2}\left(\frac{1}{4}x\sin 4x + \frac{1}{16}\cos 4x + C\right)$$
$$= -\frac{1}{4}x^2\cos 4x + \frac{1}{8}x\sin 4x + \frac{1}{32}\cos 4x + C_1$$

where C_1 is the arbitrary constant $\frac{1}{2}C$.

Notice that each time we use integration by parts, the power of x goes down by 1.

Example 6 Find $\displaystyle\int \cos^2\theta\, d\theta$.

Solution Using integration by parts with $u = \cos\theta$, $v' = \cos\theta$ gives $u' = -\sin\theta$, $v = \sin\theta$, so we get

$$\int \cos^2\theta\, d\theta = \cos\theta\sin\theta + \int \sin^2\theta\, d\theta$$

Substituting $\sin^2\theta = 1 - \cos^2\theta$ leads to

$$\int \cos^2\theta\, d\theta = \cos\theta\sin\theta + \int (1 - \cos^2\theta)\, d\theta$$
$$= \cos\theta\sin\theta + \int 1\, d\theta - \int \cos^2\theta\, d\theta.$$

Looking at the right side, we see that our original integral has 'boomeranged' back to us. If we move it to the left, we get

$$2\int \cos^2\theta\, d\theta = \cos\theta\sin\theta + \int 1\, d\theta.$$

Dividing by 2 gives

$$\int \cos^2\theta\, d\theta = \frac{1}{2}\cos\theta\sin\theta + \frac{1}{2}\int 1\, d\theta = \frac{1}{2}\cos\theta\sin\theta + \frac{1}{2}\theta + C.$$

Problems for Section 7.4

1. (a) Find formulas for the derivatives of the following functions:
   (i) $x \cos x$ (ii) $\cos 2x$ (iii) $x^2 \cos x$ (iv) $x \ln x$.
   (b) Use your answers from part (a) to find antiderivatives for
   (i) $\ln x$ (ii) $\sin 2x$ (iii) $x \sin x$ (iv) $x^2 \sin x$.
2. By writing $\arctan x = (1) \cdot (\arctan x)$, find $\int \arctan x\, dx$.

Find the indefinite integrals in Problems 3–28.

3. $\int te^{5t}\, dt$
4. $\int t^2 e^{5t}\, dt$
5. $\int pe^{-0.1p}\, dp$
6. $\int y \ln y\, dy$
7. $\int x^3 \ln x\, dx$
8. $\int t \sin t\, dt$
9. $\int t^2 \sin t\, dt$
10. $\int \theta^2 \cos 3\theta\, d\theta$
11. $\int (z+1)e^{2z}\, dz$
12. $\int \frac{z}{e^z}\, dz$
13. $\int (\theta+1)\sin(\theta+1)\, d\theta$
14. $\int \sin^2 \theta\, d\theta$
15. $\int \cos^2(3\alpha+1)\, d\alpha$
16. $\int q^5 \ln 5q\, dq$
17. $\int \frac{\ln x}{x^2}\, dx$
18. $\int y\sqrt{y+3}\, dy$
19. $\int \frac{y}{\sqrt{5-y}}\, dy$
20. $\int (t+2)\sqrt{2+3t}\, dt$
21. $\int \frac{t+7}{\sqrt{5-t}}\, dt$
22. $\int (\ln t)^2\, dt$
23. $\int x(\ln x)^4\, dx$
24. $\int \arctan 7z\, dz$
25. $\int x \arctan x^2\, dx$
26. $\int \arcsin w\, dw$
27. $\int x^3 e^{x^2}\, dx$
28. $\int x^5 \cos x^3\, dx$
29. In Problem 14, you evaluated $\int \sin^2 \theta\, d\theta$ using integration by parts. (If you didn't do it by parts, do so now!) Redo this integral by changing the form of the integrand using the identity $\sin^2 \theta = (1 - \cos 2\theta)/2$. Explain any differences in the form of the answer obtained by the two methods.
30. Compute $\int \cos^2 \theta\, d\theta$ in two different ways and explain any differences in the form of your answers. (The identity $\cos^2 \theta = (1 + \cos 2\theta)/2$ may be useful.)
31. Use integration by parts twice to find $\int e^x \sin x\, dx$.
32. Use integration by parts twice to find $\int e^\theta \cos \theta\, d\theta$.
33. Use the results of Problems 31 and 32 to find $\int xe^x \sin x\, dx$.
34. Use the results from Problems 31 and 32 to find $\int \theta e^\theta \cos \theta\, d\theta$.
35. Show that $\int x^n e^x\, dx = x^n e^x - n \int x^{n-1} e^x\, dx$.

36. Show that $\int x^n \sin ax\,dx = -\frac{1}{a}x^n \cos ax + \frac{n}{a}\int x^{n-1}\cos ax\,dx.$

37. Show that $\int x^n \cos ax\,dx = \frac{1}{a}x^n \sin ax - \frac{n}{a}\int x^{n-1}\sin ax\,dx.$

38. Show that $\int \cos^n x\,dx = \frac{1}{n}\cos^{n-1} x \sin x + \frac{n-1}{n}\int \cos^{n-2} x\,dx.$

Evaluate the integrals in Problems 39–47 both exactly [e.g., $\ln(3\pi)$] and numerically [e.g. $\ln(3\pi) \approx 2.243$]:

39. $\int_1^5 \ln t\,dt$
40. $\int_0^{10} ze^{-z}\,dz$
41. $\int_3^5 x\cos x\,dx$
42. $\int_1^3 t\ln t\,dt$
43. $\int_0^5 \ln(1+t)\,dt$
44. $\int_0^1 \arctan y\,dy$
45. $\int_0^1 x\arctan x^2\,dx$
46. $\int_0^1 \arcsin z\,dz$
47. $\int_0^1 u\arcsin u^2\,du$

48. Using Riemann sums, find an approximate value for $\int_1^2 \ln x\,dx$. Find, using antiderivatives, $\int_1^2 \ln x\,dx$. Explain in words why your answers verify the Fundamental Theorem of Calculus.

49. Integrating $e^{ax}\sin bx$ by parts twice yields a result of the form

$$\int e^{ax}\sin bx\,dx = e^{ax}(A\sin bx + B\cos bx) + C.$$

(a) Find the constants A and B in terms of a and b. [Hint: Don't actually perform the integration by parts.]

(b) Evaluate $\int e^{ax}\cos bx\,dx$ by modifying the result in part (a). [Again, it is not necessary to perform integration by parts, as the result is of the same form as that in part (a).]

50. During a surge in the demand for electrical power, the rate, r, at which power is used can be approximated by

$$r = te^{-at},$$

where t is the time in hours and a is a positive constant.

(a) Find the total energy, E, used in the first T hours. Give your answer as a function of a.

(b) What happens to E as $T \to \infty$?

7.5 TABLES OF INTEGRALS

As we have said, few functions have elementary antiderivatives. Since there are so few, it would make sense for someone to figure them all out, once and for all, and compile them in a list or table of integrals. Such integral tables are available[1] so that if you need an antiderivative, you can just look it up. There are now computer programs and some calculators that can compute antiderivatives. However, until such calculators are widely used, it is a good idea to learn to use an integral table. The key to using these tables is being able to recognize the general class of function that you are trying to integrate, so you can know what section of the table to look in. In this section, we present a short table of indefinite integrals.

Warning: This section involves long division of polynomials and completing the square. You may want to review these topics!

[1]See, for example, *CRC Standard Mathematical Tables* (Boca Raton, Fl: CRC Press).

A Short Table of Indefinite Integrals

I. Basic Functions

1. $\displaystyle\int x^n\,dx = \frac{1}{n+1}x^{n+1} + C, \quad n \neq -1$

2. $\displaystyle\int \frac{1}{x}\,dx = \ln|x| + C$

3. $\displaystyle\int a^x\,dx = \frac{1}{\ln a}a^x + C$

4. $\displaystyle\int \ln x\,dx = x\ln x - x + C, \quad x > 0$

5. $\displaystyle\int \sin x\,dx = -\cos x + C$

6. $\displaystyle\int \cos x\,dx = \sin x + C$

7. $\displaystyle\int \tan x\,dx = -\ln|\cos x| + C$

II. Products of e^x, $\cos x$, and $\sin x$

8. $\displaystyle\int e^{ax}\sin(bx)\,dx = \frac{1}{a^2+b^2}e^{ax}[a\sin(bx) - b\cos(bx)] + C$

9. $\displaystyle\int e^{ax}\cos(bx)\,dx = \frac{1}{a^2+b^2}e^{ax}[a\cos(bx) + b\sin(bx)] + C$

10. $\displaystyle\int \sin(ax)\sin(bx)\,dx = \frac{1}{b^2-a^2}[a\cos(ax)\sin(bx) - b\sin(ax)\cos(bx)] + C, \quad a \neq b$

11. $\displaystyle\int \cos(ax)\cos(bx)\,dx = \frac{1}{b^2-a^2}[b\cos(ax)\sin(bx) - a\sin(ax)\cos(bx)] + C, \quad a \neq b$

12. $\displaystyle\int \sin(ax)\cos(bx)\,dx = \frac{1}{b^2-a^2}[b\sin(ax)\sin(bx) + a\cos(ax)\cos(bx)] + C, \quad a \neq b$

III. Product of Polynomial $p(x)$ with $\ln x$, e^x, $\cos x$, $\sin x$

13. $\displaystyle\int x^n\ln x\,dx = \frac{1}{n+1}x^{n+1}\ln x - \frac{1}{(n+1)^2}x^{n+1} + C, \quad n \neq -1, \quad x > 0$

14. $$\begin{aligned}\int p(x)e^{ax}\,dx &= \frac{1}{a}p(x)e^{ax} - \frac{1}{a}\int p'(x)e^{ax}\,dx\\ &= \frac{1}{a}p(x)e^{ax} - \frac{1}{a^2}p'(x)e^{ax} + \frac{1}{a^3}p''(x)e^{ax} - \cdots\\ &\quad (+-+-\ldots) \quad \text{(signs alternate)}\end{aligned}$$

15. $$\begin{aligned}\int p(x)\sin ax\,dx &= -\frac{1}{a}p(x)\cos ax + \frac{1}{a}\int p'(x)\cos ax\,dx\\ &= -\frac{1}{a}p(x)\cos ax + \frac{1}{a^2}p'(x)\sin ax + \frac{1}{a^3}p''(x)\cos ax - \cdots\\ &\quad (-++--++\ldots) \quad \text{(signs alternate in pairs after first term)}\end{aligned}$$

16. $$\begin{aligned}\int p(x)\cos ax\,dx &= \frac{1}{a}p(x)\sin ax - \frac{1}{a}\int p'(x)\sin ax\,dx\\ &= \frac{1}{a}p(x)\sin ax + \frac{1}{a^2}p'(x)\cos ax - \frac{1}{a^3}p''(x)\sin ax - \cdots\\ &\quad (++--++--\ldots) \quad \text{(signs alternate in pairs)}\end{aligned}$$

IV. Integer Powers of $\sin x$ and $\cos x$

17. $$\int \sin^n x\,dx = -\frac{1}{n}\sin^{n-1} x\cos x + \frac{n-1}{n}\int \sin^{n-2} x\,dx, \quad n \text{ positive}$$

18. $$\int \cos^n x\,dx = \frac{1}{n}\cos^{n-1} x\sin x + \frac{n-1}{n}\int \cos^{n-2} x\,dx, \quad n \text{ positive}$$

19. $$\int \frac{1}{\sin^m x}\,dx = \frac{-1}{m-1}\frac{\cos x}{\sin^{m-1} x} + \frac{m-2}{m-1}\int \frac{1}{\sin^{m-2} x}\,dx, \quad m \neq 1, m \text{ positive}$$

20. $$\int \frac{1}{\sin x}\,dx = \frac{1}{2}\ln\left|\frac{(\cos x) - 1}{(\cos x) + 1}\right| + C$$

21. $$\int \frac{1}{\cos^m x}\,dx = \frac{1}{m-1}\frac{\sin x}{\cos^{m-1} x} + \frac{m-2}{m-1}\int \frac{1}{\cos^{m-2} x}\,dx, \quad m \neq 1, m \text{ positive}$$

22. $$\int \frac{1}{\cos x}\,dx = \frac{1}{2}\ln\left|\frac{(\sin x) + 1}{(\sin x) - 1}\right| + C$$

23. $\int \sin^m x\cos^n x\,dx$: If m is odd, let $w = \cos x$. If n is odd, let $w = \sin x$. If both m and n are even and non-negative, convert all to $\sin x$ or all to $\cos x$ (using $\sin^2 x + \cos^2 x = 1$), and use IV-17 or IV-18. If m and n are even and one of them is negative, convert to whichever function is in the denominator and use IV-19 or IV-21. The case in which both m and n are even and negative is omitted.

V. Quadratic in the Denominator

24. $$\int \frac{1}{x^2 + a^2}\,dx = \frac{1}{a}\arctan\frac{x}{a} + C, \quad a \neq 0$$

25. $$\int \frac{bx + c}{x^2 + a^2}\,dx = \frac{b}{2}\ln|x^2 + a^2| + \frac{c}{a}\arctan\frac{x}{a} + C, \quad a \neq 0$$

26. $$\int \frac{1}{(x-a)(x-b)}\,dx = \frac{1}{a-b}(\ln|x-a| - \ln|x-b|) + C, \quad a \neq b$$

27. $$\int \frac{cx + d}{(x-a)(x-b)}\,dx = \frac{1}{a-b}\left[(ac+d)\ln|x-a| - (bc+d)\ln|x-b|\right] + C, \quad a \neq b$$

VI. Integrands Involving $\sqrt{a^2 + x^2}$, $\sqrt{a^2 - x^2}$, $\sqrt{x^2 - a^2}$, $a > 0$

28. $$\int \frac{1}{\sqrt{a^2 - x^2}}\,dx = \arcsin\frac{x}{a} + C$$

29. $$\int \frac{1}{\sqrt{x^2 \pm a^2}}\,dx = \ln\left|x + \sqrt{x^2 \pm a^2}\right| + C$$

30. $$\int \sqrt{a^2 \pm x^2}\,dx = \frac{1}{2}\left(x\sqrt{a^2 \pm x^2} + a^2\int \frac{1}{\sqrt{a^2 \pm x^2}}\,dx\right) + C$$

31. $$\int \sqrt{x^2 - a^2}\,dx = \frac{1}{2}\left(x\sqrt{x^2 - a^2} - a^2\int \frac{1}{\sqrt{x^2 - a^2}}\,dx\right) + C$$

Using the Table of Integrals

Part I of the table on pages 366–367 gives the antiderivatives of the basic functions x^n, a^x, $\ln x$, $\sin x$, $\cos x$, and $\tan x$. (The antiderivative for $\ln x$ is found using integration by parts and is a special case of the more general formula III-13.) Most of these you know already.

Part II of the table contains antiderivatives of functions involving products of e^x, $\sin x$, and $\cos x$. All of these antiderivatives were obtained using integration by parts.

Example 1 Find $\int \sin 7z \sin 3z\, dz$.

Solution Since the integrand is the product of two sines, we should use II-10 in the table,

$$\int \sin 7z \sin 3z\, dz = -\frac{1}{40}(7\cos 7z \sin 3z - 3\cos 3z \sin 7z) + C.$$

Part III of the table contains antiderivatives for products of a polynomial and e^x, $\sin x$, or $\cos x$. It also has an antiderivative for $x^n \ln x$, which can easily be used to find the antiderivatives of the product of a general polynomial and $\ln x$. Each *reduction formula* is used repeatedly to reduce the degree of the polynomial until a zero degree polynomial is obtained.

Example 2 Find $\int x^3 \sin 5x\, dx$.

Solution Here we have a polynomial, x^3 multiplied by $\sin 5x$, which is in the form of III-15. There are only three successive derivatives of x^3 before 0 is reached (namely, $3x^2$, $6x$, and 6), so there will be four terms. The signs in the terms will be $-++-$, as given in III-15, so we get

$$\int x^3 \sin 5x\, dx = -\frac{1}{5}x^3\cos 5x + \frac{1}{25}\cdot 3x^2 \sin 5x + \frac{1}{125}\cdot 6x\cos 5x - \frac{1}{625}\cdot 6\sin 5x + C.$$

Example 3 Find $\int (x^5 + 2x^3 - 8)e^{3x}\, dx$.

Solution Since $p(x) = x^5 + 2x^3 - 8$ is a polynomial multiplied by e^{3x}, this is of the form in III-14. Now $p'(x) = 5x^4 + 6x^2$ and $p''(x) = 20x^3 + 12x$, and so on, giving

$$\int (x^5 + 2x^3 - 8)e^{3x}\, dx = e^{3x}\left[\frac{1}{3}(x^5 + 2x^3 - 8) - \frac{1}{9}(5x^4 + 6x^2) + \frac{1}{27}(20x^3 + 12x) - \frac{1}{81}(60x^2 + 12) + \frac{1}{243}(120x) - \frac{1}{729}\cdot 120\right] + C.$$

Here we have the successive derivatives of the original polynomial $x^5 + 2x^3 - 8$, occurring with alternating signs and multiplied by successive powers of 1/3.

Example 4 Find $\int (x^2 + 3)\ln x\, dx$.

Solution Formula III-13 applies only to functions of the form $x^n \ln x$, so we'll have to multiply out and separate into two integrals.

$$\int (x^2+3)\ln x\,dx = \int x^2 \ln x\,dx + 3\int \ln x\,dx.$$

Now we can use formula III-13 on each integral separately, to get

$$\int (x^2+3)\ln x\,dx = \frac{x^3}{3}\ln x - \frac{x^3}{9} + 3(x\ln x - x) + C.$$

Part IV of the table contains reduction formulas for the antiderivatives of $\cos^n x$ and $\sin^n x$, which can be obtained by integration by parts. When n is a positive integer, formulas IV-17 and IV-18 can be used repeatedly to reduce the power n until it is 0 or 1.

Example 5 Find $\int \sin^6\theta\,d\theta$.

Solution Use IV-17 repeatedly:

$$\int \sin^6\theta\,d\theta = -\frac{1}{6}\sin^5\theta\cos\theta + \frac{5}{6}\int \sin^4\theta\,d\theta$$
$$\int \sin^4\theta\,d\theta = -\frac{1}{4}\sin^3\theta\cos\theta + \frac{3}{4}\int \sin^2\theta\,d\theta$$
$$\int \sin^2\theta\,d\theta = -\frac{1}{2}\sin\theta\cos\theta + \frac{1}{2}\int 1\,d\theta.$$

Calculate $\int \sin^2\theta\,d\theta$ first, and use this to find $\int \sin^4\theta\,d\theta$; then calculate $\int \sin^6\theta\,d\theta$. Putting this all together, we get

$$\int \sin^6\theta\,d\theta = -\frac{1}{6}\sin^5\theta\cos\theta - \frac{5}{24}\sin^3\theta\cos\theta - \frac{15}{48}\sin\theta\cos\theta + \frac{15}{48}\theta + C.$$

There are reduction formulas that work for negative powers of $\sin x$ and $\cos x$ in IV-19 and IV-21, respectively. Here is an example that uses IV-21:

Example 6 Find $\int \frac{1}{\cos^5 x}\,dx$.

Solution Use IV-21 twice to get the exponent down to 1:

$$\int \frac{1}{\cos^5 x}\,dx = \frac{1}{4}\frac{\sin x}{\cos^4 x} + \frac{3}{4}\int \frac{1}{\cos^3 x}\,dx$$
$$\int \frac{1}{\cos^3 x}\,dx = \frac{1}{2}\frac{\sin x}{\cos^2 x} + \frac{1}{2}\int \frac{1}{\cos x}\,dx.$$

Now use IV-22 to get

$$\int \frac{1}{\cos x}\,dx = \frac{1}{2}\ln\left|\frac{(\sin x)+1}{(\sin x)-1}\right| + C.$$

Putting this all together gives

$$\int \frac{1}{\cos^5 x}\,dx = \frac{1}{4}\frac{\sin x}{\cos^4 x} + \frac{3}{8}\frac{\sin x}{\cos^2 x} + \frac{3}{16}\ln\left|\frac{(\sin x)+1}{(\sin x)-1}\right| + C.$$

The last item in **Part IV** of the table is not a formula at all: it is, instead, advice on how to antidifferentiate products of integer powers of $\sin x$ and $\cos x$. There are various techniques to choose from, depending on the nature (odd or even, positive or negative) of the exponents.

Example 7 Find $\int \cos^3 t \sin^4 t\, dt$.

Solution Here the exponent of $\cos t$ is odd, so IV-23 recommends making the substitution $w = \sin t$. Then $dw = \cos t\, dt$. To make this work, we'll have to separate off one of the cosines to be part of dw. Also, the remaining even power of $\cos t$ can be rewritten in terms of $\sin t$ by using $\cos^2 t = 1 - \sin^2 t = 1 - w^2$, so that

$$\begin{aligned} \int \cos^3 t \sin^4 t\, dt &= \int \cos^2 t \sin^4 t \cos t\, dt \\ &= \int (1 - w^2) w^4\, dw = \int (w^4 - w^6)\, dw \\ &= \frac{1}{5} w^5 - \frac{1}{7} w^7 + C = \frac{1}{5} \sin^5 t - \frac{1}{7} \sin^7 t + C. \end{aligned}$$

Example 8 Find $\int \cos^2 x \sin^4 x\, dx$.

Solution In this example, both exponents are even. The advice given in IV-23 is to convert to all sines or all cosines. We'll convert to all sines by substituting $\cos^2 x = 1 - \sin^2 x$, and then we'll multiply out the integrand:

$$\int \cos^2 x \sin^4 x\, dx = \int (1 - \sin^2 x) \sin^4 x\, dx = \int \sin^4 x\, dx - \int \sin^6 x\, dx.$$

In Example 5 you found $\int \sin^4 x\, dx$ and $\int \sin^6 x\, dx$. Put them together to get:

$$\begin{aligned} \int \cos^2 x \sin^4 x\, dx &= -\frac{1}{4} \sin^3 x \cos x - \frac{3}{8} \sin x \cos x + \frac{3}{8} x \\ &\quad - \left(-\frac{1}{6} \sin^5 x \cos x - \frac{5}{24} \sin^3 x \cos x - \frac{15}{48} \sin x \cos x + \frac{15}{48} x \right) + C \\ &= \frac{1}{6} \sin^5 x \cos x - \frac{1}{24} \sin^3 x \cos x - \frac{3}{48} \sin x \cos x + \frac{3}{48} x + C. \end{aligned}$$

The last two parts of our table are concerned with quadratic functions: **Part V** has expressions with quadratic denominators; **Part VI** contains square roots of quadratics. The quadratics that appear in these formulas are of the form $x^2 \pm a^2$ or $a^2 - x^2$, or in factored form $(x - a)(x - b)$, where a and b are constants. Many integrands with quadratics in them are either of one of these forms or can be coerced into one of them by completing the square or factoring.

Using Factoring

Example 9 Find $\displaystyle\int \frac{3x+7}{x^2+6x+8}\,dx$.

Solution In this case the denominator factors

$$x^2+6x+8=(x+2)(x+4).$$

Now in V-27 we let $a=-2$, $b=-4$, $c=3$, and $d=7$, to obtain:

$$\int \frac{3x+7}{x^2+6x+8}\,dx=\frac{1}{2}(\ln|x+2|-(-5)\ln|x+4|)+C.$$

Completing the Square to Rewrite the Quadratic in the Form w^2+a^2

Example 10 Find $\displaystyle\int \frac{1}{x^2+6x+14}\,dx$.

Solution By completing the square, we get:

$$\begin{aligned} x^2+6x+14 &= (x^2+6x+9)-9+14 \\ &= (x+3)^2+5. \end{aligned}$$

Let $w=x+3$. Then $dw=dx$ and so the substitution gives

$$\int \frac{1}{x^2+6x+14}\,dx=\int \frac{1}{w^2+5}\,dw=\frac{1}{\sqrt{5}}\arctan\frac{w}{\sqrt{5}}+C=\frac{1}{\sqrt{5}}\arctan\frac{x+3}{\sqrt{5}}+C$$

where the antidifferentiation uses V-24 with $a^2=5$.

Partial Fractions

The formulas in V-26 and V-27 are obtained by splitting the integrand into *partial fractions*. For example, to find

$$\int \frac{1}{(x-2)(x-5)}\,dx$$

the integrand is split into partial fractions with denominators $(x-2)$ and $(x-5)$. We write

$$\frac{1}{(x-2)(x-5)}=\frac{A}{x-2}+\frac{B}{x-5}.$$

Multiplying by $(x-2)(x-5)$ gives

$$1=A(x-5)+B(x-2)$$

so,

$$1=(A+B)x-5A-2B.$$

The only way that the quantity $(A+B)x-5A-2B$ can be equal to 1 for all x is if

$$\begin{aligned} A+B&=0 \\ -5A-2B&=1. \end{aligned}$$

Solving these equations gives $A = -1/3$, $B = 1/3$. Now we can rewrite the integral in such a way that it can be integrated:

$$\int \frac{1}{(x-2)(x-5)}\,dx = \int \left(\frac{-1/3}{x-2} + \frac{1/3}{x-5}\right)dx = -\frac{1}{3}\ln|x-2| + \frac{1}{3}\ln|x-5| + C.$$

You can check that using V-26 gives the same result.

Example 11 Find $\displaystyle\int \frac{x+2}{x^2+x}\,dx$.

Solution We factor the denominator and split the integrand into partial fractions:

$$\frac{x+2}{x^2+x} = \frac{x+2}{x(x+1)} = \frac{A}{x} + \frac{B}{x+1}.$$

Multiplying by $x(x+1)$ gives

$$\begin{aligned} x+2 &= A(x+1) + Bx \\ &= (A+B)x + A. \end{aligned}$$

Thus we must have $A = 2$ and $A + B = 1$, so $B = -1$. Then

$$\int \frac{x+2}{x^2+x}\,dx = \int \left(\frac{2}{x} - \frac{1}{x+1}\right)dx = 2\ln|x| - \ln|x+1| + C.$$

Preparing to Use the Table: Transforming the Integrand

Some integration problems will arise in forms that look nearly the same as the formulas in our table, but many will not. In order to use the table, you'll often need to manipulate or reshape integrands to fit entries in the table. The kinds of manipulation that tend to be useful are expansion, factoring, long division, completing the square, and substitution.

Example 12 Find $\displaystyle\int (x^3+5)^2\,dx$.

Solution Note that you can't use substitution here: letting $w = x^3 + 5$ doesn't work, since there is no $dw = 3x^2\,dx$ in the integrand. What will work is simply multiplying out the square: $(x^3+5)^2 = x^6 + 10x^3 + 25$. Then use I-1:

$$\int (x^3+5)^2\,dx = \int x^6\,dx + 10\int x^3\,dx + 25\int 1\,dx = \frac{1}{7}x^7 + 10\cdot\frac{1}{4}x^4 + 25x + C.$$

Example 13 Find $\displaystyle\int \frac{2x^2-3x+7}{2x+5}\,dx$.

Solution You might be tempted to try to factor the numerator or to complete the square. But a good rule of thumb when integrating a rational function whose numerator has a degree greater than or equal to that of the denominator is to do *long division*. This results in a polynomial plus a simpler rational function as a remainder.

If we perform long division, we get

$$\frac{2x^2-3x+7}{2x+5} = x - 4 + \frac{27}{2x+5}.$$

Then use I-1 and I-2:

$$\begin{aligned}
\int \frac{2x^2 - 3x + 7}{2x + 5}\, dx &= \int \left(x - 4 + \frac{27}{2x+5} \right) dx \\
&= \frac{x^2}{2} - 4x + 27 \int \frac{dx}{2x+5} \\
&= \frac{x^2}{2} - 4x + 27 \int \frac{1}{w} \left(\frac{1}{2} dw \right) \qquad \text{(where } w = 2x + 5\text{)} \\
&= \frac{x^2}{2} - 4x + \frac{27}{2} \ln |w| + C \\
&= \frac{x^2}{2} - 4x + \frac{27}{2} \ln |2x + 5| + C.
\end{aligned}$$

Example 14 Find $\int \frac{x^2}{x^2+4}\, dx$.

Solution Since the degree of the numerator and denominator are equal, start with long division:

$$\frac{x^2}{x^2+4} = 1 - \frac{4}{x^2+4}.$$

Then by V-24 with $a = 2$ we get:

$$\int \frac{x^2}{x^2+4}\, dx = \int 1\, dx - 4 \int \frac{1}{x^2+4}\, dx = x - 4 \cdot \frac{1}{2} \arctan \frac{x}{2} + C.$$

Example 15 Find $\int e^t \sin(5t+7)\, dt$.

Solution This looks pretty similar to II-8. To make the correspondence more complete, let's try the substitution $w = 5t + 7$. Then $dw = 5\, dt$, so $dt = \frac{1}{5}\, dw$. Also, $t = (w - 7)/5$. Then the integral becomes

$$\begin{aligned}
\int e^t \sin(5t+7)\, dt &= \int e^{(w-7)/5} \sin w \, \frac{dw}{5} \\
&= \frac{e^{-7/5}}{5} \int e^{w/5} \sin w \, dw \qquad \text{(Since } e^{(w-7)/5} = e^{w/5}e^{-7/5} \text{ and } e^{-7/5} \text{ is a constant)}
\end{aligned}$$

And now we can use II-8 with $a = \frac{1}{5}$ and $b = 1$ to write

$$\int e^{w/5} \sin w \, dw = \frac{1}{(\frac{1}{5})^2 + 1^2} e^{w/5} \left(\frac{\sin w}{5} - \cos w \right) + C,$$

so

$$\begin{aligned}
\int e^t \sin(5t+7)\, dt &= \frac{e^{-7/5}}{5} \left(\frac{25}{26} e^{(5t+7)/5} \left(\frac{\sin(5t+7)}{5} - \cos(5t+7) \right) \right) + C \\
&= \frac{5e^t}{26} \left(\frac{\sin(5t+7)}{5} - \cos(5t+7) \right) + C.
\end{aligned}$$

Problems for Section 7.5

For Problems 1–4, write down which formula in the table of integrals applies and what substitution you would use (if one is necessary). Do not do the integration.

1. The integrals in Problems 5–13.
2. The integrals in Problems 14–25.
3. The integrals in Problems 26–38.
4. The integrals in Problems 39–50.

For Problems 5–50, antidifferentiate, using the table of integrals and a substitution if necessary.

5. $\int x^3 e^{2x}\,dx$
6. $\int \sin 3\theta \cos 5\theta\,d\theta$
7. $\int \sin 3\theta \sin 5\theta\,d\theta$
8. $\int e^{-3\theta} \cos\theta\,d\theta$
9. $\int x^5 \ln x\,dx$
10. $\int \sin w \cos^4 w\,dw$
11. $\int \frac{1}{\cos^3 x}\,dx$
12. $\int \sin^4 x\,dx$
13. $\int \frac{1}{3+y^2}\,dy$
14. $\int x^2 e^{3x}\,dx$
15. $\int x^4 e^{3x}\,dx$
16. $\int x^2 e^{x^3}\,dx$
17. $\int y^2 \sin 2y\,dy$
18. $\int \cos 2y \cos 7y\,dy$
19. $\int e^{5x} \sin 3x\,dx$
20. $\int x^3 \sin x^2\,dx$
21. $\int z e^{2z^2} \cos(2z^2)\,dz$
22. $\int u^5 \ln(5u)\,du$
23. $\int y^7 \ln(y^3)\,dy$
24. $\int \cos^4 3y\,dy$
25. $\int z \sin^3(z^2)\,dz$
26. $\int \frac{1}{\sqrt{2-x^2}}\,dx$
27. $\int \frac{1}{\sqrt{1-9x^2}}\,dx$
28. $\int \frac{1}{1+4x^2}\,dx$
29. $\int \frac{1}{\sin^2 2\theta}\,d\theta$
30. $\int \frac{1}{\sin^3 3\theta}\,d\theta$
31. $\int \frac{1}{\cos^4 7x}\,dx$
32. $\int \frac{1}{\sqrt{9-4x^2}}\,dx$
33. $\int \frac{x}{\sqrt{9-4x^2}}\,dx$
34. $\int \cos^2\theta \sin^2\theta\,d\theta$
35. $\int \sin^3 3\theta \cos^2 3\theta\,d\theta$
36. $\int \tan^4 x\,dx$
37. $\int \frac{dz}{z(z-3)}$
38. $\int \frac{dy}{4-y^2}$
39. $\int \frac{1}{1+(z+2)^2}\,dz$
40. $\int \frac{dy}{\sqrt{4+(1+y)^2}}$
41. $\int \frac{1}{\sqrt{2-(x+1)^2}}\,dx$
42. $\int \frac{1}{x^2+4x+3}\,dx$
43. $\int \frac{1}{x^2+4x+4}\,dx$
44. $\int \frac{1}{y^2+4y+5}\,dy$
45. $\int \frac{x^3+3}{x^2-3x+2}\,dx$
46. $\int \frac{x^2}{x^2+6x+13}\,dx$
47. $\int \sqrt{x^2+8x+7}\,dx$
48. $\int \frac{4y+9}{3y+y^2}\,dy$
49. $\int \frac{5z-13}{6-5z+z^2}\,dz$
50. $\int \frac{t^2+1}{t^2-1}\,dt$

In Problems 51–63, find the definite integrals by using antiderivatives and the Fundamental Theorem of Calculus. Check your answers numerically, using left- and right-hand sums.

51. $\int_0^2 \frac{1}{4+x^2}\,dx$
52. $\int_\pi^{2\pi} (y^2+3)\cos 3y\,dy$
53. $\int_0^1 \frac{dx}{x^2+2x+5}$

54. $\displaystyle\int_{\pi/4}^{\pi/3} \frac{dx}{\sin^3 x}$

55. $\displaystyle\int_{-3}^{-1} \frac{dx}{\sqrt{x^2+6x+10}}$

56. $\displaystyle\int_0^3 \frac{1+x}{2+x^2}\,dx$

57. $\displaystyle\int_0^2 \cos\theta \sin 2\theta \, d\theta$

58. $\displaystyle\int_{-\pi}^{\pi} \sin 5x \cos 6x \, dx$

59. $\displaystyle\int_{-\pi}^{\pi} \sin 5x \cos 5x \, dx$

60. $\displaystyle\int_1^2 (x-2x^3)\ln x \, dx$

61. $\displaystyle\int_0^1 \sqrt{3-x^2}\,dx$

62. $\displaystyle\int_0^1 \sqrt{4-x^2}\,dx$

63. $\displaystyle\int_0^2 \sqrt{4-x^2}\,dx$. [Hint: You can evaluate this by thinking about its geometric meaning.]

64. The voltage, V, in an electrical outlet is given as a function of time, t, by the function $V = V_0 \cos(120\pi t)$, where V is in volts and t in seconds, and V_0 is a constant representing the maximum voltage.
    (a) What is the average value of the voltage over 1 second?
    (b) Engineers do not use the average voltage. They use the *root mean square* voltage defined by $\overline{V} = \sqrt{\text{average of } (V^2)}$. Find $\overline{V}$ in terms of V_0.
    (c) The standard voltage in an American house is 110 volts, meaning that $\overline{V} = 110$. What is V_0?

65. An economist studying the rate of production, $R(t)$, of oil in a new oil well has proposed the following model:

$$R(t) = A + Be^{-t}\sin(2\pi t)$$

where t is the time in years, A is the average rate (a constant), and B is the "variability" coefficient (a constant).
    (a) Find the total amount of oil produced in the first N years of operation. (Take N to be an integer.)
    (b) Find the average amount of oil produced per year over the first N years (N an integer).
    (c) From your answer to part (b), find the average amount of oil produced per year as $N \to \infty$.
    (d) Looking at the function $R(t)$, explain how you might have predicted your answer to part (c) without doing any calculations.
    (e) Do you think it is reasonable to expect this model to hold over a very long period? Why or why not?

66. (a) Show how you can use the fact that $\dfrac{1}{x^2-x} = \dfrac{1}{x-1} - \dfrac{1}{x}$ to find $\displaystyle\int \frac{1}{x^2-x}\,dx$.
    (b) Show that your answer to part (a) agrees with the answer you get by using the integral tables.

67. (a) Show how combining terms in the expression $\dfrac{2}{x} + \dfrac{1}{x+3}$ yields $\dfrac{3x+6}{x^2+3x}$, and use this to evaluate $\displaystyle\int \frac{3x+6}{x^2+3x}\,dx$.
    (b) Show that your answer to part (a) agrees with the answer you get by using the integral tables.

68. Use partial fractions to find $\displaystyle\int \frac{1}{x(L-x)}\,dx$, where L is constant.

69. (a) Differentiate the following function

$$a\ln x + b\ln(1+x) + \frac{c}{1+x},$$

where a, b, and c are constants.

(b) Collect the terms in part (a) over a common denominator, then write the numerator as a polynomial in x, and find

$$\int_1^2 \frac{1+x^2}{x(1+x)^2}\,dx$$

70. Show that for all integers m and n, with $m \neq \pm n$, $\displaystyle\int_{-\pi}^{\pi} \sin m\theta \sin n\theta \, d\theta = 0.$

71. Show that for all integers m and n, with $m \neq \pm n$, $\displaystyle\int_{-\pi}^{\pi} \cos m\theta \cos n\theta \, d\theta = 0.$

7.6 APPROXIMATING DEFINITE INTEGRALS

The methods of the last few sections allow us to get exact answers for definite integrals in a variety of special cases. In fact, most real-world applications of calculus don't require exact answers. If the problem is to predict the total amount of fuel used by the space shuttle or the length of a steel cable supporting a bridge, we may require only a certain number of decimal places of accuracy. Even when we can use the Fundamental Theorem to find an exact answer, the precision may be meaningless if the integrand is modeled on data that was measured in an inexact way.

We can approximate a definite integral numerically using left- and right-hand Riemann sums. The purpose of the next two sections is to introduce better methods for approximating definite integrals—better in the sense that they give more accurate results with less work than that required to find the left- and right-hand sums. The methods work so well that in most practical situations definite integrals are evaluated numerically instead of by antidifferentiation and the Fundamental Theorem.

The General Riemann Sum

Remember, a left-hand Riemann sum of $f(x)$ over an interval $a \leq x \leq b$ with n subdivisions is a sum of terms like $f(z_i)\Delta x$, where z_i is chosen to be the left-hand point in each subinterval. The right-hand Riemann sum is the same except that we choose z_i to be the right-hand point. There is nothing special about the left or right endpoints of the subintervals—we can get a Riemann sum that approximates the definite integral by choosing *any* point z_i in each subinterval. This is called a *general Riemann sum* and written:

$$\sum_{i=1}^{n} f(z_i)\Delta x = f(z_1)\Delta x + \cdots + f(z_n)\Delta x,$$

where, as before, Δx is the length of an individual subinterval,[2] so that $\Delta x = (b-a)/n$. Then z_1 is the point you chose from the first subinterval, z_2 is the point you chose from the second, and so on. The reason for considering Riemann sums other than left- and right-hand sums is that other choices of points within the subinterval can yield better approximation methods.

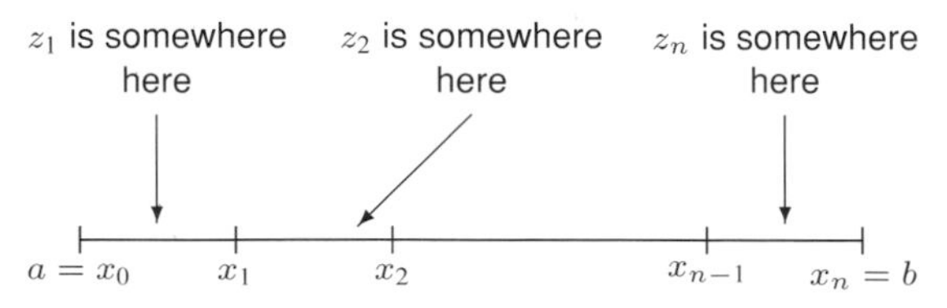

[2]It is also possible to have unequal subdivisions, although we will not consider them.

The Midpoint Rule

An important example of this more general Riemann sum is when z_i is chosen to be the *midpoint* of each interval. For example, in approximating $\int_1^2 f(x)\,dx$ by a Riemann sum with two subdivisions, you could divide the interval $1 \le x \le 2$ into two pieces and then take $z_1 = 1.25$, the midpoint of the first piece, and $z_2 = 1.75$, the midpoint of the second piece (see Figure 7.2).

The Riemann sum is then

$$f(1.25)0.5 + f(1.75)0.5.$$

Figure 7.2 shows that evaluating f at the midpoint of each subdivision usually gives a better approximation to the area under the curve than evaluating f at either end. For this particular f, for instance, you can see that each rectangle both overshoots and undercuts the exact area, and the amounts of overshoot and undercut are nearly equal. Thus, the area of the rectangles in Figure 7.2 will be closer to the area under the curve than the left- or right-hand sums. In fact, we shall see that this new midpoint Riemann sum *is* generally a better approximation to the definite integral than the left- or right-hand sum with the same number of subdivisions, n.

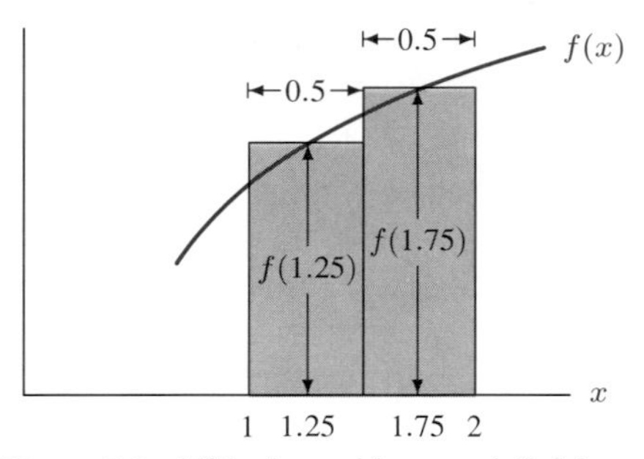

Figure 7.2: Midpoints with two subdivisions

> So far we have three rules for choosing the points $z_1, \ldots, z_n$ in a Riemann sum:
>
> 1. Use the left endpoint of each subinterval.
> 2. Use the right endpoint of each subinterval.
> 3. Use the midpoint of each subinterval.
>
> These three methods are called, not surprisingly, the *left (rectangle) rule*, the *right (rectangle) rule*, and the *midpoint (rectangle) rule*. We'll use LEFT(n), RIGHT(n), and MID(n) to denote the results obtained by using these rules with n subdivisions.

Example 1 For $\displaystyle\int_1^2 \frac{1}{x}\,dx$, compute LEFT(2), RIGHT(2) and MID(2), and compare your answers with the true value of the integral.

Solution

$$\text{LEFT}(2) = f(1)(0.5) + f(1.5)(0.5) = \frac{1}{1}(0.5) + \frac{1}{1.5}(0.5) = 0.8333\ldots$$

$$\text{RIGHT}(2) = f(1.5)(0.5) + f(2)(0.5) = \frac{1}{1.5}(0.5) + \frac{1}{2}(0.5) = 0.5833\ldots$$

$$\text{MID}(2) = f(1.25)(0.5) + f(1.75)(0.5) = \frac{1}{1.25}(0.5) + \frac{1}{1.75}(0.5) = 0.6857\ldots$$

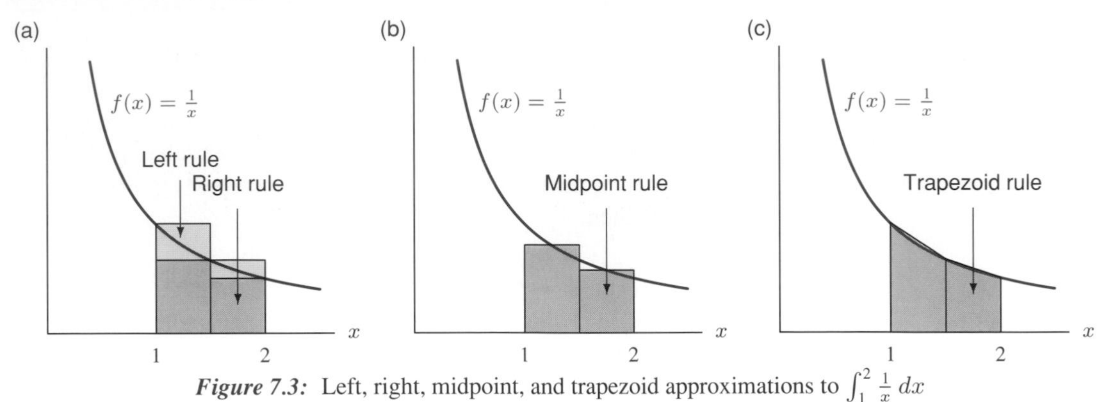

Figure 7.3: Left, right, midpoint, and trapezoid approximations to $\int_1^2 \frac{1}{x}\,dx$

All three Riemann sums in this example are approximating

$$\int_1^2 \frac{1}{x}\,dx = \ln x\Big|_1^2 = \ln 2 - \ln 1 = \ln 2 = 0.6931\ldots.$$

With only two subdivisions, the left and right rules give quite poor approximations but the midpoint rule is already fairly close to the true answer.

It is easy to see why the left and right rules are so bad; since $f(x) = 1/x$ is decreasing from 1 to 2, the left rule overestimates on each subdivision, and the right rule underestimates. (See Figure 7.3(a).) On the other hand, the midpoint rule approximates with a rectangle on each subdivision that is partly above and partly below the graph, so the errors tend to balance out. (See Figure 7.3(b).)

The Trapezoid Rule

We have just seen how the midpoint rule can have the effect of balancing out the errors of the left and right rules. There is another way of balancing these errors: why not average the results from the left and right rules? This approximation is called the *trapezoid rule*:

$$\text{TRAP}(n) = \frac{\text{LEFT}(n) + \text{RIGHT}(n)}{2}.$$

The trapezoid rule averages the values of f at the left and right endpoints of each subinterval and multiplies by Δx. This is the same as approximating the area under the graph of f in each subinterval by a trapezoid (see Figure 7.4).

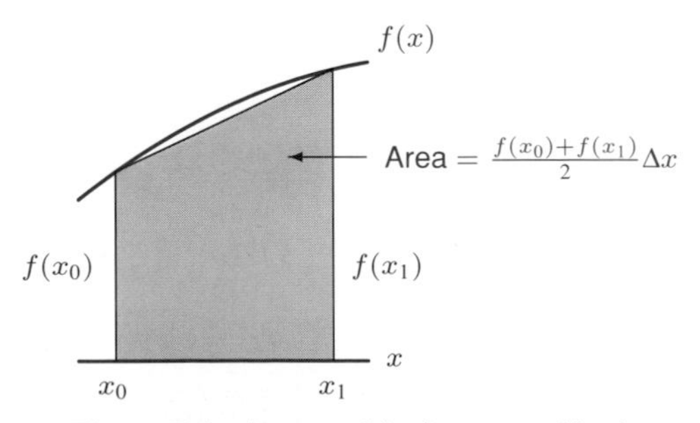

Figure 7.4: Area used in the trapezoid rule

Example 2 For $\int_1^2 \frac{1}{x}\,dx$, compare the trapezoid rule with two subdivisions with the left, right, and midpoint rules.

Solution In the previous example we got LEFT$(2) = 0.8333\ldots$ and RIGHT$(2) = 0.5833\ldots$. The trapezoid rule is the average of these, so TRAP$(2) = 0.7083\ldots$. (See Figure 7.3(c).) The exact value of the integral is $0.6931\ldots$, so the trapezoid rule is better than the left or right rules. The midpoint rule is still the best, however, since MID$(2) = 0.6857\ldots$.

Is the Approximation an Over- or Underestimate?

It is useful to know when a rule is producing an overestimate and when it is producing an underestimate. In Chapter 3 we saw that

When f is increasing,

$$\text{LEFT}(n) \le \int_a^b f(x)\,dx \le \text{RIGHT}(n).$$

When f is decreasing,

$$\text{RIGHT}(n) \le \int_a^b f(x)\,dx \le \text{LEFT}(n).$$

The Trapezoid Rule

If the graph of the function is concave down, then each trapezoid will lie below the graph, so the trapezoid rule will provide an underestimate. If the graph is concave up, the trapezoid rule will provide an overestimate. (See Figure 7.5.)

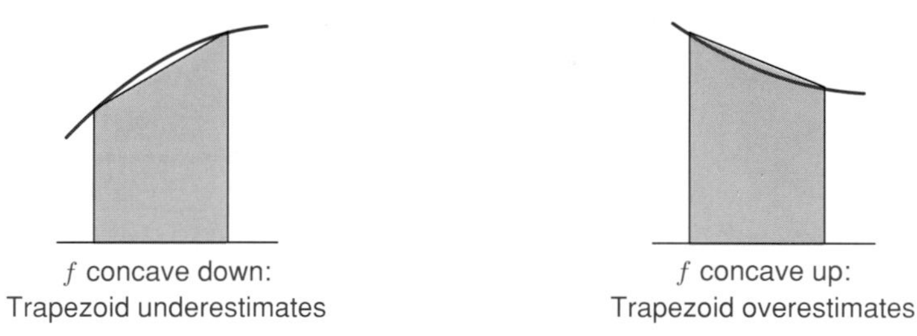

Figure 7.5: Error in the trapezoid rule

The Midpoint Rule

To understand the relationship between the midpoint rule and concavity, take a rectangle whose top intersects the curve at the midpoint of the interval. In addition, draw a tangent to the curve at the midpoint; this gives a trapezoid. See Figure 7.6. (This is *not* the same trapezoid as in the trapezoid rule; this one does not touch the curve at the endpoints of the interval.) The midpoint rectangle and the new trapezoid have the same area, because the shaded triangles in Figure 7.6 are congruent and have the same area. Hence, if the graph of the function is concave down, the midpoint rule overestimates; if the graph is concave up, the midpoint rule underestimates. (See Figure 7.7.)

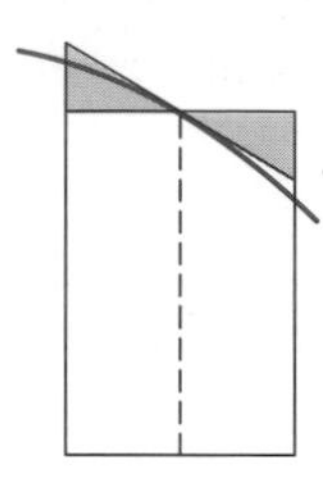

Figure 7.6: Midpoint rectangle and trapezoid with same area

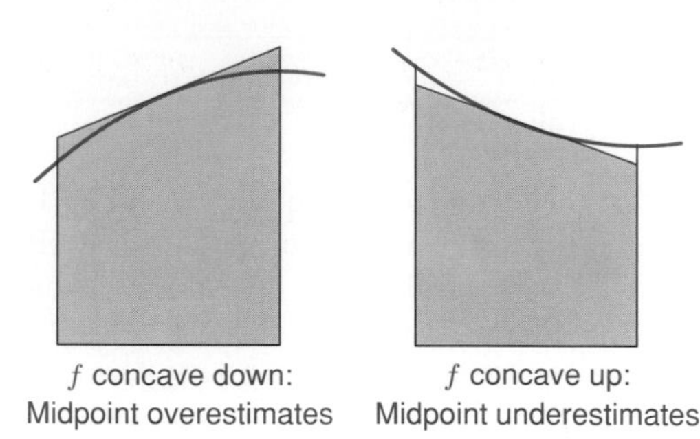

Figure 7.7: Error in the midpoint rule

If the graph of f is concave down,

$$\text{TRAP}(n) \leq \int_a^b f(x)\,dx \leq \text{MID}(n).$$

If the graph of f is concave up,

$$\text{MID}(n) \leq \int_a^b f(x)\,dx \leq \text{TRAP}(n).$$

How Do We Know If n Is Large Enough?

Any of these rules can be made more accurate by taking n larger.[3] We usually start by taking larger and larger values of n until the decimals in the answer start to stabilize, or remain the same, as n increases. It is then likely (though there's no guarantee) that we're close to the true value. For example, since the last two values in Table 7.1 agree to three decimal places, it is probably safe to conclude that $\int_0^{2.5} \sin(t^2)\,dt = 0.43$ to two decimal places.

TABLE 7.1 *Trapezoid and midpoint rules for* $\int_0^{2.5} \sin(t^2)\,dt$

n	TRAP(n)	MID(n)
2	1.2292	0.0192
10	0.4572	0.4169
50	0.4316	0.4300
250	0.4306	0.4305

If f is monotonic or doesn't change concavity, we can be certain how close our estimates are to the true value. If f is monotonic over the range of integration, we know that the true value of the integral is trapped between LEFT(n) and RIGHT(n), so we can just increase n until LEFT(n) and RIGHT(n) are sufficiently close together. Similarly, if f doesn't change concavity, the true value of the integral will be trapped between MID(n) and TRAP(n), and we can just increase n until the difference between them is very small.

[3] In practice, increasing n increases the accuracy of the approximation until n becomes so large that round-off error comes into play.

For a function that is not monotonic or does change concavity, we can often split up the interval of integration in such a way that the behavior does not change on each subinterval, as in the next example.

Example 3 Use the midpoint and trapezoid rules to find upper and lower estimates for the integral

$$\int_0^1 e^{-z^2}\,dz.$$

Solution The graph of $f(z) = e^{-z^2}$ in Figure 7.8 appears to change concavity between 0 and 1. Differentiating f shows that

$$f'(z) = -2ze^{-z^2} \quad \text{and} \quad f''(z) = 2(2z^2 - 1)e^{-z^2}$$

Thus the points of inflection are at $z = \pm 1/\sqrt{2}$. Since the concavity changes between 0 and 1, we do not know whether the midpoint and trapezoid rules give over- or underestimates for $\int_0^1 e^{-z^2}\,dz$.

If, however, we split the interval at $z = 1/\sqrt{2}$, we can be sure the midpoint rule overestimates the integral on $[0, 1/\sqrt{2}]$ (where the integrand is concave down) and the trapezoid rule overestimates on $[1/\sqrt{2}, 1]$ (where the integrand is concave up), and vice versa for the underestimate. Using $n = 20$ for $\int_0^{1/\sqrt{2}} e^{-z^2}\,dz$ gives TRAP(20) $= 0.604928$ and MID(20) $= 0.605062$, so:

$$0.604928 < \int_0^{1/\sqrt{2}} e^{-z^2}\,dz < 0.605062.$$

Similarly for $\displaystyle\int_{1/\sqrt{2}}^1 e^{-z^2}\,dz$, we have TRAP(20) $= 0.141809$ and MID(20) $= 0.141805$:

$$0.141805 < \int_{1/\sqrt{2}}^1 e^{-z^2}\,dz < 0.141809.$$

Putting these two results together

$$0.604928 + 0.141805 < \int_0^{1/\sqrt{2}} e^{-z^2}\,dz + \int_{1/\sqrt{2}}^1 e^{-z^2}\,dz < 0.605062 + 0.141809$$

giving

$$0.746733 < \int_0^1 e^{-z^2}\,dz < 0.746871.$$

Thus we can say that $\int_0^1 e^{-z^2}\,dz = 0.747$ to three decimal places. If more accuracy is needed, we simply increase the value of n and repeat the calculation.

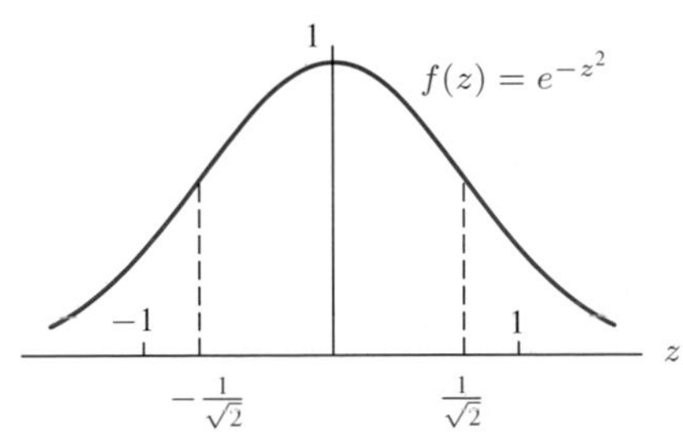

Figure 7.8

Problems for Section 7.6

1. The following table contains approximations to $\int_0^4 \sqrt{100+x^3}\,dx$. Fill in the rest of the table with values rounded to four decimal places.

	$n=1$	$n=2$	$n=4$
LEFT			
RIGHT	51.2250		
TRAP		43.5909	
MID			

Estimate the integrals in Problems 2–7 with $N = 10$, 100, and 1000 using left sums, right sums, trapezoids, and midpoint sums. State whether each sum is an over- or underestimate.

2. $\int_0^{1/2} (\sin\theta)^{3/2}\,d\theta$
3. $\int_1^2 \sqrt{x}e^x\,dx$
4. $\int_0^{\pi/2} \sqrt{3+\cos\theta}\,d\theta$
5. $\int_0^1 \sin\frac{\theta^2}{2}\,d\theta$
6. $\int_0^{\pi/2} e^{-\sin x}\,dx$
7. $\int_0^3 \sqrt{1+x^3}\,dx$

8. (a) Estimate $\int_0^1 \frac{1}{1+x^2}\,dx$ by subdividing the interval into eight parts using:
   (i) The left-hand Riemann sum. (ii) The right-hand Riemann sum.
   (iii) The trapezoidal rule.
   (b) Since the exact value of the integral is $\pi/4$, you can estimate the value of π. Explain why your first estimate is too large and your second estimate too small.

9. Consider the following integrals:

   (i) $\int_1^{10} \ln x\,dx$ (ii) $\int_0^4 e^x\,dx$

   (a) For each integral, find LEFT(32), RIGHT(32), and TRAP(32). Also find the exact value of each integral.
   (b) For each integral, arrange LEFT(32), RIGHT(32), TRAP(32), and the true value in ascending order. Explain, using diagrams, how you could predict this ordering without doing the calculations in part (a).

10. The graph of g is shown in Figure 7.9. The results from the left, right, trapezoid, and midpoint rules used to approximate $\int_0^1 g(t)\,dt$, with the same number of subdivisions for each rule, are as follows: 0.601, 0.632, 0.633, 0.664.
    (a) Match each rule with its approximation.
    (b) Between which two approximations does the true value of the integral lie?

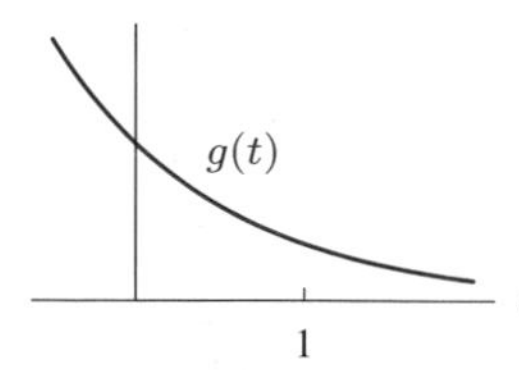

Figure 7.9

11. Given the velocity-time data in the table, estimate the total distance traveled from time $t = 0$ to time $t = 6$ using LEFT, RIGHT, and TRAP.

t	0	1	2	3	4	5	6
v	3	4	5	4	7	8	11

12. (a) Show that $\displaystyle\int_0^1 \sqrt{2-x^2}\,dx = \frac{\pi}{4}+\frac{1}{2}$. [Hint: Break up the area under $y=\sqrt{2-x^2}$ from $x=0$ to $x=1$ into two pieces: a sector of a circle and a right triangle.]
    (b) Approximate $\int_0^1 \sqrt{2-x^2}\,dx$ for $n=5$ using the left, right, trapezoid, and midpoint rules. Compute the error in each case using the answer to (a), and compare the errors.

13. (a) Find the exact value of $\displaystyle\int_0^{2\pi} \sin\theta\,d\theta$.
    (b) Explain, using pictures, why the MID(1) and MID(2) approximations to this integral give the exact value.
    (c) Does MID(3) give the exact value of this integral? How about MID(n)? Explain.

14. (a) Explain why the area of the trapezoid in Figure 7.10 is $h\cdot(l_1+l_2)/2$.
    (b) On a graph similar to Figure 7.11, sketch areas representing each of the following quantities:

$$E = h\cdot f(0), F = h\cdot f(h), R = h\cdot f\left(\frac{h}{2}\right)$$

$$C = h\cdot\frac{f(0)+f(h)}{2} = \frac{E+F}{2}$$

$$N = \frac{h}{2}\cdot\frac{f(0)+f(\frac{h}{2})}{2} + \frac{h}{2}\cdot\frac{f(\frac{h}{2})+f(h)}{2} = \frac{R+C}{2}.$$

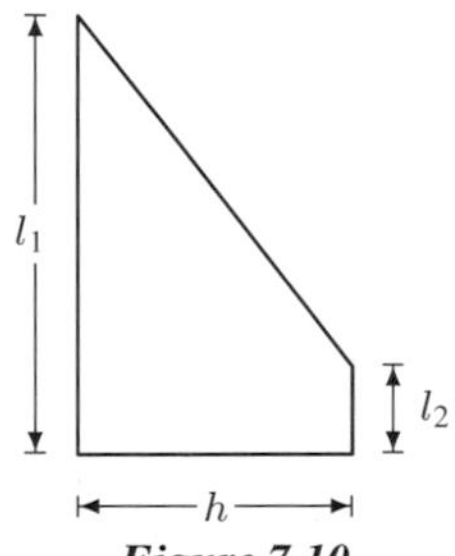

Figure 7.10

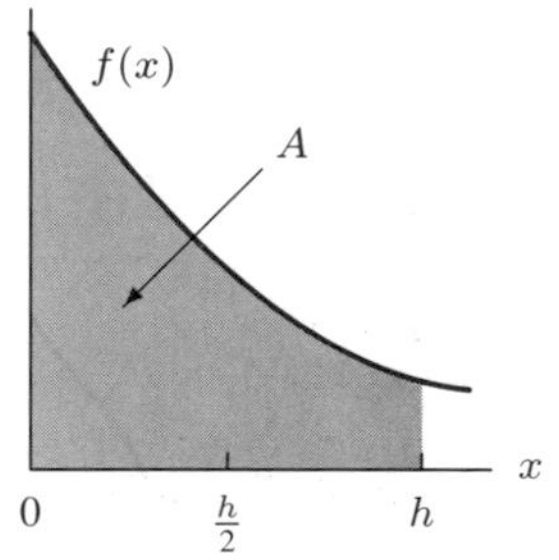

Figure 7.11

    (c) Let A be the area under the function shown in Figure 7.11. Write the values A, E, F, R, C, and N in increasing order.
    (d) Which is the better approximation to A: E or F?
    (e) Which is the better approximation to A: R or C?

Problems 15–19 involve approximating $\int_a^b f(x)\,dx$.

15. Show that RIGHT(n) $=$ LEFT(n) $+ f(b)\Delta x - f(a)\Delta x$.

16. Show that TRAP(n) $=$ LEFT(n) $+ \frac{1}{2}\left(f(b)-f(a)\right)\Delta x$.

17. Show that LEFT($2n$) $= \frac{1}{2}\left(\text{LEFT}(n)+\text{MID}(n)\right)$.

18. Using a computer or calculator, verify that the equations given in Problems 15 and 16 hold for $\int_1^2 (1/x)\,dx$, when $n=10$.

19. Suppose that $a=2$, $b=5$, $f(2)=13$, $f(5)=21$ and that LEFT(10) $=3.156$ and MID(10) $=3.242$. Use the equations given in Problems 15–17 to compute RIGHT(10), TRAP(10), LEFT(20), RIGHT(20), and TRAP(20).

20. The width, in feet, at various points along the fairway of a hole on a golf course is given in Figure 7.12. If one pound of fertilizer covers 200 square feet, estimate the amount of fertilizer needed to fertilize the fairway.

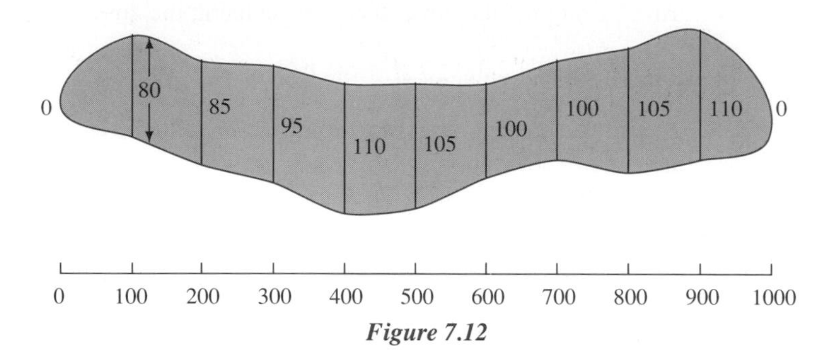

Figure 7.12

21. Estimate

$$\sum_{k=1}^{100,000} \frac{1}{k}$$

to the closest integer. [Hint: Work with left- or right-hand sums of the function $f(x) = 1/x$.]

7.7 APPROXIMATION ERRORS AND SIMPSON'S RULE

When we compute an approximation, we are always concerned about the error, namely the difference between the exact answer and the approximation. We never know the exact error; if we did, we would also know the exact answer! What we want is some bound on the error and some idea of how much work would be involved in making the error smaller. The way to do this with definite integrals is quite practical: Do the computation once for some moderate size of n, then do the computation again, increasing n by a factor of 2 or 10 and see how many decimal places agree in the two answers. Continue increasing n until the desired digits have stopped changing; we say they have stabilized. What is most interesting is that the errors are not random but follow a definite pattern as n increases. The errors for some methods are much smaller than those for others, and the errors for the midpoint and trapezoid methods are related to each other in a way that suggests an even better method, called Simpson's rule. We will work with the example $\int_1^2 \frac{1}{x}\,dx$, because we know the value of this integral to about 10 decimal places just by punching $\ln 2$ on a calculator. By studying the error in a case where we do know the exact answer, we will learn how the error behaves for other definite integrals where we may not know the answer exactly.

Error in Left and Right Rules

Let us see what happens to the error in the left and right rules as we increase n. We will increase n each time by a factor of 5 starting at $n = 2$. Since we already know what $\ln 2$ is (at least to 10 digits), we can compute the error in approximating $\ln 2$ by the left- and right-hand Riemann sums for each n. The results are in Table 7.2. A negative error indicates the Riemann sum is less than $\ln 2$.

TABLE 7.2 *Errors for the left and right rule approximation to* $\int_1^2 \frac{1}{x}\,dx = \ln 2 \approx 0.6931471806$

n	Error	
	Left rule	Right rule
2	0.1402	−0.1098
10	0.0256	−0.0244
50	0.0050	−0.0050
250	0.0010	−0.0010

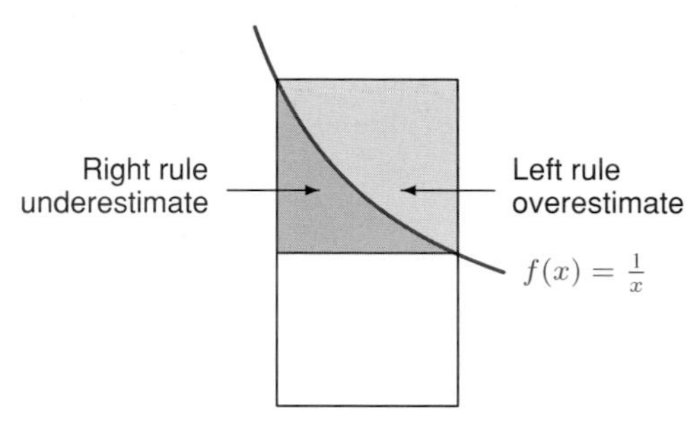

Figure 7.13: Errors in left and right sums

One thing stands out: The errors for the left and right rules have opposite signs but are approximately equal in magnitude. This should come as no surprise. Since $f(x) = 1/x$ decreases as we move from left to right in the interval $[1, 2]$, the left rule will always overestimate and the right rule will always underestimate. Moreover, since the graph of f is almost a straight line in each subinterval, the left rule overestimates by about the same amount (geometrically, a "triangle" above the curve) as the right rule underestimates (a "triangle" under the curve). See Figure 7.13. If we had not already decided to use the trapezoid rule, we might have been led to invent it by the observation that the errors in the left and right rule are approximately equal in magnitude and opposite in sign. The best way to try to get the errors to cancel is to average the left and right rule; this average is the trapezoidal rule.

There is another pattern to the errors in Table 7.2. If we compute the *ratio* of the errors for $n = 2$ versus $n = 10$, for $n = 10$ versus $n = 50$, and for $n = 50$ versus $n = 250$, as in Table 7.3,[4] we see that the error in both the left and right rules decreases by a factor of about 5 as n increases by a factor of 5.

There is nothing special about the number 5; the same holds for any factor. In particular, to get one extra digit of accuracy in any calculation, we must make the error $1/10$ as big, so we must increase n by a factor of 10. Thus, *for the left or right rules, each extra digit of accuracy requires about 10 times the work*. The calculator used to produce these tables took about 4 seconds to compute the left rule approximation for $n = 50$, and this yields ln 2 to two digits. To get three correct digits, n would need to be around 500 and the time would be 40 seconds. Four digits requires $n = 5000$ and 400 seconds. Ten digits requires $n = 5 \times 10^9$ and 4×10^8 seconds, which is more than 12 years! The fastest computers would still take 4 seconds for 10 digits, 40 seconds for 11 digits, 400 seconds for 12 digits, and 4×10^{10} seconds (or about 1200 years) for 20 digits. Clearly, the error for the left and right rules does not decrease fast enough as n increases for practical computations accurate to more than a few digits.

TABLE 7.3 *The ratio of the errors as* n *increases for* $\int_1^2 \frac{1}{x}\,dx$

	Ratio of Errors	
	Left rule	Right rule
2 vs. 10	5.47	4.51
10 vs. 50	5.10	4.90
50 vs. 250	5.02	4.98

[4]The values in Table 7.2 are rounded to 4 decimal places; those in Table 7.3 were computed using more decimal places and then rounded.

Error in Trapezoid and Midpoint Rules

Table 7.4 shows that the trapezoid and midpoint rules produce much better approximations to $\int_1^2 \frac{1}{x}\,dx$ than the left and right rules.

TABLE 7.4 *The errors for the trapezoid and midpoint rules for* $\int_1^2 \frac{1}{x}\,dx$

n	Error	
	Trapezoid rule	Midpoint rule
2	0.0152	−0.0074
10	0.00062	−0.00031
50	0.0000250	−0.0000125
250	0.0000010	−0.0000005

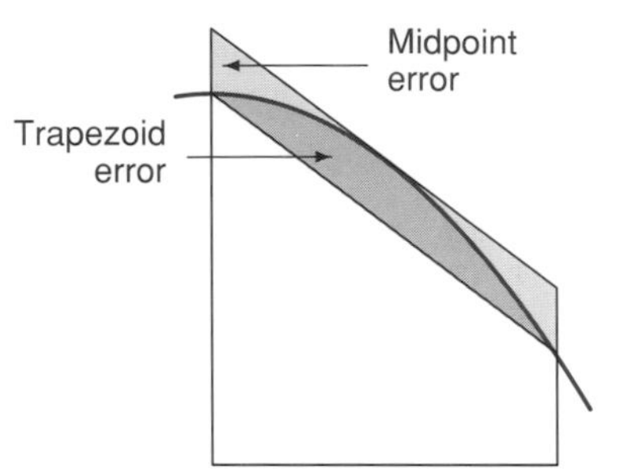

Figure 7.14: Errors in the midpoint and trapezoid rules

Again there is a pattern to the errors. The midpoint rule is noticeably better than the trapezoid rule; the error for the midpoint rule, in absolute value, seems to be about half the error of the trapezoid rule. To see why, remember from the last section that the trapezoid rule uses the line joining the endpoints of each subdivision to form the approximating trapezoid, whereas the midpoint rule uses a tangent line at the midpoint. To visualize the difference, compare the shaded areas representing the errors of the two rules in Figure 7.14. Also, notice in Table 7.4 that the errors for the two rules have opposite signs. This is due to the concavity of the curve: if the curve is concave down, the midpoint rule overestimates and the trapezoid rule underestimates. If the curve is concave up, the midpoint rule underestimates and the trapezoid rule overestimates. (See Figure 7.14.)

In addition, we are interested in how the errors behave for each individual rule as n increases. Table 7.5 gives the ratios of the errors for each rule. For both rules we see that as n increases by a factor of 5, the error decreases by a factor of about $25 = 5^2$. In fact, it can be shown that this squaring relationship holds for any factor, so increasing n by a factor of 10 will decrease the error by a factor of about $100 = 10^2$. Reducing the error by a factor of 100 is equivalent to adding two more decimal places of accuracy to the result. In other words: *In the trapezoid or midpoint rules, each extra* two *digits of accuracy requires about 10 times the work.*

TABLE 7.5 *The ratios of the errors as* n *increases for* $\int_1^2 \frac{1}{x}\,dx$

	Ratio of Errors	
	Trapezoid rule	Midpoint rule
2 vs. 10	24.33	23.84
10 vs. 50	24.97	24.95
50 vs. 250	25.00	25.00

This result shows the advantage of the midpoint and trapezoid methods over the left and right methods: less additional work needs to be done to get another decimal place of accuracy.

The calculator used to produce these tables again took about 4 seconds to compute the midpoint rule for $\int_1^2 \frac{1}{x}\,dx$ with $n = 50$, and this gets 4 digits correct. Thus to get 6 digits would take $n = 500$ and 40 seconds, to get 8 digits would take 400 seconds, and to get 10 digits would take 4000 seconds, or about one hour. That is still not great, but it is certainly better than the 12 years required by the left or right rule.

Simpson's Rule

Still more improvement is possible. If you observe that the trapezoid error has the opposite sign and about twice the magnitude of the midpoint error, you may agree that a weighted average of the two rules, with the midpoint rule weighted twice the trapezoid rule, will have a much smaller error. This approximation is called *Simpson's rule*[5]:

$$\mathrm{SIMP}(n) = \frac{2 \cdot \mathrm{MID}(n) + \mathrm{TRAP}(n)}{3}.$$

Table 7.6 gives the errors for Simpson's rule. Notice how much smaller the errors are than the previous errors. Of course, it is a little unfair to compare Simpson's rule at $n = 50$, say, with the previous rules, because Simpson's rule must compute the value of f at both the midpoint and the endpoints of each subinterval and hence involves evaluating the function at twice as many points. We know by our previous analysis, however, that even if we did compute the other rules at $n = 100$ to compare with Simpson's rule at $n = 50$, the other errors would only decrease by a factor of 2 for the left and right rules and by a factor of 4 for the trapezoid and midpoint rules.

TABLE 7.6 *The errors for Simpson's rule and the ratios of the errors*

n	Error	Ratio
2	0.0001067877	
		550.15
10	0.0000001940	
		632.27
50	0.0000000003	

Table 7.6 also gives the ratios of the errors. This time as n increases by a factor of 5, the errors decrease by a factor of about 600, or about 5^4. Again this behavior holds for any factor, so increasing n by a factor of 10 decreases the error by a factor of about 10^4. In other words: *In Simpson's rule, each extra* four *digits of accuracy requires about 10 times the work.*

This is a great improvement over either the midpoint or trapezoid rules, which only give two extra digits of accuracy when n is increased by a factor of 10. Simpson's rule is so efficient that we get 9 digits correct with $n = 50$ in about 8 seconds on our calculator. Doubling n will decrease the error by a factor of about $2^4 = 16$ and hence will give the tenth digit. The total time is 16 seconds, which is pretty good.

In general, Simpson's rule achieves a reasonable degree of accuracy when using relatively small values of n, and is a good choice for an all-purpose approximation method.

[5]You should be aware that some books and computer programs use slightly different terminology for Simpson's rule; what we call $n = 50$, they call $n = 100$.

Alternate View of the Trapezoid and Simpson's Rules

Our approach to approximating $\int_a^b f(x)\,dx$ numerically has been empirical: try something, see how the error behaves, and then try to improve it. We can also develop the various rules for numerical integration by making better and better approximations to the integrand, f. The left, right, and midpoint rectangular rules are all examples of approximating f by a constant (flat) function on each subinterval. The trapezoid rule is obtained by approximating f by a linear function on each subinterval. It turns out that Simpson's rule can, in the same spirit, be obtained by approximating f by quadratic functions. The details are given in Problems 20 and 21 on page 390.

How the Error Depends on the Integrand

Other factors besides the size of n affect the size of the error in each of the rules. Instead of looking at how the error behaves as we increase n, let's leave n fixed and imagine trying our approximation methods on different functions. We observe that the error in the left or right rule depends on how steeply the graph of f rises or falls. A steep curve makes the triangular regions missed by the left or right rectangles tall and hence large in area. Thus, the error in the left or right rules depends on the size of the derivative of f (see Figure 7.15).

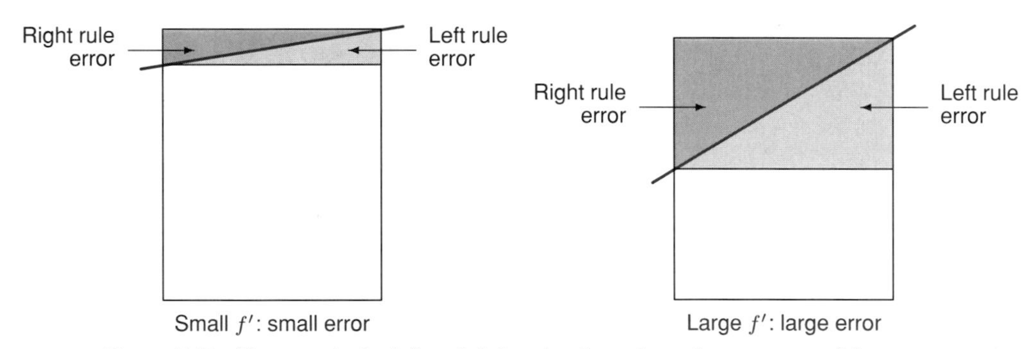

Figure 7.15: The error in the left and right rules depends on the steepness of the curve

From Figure 7.16 we can see that the errors in the trapezoid and midpoint rules depend on how much the curve is bent up or down. In other words, the concavity, and hence the size of the second derivative of f, has an effect on the errors of these two rules. Finally, it can be shown that the error in Simpson's rule depends on the size of the *fourth* derivative of f, written $f^{(4)}$.

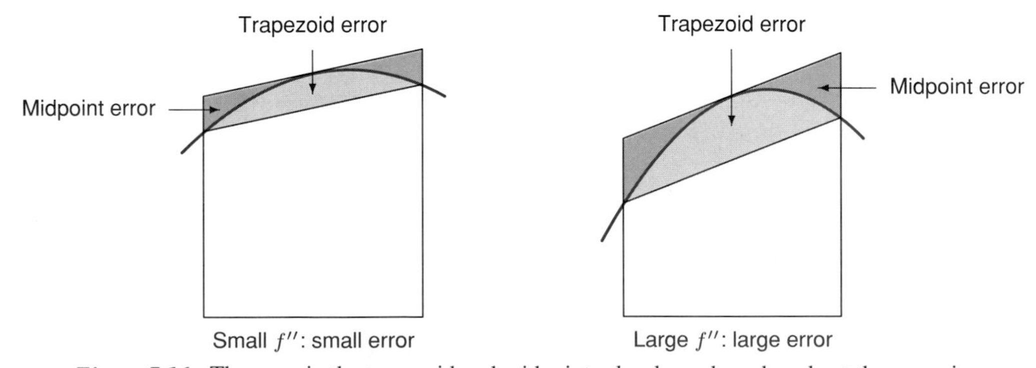

Figure 7.16: The error in the trapezoid and midpoint rules depends on how bent the curve is

Summary of Errors

Left and Right Rules

- Errors are approximately proportional to $1/n$. For example, doubling n decreases the error by a factor of $1/2$ or increasing n by a factor of 10 gives one extra digit of accuracy.
- For a given n, the errors for the left and right rules are approximately equal in absolute value and opposite in sign.
- For a given n, the size of the error depends on the size of f'.

Midpoint and Trapezoid Rules

- Errors are approximately proportional to $1/n^2$. For example, doubling n decreases the error by a factor of $1/4$ or increasing n by a factor of 10 gives two extra digits of accuracy.
- For a given n, the midpoint error is about half the size of the trapezoid error and opposite in sign.
- For a given n, the size of the error depends on the size of f''.

Simpson's Rule

- Errors are approximately proportional to $1/n^4$. For example, doubling n decreases the error by a factor of $1/16$ or increasing n by a factor of 10 gives four extra digits of accuracy.
- For a given n, the size of the error depends on the size of the fourth derivative, $f^{(4)}$.

Conclusion

The definite integrals $\int_a^b f(x)\,dx$ can be computed quickly and accurately in most cases using Simpson's rule. The only trouble occurs when f' or a higher derivative of f fails to exist or gets very large in the interval $a \le x \le b$.

Problems for Section 7.7

1. Compute SIMP(2), the value for Simpson's rule with $n = 2$, for Example 1, on page 377, and compare your value with the true value.

2. Approximate the value of $\displaystyle\int_1^2 \frac{1}{x}\,dx$ using Simpson's rule for $n = 10$, and compute the error.

For Problems 3–10, use Simpson's rule with various values of n to evaluate the definite integrals with error less than 0.001. Explain why you think you have reached a large enough value of n.

3. $\displaystyle\int_0^1 \frac{1}{4+x^2}\,dx$

4. $\displaystyle\int_0^2 e^{\sin t}\,dt$

5. $\displaystyle\int_0^3 \sin^2 x\,dx$

6. $\displaystyle\int_0^3 \cos^2 \theta\,d\theta$

7. $\displaystyle\int_0^1 \sqrt{1+x^4}\,dx$

8. $\displaystyle\int_0^{10} \ln(z^2+1)\,dz$

9. $\displaystyle\int_0^1 \cos(t^2)\,dt$

10. $\displaystyle\int_1^2 \frac{\sin x}{x}\,dx$

11. (a) What is the exact value of $\int_0^1 7x^6\,dx$?
    (b) Find LEFT(5), RIGHT(5), TRAP(5), MID(5), and SIMP(5), and compute the error for each.
    (c) Repeat part (b) with $n = 10$ (instead of $n = 5$).
    (d) For each rule in part (b), compute the ratio of the error for $n = 5$ divided by the error for $n = 10$. Are the values what you expect? Why?

12. Consider Simpson's rule approximations to $\int_0^2 (x^3 + 3x^2)\,dx$.
    (a) What is the exact value of this integral?
    (b) Find SIMP(n) for $n = 2, 4$, and 100. What do you notice?

13. (a) Suppose a certain computer takes two seconds to compute a certain definite integral accurate to 4 digits to the right of the decimal point, using the left rectangle rule. How long will it take to get 8 digits correct using the left rectangle rule? How about 12 digits? 20 digits? (Give answers in years.)
    (b) Repeat part(a) but this time assume it is the trapezoidal rule which is being used throughout.

14. Suppose that for a certain definite integral, LEFT(10) $= 0.38745$ and LEFT(20) $= 0.36517$. Estimate the actual error for LEFT(10) (and thus the actual value of the integral) by assuming that the error is reduced by a factor of 2 in going from LEFT(10) to LEFT(20).

15. Suppose for a certain definite integral that MID(10) $= 35.619$ and MID(20) $= 35.415$. Estimate the actual error for MID(10) assuming that the error for MID(10) is reduced by a factor of 4 in going to MID(20).

Are the statements in Problems 16–19 true or false? Why?

16. The midpoint rule approximation to $\int_0^1 (y^2 - 1)\,dy$ is always smaller than the exact value of the integral.

17. The trapezoid rule approximation is never exact.

18. If LEFT(2) $< \int_a^b f(x)\,dx$, then LEFT(4) $< \int_a^b f(x)\,dx$.

19. If $0 < f' < g'$ everywhere, then the error in approximating $\int_a^b f(x)\,dx$ by LEFT(n) is less than the error in approximating $\int_a^b g(x)\,dx$ by LEFT(n).

Problems 20–21 show how Simpson's rule can be obtained by approximating the integrand, f, by quadratic functions.

20. Suppose that $a < b$ and that m is the midpoint $m = (a+b)/2$. Let $h = b - a$. We wish to show that if f is a quadratic function, so that $f(x) = Ax^2 + Bx + C$, then:

$$\int_a^b f(x)\,dx = \frac{h}{3}\left(\frac{f(a)}{2} + 2f(m) + \frac{f(b)}{2}\right).$$

    (a) Prove that this equation holds for the functions $f(x) = 1$, $f(x) = x$, and $f(x) = x^2$.
    (b) Use part (a) and the "Facts about Sums and Constant Multiples of the Integrand" from Section 6.2, page 315, to prove that the equation holds for any quadratic function f.

21. Consider the following method for approximating $\int_a^b f(x)dx$. Divide the interval $[a, b]$ into n equal subintervals. On each subinterval approximate f by a quadratic function that agrees with f at both endpoints and at the midpoint of the subinterval.
   (a) Explain why the integral of f on the subinterval $[x_i, x_{i+1}]$ is approximately equal to the expression
   $$\frac{h}{3}\left(\frac{f(x_i)}{2} + 2f(m_i) + \frac{f(x_{i+1})}{2}\right),$$
   where m_i is the midpoint of the subinterval, $m_i = (x_i + x_{i+1})/2$. (See Problem 20.)
   (b) Show that if we add up these approximations for each subinterval, we get Simpson's rule:
   $$\int_a^b f(x)dx \approx \frac{2 \cdot \text{MID}(n) + \text{TRAP}(n)}{3}.$$

7.8 IMPROPER INTEGRALS

Our original definition of the definite integral $\int_a^b f(x)\,dx$ assumed that the interval $a \leq x \leq b$ was of finite length and that f was continuous, except perhaps at a few points, and bounded everywhere. Integrals that arise in applications don't necessarily have these nice properties. In this section we investigate a class of integrals, called *improper* integrals, in which one limit of integration is infinite or the integrand is unbounded. As an example, to estimate the mass of the earth's atmosphere, you might calculate an integral which sums the mass of the air up to different heights. In order to represent the fact that the atmosphere doesn't end at a specific height, you can let the upper limit of integration get larger and larger, or tend to infinity.

We will usually consider only improper integrals with positive integrands since they are the most common.

One Type of Improper Integral: When the Limit of Integration Is Infinite

Let's examine an example of an improper integral:

$$\int_1^\infty \frac{1}{x^2}\,dx.$$

This notation is shorthand for the idea of computing a definite integral over $1 \leq x \leq b$ and then taking the limit as b approaches infinity. We compute the definite integral $\int_1^b \frac{1}{x^2}\,dx$:

$$\int_1^b \frac{1}{x^2}\,dx = -x^{-1}\Big|_1^b = -\frac{1}{b} + \frac{1}{1}.$$

Now consider what happens as $b \to \infty$. Since

$$\lim_{b\to\infty} \int_1^b \frac{1}{x^2}\,dx = \lim_{b\to\infty}\left(-\frac{1}{b} + 1\right) = 1,$$

we say that the improper integral

$$\int_1^\infty \frac{1}{x^2}\,dx$$

converges to 1.

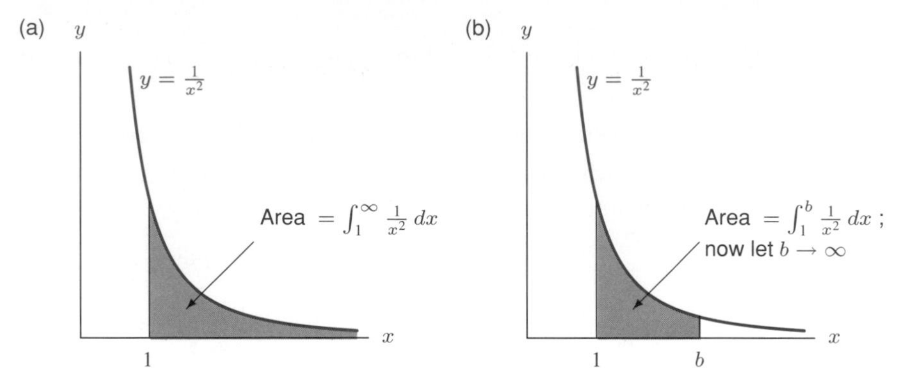

Figure 7.17: Area representation of improper integral

If we think of this in terms of areas, it may seem strange that the region whose area is computed by $\int_1^\infty \frac{1}{x^2}\,dx$ extends from $x = 1$ infinitely far to the right. How could it have finite area? (See Figure 7.17(a).) What our limit computations are saying is that

$$\text{When } b = 10: \quad \int_1^{10} \frac{1}{x^2}\,dx = -\frac{1}{x}\bigg|_1^{10} = -\frac{1}{10} + 1 = 0.9$$

$$\text{When } b = 100: \quad \int_1^{100} \frac{1}{x^2}\,dx = -\frac{1}{100} + 1 = 0.99$$

$$\text{When } b = 1000: \quad \int_1^{1000} \frac{1}{x^2}\,dx = -\frac{1}{1000} + 1 = 0.999$$

and so on. In other words, as b gets larger and larger, the area between $x = 1$ and $x = b$ tends to 1. See Figure 7.17(b). Thus, it does make sense to declare that $\int_1^\infty \frac{1}{x^2}\,dx = 1$.

Of course, in another example, we might not get a finite limit as b gets larger and larger. In this case we say the improper integral *diverges*.

Suppose $f(x)$ is positive for $x \geq a$.

If $\lim\limits_{b\to\infty} \int_a^b f(x)\,dx$ is a finite number, we say that $\int_a^\infty f(x)\,dx$ **converges** and define

$$\int_a^\infty f(x)\,dx = \lim_{b\to\infty} \int_a^b f(x)\,dx.$$

Otherwise, we say that $\int_a^\infty f(x)\,dx$ **diverges**. We define $\int_{-\infty}^b f(x)\,dx$ similarly.

Example 1 Does the improper integral $\int_1^\infty \frac{1}{\sqrt{x}}\,dx$ converge or diverge?

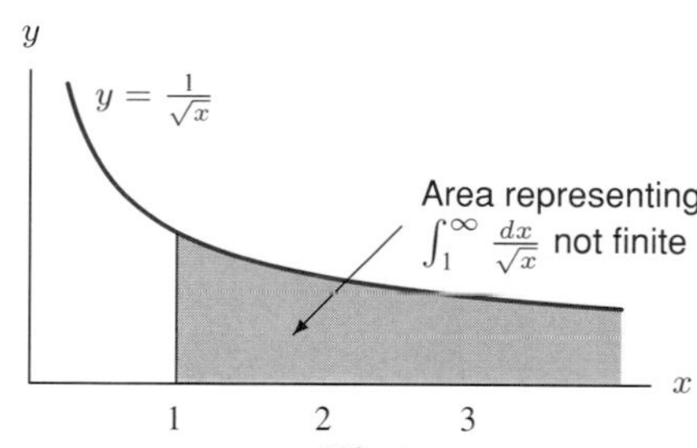

Figure 7.18: $\int_1^\infty \frac{1}{\sqrt{x}}\,dx$ diverges

Solution We consider

$$\int_1^b \frac{1}{\sqrt{x}}\,dx = \int_1^b x^{-1/2}\,dx = 2x^{1/2}\Big|_1^b = 2b^{1/2} - 2.$$

We can see that $\int_1^b \frac{1}{\sqrt{x}}\,dx$ grows without bound as $b \to \infty$. Thus we say the integral $\int_1^\infty \frac{1}{\sqrt{x}}\,dx$ *diverges*. We have shown that the area in Figure 7.18 is not finite.

What is the difference between the functions $1/x^2$ and $1/\sqrt{x}$ that makes the area under the graph of $1/x^2$ approach 1 as $x \to \infty$, whereas the area under $1/\sqrt{x}$ grows very large? Both functions approach 0 as x grows, so as b grows larger, smaller bits of area are being added to the definite integral. The difference between the functions is subtle: the values of the function $1/\sqrt{x}$ *don't shrink fast enough* for the integral to have a finite value. Of the two functions, $1/x^2$ drops to 0 much faster than $1/\sqrt{x}$, and this feature keeps the area under $1/x^2$ from growing beyond 1.

Example 2 Find $\int_0^\infty e^{-5x}\,dx$.

Solution First we consider $\int_0^b e^{-5x}\,dx$:

$$\int_0^b e^{-5x}\,dx = -\frac{1}{5}e^{-5x}\Big|_0^b = -\frac{1}{5}e^{-5b} + \frac{1}{5}.$$

Since $e^{-5b} = \frac{1}{e^{5b}}$, this term tends to 0 as b approaches infinity, so $\int_0^\infty e^{-5x}\,dx$ converges. Its value is

$$\int_0^\infty e^{-5x}\,dx = \lim_{b\to\infty}\int_0^b e^{-5x}\,dx = \lim_{b\to\infty}\left(-\frac{1}{5}e^{-5b} + \frac{1}{5}\right) = 0 + \frac{1}{5} = \frac{1}{5}.$$

Since e^{5x} grows very rapidly, we expect that e^{-5x} will approach 0 rapidly. The fact that the area approaches $\frac{1}{5}$ instead of growing without bound is a consequence of the speed with which e^{-5x} approaches 0.

Example 3 Determine for which values of the exponent, p, the improper integral $\displaystyle\int_1^\infty \frac{1}{x^p}\,dx$ diverges.

Solution For $p \neq 1$,

$$\int_1^b x^{-p}\,dx = \frac{1}{-p+1}x^{-p+1}\Big|_1^b = \left(\frac{1}{-p+1}b^{-p+1} - \frac{1}{-p+1}\right).$$

The important question is whether the exponent of b is positive or negative. If it is negative, then as b approaches infinity, b^{-p+1} approaches 0. If the exponent is positive, then b^{-p+1} grows without bound as b approaches infinity. Thus, the integral converges if $-p+1 < 0$, or $p > 1$. It diverges if $p < 1$.

What happens if $p = 1$? In this case we get

$$\int_1^\infty \frac{1}{x}\,dx = \lim_{b\to\infty} \ln x\Big|_1^b = \lim_{b\to\infty} \ln b - \ln 1.$$

Since $\ln b$ becomes arbitrarily large as b approaches infinity, the integral grows without bound. We conclude that $\int_1^\infty \frac{1}{x^p}\,dx$ diverges precisely when $p \leq 1$. For $p > 1$ the integral has the value

$$\int_1^\infty \frac{1}{x^p}dx = \lim_{b\to\infty}\int_1^b \frac{1}{x^p}dx = \lim_{b\to\infty}\left(\frac{1}{-p+1}b^{-p+1} - \frac{1}{-p+1}\right) = -\left(\frac{1}{-p+1}\right) = \frac{1}{p-1}.$$

Application of Improper Integrals to Energy

The energy, E, required to separate two charged particles, originally a distance a apart, to a distance b, is given by the integral

$$E = \int_a^b \frac{kq_1q_2}{r^2}\,dr$$

where q_1 and q_2 are the magnitudes of the charges and k is a constant. If q_1 and q_2 are in coulombs, a and b are in meters, and E is in joules, the value of the constant k is 9×10^9.

Example 4 A hydrogen atom consists of a proton and an electron, with opposite charges of magnitude 1.6×10^{-19} coulombs. Find the energy required to take a hydrogen atom apart (i.e., to move the electron from its orbit to an infinite distance from the proton). Assume that the initial distance between the electron and the proton is the Bohr radius, $R_B = 5.3 \times 10^{-11}$ meter.

Solution Since we are moving from an initial distance of R_B to a final distance of ∞, the energy is represented by the improper integral

$$\begin{aligned} E &= \int_{R_B}^\infty k\frac{q_1q_2}{r^2}\,dr = kq_1q_2 \lim_{b\to\infty}\int_{R_B}^b \frac{1}{r^2}\,dr \\ &= kq_1q_2 \lim_{b\to\infty} -\frac{1}{r}\Big|_{R_B}^b = kq_1q_2 \lim_{b\to\infty}\left(-\frac{1}{b} + \frac{1}{R_B}\right) = \frac{kq_1q_2}{R_B}. \end{aligned}$$

With the units we are using, E is measured in joules, and therefore

$$E = \frac{(9\times 10^9)(1.6\times 10^{-19})^2}{5.3\times 10^{-11}} \approx 4.35 \times 10^{-18} \text{ joules}.$$

This is about the amount of energy needed to lift a speck of dust 0.000000025 inch off the ground. (In other words, not much!) For comparison, the energy required to bring two 1-coulomb charges of the same sign from infinity to within 1 meter of one another is about equal to the work needed to lift 1 million elephants 6 inches off the ground.

An improper integral is used to model the energy required to take a hydrogen atom apart because the difference between the energy required to separate the electron and the proton by an infinite distance and by a large finite distance is negligible, and the improper integral can be calculated without knowing the final distance.

What happens if the limits of integration are $-\infty$ and ∞? In this case, we break the integral at any point and write the original integral as a sum of two new improper integrals.

> We can use any (finite) number c to define
>
> $$\int_{-\infty}^{\infty} f(x)\,dx = \int_{-\infty}^{c} f(x)\,dx + \int_{c}^{\infty} f(x)\,dx.$$
>
> If *either* of the two new improper integrals diverges, we say the original integral diverges. Only if both of the new integrals have a finite value do we add the values to get a finite value for the original integral.

It is not hard to show that the preceding definition does not depend on the choice for c.

Another Type of Improper Integral: When the Integrand Becomes Infinite

There is another way for an integral to be improper. The interval may be finite but the function may be unbounded at some points in the interval. For example, consider $\displaystyle\int_0^1 \frac{1}{\sqrt{x}}\,dx$. Since the graph of $y = 1/\sqrt{x}$ has a vertical asymptote at $x = 0$, the region between the graph, the x-axis, and the lines $x = 0$ and $x = 1$ is unbounded. Instead of extending to infinity in the horizontal direction as in the previous improper integrals, this region extends to infinity in the vertical direction. See Figure 7.19(a). We can handle this improper integral in the same way as before: we compute $\int_a^1 \frac{1}{\sqrt{x}}\,dx$ for values of a slightly larger than 0 and look at what happens as a approaches 0 from the positive side. (This is written as $a \to 0^+$.)

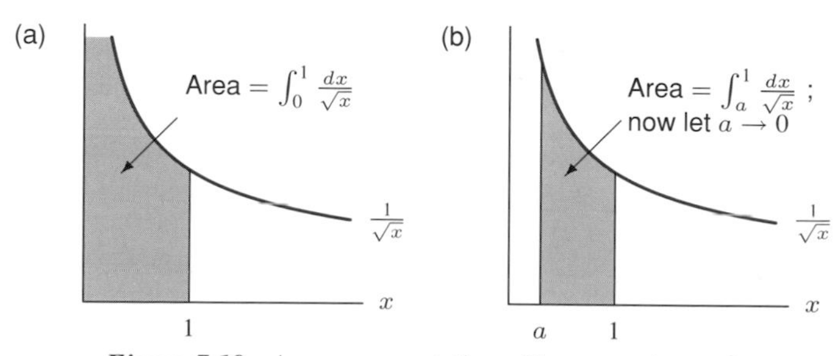

Figure 7.19: Area representation of improper integral

First we compute the integral:

$$\int_a^1 \frac{1}{\sqrt{x}}\,dx = 2x^{1/2}\Big|_a^1 = 2 - 2a^{1/2}.$$

Now we take the limit:

$$\lim_{a\to 0^+} \int_a^1 \frac{1}{\sqrt{x}}\,dx = \lim_{a\to 0^+} (2 - 2a^{1/2}) = 2.$$

Since the limit is finite, we say the improper integral converges, and that

$$\int_0^1 \frac{1}{\sqrt{x}}\,dx = 2.$$

Geometrically, what we have done is to calculate the finite area between $x = a$ and $x = 1$ and take the limit as a tends to 0 from the right. See Figure 7.19(b). Since the limit exists, we say that the integral *converges* to 2. If the limit did not exist, we would say the improper integral *diverges*.

Example 5 Investigate the convergence of $\displaystyle\int_0^2 \frac{1}{(x-2)^2}\,dx$.

Solution This is an improper integral since the integrand tends to infinity as x approaches 2, and is undefined at $x = 2$. Since the trouble is at the right endpoint, we replace the upper limit by b, and let b tend to 2 from the left. This is written $b \to 2^-$, with the "$-$" signifying that 2 is approached from below. See Figure 7.20.

$$\int_0^2 \frac{1}{(x-2)^2}\,dx = \lim_{b\to 2^-} \int_0^b \frac{1}{(x-2)^2}\,dx = \lim_{b\to 2^-} (-1)(x-2)^{-1}\Big|_0^b = \lim_{b\to 2^-} \left(-\frac{1}{(b-2)} - \frac{1}{2}\right).$$

Therefore, since $\displaystyle\lim_{b\to 2^-}\left(-\frac{1}{b-2}\right)$ does not exist, the integral diverges.

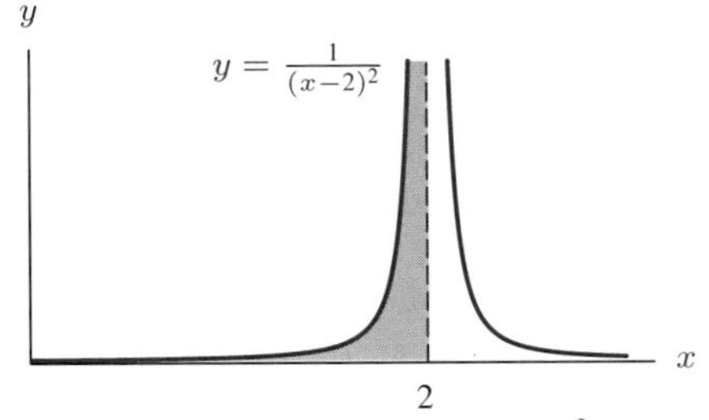

Figure 7.20: Shaded area represents $\int_0^2 \frac{1}{(x-2)^2}\,dx$

Suppose $f(x)$ is positive and tends to infinity as x approaches b.

If $\displaystyle\lim_{c\to b^-} \int_a^c f(x)\,dx$ is a finite number, we say that $\displaystyle\int_a^b f(x)\,dx$ **converges** and define

$$\int_a^b f(x)\,dx = \lim_{c\to b^-} \int_a^c f(x)\,dx.$$

Otherwise, we say that $\displaystyle\int_a^b f(x)\,dx$ **diverges.**

It can also be the case that an integral is improper because the integrand tends to infinity *inside* the interval of integration rather than at an endpoint. In this case, we break the given integral into two (or more) improper integrals so that the integrand tends to infinity only at endpoints. The following is a more precise statement of the procedure.

> If $f(x)$ is positive and tends to infinity as x approaches some point c in the interval $[a, b]$, then we define
> $$\int_a^b f(x)\,dx = \int_a^c f(x)\,dx + \int_c^b f(x)\,dx.$$
> If *either* of the two new improper integrals diverges, we say the original integral diverges. Only if *both* of the new integrals have a finite value do we add the values to get a finite value for the original integral.

Example 6 Investigate the convergence of $\displaystyle\int_{-1}^{2} \frac{1}{x^4}\,dx$.

Solution See the graph in Figure 7.21. The trouble spot is $x = 0$, rather than $x = -1$ or $x = 2$. To handle this situation, we break the given improper integral into two other improper integrals which have $x = 0$ as one of the endpoints:

$$\int_{-1}^{2} \frac{1}{x^4}\,dx = \int_{-1}^{0} \frac{1}{x^4}\,dx + \int_{0}^{2} \frac{1}{x^4}\,dx.$$

We can now use the previous technique to evaluate the new integrals, if they converge. In this example, both of the new integrals diverge. Since

$$\int_0^2 \frac{1}{x^4}\,dx = \lim_{a\to 0+} -\frac{1}{3}x^{-3}\Big|_a^2 = \lim_{a\to 0+}\left(-\frac{1}{3}\right)\left(\frac{1}{8} - \frac{1}{a^3}\right)$$

the integral $\int_0^2 \frac{1}{x^4}\,dx$ does not exist. A similar computation shows that $\int_{-1}^0 \frac{1}{x^4}\,dx$ also diverges. Thus, the original integral diverges.

It is easy to miss an integral which is improper because the integrand tends to infinity inside the interval. For example, it is seriously incorrect to say that $\int_{-1}^{2} \frac{1}{x^4}\,dx = -\frac{1}{3}x^{-3}\Big|_{-1}^{2} = -\frac{1}{24} - \frac{1}{3} = -\frac{3}{8}$.

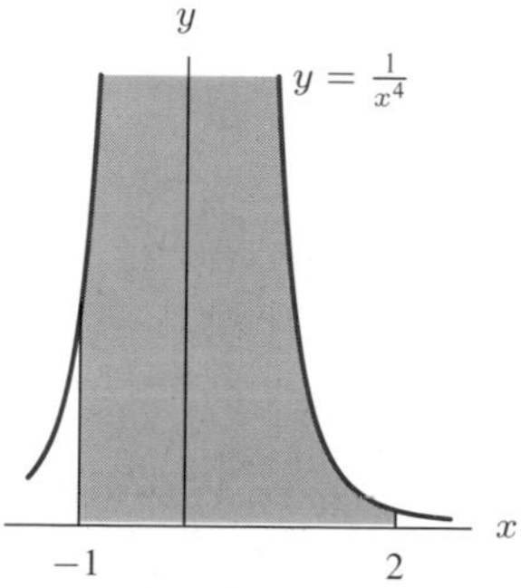

Figure 7.21: Shaded area represents $\int_{-1}^{2} \frac{1}{x^4}\,dx$

Example 7 Find $\displaystyle\int_0^6 \frac{1}{(x-4)^{2/3}}\,dx$.

Solution Figure 7.22 shows that the trouble spot is at $x = 4$, so we break the integral at $x = 4$ and consider the separate parts.

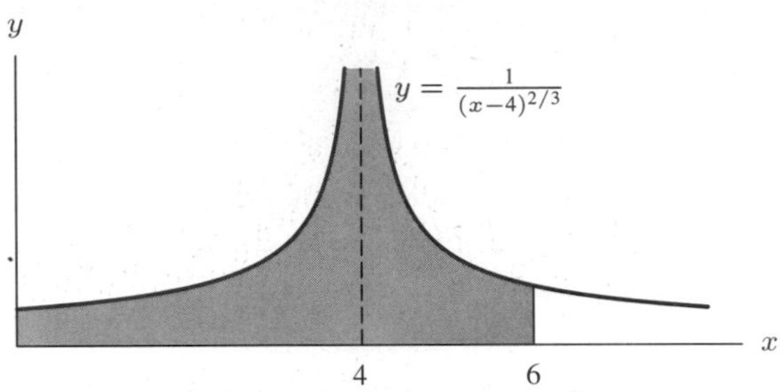

Figure 7.22: Shaded area represents $\int_0^6 \frac{1}{(x-4)^{2/3}}\,dx$

$$\int_0^4 \frac{1}{(x-4)^{2/3}}\,dx = \lim_{b\to 4-} 3(x-4)^{1/3}\Big|_0^b = \lim_{b\to 4-} \left(3(b-4)^{1/3} - 3(-4)^{1/3}\right) = 3(4)^{1/3}.$$

Similarly,

$$\int_4^6 \frac{1}{(x-4)^{2/3}}\,dx = \lim_{a\to 4+} 3(x-4)^{1/3}\Big|_a^6 = \lim_{a\to 4+} \left(3\cdot 2^{1/3} - 3(a-4)^{1/3}\right) = 3(2)^{1/3}.$$

Since both of these integrals converge, the original integral converges:

$$\int_0^6 \frac{1}{(x-4)^{2/3}}\,dx = 3(4)^{1/3} + 3(2)^{1/3} \approx 8.54.$$

Finally, there is a question of what to do when an integral is improper at both endpoints. In this case, we just break the integral at any interior point of the interval. The original integral diverges if either or both of the new integrals diverge.

Example 8 Investigate the convergence of $\displaystyle\int_0^\infty \frac{1}{x^2}\,dx$.

Solution This integral is improper both because the upper limit is ∞ and because the function is undefined at $x = 0$. We break the integral into two parts at, say, $x = 1$. We know by Example 3 that $\int_1^\infty \frac{1}{x^2}\,dx$ has a finite value, but the other part, $\int_0^1 \frac{1}{x^2}\,dx$, diverges since:

$$\int_0^1 \frac{1}{x^2}\,dx = \lim_{a\to 0+} -x^{-1}\Big|_a^1 = \lim_{a\to 0+} \left(\frac{1}{a} - 1\right).$$

Therefore $\displaystyle\int_0^\infty \frac{1}{x^2}\,dx$ diverges as well.

Problems for Section 7.8

Calculate the values of the integrals in Problems 1–24, if they converge.

1. $\int_1^\infty e^{-2x}\,dx$
2. $\int_1^\infty \frac{x}{4+x^2}\,dx$
3. $\int_0^\infty \frac{x}{e^x}\,dx$
4. $\int_{-\infty}^0 \frac{e^x}{1+e^x}\,dx$
5. $\int_\pi^\infty \sin y\,dy$
6. $\int_{-\infty}^\infty \frac{dz}{z^2+25}$
7. $\int_{\pi/4}^{\pi/2} \frac{\sin x}{\sqrt{\cos x}}\,dx$
8. $\int_0^4 \frac{dx}{\sqrt{16-x^2}}$
9. $\int_{-1}^1 \frac{1}{v}\,dv$
10. $\int_1^\infty \frac{1}{x^2+1}\,dx$
11. $\int_1^\infty \frac{1}{\sqrt{x^2+1}}\,dx$
12. $\int_0^1 \frac{x^4+1}{x}\,dx$
13. $\int_1^\infty \frac{y}{y^4+1}\,dy$
14. $\int_{16}^{20} \frac{1}{y^2-16}\,dy$
15. $\int_0^4 \frac{1}{u^2-16}\,du$
16. $\int_0^1 \frac{\ln x}{x}\,dx$
17. $\int_2^\infty \frac{dx}{x\ln x}$
18. $\int_0^2 \frac{1}{\sqrt{4-x^2}}\,dx$
19. $\int_0^\pi \frac{1}{\sqrt{x}}e^{-\sqrt{x}}\,dx$
20. $\int_3^\infty \frac{dx}{x(\ln x)^2}$
21. $\int_1^2 \frac{dx}{x\ln x}$
22. $\int_7^\infty \frac{dy}{\sqrt{y-5}}$
23. $\int_4^\infty \frac{dx}{x^2-1}$
24. $\int_4^\infty \frac{dx}{(x-1)^2}$

25. Find the area under the curve $y = 1/\cos^2 t$ between $t = 0$ and $t = \pi/2$.

26. Suppose a function h is defined by $h(x) = \dfrac{1}{x\sqrt{x}} - \dfrac{1}{16}$, $0 < x \le 4$, and $h(x) = \dfrac{1}{x^2}$, $x > 4$;
    (a) Evaluate $\int_0^\infty h(x)\,dx$.
    (b) Is h differentiable at $x = 4$? If not, why not? If so, find $h'(4)$.

27. (a) Find the indefinite integral $\displaystyle\int \frac{dz}{z^2-z}$.
    (b) Sketch a graph of the integrand $f(z) = \dfrac{1}{z^2-z}$. For what value(s) of z is $f(z)$ undefined?
    (c) Find $\displaystyle\int_3^\infty \frac{dz}{z^2-z}$, if it exists.
    (d) Find $\displaystyle\int_1^3 \frac{dz}{z^2-z}$, if it exists.
    (e) Find a formula for $\displaystyle\int_a^x \frac{dz}{z^2-z}$, and explain for what values of a and x the formula holds.

28. The gamma function is defined for all $x > 0$ by the rule

$$\Gamma(x) = \int_0^\infty t^{x-1}e^{-t}\,dt.$$

    (a) Find $\Gamma(1)$ and $\Gamma(2)$.
    (b) Integrate by parts with respect to t to show that, for positive n,

$$\Gamma(n+1) = n\Gamma(n).$$

    (c) Find a simple expression for $\Gamma(n)$ for positive integers n .

29. Find the energy required to separate opposite electric charges of magnitude 1 coulomb. Assume the charges are initially 1 meter apart and that one is moved infinitely far away from the other. (Energy is defined on page 394.)

30. The rate, r, at which people get sick during an epidemic of the flu can be approximated by

$$r = 1000te^{-0.5t}$$

where r is measured in people/day and t is measured in days since the start of the epidemic.

(a) Sketch a graph of r as a function of t. (b) When are people getting sick fastest?
(c) How many people get sick altogether?

31. For what values of p does the following integral converge?

$$\int_e^{\infty} x^p \ln x \, dx$$

What is the value of the integral when it converges? (A graph may be helpful for the limits.)

32. For what values of p does the following integral converge?

$$\int_0^{e} x^p \ln x \, dx$$

What is the value of the integral when it converges? (A graph may be helpful for the limits.)

7.9 MORE ON IMPROPER INTEGRALS

Making Comparisons

Sometimes it is too difficult to find an exact value of an improper integral by antidifferentiation. Still, it may be possible to determine whether an integral converges or diverges, and if it converges, to estimate its value. The key is to *compare* the given integral to one whose behavior you already know. As in the preceding section, we will restrict our attention to integrals with positive integrands. Let's look at an example.

Example 1 Determine whether $\displaystyle\int_1^{\infty} \frac{1}{\sqrt{x^3+5}}\, dx$ converges.

Solution First, let's see what this integrand does as $x \to \infty$. For large x, the 5 becomes insignificant compared with the x^3, so

$$\frac{1}{\sqrt{x^3+5}} \approx \frac{1}{\sqrt{x^3}} = \frac{1}{x^{3/2}}.$$

Since

$$\int_1^b \frac{1}{x^{3/2}}\, dx = -2x^{-1/2}\Big|_1^b = 2 - 2b^{-1/2},$$

the integral $\int_1^{\infty} \frac{1}{x^{3/2}}\, dx$ converges, so we expect our integral to converge too. In order to confirm this, we observe that $x^3 \leq x^3 + 5$, so

$$\frac{1}{\sqrt{x^3+5}} \leq \frac{1}{\sqrt{x^3}}.$$

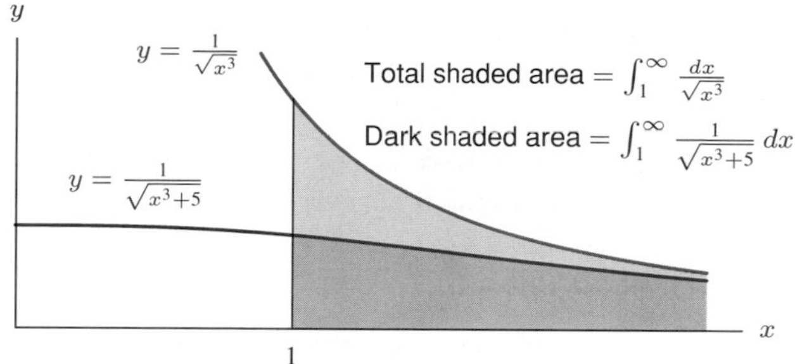

Figure 7.23: Graph showing $\int_1^\infty \frac{1}{\sqrt{x^3+5}}\,dx \le \int_1^\infty \frac{dx}{\sqrt{x^3}}$

Thus, for any b, with $b \ge 1$,

$$\int_1^b \frac{1}{\sqrt{x^3+5}}\,dx \le \int_1^b \frac{1}{\sqrt{x^3}}\,dx.$$

(See Figure 7.23.) The values of both integrals increase as b approaches infinity, since both integrands are positive. The right-hand integral converges because

$$\int_1^\infty \frac{1}{\sqrt{x^3}}\,dx = \int_1^\infty \frac{1}{x^{3/2}}\,dx = \lim_{b\to\infty}\int_1^b \frac{1}{x^{3/2}}\,dx = \lim_{b\to\infty} -2x^{-1/2}\Big|_1^b = \lim_{b\to\infty}\left(2-2b^{-1/2}\right) = 2.$$

Since $\int_1^b \frac{1}{\sqrt{x^3+5}}\,dx$ increases as b approaches infinity but is always smaller than $\int_1^\infty \frac{1}{x^{3/2}}\,dx = 2$, we know $\int_1^\infty \frac{1}{\sqrt{x^3+5}}\,dx$ must have a finite value ≤ 2. Thus,

$$\int_1^\infty \frac{dx}{\sqrt{x^3+5}} \text{ converges,}$$

and it converges to a value less than or equal to 2.

Notice that we first looked at the behavior of the integrand as $x \to \infty$. This is useful because the convergence or divergence of the integral is determined by what happens as $x \to \infty$. The integrand can do any wild thing it likes for x in a finite interval (except perhaps be unbounded) without affecting the convergence of the integral.

The Comparison Test for $\int_a^\infty f(x)\,dx$

Assume $f(x)$ is positive. Making a comparison usually involves two stages:

- Guessing, by looking at the behavior of the integrand for large x, whether the integral converges or not. (This is the "behaves like" principle.)
- Confirming the guess by comparison, usually with a power function ($g(x) = 1/x^p$) or an exponential ($g(x) = e^{-ax}$, $a > 0$).

Example 2 Decide whether $\displaystyle\int_4^\infty \frac{dt}{(\ln t)-1}$ converges or diverges.

Solution Since $\ln t$ grows without bound as $t \to \infty$, the -1 is eventually going to be insignificant in comparison to $\ln t$. Thus, as far as convergence is concerned,

$$\int_4^\infty \frac{1}{(\ln t)-1}\,dt \quad \text{will behave like} \quad \int_4^\infty \frac{1}{\ln t}\,dt.$$

Does $\int_4^\infty \frac{1}{\ln t}\,dt$ converge or diverge? Since $\ln t$ grows very slowly, $1/\ln t$ goes to zero very slowly, and so the integral probably doesn't converge. Specifically, $0 < \ln t < t$ for $t \geq 4$, so $1/\ln t > 1/t$. Since $\int_4^b \frac{1}{t}\,dt$ grows without bound as $b \to \infty$ (see Example 3, page 394), $\int_4^b \frac{1}{\ln t}\,dt$ must too.

Here is the more precise argument. We know that $\ln t < t$ for all positive t, and we know that a positive fraction is increased by making its denominator smaller, but still positive, so:

$$\frac{1}{(\ln t)-1} > \frac{1}{\ln t} > \frac{1}{t}.$$

Since these inequalities hold throughout the interval between the limits of integration, we can say for all $b \geq 4$,

$$\int_4^b \frac{1}{(\ln t)-1}\,dt \geq \int_4^b \frac{1}{t}\,dt.$$

Since

$$\int_4^b \frac{1}{t}\,dt = \ln t\Big|_4^b = \ln\frac{b}{4},$$

the values of $\int_4^b \frac{1}{t}\,dt$ increase without bound as $b \to \infty$. Therefore

$$\int_4^\infty \frac{1}{(\ln t)-1}\,dt \text{ diverges.}$$

How Do You Know What To Compare With?

In Examples 1 and 2, we investigated the convergence of an integral by comparing it with an easier integral. How did we pick the easier integral? Indeed, how do we know if the integrand we start with is to be larger or smaller than the one we use for comparison?

Which Way Should the Inequalities Go?

Assuming $f(x)$ is positive:
To show convergence, $f(x)$ needs to be *smaller* than one whose integral converges.
To show divergence, $f(x)$ needs to be *larger* than one whose integral diverges.

Finding the right integrand to compare with is a matter of trial and error, guided by any information you get by looking at your own integrand as $x \to \infty$. You want your comparison integrand to be easy and, in particular, to have a simple antiderivative.

Useful Integrals for Comparison

- $\int_1^\infty \frac{1}{x^p}\,dx$ converges for $p > 1$ and diverges for $p \le 1$.
- $\int_0^1 \frac{1}{x^p}\,dx$ converges for $p < 1$ and diverges for $p \ge 1$.
- $\int_0^\infty e^{-ax}\,dx$ converges for $a > 0$.

Of course, you can use any function for comparison, provided you can determine its behavior easily.

Example 3 Investigate the convergence of $\int_1^\infty \frac{(\sin x)+3}{\sqrt{x}}\,dx$.

Solution Since it looks difficult to find an antiderivative of this function, we try comparison. What happens to this integrand as $x \to \infty$? Since $\sin x$ oscillates between -1 and 1,

$$\frac{2}{\sqrt{x}} = \frac{-1+3}{\sqrt{x}} \le \frac{(\sin x)+3}{\sqrt{x}} \le \frac{1+3}{\sqrt{x}} = \frac{4}{\sqrt{x}},$$

our integrand oscillates between $2/\sqrt{x}$ and $4/\sqrt{x}$. (See Figure 7.24.)

What do $\int_1^\infty \frac{2}{\sqrt{x}}\,dx$ and $\int_1^\infty \frac{4}{\sqrt{x}}\,dx$ do? As far as convergence is concerned, they certainly do the same thing, and whatever that is, our integral does it too. The important thing to notice is that $\sqrt{x}$ grows very slowly, so $1/\sqrt{x}$ gets small slowly — which means convergence is unlikely. Since $\sqrt{x} = x^{1/2}$, the result in the preceding box (with $p = \frac{1}{2}$) tells us that $\int_1^\infty \frac{dx}{\sqrt{x}}$ diverges. So we believe our original integral diverges. Therefore we compare our integrand with a smaller integrand, namely $2/\sqrt{x}$. We use the fact that

$$\frac{2}{\sqrt{x}} \le \frac{(\sin x)+3}{\sqrt{x}}.$$

Since this inequality holds throughout the interval of integration, we can say for all $b \ge 1$,

$$\int_1^b \frac{2}{\sqrt{x}}\,dx \le \int_1^b \frac{(\sin x)+3}{\sqrt{x}}\,dx.$$

Now, since $\int_1^\infty \frac{2}{\sqrt{x}}\,dx$ diverges, $\int_1^\infty \frac{\sin x+3}{\sqrt{x}}\,dx$ diverges too.

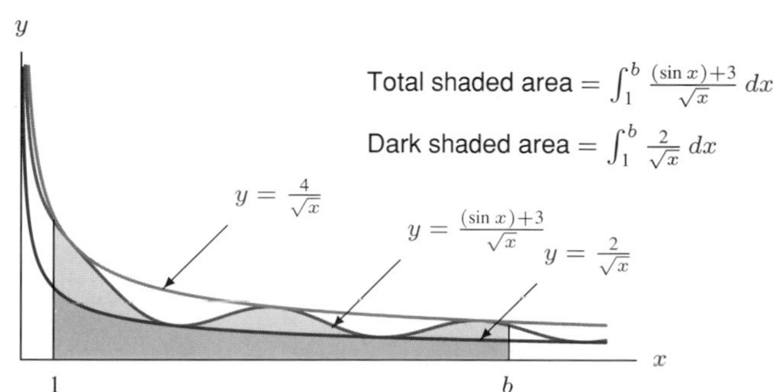

Figure 7.24: Graph showing $\int_1^b \frac{2}{\sqrt{x}}\,dx \le \int_1^b \frac{(\sin x)+3}{\sqrt{x}}\,dx$, for $b \ge 1$

Notice that there are two possible comparisons we could have made in Example 3:

$$\frac{2}{\sqrt{x}} \leq \frac{(\sin x)+3}{\sqrt{x}} \quad \text{or} \quad \frac{(\sin x)+3}{\sqrt{x}} \leq \frac{4}{\sqrt{x}}.$$

Since both $\int_1^\infty \frac{2}{\sqrt{x}}\,dx$ and $\int_1^\infty \frac{4}{\sqrt{x}}\,dx$ diverge, only the first comparison is useful. Knowing that our integral is *smaller* than a divergent integral is of no help whatsoever!

The next example shows what to do if your comparison does not hold throughout the interval of integration.

Example 4 Show $\displaystyle\int_1^\infty e^{-x^2/2}\,dx$ converges to a finite value.

Solution We know that $e^{-x^2/2}$ goes very rapidly to zero as $x \to \infty$, so we expect this integral to converge. Hence we look for some larger integrand which has a convergent integral. The first thing that comes to mind is probably $\int_1^\infty e^{-x}\,dx$, because e^{-x} has an elementary antiderivative and $\int_1^\infty e^{-x}\,dx$ converges. What is the relationship between $e^{-x^2/2}$ and e^{-x}? We know that for $x \geq 2$,

$$x \leq \frac{x^2}{2} \quad \text{so} \quad -\frac{x^2}{2} \leq -x,$$

and so, for $x \geq 2$

$$e^{-x^2/2} \leq e^{-x}.$$

But since this inequality holds only for $x \geq 2$, we split our original interval of integration into two pieces:

$$\int_1^\infty e^{-x^2/2}\,dx = \int_1^2 e^{-x^2/2}\,dx + \int_2^\infty e^{-x^2/2}\,dx.$$

Now $\int_1^2 e^{-x^2/2}\,dx$ is finite (after all, it is the integral of a bounded function over a finite interval), and $\int_2^\infty e^{-x^2/2}\,dx$ is finite by comparison with $\int_2^\infty e^{-x}\,dx$. Therefore, $\int_1^\infty e^{-x^2/2}\,dx$ is the sum of two finite pieces and therefore must be finite.

Using Numerical Methods for Improper Integrals

In the next example we will see how to use numerical methods, such as the trapezoidal or midpoint rule, for an improper integral over an infinite interval.

Example 5 Approximate the value of $\displaystyle\int_1^\infty e^{-x^2/2}\,dx$ to three decimal places.

Solution Since $f(x) = e^{-x^2/2}$ has no elementary antiderivative, we will use numerical methods to compute $\int_1^c e^{-x^2/2}\,dx$ for some value of c which is large enough to make the part we missed, $\int_c^\infty e^{-x^2/2}\,dx$, so small that we don't care about it. We estimate the part we missed by comparison with e^{-x}:

$$\int_c^\infty e^{-x^2/2}\,dx \leq \int_c^\infty e^{-x}\,dx = \lim_{b\to\infty} -e^{-x}\Big|_c^b = e^{-c}, \qquad \text{for } c \geq 2.$$

Suppose we pick $c = 10$. Then $e^{-10} < 0.00005$, so we know that $\int_{10}^{\infty} e^{-x^2/2}\,dx < 0.00005$. Therefore, $\int_1^{\infty} e^{-x^2/2}\,dx$ and $\int_1^{10} e^{-x^2/2}\,dx$ differ by less than 0.00005. If we compute $\int_1^{10} e^{-x^2/2}\,dx$ by the midpoint rule, the approximate value will be an underestimate, while the trapezoidal approximation will be an overestimate because the graph of $e^{-x^2/2}$ is concave up for $x > 1$. Using 200 subdivisions, we obtain

$$\text{MID}(200) = 0.39764$$
$$\text{TRAP}(200) = 0.39779.$$

Thus,

$$0.39764 \le \int_1^{10} e^{-x^2/2} dx \le 0.39779.$$

Since $0 < \int_{10}^{\infty} e^{-x^2/2} dx \le e^{-10} < 0.00005$, we have

$$0.39764 + 0 < \int_1^{10} e^{-x^2/2} dx + \int_{10}^{\infty} e^{-x^2/2} dx < 0.39779 + 0.00005.$$

Therefore,

$$0.39764 < \int_1^{\infty} e^{-x^2/2} dx < 0.39784.$$

Thus, we have $\int_1^{\infty} e^{-x^2/2} dx \approx 0.398$ to three decimal places.

Notice that we didn't begin to do numerical approximation in this example until after we had ascertained that the integral converged.

Problems for Section 7.9

In Problems 1–15 decide if the improper integral converges or diverges. Explain your reasoning.

1. $\displaystyle\int_{50}^{\infty} \frac{dz}{z^3}$
2. $\displaystyle\int_{0.5}^{1} \frac{1}{x^{(19/20)}}\,dx$
3. $\displaystyle\int_{1}^{\infty} \frac{dx}{x^3+1}$
4. $\displaystyle\int_{2}^{\infty} \frac{d\theta}{\sqrt{\theta^3+1}}$
5. $\displaystyle\int_{1}^{\infty} \frac{dx}{1+x}$
6. $\displaystyle\int_{0}^{\infty} \frac{dy}{1+e^y}$
7. $\displaystyle\int_{-1}^{5} \frac{dt}{(t+1)^2}$
8. $\displaystyle\int_{1}^{\infty} \frac{2+\cos\phi}{\phi^2}\,d\phi$
9. $\displaystyle\int_{0}^{\infty} \frac{dz}{e^z+2^z}$
10. $\displaystyle\int_{0}^{\pi} \frac{2-\sin\phi}{\phi^2}\,d\phi$
11. $\displaystyle\int_{-\infty}^{\infty} \frac{du}{1+u^2}$
12. $\displaystyle\int_{1}^{\infty} \frac{du}{u+u^2}$
13. $\displaystyle\int_{4}^{\infty} \frac{3+\sin\alpha}{\alpha}\,d\alpha$
14. $\displaystyle\int_{1}^{\infty} \frac{d\theta}{\sqrt{\theta^2+1}}$
15. $\displaystyle\int_{0}^{1} \frac{d\theta}{\sqrt{\theta^3+\theta}}$

16. Does the integral $\displaystyle\int_1^{\infty} \frac{x}{e^{-x}+x}\,dx$ converge? Why or why not?

Estimate the values of the integrals in Problems 17–18 correct to two decimal places by integrating the functions on your calculator or computer for large values of the upper bound of integration.

17. $\displaystyle\int_1^{\infty} e^{-x^2}\,dx$
18. $\displaystyle\int_0^{\infty} e^{-x^2}\cos^2 x\,dx$

19. The bell-shaped curve of statistics has the equation

$$f(x) = ae^{-x^2/2}.$$

Find the value of a (to three decimal places) that makes

$$\int_{-\infty}^{\infty} f(x)\,dx = 1.$$

20. Statisticians often use the function

$$g(x) = ae^{-(x-k)^2/2}.$$

(a) To three decimal places, what value of a should be chosen to ensure that

$$\int_{-\infty}^{\infty} g(x)\,dx = 1?$$

(b) Is your answer the same as or different from your answer to Problem 19? Why?

21. (a) Find an upper bound for

$$\int_{3}^{\infty} e^{-x^2}\,dx.$$

[Hint: $e^{-x^2} \le e^{-3x}$ for $x \ge 3$.]

(b) For any positive n, generalize the result of part (a) to find an upper bound for

$$\int_{n}^{\infty} e^{-x^2}\,dx$$

by noting that $nx \le x^2$ for $x \ge n$.

22. Do the following integrals converge? If so, give an upper bound for the value of the integral.

(a) $\displaystyle\int_1^{\infty} \frac{2x^2+1}{4x^4+4x^2-2}\,dx$

(b) $\displaystyle\int_1^{\infty} \left(\frac{2x^2+1}{4x^4+4x^2-2}\right)^{1/4} dx$

23. For what values of p does the integral $\displaystyle\int_2^{\infty} \frac{dx}{x(\ln x)^p}$ converge?

24. For what values of p does the integral $\displaystyle\int_1^{2} \frac{dx}{x(\ln x)^p}$ converge?

25. In Planck's Radiation Law, we encounter the integral

$$\int_1^{\infty} \frac{dx}{x^5(e^{1/x}-1)}.$$

(a) Explain why the tangent line approximation to e^t at $t = 0$ tells us that for all t

$$1 + t \le e^t.$$

(b) Substituting $t = 1/x$, show that for all x

$$e^{1/x} - 1 > \frac{1}{x}.$$

(c) Use the comparison test to show that the original integral converges.

26. In this problem you will look at two possible ways to approximate (to four decimal places) the value of the integral $\int_1^\infty e^{-x^2/2}\,dx$. You will compute $\int_1^b e^{-x^2/2}\,dx$ using midpoint rectangles with n subdivisions. Since the original integral is improper, you need to let $b \to \infty$. In addition, you need to let $n \to \infty$. The question is: Which should go to infinity first?
   (a) First fix b and increase n. Start with $b = 3$ and let $n = 20, 50, 100, \ldots$ until your result stabilizes. Take the value to which your result has stabilized as the value of the integral with $b = 3$. Now increase $b = 4, 5, \ldots$ and for each new b value increase n as before. Continue until successive values of b give the same answer. Take the value you get as the value of $\int_1^\infty e^{-x^2/2}\,dx$.
   (b) In this part, fix n and increase b. Start with $n = 20$ and increase $b = 3, 4, 5, \ldots$. What happens? You may have to use very large values of b before the results stabilize.
   (c) Explain why the method in part (a) gives a reasonable way of approximating the improper integral, whereas the method in part (b) does not.

7.10 NOTES ON CONSTRUCTING ANTIDERIVATIVES

We have seen several techniques for finding an antiderivative F of a function f. The fact remains, however, that most of the time F cannot be found easily. Suppose f is an elementary function, that is, a combination of constants, powers of x, $\sin x$, $\cos x$, e^x, and $\ln x$. Then you have to be lucky to find an antiderivative F which is also an elementary function. But if F is not an elementary function, how can we be sure that it exists at all? After all, elementary functions are the only functions for which we have formulas. In this section we will learn to use the definite integral to construct antiderivatives, and, along the way, we will learn how to create genuinely new functions.

Construction of Antiderivatives Using a Slope Field

In Chapter 6 we saw how to graph an antiderivative, F, of the function f in Figure 7.25(a) with the property that $F(0) = 0$. Since $F' = f$, the graph of F will be increasing where f is positive, and decreasing where f is negative. Also, it will slope steeply when f is large (positive or negative) and gently when f is small, and it will be horizontal when f is zero. Using this information, we can generate the lower graph in Figure 7.25(b). This graph starts at $(0, 0)$, since $F(0) = 0$.

What about other antiderivatives? They all have the same slope for a particular value of x and so are the same shape as F, but shifted up or down. Thus, they all have the form $F(x) + C$, for some constant C. If you imagine the whole family of functions $F(x) + C$ as C varies, you get a pile of graphs stacked on top of each other, all of the same shape but at different heights above the x-axis. Figure 7.25(b) shows $F(x) + 1$ as well as $F(x)$.

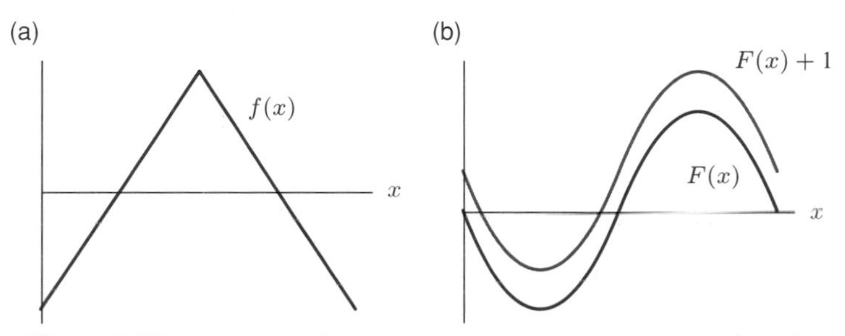

Figure 7.25: A function f and two antiderivatives (Note that $F' = f$.)

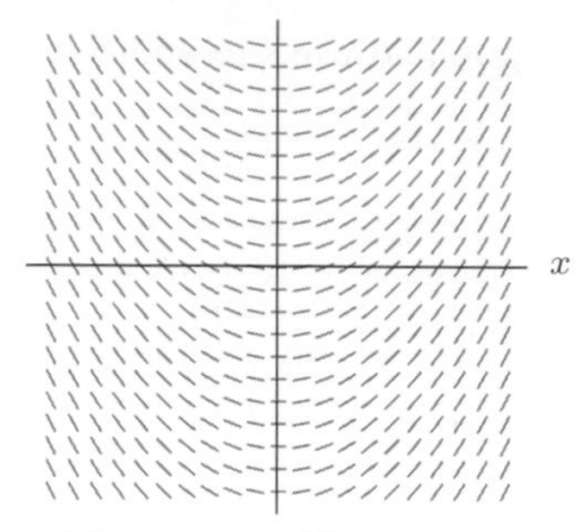

Figure 7.26: Slope field of $f(x) = x$

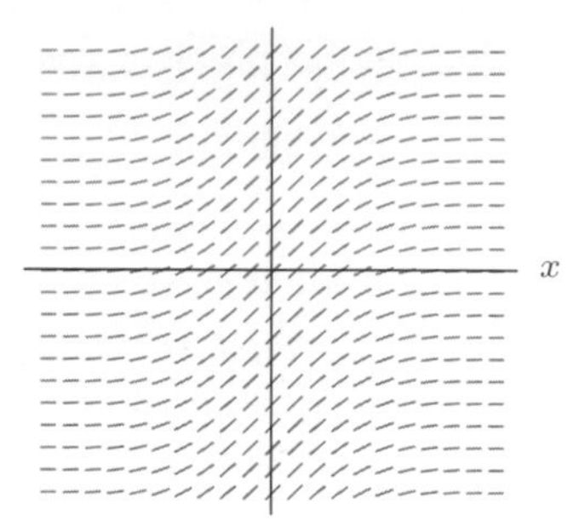

Figure 7.27: Slope field of $f(x) = e^{-x^2}$

So far, all we have is a rough sketch of the antiderivative. To get a more accurate graph of F, we should be more careful about making F have the right slope at every point. The slope of F at any point (x, y) on its graph should be $f(x)$, since $F'(x) = f(x)$. We can arrange this as follows: at the point (x, y) in the plane, draw a small line segment with slope $f(x)$. Do this at many points. We call such a diagram a *slope field* (or a *direction field* or *tangent field*).

If $f(x) = x$, you get Figure 7.26. Notice how the lines in Figure 7.26 seem to be arranged in a parabolic pattern. This is because the general antiderivative of x is $x^2/2 + C$, so the lines are all the tangent lines to the family of parabolas $y = x^2/2 + C$. This suggests a way of finding antiderivatives graphically even if we can't write down a formula for them: plot the slopes, and see if they suggest the graph of an antiderivative. For example, if you do this with $f(x) = e^{-x^2}$, which is one of the functions that does not have an elementary antiderivative, you get Figure 7.27.

You can see the ghost of the graph of a function lurking behind the slopes in Figure 7.27; in fact there is a whole stack of them. If you move across the plane in the direction suggested by the slope field at every point, you will trace out a curve. The slope field will be tangent to the curve everywhere, so this is the graph of an antiderivative of e^{-x^2}.

This construction is important because it shows that the antiderivative of e^{-x^2} does exist—the graph defines it as well as showing us its qualitative behavior.

Construction of Antiderivatives Using the Definite Integral

The previous discussion may have convinced you that an antiderivative of e^{-x^2} exists, but it does not give you an easy way of evaluating the antiderivative. If it were an elementary function, we could evaluate it using a calculator, but it isn't. However, we know from the Fundamental Theorem of Calculus that if F is an antiderivative of e^{-x^2}, then

$$F(b) - F(a) = \int_a^b e^{-t^2}\,dt$$

so

$$F(x) - F(0) = \int_0^x e^{-t^2}\,dt.$$

All we have done here is to replace the usual b in the upper limit with an x and set $a = 0$. Suppose we are looking for the antiderivative that satisfies $F(0) = 0$. Then $F(0)$ drops out of the equation, and we get

$$F(x) = \int_0^x e^{-t^2}\,dt.$$

This is a formula for F. For any value of x, there is a unique value for $F(x)$, so F is a function. For any fixed x, we can calculate $F(x)$ to any accuracy we like by using numerical methods such as the trapezoid rule or Simpson's rule. For example, we can show that

$$F(2) = \int_0^2 e^{-t^2}\,dt = 0.88208\ldots.$$

Notice that our expression for F is not an elementary function; we have *created* a new function using the definite integral. Our next theorem says that this method of constructing antiderivatives works in general. This means that if we define F by

$$F(x) = \int_a^x f(t)\,dt$$

then F must be an antiderivative of f. This can happen in one of two ways. Either we already know what the antiderivatives of f are, in which case F turns out to be one of them. Or, if f has no elementary antiderivatives, F defines an antiderivative of f for you.

Construction Theorem for Antiderivatives

If f is a continuous function on an interval, and if a is any number in that interval, then the function F defined by

$$F(x) = \int_a^x f(t)\,dt$$

is an antiderivative of f.

Notice that if we already know an antiderivative of f, which we may call G, then the Fundamental Theorem of Calculus tells us that

$$G(x) - G(a) = \int_a^x f(t)\,dt.$$

By the definition of F, we have

$$F(x) = \int_a^x f(t)\,dt.$$

Therefore,

$$G(x) - G(a) = F(x),$$

so F and G differ by a constant [namely $G(a)$], and so F and G are *both* antiderivatives of f.

Justification of the Construction Theorem for Antiderivatives To justify the theorem, we cannot assume that we know any antiderivatives of f. Our task is to show that F, defined by this integral, is an antiderivative of f. We want to show that $F'(x) = f(x)$. In fact, the argument we will use works even if we already know an antiderivative. By the definition of the derivative,

$$F'(x) = \lim_{h\to 0} \frac{F(x+h) - F(x)}{h}.$$

Let's suppose f is positive and h is positive. Then we can visualize

$$F(x) = \int_a^x f(t)\,dt$$

and

$$F(x+h)=\int_a^{x+h} f(t)\,dt$$

as areas, which leads to representing

$$F(x+h)-F(x)=\int_x^{x+h} f(t)\,dt$$

as a difference of two areas. From Figure 7.28, we see that $F(x+h)-F(x)$ is roughly a rectangle of height $f(x)$ and width h (shaded darker in Figure 7.28), so we have

$$F(x+h)-F(x)\approx f(x)h.$$

Because f is continuous, as h approaches 0, the height of this rough rectangle approaches $f(x)$. That is, as $h \rightarrow 0$

$$\frac{F(x+h)-F(x)}{h} \rightarrow f(x).$$

A similar argument holds if h is negative and $h \rightarrow 0$, so

$$F'(x)=\lim_{h\rightarrow 0}\frac{F(x+h)-F(x)}{h}=f(x).$$

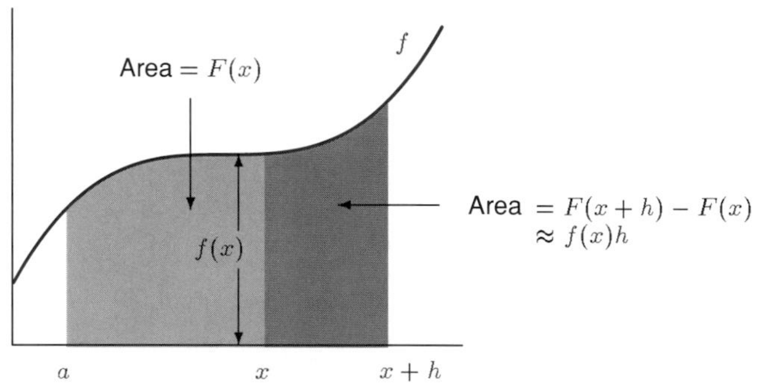

Figure 7.28: Justification of the Construction Theorem for Antiderivatives

Using the Construction Theorem for Antiderivatives

The construction theorem enables us to write down antiderivatives of functions that do not have elementary antiderivatives. For example, an antiderivative of $(\sin x)/x$ is

$$F(x)=\int_0^x \frac{\sin t}{t}\,dt.$$

This is an honest-to-goodness function, whose values you can calculate to any degree of accuracy using left- or right-hand sums or the more efficient methods of Section 7.6. If you were a manufacturer of calculators, you might even put a button for it on your calculator. In fact, this particular function already has a name: it is called the *sine-integral*, and it is denoted $\mathrm{Si}(x)$.

Example 1 Construct a table of values of $\text{Si}(x)$ for $x = 0, 1, 2, 3$.

Solution Using Riemann sums, we calculate the values of $\text{Si}(x) = \int_0^x \frac{\sin t}{t}\,dt$ given in Table 7.7. Since the integrand is undefined at $t = 0$, we took the lower limit as 0.00001 instead of 0.

TABLE 7.7 *Values of* $\text{Si}(x)$

x	0	1	2	3
$\text{Si}(x)$	0	0.95	1.61	1.85

The reason the sine-integral has a name is that scientists have found a use for it in optics. For people who use it all the time, it is just another elementary function like sine or cosine, and tables of its values have been compiled. In addition to knowing all the derivatives that we know, (i.e., the derivatives of e^x, $\sin x$, $\cos x$, and $\ln x$), such people also know the formula

$$\frac{d}{dx}\,\text{Si}(x) = \frac{\sin x}{x},$$

and can use this fact with the other differentiation rules.

Example 2 Find the derivative of $x\,\text{Si}(x)$.

Solution Using the product rule,

$$\begin{aligned}\frac{d}{dx}(x\,\text{Si}(x)) &= \left(\frac{d}{dx}x\right)\text{Si}(x) + \left(\frac{d}{dx}\,\text{Si}(x)\right)x\\ &= 1\cdot\text{Si}(x) + \frac{\sin x}{x}x\\ &= \text{Si}(x) + \sin x.\end{aligned}$$

If we were to add $\text{Si}(x)$ to our list of elementary functions we would be able to do more indefinite integrals using the methods of this chapter.

Example 3 Find $\displaystyle\int \frac{\sin(t^2)}{t}\,dt$.

Solution Make the substitution $t = \sqrt{w}$, giving $dt = \dfrac{1}{2\sqrt{w}}\,dw$ and $w = t^2$, so

$$\int \frac{\sin(t^2)}{t}\,dt = \int \frac{\sin w}{\sqrt{w}}\frac{1}{2\sqrt{w}}\,dw = \frac{1}{2}\int \frac{\sin w}{w}\,dw = \frac{1}{2}\,\text{Si}(w) + C = \frac{1}{2}\,\text{Si}(t^2) + C.$$

Problems for Section 7.10

For each of the functions in Problems 1-3, let $F(x) = \int_0^x f(t)\,dt$. Draw a graph showing $F(x)$ as a function of x.

1.

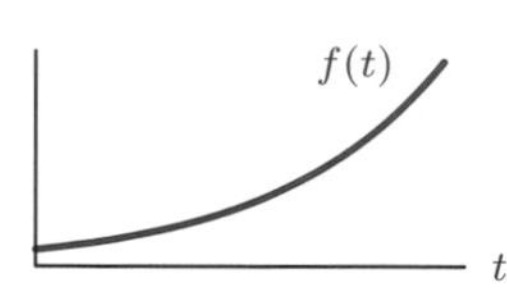

2.

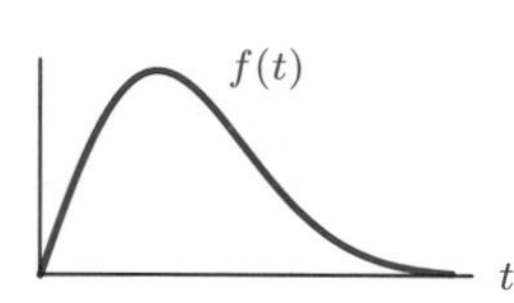

3.

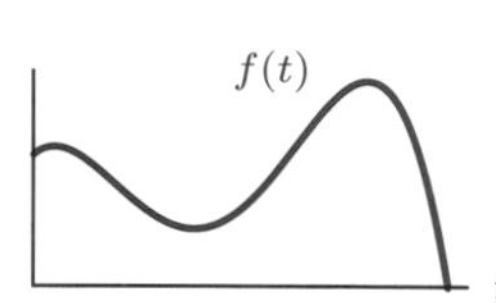

4. Make a table of values for the function

$$I(x) = \int_0^x \sqrt{t^4 + 1}\,dt$$

for $x = 0, 0.5, 1.0, 1.5, 2.0$.

5. (a) Continue the table of values for $\mathrm{Si}(x) = \int_0^x \frac{\sin t}{t}\,dt$, in Example 1, page 411 for $x = 4, 5$.
   (b) Why is $\mathrm{Si}(x)$ decreasing between $x = 4$ and $x = 5$?

6. Match the following functions with the slope fields shown in Figure 7.29.
   (a) $f(x) = e^{x^2}$ (b) $f(x) = e^{-2x^2}$ (c) $f(x) = e^{-x^2/2}$
   (d) $f(x) = e^{-0.5x}\cos x$ (e) $f(x) = \frac{1}{(1 + 0.5\cos x)^2}$ (f) $f(x) = -e^{-x^2}$

(I)

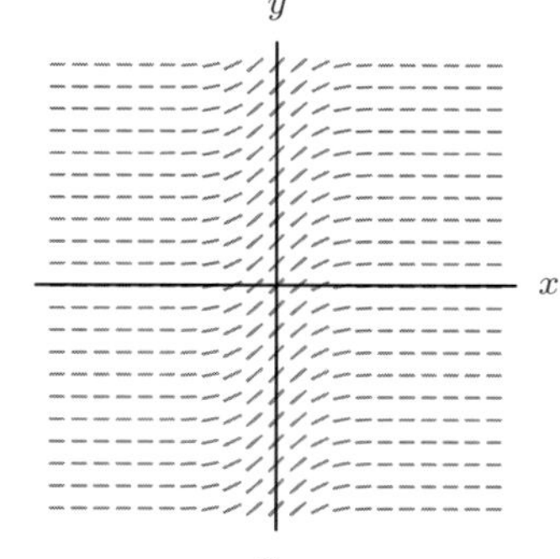

(II)

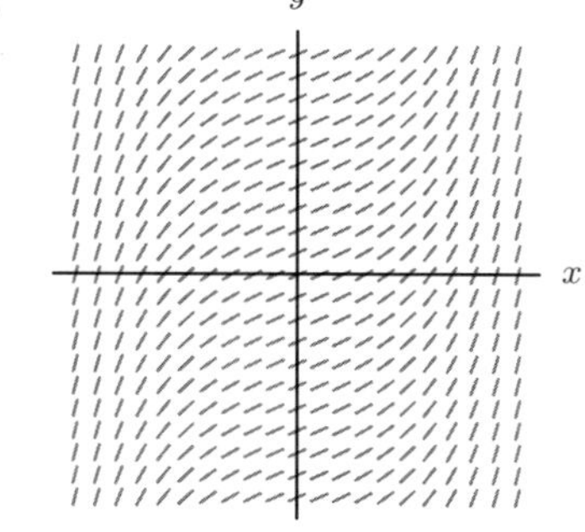

(III)

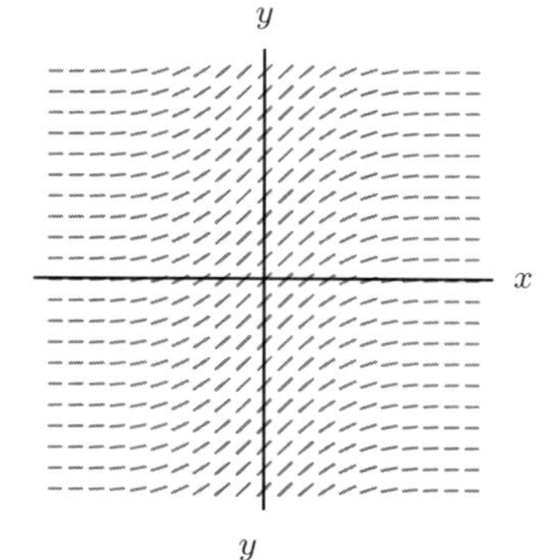

(IV)

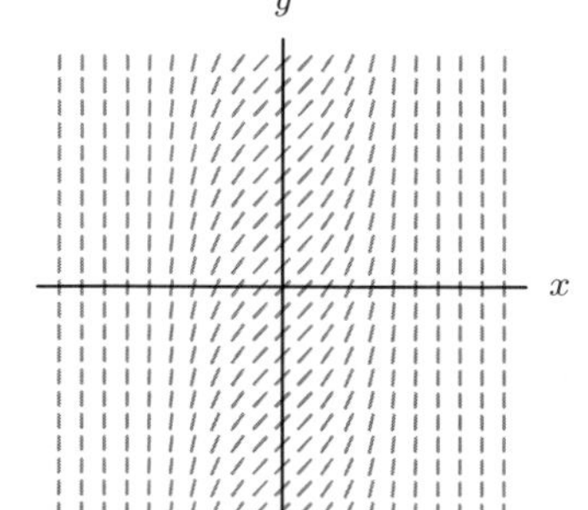

(V)

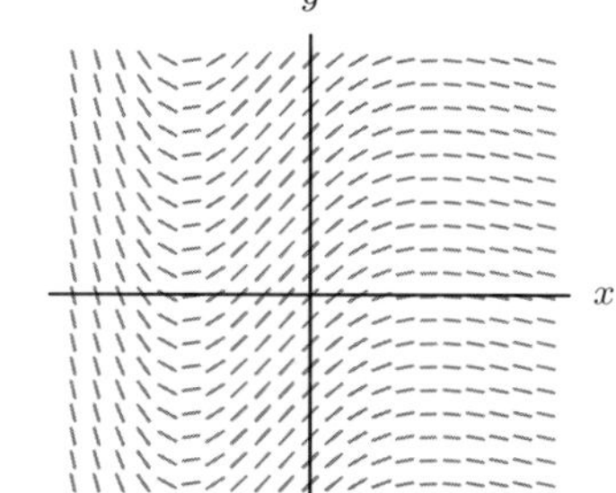

(VI)

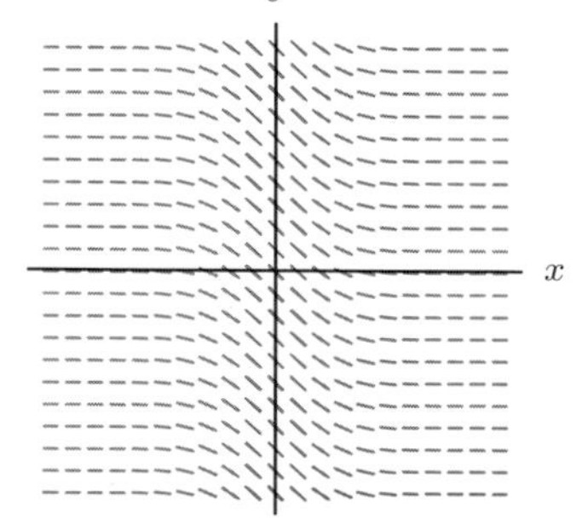

Figure 7.29: All slope fields are for $-3 \le x \le 3$ and $-3 \le y \le 3$

7. On each slope field in Figure 7.29, sketch the antiderivative, F, with $F(0) = 0$.

Find the derivatives in Problems 8–13 .

8. $\displaystyle \frac{d}{dx}\int_0^x \sqrt{3+\cos(t^2)}\,dt$

9. $\displaystyle \frac{d}{dx}\int_1^x (1+t)^{200}\,dt$

10. $\displaystyle \frac{d}{dx}\int_{0.5}^x \arctan(t^2)\,dt$

11. $\displaystyle \frac{d}{dt}\int_t^\pi \cos(z^3)\,dz$

12. $\displaystyle \frac{d}{dx}\int_x^1 \ln t\,dt$

13. $\displaystyle \frac{d}{dx}\left[\mathrm{Si}(x^2)\right]$

14. Using a calculator or computer, draw the slope field of e^{-x^2}. Use it to sketch the antiderivative $F(x)$ of e^{-x^2} that satisfies $F(0)=0$. Estimate $\lim_{x\to\infty} F(x)$.

15. (a) Sketch a graph of $f(t) = \dfrac{\sin t}{t}$.
    (b) What does your graph tell you about the behavior of $\mathrm{Si}(x)$ for $x > 0$? Is $\mathrm{Si}(x)$ always increasing or always decreasing? Does $\mathrm{Si}(x)$ cross the x-axis for $x > 0$?
    (c) By drawing the slope field for $f(t) = \dfrac{\sin t}{t}$, decide whether $\lim_{x\to\infty} \mathrm{Si}(x)$ exists.

16. (a) Use your calculator or computer to sketch a graph of $y = x^{\sin x}$ for $0 < x \le 20$.
    (b) Using your answer to part (a), sketch by hand a graph of the function F, where

$$F(x) = \int_0^x t^{\sin t}\,dt.$$

    (c) Use a slope field program to check your answer to part (b).

17. Let $F(x)$ be the antiderivative of $\sin(x^2)$ satisfying $F(0)=0$.
    (a) Describe any general features of the graph of F that you can deduce by looking at the graph of $\sin(x^2)$ in Figure 7.30.
    (b) By drawing a slope field (using a calculator or computer), sketch a graph of F. Does F ever cross the x-axis in the region $x > 0$? Does $\lim_{x\to\infty} F(x)$ exist?

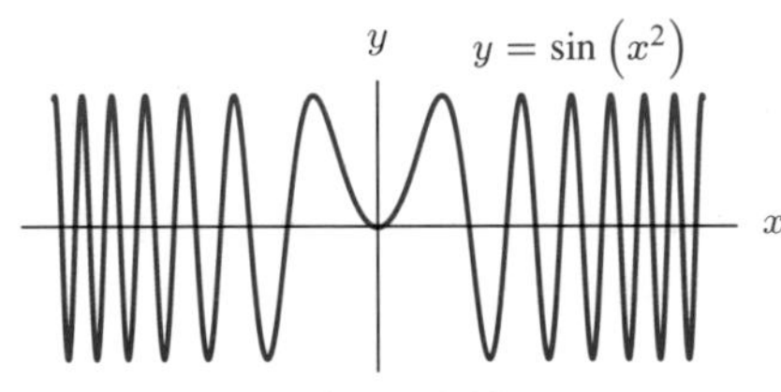

Figure 7.30

18. Consider the integral

$$F(x) = \int_1^x (t+1)^2 \cos 2t\,dt.$$

    (a) Evaluate the definite integral using the table of integrals and the Fundamental Theorem of Calculus.
    (b) Verify the Construction Theorem by finding $F'(x)$ from your answer to part (a).

Problems 19–22 concern the *error function*, $\text{erf}(x)$, defined by

$$\text{erf}(x) = \frac{2}{\sqrt{\pi}} \int_0^x e^{-t^2}\, dt.$$

19. Find the derivative of $x\,\text{erf}(x)$.

20. Find the derivative of $\text{erf}(\sqrt{x})$.

21. Find an expression for the cumulative probability distribution function,

$$F(x) = \sqrt{\frac{2}{\pi}} \int_0^x e^{-t^2/2} dt$$

in terms of $\text{erf}(x)$.

22. Find an expression for

$$\sqrt{\frac{2}{\pi}} \int_{x_1}^{x_2} e^{-t^2/2} dt$$

in terms of $\text{erf}(x)$.

REVIEW PROBLEMS FOR CHAPTER SEVEN

For each region in Problems 1–3, write a definite integral which represents its area. Evaluate the integral to derive a formula for the area.

1. A rectangle with base b and height h:

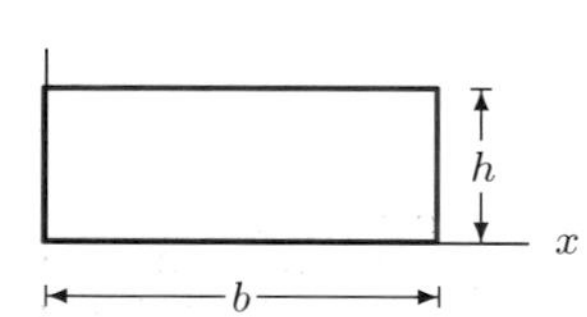

2. A right triangle of base b and height h:

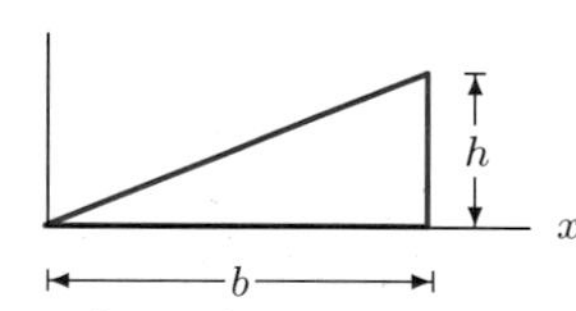

3. A circle of radius r:

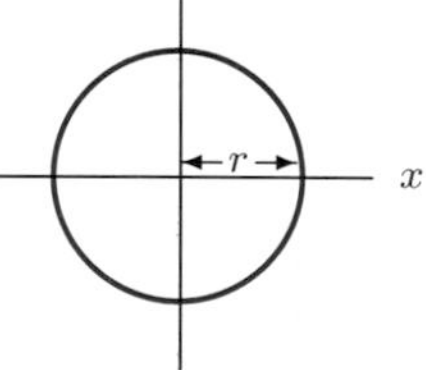

4. (a) Using a substitution, find $\displaystyle\int \frac{dx}{\sqrt{1-4x^2}}$ from the table of integrals.

(b) Use your answer to part (a) to find $\displaystyle\int_0^{\pi/8} \frac{dx}{\sqrt{1-4x^2}}$.

(c) Check your answer to part (b) by numerical integration.

5. (a) Explain why you can rewrite x^x as $x^x = e^{x \ln x}$.

(b) Use your answer to part (a) to find $\dfrac{d}{dx}(x^x)$.

(c) Find $\displaystyle\int x^x(1 + \ln x)dx$.

(d) Find $\displaystyle\int_1^2 x^x(1 + \ln x)dx$ using the Fundamental Theorem of Calculus and check your answer using numerical methods.

6. Suppose the function f is defined by $f(t) = t^2$ for $0 \le t \le 1$ and $f(t) = 2 - t$ for $1 < t \le 2$. Compute $\displaystyle\int_0^2 f(t)\, dt$.

For Problems 7–30, evaluate the following integrals. For Problems 27–30, find the definite integrals using the Fundamental Theorem of Calculus, and check your answers numerically.

7. $\int te^{t^2}\,dt$

8. $\int x\cos x\,dx$

9. $\int \sin x\left(\sqrt{2+3\cos x}\right)dx$

10. $\int x^2e^{2x}\,dx$

11. $\int x\sqrt{1-x}\,dx$

12. $\int x\ln x\,dx$

13. $\int y\sin y\,dy$

14. $\int \frac{dz}{z^2+z}$

15. $\int \frac{\cos\sqrt{y}}{\sqrt{y}}\,dy$

16. $\int (\ln x)^2\,dx$

17. $\int \ln(x^2)\,dx$

18. $\int e^{0.5-0.3t}\,dt$

19. $\int \cos^3 2\theta\sin 2\theta\,d\theta$

20. $\int \frac{5x+6}{x^2+4}\,dx$

21. $\int x\sqrt{4-x^2}\,dx$

22. $\int \sqrt{4-x^2}\,dx$

23. $\int \frac{1}{\cos^2 z}\,dz$

24. $\int \cos^2\theta\,d\theta$

25. $\int \frac{(u+1)^3}{u^2}\,du$

26. $\int \tan(2x-6)\,dx$

27. $\int_1^3 \ln(x^3)\,dx$

28. $\int_1^e (\ln x)^2\,dx$

29. $\int_{-\pi}^{\pi} e^{2x}\sin 2x\,dx$

30. $\int_{-\pi/4}^{\pi/4} x^3\cos x^2\,dx$

In Problems 31–34, explain why the following pairs of antiderivatives are really, despite their apparent dissimilarity, different expressions of the same problem. You do not need to evaluate the integrals.

31. $\int \frac{dx}{x^2+4x+4}$ and $\int \frac{x}{(x^2+1)^2}\,dx$

32. $\int \frac{1}{\sqrt{1-x^2}}\,dx$ and $\int \frac{x\,dx}{\sqrt{1-x^4}}$

33. $\int \frac{x}{1-x^2}\,dx$ and $\int \frac{1}{x\ln x}\,dx$

34. $\int \frac{x}{x+1}\,dx$ and $\int \frac{1}{x+1}\,dx$

35. Integrate

(a) $\int \frac{e^x}{1+e^x}\,dx$ (b) $\int \frac{e^x}{1+e^{2x}}\,dx$ (c) $\int \frac{1}{1+e^x}\,dx$

36. (a) Find the average value of the following functions over one cycle:

(i) $f(t)=\cos t$ (ii) $g(t)=|\cos t|$ (iii) $k(t)=(\cos t)^2$

(b) Write the averages you have just found in ascending order. Explain clearly, using words and graphs, why the averages should be expected to come out in the order they did.

37. In 1987 the average per capita income in the US was \$26,000. Suppose that average per capita income is increasing at a rate in dollars per year given by

$$r(t)=480(1.024)^t,$$

where t is the number of years since 1987.

(a) Estimate the average per capita income in 1995.

(b) Find a formula for the average per capita income as a function of time after 1987.

38. In 1990 humans generated 1.4×10^{20} joules of energy through the combustion of petroleum. All of the earth's petroleum would generate approximately 10^{22} joules. Assuming the use of energy generated by petroleum combustion will increase by 2% each year, how long will it be before all of our petroleum resources are used up?

39. A store has an inventory of Q units of a certain product at time $t = 0$. The store sells the product at the steady rate of Q/A units per week, and it exhausts the inventory in A weeks.

(a) Find a formula $f(t)$ for the amount of product in inventory at time t. Graph $f(t)$.

(b) Find the average inventory level during the period $0 \leq t \leq A$ in two ways: graphically and by antidifferentiation. Check your answer by common sense.

40. Find the average (vertical) height of the shaded area in Figure 7.31.

41. Find the average (horizontal) width of the shaded area in Figure 7.31.

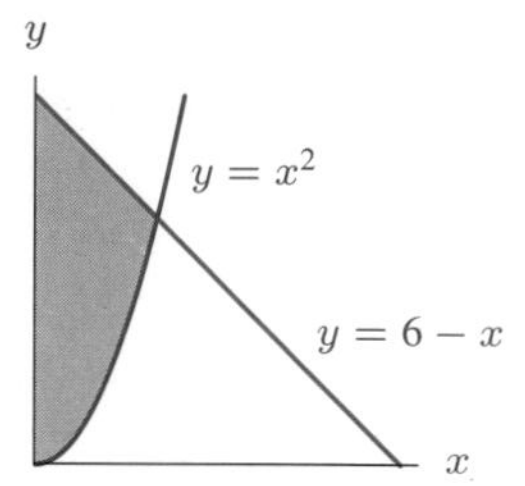

Figure 7.31

42. The curves $y = \sin x$ and $y = \cos x$ cross each other infinitely often. What is the area of the region bounded by these two curves between two consecutive crossings?

For Problems 43–51, decide if the integral converges or diverges. If the integral converges, find its value.

43. $\displaystyle\int_4^{\infty} \frac{dt}{t^{3/2}}$ **44.** $\displaystyle\int_0^{\infty} we^{-w}\, dw$ **45.** $\displaystyle\int_{10}^{\infty} \frac{dx}{x \ln x}$

46. $\displaystyle\int_2^{\infty} \frac{1}{4+z^2}\, dz$ **47.** $\displaystyle\int_{10}^{\infty} \frac{1}{z^2-4}\, dz$ **48.** $\displaystyle\int_{-1}^{1} \frac{1}{x^4}\, dx$

49. $\displaystyle\int_{-\pi/4}^{\pi/4} \tan\theta\, d\theta$ **50.** $\displaystyle\int_0^{\pi/2} \frac{1}{\sin\phi}\, d\phi$ **51.** $\displaystyle\int_{-5}^{10} \frac{dt}{\sqrt{t+5}}$

For Problems 52–56, decide if the integral converges or diverges. If the integral converges, find its value, or give a bound for the value.

52. $\displaystyle\int_0^{\pi/4} \tan 2\theta\, d\theta$ **53.** $\displaystyle\int_0^{\infty} \frac{\sin^2\theta}{\theta^2+1}\, d\theta$ **54.** $\displaystyle\int_0^{\pi} \tan^2\theta d\theta$

55. $\displaystyle\int_0^{1} (\sin x)^{-3/2} dx$ **56.** $\displaystyle\int_1^{\infty} \frac{x}{x+1}\, dx$

57. Estimate the values of the following integrals correct to two decimal places by integrating the functions on your calculator for larger and larger values of the upper bound of integration.

(a) $\displaystyle\int_0^{\infty} \frac{\sqrt{x}}{e^x}\, dx$ (b) $\displaystyle\int_1^{\infty} \ln\left(\frac{e^x+1}{e^x-1}\right) dx$

58. (a) Find the following indefinite integrals.

(i) $\displaystyle\int \ln x\, dx$ (ii) $\displaystyle\int (\ln x)^2\, dx$ (iii) $\displaystyle\int (\ln x)^3\, dx$ (iv) $\displaystyle\int (\ln x)^4\, dx$

(b) Find the following improper definite integrals.

(i) $\displaystyle\int_0^1 \ln x\, dx$ (ii) $\displaystyle\int_0^1 (\ln x)^2\, dx$ (iii) $\displaystyle\int_0^1 (\ln x)^3\, dx$ (iv) $\displaystyle\int_0^1 (\ln x)^4\, dx$

(c) Guess a formula for $\int_0^1 (\ln x)^n\, dx$, where $n \geq 0$.

59. Find each of the integrals

$$\int_0^\infty e^{-x}\,dx, \qquad \int_0^\infty xe^{-x}\,dx, \qquad \int_0^\infty x^2e^{-x}\,dx, \qquad \int_0^\infty x^3e^{-x}\,dx,$$

and hence guess the value of $\displaystyle\int_0^\infty x^n e^{-x}\,dx$.

60. Suppose you estimate $\int_0^{0.5} f(x)dx$ by the trapezoid and midpoint rules with 100 steps. Use a sketch to explain the relation between the two estimates and the true value of the integral if

(a) $f(x) = 1 + e^{-x}$
(b) $f(x) = e^{-x^2}$
(c) $f(x)$ is a straight line.

61. Suppose for a certain definite integral that TRAP(10) = 4.6891 and TRAP(50) = 4.6966. Estimate the actual error for TRAP(10) (and thus the actual value of the integral) by assuming that the error is reduced by a factor of roughly 25 in going from TRAP(10) to TRAP(50).

62. Suppose for a definite integral that SIMP(5) = 7.41562 and SIMP(10) = 7.41738. Estimate the actual value of the integral by using the fact that the error is reduced by a factor of roughly 16 in going from SIMP(5) to SIMP(10).

63. Let $F(x) = \int_0^x \sin 2t\,dt$.

(a) Evaluate $F(\pi)$.
(b) Draw a sketch to explain geometrically why the answer to part (a) is correct.
(c) For what values of x is $F(x)$ positive? negative?

64. Let

$$F(x) = \int_2^x \frac{1}{\ln t}\,dt \text{ for } x \geq 2.$$

(a) Find $F'(x)$.
(b) Is F increasing or decreasing? What can you say about the concavity of its graph?
(c) Sketch a graph of $F(x)$.

65. Let $F(x) = \displaystyle\int_{\pi/2}^x \frac{\sin t}{t}\,dt$. Find the value(s) of x between $\pi/2$ and $3\pi/2$ for which $F(x)$ is at a global maximum and global minimum. Show your reasoning.

66. Use the fact that $e^x \geq 1 + x$ for all values of x and the formula

$$e^x = 1 + \int_0^x e^t\,dt$$

to show that

$$e^x \geq 1 + x + \frac{x^2}{2}$$

for all positive values of x. Generalize this idea to get inequalities involving higher-degree polynomials.

67. Use the fact that $\cos x \leq 1$ for all x and repeated integration to show that

$$\cos x \leq 1 - \frac{x^2}{2!} + \frac{x^4}{4!}.$$

CHAPTER EIGHT

USING THE DEFINITE INTEGRAL

In Chapter 3 we saw how a definite integral can represent an area, an average value, or a total change. In Chapter 7 we focused on different ways of evaluating integrals. In this chapter we will use definite integrals to solve problems in geometry, physics, economics, and probability. In each section, we use the same method to represent a quantity as a definite integral: we chop up the quantity and approximate it by a Riemann sum.

8.1 SETTING UP RIEMANN SUMS

Using the Notation

There are many situations where a quantity we want to calculate can be expressed as a definite integral. Typically this will happen when the quantity can be approximated by dividing it into little pieces, solving the problem approximately for each little piece, and then adding the resulting approximations. For example, in Section 3.1, the way we learned that distance traveled is the definite integral of velocity was by adding up the distances traveled over small time intervals; this yielded a Riemann sum, and the definite integral came about as the limit of that Riemann sum as subdivisions got smaller and smaller. In many of the problems we will encounter in this chapter, however, we will not immediately recognize the quantity we want as a Riemann sum. The key, then, is to understand how the desired quantity can be chopped up into lots of small pieces in such a way that you can calculate the desired quantity for each small piece. If adding up all these pieces gives a Riemann sum, then taking the limit as the subdivisions get smaller and smaller gives a definite integral.

Recall how we do this for a particle moving with velocity $v = f(t)$ at time t. We find the net distance traveled between $t = a$ and $t = b$ by cutting the time interval into subintervals of length Δt. Provided Δt is small, we can approximate the distance traveled during this time by assuming that the velocity is constant over each subinterval $t_i \le t \le t_i + \Delta t$, where $i = 0, 1, \ldots n$. Let's suppose the velocity throughout the ith interval is $v = f(t_i)$, the velocity at the start of the interval. The distance moved during that time is approximately

$$\text{Velocity} \times \text{Time} \quad \text{or} \quad f(t_i)\,\Delta t.$$

Therefore, the net change in position is approximated by the sum of all of these small distances, so

$$\text{Change in position} \approx \sum_{i=0}^{n-1} f(t_i)\,\Delta t.$$

In the limit as Δt tends to zero, this sum becomes an integral:

$$\text{Change in position} = \lim_{\Delta t \to 0} \sum_{i=0}^{n-1} f(t_i)\,\Delta t = \int_a^b f(t)\,dt.$$

Although the dt in the definite integral is not a number, it behaves as if it were a very small version of Δt. Informally, $f(t)\,dt$ may be thought of as the distance traveled by the particle over a very small time interval, since it is the velocity, $f(t)$, multiplied by the time, dt. Thus, the integral may be thought of as the sum of all these small distances.

Density and How to Slice

Most of the examples in this section involve the idea of *density*. For example,

- A population density is measured in, say, people per mile (along the edge of a road), or people per unit area (in a city), or bacteria per cubic centimeter (in a test tube).
- The density of a substance (e.g. air, wood, or metal), is the mass of a unit volume of the substance and is measured in, say, grams per cubic centimeter.

Suppose we want to calculate the total mass or total population, but the density is not constant over the region. Then to find the total quantity:

> Divide the region into small pieces in such a way that the density is approximately constant on each piece, and add up the contributions of all the pieces.

Example 1 The Massachusetts Turnpike ("the Pike") starts in the middle of Boston and heads west. The number of people living next to it varies as it gets further and further from town. Suppose that, x miles out of town, the population density adjacent to the Pike is $P = f(x)$ people/mile. Express the total population living next to the Pike within 5 miles of Boston as a definite integral.

Solution Think about how you would calculate the population. First, you could get a rough figure for each mile of the Pike. The population density at the center of Boston is $f(0)$; let's use that density for the whole first mile, giving an estimate of $(f(0)\text{ people/mile}) \cdot (1\text{ mile}) = f(0)$ people living in the first mile. The density 1 mile out is $f(1)$. Using that figure for the portion out to the second mile gives an estimate of $f(1) \cdot 1 = f(1)$ people living in the second mile. Continuing in this way, we get estimates of $f(2)$, $f(3)$, and $f(4)$ for the remaining 3 miles. Our estimate of the total population is

$$f(0) + f(1) + f(2) + f(3) + f(4).$$

Now, to get a more accurate estimate, you could break the distance up into tenths of a mile. The population in the first $\frac{1}{10}$ mile would be roughly the population density at the beginning of that $\frac{1}{10}$ mile times $\frac{1}{10}$ (since the density is in people per *mile*), or $f(0) \cdot \frac{1}{10}$. The population in the second $\frac{1}{10}$ mile would be roughly the population density at the beginning of that $\frac{1}{10}$ mile times $\frac{1}{10}$, or $f(0.1) \cdot \frac{1}{10}$, and so on. The total estimate would be

$$\underbrace{f(0) \cdot 0.1 + f(0.1) \cdot 0.1 + f(0.2) \cdot 0.1 + \cdots + f(4.9) \cdot 0.1}_{50\text{ terms}}.$$

This is a Riemann sum approximating the total population, with subdivision of the interval $0 \le x \le 5$ into 50 pieces and with $\Delta x = 0.1$. The population in each piece $x_i \le x \le x_i + \Delta x$ is roughly $f(x_i)\,\Delta x$, that is, the population density times the length of the interval. (See Figure 8.1.)

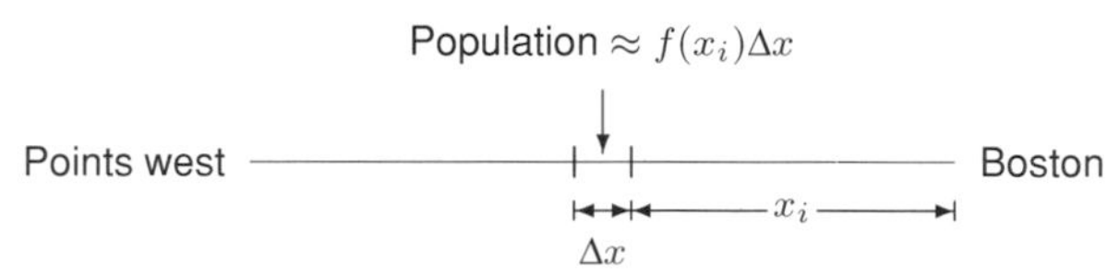

Figure 8.1: Population along the Massachusetts Turnpike

The sum of all these estimates gives the estimate

$$\text{Total population} \approx \sum_{i=0}^{49} f(x_i)\,\Delta x.$$

With n subdivisions, we get

$$\text{Total population} \approx \sum_{i=0}^{n-1} f(x_i)\,\Delta x.$$

Letting n tend to ∞ shows that

$$\text{Total population} = \lim_{n\to\infty}\sum_{i=0}^{n-1} f(x_i)\,\Delta x = \int_0^5 f(x)\,dx.$$

The 5 and 0 in the limits of the integral are the upper and lower limits of the interval over which we are integrating.

Example 2 The air density (in kg/m^3) h meters above the earth's surface is $P = f(h)$. Find the mass of a cylindrical column of air 2 meters in diameter and 25 kilometers high.

Solution Our column of air is a circular cylinder 2 m in diameter and 25 km, or 25,000 m, high. First we should decide how we are going to slice this column into small pieces. Since the air density varies with altitude but remains constant horizontally, it makes sense to take horizontal slices of air. That way the density will be more or less constant over the whole slice, being close to its value at the bottom of the slice. (See Figure 8.2.)

A slice will be a cylinder of height Δh and diameter 2 m, so its radius will be 1 m. We can find the approximate mass of the slice by multiplying its volume and the density together. If the thickness of the slice is Δh, then its volume is $\pi r^2 \cdot \Delta h = \pi 1^2 \cdot \Delta h = \pi\,\Delta h$ m^3 (where m^3 = cubic meter). The density of the slice between h_i and $h_i + \Delta h$ is roughly $f(h_i)$. Thus

$$\text{Mass of slice} \approx \text{Volume} \cdot \text{Approximate density} = (\pi \Delta h \text{m}^3)(f(h_i)\ \text{kg/m}^3) = \pi\,\Delta h \cdot f(h_i)\ \text{kg}.$$

If there are n slices, then adding all these up will yield a Riemann sum:

$$\text{Total mass} \approx \sum_{i=0}^{n-1} \pi f(h_i)\,\Delta h\ \text{kg}.$$

As $n \to \infty$, this sum approximates the definite integral:

$$\text{Total mass} = \int_0^{25{,}000} \pi f(h)\,dh\ \text{kg}.$$

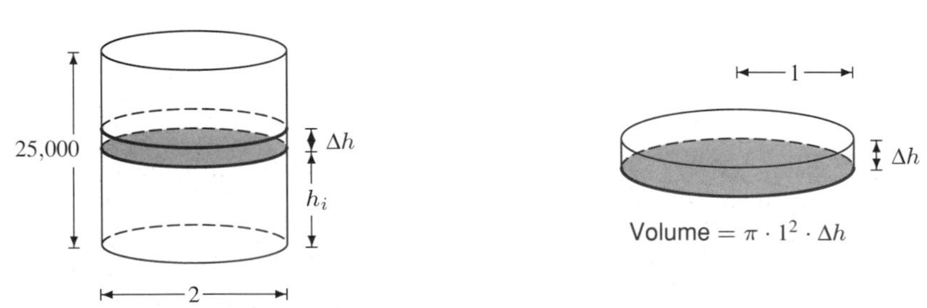

Figure 8.2: How to slice a column of air.

In order to get a numerical value for the mass of air, we need an explicit formula for the density as a function of height, as in the next example.

Example 3 Find the mass of the column of air in Example 2 if the density of air at height h is given by

$$P = f(h) = 1.28e^{-0.000124h} \text{ kg/m}^3.$$

Solution Using the result of the previous example, we have

$$\text{Mass} = \int_0^{25{,}000} \pi 1.28e^{-0.000124h}\,dh = \frac{1.28\pi}{0.000124}\left(-e^{-0.000124h}\Big|_0^{25{,}000}\right)$$

$$\approx 32{,}429(e^0 - e^{-0.000124(25{,}000)}) \approx 31{,}000 \text{ kg}.$$

In Example 2, we did not write down any specific Riemann sums, say, with 5 or 50 subdivisions, as we did in Example 1. Rather, we thought about the general Riemann sum for the problem, in order to be able to write down the definite integral. The Δh became dh, and the limits in the definite integral were obtained by considering the limits for the variable h. Once we decided to slice the column horizontally, we knew that a typical slice has thickness Δh, and so h must be the variable in our definite integral. The limits in our definite integral must be values of h.

Example 4 Water leaks out of a tank through a square hole with 1-inch sides. At time t (in seconds) the velocity of water flowing through the hole is $v = g(t)$ ft/sec. Write down a definite integral that represents the total amount of water lost in the first minute.

Solution In this problem we slice time into small intervals, because we are trying to get from a rate of change to a total amount. Since t is given in seconds, we will convert the minute to 60 seconds. Thus, we are considering water loss over the time interval $0 \le t \le 60$. We also need to convert inches into feet since the velocity is given in ft/sec. Since 1 inch = $1/12$ foot, the square hole has area $1/144$ square feet. For water flowing through a hole with constant velocity v, the amount of water which has passed through in some time, Δt, can be pictured as the rectangular solid in Figure 8.3, which has volume

$$\text{Area} \cdot \text{Height} = \text{Area} \cdot \text{Velocity} \cdot \text{Time}.$$

Over a small time interval of length Δt, starting at time t, water flows with a nearly constant velocity $v = g(t)$ through a hole $1/144$ square feet in area. In Δt seconds, we know that

$$\text{Water lost} \approx \left(\frac{1}{144}\text{ ft}^2\right)(g(t)\text{ ft/sec})(\Delta t\text{ sec}) = \left(\frac{1}{144}\right) g(t)\,\Delta t \text{ ft}^3.$$

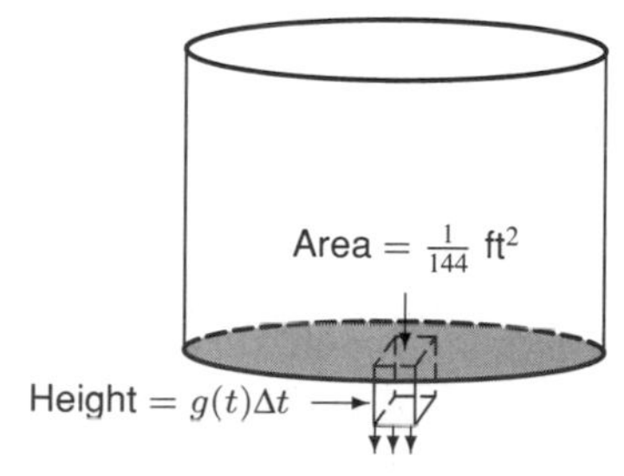

Figure 8.3: Volume of water passing through hole

Adding the water from all subintervals gives

$$\text{Total water lost} \approx \sum \frac{1}{144} g(t)\,\Delta t \text{ ft}^3.$$

As Δt tends to 0, the sum tends to the definite integral:

$$\text{Total water lost} = \int_0^{60} \frac{1}{144} g(t)\,dt \text{ ft}^3.$$

Note that the sum represented by the $\sum$ sign is over all the small time intervals. In this example we did not bother to write down the points t_0, t_1, and so on, that are needed to write the Riemann sum precisely. This is unnecessary if all you want is the final expression for the definite integral.

Sometimes it requires some thought to figure out how to slice a quantity up. The key point to remember is that you want the density, rate, or whatever to be nearly constant within each piece.

Example 5 The population density in Ringsburg is a function of the distance from the city center: at r miles from the center, the density is $P = f(r)$ people per square mile. Ringsburg has a radius of 5 miles. Write down a definite integral that expresses the total population of Ringsburg.

Solution We want to slice Ringsburg up and estimate the population on each slice. If we were to take straight-line slices, the population density would vary on each slice, since it depends on the distance from the city center. We want the population density to be pretty close to constant on each slice, so that we will be able to estimate the population by multiplying the density and the area together. We therefore take slices that are thin rings around the center, at a constant distance from the center (see Figure 8.4), and since the ring is very thin, we can approximate its area quite closely by straightening it into a thin rectangle. (See Figure 8.5.) The width of the rectangle is Δr miles, and its length is approximately equal to the circumference of the ring, $2\pi r$ miles, so its area is about $2\pi r\,\Delta r$ mi^2. Thus,

$$\text{Population on ring} \approx \text{Density} \cdot \text{Area},$$

so

$$\text{Population on ring} \approx (f(r) \text{ people/mi}^2)(2\pi r \Delta r \text{ mi}^2) = f(r) \cdot 2\pi r\,\Delta r \text{ people}.$$

Adding the contributions from each ring, we get

$$\text{Total population} \approx \sum 2\pi r f(r)\,\Delta r \text{ people}.$$

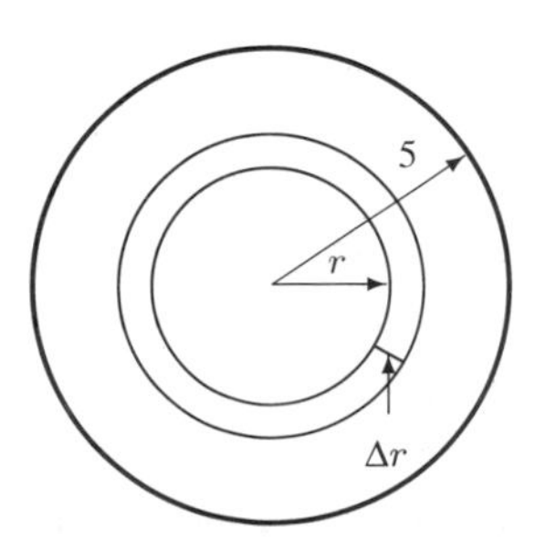

Figure 8.4: Ringsburg

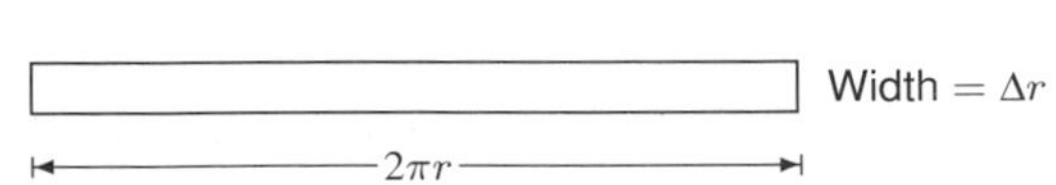

Figure 8.5: Ring from Ringsburg (straightened out)

So

$$\text{Total population} = \int_0^5 2\pi r f(r)\, dr \text{ people.}$$

Note: You may wonder what happens if we calculate the area of the ring by subtracting the area of the inner circle (πr^2) from the area of the outer circle [$\pi(r + \Delta r)^2$], giving

$$\text{Area} = \pi(r + \Delta r)^2 - \pi r^2.$$

Multiplying out and subtracting, we get

$$\begin{aligned}\text{Area} &= \pi[r^2 + 2r\,\Delta r + (\Delta r)^2] - \pi r^2 \\ &= 2\pi r\,\Delta r + \pi(\Delta r)^2.\end{aligned}$$

This expression differs from the one we used before by the $\pi(\Delta r)^2$ term. However, as Δr becomes very small, $\pi(\Delta r)^2$ becomes much, much smaller. We say it is *second order* of smallness, since the power of the small factor, Δr, is 2. In the limit as $\Delta r \to 0$, we can ignore $\pi(\Delta r)^2$.

Problems for Section 8.1

1. A rod has length 2 meters. At a distance x meters from its left end, the density of the rod is given by
$$\rho(x) = 2 + 6x \text{ g/m}.$$
   (a) Write a Riemann sum approximating the total mass of the rod.
   (b) Find the exact mass by converting the sum into an integral.
2. The *center of mass* of a rod is the point at which you would place a pivot so that the rod balances. To find the position of the center of mass of a rod, you first calculate its *moment*, and then divide the moment by the total mass. If the rod lies along the x-axis between a and b, the moment of the rod is $\int_a^b x\rho(x)\,dx$, where $\rho(x)$ is its density at a position x. Find the moment of the rod in Problem 1.
3. The density of cars (in cars per mile) down a 20-mile stretch of the Pennsylvania Turnpike can be approximated by
$$\rho(x) = 300\left(2 + \sin\left(4\sqrt{x + 0.15}\right)\right),$$
where x is the distance in miles from the Breezewood toll plaza.
   (a) Sketch a graph of this function for $0 \le x \le 20$.
   (b) Write a sum that approximates the total number of cars on this 20-mile stretch.
   (c) Find the total number of cars on the 20-mile stretch.
4. Circle City, a typical metropolis, is very densely populated near its center, and its population gradually thins out toward the city limits. In fact, its population density is 10,000($3 - r$) people/square mile at distance r miles from the center.
   (a) Assuming that the population density at the city limits is zero, find the radius of the city.
   (b) What is the total population of the city?

5. Greater Boston can be approximated by a semicircle of radius 8 miles with its center on the coast. Moving away from the center along a radius, the population density is constant for the first mile. Beyond that, the density starts to decrease according to the data given in the table, where $\rho(r)$ is the population density at a distance r miles from the center.

r (mi)	0	1	2	3	4	5	6	7	8
$\rho(r)$ (thousand people/mi^2)	75	75	67.5	60	52.5	45	37.5	30	22.5

   (a) Using this data and Riemann sums, estimate the total population of Boston living within the 8-mile limit.
   (b) Looking at the data, do you think your answer to part (a) is an overestimate or an underestimate? Why?
   (c) Find a possible formula for $\rho(r)$. Use this to make another estimate of the population.

6. The density of oil in a circular oil slick on the surface of the ocean at a distance r meters from the center of the slick is given by $\rho(r) = 50/(1 + r)$ kg/m^2.
   (a) If the slick extends from $r = 0$ to $r = 10{,}000$ m, find a Riemann sum approximating the total mass of oil in the slick.
   (b) Find the exact value of the mass of oil in the slick by turning your sum into an integral and evaluating it.
   (c) Within what distance r is half the oil of the slick contained?

7. Suppose you want to find the total mass of a 3×5 rectangular sheet, whose density per unit area at a distance x from one of the sides of length 5 is $1/(1 + x^4)$.
   (a) Find a Riemann sum which approximates the total mass.
   (b) Find the mass to one decimal place. How do you know your answer is accurate enough?

8. You are given a cardboard figure in the shape shown in Figure 8.6. The region is bounded on the left by the line $x = a$, on the right by the line $x = b$, above by $f(x)$, and below by $g(x)$. If the density $\rho(x)$ gm/cm^2 varies only with x, find an expression for the total mass of the figure, in terms of $f(x)$, $g(x)$, and $\rho(x)$.

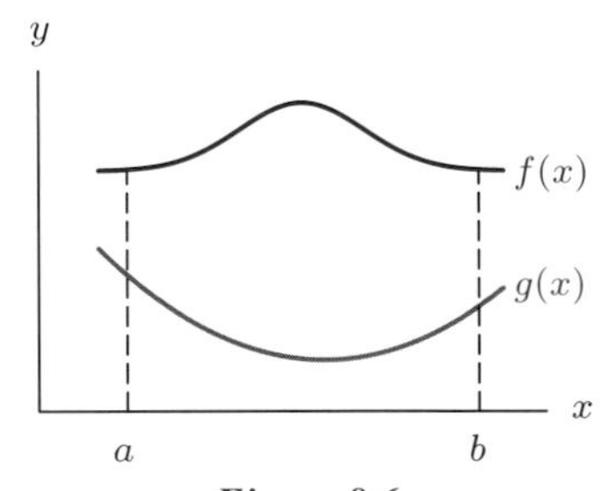

Figure 8.6

9. In the table below we list the density D (gm/cm^3) of the earth at a depth x km below the earth's surface. The radius of the earth is about 6370 km. Find an upper and a lower bound for the earth's mass such that the upper bound is less than twice the lower bound. Explain your reasoning; in particular, what assumptions have you made about the density?

x	0	1000	2000	2900	3000	4000	5000	6000	6370
D	3.3	4.5	5.1	5.6	10.1	11.4	12.6	13.0	13.0

10. An exponential model for the density of the earth's atmosphere says that if the temperature of the atmosphere were constant, then the density of the atmosphere as a function of height, h (in meters), above the surface of the earth would be given by

$$\rho(h) = 1.28e^{-0.000124h} \text{ kg/m}^3.$$

(a) Write (but do not evaluate) a sum that approximates the mass of the portion of the atmosphere from $h = 0$ to $h = 100$ m (i.e., the first 100 meters above sea level). Assume the radius of the earth is 6370 km.

(b) Find the exact answer by turning your sum in part (a) into an integral. Evaluate the integral.

11. Water is flowing in a cylindrical pipe of radius 1 inch. Because water is viscous and sticks to the pipe, the rate of flow varies with the distance from the center. The speed of the water at a distance r inches from the center is $10(1 - r^2)$ inches per second. What is the rate (in cubic inches per second) at which water is flowing through the pipe?

8.2 APPLICATIONS TO GEOMETRY

We have already seen how the definite integral can be used to compute the areas of certain regions. In this section we will show how the definite integral can be used to compute the volume of a solid and the length of a curve. The strategy we will employ to make these computations is to divide the solid (or curve) into small pieces whose volume (or length) we can easily approximate. We then add the contributions of all the pieces, giving a Riemann sum that approximates the total volume (or length). As the number of terms in the sum tends to infinity, the approximation gets better, and the sum approaches a definite integral. The value of the definite integral is therefore the total volume of the solid (or the total length of the curve).

Volumes of Given Cross-Section

When calculating the volume of a solid using Riemann sums, we chop the solid up into smaller pieces whose volumes we can estimate.

How to Slice a Cone

Let's see how we might slice up a cone which is standing with the vertex uppermost. We'll consider two different types of slices. We could divide the cone into vertical slices, each parallel to a vertical plane that passes through its tip. The slices will be arch-shaped (their cross-sections are actually hyperbolas); see Figure 8.7. We could also divide the cone horizontally, giving coin-shaped slices; see Figure 8.8.

To construct a Riemann sum, we probably want to cut the cone into the horizontal (circular) slices, because it would be difficult to estimate the volumes of the arch-shaped slices.

Figure 8.7: Cone cut into vertical slices

Figure 8.8: Cone cut into horizontal slices

Finding Volumes by Slicing

Example 1 Compute the volume, in cubic feet, of the Great Pyramid of Egypt, whose base is a square 755 feet by 755 feet and whose height is 410 feet.

Solution You may already know the equation $V = \frac{1}{3}b^2 \cdot h$ for the volume V of a pyramid with height h and base length b. We will in effect derive this formula in solving this problem.

We will view the pyramid as built up in layers starting from the base. Each layer will be square with a little bit of thickness, which we denote Δh (think of Δh as meaning "a little bit of h"). The bottom layer is a square slab 755 feet by 755 feet by Δh feet. As we move up the pyramid, the layers have shorter side length. If we let s denote this side length, then the volume of each layer is approximately $s^2\,\Delta h$ ft^3, where s varies from 755 feet for the bottom layer to 0 feet for the top layer. (The volume is only approximately $s^2\,\Delta h$ because the sides of the slab are not vertical.) See Figure 8.9.

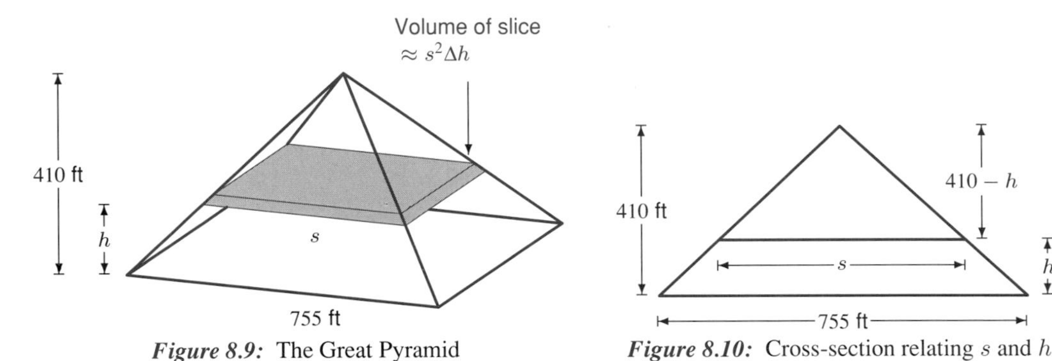

Figure 8.9: The Great Pyramid

Figure 8.10: Cross-section relating s and h

The total volume of the pyramid is the sum of the volumes, $s^2\,\Delta h$, of each layer. Since each layer is at a different height h, we must express s as a function of h so that each term in the sum depends on h alone. If we cut open the pyramid along a plane through the top point of the pyramid and perpendicular to a side of the base, we obtain the triangular cross-section drawn in Figure 8.10. By similar triangles, we get $s/755 = (410-h)/410$. Thus $s = \left(755/410\right)(410-h)$, and the total volume, V, is approximated by

$$V \approx \sum s^2\,\Delta h = \sum \left[\left(\frac{755}{410}\right)(410-h)\right]^2 \Delta h \text{ ft}^3.$$

As the thickness of each slice tends to zero, the sum becomes a definite integral. Finally, since h varies from 0 to 410, the height of the pyramid, we have

$$V = \int_{h=0}^{h=410} \left[\left(\frac{755}{410}\right)(410-h)\right]^2 dh = \left(\frac{755}{410}\right)^2 \int_0^{410} (410-h)^2\,dh$$

$$= \left(\frac{755}{410}\right)^2 \left[-\frac{(410-h)^3}{3}\right]\Bigg|_0^{410} = \frac{1}{3}\left(\frac{755}{410}\right)^2 (410)^3 = \frac{1}{3}(755)^2(410)$$

$$\approx 78 \text{ million cubic feet}$$

Note that $V = \frac{1}{3}(755)^2(410) = \frac{1}{3}b^2 \cdot h$, as expected.

Example 2 Find the volume of a hemisphere of radius 7 inches.

Solution Again, you probably know the formula $\frac{4}{3}\pi R^3$ for the volume of a sphere of radius R. We will derive this using a Riemann sum and a definite integral.

We imagine the hemisphere sitting flat on a plane and divide it into horizontal layers of thickness Δh inches. (See Figure 8.11.) Each layer is a circular slab with radius r, where r varies from 7 inches for the bottom layer to 0 inches for the top layer. Thus, the volume of each layer is approximately $\pi r^2 \Delta h$ in^3, and our problem is to express r in terms of h so that we can write a Riemann sum (all of whose terms depend only on h) representing the sum of the volumes of all the separate layers. From Figure 8.12, showing a vertical slice through the center of the hemisphere, we find that $r^2 = 7^2 - h^2$. So at height h,

$$\text{Volume of slice} \approx \pi r^2 \Delta h = \pi(7^2 - h^2)\,\Delta h \text{ in}^3.$$

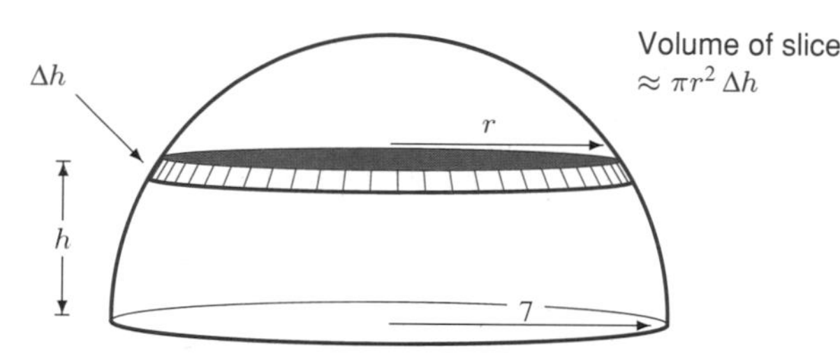

Figure 8.11: Slicing to find the volume of a hemisphere

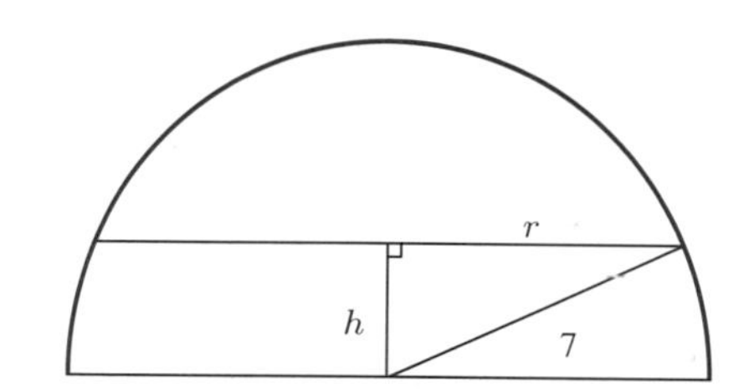

Figure 8.12: Vertical slice through center of hemisphere showing relation between r and h

Summing the volumes of all slices, we have

$$\text{Volume} = V \approx \sum \pi r^2 \Delta h = \sum \pi(7^2 - h^2)\,\Delta h \text{ in}^3.$$

As the thickness of each slab tends to zero, the sum becomes a definite integral. Since h varies from 0 to 7, the radius of the hemisphere, we have

$$V = \int_{h=0}^{h=7} \pi r^2\,dh = \pi \int_0^7 (7^2 - h^2)\,dh = \pi\left(7^2 h - \frac{1}{3}h^3\right)\Bigg|_0^7 = \pi\left(7^3 - \frac{1}{3}\cdot 7^3\right) = \frac{2}{3}\pi 7^3 \text{ in}^3.$$

The volume of the whole sphere is $2(\frac{2}{3}\pi 7^3) = \frac{4}{3}\pi 7^3$ in^3, as we expected.

In Examples 1 and 2 all cross-sections of the solid perpendicular to a certain direction were the same geometric figure — a square in the case of the pyramid and a circle in the case of the hemisphere. The volumes of other solids with known cross-sections can be calculated by a similar method.

Example 3 Find the volume of the solid whose base is the region in the xy-plane bounded by the curves $y = x^2$ and $y = 8 - x^2$ and whose cross-sections perpendicular to the x-axis are squares with one side in the xy-plane. (See Figure 8.13.)

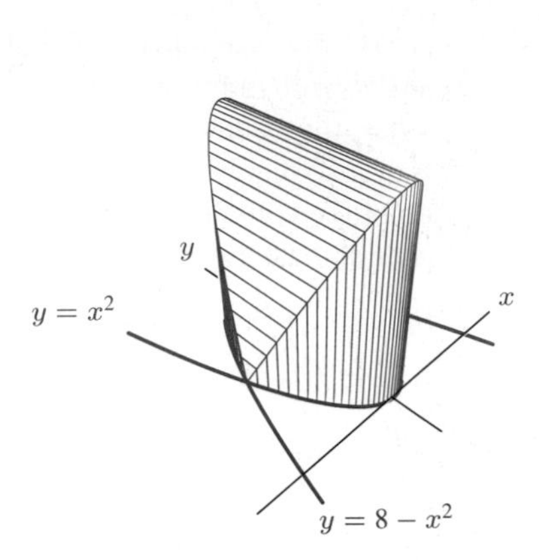

Figure 8.13: The solid for Example 3

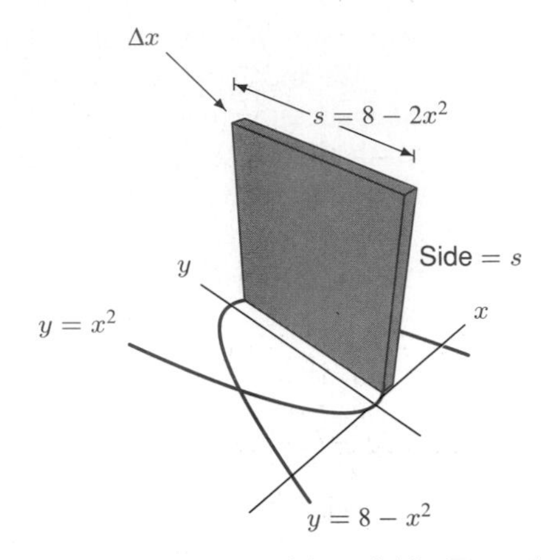

Figure 8.14: A slice of the solid for Example 3

Solution We view the solid as a loaf of bread sitting on the xy-plane and made up of square slices. These slices are analogous to the square layers making up the pyramid in our first example. A typical slice is shown in Figure 8.14. The thickness of each slice is a little bit of distance along the x-axis, and hence we denote it by Δx. The side length s of the square face of each slice is just the distance (in the y direction) between the two curves. Therefore, in terms of x we have $s = (8 - x^2) - x^2 = 8 - 2x^2$.

Notice that this side length grows from $s = 0$ at $x = -2$ to $s = 8$ at $x = 0$, and then it shrinks back to $s = 0$ at $x = 2$. The volume of the solid is obtained by summing up the volumes of all the slices. Each slice has volume $s^2 \,\Delta x = (8 - 2x^2)^2 \,\Delta x$, so the total volume is

$$V \approx \sum s^2 \,\Delta x = \sum (8 - 2x^2)^2 \,\Delta x.$$

As the thickness Δx of each slice tends to zero, the sum becomes a definite integral. Since x varies between -2 and 2, we have

$$\begin{aligned}
V &= \int_{-2}^{2} (8 - 2x^2)^2 \,dx \\
&= \int_{-2}^{2} (64 - 32x^2 + 4x^4) \,dx \\
&= 64x - \frac{32}{3}x^3 + 4\frac{x^5}{5}\Bigg|_{-2}^{2} \\
&= 256\left(1 - \frac{2}{3} + \frac{1}{5}\right) = 256\left(\frac{8}{15}\right) \\
&\approx 136.5.
\end{aligned}$$

We can check the plausibility of this answer by observing that the given solid is somewhat larger than the solid obtained by putting together two pyramids with 8×8 bases at $x = 0$ and apexes (highest points) at $x = 2$ and $x = -2$. By our first example, the total volume of these two pyramids is $2(\frac{1}{3})8^2 \cdot 2 = 256(\frac{1}{3})$, which is, as expected, less than our answer, $256(\frac{8}{15})$.

Volumes of Revolution

Another way to create a solid having known cross-sections is to revolve a region in the plane around some line in 3-space, giving a *solid of revolution,* as in the following examples.

Example 4 The region bounded by the curve $y = e^{-x}$ and the x-axis between $x = 0$ and $x = 1$ is revolved around the x-axis. Find the volume of this solid of revolution.

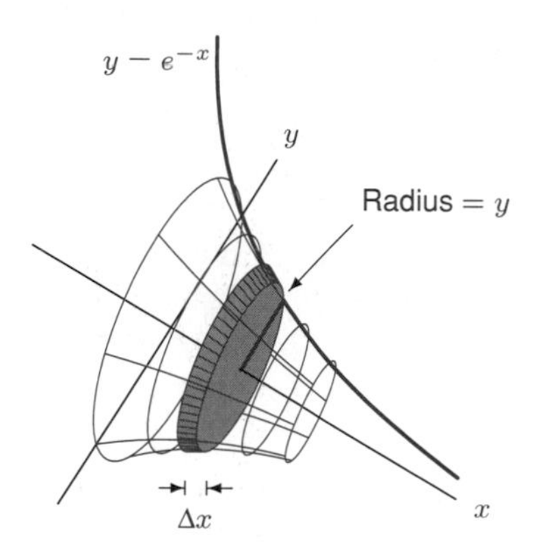

Figure 8.15: A thin strip rotated around the x-axis to form a circular slice

Solution Imagine the plane region divided into thin strips perpendicular to the x-axis. Each strip has width Δx. As we rotate the whole region about the x-axis, each strip sweeps out a circular slice or disk of thickness Δx (see Figure 8.15). The radius of the slice, as can be seen in Figure 8.15, is the distance from the x-axis to the curve, which is just the value of $y = e^{-x}$ on the slice. Thus the volume of the slice is $\pi y^2 \Delta x = \pi(e^{-x})^2 \Delta x$, and the total volume, V, of the solid is given by

$$V \approx \sum \pi y^2 \Delta x = \sum \pi \left(e^{-x}\right)^2 \Delta x.$$

As the thickness of each slice tends to zero, we get

$$V = \int_0^1 \pi(e^{-x})^2\, dx = \pi \int_0^1 e^{-2x}\, dx = \pi \left(-\frac{1}{2}\right) e^{-2x}\Big|_0^1$$
$$= \pi \left(-\frac{1}{2}\right)(e^{-2} - e^0) = \frac{\pi}{2}(1 - e^{-2}) \approx 1.36.$$

Example 5 The region bounded by the curves $y = x$ and $y = x^2$ is rotated about the line $y = 3$. Compute the volume of the resulting solid.

Solution Again imagine the region divided into thin vertical strips of thickness Δx, as in Figure 8.16. As each strip is rotated around the line $y = 3$, it sweeps out a slice of the total solid shaped like a disk with a hole in it (see Figures 8.17 and 8.18). This disk-with-a-hole has two radii: an inner radius, r_{in}, which

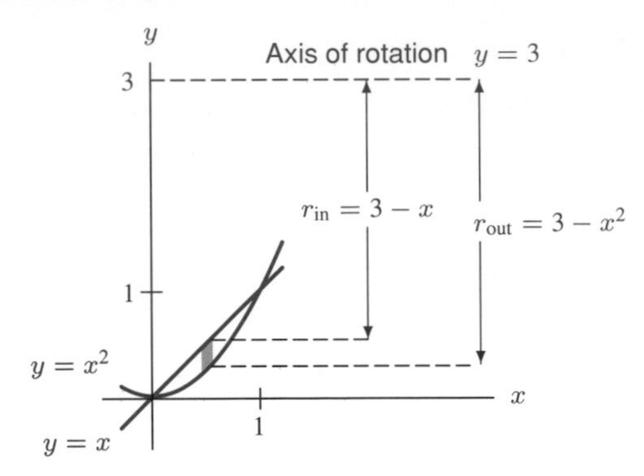

Figure 8.16: The region for Example 5

is the distance from the line $y = 3$ to the nearer curve $y = x$, and an outer radius, r_{out}, which is the distance from the line $y = 3$ to the more distant curve, $y = x^2$. So the slice at x has inner radius $r_{\text{in}} = 3 - x$ and outer radius $r_{\text{out}} = 3 - x^2$. Think of the slice as a circular disk of radius r_{out} from which has been removed a smaller circular disk of radius r_{in}. Thus, the volume of the slice is

$$\pi r_{\text{out}}^2 \,\Delta x - \pi r_{\text{in}}^2 \,\Delta x = \pi(3 - x^2)^2 \,\Delta x - \pi(3 - x)^2 \,\Delta x.$$

Adding up the volumes of all the slices, we have

$$V \approx \sum \left(\pi r_{\text{out}}^2 - \pi r_{\text{in}}^2\right) \Delta x = \sum \left[\pi(3 - x^2)^2 - \pi(3 - x)^2\right] \Delta x.$$

We let Δx, the thickness of each slice, tend to zero to obtain the definite integral. Since the curves $y = x$ and $y = x^2$ intersect at $x = 0$ and $x = 1$, we have

$$\begin{aligned} V &= \int_0^1 \left[\pi(3 - x^2)^2 - \pi(3 - x)^2\right] dx = \pi \int_0^1 \left[(9 - 6x^2 + x^4) - (9 - 6x + x^2)\right] dx \\ &= \pi \int_0^1 (6x - 7x^2 + x^4) \, dx = \pi \left(3x^2 - \frac{7x^3}{3} + \frac{x^5}{5}\right) \bigg|_0^1 \approx 2.72. \end{aligned}$$

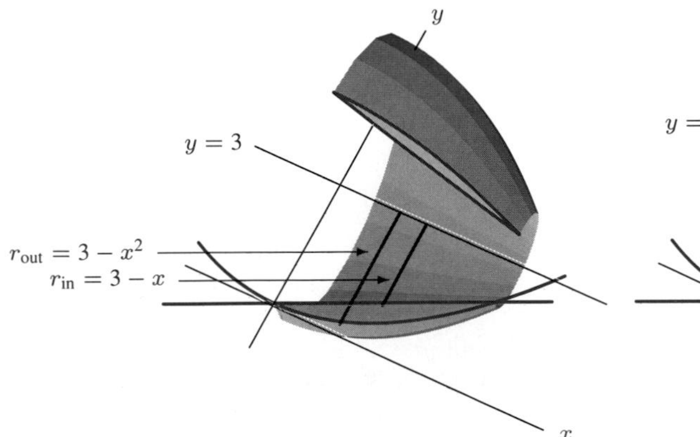

Figure 8.17: Cutaway view of volume (note inner and outer radii)

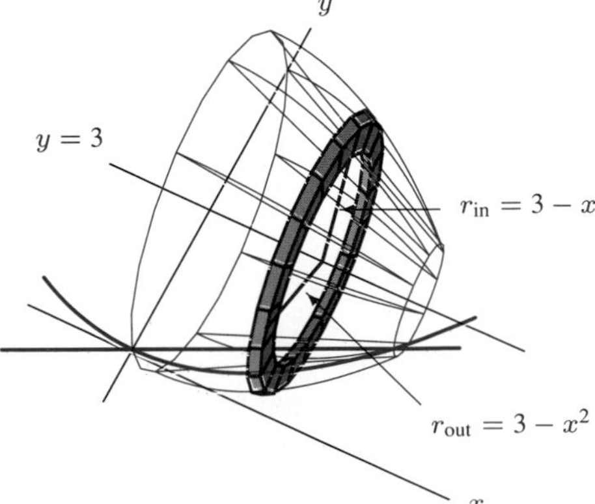

Figure 8.18: One slice (a disk-with-a-hole)

Arc Length

Just as the definite integral can be used to compute the volumes of various solids, it can also be used to compute the lengths of various curves. Suppose we wish to compute the length, or the *arc length*, as it is usually called, of a curve $y = f(x)$ from $x = a$ to $x = b$, where $a < b$. As usual, we divide the curve into many little pieces; each one is approximately straight.

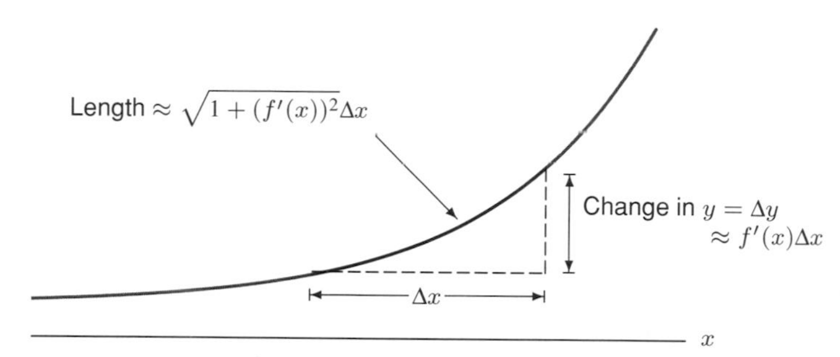

Figure 8.19: A little bit of arc length

Figure 8.19 shows that a small change Δx in the x coordinate produces a corresponding small change in the y coordinate of approximately $\Delta y \approx f'(x)\,\Delta x$. The length of the little piece of the curve is then approximately

$$\sqrt{(\Delta x)^2 + (\Delta y)^2} \approx \sqrt{(\Delta x)^2 + \left(f'(x)\,\Delta x\right)^2} = \sqrt{1 + (f'(x))^2}\,\Delta x.$$

Thus, for $a < b$, the arc length of the curve $y = f(x)$ from $x = a$ to $x = b$ is approximated by a Riemann sum:

$$\text{Arc length} = L \approx \sum \sqrt{1 + \left(f'(x)\right)^2}\,\Delta x.$$

As we let Δx tend to zero, the sum becomes the definite integral. Since x is to vary between a and b, we have the following expression for the arc length.

$$\text{Arc length} = L = \int_a^b \sqrt{1 + (f'(x))^2}\,dx.$$

Example 6 Set up and evaluate an integral to compute the length of the curve $y = x^3$ from $x = 0$ to $x = 5$.

Solution Since if $f(x) = x^3$, then $f'(x) = 3x^2$, we have

$$L = \int_0^5 \sqrt{1 + (3x^2)^2}\,dx.$$

Although the formula for the arc length of a curve is easy to apply, the integrands it generates often do not have elementary antiderivatives. As a result, the only way to evaluate an arc length integral such as this one is by numerical methods. In this case, Simpson's rule for $\int_0^5 \sqrt{1 + 9x^4}\,dx$ gives 125.680669093 with 10 steps, and 125.680300379 with 20 steps. An answer of 125.68 looks reasonable. The curve starts at $(0, 0)$ and goes to $(5, 125)$, so its length must be at least the length of

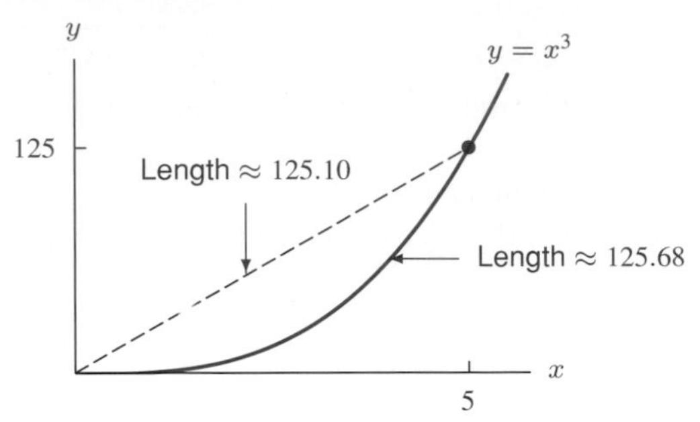

Figure 8.20: Arc length of $y = x^3$ (note that the picture is distorted because the scales on the two axes are so different)

a straight line between these points, or $\sqrt{5^2 + 125^2} = 125.10$. (See Figure 8.20.) Are you surprised that the lengths along the line and the curve are so close?

Problems for Section 8.2

1. Imagine a hard-boiled egg lying on its side cut into thin slices. First think about vertical slices and then horizontal ones. What would these slices look like? Sketch them.
2. Find the volume of a sphere of radius r by slicing up the sphere and setting up a Riemann sum that approximates the volume, and then finding the definite integral obtained when the thickness of each slice tends to zero.
3. Find, by slicing, the volume of a cone whose height is 3 cm and whose base radius is 1 cm. Slice the cone as shown in Figure 8.8, page 427.
4. Find, by slicing, a formula for the volume of a cone of height h and base radius r.

For Problems 5–7, sketch the solid obtained by rotating each region around the indicated axis. Using the sketch, show how to approximate the volume of the solid by a Riemann sum, and hence find the volume.

5. Region bounded by the first arch of $y = \sin x$, $y = 0$. Axis: x axis.
6. Region bounded by $y = x^3$, $x = 1$, $y = -1$. Axis: $y = -1$.
7. Region bounded by $y = \sqrt{x}$, $x = 1$, $y = 0$. Axis: $x = 1$.
8. When the ellipse $x^2/a^2 + y^2/b^2 = 1$ is rotated about the x-axis, it generates an *ellipsoid*. Compute its volume.

For Problems 9–13 consider the region bounded by $y = e^x$, the x-axis, and the lines $x = 0$ and $x = 1$. Find the volume of the following solids.

9. The solid obtained by rotating the region about the x-axis.
10. The solid obtained by rotating the region about the horizontal line $y = -3$.
11. The solid obtained by rotating the region about the horizontal line $y = 7$.
12. The solid whose base is the given region and whose cross-sections perpendicular to the x-axis are squares.
13. The solid whose base is the given region and whose cross-sections perpendicular to the x-axis are semicircles.

14. Rotate the bell-shaped curve $y = e^{-x^2/2}$ shown in Figure 8.21 around the y-axis, forming a hill-shaped solid of revolution. By slicing horizontally, find the volume of this hill.

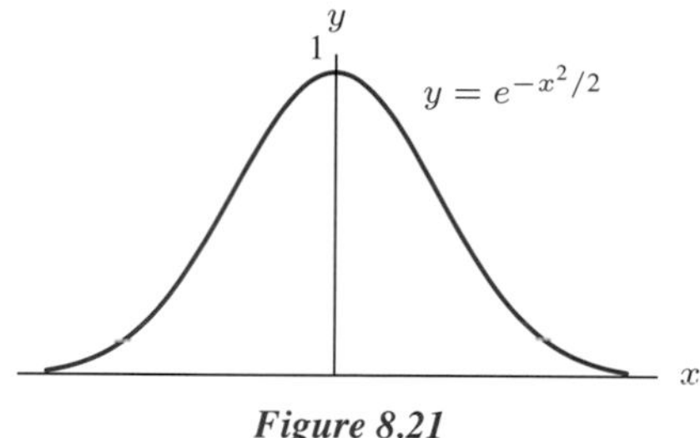

Figure 8.21

15. The circumference of the trunk of a certain tree at different heights above the ground is given in the following table.

Height (feet)	0	20	40	60	80	100	120
Circumference (feet)	26	22	19	14	6	3	1

Assume all horizontal cross-sections of the trunk are circles. Estimate the volume of the tree trunk using the trapezoid rule.

16. Most states expect to run out of space for their garbage soon. In New York, solid garbage is packed into pyramid-shaped dumps with square bases. (The largest such dump is on Staten Island.) A small community has a dump with a base length of 100 yards. One yard vertically above the base, the length of the side parallel to the base is 99 yards; the dump can be built up to a vertical height of 20 yards. (The top of the pyramid is never reached.) If 65 cubic yards of garbage arrive at the dump every day, how long will it be before the dump is full?

17. The hull of a certain boat has widths given by the following table. Reading across a row of the table gives widths at points $0, 10, \ldots, 60$ feet from the front to the back at a certain level below waterline. Reading down a column of the table gives widths at levels 0, 2, 4, 6, 8 feet below waterline at a certain point from the front to the back. Use the trapezoidal rule to estimate the volume of the hull below waterline.

		Front of boat $\longrightarrow$ Back of boat						
		0	10	20	30	40	50	60
	0	2	8	13	16	17	16	10
Depth	2	1	4	8	10	11	10	8
below	4	0	3	4	6	7	6	4
waterline	6	0	1	2	3	4	3	2
(in feet)	8	0	0	1	1	1	1	1

For Problems 18 and 19, find the arc length of the given function from $x = 0$ to $x = 2$.

18. $f(x) = \sqrt{4 - x^2}$

19. $f(x) = \sqrt{x^3}$

20. (a) Write an integral which represents the circumference of a circle of radius r.
(b) Evaluate the integral, and show that you get the answer you expect.

21. Figure 8.22 is a graph of the function

$$y = \frac{1}{2}(e^x + e^{-x}).$$

It is called a *catenary* and represents the shape in which a cable hangs. Find the length of this catenary between $x = -1$ and $x = 1$.

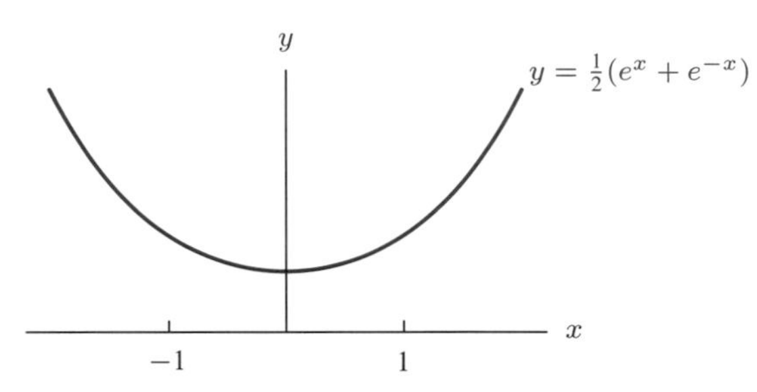

Figure 8.22

22. Compute the perimeter of the region used in Problems 9–13 with error at most 0.001.

23. Set up an integral for the circumference of an ellipse with semimajor axis $a = 2$ and semiminor axis $b = 1$. As you try to evaluate this integral numerically (i.e., using a calculator or computer), you are faced with a problem. Describe this problem. Formulate a way to solve it.

24. There are very few elementary functions $y = f(x)$ for which arc length can be computed in elementary terms using the formula

$$\int_a^b \sqrt{1 + \left(\frac{dy}{dx}\right)^2}\, dx.$$

You have seen some such functions f in Problems 18, 19, and 21, namely, $f(x) = \sqrt{4 - x^2}$, $f(x) = \sqrt{x^3}$, and $f(x) = \frac{1}{2}(e^x + e^{-x})$. Try to find some other function that "works," that is, a function whose arc length you can find using this formula and antidifferentiation.

25. After doing Problem 24 you may wonder what sort of functions can represent arc length. If $g(0) = 0$ and g is differentiable and increasing, then can $g(x)$, $x \geq 0$, represent arc length? That is, can we find a function $f(t)$ such that

$$\int_0^x \sqrt{1 + (f'(t))^2}\, dt = g(x)?$$

(a) Show that $f(x) = \int_0^x \sqrt{(g'(t))^2 - 1}\, dt$ works, namely that the arc length of the graph of f from 0 to x is $g(x)$, as long as $g'(x) \geq 1$.
(b) Show that if $g'(x) < 1$ for some x, then $g(x)$ cannot represent the arc length of the graph of any function.
(c) Find a function f whose arc length from 0 to x is $2x$.

8.3 APPLICATIONS TO PHYSICS

Although geometric problems of area, length, and volume were a driving force for the development of the calculus in the seventeenth century, it was Newton's spectacularly successful applications of the calculus to physics that most clearly demonstrated the power of this new mathematics.

Work

Our first application involves the concept of work. In physics the word "work" has a technical meaning which is different from its everyday meaning. Physicists say that if a constant force, F, is applied to some object to move it a distance, d, then the force has done work on the object. We make the following definition:

$$\text{Work done} = \text{Force} \times \text{Distance}$$
$$W = F \cdot d$$

Notice that by the definition, if an object does not move, then no work is done. Thus, holding a 5-pound book at the same height for an hour accomplishes no work, even though it might make you tired. You are exerting a force on the book to keep the book from falling (that is, to counteract the force of gravity on the book), but you are not moving that force through a distance. This does make sense, since the book could just as easily have been sitting on a table for an hour and we wouldn't say that the table had accomplished any work. Also, if we walked across the room holding the book, we would not accomplish any work on the book, since the only force we exert on the book is vertical but the motion of the book is horizontal; the distance, d, in the equation $W = Fd$ must be in the same direction as the force, F. On the other hand, if you lift the book from the floor to a height of 3 feet off the floor, this process of lifting accomplishes work. Note that time has nothing to do with work. Whether the object is moved quickly or slowly is irrelevant.

Note on Units: Mass versus Weight

One further complication should be mentioned. When we talk about how much something weighs, we're talking about the force that gravity exerts on that object. A lump of iron, for example, might weigh 10 pounds on the surface of the earth. This is because the earth's gravitational field is busily exerting 10 pounds of force on that lump of iron. If the same lump, however, is adrift in interstellar space, it is weightless because there is no gravitational field in its vicinity to exert a force on it. Of course, the quantity of iron present hasn't changed; there's no less iron there than there was before. So we make a distinction between the *mass* of the lump (the quantity of matter it comprises) and the *weight* of that mass (the force exerted on it due to gravity).

In the British system of units, a *pound* is a unit of weight. In the metric system, however, a *kilogram* is a unit of mass, not weight. Thus knowing we have 1 kilogram of iron means that we know the mass of iron present. To figure out how much our iron weighs, we have to figure out how much force is being exerted on it by gravity. Since Newton's second law tells us that Force = Mass × Acceleration, we see that we have to multiply our iron's mass by the acceleration due to gravity, which (on the surface of the earth) is 9.8 m/sec^2. Thus our kilogram of iron weighs $1 \text{ kg} \cdot 9.8 \text{ m/sec}^2 = 9.8 \text{ kg m/sec}^2$, which is almost always written 9.8 newtons. (A newton $= 1 \text{ kg} \cdot \text{m/sec}^2$ is a unit of force or weight. It is abbreviated nt.)

Example 1 How much work is done by lifting

(a) A 5-pound book 3 feet off the floor? (b) A 1.5-kilogram book 2 meters off the floor?

Solution (a) $W = F \cdot d = (5 \text{ lb})(3 \text{ ft}) = 15$ foot-pounds.

(b) In this case we have not done $(1.5)(2) = 3$ units of work. We need to use the book's weight (not its mass), namely $(1.5 \text{ kg})(g \text{ m/sec}^2) = (1.5)(9.8)$ newtons. Thus, the work done is

$$W = F \cdot d = [(1.5 \text{ kg})(9.8 \text{ m/sec}^2)] \cdot (2 \text{ m}) = 29.4 \text{ newton-meters.}$$

A newton-meter is usually called a joule, so you did 29.4 joules of work in the metric system of units. In the British system the unit of work is the foot-pound, the work done when a 1-pound force moves an object a distance of 1 foot.

Calculating the Work Done

In the following examples, we calculate the work done where both the force and distance moved may vary. To find the work done on a complicated object, *we slice the object up in such a way that we can find the work done on each piece.* We will calculate the work for each piece using $W = Fd$, and we will sum these pieces to approximate the total work as a Riemann sum. Letting the size of each piece tend to zero, we will obtain a definite integral that represents the total work.

Example 2 A 28-meter uniform chain with a mass of 20 kilograms is dangling from the roof of a building. How much work is needed to pull the chain up to the top of the building?

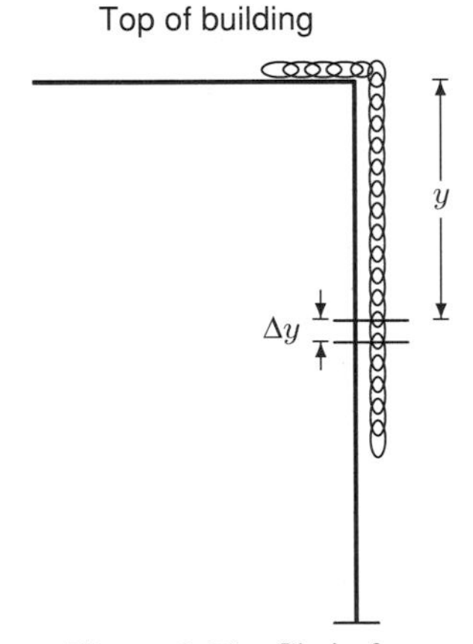

Figure 8.23: Chain for Example 2

Solution Since a 20-kilogram chain weighs $(20 \text{ kg})(9.8 \text{ m/ sec}^2) = 196$ newtons, it might seem that the answer should be $(196 \text{ nt})(28 \text{ m}) = 5488$ joules. But remember that not all of the chain has to move 28 meters—links near the top move less.

Let's divide the chain into small sections of length Δy, each weighing $7\,\Delta y$ newtons. (A length of 28 meters weighs 196 newtons, so 1 meter weighs 7 newtons.) See Figure 8.23. If Δy is small,

all of this piece is hauled up approximately the same distance, namely y, against the force due to gravity of $7\,\Delta y$ newtons. Thus, the work done on the small piece is approximately

$$(7\,\Delta y \text{ newtons})(y \text{ meters}) = 7y\,\Delta y \text{ joules.}$$

The work done on the entire chain is given by the total of the work done on each piece:

$$W \approx \sum 7y\,\Delta y \text{ joules.}$$

As Δy tends to zero, we obtain a definite integral. Since y varies from 0 to 28 meters, the total work is

$$W = \int_0^{28} (7y)\,dy = \frac{7}{2}y^2 \Big|_0^{28} = 2744 \text{ joules.}$$

Example 3 It is reported that the Great Pyramid of Egypt was built in 20 years. If the stone making up the pyramid has density 200 pounds per cubic foot, find the total amount of work done in building the pyramid. Estimate how many workers were needed to build the pyramid.

Solution We assume that the stones were originally located at the approximate height of the construction site. Imagine the pyramid built in layers as we did in Example 1, page 428, of the previous section. Recall that the pyramid is 410 feet high and has a square base 755 feet by 755 feet.

We found by similar triangles that the layer at height h has a side length $s = \frac{755}{410}(410 - h)$ ft. The layer at height h has a volume of $s^2\,\Delta h$ ft^3 and hence a weight of $200s^2\,\Delta h$ lb. Constructing this layer involved lifting a weight of $200s^2\,\Delta h$ through a height of h, so the work required was $(200s^2\,\Delta h \text{ lb}))(h \text{ ft}) = 200s^2 h\,\Delta h$ ft-lb. (See Figure 8.24.) The total work, W, of constructing all the layers is then

$$W \approx \sum 200s^2\,h\Delta h = \sum 200\left(\frac{755}{410}\right)^2 (410 - h)^2 h\,\Delta h \text{ ft-lb.}$$

As the thickness of each layer, Δh, tends to zero, we obtain a definite integral. Since h varies from 0 to 410, we have

$$W = \int_0^{410} 200s^2 h\,dh = 200\int_0^{410} \left(\frac{755}{410}\right)^2 (410 - h)^2 h\,dh = 200\left(\frac{755}{410}\right)^2 \int_0^{410} (410 - h)^2 h\,dh$$
$$= 200\left(\frac{755}{410}\right)^2 \left(410^2\frac{h^2}{2} - 2(410)\frac{h^3}{3} + \frac{h^4}{4}\right)\Big|_0^{410} = 200\left(\frac{755}{410}\right)^2 (410)^4\left(\frac{1}{12}\right)$$
$$\approx 1{,}597{,}020{,}000{,}000 \approx 1.6 \times 10^{12} \text{ foot-pounds.}$$

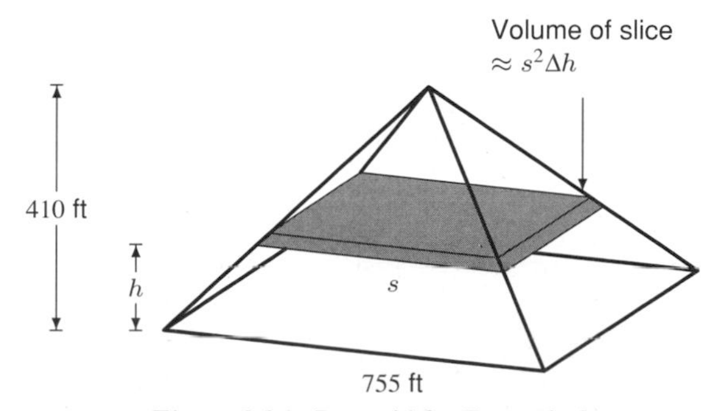

Figure 8.24: Pyramid for Example 3

We have calculated the total work done in building the pyramid; now we want to estimate the total number of workers needed. Let's assume every laborer worked 10 hours a day, 300 days a year, for 20 years. Assume that a typical worker lifted ten 50 pound blocks a distance of 4 feet every hour, thus performing 2000 foot-pounds of work per hour (this is a very rough estimate). Then each laborer performed $(10)(300)(20)(2000) = 1.2 \times 10^8$ foot-pounds of work over a twenty year period. Thus, the number of workers needed was about $(1.6 \times 10^{12})/(12 \times 10^7)$, or about 13,000.

Escape from the Earth's Gravitational Field

If you throw an object into the air, you expect gravity to bring it back to earth. However, the harder you throw it, the longer you expect to wait before it returns. Have you ever wondered if it is possible to throw something hard enough so that it never returns to earth? We will calculate the total amount of work done to move an object infinitely far away from the earth. Amazingly enough, this total amount of work is finite, and it *is* possible to throw something hard enough that it never returns to earth.

Previously, when we analyzed the motion of a freely falling body, we assumed that the force of gravity is constant. Now, since we are thinking of a body moving into outer space, we have to take into account the fact that, as we move away from the earth, its gravitational force gets weaker. Newton's Law of Gravitation says that the force, F, exerted on a mass m at a distance, r, from the center of the earth is given by

$$F = \frac{GMm}{r^2}$$

where $r > R$, the earth's radius, M is the mass of the earth, and G is called the gravitational constant, whose value is about 6.67×10^{-11} if mass is measured in kilograms, distance in meters, and force in newtons.

Example 4 Find the total work done to move an object of mass m infinitely far from the earth.

Solution The force pulling the object back to the earth is GMm/r^2. The work needed to move the object a distance Δr further away is

$$\left(\frac{GMm}{r^2}\right)\Delta r.$$

Therefore the total work required to move the object from the surface of the earth $r = R$ to a point "infinitely far away" is approximated by

$$W \approx \sum \frac{GMm}{r^2}\Delta r$$

where the value of r runs from $r = R$ (at the earth's surface) to $r = \infty$ (very far away). As Δr tends to zero, we get the definite integral

$$W = \int_R^\infty \frac{GMm}{r^2}\,dr = \lim_{b\to\infty} -\left.\frac{GMm}{r}\right|_R^b = \frac{GMm}{R}.$$

Imagine an object shot upward with initial velocity v. The energy of the body by virtue of its motion is called its *kinetic energy* and is equal to $\frac{1}{2}mv^2$. When the object moves upward, it slows down, losing kinetic energy as work is done against gravity. The body moves upward until all the kinetic energy is expended: it will stop at the height at which

$$\text{Initial kinetic energy} = \text{Work done against gravity}.$$

Since the work done to move an object infinitely far from the earth is finite, a finite starting velocity can give the object enough kinetic energy to move infinitely far away. The *escape velocity* from the earth is defined to be the minimum vertical velocity that must be given to an object so that it is never drawn back to earth by gravity.

Example 5 Find the escape velocity of an object from the earth.

Solution If v is the initial velocity, then setting the initial kinetic energy equal to the work done to escape infinitely far from the earth, we get

$$\frac{1}{2}mv^2 = \frac{GMm}{R}.$$

Solving for v gives

$$v = \sqrt{\frac{2GM}{R}}.$$

Since the mass of the earth is $M \approx 6 \times 10^{24}$ kg and its radius is $R \approx 6.4 \times 10^6$ meters, the escape velocity is about 11,000 meters/sec $\approx$ 25,000 mph, or about 30 times the speed of sound. Thus, the escape velocity is finite, though it is much higher than you can achieve by throwing or by shooting a gun.

Force and Pressure

We can use the definite integral to compute the force exerted by a liquid on a surface, for example, the force of water on a dam. The idea is to get the force from the *pressure*. The pressure in a liquid is the force per unit area exerted on a small piece of area in the liquid. Two things you need to know about pressure are:

- At any point, pressure is exerted equally in all directions — up, down, sideways.
- Pressure increases with depth. (That is one of the reasons why deep sea divers have to take much greater precautions than scuba divers.)

At a depth of h feet, the pressure exerted by the liquid measured in pounds per square foot, is given by computing the total weight of a column of liquid h feet high with a base of 1 square foot. The volume of such a column of liquid is just h cubic feet.

If the liquid has density ρ (mass per unit volume), then its weight per unit volume is ρg (where g is the acceleration due to gravity). Thus the weight of the column of liquid is $\rho g h$, so

$$\text{Pressure} = \text{Density} \times g \times \text{Depth}$$
$$p = \rho g h$$

Sometimes you may be given the density of the liquid as a weight per unit volume, rather than a mass per unit volume. In that case, you do not need to multiply by g because it has already been

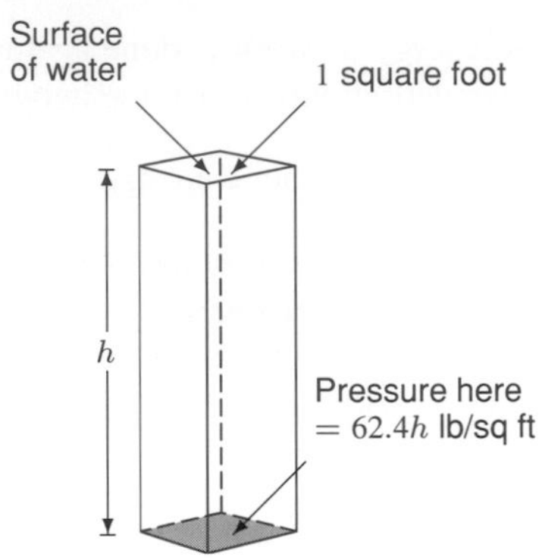

Figure 8.25: Pressure exerted by a column of water

done. Since pounds are units of weight, and not mass, knowing that water weighs 62.4 lb/ft^3 tells us that for water, $\rho g = 62.4$ lb/ft^3, and pressure at depth h is $62.4h$ lb/ft^2. (See Figure 8.25.)

Next you need to know the relation between force and pressure. Provided the pressure is constant over a given area, we have the following relation:

$$\text{Force} = \text{Pressure} \times \text{Area}$$

When the pressure isn't constant over a surface, we divide the surface into small pieces *in such a way that the pressure is nearly constant on each one*. Then we can use this formula on each piece and obtain a definite integral to get the force over an entire surface. Since the pressure varies with depth, we should divide the surface into horizontal strips, each of which is at an approximately constant depth.

Example 6 A trough 14 feet long has one rectangular vertical side 14 feet by 3 feet, one rectangular inclined side 14 feet by 5 feet, and two triangular ends with sides 3 feet, 4 feet, and 5 feet (see Figure 8.26). If the trough is filled to the top with water, compute the force on each end and each side.

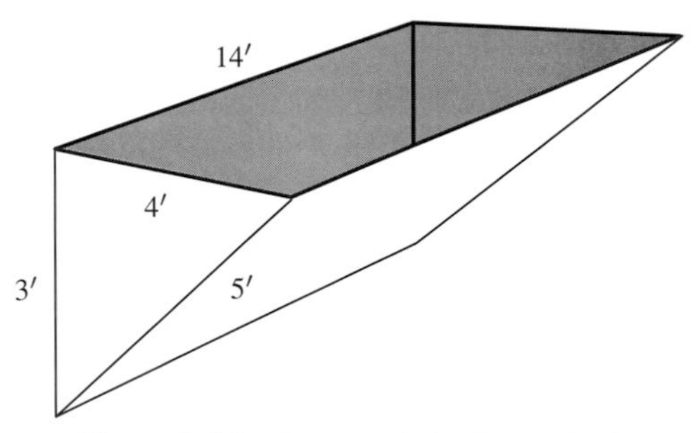

Figure 8.26: The trough for Example 6

Solution We will do the long vertical side first. Divide the side into thin horizontal strips of width Δh. Since pressure depends only on the depth h and the strips are thin and horizontal, we can assume the pressure is the same at every point in the strip. The area of the strip is $14\,\Delta h$ ft^2, and the force on the strip at depth h ft is the pressure at that depth times the area of the strip:

$$\text{Force on strip} = \text{Pressure} \times \text{Area} \approx (62.4h \text{ lb/ft}^2)(14\,\Delta h \text{ ft}^2) = (62.4h)14\,\Delta h \text{ lb}.$$

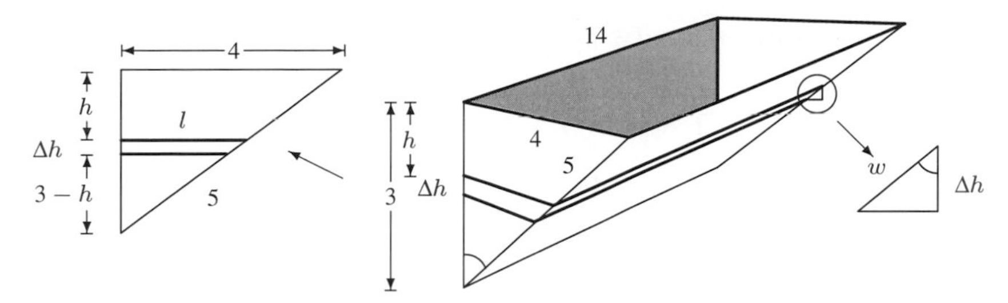

Figure 8.27: Calculating the widths of strips on the end and the inclined side

Summing the forces on these strips, we get an expression for the total force on the vertical side:

$$F_{\substack{\text{vertical}\\\text{side}}} \approx \sum (62.4h)14\,\Delta h \text{ lb.}$$

Letting $\Delta h \to 0$, we get a definite integral. Since h varies from 0 to 3 feet, we have

$$F_{\substack{\text{vertical}\\\text{side}}} = \int_0^3 (62.4h)14\,dh = (62.4)(14)\frac{h^2}{2}\bigg|_0^3 = (62.4)(14)\left(\frac{9}{2}\right) \approx 3930 \text{ pounds.}$$

To compute the force on the inclined side, we use an inclined horizontal strip. Each strip has length 14, but its width is not Δh. By similar triangles (see Figure 8.27) the width w of the strip satisfies

$$\frac{\Delta h}{w} = \frac{3}{5}, \quad \text{so} \quad w = \frac{5}{3}\Delta h.$$

Therefore,

$$\text{Area of strip} = 14w = 14\left(\frac{5}{3}\right)\Delta h \text{ ft}^2,$$

so

$$\text{Force on strip} = (62.4h)(14)\left(\frac{5}{3}\right)\Delta h \text{ lb.}$$

The total force on the inclined side is obtained by adding the forces on the strips, giving

$$F_{\substack{\text{inclined}\\\text{side}}} \approx \sum (62.4h)(14)\left(\frac{5}{3}\right)\Delta h \text{ lb.}$$

We let $\Delta h \to 0$ to arrive at a definite integral. Again, h varies between 0 and 3 feet, so

$$F_{\substack{\text{inclined}\\\text{side}}} = \int_0^3 (62.4h)(14)\left(\frac{5}{3}\right) dh = \left(\frac{5}{3}\right)(62.4)(14)\left(\frac{9}{2}\right) \approx 6550 \text{ pounds.}$$

Notice that this is just $\frac{5}{3}$ the force on the vertical side.

Finally, to compute the force on each triangular end, we again divide the end into horizontal strips. The width of the strip is Δh as with the vertical side, but this time the length of the strip varies from 4 feet at the top of the trough to 0 feet at the bottom. By similar triangles (see Figure 8.27) the length l of a strip is related to the depth h by

$$\frac{l}{4} = \frac{3-h}{3}, \quad \text{so} \quad l = \frac{4}{3}(3-h).$$

Thus, the area of the strip is $l\,\Delta h = \frac{4}{3}(3-h)\Delta h$ ft^2. As before, this means the total force is

$$F_{\text{end}} \approx \sum (62.4h)\left(\frac{4}{3}(3-h)\right)\Delta h \text{ lb.}$$

Since h varies between 0 and 3, as $\Delta h \to 0$ we have

$$F_{\text{end}} = \int_0^3 (62.4h)\left(\frac{4}{3}(3-h)\right)dh = (62.4)\left(\frac{4}{3}\right)\left(3\frac{h^2}{2} - \frac{h^3}{3}\right)\Bigg|_0^3$$

$$= (62.4)\left(\frac{4}{3}\right)\left(\frac{27}{6}\right) \approx 374 \text{ pounds.}$$

Problems for Section 8.3

1. A water tank is in the form of a right circular cylinder with height 20 ft and radius 6 ft. If the tank is half full of water, find the work required to pump all of it over the top rim. (Note that 1 cubic foot of water weighs 62.4 lb.)
2. Suppose the tank in Problem 1 is full of water. Find the work required to pump all of it to a point 10 ft above the top of the tank.
3. A rectangular water tank has length 20 ft, width 10 ft, and depth 15 ft. If the tank is full, how much work does it take to pump all the water out? (Note that 1 cubic foot of water weighs 62.4 lb.)
4. At a construction site a crane lifts a bucket of cement from the ground to a point 30 ft above the ground. The distance from the end of the crane arm to the ground is 75 ft. If the mass of the bucket of cement is 500 lb and that of the cable is 5 lb/ft, find the work required to lift the cement.
5. A 2000-lb cube of ice must be lifted 100 ft, and it is melting at a rate of 4 lb per minute. Assume that it can be lifted at a rate of one foot every minute. Find the work needed to get the block of ice to the desired height.
6. A gas station stores its gasoline in a tank under the ground. The tank is a cylinder lying horizontally on its side. (In other words, the tank is not standing vertically on one of its flat ends.) If the radius of the cylinder is 4 feet, its length is 12 feet, and its top is 10 feet under the ground, find the total amount of work needed to pump the gasoline out of the tank. (Gasoline weighs 42 lb/ft^3.)
7. On a hot afternoon in Pétionville, Haiti, Lubonga is trying to cool off by sipping the glass of palm wine shown in Figure 8.28. How much work does she have to do to empty the glass? (The density of the palm wine is 1.2 gm/cm^3. Initially, the glass is filled to a vertical depth of 8 cm. Note that the force due to gravity acting on 1 gram is 980 dynes, and that in these units work is measured in ergs.)

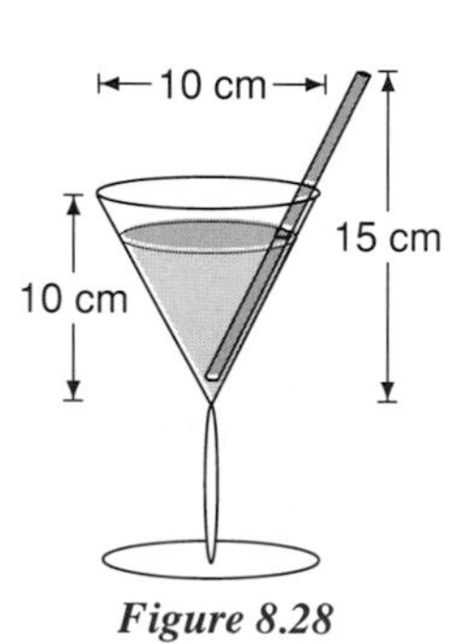

Figure 8.28

8. How much work is required to lift a 1000-kg satellite to an altitude of 2×10^6 m above the surface of the earth? (The radius of the earth is 6.4×10^6 m, its mass is 6×10^{24} kg, and in these units the gravitational constant G, is 6.67×10^{-11}.)

9. Show that the escape velocity needed by an object to escape the gravitational influence of a spherical planet of density ρ and radius R is proportional to R and to $\sqrt{\rho}$.

10. Calculate the escape velocity of an object from the moon. The acceleration due to gravity on the moon is 1.6 m/sec^2, whereas on earth it is 9.8 m/sec^2; the radius of the moon is approximately 1740 km. [Hint: If g is the acceleration due to gravity, the force on an object is $F = mg = GMm/R^2$. See page 440.]

11. A reservoir has a dam at one end. The dam is a rectangular wall, 1000 feet long and 50 feet high. The problem is to find the total force of the water acting on the dam.
    (a) Show how to approximate this force by a Riemann sum, explaining your reasoning clearly.
    (b) Derive an integral which represents the force, and evaluate it.

12. Before you pumped out any water, what was the total force on the bottom and each side of the tank in Problem 3?

13. We define the electric potential at a distance r from an electric charge q by q/r. The electric potential of a charge distribution is obtained by adding up the potential from each point. Suppose electric charge is sprayed (with constant density σ in units of charge/unit area) on to a circular disk of radius a. Consider the axis perpendicular to the disk and through its center. Find the electric potential at the point P on this axis at a distance R from the center. (See Figure 8.29.)

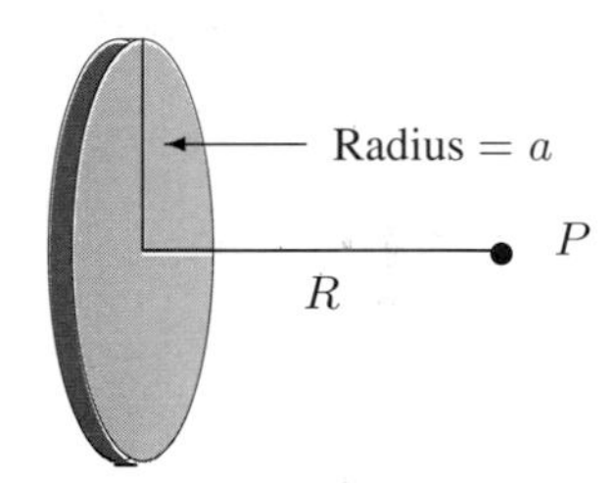

Figure 8.29

14. A car moving at a speed of v mph achieves $25 + 0.1v$ mpg (miles per gallon) for v between 20 and 60 mph. Suppose your speed as a function of time is given by

$$v = 50\frac{t}{t+1}.$$

How many gallons do you consume between $t = 2$ and $t = 3$ (i.e., in the third hour)?

For Problems 15-16, find the kinetic energy of a rotating body, using the fact that the kinetic energy of a particle of mass m moving at a speed v is $\frac{1}{2}mv^2$. Slice the object into pieces in such a way that the velocity is approximately constant across each piece.

15. Find the kinetic energy of a rod of mass 10 kg and length 6 m rotating about an axis perpendicular to the rod at its midpoint, with an angular velocity of 2 radians per second. (Imagine a helicopter blade of uniform thickness.)

16. Find the kinetic energy of a phonograph record of mass 50 gm and radius 10 cm rotating at $33\frac{1}{3}$ revolutions per minute.

For Problems 17-18, find the gravitational force (see page 440) between two objects. Use the fact that the gravitational attraction between particles of mass m_1 and m_2 at a distance r apart is Gm_1m_2/r^2. For each problem, slice the objects into pieces, use this formula for the pieces, and sum using a definite integral.

17. What is the force of gravitational attraction between a thin uniform rod of mass M and length l and a particle of mass m lying in the same line as the rod at a distance a from one end?

18. Two long thin rods of lengths l_1 and l_2 lie on a straight line with a gap between them of length a. Suppose their masses are M_1 and M_2, respectively, and the constant of the gravitation is G. What is the force of attraction between the rods? (Use the result of Problem 17.)

8.4 APPLICATIONS TO ECONOMICS

Present and Future Value

Many business deals involve payments in the future. If you buy a car or furniture, for example, you may buy it on credit and pay over a period of time. If you are going to accept payment in the future under such a deal, you obviously need to know how much you should be paid. Being paid $100 in the future is clearly worse than being paid $100 today for many reasons. One is inflation: $100 in the future may well buy less than $100 today because prices are likely to rise. Perhaps more important, if you are given the money today, you can do something else with it—for example, put it in the bank, invest it somewhere, or spend it. Thus, even without inflation, if you are to accept payment in the future, you would expect to be paid more to compensate for this loss of potential earnings. The question we will consider now is, how much more?

To simplify matters, we will only consider what you would lose by not earning interest; we will not consider the effect of inflation. Let's look at some specific numbers. Suppose you deposit $100 in an account which earns 7% interest compounded annually, so that in a year's time you will have $107. Thus, $100 today will be worth the same as $107 a year from now. We say that the $107 is the *future value* of the $100, and that the $100 is the *present value* of the $107. In general, we say the following:

- The **future value**, $\$B$, of a payment, $\$P$, is the amount to which the $\$P$ would have grown if deposited in an interest bearing bank account.
- The **present value**, $\$P$, of a future payment, $\$B$, is the amount which would have to be deposited in a bank account today to produce exactly $\$B$ in the account at the relevant time in the future.

As you might expect, the present value will always be smaller than the future value. In our work on compound interest in Section 1.8, we saw that with an interest rate of r, compounded annually, and a time period of t years, a deposit of $\$P$ grows to a future balance of $\$B$, where

$$B = P(1+r)^t, \quad \text{or equivalently,} \quad P = \frac{B}{(1+r)^t}.$$

If the interest is compounded n times a year for t years at rate r, and if $\$B$ is the *future value* of $\$P$ after t years and $\$P$ is the *present value* of $\$B$, then

$$B = P\left(1+\frac{r}{n}\right)^{nt}, \quad \text{or equivalently,} \quad P = \frac{B}{(1+r/n)^{nt}}.$$

Note that for a 7% interest rate, $r = 0.07$. In addition, as we saw in Section 1.8, when n tends towards infinity, we say that the compounding becomes continuous, and we get the following result:

$$B = Pe^{rt}, \quad \text{or equivalently,} \quad P = \frac{B}{e^{rt}} = Be^{-rt}.$$

Example 1 You win the lottery and are offered the choice between $1 million in four yearly installments of $250,000 each, starting now, and a lump-sum payment of $920,000 now. Assuming a 6% interest rate, compounded continuously, and ignoring taxes, which should you choose?

Solution We will do the problem in two ways. First, we assume that you pick the option with the largest present value. The first of the four $250,000 payments is made now, so

$$\text{Present value of first payment} = \$250{,}000.$$

The second payment is made one year from now and so

$$\text{Present value of second payment} = \$250{,}000e^{-0.06(1)}.$$

Calculating the present value of the third and fourth payments similarly, we find:

$$\begin{aligned}\text{Total present value} &= \$250{,}000 + \$250{,}000e^{-0.06(1)} + \$250{,}000e^{-0.06(2)} + \$250{,}000e^{-0.06(3)}\\ &\approx \$250{,}000 + \$235{,}441 + \$221{,}730 + \$208{,}818\\ &= \$915{,}989.\end{aligned}$$

Since the present value of the four payments is less than $920,000, you are better off taking the $920,000 right now.

Alternatively, we can do the problem by comparing the future values of the two pay schemes. The scheme with the highest future value is the best from a purely financial point of view. We calculate the future value of both schemes three years from now, on the date of the last $250,000 payment. At that time,

$$\text{Future value of the lump sum payment} = \$920{,}000e^{0.06(3)} \approx \$1{,}101{,}440.$$

Now we calculate the future value of the first $250,000 payment:

$$\text{Future value of the first payment} = \$250{,}000e^{0.06(3)}.$$

Calculating the future value of the other payments similarly, we find:

$$\begin{aligned}\text{Total future value} &= \$250{,}000e^{0.06(3)} + \$250{,}000e^{0.06(2)} + \$250{,}000e^{0.06(1)} + \$250{,}000\\ &\approx \$299{,}304 + \$281{,}874 + \$265{,}459 + \$250{,}000\\ &= \$1{,}096{,}637.\end{aligned}$$

The future value of the $920,000 payment is greater, so you are better off taking the $920,000 right now. Of course, since the present value of the $920,000 payment is greater than the present value of the four separate payments, you would expect the future value of the $920,000 payment to be greater than the future value of the four separate payments.

(**Note:** If you read the fine print, you will find that many lotteries do not make their payments right away, but often spread them out, sometimes far into the future. This is to reduce the present value of the payments made, so that the value of the prizes is much less than it might first appear!)

Income Stream

When we consider payments made to or by an individual, we usually think of *discrete* payments, that is, payments made at specific moments in time. However, we may think of payments made by a company as being *continuous*. The revenues earned by a huge corporation, for example, come in essentially all the time, and therefore they can be represented by a continuous *income stream*. Since the rate at which revenue is earned may vary from time to time, the income stream is described by

$$P(t) \text{ dollars/year.}$$

Notice that $P(t)$ is a *rate* (its units are dollars per year, for example) and that the rate depends on the time, t, usually measured in years from the present.

Present and Future Values of an Income Stream

Just as we can find the present and future values of a single payment, so we can find the present and future values of a stream of payments. As before, the future value represents the total amount of money obtained by depositing the income stream into a bank account and letting it gather interest. The present value represents the amount of money you would have to deposit today (in an interest-bearing bank account) in order to match what you would get from the income stream.

When we are working with a continuous income stream, we will assume that interest is compounded continuously. The reason for this is that the approximations we are going to make (of sums by integrals) are much simpler if both payments and interest are continuous.

Suppose that we want to calculate the present value of the income stream described by a rate of $P(t)$ dollars per year, and that we are interested in the period from now until T years in the future. In order to use what we know about single deposits to calculate the present values of an income stream, we must first divide the stream into many small deposits, each of which is made at approximately one instant. We divide the interval $0 \leq t \leq T$ into subintervals, each of length Δt:

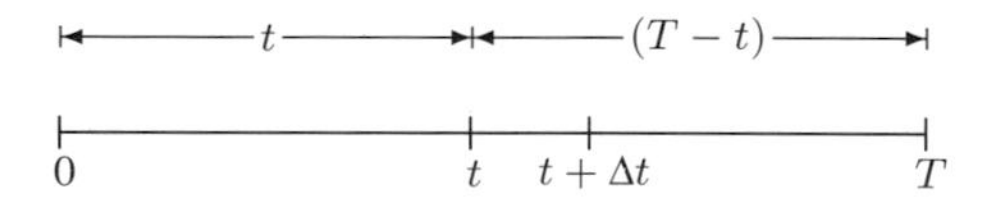

Assuming Δt is small, the rate, $P(t)$, at which deposits are being made will not vary much within one subinterval. Thus, between t and $t + \Delta t$:

$$\begin{aligned} \text{Amount paid} &\approx \text{Rate of deposits} \times \text{Time} \\ &\approx (P(t) \text{ dollars/year})(\Delta t \text{ years}) \\ &= P(t)\Delta t \text{ dollars.} \end{aligned}$$

Measured from the present, the deposit of $P(t)\Delta t$ is made t years in the future. Thus,

$$\begin{matrix} \text{Present value of money} \\ \text{deposited in interval } t \text{ to } t + \Delta t \end{matrix} \approx P(t)\Delta t e^{-rt}.$$

Summing over all subintervals gives

$$\text{Total present value} \approx \sum P(t)e^{-rt}\Delta t \text{ dollars.}$$

In the limit as $\Delta t \to 0$, we get the following integral:

$$\text{Present value} = \int_0^T P(t)e^{-rt}\,dt.$$

In computing future value, the deposit of $P(t)\Delta t$ has a period of $(T-t)$ years to earn interest, and therefore

$$\text{Future value of money deposited in interval } t \text{ to } t+\Delta t \approx \left[P(t)\Delta t\right] e^{r(T-t)}.$$

Summing over all subintervals, we get:

$$\text{Total future value} \approx \sum P(t)\Delta t e^{r(T-t)} \text{ dollars.}$$

As the length of the subdivisions tends toward zero, the sum becomes an integral:

$$\text{Future value} = \int_0^T P(t)e^{r(T-t)}\,dt.$$

Example 2 Find the present and future values of a constant income stream of \$100 per year over a period of 20 years, assuming an interest rate of 10% compounded continuously.

Solution

$$\text{Present value} = \int_0^{20} 100e^{-0.1t}\,dt = 100\left(-\frac{e^{-0.1t}}{0.1}\right)\Bigg|_0^{20} = 1000(1-e^{-2}) \approx \$864.66.$$

$$\begin{aligned}\text{Future value} &= \int_0^{20} 100e^{0.1(20-t)}\,dt = \int_0^{20} 100e^2e^{-0.1t}\,dt \\ &= 100e^2\left(-\frac{e^{-0.1t}}{0.1}\right)\Bigg|_0^{20} = 1000e^2(1-e^{-2}) \approx \$6389.06.\end{aligned}$$

Example 3 What is the relationship between the present and future values in the previous example? Explain this relationship.

Solution Since

$$\begin{aligned}\text{Present value} &= 1000(1-e^{-2}) \approx \$864.66 \\ \text{Future value} &= 1000c^2(1-e^{-2}) \approx \$6389.06\end{aligned}$$

you can see that

$$\text{Future value} = (\text{Present value})e^2.$$

The reason for this is that the stream of payments is equivalent to a single payment of \$864.66 at time $t = 0$. With an interest rate of 10%, in 20 years that single payment will have grown to a future value of

$$864.66e^{0.1(20)} = 864.66e^2 \approx \$6389.02.$$

To within rounding error, this is the future value calculated previously.

Supply and Demand Curves

In a free market, the quantity of a certain item produced and sold can be described by the supply and demand curves of the item. The *supply curve* shows what quantity of the item the producers will supply at different price levels. It is usually assumed that as the price increases, the quantity supplied will increase. The consumers' behavior is reflected in the *demand curve*, which shows what quantity of goods are bought at various prices. Here, an increase in price is assumed to cause a decrease in the quantity purchased.

For most producers and consumers, the price of an item has to be thought of as the independent variable, because it is set by the market as a whole, and there is nothing they can do to alter it. The quantity produced or consumed is then thought of as depending on the price. In other words, quantity is considered a function of price. However, although in this context we think of price as the independent variable, economic tradition has placed price along the vertical axis. (The reason for this state of affairs is that economists originally took price to be the dependent variable and put it on the vertical axis. Unfortunately, when the point of view changed, the axes did not.) The higher the price, the more suppliers will produce but the less consumers will demand. Thus, typical supply and demand curves are monotonic (either increasing or decreasing) and so it is in fact possible to think of either variable as a function of the other. (See Figure 8.30.)

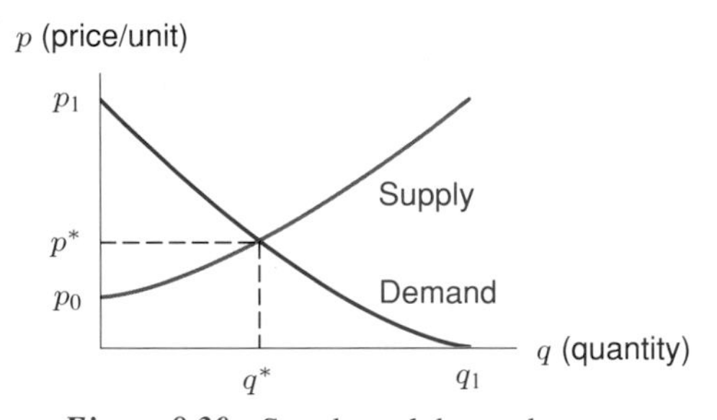

Figure 8.30: Supply and demand curves

Example 4 What is the economic meaning of the prices p_0 and p_1 and the quantity q_1 in Figure 8.30?

Solution The price p_0 is the vertical intercept on the supply curve, so at this price the quantity supplied is zero. In other words, unless the price is above p_0, the supplier will not produce anything.

The price p_1 is the vertical intercept of the demand curve. At this price, the quantity demanded is zero. In other words, unless the price is below p_1, consumers won't buy any of the product.

The quantity q_1 is the quantity that would be demanded if the price were zero—or the quantity which could be given away if the item were free.

It is assumed that the market will settle to the *equilibrium price* and *quantity*, p^* and q^*, where the graphs cross. This means that at the equilibrium point, a quantity q^* of an item will be produced and sold for a price of p^* each.

Consumer and Producer Surplus

Notice that at equilibrium, a number of consumers have bought the item at a lower price than they would have been willing to pay. (For example, there are some consumers who would have been willing to pay prices up to p_1.) Similarly, there are some suppliers who would have been willing to produce the item at a lower price (down to p_0, in fact). We define the following terms:

- The **consumer surplus** measures the total amount saved by consumers by buying the good at the equilibrium price, rather than at the price they would have been willing to pay.
- The **producer surplus** measures the additional revenue earned by suppliers by selling at the equilibrium price, rather than at the price they would have been willing to accept.

Suppose that all consumers buy the good at the maximum price they are willing to pay. Divide the interval from 0 to q^* into subintervals of length Δq. Figure 8.31 shows that a quantity Δq of items are sold at a price of about p_1, another Δq are sold for a slightly lower price of about p_2, the next Δq for a price of about p_3, and so on. Thus, the consumers' total expenditure is about

$$p_1\Delta q + p_2\Delta q + p_3\Delta q + \cdots = \sum p_i \Delta q.$$

If D is the demand function given by $p = D(q)$, and if all consumers who were willing to pay more than p^* paid as much as they were willing, then as $\Delta q \to 0$, we would have

$$\text{Consumer expenditure} = \int_0^{q^*} D(q)\,dq = \text{Area under demand curve from 0 to } q^*.$$

Now if all goods are sold at the equilibrium price, the consumers' actual expenditure is p^*q^*, the area of the rectangle between the axes and the lines $q = q^*$ and $p = p^*$. Thus the consumer surplus may be calculated as follows:

$$\text{Consumer surplus} = \left(\int_0^{q^*} D(q)\,dq\right) - p^*q^* = \text{Area under demand curve above } p = p^*.$$

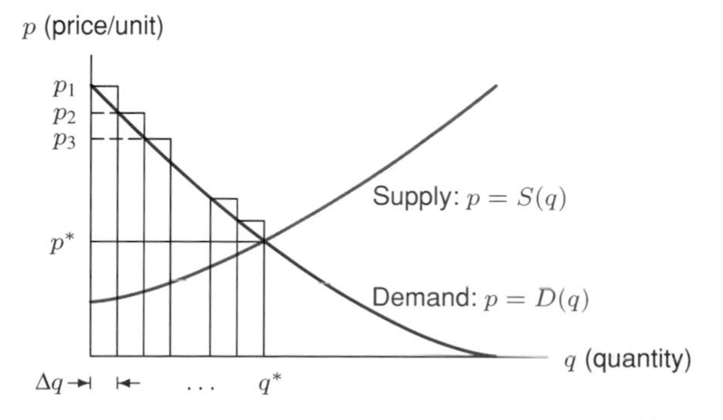

Figure 8.31: Calculation of consumer surplus

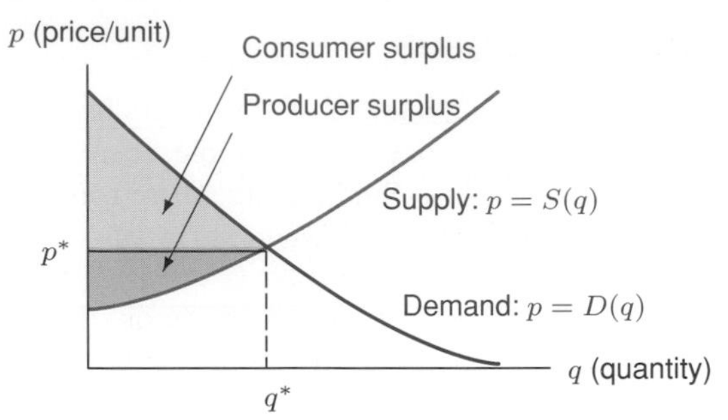

Figure 8.32: Consumer and producer surplus

See Figure 8.32. Similarly, if the supply curve is given by the function $p = S(q)$, we define the producer surplus as follows:

$$\text{Producer Surplus} = p^*q^* - \left(\int_0^{q^*} S(q)dq\right) = \begin{array}{c}\text{Area between supply}\\ \text{curve and line } p = p^*.\end{array}$$

Problems for Section 8.4

1. Draw a graph, with time in years on the horizontal axis, of what an income stream might look like for a company that sells sunscreen in the Northeast.
2. A company that owes your company money offers to begin repaying the debt either by making four annual payments of \$5000, spread out over the next three years (one now, one in a year, one in two years, and one in three years), or by waiting and making a lump-sum payment of \$25,000 at the end of the three years. If you can assume an 8% annual return on the money, compounded continuously, and you use only financial considerations, which option for repayment should you choose? Justify your answer. What other considerations might you want to consider in making your decision?
3. Find the present and future values of a constant income stream of \$3000 per year over a period of 15 years, assuming a 6% annual interest rate compounded continuously.
4. A certain bond is guaranteed to pay $(100 + 10t)$ dollars per year for 10 years, where t is the number of years from the present. Find the present value of this income stream, given an interest rate of 5%, compounded continuously.
5. (a) A bank account earns 10% interest compounded continuously. At what (constant, continuous) rate must a parent deposit money into such an account in order to save \$100,000 in 10 years for a child's college expenses?
   (b) If the parent decides instead to deposit a lump sum now in order to attain the goal of \$100,000 in 10 years, how much must be deposited now?
6. The value of good wine increases with age. Thus, if you are a wine dealer, you have the problem of deciding whether to sell your wine now, at a price of \$$P$ a bottle, or to sell it later at a higher price. Suppose you know that the price of your wine t years from now is well approximated by \$$P(1 + 20\sqrt{t})$. Assuming continuous compounding and a prevailing interest rate of 5% per year, when is the best time to sell your wine?

7. (a) If you deposit money continuously at a constant rate of \$1000 per year into a bank account that earns 5% interest, how many years will it take for the balance to reach \$10,000?
   (b) How many years would it take if the account had \$2000 in it initially?

8. (a) During the 1970s, ACME Widgets sold at a yearly rate, in widgets per year, given by $R = R_0e^{0.15t}$, where t is time in years since January 1, 1970. Suppose they were selling widgets at a rate of 1000 per year on the first day of the decade. How many widgets did they sell during the decade? How many did they sell if the rate on January 1, 1970 was 150,000,000 widgets per year?
   (b) In the first case above (1000 widgets per year on January 1, 1970), how long did it take for half the widgets in the 1970s to be sold? In the second case (150,000,000 widgets per year on January 1, 1970), when had half the widgets in the 1970s been sold?
   (c) In 1980 ACME began an advertising campaign claiming that half the widgets it had sold in the previous ten years were still in use. Comment in light of your answer to part (b).

9. Suppose you are manufacturing a particular item, and that the rate at which you earn a profit on the item is decreasing with time according to the formula

$$\text{Rate profit earned} = (2 - 0.1t) \text{ thousand dollars per year}$$

after t years. (A negative profit represents a loss.) If interest is 10%, compounded continuously:
   (a) Write a Riemann sum approximating the present value of the total profit earned up to a time T years in the future.
   (b) Write an integral representing the present value of the total profit earned T years in the future. (You need not evaluate this integral.)
   (c) At what time is the present value of the stream of profits on this item maximized? What is the present value of the total profit earned up to that time?

10. In 1980 West Germany made a loan of 20 billion deutsche marks to the Soviet Union, to be used for the construction of a natural gas pipeline connecting Siberia to Western Russia, and continuing to West Germany (Urengoi – Uschgorod – Berlin). The deal was as follows: In 1985, upon completion of the pipeline, the Soviet Union would deliver natural gas to West Germany, at a constant rate, for all future times. Assuming a constant price of natural gas of 10 pfennig (= 0.10 deutsche mark) per cubic meter, and assuming West Germany expects 10% annual interest on its investment (compounded continuously), at what rate does the Soviet Union have to deliver the gas, in billions of cubic meters per year? Keep in mind that delivery of gas could not begin until the pipeline was completed. Thus, West Germany received no return on its investment until after five years had passed. (Note: A deal of this type was actually made between the two countries; obviously, the terms of the deal were much more complex than presented here.)

11. In May 1991 *Car and Driver* described a model Jaguar that sells for \$980,000. At that price only 50 have been sold. It is estimated that 350 could have been sold if the price had been \$560,000. Assuming that the demand curve is a straight line, and that \$560,000 and 350 are the equilibrium price and quantity, find the consumer surplus.

12. Using Riemann sums, give an interpretation of producer surplus,

$$\int_0^{q^*} (p^* - S(q))\, dq$$

analogous to the interpretation of consumer surplus.

13. Using Riemann sums, explain the economic significance of $\int_0^{q^*} S(q)\,dq$ to the producers.

14. In Figure 8.30, page 450, find the regions with the following areas:

(a) p^*q^* (b) $\int_0^{q^*} D(q)\,dq$ (c) $\int_0^{q^*} S(q)\,dq$

(d) $\left(\int_0^{q^*} D(q)\,dq\right) - p^*q^*$ (e) $p^*q^* - \int_0^{q^*} S(q)\,dq$ (f) $\int_0^{q^*} (D(q) - S(q))\,dq$

8.5 APPLICATIONS TO DISTRIBUTION FUNCTIONS

Understanding the distribution of various quantities through the population can be important to decision makers. For example, the income distribution gives useful information about the economic structure of a society. In this section we will look at the distribution of ages in the US. To allocate funding for education, health care, and social security, the government needs to know how many people are in each age group. We will see how to represent such information by a density function.

US Age Distribution

TABLE 8.1 *Distribution of ages in the US in 1990*

Age group	Percentage of total population
0–20	30%
20–40	31%
40–60	24%
60–80	14%
Over 80	1%

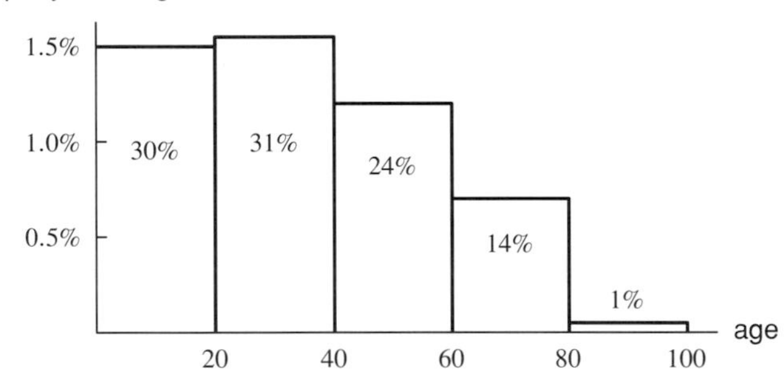

Figure 8.33: How ages were distributed in the US in 1990

Suppose we have the data in Table 8.1 showing how the ages of the US population were distributed in 1990. To represent this information graphically we use a *histogram*, putting a vertical bar above each age group in such a way that the *area* of each bar represents the percentage in that age group. The total area of all the rectangles is $100\% = 1$. We will assume that there is nobody over 100 years old, so that the last age group is 80–100. For the 0–20 age group, the base of the rectangle is 20, and we want the area to be 30%, so the height must be $30\%/20 = 1.5\%$. Notice that the vertical axis is measured in percent/year. (See Figure 8.33.)

Example 1 In 1990, what percentage of the US population was:

(a) Between 20 and 60 years old?

(b) Less than 10 years old?

(c) Between 75 and 80 or between 80 and 85 years old?

Solution (a) We add the percentages, so $31\% + 24\% = 55\%$.

(b) To find the percentage less than 10 years old, we could assume, for example, that the population was distributed evenly over the 0–20 group. (This means we are assuming that babies were born at a fairly constant rate over the last 20 years, which is probably reasonable.) If we make this assumption, then we can say that the population less than 10 years old was about half that in that 0–20 group, that is, 15%. Notice that we get the same result by computing the area of the rectangle from 0 to 10. (See Figure 8.34.)

(c) To find the population between 75 and 80 years old, since 14% of Americans in 1990 were in the 60-80 group, we might apply the same reasoning and say that $\frac{1}{4}(14\%) = 3.5\%$ of the population was in this age group. This result is represented as an area in Figure 8.34. The assumption that the population was evenly distributed is not a good one here; certainly there were more people between the ages of 60 and 65 than between 75 and 80. Thus, the estimate of 3.5% is certainly too high.

Again using the (faulty) assumption that ages in each group were distributed uniformly, we would find that the percentage between 80 and 85 was $\frac{1}{4}(1\%) = 0.25\%$. (See Figure 8.34.) This estimate is also poor — there were certainly more people in the 80–85 group than, say, the 95–100 group, and so the 0.25% is too low. In addition, although the percentage of 80–85-year-olds was certainly smaller than the percentage of 75–80-year-olds, the difference between 0.25% and 3.5% (a factor of 14) is unreasonably large. We can expect the transition from one age group to the next to be smoother and more gradual.

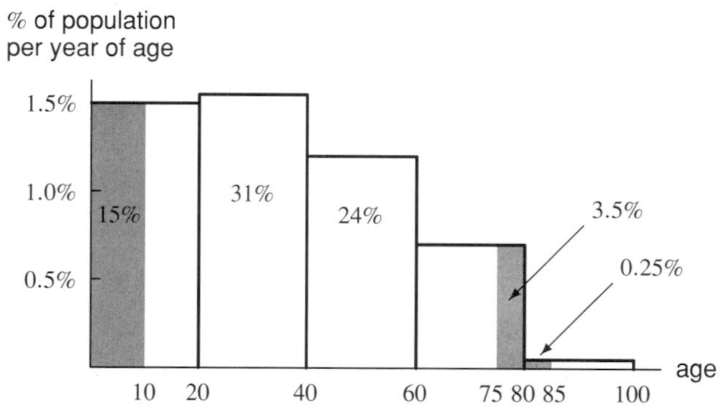

Figure 8.34: Ages in the US in 1990 — various subgroups (for Example 1)

Smoothing Out the Histogram

We could get better estimates if we had smaller age groups (each age group in Figure 8.33 is 20 years, which is quite large) or if the histogram were smoother. Suppose we have the more detailed data in Table 8.2, which leads to the new histogram in Figure 8.35. As we get more detailed information, the upper silhouette of the histogram becomes smoother, but the area of any of the bars still represents the percentage of the population in that age group. Imagine, in the limit, replacing the upper silhouette of the histogram by a smooth curve in such a way that area under the curve above one age group is the same as the area in the corresponding rectangle. The total area under the whole curve is again $100\% = 1$. (See Figure 8.35.)

TABLE 8.2 *Ages in the US in 1990 (more detailed)*

Age group	Percentage of total population
0–10	15%
10–20	15%
20–30	16%
30–40	15%
40–50	13%
50–60	11%
60–70	9%
70–80	5%
80–90	1%

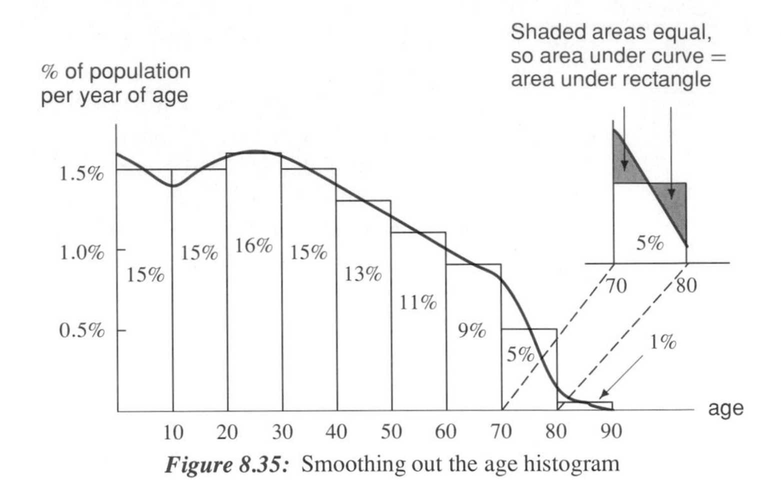

Figure 8.35: Smoothing out the age histogram

The Age Density Function

If t is age in years, we define $p(t)$, the age *density function*, to be a function which "smoothes out" the age histogram. This function has the property that

$$\begin{array}{c}\text{Fraction of population}\\ \text{between ages } a \text{ and } b\end{array} = \begin{array}{c}\text{Area under}\\ \text{graph of } p\\ \text{between } a \text{ and } b\end{array} = \int_a^b p(t)dt.$$

If a and b are the smallest and largest possible ages (say, $a = 0$ and $b = 100$), so that the ages of all of the population are between a and b, then

$$\int_a^b p(t)dt = \int_0^{100} p(t)dt = 1.$$

What does the age density function p tell us? Notice that we have not talked about the meaning of $p(t)$ itself, but *only* of the integral $\int_a^b p(t)\,dt$. Let's look at this in a bit more detail. Suppose, for example, that $p(10) = 0.015 = 1.5\%$ per year. This is *not* telling us that 1.5% of the population is precisely 10 years old (where 10 years old means exactly 10, not $10\frac{1}{2}$, not $10\frac{1}{4}$, not 10.1). However, $p(10) = 0.015$ does tell us that for some small interval Δt around 10, the fraction of the population with ages in this interval is approximately $p(10)\,\Delta t = 0.015\,\Delta t$. Notice also that the units of $p(t)$ are *% per year*, so $p(t)$ must be multiplied by years to give a percentage of the population.

The Density Function

In order to generalize the idea of the age distribution, let us look at a general density function. Suppose we are interested in how a certain characteristic, x, is distributed through a population. For example, x might be height, age, or wattage, and the population might be people, or any set of objects such as light bulbs. Then we define a general density function with the following properties:

The function, $p(x)$, is a **density function** if

$$\begin{array}{c}\text{Fraction of population}\\ \text{for which } x \text{ is}\\ \text{between } a \text{ and } b\end{array} = \begin{array}{c}\text{Area under}\\ \text{graph of } p\\ \text{between } a \text{ and } b\end{array} = \int_a^b p(x)dx.$$

$$\int_{-\infty}^{\infty} p(x)\,dx = 1 \quad \text{and} \quad p(x) \geq 0 \quad \text{for all } x.$$

The density function must be nonnegative if its integral always gives a fraction of the population. Also, the fraction of the population with x between $-\infty$ and ∞ is 1 because the entire population has the characteristic x between $-\infty$ and ∞. The function $p(t)$ used to smooth out the age histogram satisfies this definition of a density function. We do not assign a meaning to the value of $p(x)$ alone, but rather interpret $p(x)\,\Delta x$ as the fraction of the population with the characteristic in a short interval of length Δx around x.

Example 2 The graph in Figure 8.36 shows the distribution of the number of years of education completed by adults in a population. What does the graph tell us?

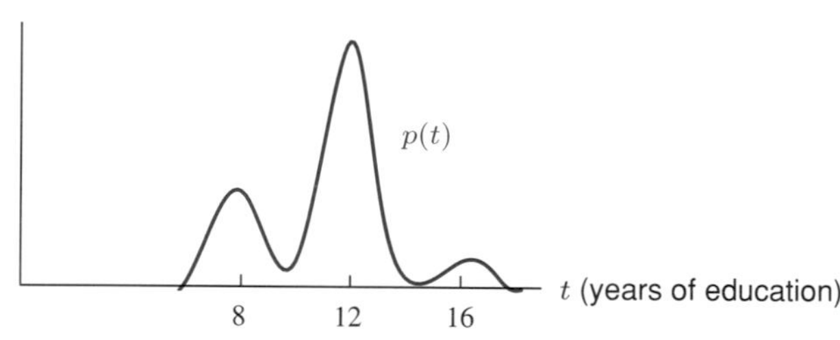

Figure 8.36: Distribution of years of education

Solution The fact that most of the area under the graph of the density function is concentrated in two humps, centered at 8 and 12 years, indicates that most of the population belong to one of two groups, those who leave school after finishing approximately 8 years and those who finish about 12 years. There is a smaller group of people who finish approximately 16 years of school.

The density function is often approximated by formulas, as in the next example.

Example 3 Find reasonable formulas representing the density function for the US age distribution, assuming the function is constant at 1.5% up to age 40 and then drops linearly.

Solution We need to construct a linear function sloping downward from age 40 in such a way that $p(40) =$ 1.5% per year $= 0.015$ and that $\int_0^{100} p(t)dt = 1$. Suppose b is as in Figure 8.37. Since

$$\int_0^{100} p(t)dt = \int_0^{40} p(t)dt + \int_{40}^{100} p(t)dt = 40(0.015) + \frac{1}{2}(0.015)b = 1,$$

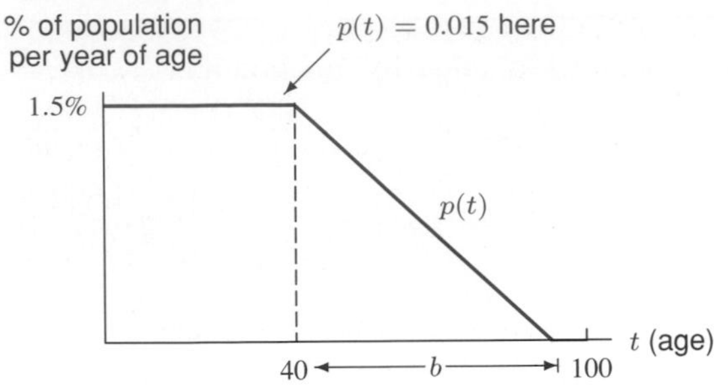

Figure 8.37: Age density function

we have

$$\frac{0.015}{2}b = 0.4, \quad \text{giving} \quad b \approx 53.3.$$

Thus the slope of the line is $-0.015/53.3 \approx -0.00028$, so for $40 \leq t \leq 93.3$,

$$p(t) - 0.015 = -0.00028(t - 40),$$
$$p(t) = 0.0263 - 0.00028t.$$

According to this way of smoothing the data, there is no one over 93.3 years old.

Cumulative Distribution Function for Ages

Another way of showing how ages are distributed in the US is by using the *cumulative distribution function* $P(t)$, defined by

$$P(t) = \begin{array}{c}\text{Fraction of population} \\ \text{of age less than } t\end{array} = \int_0^t p(x)dx.$$

Thus, P is the antiderivative of p with $P(0) = 0$.

Notice that the cumulative distribution function is nonnegative and increasing (or at least nondecreasing), since the number of people younger than age t increases as t increases. Another way of seeing this is to notice that $P' = p$, and p is positive (or nonnegative). Thus the cumulative age distribution is a function which starts with $P(0) = 0$ and increases as t increases. $P(t) = 0$ for $t < 0$ because, when $t < 0$, there is no one whose age is less than t. The limiting value of P, as $t \to \infty$, is 1 since as t becomes very large (100 say), everyone will be younger than age t, so the fraction of people with age less than t tends towards 1. (See Figure 8.38.) Notice that for t less than 40, the graph of P is a straight line, because p is constant there. For $t > 40$, the graph of P levels off as p tends to 0.

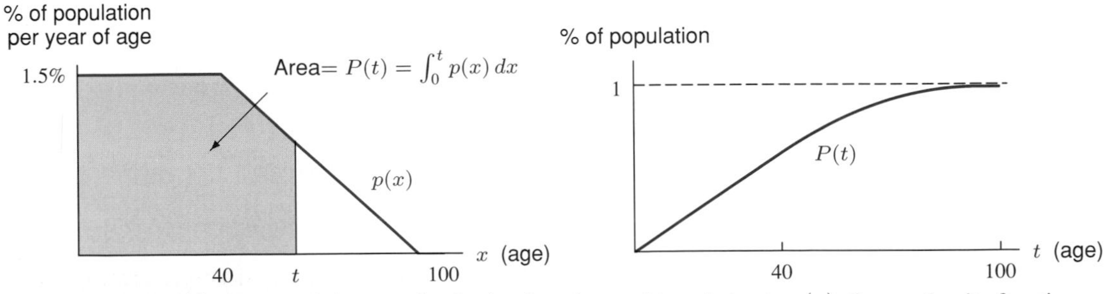

Figure 8.38: $P(t)$, the cumulative age distribution function, and its relation to $p(x)$, the age density function

Cumulative Distribution Function

A **cumulative distribution function**, $P(t)$, of a density function p, is defined by

$$P(t) = \int_{-\infty}^{t} p(x)\,dx = \begin{array}{c}\text{Fraction of population having}\\ \text{values of } x \text{ below } t.\end{array}$$

Thus, P is an antiderivative of p, that is, $P' = p$.
Any cumulative distribution has the following properties:

- P is increasing (or nondecreasing).
- $\lim_{t\to\infty} P(t) = 1$ and $\lim_{t\to-\infty} P(t) = 0$.
- By the Fundamental Theorem of Calculus,

$$\begin{array}{c}\text{Fraction of population}\\ \text{having values of } x\\ \text{between } a \text{ and } b\end{array} = \int_a^b p(x)\,dx = P(b) - P(a).$$

Problems for Section 8.5

In Problems 1–3, sketch graphs of a density function and a cumulative distribution function which could represent the distribution of income through a population with the given characteristics.

1. A large middle class.
2. Small middle and upper classes and many poor people.
3. Small middle class, many poor and many rich people.
4. A large number of people take a standardized test, receiving scores described by the density function p graphed in Figure 8.39. Does the density function imply that most people receive a score near 50? Explain why or why not.

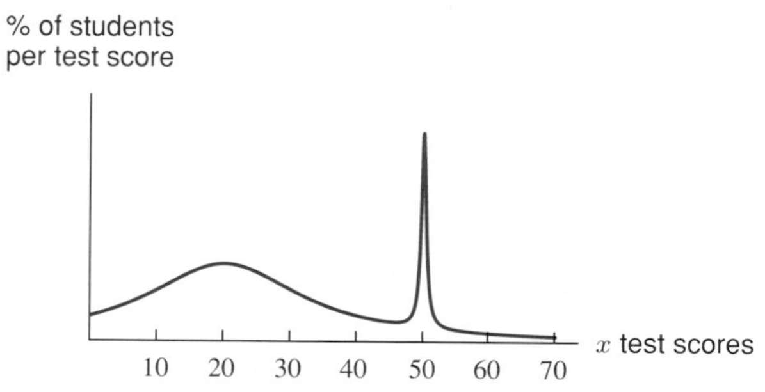

Figure 8.39: Density function of test scores

5. An experiment is done to determine the effect of two new fertilizers A and B on the growth of a species of peas. The cumulative distribution functions of the heights of the mature peas without treatment and treated with each of A and B are graphed in Figure 8.40.

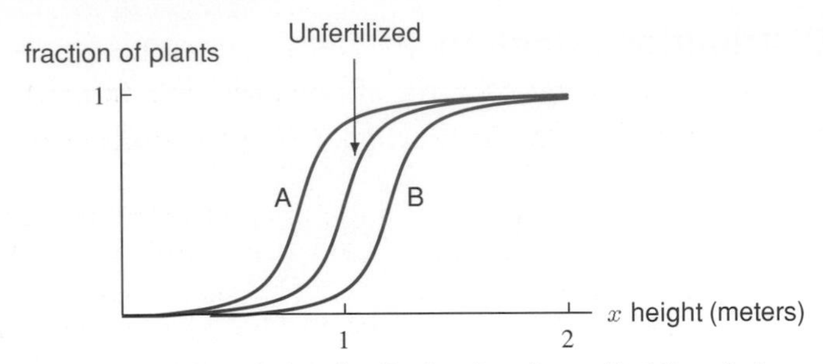

Figure 8.40: Cumulative distribution functions of heights of plants

(a) About what height are most of the unfertilized plants?
(b) Explain in words the effect of the fertilizers A and B on the mature height of the plants.

6. The density function for the duration of telephone calls within a certain city is $p(x) = 0.4e^{-0.4x}$, where x denotes the duration in minutes of a randomly selected call.

(a) What percentage of calls last between 1 and 2 minutes?
(b) What percentage of calls last 1 minute or less?
(c) What percentage of calls last 3 minutes or more?
(d) Find the cumulative distribution function.

7. A congressional committee is investigating a defense contractor. It seems that by the time this contractor's projects are completed, there are almost always cost overruns. The data in Table 8.3 were presented to the committee.

TABLE 8.3 *Defense contractor data*

Cost overrun (C)	Fraction of cost overruns that are at most C
−20%	0.01
−15%	0.03
−10%	0.08
− 5%	0.13
0%	0.19
10%	0.32
20%	0.50
30%	0.80
40%	0.94
50%	0.99

(a) Plot the data with C on the horizontal axis. Is this a density function or a cumulative distribution function? Sketch the best curve you can through these points.
(b) If you think you drew a density function in part (a), sketch the corresponding cumulative distribution function on another set of axes. If you think you drew a cumulative distribution function in part (a), sketch the corresponding density function.
(c) Based on the table, what is the probability that there will be a cost overrun of 50% or more? Between 20% and 50%? What appears to be the most likely cost overrun?

8. A person who travels regularly on the 9:00 am bus from Oakland to San Francisco reports that the bus is almost always a few minutes late but rarely more than five minutes late. We are told further that the bus is never more than two minutes early, although it is on very rare occasions a little early.
   (a) Sketch a possible density function, $p(t)$, for this situation. Shade the region under the graph between $t = 2$ minutes and $t = 4$ minutes. Explain what this region represents.
   (b) Now sketch the cumulative distribution function $P(t)$ for this situation. What measurement(s) on this graph correspond to the area shaded? What do the inflection point(s) on your graph of P correspond to on the graph of p? How can you interpret the inflection points on the graph of P without referring to the graph of p?

9. Students at the University of California were surveyed and asked their grade point average. (The GPA ranges from 0 to 4, where 2 is just passing.) The distribution of GPAs is shown in Figure 8.41.

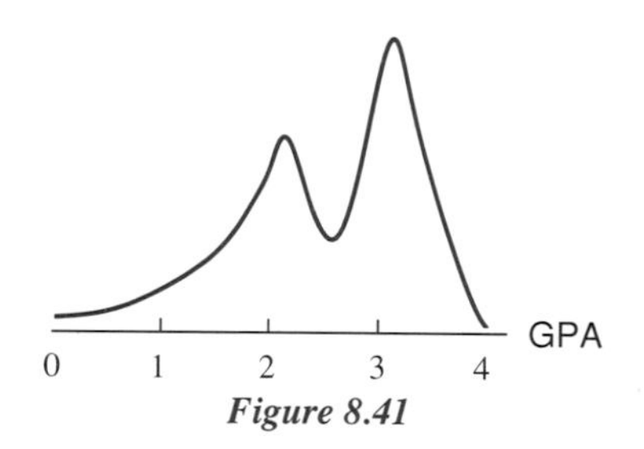

Figure 8.41

   (a) Roughly what fraction of students are passing?
   (b) Roughly what fraction of the students have honor grades (GPAs above 3)?
   (c) Why do you think there is a peak around 2?
   (d) Sketch a graph of the cumulative distribution function.

10. Figure 8.42 shows the distribution of elevation, in miles, across the earth's surface. Positive elevation denotes land above sea level; negative elevation shows land below sea level (i.e., the ocean floor).
   (a) Describe in words the elevation of most of the earth's surface.
   (b) Approximately what fraction of the earth's surface is below sea level?

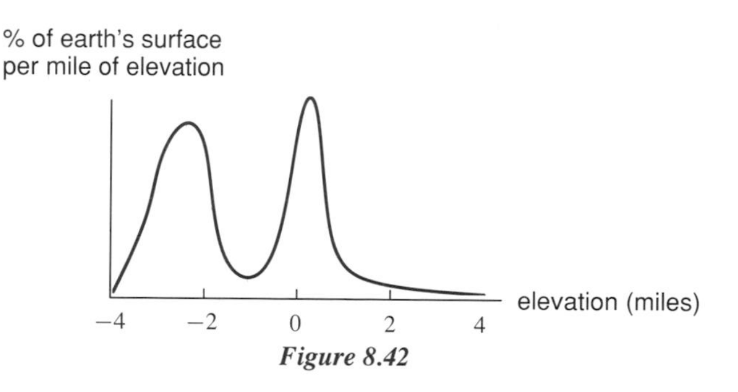

Figure 8.42

11. In southern Switzerland, most of the rain falls in the spring and fall; summers and winters are relatively dry. Sketch possible graphs for the density function and the cumulative distribution function of the rain distribution over the course of one year. Put the date on the horizontal axis and percentage of the year's rainfall on the vertical axis.

12. Consider a pendulum swinging through a small angle. The x-coordinate of the bob moves between $-a$ and a, as shown in Figure 8.43.

    (a) Draw the density function for the location of the x-coordinate of the pendulum bob (i.e., neglect up-and-down motion). In order to do this, imagine a camera taking pictures of the pendulum at random instants. Where is the bob most likely to be found? Least likely? [Hint: Consider the speed of the pendulum at different points in its path. Is the camera more likely to take a photograph of the bob at a point on its path where the bob is moving quickly, or where it is moving slowly?]

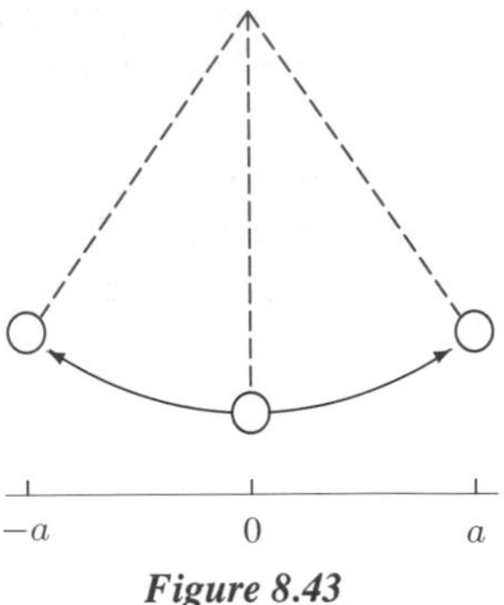

Figure 8.43

    (b) Now sketch the function

$$f(x) = \begin{cases} \dfrac{1}{\pi\sqrt{a^2 - x^2}} & -a < x < a; \\ 0 & |x| \geq a. \end{cases}$$

    How does this graph compare with the one you drew in part (a)?

    (c) Assuming that the function given in part (b) is the density function for the pendulum, what do you expect

$$\int_{-a}^{a} \frac{1}{\pi\sqrt{a^2 - x^2}}\, dx$$

    to be? Check this by computing the integral.

    (d) Does it seem reasonable, physically speaking, that $f(x)$ "blows up" at a and $-a$? Explain your answer.

8.6 PROBABILITY AND MORE ON DISTRIBUTIONS

Probability

Suppose we pick a member of the US population at random and ask what is the probability that the person is between, say, the ages of 60 and 65. The probability, or chance, that the person is in a certain age group is equal to the fraction of the population in that age group. Consider the density function $p(t)$ defined in Section 8.5 to describe the distribution of ages in the US. We can use the density function to calculate probabilities as follows:

$$\begin{array}{c}\text{Probability that}\\ \text{a person is between}\\ \text{ages } a \text{ and } b\end{array} = \begin{array}{c}\text{Fraction of population}\\ \text{between ages } a \text{ and } b\end{array} = \int_a^b p(t)\, dt.$$

Since the cumulative distribution function gives the fraction of the population younger than age t, the cumulative distribution can also be used to calculate the probability that a randomly selected person is in a given age group.

$$\begin{array}{c}\text{Probability that} \\ \text{a person is younger} \\ \text{than age } t\end{array} = \begin{array}{c}\text{Fraction of population} \\ \text{younger than age } t\end{array} = P(t) = \int_0^t p(x)\,dx.$$

In the next example, both a density function and a cumulative distribution function are used to describe the same situation.

Example 1 Suppose you want to analyze the fishing industry in a small town. Each day, the boats bring back at least 2 tons of fish, and never more than 8 tons.

(a) Using the density function describing the daily catch in Figure 8.44, find and graph the corresponding cumulative distribution function and explain its meaning.

(b) What is the probability that the catch will be between 5 and 7 tons?

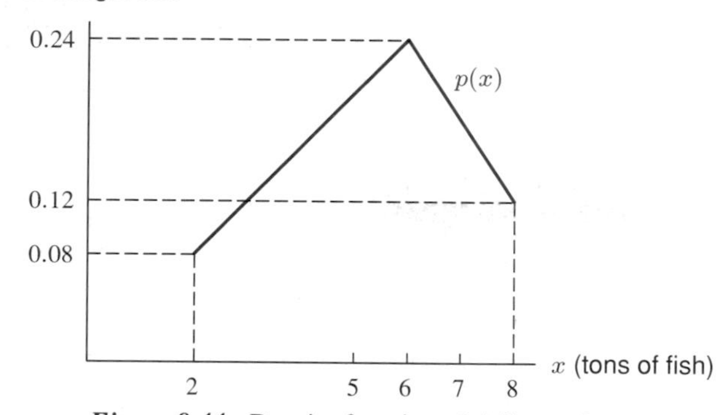

Figure 8.44: Density function of daily catch

Solution (a) The equations of the lines that determine the density function are

$$p(x) = \begin{cases} 0.04x & \text{for } 2 \le x \le 6 \\ -0.06x + 0.6 & \text{for } 6 < x \le 8 \end{cases}$$

and $p(x) = 0$ for $x < 2$ or $x > 8$. Since the catch is at least 2 tons, the cumulative distribution function is given by $P(t) = \int_2^t p(x)\,dx$. Thus, for $2 \le t \le 6$,

$$P(t) = \int_2^t 0.04x\,dx = 0.04\frac{x^2}{2}\bigg|_2^t = 0.02t^2 - 0.08.$$

And for $6 \le t \le 8$,

$$\begin{aligned} P(t) &= \int_2^t p(x)\,dx = \int_2^6 p(x)\,dx + \int_6^t p(x)\,dx \\ &= 0.64 + \int_6^t (-0.06x + 0.6)\,dx = 0.64 + \left(-0.06\frac{x^2}{2} + 0.6x\right)\bigg|_6^t \\ &= -0.03t^2 + 0.6t - 1.88. \end{aligned}$$

Thus

$$P(t) = \begin{cases} 0.02t^2 - 0.08 & \text{for } 2 \le t \le 6 \\ -0.03t^2 + 0.6t - 1.88 & \text{for } 6 < t \le 8. \end{cases}$$

In addition $P(t) = 0$ for $t < 2$ and $P(t) = 1$ for $8 < t$. (See Figure 8.45.)

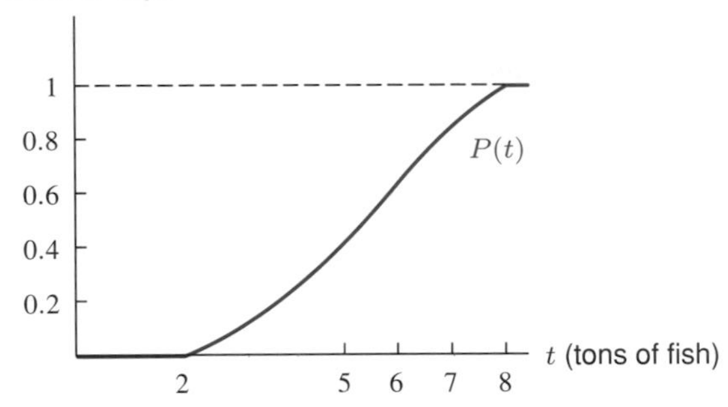

Figure 8.45: Cumulative distribution of daily catch

(b) The probability that the catch is between 5 and 7 tons can be found using either p or P. This probability can be represented by the shaded area in Figure 8.46, which shows that the probability is about 0.43.

$$\begin{array}{c}\text{Probability catch is} \\ \text{between 5 and 7 tons}\end{array} = \int_5^7 p(x)\,dx = 0.43.$$

The probability can also be found from the cumulative distribution as follows:

$$\begin{array}{c}\text{Probability catch is} \\ \text{between 5 and 7 tons}\end{array} = P(7) - P(5) = 0.85 - 0.42 = 0.43.$$

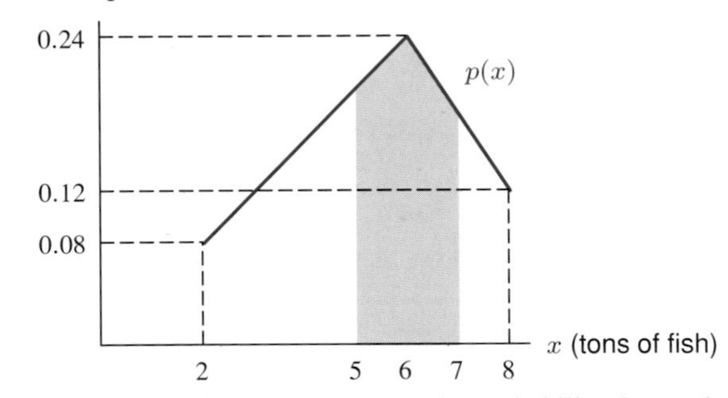

Figure 8.46: Shaded area represents the probability the catch is between 5 and 7 tons

The Median and Mean

It is often useful to be able to give an "average" value for a distribution. Two measures that are in common use are the *median* and the *mean*.

The Median

A **median** is a value T such that half the population has values of x less than (or equal to) T, and half the population has values of x greater than (or equal to) T. Thus, a median T satisfies:

$$\int_{-\infty}^{T} p(x)\,dx = 0.5,$$

where p is the density function. In other words, half the area under the graph of p lies to the left of T.

Example 2 Find the median age in the US in 1990, using the age density function given by

$$p(t) = \begin{cases} 0.015 & \text{for } 0 \le t \le 40 \\ 0.0263 - 0.00028t & \text{for } 40 < t \le 93.3. \end{cases}$$

Solution We want to find the value of T such that

$$\int_{-\infty}^{T} p(t)\,dt = \int_{0}^{T} p(t)\,dt = 0.5.$$

Since $p(t) = 1.5\%$ up to age 40, we have

$$\text{Median} = T = \frac{50\%}{1.5\%} \approx 33 \text{ years.}$$

(See Figure 8.47.)

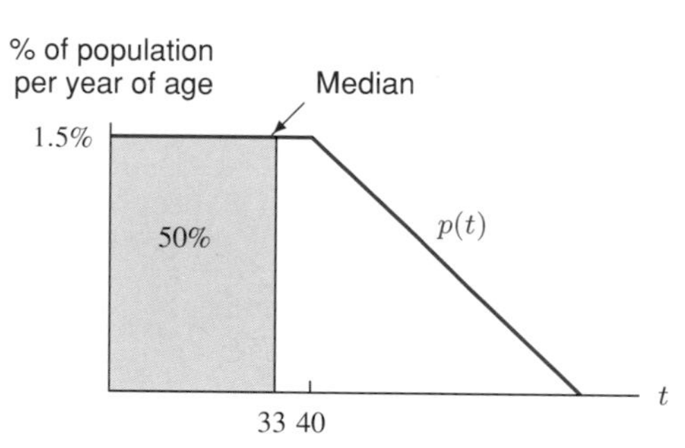

Figure 8.47: Median of age distribution

The Mean

Another commonly used average value is the *mean*. To find the mean of N numbers, you add the numbers and divide the sum by N. For example, the mean of the numbers 1, 2, 7, and 10 is $(1 + 2 + 7 + 10)/4 = 5$. The mean age of the entire US population is therefore defined as

$$\frac{\sum \text{Ages of all people in the US}}{\text{Total number of people in the US}}.$$

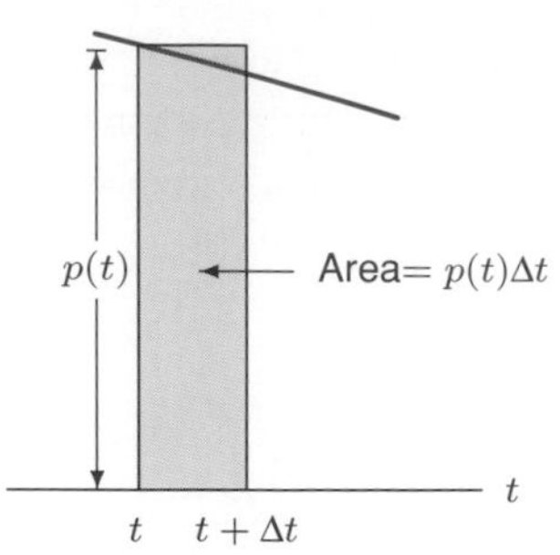

Figure 8.48: Shaded area is percentage of population with age between t and $t+\Delta t$

Calculating the sum of all the ages directly would be an enormous task; we will approximate the sum by an integral. The idea is to "slice up" the age axis and consider the people whose age is between t and $t+\Delta t$. How many are there?

The percentage of the population between t and $t+\Delta t$ is the area under the graph of p between these points, which is well approximated by the area of the rectangle, $p(t)\Delta t$. (See Figure 8.48.) If the total number of people in the population is N, then

$$\begin{array}{c}\text{Number of people with age}\\ \text{between } t \text{ and } t+\Delta t\end{array} \approx p(t)\Delta t N.$$

The age of all of these people is approximately t:

$$\begin{array}{c}\text{Sum of ages of people}\\ \text{between age } t \text{ and } t+\Delta t\end{array} \approx tp(t)\Delta t N.$$

Therefore, adding and factoring out an N gives us

$$\text{Sum of ages of all people} \approx \left(\sum tp(t)\Delta t\right) N.$$

In the limit, as we allow Δt to shrink to 0, the sum becomes an integral, so as an approximation:

$$\text{Sum of ages of all people} = \left(\int_0^{100} tp(t)dt\right) N.$$

Therefore, with N equal to the total number of people in the US, and assuming no person is over 100 years old,

$$\text{Mean age} = \frac{\text{Sum of ages of all people in US}}{N} = \int_0^{100} tp(t)dt.$$

We can give the same argument for any[1] density function $p(x)$.

If a quantity has density function $p(x)$,

$$\textbf{Mean value} \text{ of the quantity} = \int_{-\infty}^{\infty} xp(x)\,dx.$$

[1]Provided all the relevant improper integrals converge.

It can be shown that the mean is the point on the horizontal axis where the region under the graph of the density function, if it were made out of cardboard, would balance.

Example 3 Find the mean age of the US population, using the density function of Example 2.

Solution The approximate formulas for p are

$$p(t) = \begin{cases} 0.015 & \text{for } 0 \le t \le 40 \\ 0.0263 - 0.00028t & \text{for } 40 < t \le 93.3. \end{cases}$$

Using these formulas, we compute:

$$\text{Mean age} = \int_0^{100} tp(t)dt = \int_0^{40} t(0.015)dt + \int_{40}^{93.3} t(0.0263 - 0.00028t)dt$$

$$= 0.015\frac{t^2}{2}\bigg|_0^{40} + 0.0263\frac{t^2}{2}\bigg|_{40}^{93.3} - 0.00028\frac{t^3}{3}\bigg|_{40}^{93.3} \approx 36 \text{ years.}$$

The mean is shown is Figure 8.49..

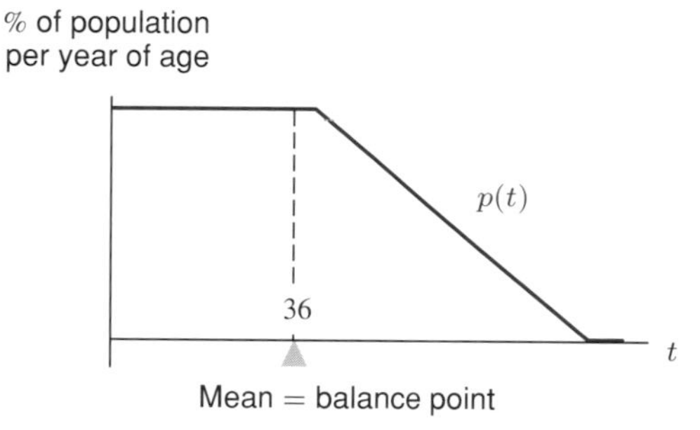

Figure 8.49: Mean of age distribution

Normal Distributions

How much rain do you expect will fall in your home town this year? If you live in Anchorage, Alaska, the answer would be something close to 15 inches (including the snow). Of course, you don't expect exactly 15 inches. Some years there will be more than 15 inches, and some years there will be less. Most years, however, the amount of rainfall will be close to 15 inches; only rarely will it be well above or well below 15 inches. What does the density function for the rainfall look like? To answer this question, we look at rainfall data over many years. It lies on a bell-shaped curve which peaks at 15 inches and slopes downward approximately symmetrically on either side. This is an example of a normal distribution.

Normal distributions are frequently used to model real phenomena, from grades on an exam to the number of airline passengers on a particular flight. A normal distribution is characterized by its *mean*, μ, and its *standard deviation*, σ. The mean tells us where the data is clustered: the location of the central peak. The standard deviation tells us how closely the data is clustered around the mean. A small value of σ tells us that the data is close to the mean; a large σ tells us the data is spread out. The formula for a normal distribution is as follows.

A **normal distribution** has a density function of the form

$$p(x) = \frac{1}{\sigma\sqrt{2\pi}} e^{-(x-\mu)^2/(2\sigma^2)},$$

where μ is the mean of the distribution and σ is the standard deviation, with $\sigma > 0$.

The factor of $1/(\sigma\sqrt{2\pi})$ in front of the function is there to make the area under its graph equal to 1. That the factor $\sqrt{2\pi}$ is involved is one of the truly remarkable discoveries of mathematics.

To model the rainfall in Anchorage, we use a normal distribution with $\mu = 15$. The standard deviation can be estimated by looking at the data; we will take it to be 1. (See Figure 8.50.)

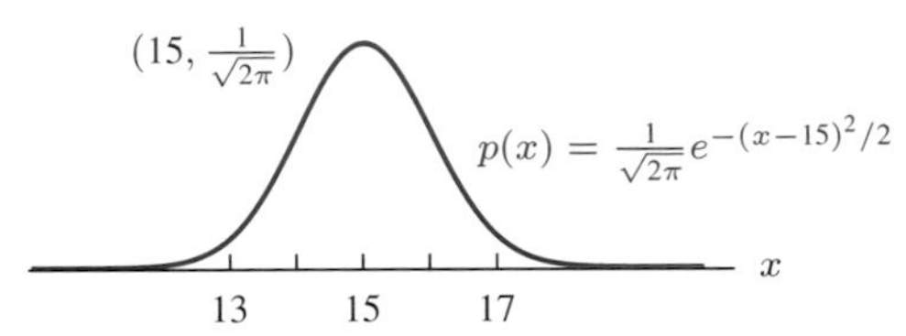

Figure 8.50: Normal distribution with $\mu = 15$ and $\sigma = 1$

In the following example, we will verify that for a normal distribution a certain percentage of the data always lies within a certain number of standard deviations from the mean.

Example 4 For Anchorage's rainfall, use the normal distribution with the density function

$$p(x) = \frac{1}{\sqrt{2\pi}} e^{-(x-15)^2/2},$$

to compute the fraction of the years with rainfall between
(a) 14 and 16 inches, (b) 13 and 17 inches, (c) 12 and 18 inches.

Solution (a) The fraction of the years with annual rainfall between 14 and 16 inches is $\int_{14}^{16} \frac{1}{\sqrt{2\pi}} e^{-(x-15)^2/2}\,dx$. Since there is no elementary antiderivative for $e^{-(x-15)^2/2}$, we find the integral numerically. Its value is about 0.68.

$$\text{Fraction of years with rainfall between 14 and 16 inches} = \int_{14}^{16} \frac{1}{\sqrt{2\pi}} e^{-(x-15)^2/2}\,dx \approx 0.68.$$

(b) Finding the integral numerically again:

$$\text{Fraction of years with rainfall between 13 and 17 inches} = \int_{13}^{17} \frac{1}{\sqrt{2\pi}} e^{-(x-15)^2/2}\,dx \approx 0.95.$$

(c)

$$\text{Fraction of years with rainfall between 12 and 18 inches} = \int_{12}^{18} \frac{1}{\sqrt{2\pi}} e^{-(x-15)^2/2}\,dx \approx 0.997.$$

Since 0.95 is so close to 1, we expect that most of the time the rainfall will be between 13 and 17 inches a year.

Notice that in the preceding example, the standard deviation is 1 inch, so rainfall between 14 and 16 inches a year is within one standard deviation of the mean. Similarly, rainfall between 13 and 17 inches is within 2 standard deviations of the mean, and rainfall between 12 and 18 inches is within three standard deviations of the mean. The fractions of the observations within one, two, and three standard deviations of the mean calculated in the previous example hold for any normal distribution.

Rules of Thumb for Any Normal Distribution

- About 68% of the observations are within one standard deviation of the mean.
- About 95% of the observations are within two standard deviations of the mean.
- Over 99% of the observations are within three standard deviations of the mean.

Problems for Section 8.6

1. Consider the fishing data given in Example 1 on page 463. Show that the area under the density function in Figure 8.44 is 1. Why is this to be expected?
2. Find the mean daily catch for the fishing data in Figure 8.44, page 463.
3. IQ scores are believed to be normally distributed with mean 100 and standard deviation 15.
   (a) Write a formula for the density distribution of IQ scores.
   (b) Estimate the fraction of the population with IQ between 115 and 120.
4. Show that the area under the graph of the density function of the normal distribution

$$p(x) = \frac{1}{\sqrt{2\pi}} e^{-(x-15)^2/2}$$

   is 1. This function has no elementary antiderivative, so you must do it numerically. Make it clear in your solution what limits of integration you used.
5. (a) Using a calculator or computer, sketch graphs of the density function of the normal distribution

$$p(x) = \frac{1}{\sigma\sqrt{2\pi}} e^{-(x-\mu)^2/(2\sigma^2)}$$

   (i) For fixed μ (say, $\mu = 5$) and varying σ (say, $\sigma = 1, 2, 3$).
   (ii) For varying μ (say, $\mu = 4, 5, 6$) and fixed σ (say, $\sigma = 1$).

   (b) Explain how the graphs confirm that μ is the mean of the distribution and that σ shows how closely the data is clustered around the mean.
6. Let $P(x)$ be the cumulative function for the income distribution in the US in 1973 (income is measured in thousands of dollars). Some values of $P(x)$ are in the following table:

Income x (thousands)	1	4.4	7.8	12.6	20	50
$P(x)$ (%)	1	10	25	50	75	99

   (a) What fraction of the population made between \$20,000 and \$50,000?

(b) What was the median income?
(c) Sketch a density function for this distribution. Where, approximately, does your density function have a maximum? What is the significance of this point, in terms of income distribution? How can you recognize this point on the graph of the density function and on the graph of the cumulative distribution?

7. In 1950 an experiment was done observing the time gaps between successive cars on the Arroyo Seco Freeway. The data[2] show that the density function of these time gaps was given approximately by

$$p(x) = ae^{-0.122x}$$

where x is the time in seconds and a is a constant.

(a) Find a.
(b) Find P, the cumulative distribution function.
(c) Find the median and mean time gap.
(d) Sketch rough graphs of p and P.

8. Consider a group of people who have received treatment for a disease such as cancer. Let t be the *survival time*, the number of years a person lives after receiving treatment. The density function giving the distribution of t is $p(t) = Ce^{-Ct}$ for some positive constant C.

(a) What is the practical meaning for the cumulative distribution function $P(t) = \int_0^t p(x)\, dx$?
(b) The survival function, $S(t)$, is the probability that a randomly selected person survives for at least t years. Find $S(t)$.
(c) Suppose a patient has a 70% probability of surviving at least two years. Find C.

9. Let v be the speed, in meters/second, of an oxygen molecule, and let $p(v)$ be the density function of the speed distribution of oxygen molecules at room temperature. Maxwell showed that

$$p(v) = av^2e^{-mv^2/(2kT)},$$

where $k = 1.4 \times 10^{-23}$ is the Boltzmann constant, T is the temperature in degrees Kelvin (at room temperature, $T = 293$), and $m = 5 \times 10^{-26}$ is the mass of the oxygen molecule in kilograms.

(a) Find the value of a.
(b) Estimate the median and the mean speed. Find the maximum of $p(v)$.
(c) How do your answers in part (b) for the mean and the maximum of $p(v)$ change as T changes?

10. If we think of an electron as a particle, the function

$$P(r) = 1 - (2r^2 + 2r + 1)e^{-2r}$$

is the cumulative distribution function of the radial distribution of the sole electron in a hydrogen atom (at ground state), where the distance r of the electron from the center of the atom is measured in Bohr radii. (1 Bohr radius = 5.29×10^{-11} m. Niels Bohr (1885–1962) was a Danish physicist.)

For example, $P(1) = 1 - 5e^{-2} \approx 0.32$ means that the sole electron is within 1 Bohr radius from the center of the atom 32% of the time.

(a) Find a formula for the density function of this distribution. Sketch the density function and the cumulative distribution function.

[2] Reported by Daniel Furlough and Frank Barnes

(b) Find the median distance, the mean distance, and the most likely distance of the electron from the center of the atom.

(c) The Bohr radius is sometimes called the "radius of the hydrogen atom." Why?

REVIEW PROBLEMS FOR CHAPTER EIGHT

1. (a) Sketch the solid obtained by rotating the region bounded by $y = \sqrt{x}$, $x = 1$, and $y = 0$ around the line $y = 0$.
   (b) Approximate its volume by Riemann sums, showing the volume represented by each term in your sum on the sketch.
   (c) Now find the volume of this solid using an integral.

2. Using the region of Problem 1, find the volume when it is rotated around
   (a) The line $y = 1$. (b) The y-axis.

3. In this problem, you will derive the formula for the volume of a right circular cone with height l and base radius b by rotating the line $y = ax$ from $x = 0$ to $x = l$ around the x-axis. See Figure 8.51.
   (a) What value should you choose for a such that the cone will have height l and base radius b?
   (b) Given this value of a, find the volume of the cone.

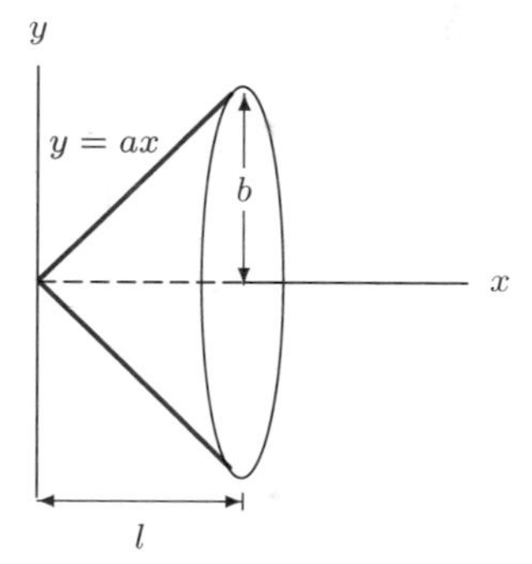

Figure 8.51

4. Mt. Shasta is a cone-like volcano whose radius at an elevation of h feet above sea level is approximately $(3.5 \cdot 10^5)/\sqrt{h + 600}$ feet. Its bottom is 400 feet above sea level, and its top is 14,400 feet above sea level. See Figure 8.52.
   (a) Give a Riemann sum approximating the volume of Mt. Shasta.
   (b) Find the volume in cubic feet. (Note: This data about Mt. Shasta is more or less accurate. Mt. Shasta lies in northern California, and for some time was thought to be the highest point in the US outside Alaska.)

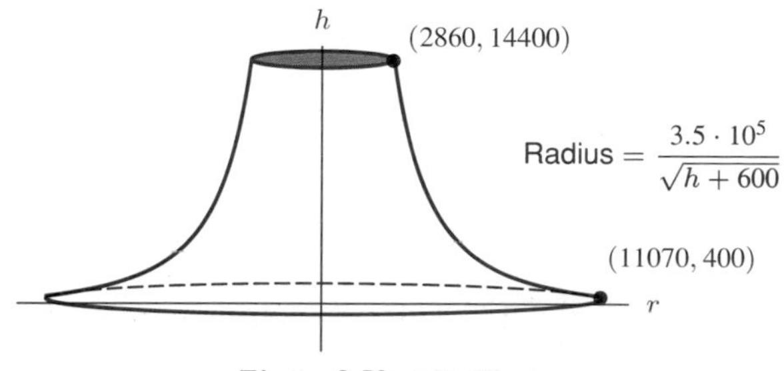

Figure 8.52: Mt. Shasta

5. The reflector behind a car headlight is made in the shape of the parabola, $x = \frac{4}{9}y^2$, with a circular cross-section, as shown in Figure 8.53.

(a) Find a Riemann sum approximating the volume contained by this headlight.
(b) Find the volume exactly.

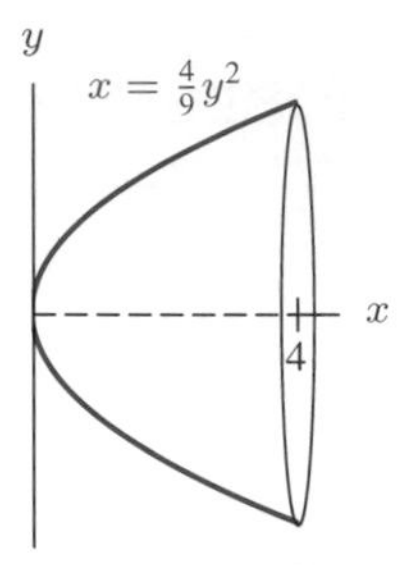

Figure 8.53

6. If the circle $x^2 + y^2 = 1$ is rotated about the line $y = 3$, it forms a torus (a doughnut-shaped figure). Find the volume of this torus.

7. Find the volume of a generalized pyramid, whose base is a planar region of area A and whose apex is a height h above the plane containing the base. [Hint: The cross-section of the pyramid parallel to the base is of the same shape as the base, but is scaled down in area.]

For Problems 8–10, find the arc lengths.

8. $f(x) = \sqrt{1 - x^2}$ from $x = 0$ to $x = 1$

9. $f(x) = e^x$ from $x = 1$ to $x = 2$

10. $f(x) = \frac{1}{3}x^3 + \frac{1}{4x}$ from $x = 1$ to $x = 2$.

11. Find a function $f(x)$ such that its length L from $(0, 0)$ to $(x, f(x))$, $x > 0$, is given by
(a) $L = x$ (b) $L = ax$, $a > 0$

For the curves described in Problems 12 and 13, write the integral that gives the exact length of the curve; you need not evaluate the integral.

12. One arch of the sine curve, from $x = 0$ to $x = \pi$.

13. The ellipse with equation $(x^2/a^2) + (y^2/b^2) = 1$.

14. Find the average horizontal width of A in Figure 8.54.

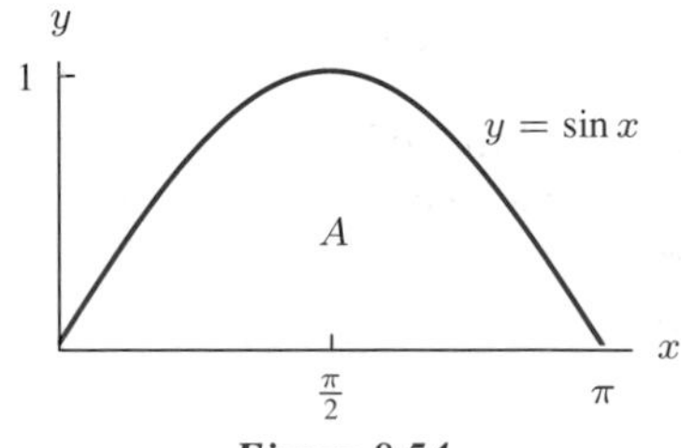

Figure 8.54

15. (a) Find the present and future values of a constant income stream of \$100 per year over a period of 20 years, assuming a 10% annual interest rate compounded continuously.
(b) How many years will it take for the balance to reach \$5000?

16. Marie-Jolaine walks about the xy-plane. She starts at the origin, and at time t she finds herself at the point (t, t^2).

(a) What is her speed v at time t?
(Note: The speed v is defined as $v = ds/dt = \lim_{\Delta t \to 0} \Delta s/\Delta t$, where s is the distance traveled; see Figure 8.55.)

(b) Find integrals representing s, the distance traveled from the origin, two ways: using arc length and using the relation $v = ds/dt$.

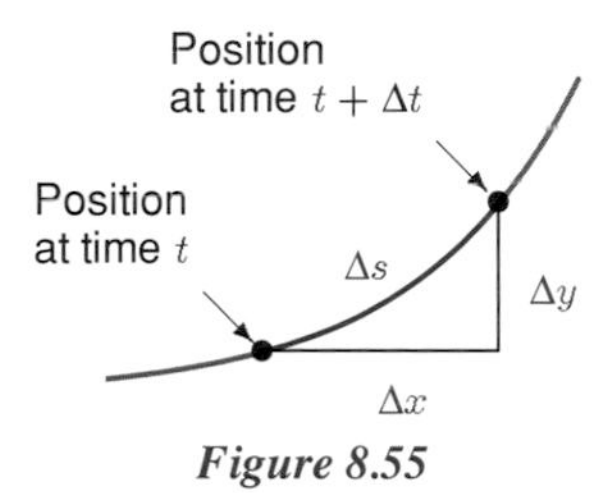

Figure 8.55

17. The heights of grass plants in a meadow were measured and the density function and cumulative distribution function are graphed in Figure 8.56 and Figure 8.57.

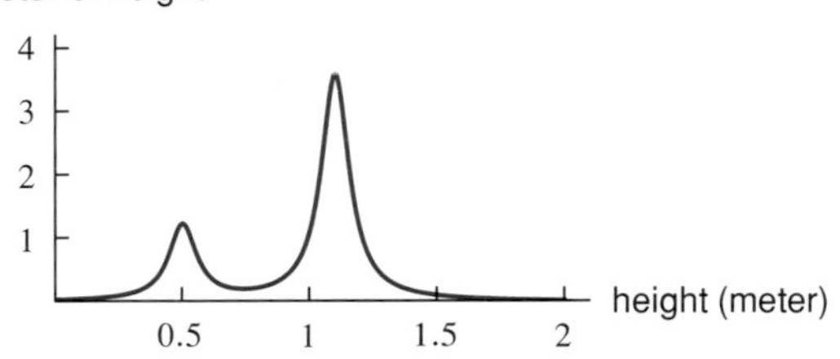

Figure 8.56: Height density of meadow grass

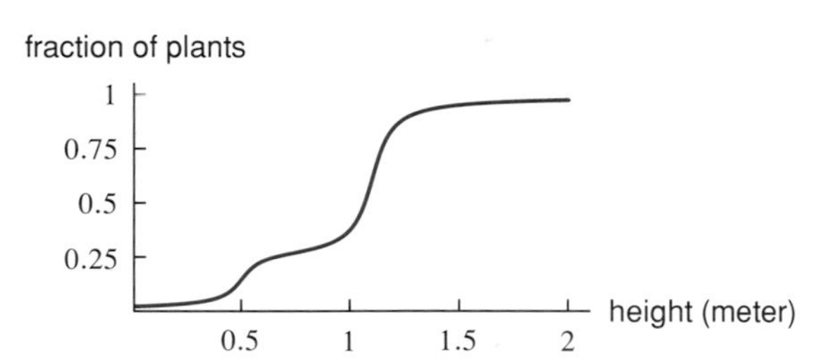

Figure 8.57: Cumulative distribution of meadow grass

(a) You know that there are two species of grass in the meadow, a short grass and a tall grass. Explain how the graph of the density function reflects this fact.

(b) Explain how the graph of the cumulative distribution functions reflects the fact that there are two species of grass in the meadow.

(c) About what percentage of the grasses in the meadow belong to the short grass species?

18. The probability of a transistor failing between $t = a$ months and $t = b$ months is given by $c \int_a^b e^{-ct}\, dt$ for some constant c.

(a) If the probability of failure within the first six months is 10%, what is c?

(b) Given the value of c in part (a), what is the probability the transistor will fail within the second six months?

19. Consider the normal distribution

$$p(x) = \frac{1}{\sigma\sqrt{2\pi}} e^{-(x-\mu)^2/2\sigma^2}.$$

(a) Show that $p(x)$ is a maximum when $x = \mu$. What is that maximum value?

(b) Show that $p(x)$ has points of inflection where $x = \mu + \sigma$ and $x = \mu - \sigma$.

(c) Describe in your own words what μ and σ tell you about the distribution.

20. A nuclear power plant produces strontium-90 at a rate of 1 kg/yr. How much of the strontium produced since 1971 (when the plant opened) was still around in 1992? (The half–life of strontium-90 is 28 years.)

21. A 200-lb weight is to be hoisted from the ground to a point 20 ft high by a chain passing over a pulley. If the chain weighs 2 lb/ft, find the work done in lifting the weight.

22. Water is to be raised from a well 40 ft deep by means of a bucket attached to a rope. When the bucket is full of water it weighs 30 lb, but the bucket has a leak that causes it to lose water at the constant rate of 1/4 lb for each foot that the bucket is raised. Neglecting the weight of the rope, find the work done in raising the bucket to the top.

23. A cylindrical garbage can of depth 3 ft and radius 1 ft fills with rainwater up to a depth of 2 ft. How much work would be done in pumping the water up to the top edge of the can? (Recall that water weighs 62.4 lb/ft^3.)

24. A water tank is in the shape of a right circular cone with height 18 ft and radius 12 ft at the top. If it is filled with water to a depth of 15 ft, find the work done in pumping all of the water over the top of the tank. (Note: Water weighs 62.4 lb/ft^3.)

25. What is the total force on the bottom and sides of the garbage can in Problem 23?

26. In Figure 8.58 you see a cross section through an apple (Scale: One division = 1/2 inch.)

(a) Give a rough estimate for the volume of this apple (in cubic inches).

(b) The density of these apples is about 0.03 lb/in^3 (a little less than the density of water—as you might expect, since apples float). Estimate how much this apple would cost. (They go for 80 cents a pound.)

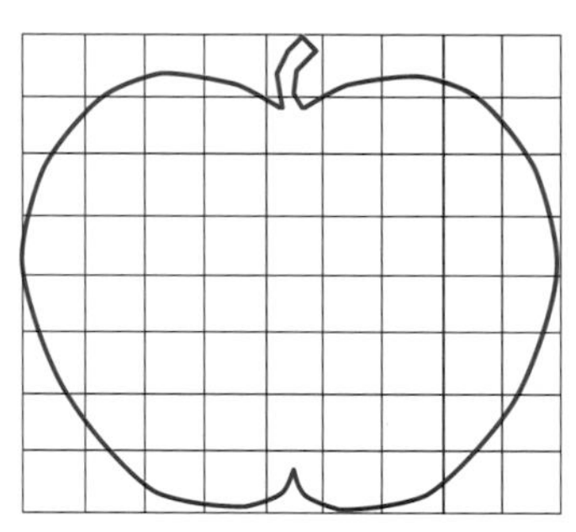

Figure 8.58

27. A cylindrical centrifuge of radius 1 m and height 2 m is filled with water to a depth of 1 meter (see Figure 8.59(a)). As the centrifuge accelerates, the water level rises along the wall and drops in the center; the cross-section will be a parabola. (See Figure 8.59(b).)

(a) Find the equation of the parabola in Figure 8.59(b) in terms of h, the depth of the water at its lowest point.

(b) As the centrifuge rotates faster and faster, either water will be spilled out the top, as in Figure 8.59(c), or the bottom of the centrifuge will be exposed, as in Figure 8.59(d). Which happens first?

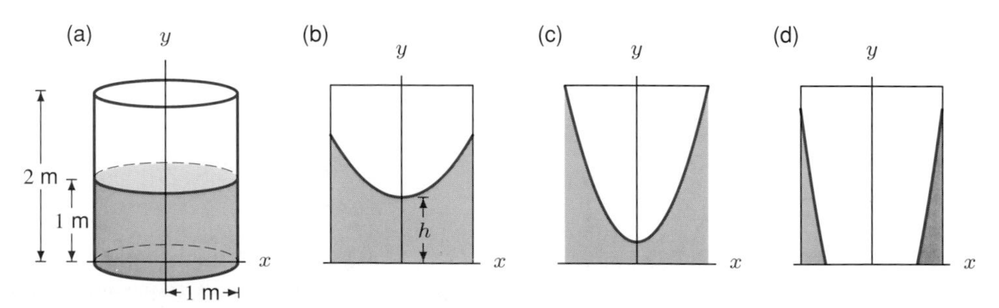

Figure 8.59

28. Figure 8.60 shows an ancient Greek water clock called a clepsydra, which is designed so that the depth of the water decreases at a constant rate as the water runs out a hole in the bottom. This design allows the hours to be marked by a uniform scale. The tank of the clepsydra is a volume of revolution about a vertical axis. According to Torricelli's law, the exit speed of the water flowing through the hole is proportional to the square root of the depth of the water. Use this to find the formula $y = f(x)$ for this profile, assuming that $f(1) = 1$.

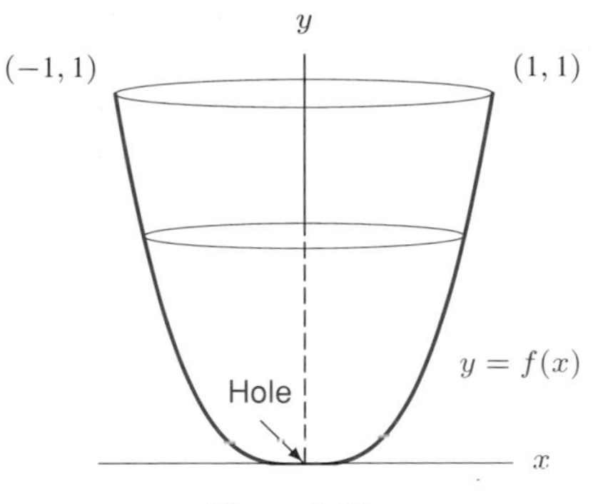

Figure 8.60

Problems 29 and 30 concern Figures 8.61 and 8.62. In each problem, you are given two objects which have the same mass M, the same radius R, and the same angular velocity about the indicated axes (say, for the sake of definiteness, one revolution per minute). For each problem, say which object has the greater kinetic energy. (Recall that the kinetic energy of a particle of mass m with speed v is $\frac{1}{2}mv^2$.) Don't attempt to compute the kinetic energy of the objects to do this; just use reasoning.

29. **30.**

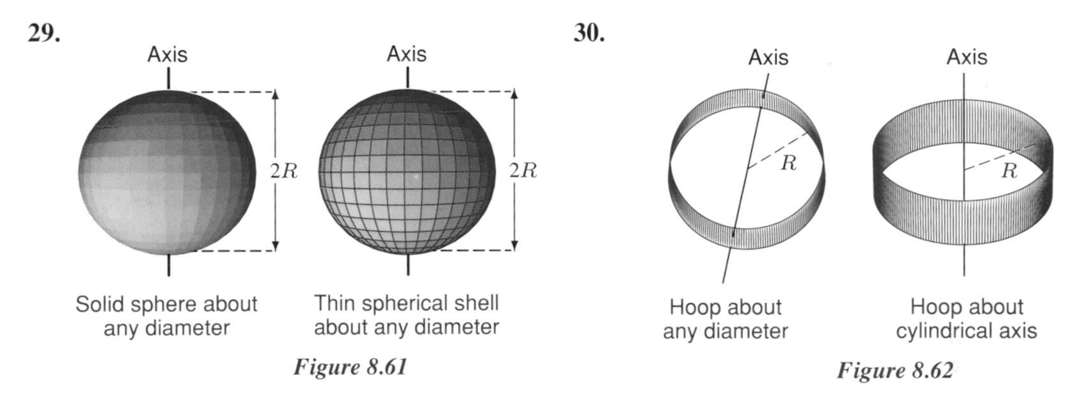

Figure 8.61

Figure 8.62

31. The distance between the towers of the main span of the Golden Gate Bridge is about 1280 m; the sag of the cable halfway between the towers on a cold winter day is about 143 m. See Figure 8.63.

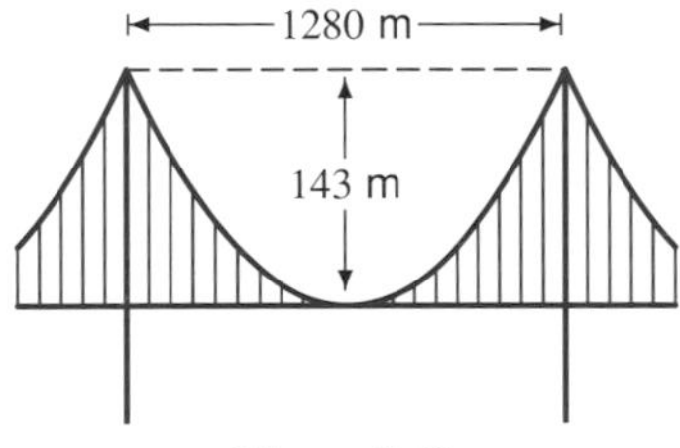

Figure 8.63

(a) How long is the cable, assuming it has a parabolic shape? (Represent the cable as a parabola of the form $y = kx^2$ and determine k to at least 1 decimal place.)

(b) On a hot summer day the cable is about 0.05% longer, due to thermal expansion. By how much does the sag increase? Assume no movement of the towers.

32. Two cylinders are inscribed in a cube of side length 2, as shown in Figure 8.64. What is the volume of the solid that the two cylinders enclose? [Hint: Use horizontal slices.] Note: The solution was known to Archimedes. The Chinese mathematician Liu Hui (third century A.D.) tried to find this volume, but he failed; he wrote a poem about his efforts calling the enclosed volume a "box-lid":

Look inside the cube
And outside the box-lid;
Though the dimension increases,
It doesn't quite fit.
The marriage preparations are complete;
But square and circle wrangle,
Thick and thin are treacherous plots,
They are incompatible.
I wish to give my humble reflections,
But fear that I will miss the correct principle;
I dare to let the doubtful points stand,
Waiting
For one who can expound them.

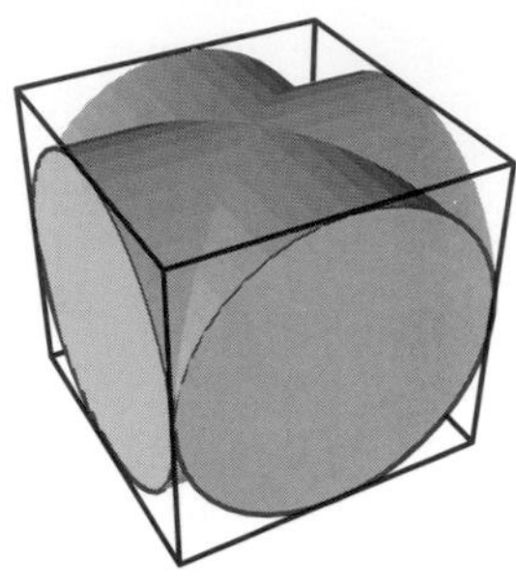

Figure 8.64

33. Find a formula for the volume of a bagel in terms of the diameter D of the bagel and the diameter d of the hole.

34. Triangular probability distributions, such as the one graphed in Figure 8.65, are used in business to model uncertainty. Such a distribution can be used to model a variable where only three pieces of information are available: a lower bound ($x = a$), a most likely value ($x = c$), and an upper bound ($x = b$).

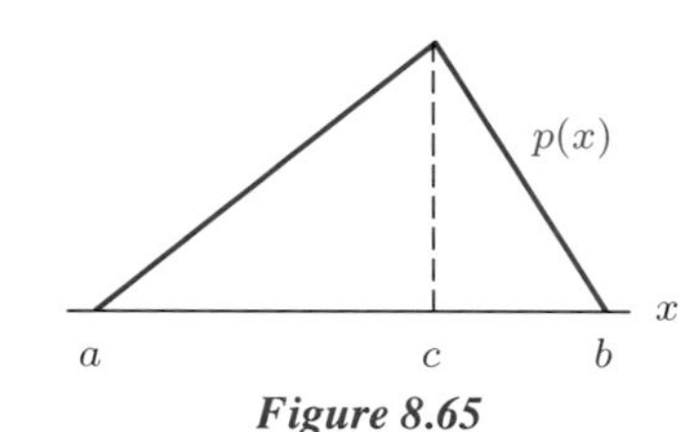

Figure 8.65

Thus one can write the function $p(x)$ as two lines:

$$p(x) = \begin{cases} m_1x + b_1 & a \le x \le c \\ m_2x + b_2 & c < x \le b. \end{cases}$$

(a) Find the value of $p(c)$ geometrically, using the criterion that the probability that x takes on some value between a and b is 1.

Suppose a new product costs between \$6 and \$10 per unit to produce, with a most likely cost of \$9.

(b) Find $p(9)$.
(c) Use the fact that $p(6) = p(10) = 0$ and the value of $p(9)$ you found in part (b) to find m_1, m_2, b_1, and b_2.
(d) What is the probability that the production cost per unit will be less than \$8?
(e) What is the median cost?
(f) Write a formula for the cumulative probability distribution function $P(x)$ for

(i) $6 \le x \le 9$, (ii) $9 < x \le 10$.

Sketch the graph of $P(x)$.

CHAPTER NINE

DIFFERENTIAL EQUATIONS

The derivative was introduced as the rate of change of a function in Chapter 2. In Chapter 3 we saw how to use the definite integral to compute changes in the original function from the derivative. Chapter 6 introduced antiderivatives and 'going backward' from the derivative to the original function.

In this chapter we will start with an equation involving the derivative of an unknown function, and 'go backward' to the original function. Such an equation is called a *differential equation.*

Although you may not have realized it, you have been solving differential equations for some time. Every time you antidifferentiate to find $\int f(x)\,dx$, you are actually solving the differential equation

$$\frac{dy}{dx} = f(x).$$

The new feature of this chapter is that the right-hand side of the differential equation may now be a function of y, or of both x and y, instead of just x.

We will see how to solve differential equations graphically first, then numerically, and, finally, in some of the special cases when it can be done, analytically. In the course of this chapter, we will treat first-order differential equations (involving first derivatives only), systems of differential equations, and some second-order differential equations (involving first and second derivatives).

9.1 WHAT IS A DIFFERENTIAL EQUATION?

How Fast Does a Person Learn?

Suppose we are interested in how fast an employee learns a new task. One theory claims that the more the employee already knows of the task, the slower he or she learns. In other words, if $y\%$ represents the percentage of the task that has already been mastered, and dy/dt the rate at which the employee is learning, then dy/dt decreases as y increases.

What can we say about y as a function of time, t? Figure 9.1 shows three graphs whose slope, dy/dt, decreases as y increases. Figure 9.1(a) represents an employee who starts learning at $t = 0$ and who eventually masters 100% of the task. Figure 9.1(b) represents an employee who starts later but eventually masters 100% of the task. Figure 9.1(c), on the other hand, represents an employee who starts learning at $t = 0$, but who does not master the whole task (since y levels off at some percentage below 100%).

Notice the fact that y levels off does not follow from the fact that dy/dt decreases as y increases. However, common sense tells us that a person's knowledge of the task must level off at or below 100%, and all the graphs in Figure 9.1 have been drawn to reflect this fact.

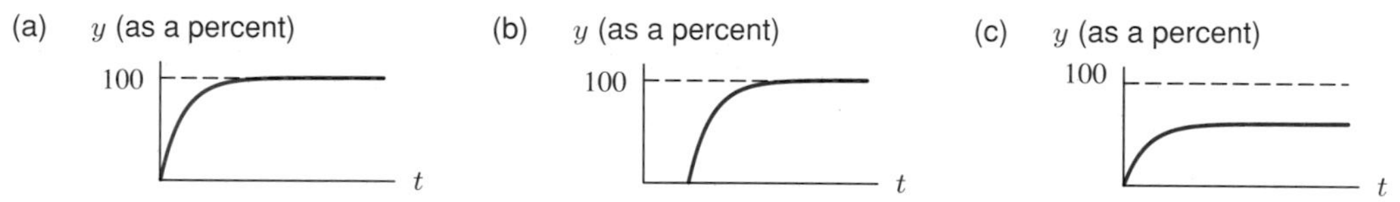

Figure 9.1: Possible graphs showing percentage of task learned, y, as a function of time, t

Setting up a Differential Equation to Model How a Person Learns

To describe more precisely how a person learns, we need more exact information about how dy/dt depends on y. Suppose we are told that if time, t, is measured in weeks,

$$\text{Rate a person learns} = \text{Percentage of task not yet learned.}$$

If y is the percentage learned by time t (in weeks), the percentage not yet learned by that time is $100 - y$. Thus we have

$$\frac{dy}{dt} = 100 - y.$$

Such an equation, which gives information about the rate of change of an unknown function, is called a *differential equation.*

Solving the Differential Equation Numerically

Suppose that the person starts learning at time zero, so $y = 0$ when $t = 0$. Then initially the person is learning at a rate

$$\frac{dy}{dt} = 100 - 0 = 100\% \text{ per week.}$$

In other words, if the person were to continue learning at this rate, the task would be mastered in a week. In fact, however, the rate at which the person learns decreases, so it takes more than a week to get close to mastering the task. Let's assume a five-day work week and that the 100% per week learning rate holds for the whole first day. (It doesn't, but we will assume this for now.) One day is 1/5 of a week, so during the first day the person learns $100(1/5) = 20\%$ of the task. By the end of the first day the rate at which the person learns has therefore been reduced to

$$\frac{dy}{dt} = 100 - 20 = 80\% \text{ per week.}$$

Thus, during the second day the person learns $80(1/5) = 16\%$, so by the end of the second day the person knows $20 + 16 = 36\%$ of the task. Continuing in this fashion, we compute the approximate values of y in Table 9.1.[1] These values for y represent an approximate numerical solution to the differential equation.

TABLE 9.1 *Approximate percentage of task learned as a function of time*

Time (working days)	0	1	2	3	4	5	10	20
Percentage learned	0	20	36	48.8	59.0	67.2	89.3	98.8

A Formula for the Solution to the Differential Equation

The differential equation

$$\frac{dy}{dt} = 100 - y$$

gives information about the unknown function y. A function which satisfies it is called a *solution*. Figure 9.1 contains graphs of possible solutions, and Table 9.1 shows approximate numerical values of a solution. It is also sometimes (but not always) possible to find a formula for the solution. In this particular case there is a formula, which we will derive in Section 9.4. For now, we will simply check that the formula is correct, and look at its general form. Consider the function

$$y = 100 + Ce^{-t},$$

where C is any *arbitrary constant*. Substituting this into the differential equation $dy/dt = 100 - y$ gives:

$$\text{Left side} = \frac{dy}{dt} = -Ce^{-t}$$
$$\text{Right side} = 100 - y = 100 - (100 + Ce^{-t}) = -Ce^{-t}.$$

Since we get the same thing on both sides, we say $y = 100 + Ce^{-t}$ *satisfies* this differential equation.

Finding the Arbitrary Constant: Initial Conditions

To find a value for the arbitrary constant C, we need an additional piece of information — usually the initial value of y. If, for example, we are told that $y = 0$ when $t = 0$, then substituting into

$$y = 100 + Ce^{-t}$$

shows us that

$$0 = 100 + Ce^{0}, \quad \text{so} \quad C = -100.$$

So the function $y = 100 - 100e^{-t}$ satisfies the differential equation *and* the condition that $y = 0$ when $t = 0$.

[1]The values of y after 6, 7, ..., 9, 11, ..., 19 days were computed by the same method, but omitted from the table.

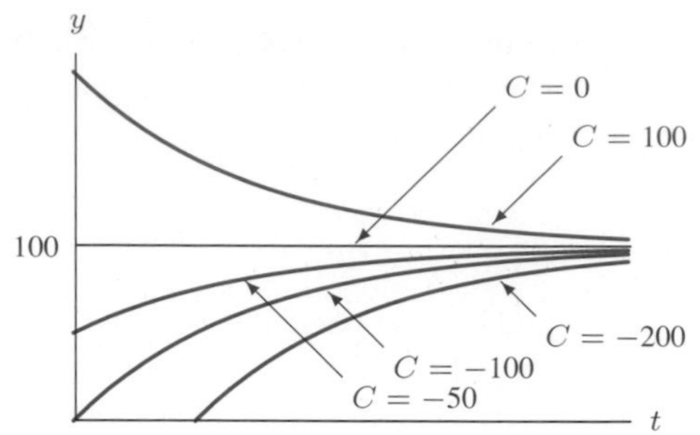

Figure 9.2: Solution curves for $dy/dt = 100 - y$: Members of the family $y = 100 + Ce^{-t}$

The Family of Solutions

Later in this chapter we will show that any solution to this differential equation is of the form $y = 100 + Ce^{-t}$ for some constant C. Since any value of C gives a solution, we say that the *general solution* to the differential equation $dy/dt = 100 - y$ is the family of functions $y = 100 + Ce^{-t}$. The solution $y = 100 - 100e^{-t}$ that satisfies the differential equation together with the *initial condition* that $y = 0$ when $t = 0$ is called a *particular solution.* The differential equation and the initial condition together are called an *initial-value problem.* Several members of the family of solutions are graphed in Figure 9.2. Notice that, like a family of antiderivatives, this family contains one arbitrary constant, C. However, this family has a different form: in a family of antiderivatives, the arbitrary constant is added, and the graph of any antiderivative can be obtained from another by shifting vertically. In this family, the arbitrary constant affects the shape of the solution, and a change in its sign changes the concavity.

First- and Second-Order Differential Equations

First, some more definitions. The differential equation

$$\frac{dy}{dt} = 100 - y$$

is called *first-order* because it involves the first derivative, but no higher derivatives. By contrast, if s is the height (in feet) of a body moving under the force of gravity and t is time (in seconds), then

$$\frac{d^2s}{dt^2} = -32.$$

This is a *second-order* differential equation because it involves the second derivative of the unknown function, $s = f(t)$, but no higher derivatives.

Example 1 Show that $y = e^{2t}$ is not a solution to the second-order differential equation

$$\frac{d^2y}{dt^2} + 4y = 0.$$

Solution To decide whether the function $y = e^{2t}$ is a solution, substitute it into the differential equation:

$$\frac{d^2y}{dt^2} + 4y = 2(2e^{2t}) + 4e^{2t} = 8e^{2t}.$$

Since $8e^{2t}$ is not identically zero, $y = e^{2t}$ is not a solution.

How Many Arbitrary Constants Should We Expect in the Family of Solutions?

Since a differential equation involves the derivative of an unknown function, solving it usually involves antidifferentiation, which introduces arbitrary constants, or parameters. The solution to a first-order differential equation usually involves one antidifferentiation and one arbitrary constant. (For example, the C in $y = 100 + Ce^{-t}$). Picking out one particular solution involves knowing an additional piece of information, such as an initial condition. Solving a second-order differential equation generally involves two antidifferentiations and so two arbitrary constants. Consequently, finding a particular solution usually involves two initial conditions.

For example, if s is the height (in feet) of a body above the surface of the earth at time t in seconds,

$$\frac{d^2s}{dt^2} = -32.$$

Integrating gives

$$\frac{ds}{dt} = -32t + C_1,$$

and integrating again gives

$$s = -16t^2 + C_1t + C_2.$$

Thus the general solution for s involves the two arbitrary constants C_1 and C_2. We can find C_1 and C_2 if we are told, for example, that the initial velocity is 100 feet per second upward and that the initial position is 15 feet above the ground. In that case, $C_1 = 100$ and $C_2 = 15$, so

$$s = -16t^2 + 100t + 15.$$

Problems for Section 9.1

1. Match the graphs in Figure 9.3 with the descriptions below.

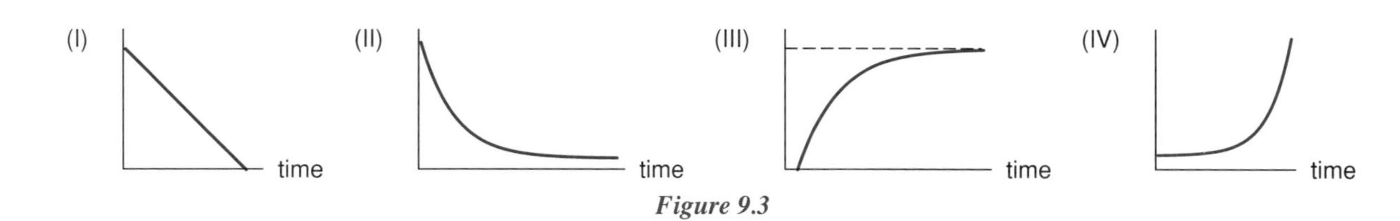

Figure 9.3

(a) The temperature of a glass of ice water left on the kitchen table.
(b) The amount of money in an interest-bearing bank account into which $50 is deposited.
(c) The speed of a constantly decelerating car.
(d) The temperature of a piece of steel heated in a furnace and left outside to cool.

2. Show that, for any constant P_0, the function $P = P_0e^t$ satisfies the differential equation

$$\frac{dP}{dt} = P.$$

3. Show that $y = \sin 2t$ satisfies
$$\frac{d^2y}{dt^2} + 4y = 0.$$

4. Suppose you know that $Q = Ce^{kt}$ satisfies the differential equation
$$\frac{dQ}{dt} = -0.03Q.$$
What (if anything) does this tell you about the values of C and k?

5. Find the value(s) of ω for which $y = \cos \omega t$ satisfies
$$\frac{d^2y}{dt^2} + 9y = 0.$$

6. (a) Show that $P = \dfrac{1}{1 + e^{-t}}$ satisfies the *logistic equation*
$$\frac{dP}{dt} = P(1 - P).$$
(b) What is the limiting value of P as time, t, tends to infinity?

7. Use the method which generated the data in Table 9.1 on page 479 to fill in the missing y values for $t = 6, 7, \cdots, 19$ days.

8. Suppose y is the height of a hanging cable above a fixed horizontal line, as in Figure 9.4. It can be shown that

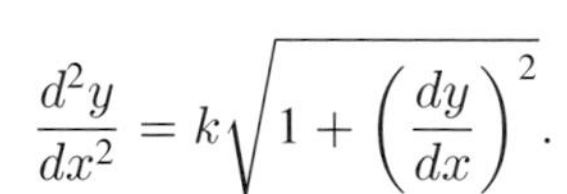

$$\frac{d^2y}{dx^2} = k\sqrt{1 + \left(\frac{dy}{dx}\right)^2}.$$

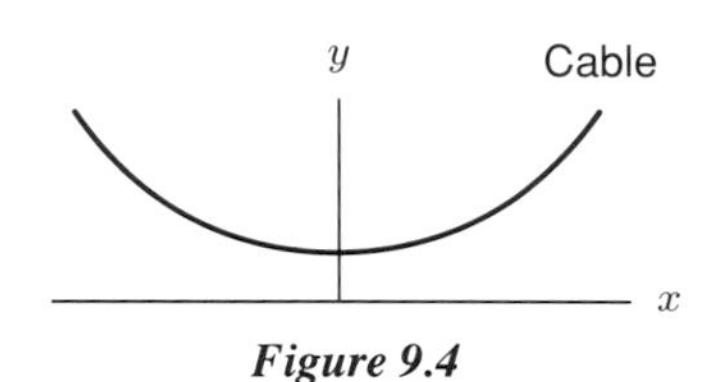

Figure 9.4

(a) Show that $y = \dfrac{e^x + e^{-x}}{2}$ satisfies this differential equation if $k = 1$.
(b) For general k, one solution to this differential equation is of the form
$$y = \frac{e^{Ax} + e^{-Ax}}{2A}.$$
Substitute this expression for y into the differential equation to find A in terms of k.

9. Pick out which functions are solutions to which differential equations. (Note: Functions may be solutions to more than one equation or to none; an equation may have more than one solution.)

(a)	$\dfrac{dy}{dx} = -2y$	(I)	$y = 2\sin x$
(b)	$\dfrac{dy}{dx} = 2y$	(II)	$y = \sin 2x$
(c)	$\dfrac{d^2y}{dx^2} = 4y$	(III)	$y = e^{2x}$
(d)	$\dfrac{d^2y}{dx^2} = -4y$	(IV)	$y = e^{-2x}$

10. Decide which of the differential equations are satisfied by the functions on the right.

(a)	$y'' + 2y = 0$	(I)	$y = xe^x$
(b)	$y'' - 2y = 0$	(II)	$y = xe^{-x}$
(c)	$y'' + 2y' + y = 0$		
(d)	$y'' - 2y' + y = 0$		

11. Match solutions and differential equations. (Note: Each equation may have more than one solution.)

(a)	$y'' - y = 0$	(I)	$y = e^x$
(b)	$x^2y'' + 2xy' - 2y = 0$	(II)	$y = x^3$
(c)	$x^2y'' - 6y = 0$	(III)	$y = e^{-x}$
		(IV)	$y = x^{-2}$

12. Families of curves often arise as solutions of differential equations. Match the families of curves with the differential equations of which they are solutions.

(a)	$\frac{dy}{dx} = \frac{y}{x}$	(I)	$y = xe^{kx}$
(b)	$\frac{dy}{dx} = \frac{y \ln y}{x}$	(II)	$y = x^p$
(c)	$\frac{dy}{dx} = \frac{y}{x}\left(1 + \ln\left(\frac{y}{x}\right)\right)$	(III)	$y = e^{kx}$
(d)	$\frac{dy}{dx} = \frac{y \ln y}{x \ln x}$	(IV)	$y = mx$

9.2 SLOPE FIELDS

In this section, you will see how to visualize a first-order differential equation. Let's start with the equation

$$\frac{dy}{dx} = y.$$

You can check that each curve in the family of exponentials, $y = Ce^x$, is a solution to this differential equation.

Any solution to this differential equation has the property that at any point in the plane, the slope of its graph is equal to its y coordinate. (That's what the equation $dy/dx = y$ is telling you!) This means that if the solution goes through the point $(0, 1)$, its slope there is 1; where it goes through a point with $y = 3$ its slope is 3. Another solution goes through $(0, 2)$ and has slope 2 there; at the point where $y = 4$ the slope of this solution is 4. (See Figure 9.5.)

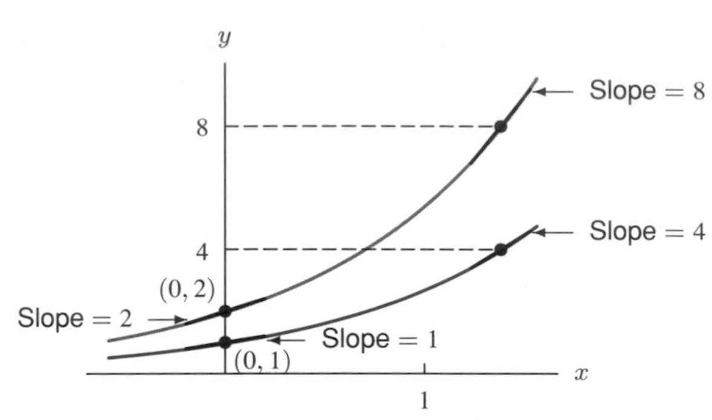

Figure 9.5: Solutions to $\frac{dy}{dx} = y$

Figure 9.6: Slope field for $\frac{dy}{dx} = y$

In Figure 9.5 a small line segment is drawn at each of the marked points showing the slope of the curve there. Imagine drawing many of these line segments, but leaving out the curves; you'll have the *slope field* for the equation $dy/dx = y$ shown in Figure 9.6. From this picture, you can see that above the x-axis, the slopes are all positive (because y is positive there), and they increase as you move upward (as y increases). Below the x-axis, the slopes are all negative, and get more so as you move downward. Notice that on any horizontal line (where y is constant) the slopes are the same.

In the slope field you can see the ghost of the solution curve lurking. Start anywhere on the plane and move so that the slope lines are tangent to your path; you will trace out one of the solution curves. Try penciling in some solution curves on Figure 9.6, some above the x-axis and some below. The curves you draw should have the shape of exponential functions.

In most problems, we will be interested in getting the solution curves from the slope field. It may be helpful to think of the slope field as a set of signposts pointing in the direction you should go at each point. Imagine starting anywhere in the plane: look at the slope field at that point and start to move in that direction. After a small step, look at the slope field again, and alter your direction if necessary. Continue to move across the plane in the direction the slope field points, and you'll trace out a solution curve. Notice that the solution curve is not necessarily the graph of a function, and even if it is, we may not have a formula for the function. Geometrically, solving a differential equation means finding the family of solution curves.

Example 1 Figure 9.7 shows the slope field of the differential equation $\dfrac{dy}{dx} = 2x$.

(a) What do you notice about the slope field?

(b) Compare the solution curves sketched on the slope field with the formula for the solutions to this differential equation.

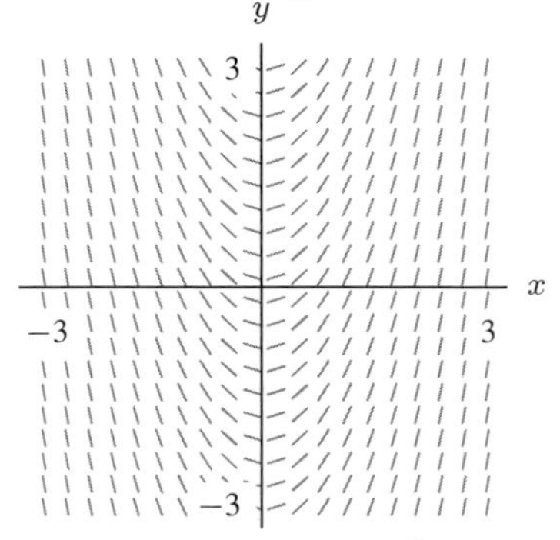

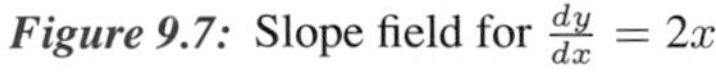

Figure 9.7: Slope field for $\frac{dy}{dx} = 2x$

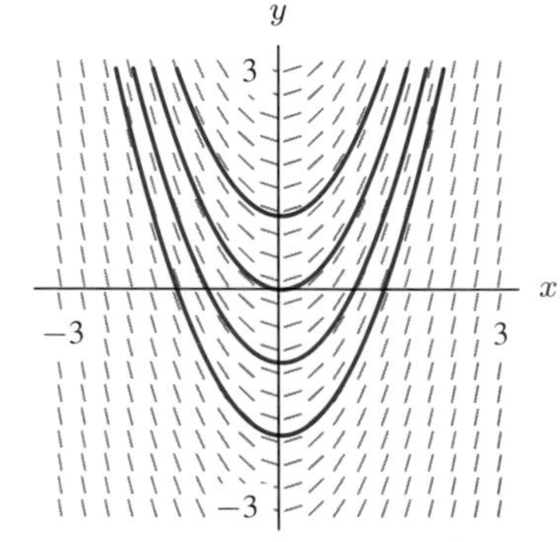

Figure 9.8: Some solutions to $\frac{dy}{dx} = 2x$

Solution (a) In Figure 9.7 you will notice that on a vertical line (where x is constant) the slopes are all the same. This is because in this differential equation dy/dx depends on x only. (In the previous example, $dy/dx = y$, the slopes depended on y only.)

(b) The solution curves in Figure 9.8 look like parabolas. Since we already know how to find a formula for the solutions to the differential equation

$$\frac{dy}{dx} = 2x$$

by antidifferentiation

$$y = \int 2x dx = x^2 + C,$$

we can confirm that the solution curves really are parabolas.

Let's look at another example where we do not (yet) have a formula for the solution of the differential equation.

Example 2 Using the slope field, guess the form of the solution curves of the differential equation

$$\frac{dy}{dx} = -\frac{x}{y}.$$

Solution The slope field is shown in Figure 9.9. Notice that on the y-axis, where x is 0, the slope is 0. On the x-axis, where y is 0, the line segments are vertical and the slope is infinite. At the origin the slope is undefined, and there is no line segment.

What do the solution curves of this differential equation look like? The slope field suggests that they look like circles centered at the origin. Later we'll see how to obtain the solution analytically, but even without this, we can still check that the circle is a solution. Let's take the circle of radius r:

$$x^2 + y^2 = r^2$$

and differentiate implicitly, thinking of y as a function of x. Using the chain rule, we get

$$2x + 2y \cdot \frac{dy}{dx} = 0.$$

Solving for dy/dx gives our differential equation,

$$\frac{dy}{dx} = -\frac{x}{y}.$$

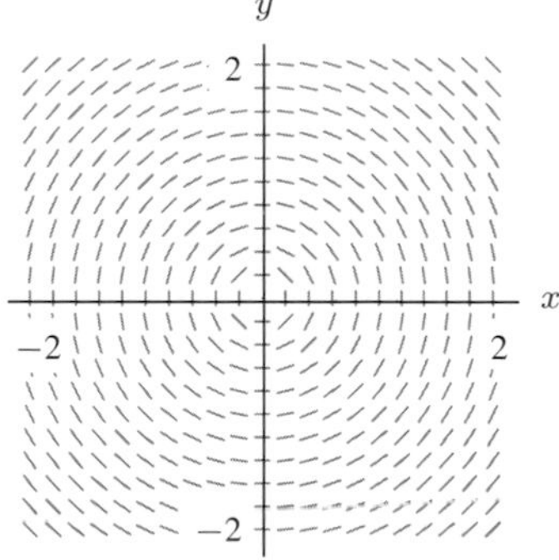

Figure 9.9: Slope field for $\frac{dy}{dx} = -\frac{x}{y}$

The previous example shows that the solutions to differential equations may often be expressed as implicit functions. (Recall that implicit functions are ones which have not been "solved" for y; in other words, the dependent variable is not expressed as an explicit function of x.)

Example 3 The slope fields for $\frac{dy}{dt} = 2 - y$ and $\frac{dy}{dt} = \frac{t}{y}$ are shown in Figures 9.10 and 9.11. For each slope field:

(a) Sketch solution curves with initial conditions
(i) $y = 1$ when $t = 0$ (ii) $y = 0$ when $t = 1$ (iii) $y = 3$ when $t = 0$

(b) For each solution curve, can you say anything about the long-run behavior of y? For example, does $\lim\limits_{t\to\infty} y$ exist? If so, what is its value?

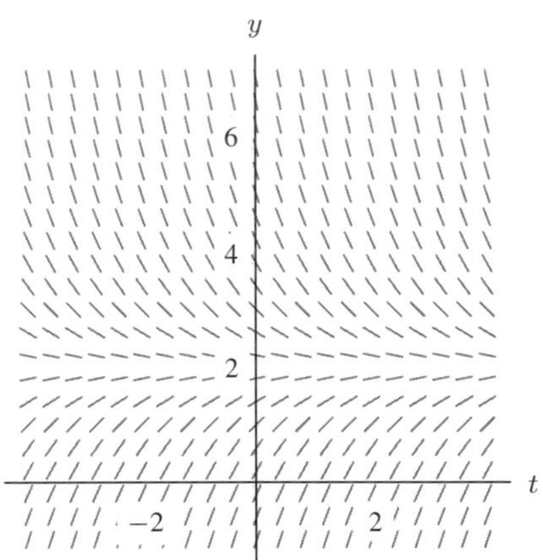

Figure 9.10: Slope field for $\frac{dy}{dt} = 2 - y$

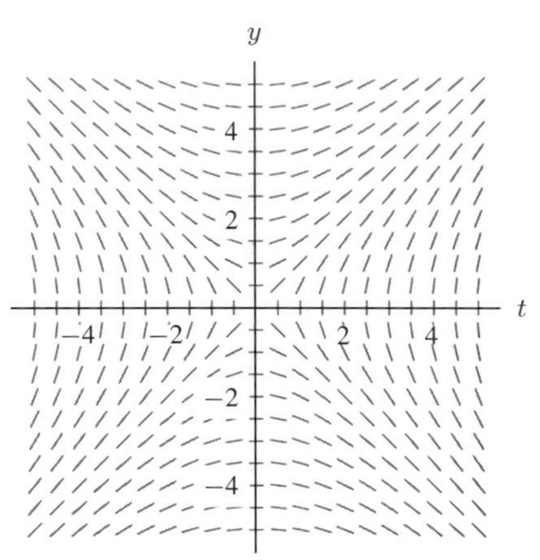

Figure 9.11: Slope field for $\frac{dy}{dt} = \frac{t}{y}$

Solution (a) See Figures 9.12 and 9.13.

(b) For $dy/dt = 2 - y$, all solution curves have $y = 2$ as a horizontal asymptote, so $\lim\limits_{t\to\infty} y = 2$. For $dy/dt = t/y$, as $t \to \infty$, it appears that either $y \to t$ or $y \to -t$.

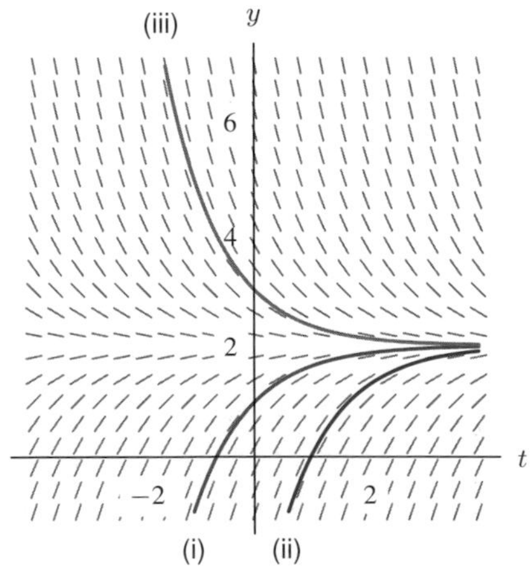

Figure 9.12: Solution curves for $\frac{dy}{dt} = 2 - y$

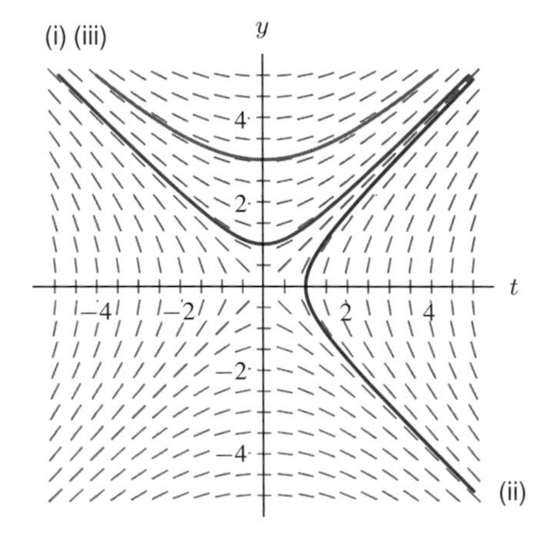

Figure 9.13: Solution curves for $\frac{dy}{dt} = \frac{t}{y}$

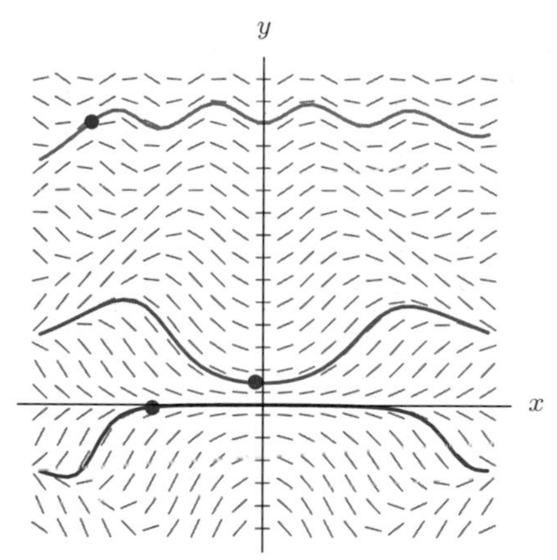

Figure 9.14: There's one and only one solution curve through each point in the plane for this slope field (Dots represent initial conditions)

Existence and Uniqueness of Solutions

Since differential equations are used to model many real situations, the question of whether a solution is unique can have great practical importance. If we know how the velocity of a satellite is changing, can we know its velocity for all future time? If we know the initial population of a city, and we know how the population is changing, can we predict the population in the future? Common sense says yes: if we know the initial value of some quantity and we know exactly how it is changing, we should be able to figure out the future value of the quantity.

In the language of differential equations, an initial-value problem (i.e., a differential equation and an initial condition) almost always has a unique solution. One way to see this is by looking at the slope field. Imagine starting at the point representing the initial condition. Through that point there will usually be a line segment pointing in the direction the solution curve must go. By following these line segments, you trace out the solution curve. Several examples with different starting points are shown in Figure 9.14. In general, at each point there is one line segment and therefore only one direction for the solution curve to go. Thus the solution curve *exists* and is *unique* provided you are given an initial point.

It can be shown that if the slope field is continuous as you move from point to point in the plane, you can be sure that the solution curve exists everywhere. Ensuring that each point has only one solution curve through it requires a slightly stronger condition.

Problems for Section 9.2

1. The slope field for the equation $y' = x + y$ is shown in Figure 9.15.
   (a) Carefully sketch the solutions that pass through the points
   (i) $(0, 0)$ (ii) $(-3, 1)$ (iii) $(-1, 0)$.
   (b) From your sketch, write the equation of the solution passing through $(-1, 0)$.
   (c) Verify your solution to part (b) by substituting it into the differential equation.
2. Figure 9.16 shows the slope field for the equation $y' = (\sin x)(\sin y)$.
   (a) Carefully sketch the solutions that pass through the points: (i) $(-2, -2)$ (ii) $(0, \pi)$.
   (b) What is the equation of the solution that passes through $(0, n\pi)$, where n is any integer?

3. (a) Sketch the slope field for the equation $y' = x - y$ in Figure 9.17 at the points indicated.
   (b) Find the equation for the solution that passes through the point $(1, 0)$.

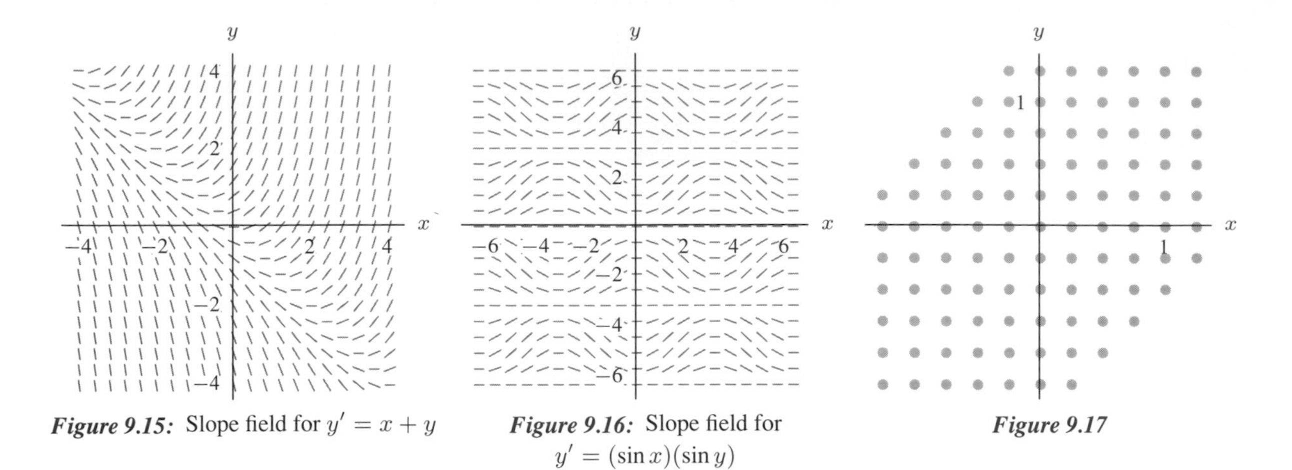

Figure 9.15: Slope field for $y' = x + y$

Figure 9.16: Slope field for $y' = (\sin x)(\sin y)$

Figure 9.17

4. Figure 9.18 displays sketches of two slope fields. Sketch three solution curves for each of these fields.

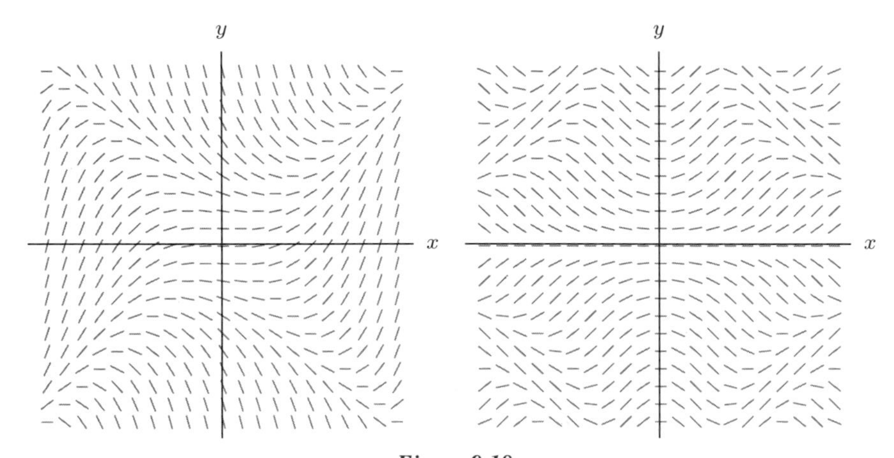

Figure 9.18

5. One of the slope fields shown in Figure 9.18 has the equation $y' = x^2 - y^2$. Which one? On this field, where is the point $(0, 1)$? The point $(1, 0)$? (Assume that the x and y scales are the same.) Now sketch in the line $x = 1$. Then sketch, as carefully as you can, the solution curve that passes through $(0, 1)$ until it crosses $x = 1$.
6. The slope field for $y' = 0.5(1 + y)(2 - y)$ is shown in Figure 9.19.
   (a) Plot the following points on the slope field:
   (i) the origin (ii) $(0, 1)$ (iii) $(1, 0)$
   (iv) $(0, -1)$ (v) $(0, -5/2)$ (vi) $(0, 5/2)$
   (b) Plot solution curves through the points in part (a).

(c) For which regions are all solution curves increasing? For which regions are all solution curves decreasing? When can the solution curves have horizontal tangents? Explain why, using both the slope field and the differential equation.

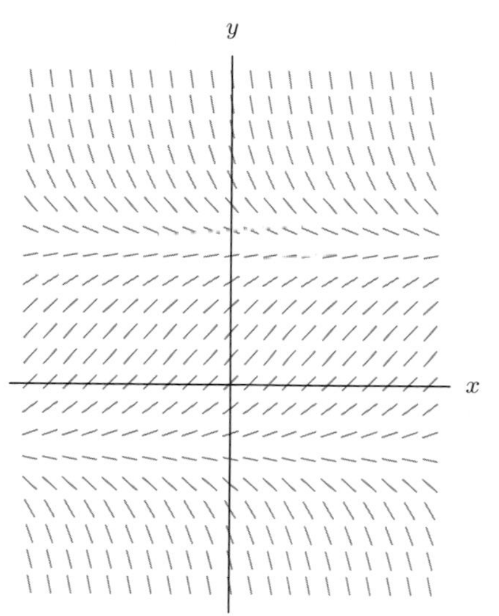

Figure 9.19: Slope field for $y' = 0.5(1+y)(2-y)$ (The x and y scales are the same)

7. The Gompertz equation, which models growth of animal tumors, is $y' = -ay\ln(y/b)$, where a and b are positive constants. Write a paragraph describing the similarities and/or differences between solutions to this differential equation with $a = 1$ and $b = 2$ and solutions to the equation $y' = y(2-y)$. Use the slope fields in Figures 9.20 and 9.21.

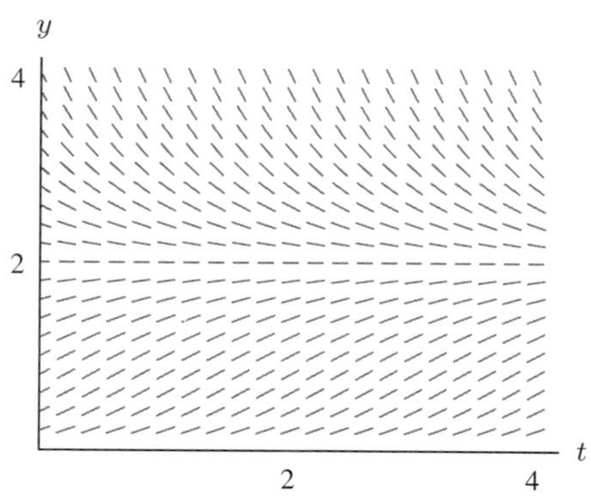

Figure 9.20: Slope field for $y' = -y\ln(y/2)$

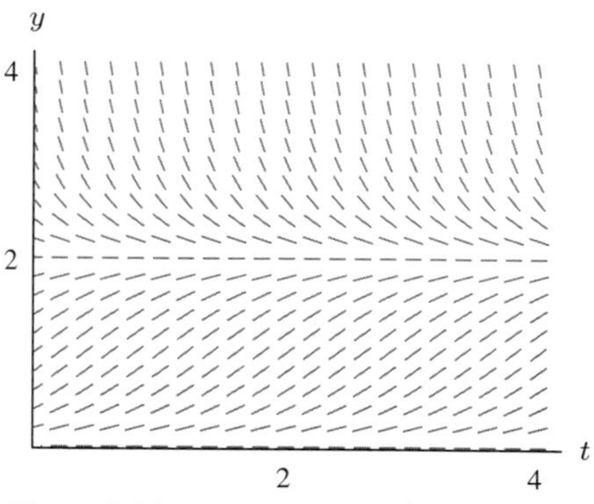

Figure 9.21: Slope field for $y' = y(2-y)$

8. Match the slope fields in Figure 9.22 with their differential equations:

(a) $y' = 1 + y^2$ (b) $y' = x$ (c) $y' = \sin x$
(d) $y' = y$ (e) $y' = x - y$ (f) $y' = 4 - y$

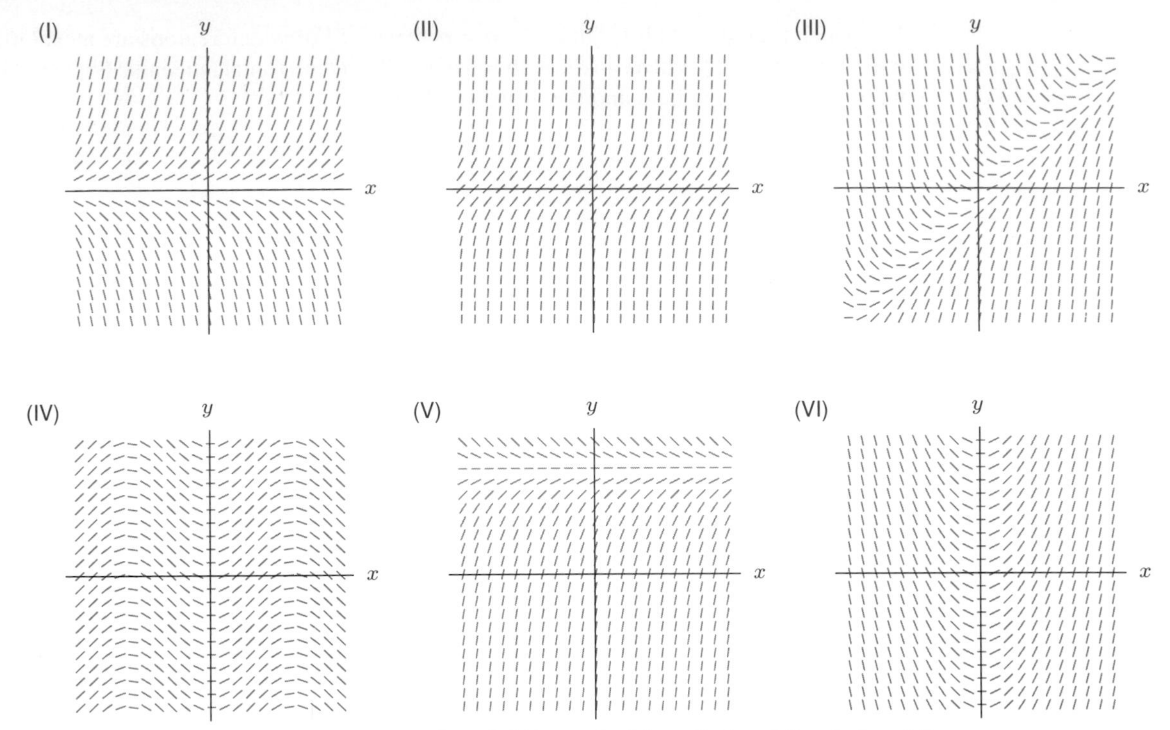

Figure 9.22: Each slope field is graphed for $-5 \leq x \leq 5$, $-5 \leq y \leq 5$

For Problems 9–14, consider a solution curve for each of the slope fields in Problem 8. Write one or two sentences describing qualitatively the long-run behavior of y. For example, as x increases, does $y \to \infty$, or does y remain finite? You may get different limiting behavior for different starting points. In each case, your answer should discuss how the limiting behavior depends on the starting point.

9. Slope field (I)
10. Slope field (II)
11. Slope field (III)
12. Slope field (IV)
13. Slope field (V)
14. Slope field (VI)

15. In Section 4.8 on implicit differentiation, the equation $y^3 - xy = -6$ was analyzed using the derivative $dy/dx = y/(3y^2 - x)$. (See page 231.) View this equation for dy/dx as a differential equation and give a rough sketch of its slope field, indicating where the slope is positive, negative, zero, or undefined (vertical). Then give a rough sketch of the solution curve through the point $(7, 2)$, the solution curve through the point $(7, 1)$, and the solution curve through the point $(7, -3)$. Notice that each of these points lies on the curve $y^3 - xy = -6$.

9.3 EULER'S METHOD

In the preceding section we saw how to sketch a solution curve to a differential equation using its slope field, whose line segments are everywhere tangent to the solution. In this section we will do the same thing numerically, by computing points on the solution curves using *Euler's method*. (Leonhard Euler was an eighteenth-century Swiss mathematician.) In some cases—but not many—we can find an equation for the solution curve. That will be the subject of the next section.

Here's the concept behind Euler's method. Think of the slope field as a set of signposts directing you across the plane. Pick a starting point (corresponding to the initial value), and calculate the slope at that point using the differential equation. This slope is a signpost telling you the direction to take. Head off a small distance in that direction. Stop and look at the new signpost. Recalculate the slope from the differential equation, using the coordinates of the new point. Change direction to correspond to the new slope, and move another small distance, and so on.

Example 1 Use Euler's method to construct a solution for $\frac{dy}{dx} = y$. Start at the point $P_0 = (0, 1)$ and take $\Delta x = 0.1$.

Solution The slope at the point $P_0 = (0, 1)$ is $dy/dx = 1$. (See Figure 9.23.) As you move from P_0 to P_1, y will increase by Δy, where

$$\Delta y = (\text{slope at } P_0)\Delta x = 1(0.1) = 0.1,$$

TABLE 9.2 *Euler's method for $dy/dx = y$, starting at $(0, 1)$*

	x	y	$\Delta y = (\text{Slope})\Delta x$
P_0	0	1	$0.1 = (1)(0.1)$
P_1	0.1	1.1	$0.11 = (1.1)(0.1)$
P_2	0.2	1.21	$0.121 = (1.21)(0.1)$
P_3	0.3	1.331	$0.1331 = (1.331)(0.1)$
P_4	0.4	1.4641	$0.14641 = (1.4641)(0.1)$
P_5	0.5	1.61051	$0.161051 = (1.61051)(0.1)$

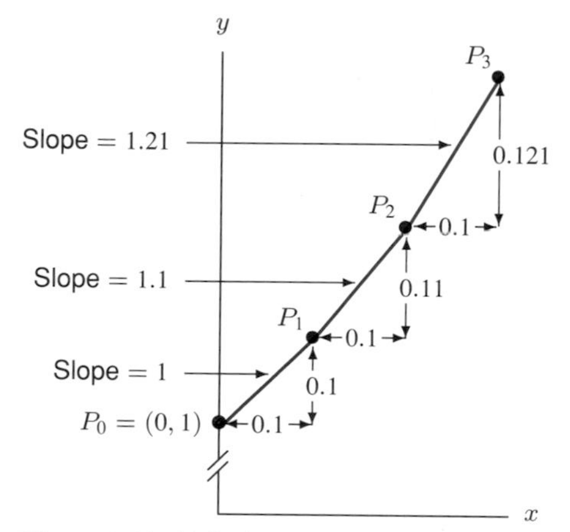

Figure 9.23: Euler's approximate solution to $\frac{dy}{dx} = y$

so

$$y \text{ value at } P_1 = (y \text{ value at } P_0) + \Delta y = 1 + 0.1 = 1.1.$$

Thus the point P_1 is $(0.1, 1.1)$. Now, using the differential equation again, we see that

$$\text{slope at } P_1 = 1.1,$$

so if we move to P_2, then y will change by

$$\Delta y = (\text{slope at } P_1)\Delta x = (1.1)(0.1) = 0.11,$$

so

$$y \text{ value at } P_2 = (y \text{ value at } P_1) + \Delta y = 1.1 + 0.11 = 1.21.$$

Thus P_2 is (0.2, 1.21). Continuing gives the results in Table 9.2. Since the solution curves of $dy/dx = y$ are exponentials, they are concave up, and they curve upward away from the line segments of the slope field. Therefore, in this case Euler's method produces y values which are too small.

Notice that Euler's method calculates approximate values of y for points on a solution curve; it does not give a general formula for y in terms of x. The graph in Figure 9.23 approximates the solution curve.

Example 2 Show that Euler's method for $\frac{dy}{dx} = y$ starting at $(0, 1)$ and using two steps with $\Delta x = 0.05$ gives $y \approx 1.1025$ when $x = 0.1$.

Solution At $(0, 1)$, the slope is 1 and $\Delta y = (1)(0.05) = 0.05$, so new $y = 1 + 0.05 = 1.05$. At $(0.05, 1.05)$, the slope is 1.05 and $\Delta y = (1.05)(0.05) = 0.0525$, so new $y = 1.05 + 0.0525 = 1.1025$.

In general, dy/dx may be a function of both x and y. Euler's method still works then, as the next example shows.

Example 3 Approximate four points on the solution curve to $\frac{dy}{dx} = -\frac{x}{y}$, starting at $(0, 1)$ and using $\Delta x = 0.1$.

Solution The results from Euler's method are in Table 9.3, along with the true y values (to two decimals) calculated from the equation of the circle $x^2 + y^2 = 1$, which is the solution curve through $(0, 1)$. The fact that the curve is concave down means that the approximate y values are above the true ones. (See Figure 9.24.)

TABLE 9.3 *Euler's method for $dy/dx = -x/y$, starting at $(0, 1)$*

x	Approximate y value	$\Delta y = (\text{Slope})\Delta x$	True y value
0	1	$0 = (0)(0.1)$	1
0.1	1	$-0.01 = (-0.1/1)(0.1)$	0.99
0.2	0.99	$-0.02 = (-0.2/0.99)(0.1)$	0.98
0.3	0.97		0.95

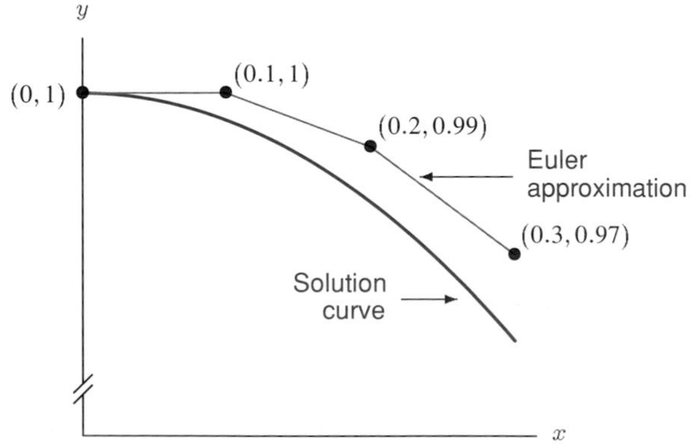

Figure 9.24: Euler's approximate solution to $\frac{dy}{dx} = -\frac{x}{y}$

The Accuracy of Euler's Method

Let's go back to the differential equation $dy/dx = y$. To improve the accuracy of our Euler's method approximation, choose Δx smaller. Let's compare the true and approximate values. The exact solution to the differential equation $dy/dx = y$ going through the point $(0, 1)$ is $y = e^x$, so the true values will be calculated using this function. (See Figure 9.25.) Where $x = 0.1$,

$$\text{True } y \text{ value } = e^{0.1} \approx 1.1051709.$$

In Example 1 we had $\Delta x = 0.1$, and where $x = 0.1$,

$$\text{Approximate } y \text{ value} = 1.1, \text{ so the error} \approx 0.005.$$

In Example 2 we decreased Δx to 0.05. After two steps, $x = 0.1$, and we had

$$\text{Approximate } y \text{ value} = 1.1025, \text{ so error} \approx 0.00267.$$

Thus halving the step size has approximately halved the error.

> The *error* in using Euler's method is the difference between the approximate value and the true value. If the number of steps used is n, the error is approximately proportional to $1/n$.

Just as there are more accurate numerical integration methods than left and right Riemann sums, there are more accurate methods than Euler's for approximating solution curves. However, Euler's method will be all we need.

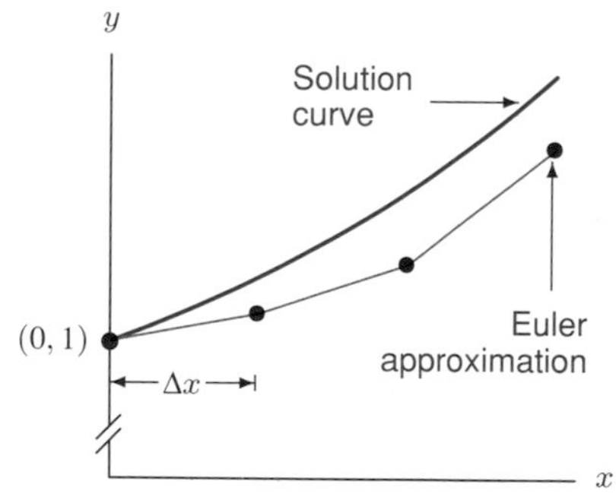

Figure 9.25: Euler's approximate solution to $\frac{dy}{dx} = y$

Problems for Section 9.3

1. Consider the differential equation $y' = (\sin x)(\sin y)$.
   (a) Calculate approximate y values using Euler's method with three steps and $\Delta x = 0.1$, starting at each of the following points: (i) $(0, 2)$ (ii) $(0, \pi)$.
   (b) Use the slope field in Figure 9.16 on page 488 to explain your solution to part (a)(ii).

2. Consider the differential equation $y' = x + y$ whose slope field is in Figure 9.15 on page 488. Use Euler's method with $\Delta x = 0.1$ to estimate y when $x = 0.4$ for the solution curves satisfying

   (a) $y(0) = 1$ (b) $y(-1) = 0$

3. (a) Use ten steps of Euler's method to approximate y-values for $dy/dt = 1/t$, starting at $(1, 0)$ and using $\Delta t = 0.1$.
   (b) Using integration, solve the differential equation to find the exact value of y at the end of these ten steps.
   (c) Is your approximate value of y at the end of ten steps bigger or smaller than the exact value? Use a slope field to explain your answer.

4. (a) Use Euler's method to approximate the value of y at $x = 1$ on the solution curve to the differential equation

$$\frac{dy}{dx} = x^3 - y^3$$

   that passes through $(0, 0)$. Use $\Delta x = 1/5$ (i.e., 5 steps).
   (b) Figure 9.26 shows the slope field for the differential equation. Sketch the solution that passes through $(0, 0)$. Show on the graph the approximation you made in part (a).
   (c) From the graph of the slope field, can you say whether your answer to part (a) is an overestimate or an underestimate?

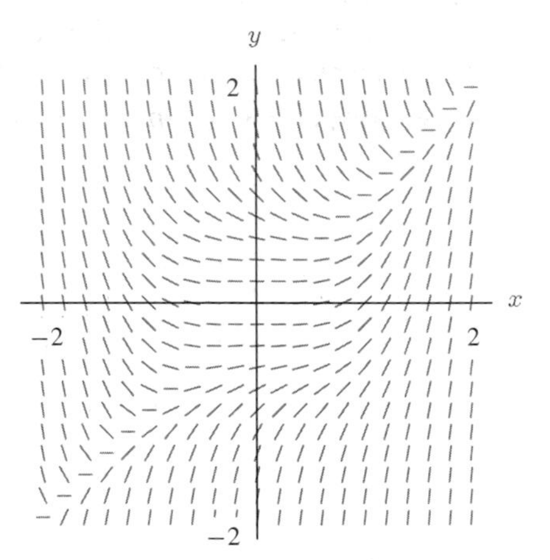

Figure 9.26: Slope field for $\frac{dy}{dx} = x^3 - y^3$

5. Consider the differential equation

$$\frac{dy}{dx} = 2x, \quad \text{with initial condition } y(0) = 1.$$

   (a) Use Euler's method with two steps to estimate y when $x = 1$.
   (b) Use Euler's method with four steps to estimate y when $x = 1$.
   (c) What is the formula for the exact value of y?
   (d) Does the error in Euler's approximation behave as predicted in the box on page 493?

6. (a) Use Euler's method with five subintervals to approximate the solution curve to the differential equation $dy/dx = x^2 - y^2$ passing through the point $(0, 1)$ and ending at $x = 1$. (Keep the approximate function values to three decimal places.)
   (b) Repeat this computation using ten subintervals, again ending with $x = 1$.

7. Why are the approximate results you obtained in Problem 6 smaller than the true values? (Note: The slope field for this differential equation is one of those in Figure 9.18 on page 488.)

8. Using the fact that the error incurred using Euler's method is roughly proportional to one over the number of subintervals used, how should the errors of the five-step calculation and the ten-step calculation in Problem 6 compare? Use your answer to estimate the true value of y when $x = 1$.

9. Use Euler's method to solve
$$\frac{dB}{dt} = 0.05B$$
with initial value $B = 1000$ when $t = 0$. Take:
(a) $\Delta t = 1$ and 1 step. (b) $\Delta t = 0.5$ and 2 steps. (c) $\Delta t = 0.25$ and 4 steps.
(d) Suppose B is the balance in a bank account earning interest. Explain why the result of your calculation in part (a) is equivalent to compounding the interest once a year instead of continuously.
(e) Interpret the result of your calculations in parts (b) and (c) in terms of compound interest.

10. Consider the differential equation $dy/dx = f(x)$ with initial value $y(0) = 0$. Explain why using Euler's method to approximate the solution curve gives the same results as using left Riemann sums to approximate $\int_0^x f(t)\,dt$.

11. In Section 4.8 on implicit differentiation, a table of values was computed for the equation $y^3 - xy = -6$ near $x = 7, y = 2$. On page 231, the implicit derivative $dy/dx = y/(3y^2 - x)$ was evaluated at $x = 7$, $y = 2$, and then the y-values at $x = 6.8, 6.9, 7.1, 7.2$ were approximated by local linearity. Now compute the y-values for $x = 7.1, 7.2$ by Euler's method using $\Delta x = 0.05$. Compare your answer for $x = 7.2$ to the actual answer, accurate to 6 decimal places, of 2.076018.

9.4 SEPARATION OF VARIABLES

We have seen how to sketch solution curves of a differential equation using a slope field and how to calculate approximate numerical solutions. Now we'll see how to find the equation of a solution curve exactly for certain differential equations.

First, we'll look at a familiar example, the differential equation
$$\frac{dy}{dx} = -\frac{x}{y},$$
whose solution curves are the circles
$$x^2 + y^2 = C.$$
We can check that these circles are solutions by differentiation; the question now is how they were obtained. The method of *separation of variables* works by putting all the x's on one side of the equation and all the y's on the other, giving
$$y\,dy = -x\,dx,$$
and then integrating each side separately:
$$\int y\,dy = -\int x\,dx,$$
$$\frac{y^2}{2} = \frac{-x^2}{2} + k,$$
thereby giving the circles we were expecting:
$$x^2 + y^2 = C \qquad \text{where } C = 2k.$$

You might worry about whether it is legitimate to integrate one side of the equation with respect to x and the other with respect to y. The reason it can be done is explained at the end of this section.

The Exponential Growth and Decay Equations

Let's use separation of variables on an equation that occurs frequently in practice:

$$\frac{dy}{dx} = ky.$$

Separating variables,

$$\frac{1}{y} dy = k\, dx$$

and integrating

$$\int \frac{1}{y}\, dy = \int k\, dx$$

gives

$$\ln|y| = kx + C \quad \text{for some constant } C.$$

Solving for $|y|$ leads to

$$|y| = e^{kx+C} = e^{kx}e^{C} = Ae^{kx}$$

where $A = e^C$, so A is positive. Thus

$$y = (\pm A)e^{kx} = Be^{kx}$$

where $B = \pm A$, so B is any nonzero constant. Even though there's no C leading to $B = 0$, B can be 0 because $y = 0$ is a solution to the differential equation. We lost this solution when we divided through by y at the first step.

The general solution to $\dfrac{dy}{dx} = ky$ is $y = Be^{kx}$ for any B.

Thus, the differential equation $dy/dx = ky$ always leads to exponential growth (if $k > 0$) or exponential decay (if $k < 0$). Graphs of solution curves for some $k > 0$ are in Figure 9.27. For $k < 0$, the graphs are reflected in the y-axis.

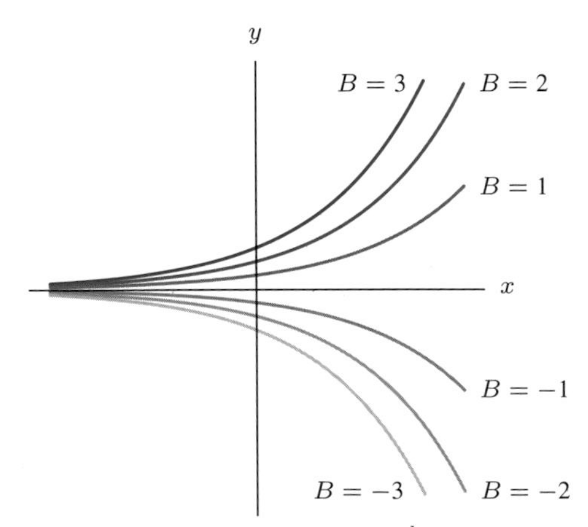

Figure 9.27: Graphs of $y = Be^{kx}$, which are solutions to $\frac{dy}{dx} = ky$, for some fixed $k > 0$

Example 1 For $k > 0$, find and graph solutions of

$$\frac{dH}{dt} = -k(H - 20).$$

Solution The slope field in Figure 9.28 shows the qualitative behavior of the solutions. To find the equation of the solution curves, we separate variables and integrate.

$$\int \frac{1}{H - 20}\, dH = -\int k\, dt$$

gives

$$\ln |H - 20| = -kt + C.$$

Solving for H leads to:

$$|H - 20| = e^{-kt+C} = e^{-kt}e^{C} = Ae^{-kt}$$

or

$$H - 20 = (\pm A)e^{-kt} = Be^{-kt}$$
$$H = 20 + Be^{-kt}.$$

Graphs for $k = 1$ and $B = -10, 0, 10$, $t \geq 0$, are in Figure 9.28.

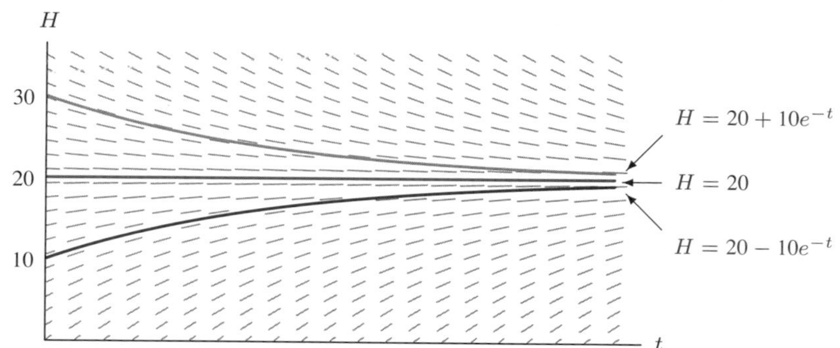

Figure 9.28: Slope field and some solution curves for $\frac{dH}{dt} = -k(H - 20)$, with $k = 1$

Example 2 Find and sketch the solution to

$$\frac{dP}{dt} = 2P - 2Pt$$

satisfying $P = 5$ when $t = 0$.

Solution Factoring the right-hand side gives

$$\frac{dP}{dt} = P(2 - 2t).$$

Separating variables, we get

$$\int \frac{dP}{P} = \int (2 - 2t)\, dt,$$

so

$$\ln |P| = 2t - t^2 + C.$$

Solving for P leads to

$$|P| = e^{2t-t^2+C} = e^C e^{2t-t^2} = Ae^{2t-t^2}$$

with $A = e^C$, so $A > 0$. Thus the general solution to the differential equation is

$$P = Be^{2t-t^2} \quad \text{for any } B.$$

As before, $B = 0$ gives a solution although there is no C which leads to it. To find the value of B for this example, substitute $P = 5$ and $t = 0$ into the general solution, giving

$$5 = Be^{2(0)-0^2} = B$$

so

$$P = 5e^{2t-t^2}.$$

The graph of this function is in Figure 9.29.

Since the solution can be rewritten as

$$P = 5e^{1-1+2t-t^2} = 5e^1 e^{-1+2t-t^2} = (5e)e^{-(1-t)^2},$$

the graph has the same shape as the graph of $y = e^{-t^2}$, the bell-shaped curve of statistics. Here, however, the maximum, normally at $t = 0$, is shifted one unit to the right and occurs at $t = 1$.

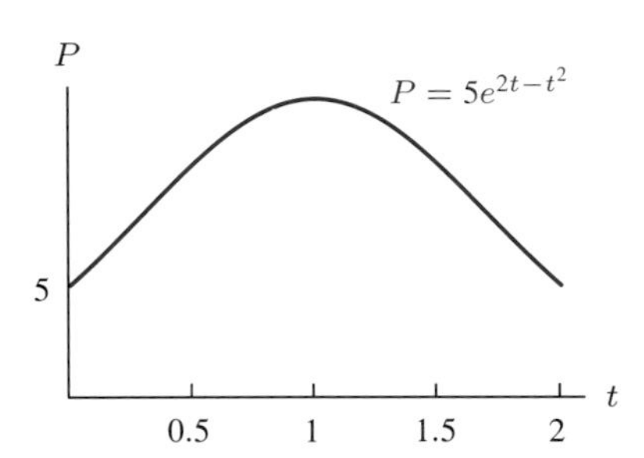

Figure 9.29: Bell-shaped solution curve

Justification of Method of Separation of Variables

Suppose a differential equation can be written in the form

$$\frac{dy}{dx} = g(x)f(y).$$

Then, provided $f(y) \neq 0$, we write $f(y) = 1/h(y)$ so that the right-hand side can be thought of as a fraction,

$$\frac{dy}{dx} = \frac{g(x)}{h(y)}.$$

If we multiply through by $h(y)$ we get

$$h(y)\frac{dy}{dx} = g(x).$$

Thinking of y as a function of x, so $y = y(x)$, and $dy/dx = y'(x)$, we can write the line above as

$$h(y(x)) \cdot y'(x) = g(x).$$

Now integrate both sides with respect to x:

$$\int h(y(x)) \cdot y'(x)\, dx = \int g(x)\, dx.$$

The form of the integral on the left suggests that we use the substitution $y = y(x)$. Since $dy = y'(x)\, dx$, we get

$$\int h(y)\, dy = \int g(x)\, dx.$$

If antiderivatives of h and g can be found, then this will give the equation of the solution curve.

Note that transforming the original differential equation,

$$\frac{dy}{dx} = \frac{g(x)}{h(y)},$$

into

$$\int h(y)\, dy = \int g(x)\, dx$$

looks as though we have treated dy/dx as a fraction, cross-multiplied and then integrated. Although that's not exactly what we've done, you may find this a helpful way of remembering the method. In fact, the dy, dx notation was introduced by Leibniz to allow shortcuts like this (more specifically, to make the chain rule look like cancellation).

Problems for Section 9.4

Find the solutions to the differential equations in Problems 1–16, subject to the given initial condition.

1. $\dfrac{dP}{dt} = 0.02P, \quad P(0) = 20.$

2. $\dfrac{dQ}{dt} = \dfrac{Q}{5}, \quad Q = 50$ when $t = 0$.

3. $\dfrac{dm}{dt} = 3m, \quad m = 5$ when $t = 1$.

4. $\dfrac{dI}{dx} = 0.2I, \quad I = 6$ where $x = -1$.

5. $\dfrac{dy}{dx} + \dfrac{y}{3} = 0, \quad y(0) = 10.$

6. $\dfrac{1}{z}\dfrac{dz}{dt} = 5, \quad z(1) = 5.$

7. $\dfrac{dP}{dt} = P + 4, \quad P = 100$ when $t = 0$.

8. $\dfrac{dy}{dx} = 2y - 4,$ through $(2, 5)$.

9. $\dfrac{dm}{dt} = 0.1m + 200, \quad m(0) = 1000.$

10. $\dfrac{dB}{dt} + 2B = 50, \quad B(1) = 100.$

11. $\dfrac{dz}{dt} = te^z,$ through the origin.

12. $\dfrac{dy}{dx} = \dfrac{5y}{x}, \quad x > 0, y = 3$ where $x = 1$.

13. $\dfrac{dy}{dt} = y^2(1+t), \quad y = 2$ when $t = 1$.

14. $\dfrac{dz}{dt} = z + zt^2, \quad z = 5$ when $t = 0$.

15. $\dfrac{dw}{d\theta} = \theta w^2 \sin\theta^2, \quad w(0) = 1$.

16. $x(x+1)\dfrac{du}{dx} = u^2, \quad u(1) = 1$.

17. (a) Find the general solution to the differential equation introduced on page 478 to describe the rate at which a person learns,

$$\frac{dy}{dt} = 100 - y.$$

(b) Sketch several solutions.
(c) Find the particular solution subject to the initial condition that $y = 0$ when $t = 0$.

Solve the differential equations in Problems 18–23. Assume a, b, and k are constants.

18. $\dfrac{dR}{dt} = kR$

19. $\dfrac{dQ}{dt} - \dfrac{Q}{k} = 0$

20. $\dfrac{dP}{dt} = P - a$

21. $\dfrac{dQ}{dt} = b - Q$

22. $\dfrac{dP}{dt} = k(P - a)$

23. $\dfrac{dR}{dt} = aR + b$

Solve the differential equations in Problems 24–27. Assume $x \geq 0$, $y \geq 0$.

24. $\dfrac{dy}{dt} = y(2-y), \quad y(0) = 1$.

25. $t\dfrac{dx}{dt} = (1 + 2\ln t)\tan x$, where $t > 0$.

26. $\dfrac{dx}{dt} = \dfrac{x \ln x}{t}$.

27. $\dfrac{dy}{dt} = -y \ln\left(\dfrac{y}{2}\right), \quad y(0) = 1$.

28. (a) Sketch the slope field for $y' = x/y$.
(b) Sketch several solution curves.
(c) Solve the differential equation analytically.

29. (a) Sketch the slope field for $y' = -y/x$.
(b) Sketch several solution curves.
(c) Solve the differential equation analytically.

30. Compare the slope field for $y' = x/y$, Problem 28, with that for $y' = -y/x$, Problem 29. Show that the solution curves of Problem 28 intersect the solution curves of Problem 29 at right angles.

9.5 GROWTH AND DECAY

In this section we will look at exponential growth and decay equations which occur frequently in practice. Consider the population of a region. If there is no immigration or emigration, the rate at which the population is changing is often proportional to the population. In other words, the larger the population, the faster it is growing, because there are more people to have babies. If the population at time t is P, and its continuous growth rate is 2% per unit time, then we know

$$\text{Rate of growth of population} = 2\%(\text{Current population})$$

and we can write this as

$$\frac{dP}{dt} = 0.02P.$$

The 2% growth rate is called the *relative growth rate* to distinguish it from the absolute growth rate, or rate of change of the population, dP/dt. Notice they are in different units. Since

$$\text{Relative growth rate} = 2\% = \frac{1}{P}\frac{dP}{dt}$$

the relative growth rate is a number (or percent) per unit time, while

$$\text{Absolute growth rate} = \text{Rate of change of population} = \frac{dP}{dt}$$

is in people per unit time.

The equation $dP/dt = 0.02P$ is of the form $dP/dt = kP$ for $k = 0.02$ and therefore has the solution

$$P = P_0 e^{0.02t}.$$

Other processes are described by differential equations similar to that for population growth, but with negative values for k. In summary, we have the following result from the preceding section:

> Every solution to the equation
>
> $$\frac{dP}{dt} = kP$$
>
> can be written in the form
>
> $$P = P_0 e^{kt},$$
>
> where P_0 is the initial value of P, and $k > 0$ represents growth, while $k < 0$ represents decay.

Recall that the *doubling time* of an exponentially growing quantity is the time required for it to double. The *half-life* of an exponentially decaying quantity is the time for half of it to decay.

Continuously Compounded Interest

In Chapter 1 we introduced continuous compounding as the limiting case in which interest was added more and more often. Here we will approach continuous compounding from a different point of view, and say that continuous compounding means that, at any time, interest is being accrued at a rate that is a fixed percentage of the balance at that moment. Thus, the larger the balance the faster interest is being earned and the faster the balance grows.

Example 1 A bank account earns interest continuously at a rate of 5% of the current balance per year. Assume that the initial deposit is $1000, and that no other deposits or withdrawals are made.

(a) Write the differential equation satisfied by the balance in the account.

(b) Solve the differential equation and graph the solution.

Solution (a) In this problem we are looking for B, the balance in the account in dollars, as a function of t, time in years. The fact that interest is being earned continuously means that interest is continuously being added to the account at a rate of 5% of the balance at that moment. Thus, at any instant,

$$\text{Rate at which interest is earned} = 5\%(\text{Current balance}).$$

If B is the balance at time t, and if no deposits or withdrawals are made, then the rate of change of the balance is exactly the rate the interest is earned, so

$$\text{Rate at which balance is increasing} = 5\%(\text{Current balance})$$

so

$$\frac{dB}{dt} = 0.05B.$$

This is the differential equation that describes the process. Notice that it does not involve the \$1000, because the initial deposit does not affect the process by which interest is earned. The \$1000 is the initial condition.

(b) Solving the differential equation using separation of variables gives

$$B = B_0 e^{0.05t}$$

where B_0 is the initial value of B, so $B_0 = 1000$. Thus

$$B = 1000e^{0.05t}$$

and this function is graphed in Figure 9.30.

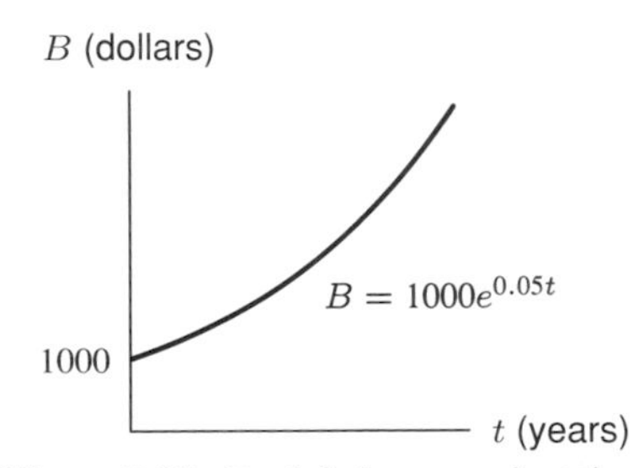

Figure 9.30: Bank balance against time

You may wonder how we can represent an amount of money by a differential equation, since money can only take on discrete values (you can't have fractions of a cent). In fact, the differential equation is only an approximation, but for large amounts of money, it is a pretty good approximation.

The Difference between Continuous and Annual Percentage Growth Rates

If $P = P_0(1+r)^t$ with t in years, we say that r is the *annual* growth rate, while if $P = P_0e^{kt}$, we say that k is the *continuous* or *instantaneous* growth rate.

It is important to note that the constant k in the differential equation $dP/dt = kP$ is not the annual growth rate, but rather the *continuous* growth rate. In Example 1, with a continuous interest rate of 5%, we obtain a balance at time t of $B = B_0e^{0.05t}$, where t is measured in years. At the end of one year the balance is $B_0e^{0.05}$. In that one year, our balance has changed from B_0 to $B_0e^{0.05}$; that is, it has changed by the factor $e^{0.05} = 1.0513$. Thus the *annual* growth rate is 5.13%. This is what the bank means when it says "5% compounded continuously for an effective annual yield of 5.13%." Since $P_0e^{0.05t} = P_0(1.0513)^t$, we are talking about two different ways to represent the same function of t.

To be honest, most growth is measured over discrete time intervals and hence a continuous growth rate is an idealized concept. A demographer who says the population of some country is growing at the rate of 2% per year usually means just what you think: after one year the population will have increased by a factor of 1.02; and after t years the population will be given by $P = P_0(1.02)^t$. If we want to find the continuous growth rate k for the population we proceed as follows. Suppose the population is expressed as $P = P_0e^{kt}$. At the end of one year $P = P_0e^k$. To get 2% growth in that year we need $e^k = 1.02$. Thus $k = \ln 1.02 \approx 0.0198$. The continuous growth rate $k = 1.98\%$ is close to the annual growth rate of 2%, but it is not the same. Again, we simply have two different representations of the same function, since $P_0(1.02)^t = P_0e^{0.0198t}$.

Pollution in the Great Lakes

In the 1960s pollution in the Great Lakes became an issue of public concern. We will set up a model for how long it would take for the lakes to flush themselves clean, assuming no further pollutants are being dumped in the lake.

Suppose Q is the total quantity of pollutant in a lake of volume V at time t. Suppose that clean water is flowing into the lake at a constant rate r and that water flows out at the same rate. Assume that the pollutant is evenly spread throughout the lake, and that the clean water coming into the lake immediately mixes with the rest of the water.

We want to investigate how Q varies with time. First, notice that since pollutants are being taken out of the lake but not added, Q decreases, and the water leaving the lake becomes less polluted, so the rate at which the pollutants leave decreases. This tells us that Q is decreasing and concave up. In addition, the pollutants will never be completely removed from the lake though the quantity remaining will become arbitrarily small: in other words, Q is asymptotic to the t-axis. (See Figure 9.31.)

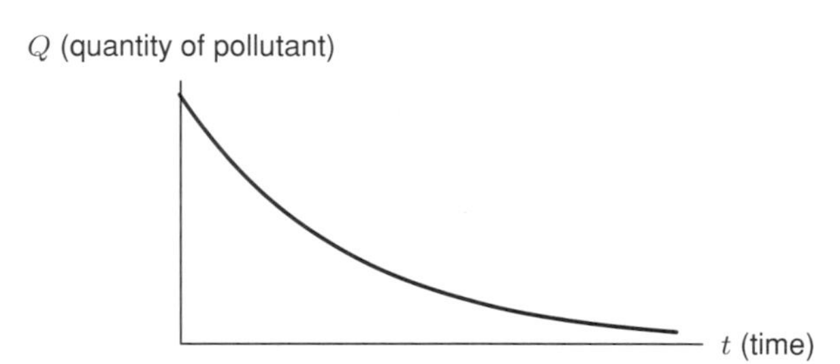

Figure 9.31: Pollutant in lake versus time

Setting Up a Differential Equation for the Pollution

To understand exactly how Q changes with time, we write an equation for the rate at which Q changes. We know that

$$\begin{matrix}\text{Rate } Q \\ \text{changes}\end{matrix} = -\begin{pmatrix}\text{Rate pollutants} \\ \text{leave in outflow}\end{pmatrix}.$$

where the minus sign represents the fact that Q is decreasing. At time t, the concentration of pollutants is Q/V and water containing this concentration is leaving at rate r. Thus

$$\begin{matrix}\text{Rate pollutants} \\ \text{leave in outflow}\end{matrix} = \begin{matrix}\text{Rate of} \\ \text{outflow}\end{matrix} \times \text{Concentration} = r \cdot \frac{Q}{V}.$$

Thus the differential equation is

$$\frac{dQ}{dt} = -\frac{r}{V}Q$$

and its solution is

$$Q = Q_0 e^{-rt/V}.$$

Table 9.4 contains values of r and V for four of the Great Lakes.[2] We will use this data to calculate how long it would take for certain fractions of the pollution to be removed.

TABLE 9.4 *Volume and outflow in Great Lakes*

	V (thousands of km^3)	r (km^3/year)
Superior	12.2	65.2
Michigan	4.9	158
Erie	0.46	175
Ontario	1.6	209

Example 2 How long will it take for 90% of the pollution to be removed from Lake Erie? For 99% to be removed?

Solution Substituting r and V for Lake Erie into the differential equation for Q gives

$$\frac{dQ}{dt} = -\frac{r}{V}Q = \frac{-175}{0.46 \times 10^3}Q = -0.38Q$$

where t is measured in years. Thus Q is given by

$$Q = Q_0 e^{-0.38t}.$$

When 90% of the pollution has been removed, 10% remains, so $Q = 0.1Q_0$. Substituting gives

$$0.1Q_0 = Q_0 e^{-0.38t}.$$

Canceling Q_0 and solving for t:

$$t = \frac{-\ln(0.1)}{0.38} \approx 6 \text{ years.}$$

When 99% of the pollution has been removed, $Q = 0.01Q_0$, so t satisfies

$$0.01Q_0 = Q_0 e^{-0.38t}.$$

Solving for t gives

$$t = \frac{-\ln(0.01)}{0.38} \approx 12 \text{ years.}$$

[2] Data from William E. Boyce and Richard C. DiPrima, *Elementary Differential Equations* (New York: Wiley, 1977).

Newton's Law of Heating and Cooling

Newton proposed that the temperature of a hot object decreases at a rate proportional to the difference between its temperature and that of its surroundings. Similarly, a cold object heats up at a rate proportional to the temperature difference between the object and its surroundings.

For example, a hot cup of coffee standing on the kitchen table cools at a rate proportional to the temperature difference between the coffee and the surrounding air. As the coffee cools down, the rate at which it cools decreases, because the temperature difference between the coffee and the air decreases. In the long run, the rate of cooling tends to zero, and the temperature of the coffee approaches room temperature — as you would expect. Figure 9.32 shows the temperature of two cups of coffee against time, one starting at a higher temperature than the other, but both tending toward room temperature in the long run.

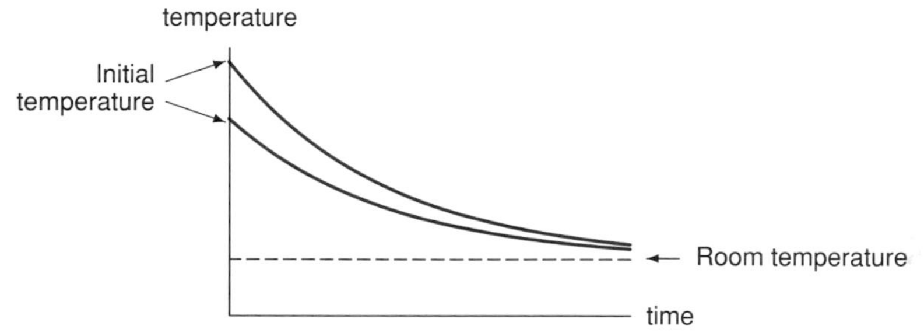

Figure 9.32: Temperature of coffee versus time

Example 3 When a murder is committed, the body, originally at 37°C, cools according to Newton's Law of Cooling. Suppose that after two hours the temperature is 35°C, and that the temperature of the surrounding air is a constant 20°C.

(a) Find the temperature, H, of the body as a function of t, the time in hours since the murder was committed.

(b) Sketch a graph of temperature against time.

(c) What happens to the temperature in the long run? Show this on the graph and algebraically.

(d) If the body is found at 4 pm at a temperature of 30°C, when was the murder committed?

Solution (a) We will first find the temperature of the body as a function of time, using the law of cooling to set up a differential equation. Newton's Law of Cooling says that for some constant α

$$\text{Rate of change of temperature} = \alpha(\text{Temperature difference}).$$

If H is the temperature of the body, then

$$\text{Temperature difference} = H - 20$$

so

$$\frac{dH}{dt} = \alpha(H - 20).$$

What about the sign of α? If the temperature difference is positive (i.e., $H > 20$), then H is falling, so the rate of change must be negative. Thus α should be negative, so we'll write:

$$\frac{dH}{dt} = -k(H - 20), \qquad \text{for some } k > 0.$$

Separating variables and solving, as in Example 1 on page 497, gives:

$$H - 20 = Be^{-kt}.$$

To find B, substitute the initial condition that $H = 37$ when $t = 0$:

$$37 - 20 = Be^{-k(0)} = B$$

so $B = 17$, the initial value of $(H - 20)$. Thus,

$$H - 20 = 17e^{-kt}.$$

To find k, we use the fact that after 2 hours, the temperature is 35°C, so

$$35 - 20 = 17e^{-k(2)}.$$

Dividing by 17 and taking natural logs, we get:

$$\ln\left(\frac{15}{17}\right) = \ln(e^{-2k})$$
$$-0.125 = -2k$$
$$k \approx 0.063.$$

Therefore, the temperature is given by

$$H - 20 = 17e^{-0.063t}$$

or

$$H = 20 + 17e^{-0.063t}.$$

(b) The graph of $H = 20 + 17e^{-0.063t}$ has a vertical intercept of $H = 37$, because the temperature of the body starts at 37°C. The temperature decays exponentially with $H = 20$ as the horizontal asymptote. (See Figure 9.33.)

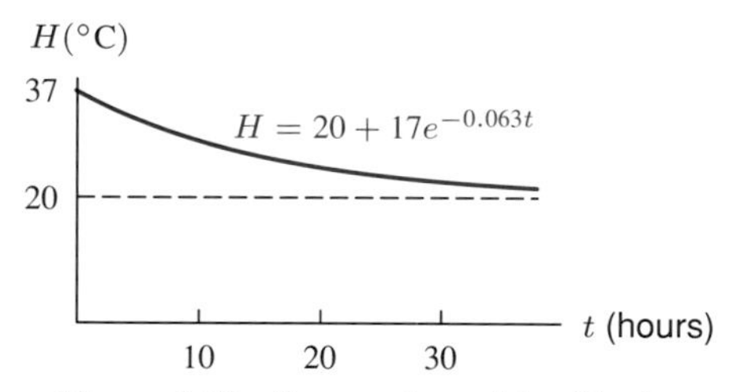

Figure 9.33: Temperature of dead body

(c) 'In the long run' means as $t \to \infty$. The graph shows that as $t \to \infty$, $H \to 20$. Algebraically, since $e^{-0.063t} \to 0$ as $t \to \infty$,

$$H = 20 + \underbrace{17e^{-0.063t}}_{\text{goes to 0 as } t \to \infty} \longrightarrow 20 \quad \text{as } t \to \infty$$

(d) We want to know when the temperature reaches 30°C. Substitute $H = 30$ and solve for t:

$$30 = 20 + 17e^{-0.063t}$$
$$\frac{10}{17} = e^{-0.063t}.$$

Taking natural logs:

$$-0.531 = -0.063t$$

gives

$$t \approx 8.4 \text{ hours}.$$

Thus the murder must have been committed about 8.4 hours before 4 pm. Since 8.4 hours = 8 hours 24 minutes, the murder was committed at about 7:36 am.

Equilibrium Solutions

Figure 9.34 shows the temperature of several objects in a 20°C room. One is initially hotter than 20°C and cools down toward 20°C; another is initially cooler and warms up toward 20°C. All these curves are solutions to the differential equation

$$\frac{dH}{dt} = -k(H - 20)$$

for some fixed $k > 0$, and all the solutions have the form

$$H = 20 + Ae^{-kt}$$

for some A. Notice that $H \to 20$ as $t \to \infty$ because $e^{-kt} \to 0$ as $t \to \infty$. In other words, in the long run, the temperature of the object always tends toward 20°C, the temperature of the room, no matter what the initial temperature. This means that what happens in the long run is independent of the initial condition.

In the special case when $A = 0$, we have the *equilibrium solution*

$$H = 20$$

for all t. This means that if the object starts at 20°C, it remains at 20°C for all time. Notice that such a solution can be found directly from the differential equation by solving $dH/dt = 0$:

$$\frac{dH}{dt} = -k(H - 20) = 0$$

giving

$$H = 20.$$

Regardless of the initial temperature, H always gets closer and closer to 20 as $t \to \infty$. As a result, $H = 20$ is called a *stable* equilibrium for H.[3]

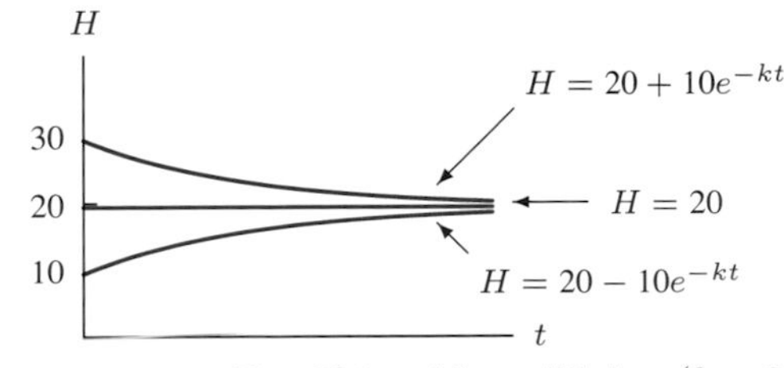

Figure 9.34: $H = 20$ is stable equilibrium $(k > 0)$

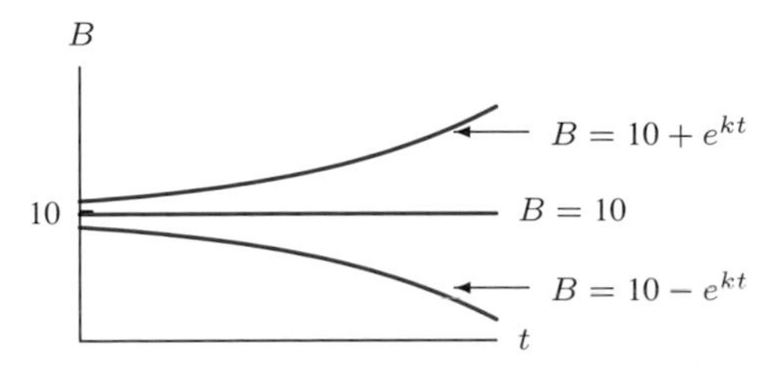

Figure 9.35: $B = 10$ is unstable equilibrium $(k > 0)$

[3]In more advanced work, this behavior is described as asymptotic stability.

A different situation is displayed in Figure 9.35, which shows solutions to the differential equation

$$\frac{dB}{dt} = k(B - 10)$$

for some fixed $k > 0$. Solving $dB/dt = 0$ gives the equilibrium $B = 10$, which is *unstable* because if B starts near 10, it moves away as $t \to \infty$.

In general, we have the following definitions.

- An **equilibrium solution** is constant for all values of the independent variable. The graph is a horizontal line. Equilibrium solutions can be identified by setting the derivative of the function to zero.
- An equilibrium is **stable** if a small change in the initial conditions gives a solution which tends toward the equilibrium as the independent variable tends to positive infinity.
- An equilibrium is **unstable** if a small change in the initial conditions gives a solution curve which veers away from the equilibrium as the independent variable tends to positive infinity.

Problems for Section 9.5

1. The amount of arable land (land that can be used for growing crops) that we use increases as the world's population increases. Suppose $A(t)$ represents the total number of hectares of arable land in use in year t. (A hectare is about $2\frac{1}{2}$ acres.)
   (a) Explain why it is reasonable to expect $A(t)$ to satisfy the equation
   $$\frac{dA}{dt} = kA.$$
   What assumptions are you making about the world's population and its relation to the amount of arable land used?
   (b) In 1950 about 1×10^9 hectares of arable land were in use; in 1980 the figure was 2×10^9. If the total amount of arable land available is thought to be 3.2×10^9 hectares, when will it be exhausted? (Let $t = 0$ in 1950.)
2. A yam is put in a 200°C oven and heats up according to the differential equation
   $$\frac{dH}{dt} = -k(H - 200),$$
   where k is positive.
   (a) If the yam is at 20°C when it is put in the oven, solve the differential equation.
   (b) Find k using the fact that after 30 minutes the temperature of the yam is 120°C.

3. Each of the curves in Figure 9.36 represents the balance in a bank account into which a single deposit was made at time zero. Assuming interest is compounded continuously, find:

   (a) The curve representing the largest initial deposit.
   (b) The curve representing the largest interest rate.
   (c) Two curves representing the same initial deposit.
   (d) Two curves representing the same interest rate.

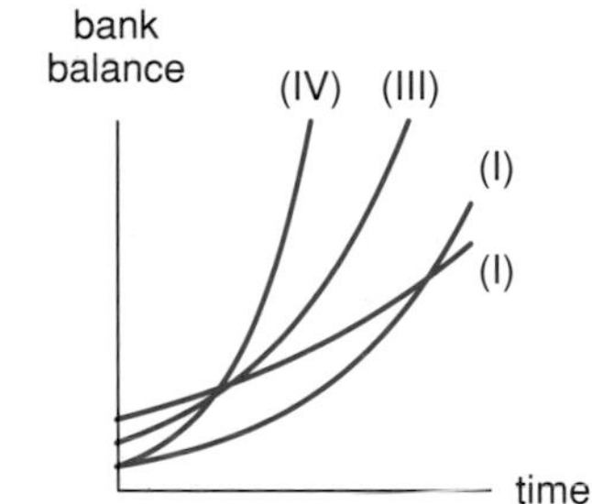

Figure 9.36

4. The graphs in Figure 9.37 represent the temperature, H°C, of four eggs as a function of time, t, measured in minutes. Match three of the graphs with the descriptions below. Write a similar description for the fourth graph, including an interpretation of any intercepts and asymptotes.

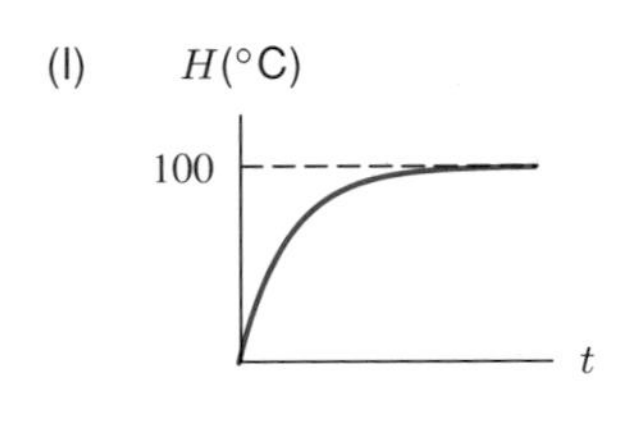

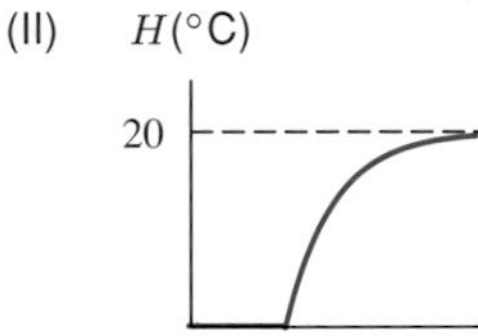

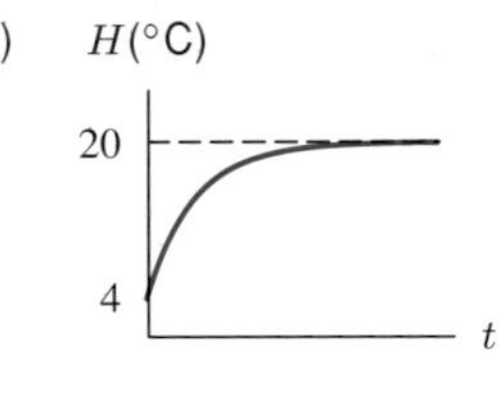

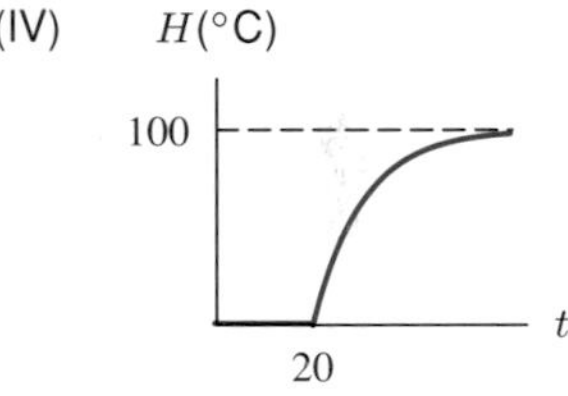

Figure 9.37

   (a) An egg is taken out of the fridge (just above 0°C) and put into boiling water.
   (b) Twenty minutes after the egg in part (a) is taken out of the fridge and put into boiling water, the same thing is done with another egg.
   (c) An egg is taken out of the fridge (just above 0°C) and left to sit on the kitchen table.

5. Money in a bank account grows continuously at a rate of r per year (when the interest rate is 5%, $r = 0.05$, and so on). Suppose \$1000 is put into the account in 1970.

   (a) Write a differential equation satisfied by M, the amount of money in the account at time t, measured in years since 1970.
   (b) Solve the equation.
   (c) Sketch the solution until the year 2000 for interest rates of 5% and 10%.

6. (a) If $B = f(t)$ is the balance of a bank account at time t that earns interest at a rate of $r\%$, compounded continuously, what is the differential equation that describes the rate at which the balance changes? What is the constant of proportionality, in terms of r?
   (b) What is the solution to this differential equation?
   (c) Sketch the graph of $B = f(t)$ for an account that starts with \$1000 and earns interest at the following rates:

   (i) 4% (ii) 10% (iii) 15%

7. In Example 2 on page 504 we saw that it would take about 6 years for 90% of the pollution in Lake Erie to be removed, and about 12 years for 99% to be removed. Explain why one time is double the other.

8. Using the model in the text and data in Table 9.4 on page 504, find how long it would take for 90% of the pollution to be removed from Lake Michigan and from Lake Ontario, assuming no

new pollutants are added. Explain how you can tell which lake will take longer to be purified just by looking at the data.

9. Use the model in the text and the data in Table 9.4 on page 504 to determine which of the Great Lakes would require the longest time and which would require the shortest time for 80% of the pollution to be removed, assuming no new pollutants are being added. Find the ratio of these two times.

10. Warfarin is a drug used as an anticoagulant. After stopping administration of the drug, the quantity remaining in a patient's body decreases at a rate proportional to the quantity remaining. The half-life of Warfarin is 37 hours.
    (a) Sketch a rough graph of the quantity, Q, of warfarin in a patient's body as function of the time, t, since stopping administration of the drug. Mark the 37 hours on your graph.
    (b) Write a differential equation satisfied by Q.
    (c) How many days does it take for the drug level in the body to be reduced to 25% of the original level?

11. In some chemical reactions, the rate at which the amount of a substance changes with time is proportional to the amount present. For example, this is the case as δ-glucono-lactone changes into gluconic acid.
    (a) Write a differential equation satisfied by y, the quantity of δ-glucono-lactone present at time t.
    (b) If 100 grams of δ-glucono-lactone is reduced to 54.9 grams in one hour, how many grams will remain after 10 hours?

12. The rate at which barometric pressure decreases with altitude is proportional to the barometric pressure at that altitude. If the barometric pressure is measured in inches of mercury, and the altitude in feet, then the constant of proportionality is 3.7×10^{-5}. Suppose the barometric pressure at sea level is 29.92 inches of mercury.
    (a) Calculate the barometric pressure at the top of Mount Whitney, 14,500 feet (the highest mountain in the US outside Alaska), and at the top of Mount Everest, 29,000 feet (the highest mountain in the world).
    (b) People cannot easily survive at a pressure below 15 inches of mercury. What is the highest altitude to which people can safely go?

13. The rate at which light is absorbed as it passes through water is proportional to the intensity, or brightness, at that point.
    (a) Find the intensity as a function of the distance the light has traveled through the water.
    (b) If 50% of the light is absorbed in 10 feet, how much is absorbed in 20 feet? 25 feet?

14. Suppose that at 1:00 pm one winter afternoon, there is a power failure at your house in Wisconsin, and your heat does not work without electricity. When the power goes out, it is 68°F in your house. At 10:00 pm, it is 57°F in the house, and you notice that it is 10°F outside.
    (a) Assuming that the temperature, T, in your home obeys Newton's Law of Cooling, write the differential equation satisfied by T.
    (b) Solve the differential equation to estimate the temperature in the house when you get up at 7:00 am the next morning. Should you worry about your water pipes freezing?
    (c) What assumption did you have to make in part (a) about the temperature outside? Given this (probably incorrect) assumption, would you revise your estimate up or down? Why?

15. Write a differential equation describing the temperature as a function of time of a bottle of orange juice taken out of a 40°F refrigerator and left in a 65°F room. Solve the equation and sketch an approximate graph of the solution.

16. A detective finds a murder victim at 9 am. The temperature of the body is measured at 90.3°F. One hour later, the temperature of the body is 89.0°F. The temperature of the room has been maintained at a constant 68°F.
    (a) Assuming the temperature, T, of the body obeys Newton's Law of Cooling, write a differential equation for T.
    (b) Solve the differential equation to estimate the time the murder occurred.
17. Radioactive carbon (carbon-14) decays at a rate of approximately 1 part in 10,000 a year. Write and solve a differential equation for the quantity of carbon-14 as a function of time. Sketch a graph of the solution.
18. The radioactive isotope carbon-14 is present in small quantities in all life forms, and it is constantly replenished until the organism dies, after which it decays to stable carbon-12 at a rate proportional to the amount of carbon-14 present, with a half-life of 5730 years. Suppose $C(t)$ is the amount of carbon-14 present at time t.
    (a) Find the value of the constant k in the differential equation $C' = -kC$.
    (b) In 1988 three teams of scientists found that the Shroud of Turin, which was reputed to be the burial cloth of Jesus, contained 91% of the amount of carbon-14 contained in freshly made cloth of the same material.[4] How old is the Shroud of Turin, according to these data?
19. The amount of radioactive carbon-14 in a sample is measured using a Geiger counter, which records each disintegration of an atom. Living tissue disintegrates at a rate of about 13.5 atoms per minute per gram of carbon. In 1977 a charcoal fragment found at Stonehenge, England, recorded 8.2 disintegrations per minute per gram of carbon. Assuming that the half-life of carbon-14 is 5730 years and that the charcoal was formed during the building of the site, estimate the date at which Stonehenge was built.
20. Before Galileo discovered that the speed of a falling body with no air resistance is proportional to the time since it was dropped, he mistakenly conjectured that the speed was proportional to the distance it had fallen.
    (a) Assume the mistaken conjecture to be true and write an equation relating the distance fallen, $D(t)$, at time t, and its derivative.
    (b) Using your answer to part (a) and the correct initial conditions, show that D would have to be equal to 0 for all t, and therefore the conjecture must be wrong.

9.6 APPLICATIONS AND MODELING

Much of this book involves functions that represent real processes, such as how the temperature of a yam or the population of Mexico is changing with time. You may wonder where such functions come from. In some cases, we fit functions to experimental data by trial and error. In other cases, we take a more theoretical approach, leading to a differential equation whose solution is the function we want. In this section we give examples of the more theoretical approach.

How a Layer of Ice Forms

When ice forms on a lake, the water on the surface freezes first. As heat from the water travels up through the ice and is lost to the air, more ice is formed. The question we will consider is: How thick is the layer of ice as a function of time? We can learn a certain amount about this function from

[4] *The New York Times*, October 18, 1988

common sense. Since the thickness of the ice increases with time, the thickness function must be increasing. In addition, as the ice gets thicker, the escaping heat has further to travel before reaching the air, therefore we expect the layer of ice to form more slowly as time goes on. This tells us that the thickness function is increasing at a decreasing rate, so its graph is concave down.

A Differential Equation for the Thickness of the Ice

To get more detailed information about the thickness function, we have to make some assumptions. Suppose y represents the thickness of the ice as a function of time, t. Since the thicker the ice, the longer it takes the heat to get through it, we will assume that the rate at which ice is formed is inversely proportional to the thickness. In other words, we assume that for some constant k,

$$\begin{array}{c}\text{Rate thickness}\\ \text{is increasing}\end{array} = \frac{k}{\text{Thickness}},$$

so

$$\frac{dy}{dt} = \frac{k}{y} \quad \text{where} \quad k > 0.$$

This differential equation enables us to find a formula for the function y. Using separation of variables:

$$\int y\,dy = \int k\,dt$$
$$\frac{y^2}{2} = kt + C.$$

If we measure time so that $y = 0$ when $t = 0$, then $C = 0$. Since y must be non-negative, we have

$$y = \sqrt{2kt}.$$

Graphs of y against t are in Figure 9.38. Notice that the larger y is, the more slowly y increases. In addition, this model suggests that y increases indefinitely as time passes. (Of course, in practice the value of y cannot increase beyond the depth of the lake.)

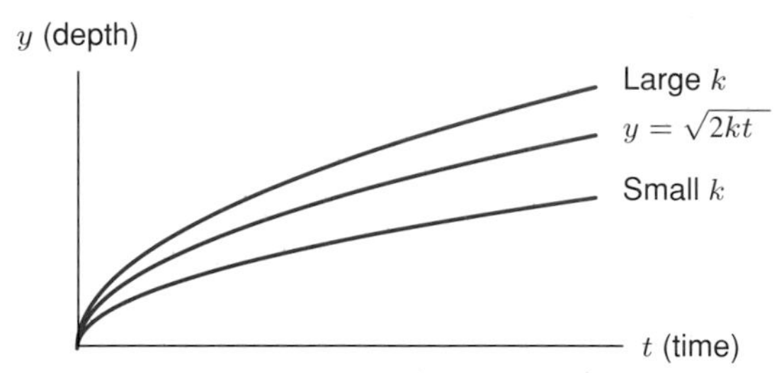

Figure 9.38: Thickness of ice as a function of time

The Net Worth of a Company

In the preceding section, we saw an example in which money in a bank account was earning interest (Example 1, page 501). Consider a company whose assets earn interest (like the money in a bank account) but which must also make payroll payments. The question is: under what circumstances does the company make money and under what circumstances does it go bankrupt?

Common sense says that if the payroll exceeds the rate at which interest is earned, the company will eventually be in trouble, whereas if interest exceeds payroll, the company should do well. In order to make this more precise, we have to make several assumptions. We assume that interest is earned continuously and that the payments are made continuously. The payments are not in practice made continuously, but for a large company this is a good approximation.

Example 1 Suppose a company's net worth increases at a continuous rate of 5% per year due to interest on its assets. At the same time, the company's payroll obligations amount to $200 million a year, paid out continuously.

(a) Write a differential equation that governs the net worth of the company, W million dollars.

(b) Solve the differential equation, assuming an initial net worth of W_0 million dollars.

(c) Sketch the solution for $W_0 = 3000$, 4000, and 5000.

Solution First, let's see what we can learn without writing a differential equation. For example, we can ask if there is any initial net worth W_0 which will exactly keep the net worth constant. If there's such an equilibrium, the rate at which interest is earned must exactly balance the payments made, so

$$\begin{matrix}\text{Rate interest} \\ \text{is earned}\end{matrix} = \begin{matrix}\text{Rate payments} \\ \text{are made}\end{matrix}.$$

We know that interest is earned continuously at a rate of 5% of whatever the net worth is at the moment. If we have equilibrium, the net worth starts as W_0 and remains W_0, so interest is always earned at a rate of $0.05W_0$ per year. Thus, for equilibrium we must have

$$0.05W_0 = 200$$

giving

$$W_0 = 4000.$$

Therefore, if the net worth starts at $4000 million, the interest and the payments are equal, and the net worth will remain constant. Therefore $4000 million is an equilibrium solution.

Suppose, however, the initial deposit is over $4000 million, the interest earned will be more than the payroll expenses, and the net worth of the company will increase, thereby increasing the interest still further. Thus the net worth will increase faster and faster. On the other hand, if the initial deposit is below $4000 million, the interest will not be enough to meet the payments, and the net worth of the company will decline, thereby decreasing the interest and making the net worth decrease still faster. Hence the net worth will eventually go to zero, and the company will go bankrupt.

(a) Now we will solve the problem analytically. To set up a differential equation for the net worth, we will use the fact that

$$\begin{matrix}\text{Rate at which} \\ \text{net worth is increasing}\end{matrix} = \begin{matrix}\text{Rate interest} \\ \text{is earned}\end{matrix} - \begin{matrix}\text{Rate payments} \\ \text{are made}\end{matrix}.$$

In millions of dollars per year, interest is earned at a rate of $0.05W$, and payments are made at a continuous rate of 200 per year, so we have

$$\frac{dW}{dt} = 0.05W - 200$$

where t is in years. Notice that the initial net worth does not enter into the differential equation. The equilibrium solution, $W = 4000$, is obtained by setting $dW/dt = 0$.

(b) We will solve this equation by separation of variables. It is helpful to factor 0.05 out of the right-hand side before separating, so that the W moves over to the left-hand side without a coefficient. This makes the manipulations less messy. So we write

$$\frac{dW}{dt} = 0.05(W - 4000).$$

Separating and integrating

$$\int \frac{dW}{W - 4000} = \int 0.05\, dt$$

gives

$$\ln|W - 4000| = 0.05t + C$$

so

$$|W - 4000| = e^{0.05t+C} = e^C e^{0.05t}$$

or

$$W - 4000 = Ae^{0.05t} \qquad \text{where } A = \pm e^C.$$

To find A we use the initial condition that $W = W_0$ when $t = 0$.

$$W_0 - 4000 = Ae^0 = A.$$

Substituting this value for A gives

$$W - 4000 = (W_0 - 4000)e^{0.05t}$$

or

$$W = 4000 + (W_0 - 4000)e^{0.05t}.$$

(c) If $W_0 = 4000$, then $W = 4000$, the equilibrium solution.
If $W_0 = 5000$, then $W = 4000 + 1000e^{0.05t}$.
If $W_0 = 3000$, then $W = 4000 - 1000e^{0.05t}$. Notice that when $t \approx 27.7$, $W = 0$, so this solution means that the company goes bankrupt in its twenty-eighth year.

The graphs of these functions are in Figure 9.39. Here $W = 4000$ is an equilibrium solution. However, if the net worth starts with W_0 near, but not equal to, \$4000 million, then W moves further away. Thus $W = 4000$ is an unstable equilibrium.

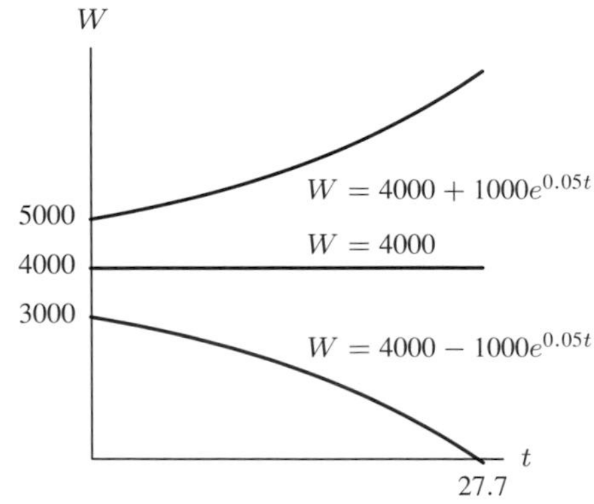

Figure 9.39: Solutions to $\frac{dW}{dt} = 0.05W - 200$

The Velocity of a Falling Body

Imagine a sky-diver jumping out of a plane. Think about how the person's velocity varies during the time before his parachute opens. When the sky-diver first jumps, his velocity is zero. Then the pull of gravity comes into play, making the velocity increase. But as the sky-diver speeds up, the air resistance also increases. Since the air resistance partly balances the pull of gravity, as the sky-diver speeds up, the force causing him to accelerate decreases. Thus the velocity must be an increasing function of time, and it must be increasing at a decreasing rate. If the air resistance increases until it balances gravity, then the sky-diver's velocity will stop increasing – the velocity levels off. Thus we expect the velocity to be increasing, and to have a graph that is concave down with a horizontal asymptote. Some possible velocity data for a sky-diver is given in Table 9.5.

TABLE 9.5 *Velocity of a sky-diver*

t (sec)	0	5	10	15	20	25	30	35	40
v (m/sec)	0	31.6	43.2	47.5	49.1	49.7	49.9	50.0	50.0

Can we find a formula for the velocity function which fits this data? The horizontal asymptote is about 50 m/sec, but that doesn't tell us what form the velocity function has. In order to compute the form of the velocity function, we must make some assumptions about the forces acting on the sky-diver and the accelerations they produce.

A Differential Equation: Air Resistance Proportional to Velocity

Common sense tells us that the faster the sky-diver is traveling, the larger the air resistance. However, to decide whether air resistance is proportional to the velocity, say, or is some other function of velocity, requires either lab experiments or some theoretical idea of how the air resistance is created. We will make an assumption – namely, that air resistance is proportional to velocity – and use that assumption to predict the velocity at different times. Checking our predictions against the data gives us some idea how good the assumptions were.

Newton's Second Law of Motion, which states that

$$\text{Force} = \text{Mass} \times \text{Acceleration}$$

relates the force to the velocity, since the acceleration is dv/dt. The force has two components: the gravitational force, mg, which acts downward, and the air resistance, kv, which acts upward, in the opposite direction to the motion (we assume $k > 0$). (See Figure 9.40.) Thus

$$\text{Gravity} - \text{Air resistance} = \text{Mass} \times \text{Acceleration},$$

so

$$mg - kv = m\frac{dv}{dt}.$$

This differential equation can be solved by separation of variables. It is easier if you factor out $-k/m$ before separating, giving

$$\frac{dv}{dt} = -\frac{k}{m}\left(v - \frac{mg}{k}\right).$$

Separating and integrating gives

$$\int \frac{dv}{v - mg/k} = -\frac{k}{m}\int dt$$

$$\ln\left|v - \frac{mg}{k}\right| = -\frac{k}{m}t + C.$$

Solving for v:

$$\left|v - \frac{mg}{k}\right| = e^{-kt/m+C} = e^{C}e^{-kt/m}$$

$$v - \frac{mg}{k} = Ae^{-kt/m},$$

where A is an arbitrary constant. We find A from the initial condition that the body starts from rest, so $v = 0$ when $t = 0$. Substituting

$$0 - \frac{mg}{k} = Ae^{0}$$

gives

$$A = -\frac{mg}{k}.$$

Thus

$$v = \frac{mg}{k} - \frac{mg}{k}e^{-kt/m} = \frac{mg}{k}(1 - e^{-kt/m}).$$

The graph of this function is in Figure 9.41. As expected, at the start the sky-diver is moving slowly, which means the air resistance is small, so he speeds up a lot. As the sky-diver starts to move faster, the air resistance builds until it balances the gravitational force. At this point the sky-diver has reached his *terminal velocity*, mg/k. In this model, the sky-diver never actually reaches terminal velocity (because $e^{-kt/m}$ is never actually equal to zero). However, the velocity is close to the terminal value after only a few seconds.

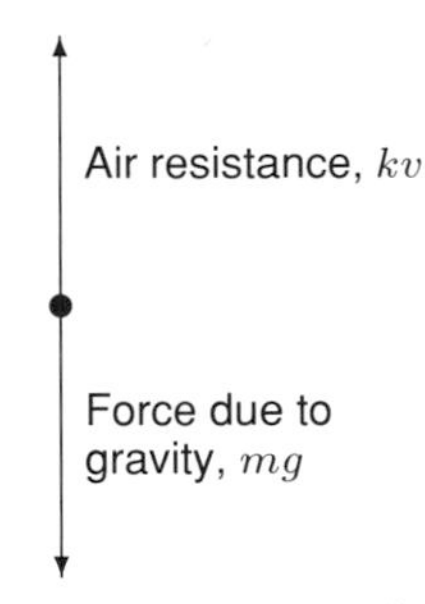

Figure 9.40: Forces acting on a sky-diver

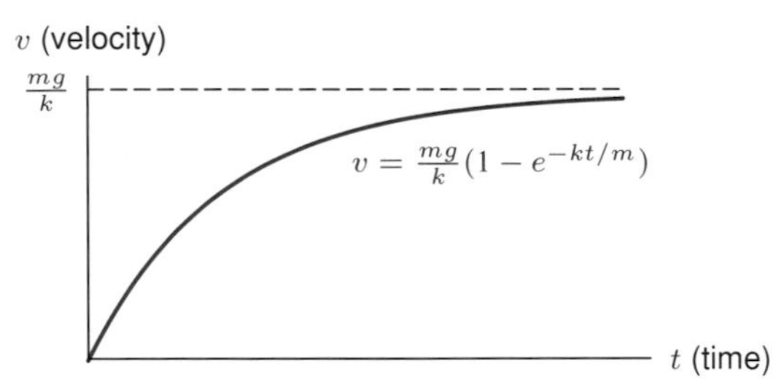

Figure 9.41: Velocity under the assumption that air resistance $= kv$

Alternative Method of Finding the Terminal Velocity

Notice that the terminal velocity can also be obtained from the differential equation by setting $dv/dt = 0$ and solving for v:

$$m\frac{dv}{dt} = mg - kv = 0$$

so

$$v = \frac{mg}{k}.$$

Comparing the Model with the Data

Does this formula for the velocity function fit the data in Table 9.5? First we need to decide the value of the parameters m, g, and k. The data shows that we can take the terminal velocity to be 50.0 m/sec, giving

$$\frac{mg}{k} = 50.0.$$

Since g is the acceleration due to gravity, $g = 9.8$ m/sec^2, we have

$$\frac{m(9.8)}{k} = 50.0$$

so

$$\frac{k}{m} = 0.196.$$

Thus we are modeling free fall with the differential equation

$$\frac{dv}{dt} = 9.8 - 0.196v$$

where t is in seconds and v is in m/sec. To compute v, we do not need to calculate values of k and m separately, since we know

$$v = \frac{mg}{k}(1 - e^{-kt/m}) = 50\left(1 - e^{-0.196t}\right).$$

Using this formula for v, we get the predicted values of v in the bottom row in Table 9.6.

TABLE 9.6 *Predicted and observed velocity*

t (sec)	0	5	10	15	20	25	30	35	40
Observed v (m/sec)	0	31.6	43.2	47.5	49.1	49.7	49.9	50.0	50.0
Predicted v (m/sec)	0	31.2	43.0	47.4	49.0	49.6	49.9	49.9	50.0

The fact that there is good agreement between the observed and predicted data suggests that our assumption about the air resistance is reasonable.

An Alternative Differential Equation: Air Resistance Proportional to a Power of the Velocity

Another common model assumes that air resistance is proportional to $v^{3/2}$. This tells us that the net force acting on the sky-diver is $mg - kv^{3/2}$, so Newton's Law tells us that

$$m\frac{dv}{dt} = mg - kv^{3/2}.$$

This small change in the differential equation makes it much more difficult to solve analytically. If you try to proceed by separation of variables, you get

$$\int \frac{dv}{g - kv^{3/2}/m} = t + C.$$

This integral can be evaluated in closed-form, but the result is such a mess that you'd be happier without it. Even if we did use this integral, we would have t in terms of v, whereas we usually want v as a function of t. This is a case in which using a slope field or Euler's method are the only reasonable methods we know to solve a differential equation.

To get a qualitative solution in a special case, we need to decide on a value for k/m. Assuming the terminal velocity remains the same, 50 m/sec, then $v = 50$ must be a solution to the following equation:

$$\frac{dv}{dt} = g - \frac{k}{m}v^{3/2} = 0.$$

Thus we can solve for k/m, giving

$$\frac{k}{m} = \frac{g}{v^{3/2}} = \frac{9.8}{50^{3/2}} = 0.0277.$$

Thus

$$\frac{dv}{dt} = 9.8 - 0.0277v^{3/2}$$

with v in m/sec. The slope field in Figure 9.42 shows that the solution curve starting at the origin has a horizontal asymptote at about $v = 50$. Thus even without solving the differential equation explicitly, we can say that, under these assumptions, the speed of the sky-diver increases towards a terminal velocity of about 50 m/sec. You may wonder how we know whether to assume the air resistance is proportional to the velocity, to the $3/2$ power of the velocity, or to some other function. This is largely decided experimentally—by dragging different objects through various media and measuring the air resistance.

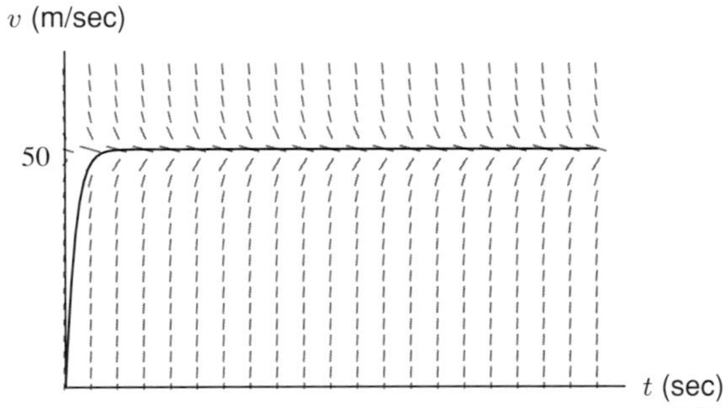

Figure 9.42: Slope field of $dv/dt = 9.8 - 0.0277v^{3/2}$

Compartmental Analysis: A Reservoir

Many processes can be modeled as a container with various solutions flowing in and out — for example, drugs given intravenously or the discharge of pollutants into a lake. We will consider a city's water reservoir, fed partly by clean water from a spring and partly by run-off from the surrounding land. In New England or any other area of the country with much snow in the winter, the run-off contains salt which has been put on the roads to make them safe for driving. Our problem will be to consider the concentration of salt in the reservoir. If there is no salt in the reservoir initially, we expect the concentration to build up until the rate at which the salt is being brought into the reservoir by the run-off is balanced by the rate at which salt flows out. If, on the other hand, the reservoir starts with a great deal of salt in it, then initially the rate at which the salt is flowing in is less than the rate at which it is flowing out, and the quantity of salt in the lake should decrease. In either case, the salt concentration levels off at an equilibrium value.

A Differential Equation for Salt Concentration

Suppose the water reservoir holds 100 million gallons of water and supplies a city with 1 million gallons a day. The reservoir is partly refilled by a spring which provides 0.9 million gallons a day, and the rest of the water, 0.1 million gallons a day, comes from run-off from the surrounding land. The spring is clean, but the run-off contains salt with a concentration of 0.0001 pound per gallon. Assume that there was no salt in the reservoir initially and that the reservoir is well mixed (i.e., that the water taken by the city residents contains the concentration of salt in the tank at that instant). We will find the concentration of salt in the reservoir as a function of time, assuming the reservoir remains full throughout.

In this type of problem, it is very important to distinguish between the total quantity and the concentration of salt, where

$$\text{Concentration} = \frac{\text{Quantity of salt}}{\text{Volume of water}}.$$

If C is the concentration of salt (in pounds/gallon) and Q is the quantity (in pounds), then since the volume of the reservoir is 100 million gallons:

$$C = \frac{Q}{100\,\text{million}} \left(\frac{\text{lb}}{\text{gal}}\right).$$

The differential equation we will use is based on the relation

$$\begin{array}{c}\text{Rate of change of}\\ \text{quantity of salt in reservoir}\end{array} = \text{Rate salt entering} - \text{Rate salt leaving},$$

which describes how the quantity is changing with time. We will find Q first, and then C.

To find the rate salt is entering, notice that all the salt is entering through the run-off. The run-off is 0.1 million gallons per day, with each gallon containing 0.0001 pound of salt. Therefore

$$\begin{aligned}\text{Rate salt entering} &= \text{Concentration} \times \text{Volume per day}\\ &= 0.0001 \left(\frac{\text{lb}}{\text{gal}}\right) \times 0.1 \left(\frac{\text{million gal}}{\text{day}}\right)\\ &= 0.00001 \left(\frac{\text{million lb}}{\text{day}}\right) = 10 \left(\frac{\text{lb}}{\text{day}}\right).\end{aligned}$$

Thus

$$\text{Rate salt entering} = 10 \text{ lb/day}.$$

Salt is leaving in the water used by the city. The city uses a million gallons of water each day with a concentration of $\dfrac{Q}{100\,\text{million}} \dfrac{\text{lb}}{\text{gal}}$. Thus

$$\begin{aligned}\text{Rate salt leaving} &= \text{Concentration} \times \text{Volume per day}\\ &= \frac{Q}{100\text{ million}} \left(\frac{\text{lb}}{\text{gal}}\right) \times 1 \left(\frac{\text{million gal}}{\text{day}}\right)\\ &= \frac{Q}{100} \left(\frac{\text{lb}}{\text{day}}\right).\end{aligned}$$

So

$$\text{Rate salt leaving} = \frac{Q}{100} \text{ lb/day}.$$

Therefore Q must satisfy

$$\frac{dQ}{dt} = 10 - \frac{Q}{100}.$$

Solving this equation by factoring out $-1/100$ and separating variables:

$$\frac{dQ}{dt} = -\frac{1}{100}(Q - 1000) = -0.01(Q - 1000)$$

gives

$$\int \frac{dQ}{Q - 1000} = -\int 0.01\, dt$$

$$\ln|Q - 1000| = -0.01t + k$$

so

$$Q - 1000 = Ae^{-0.01t}.$$

There is no salt initially, so we substitute $Q = 0$ when $t = 0$:

$$0 - 1000 = Ae^{0} \qquad \text{giving} \qquad A = -1000.$$

Thus

$$Q - 1000 = -1000e^{-0.01t}$$

so

$$Q = 1000(1 - e^{-0.01t}) \quad \text{pounds.}$$

Therefore

$$\text{Concentration } = C = \frac{Q}{100 \text{ million}} = \frac{1000}{10^8}(1 - e^{-0.01t}) = 10^{-5}(1 - e^{-0.01t}) \text{ lbs/gal.}$$

A sketch of concentration against time is in Figure 9.43.

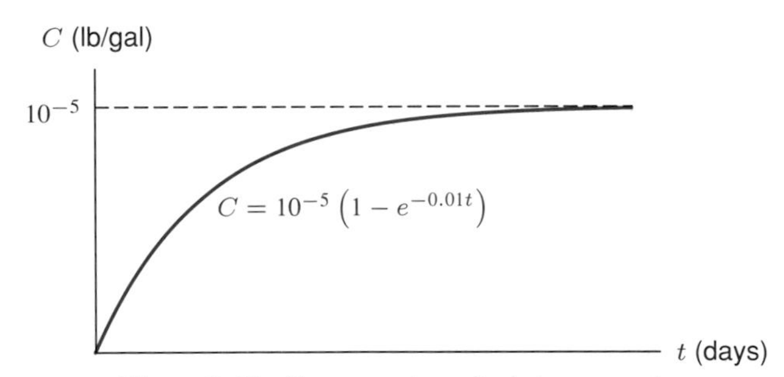

Figure 9.43: Concentration of salt in reservoir

Qualitative Information from a Differential Equation

In the last part of this section we see how information can sometimes be obtained from a differential equation without solving it. The example we take is a second-order differential equation. We first integrate to get a first-order equation, and then investigate qualitative behavior of the solution.

The Expansion of the Universe

Observations show that the universe is currently expanding, so astronomers are interested in investigating exactly how fast it is expanding. One model[5] assumes that the universe is a giant sphere of radius $R(t)$, where t is time, and that

$$R'' = -\frac{GM_0}{R^2}$$

where G is the universal gravitational constant and M_0 is the mass of the universe. This is a second-order differential equation for R. Even without solving the equation, we can get information about the expansion of the universe.

The first step is to get a first-order equation from this second-order equation. In this case we multiply both sides of the equation for R'' by R', giving

$$R''R' = -\frac{GM_0}{R^2}R'.$$

Since $\frac{d}{dt}\left(R'\right)^2 = 2R''R'$ and $\frac{d}{dt}\left(R^{-1}\right) = -R^{-2}R'$, we can write the preceding equation as

$$\frac{1}{2}\frac{d}{dt}\left(R'\right)^2 = GM_0\frac{d}{dt}\left(\frac{1}{R}\right)$$

Multiplying through by 2 and integrating with respect to t gives

$$(R')^2 = \frac{2GM_0}{R} + C$$

We will use this relation between R' and R to determine the behavior of R over time. We will look only at the case in which $C > 0$. The cases where $C < 0$ and $C = 0$ are investigated in Problems 19 and 20 on page 525. Since

$$R' = \pm\sqrt{\frac{2GM_0}{R} + C}$$

and since the universe is currently expanding, we know that R' must be positive, so R' must be given by the positive root:

$$R' = \sqrt{\frac{2GM_0}{R} + C}.$$

As R gets larger R' continues to be given by the positive root, meaning the universe continues to expand. However, R', the rate of expansion, decreases to $\sqrt{C}$ as $R \to \infty$. Therefore this model predicts that the universe will expand forever, but that the rate of the expansion is slowing down to $\sqrt{C}$.

---

[5]From M. Rowan-Robinson, *Cosmology*, Oxford University Press (1977).

Problems for Section 9.6

1. Dead leaves accumulate on the ground in a forest at a rate of 3 grams per square centimeter per year. At the same time, these leaves decompose at a continuous rate of 75% a year. Write a differential equation for the total quantity of dead leaves (per square centimeter) at time t. Sketch a solution showing that the quantity of dead leaves tends towards an equilibrium level. What is that equilibrium level?

2. According to a simple physiological model, an adult needs 20 calories per day per pound of body weight to maintain his weight. If he consumes more or fewer calories than those required to maintain his weight, his weight will change at a rate proportional to the difference between the number of calories consumed and the number needed to maintain his current weight; the constant of proportionality is $1/3500$ pounds per calorie. Suppose that a particular person has a constant caloric intake of I calories per day. Let $W(t)$ be the person's weight in pounds at time t (measured in days).
   (a) What differential equation has solution $W(t)$?
   (b) Solve this differential equation.
   (c) Draw a graph of $W(t)$ if the person starts out weighing 160 pounds and consumes 3000 calories a day. Label your axes and any intercepts and asymptotes clearly.

3. Suppose a cell contains a chemical (the solute) dissolved in it at a concentration $c(t)$, and the concentration of the same substance outside the cell is a constant k. By Fick's law, if $c(t)$ and k are unequal, solute moves across the cell wall at a rate proportional to the difference between $c(t)$ and k, toward the region of lower concentration.
   (a) Write a differential equation satisfied by $c(t)$.
   (b) Solve the differential equation with the initial condition $c(0) = c_0$.
   (c) Sketch the solution for $c_0 = 0$.

4. When interest is compounded continuously at a constant rate r, the balance B in a bank account satisfies the differential equation
$$\frac{dB}{dt} = rB$$
where t is in years. If the interest rate varies with time, the differential equation can be written as
$$\frac{dB}{dt} = r(t)B.$$
   (a) Suppose the interest rate oscillates between 7% and 9% over a period of 6 years. Write a possible formula for $r(t)$, assuming the interest rate is modeled by a trigonometric function of time.
   (b) Solve the differential equation using your answer to part (a).

5. Rainwater pours off a roof into a cylindrical barrel until the water in the barrel is 36 inches deep. When it stops raining, water leaks out of the barrel at a rate proportional to the square root of the depth of the water at that time. If the water level drops from 36 inches to 35 inches in 1 hour, how long will it take for all of the water to leak out of the barrel?

6. Suppose a chemical reaction involves one molecule of a substance A combining with one molecule of substance B to form one molecule of substance C, written $A+B \to C$. The Law of Mass Action states that the rate at which C is formed is proportional to the product of the

quantities of A and B present. Assume a and b are the initial quantities of A and B, and x is the quantity of C present at time t.

(a) Write a differential equation for x.
(b) Solve the equation with $x(0) = 0$.

7. If the substances A and B in Problem 6 are the same, write and solve a differential equation for x, with $x(0) = 0$.

8. A bank account earns 5% annual interest compounded continuously. You wish to make payments out of the account at a rate of \$12,000 per year (in a continuous cash flow) for 20 years.

   (a) Write a differential equation describing the balance $B = f(t)$, where t is in years.
   (b) Find the solution $B = f(t)$ to the differential equation given an initial balance of B_0 in the account.
   (c) What should the initial balance be such that the account has zero balance after precisely 20 years?

9. \$1000 is put into a bank account and earns interest continuously at a rate of i per year, and in addition, continuous payments are made out of the account at a rate of \$100 a year. Sketch the amount of money in the account as a function of time if the interest rate is
   (a) 5% (b) 10% (c) 15%
   In each case, you should first find an expression for the amount of money in the account at time t (in years).

10. A certain commodity is currently being sold at a price of $\$p$ per unit. Over a period of time market forces will make this price tend toward the equilibrium price, which we call $\$p_0$, at which supply exactly balances demand. The rate at which the price changes is described by the Evans Price Adjustment model, which says that the rate of change in the actual market price ($\$p$) is proportional to the difference between the actual market price and the equilibrium price.

    (a) Write a differential equation for p as a function of t.
    (b) Solve for p.
    (c) Sketch solutions for various different initial prices, both above and below the equilibrium price.
    (d) What happens to p as $t \to \infty$?

11. As you know, when a course ends, students start to forget the material they have learned. One model (called the Ebbinghaus model) assumes that the rate at which a student forgets material is proportional to the difference between the material he or she currently remembers and some positive constant, a.

    (a) Let $y = f(t)$ be the fraction of the original material remembered t weeks after the course has ended. Set up a differential equation for y. Your equation will contain two constants; the constant, a, is less than y for all t.
    (b) Solve the equation and find the arbitrary constant of integration.
    (c) Describe the practical meaning (in terms of the amount remembered) of the constants in the solution $y = f(t)$.

12. A patient is given the drug theophylline intravenously at a rate of 43.2 mg/hour to relieve acute asthma. You can imagine the drug as entering a compartment of volume 35,000 ml. (This is the volume of the part of the body through which the drug circulates.) The rate at which the drug leaves the patient is proportional to the quantity there, with proportionality constant 0.082. Assume the patient's body contains none of the drug initially.

(a) Describe in words how you would expect the concentration of theophylline in the patient to vary with time and sketch an approximate graph.
(b) Write a differential equation satisfied by the concentration of the drug, $c(t)$.
(c) Solve the differential equation and graph the solution. What happens to the concentration in the long run?

13. A drug is administered intravenously at a constant rate of r mg/hr and is excreted at a rate proportional to the quantity present, with constant of proportionality α.
    (a) Solve a differential equation for quantity, Q, in milligrams, of the drug in the body at time t hours. Your answer will contain r and α. Sketch a graph of Q against t. What is Q_∞, the limiting long-run value of Q?
    (b) What effect does doubling r have on Q_∞? What effect does doubling r have on the time to reach half the limiting value, $\frac{1}{2}Q_\infty$?
    (c) What effect does doubling α have on Q_∞? On the time to reach $\frac{1}{2}Q_\infty$?

14. Suppose people start smoking in a room of volume 60 m^3, thereby introducing air containing 5% carbon monoxide at a rate of 0.002 m^3/min into the room. (This means that 5% of the volume of the incoming air is carbon monoxide.) Assume that the smoky air mixes immediately with the rest of the air, and that the mixture leaves the room at the same rate as it enters.
    (a) Write a differential equation for $c(t)$, the concentration of carbon monoxide at time t, in minutes.
    (b) Solve the differential equation, assuming there was no carbon monoxide in the room initially.
    (c) What happens to the value of $c(t)$ in the long run?

15. (Continuation of Problem 14.) Medical texts[6] warn that exposure to air containing 0.1% carbon monoxide for some time can lead to a coma. How long does it take for the concentration of carbon monoxide in the room in Problem 14 to reach this level?

16. A rectangular swimming pool has dimensions, in meters, of 20 by 10 by 10; hence it has volume 2000 cubic meters, or 2×10^6 liters. The pool initially contains pure fresh water. At $t = 0$ minutes, water containing 10 grams/liter of salt is poured into the pool at a rate of 60 liters/minute. The salt water is instantly and totally mixed with the fresh water, and the excess mixture is drained out of the bottom of the pool at the same rate (60 liters/minute). Let $S(t)$ = the mass of salt in the pool at time t.
    (a) Write a differential equation for the amount of salt in the pool.
    (b) Solve the differential equation to find $S(t)$.
    (c) What happens to $S(t)$ as $t \to \infty$?

17. A bank account that earns 10% interest compounded continuously has an initial balance of zero. Money is deposited into the account at a constant rate of \$1000 per year.
    (a) Write a differential equation that describes the rate of change of the balance $B = f(t)$.
    (b) Solve the differential equation to find the balance as a function of time. Be sure to solve for any constants in your equation.
    (c) Using a Riemann sum, derive the integral that gives the balance at time t.
    (d) Evaluate the integral and show that you get the same formula as in part (b).
    (e) From your solution to the differential equation in part (a), what is the formula for the balance $B = f(t)$ if the initial balance is $B_0 \neq 0$?
    (f) Show how to modify your integral in part (d) if the initial balance is B_0, and show that you get the same answer in part (e) above.

[6]See, for example, R.G. Petersdorf et al., *Harrison's Principles of Internal Medicine* (New York: McGraw Hill, 1983).

18. Imagine an object of mass m thrown vertically upward from the surface of the earth with initial velocity v_0. In this problem, we will calculate the value of v_0, called the *escape velocity*, with which the object can escape the pull of the gravity and never return to earth. Suppose v is the velocity of the object (measured upward) at time t.

    Since the object is moving far from the surface of the earth, we must take into account the variation of gravity with altitude. If the acceleration due to gravity at sea level is g, the gravitational force, F, on the object of mass m at an altitude h above the surface of the earth is given by

$$F = \frac{mgR^2}{(R+h)^2},$$

    where R is the radius of the earth.

    (a) Use Newton's Law of Motion to show that

$$\frac{dv}{dt} = -\frac{gR^2}{(R+h)^2}.$$

    (b) Rewrite this equation with h instead of t as the independent variable using the chain rule $\frac{dv}{dt} = \frac{dv}{dh} \cdot \frac{dh}{dt}$. Hence show that

$$v\frac{dv}{dh} = -\frac{gR^2}{(R+h)^2}.$$

    (c) Solve the differential equation in part (b).
    (d) Find the escape velocity, the value of v_0 such that v is never zero.

Problems 19–21 refer to the model of the expansion of the universe given on page 521 by the equations

$$R'' = -\frac{GM_0}{R^2} \quad \text{and} \quad (R')^2 = \frac{2GM_0}{R} + C.$$

where $R(t)$ is the radius of the universe (assumed spherical), t is time, G is the universal gravitational constant, and M_0 is the mass of the universe. In the text we investigated the predictions of this model in the case where $C > 0$. Problems 19–21 investigate this model further.

19. In this problem we look at the case where $C < 0$. Writing $C = -K$, where $K > 0$, we have

$$R' = \pm\sqrt{\frac{2GM_0}{R} - K}.$$

    Since the universe is currently expanding, at this time $R' > 0$, so R' is currently given by the positive root. Show that R increases to some value R_{max}, and then decreases again. In addition, show that when R gets small, R' gets very large negatively; in other words, a "big crunch" happens.

20. In this problem we look at the case where $C = 0$. Then we know that

$$R' = \sqrt{\frac{2GM_0}{R}}.$$

    Solve this differential equation, assuming $R = 0$ when $t = 0$. Give R as a function of t. The resulting formula for R is called the "flat universe model." What does this model predict about the expansion of the universe?

21. (a) Einstein, who formulated the model of the universe described before Problem 19, wanted the universe to be stable — neither expanding nor shrinking. Why don't these differential equations allow for a stable universe?
    (b) One estimate for the age of the universe is $R(t_0)/R'(t_0)$ where t_0 is the current time. (This number is called the *Hubble constant*.) Why is this a reasonable estimate for the age of the universe? Is it an overestimate or an underestimate?

22. A globular cluster is a ball of gas, held together by gravity but torn apart by internal pressure. Let $p(r)$ be the pressure at a distance r from the center of the cluster, and let $f(r)$ be the density of the gas at the same point. Then if the globular cluster is in equilibrium[7]

$$\frac{dp}{dr} = \frac{Gf(r)}{r^2} \int_0^r 4\pi t^2 f(t)\,dt$$

where G is a constant. Assuming that pressure is proportional to density and that $f(r)$ is proportional to some power of r, find that power.

23. In analyzing a society, sociologists are often interested in how incomes are distributed through the society. Pareto's law asserts that each society has a constant k such that the average income of all people wealthier than you is k times your income $(k > 1)$. If $p(x)$ is the number of people in the society with an income of x or above, we define $\Delta p = p(x + \Delta x) - p(x)$, for small Δx.
    (a) Explain why the number of people with incomes between x and $x + \Delta x$ is represented by $-\Delta p$. Then show that the total amount of money earned by people with incomes between x and $x + \Delta x$ is approximated by $-x\Delta p$.
    (b) Use Pareto's law to show that the total amount of money earned by people with incomes of x or above is $kxp(x)$. Then show that the total amount of money earned by people with incomes between x and $x + \Delta x$ is approximated by $-kp\Delta x - kx\Delta p$.
    (c) Using your answers to parts (a) and (b), show that p satisfies the differential equation

$$(1 - k)xp' = kp.$$

    (d) Solve this differential equation for $p(x)$.
    (e) Sketch a graph of $p(x)$ for various values of k, some large and some near 1. How does the value of k alter the shape of the graph?

9.7 MODELS OF POPULATION GROWTH

Population projections first became important to political philosophers in the late eighteenth century. As concern for scarce resources has grown, so has interest in accurate population projections. In this section we will look at two differential equations which are used to model both human and animal population growth as well as growth processes in economics. These differential equations also have applications in epidemiology and medicine, such as modeling the spread of an infectious disease or the growth of a tumor.

Relative versus Absolute Growth Rates

When describing population growth, we often use percentages rather than absolute numbers. For example, we say the population of the world is now about 5.3 billion people and growing at a

[7]From J. Heidman, *Relativistic Cosmology*, translated by S.J. Mitton (Berlin: Springer-Verlag, 1980).

continuous rate of about 2% a year, meaning that the relative growth rate is 2%:

$$\frac{dP/dt}{P} = \frac{1}{P}\frac{dP}{dt} = 0.02.$$

The quantity dP/dt is called the *absolute growth rate* and measures the growth rate in, say, people per year. The quantity $(dP/dt)/P$ is called the *relative growth rate* and represents the absolute growth rate as a fraction of the whole population. Its units are, say, % per year. When talking about populations we will often use the relative growth rate.

The US Population: 1790--1860

Every ten years the population of the United States is recorded by a census. The first such census was in 1790. Table 9.7 contains the census data from 1790 to 1940.

TABLE 9.7 *US Population in millions, 1790–1940*

Year	Population	Year	Population	Year	Population
1790	3.9	1850	23.2	1910	92.0
1800	5.3	1860	31.4	1920	105.7
1810	7.2	1870	38.6	1930	122.8
1820	9.6	1880	50.2	1940	131.7
1830	12.9	1890	62.9		
1840	17.1	1900	76.0		

You might note that the population is given only to the nearest 0.1 million, or 100,000. The reason is that although the population is reported by the US Census Bureau down to the last digit (for example, 131,669,275 in 1940), census figures are notoriously inaccurate. For example, in the census of 1990, New York City claimed the census had missed a million people in that city alone. Thus, giving more digits in the population does not necessarily give more accuracy.

Let us concentrate first on the relative growth rate, $(dP/dt)/P$, of the US population from 1790 to 1860. If we want to estimate $(dP/dt)/P$ in 1830, we take the rate of change in the population and divide it by the population itself:[8]

$$\begin{aligned}\frac{1}{P}\frac{dP}{dt} &\approx \frac{1}{\text{Population in 1830}} \cdot \frac{\text{Population in 1840} - \text{Population in 1830}}{10 \text{ years}}\\ &= \frac{1}{12.9} \cdot \frac{17.1 - 12.9}{10} = 0.0326 = 3.26\%.\end{aligned}$$

Similar calculations for 1790, 1800, . . . , 1850 give the percentages in Table 9.8:

TABLE 9.8 *Estimated yearly growth rate of US population*

Year	1790	1800	1810	1820	1830	1840	1850
Relative growth rate	3.59%	3.58%	3.33%	3.44%	3.26%	3.57%	3.53%

These percentages are pretty close. There is no clear pattern of increase or decrease; the fluctuations appear random. Thus the relative growth rate, while not precisely constant, is nearly so. In fact,

[8]In Problem 11 at the end of this section we look at an alternative way of estimating the relative growth rate in 1830, using data from 1820 as well as from 1840.

political and economic events such as war or recession affect the population, so we don't expect the growth rate to be exactly constant.

The simplest model for population growth is to assume that the relative growth rate *is* constant, in other words

$$\frac{1}{P}\frac{dP}{dt} = k,$$

where k is the continuous growth rate. This is equivalent to assuming that the population grows exponentially:

$$P = P_0 e^{kt}.$$

What should we take for the value of k? One possibility would be the average of the percentages we just calculated, namely 3.47%. However, there is a serious objection to using this percentage as an estimate for k. Remember that k is a continuous growth rate, but the populations are given at 10-year intervals. A 10-year population growth of 34.7% doesn't come from a continuous yearly rate of 3.47%. (This would be ignoring the effects of compounding.) If the population increases by a factor of 34.7% in 10 years, then $P(10)/P_0 = 1.347$, where $P(10)$ is the population when $t = 10$. Assuming that $P = P_0 e^{kt}$, we need k to satisfy

$$\frac{P}{P_0} = e^{k\cdot 10} = 1.347.$$

Now we can find the continuous growth rate, k:

$$k = \frac{\ln(1.347)}{10} = 0.0298.$$

Let's compare predicted and actual values if we model the US population by the differential equation

$$\frac{dP}{dt} = 0.0298P.$$

We start with initial population $P_0 = 3.9$ in 1790. Notice that this says we will consider 1790 as time $t = 0$, so 1800 is $t = 10$, and 1810 is $t = 20$, etc. The solution to the differential equation is

$$P = 3.9e^{0.0298t}.$$

If we put $t = 0, 10, 20, \ldots, 70$ into this function we get the populations predicted by our model for the years 1790, 1800, ..., 1860. Table 9.9 contains the comparison to the actual populations.

TABLE 9.9 *Predicted versus actual US population 1790–1860 (exponential model)*

Year	Actual	Predicted	Year	Actual	Predicted
1790	3.9	3.9	1830	12.9	12.8
1800	5.3	5.3	1840	17.1	17.3
1810	7.2	7.1	1850	23.1	23.3
1820	9.6	9.5	1860	31.4	31.4

The agreement is remarkable. Of course, since we used the data from throughout the 70 year period to estimate k, we should expect good agreement throughout that period. What is surprising is that if we had used only the populations in 1790 and 1800 to estimate k, the predictions are still quite good. Let's find a new value of k using only the 1790 and 1800 data and compare predictions. The 10 year growth from 1790 to 1800 is 35.9% so $k = \ln(1.359)/10 = 0.0307$. We would then predict the population in 1860 to be

$$3.9e^{0.0307(70)} = 33.4,$$

which is within about 6% of the actual population of 31.4. It is remarkable that a person in 1800 could predict the population of the US 60 years later this accurately, especially when one considers all the wars, recessions, epidemics, additions of new territory, and immigration that took place from 1800 to 1860.

The Exponential Model

For the years 1790–1860, the relative growth rate of the US population was more or less constant. Assuming that the relative growth rate of a population, P, is exactly constant gives us the *exponential model*

$$\frac{1}{P}\frac{dP}{dt} = k.$$

Rewritten as

$$\frac{dP}{dt} = kP,$$

this is a familiar differential equation whose solution is the family of exponential functions

$$P = P_0 e^{kt},$$

where P_0 is the population when t is zero.

The assumption that the relative growth rate is constant is usually a good one over a small time interval. Hence, for a suitable k, an exponential function models almost any population for a short period.

The grim predictions of this model are reflected in the ideas of Thomas Malthus, an early nineteenth-century clergyman and political philosopher, who believed that, if unchecked, a population would grow geometrically (that is, exponentially) whereas the food supply would grow arithmetically (that is, linearly), and therefore that population growth would eventually outstrip increases in food supply.

Interestingly enough, an exponential model has fit the growth of world population and the population of many regions remarkably well for decades, even centuries. However, the model must break down at some point because it predicts that the population will continue to grow without bound as time goes on — and this cannot be true forever. Eventually the effects of crowding, emigration, disease, war, and lack of food will have to curb growth.

However, for small populations the exponential model is often quite accurate. It is only for large populations that the effects of crowding make themselves felt. In searching for an improvement, then, we should look for a model whose solution is approximately an exponential function for small values of the population, but which levels off later. That's what the next model does.

How to Estimate dP/dt from Data

If, as is often the case, all we know about a population, P, is its values at certain points in time, we have to approximate dP/dt by $\Delta P/\Delta t$. However there are several different ways this approximation can be made. As in the example above, we can say

$$\frac{dP}{dt} \text{ at } 1830 \approx \frac{\text{Population in 1840} - \text{Population in 1830}}{10}.$$

However we could equally well have said

$$\frac{dP}{dt} \text{ at } 1830 \approx \frac{\text{Population in } 1830 - \text{Population in } 1820}{10}.$$

Both of these are called *one-sided estimates* because they involve using the population to one side of 1830 but not the other. The first one involving 1840 is the forward, or right-hand, estimate; the second one involving 1820 is the backward, or left-hand estimate. In general, both are equally good or bad estimates for dP/dt. A more accurate approximation can be obtained by averaging the two one-sided estimates, giving

$$\frac{dP}{dt} \text{ at } 1830 \approx \frac{1}{2}\left(\frac{\text{Population in } 1840 - \text{Population in } 1830}{10} + \frac{\text{Population in } 1830 - \text{Population in } 1820}{10}\right).$$

When we add the two fractions the population in 1830 cancels and we get the two-sided estimate, so called because it involves data on both sides of 1830.

$$\frac{dP}{dt} \text{ at } 1830 \approx \frac{1}{2}\left(\frac{\text{Population } 1840 - \text{Population } 1820}{10}\right) = \left(\frac{\text{Population } 1840 - \text{Population } 1820}{20}\right).$$

In the exponential model above we used a one-sided estimate. This turned out to give good results after adjusting from 10-year to continuous growth rates. If we had been unable to get accurate predictions using one-sided estimates, we might have tried two-sided estimates instead. In the logistic model below, we use two-sided estimates all the way through, as they turn out to give noticeably better predictions.

The US Population: 1790--1940

The exponential model for the US population works very well for reasonable periods of time, but exponential growth cannot go on forever. For example, if we tried to predict the population of the US in 1990 using the exponential model we developed above with $k = 0.0298$, we would get

$$\text{US Population in } 1990 = 3.9e^{(0.0298)(200)} = 1{,}512 \text{ million},$$

which is far from the actual figure of around 250 million.

The problem is that percentage growth rate in the US population between 1790 and 1860 did not stay constant in later decades. The 10-year percentage growths from 1860 to 1930 are listed in Table 9.10[9]:

TABLE 9.10 *Estimated 10-year growth rate of US population, 1860–1930*

Year	1860	1870	1880	1890	1900	1910	1920	1930
Relative growth rate	22.9%	30.1%	25.3%	20.8%	21.1%	14.9%	16.2%	7.2%

These figures are nothing like those during the period 1790 to 1860, where they hovered around 34%. The dramatic drop to 22.9% for the decade 1860–1870 is explainable by the Civil War (but don't ascribe the entire drop to deaths of the war — see Problem 14, page 539). The growth rate goes back up during the decade 1870–1880, but by 1890–1900 it has dropped below the rate during the Civil War. There is a slight increase in the rate during the immigrations of 1900 to 1910, another

[9]Calculated using one-sided forward estimates, so that $\frac{1}{P}\frac{dP}{dt}$ at 1860 $\approx \frac{\text{Population in } 1870 - \text{Population in } 1860}{10(\text{Population in } 1860)}$.

drop during World War I, a small bounce back, and finally the rate plummets during the recession of the 1930s. Notice that the effect of the recession is more dramatic than the effect of wars, which suggests that it is a decrease in birth rate rather than an increase in death rate that is a major factor in declines in the relative population growth rate.

Our exponential model gives accurate predictions up to 1860. But for the years following 1860, the exponential model is inadequate. We look for a new model that will take into account the effects of overcrowding. Because of the effects of crowding, we expect the relative growth rate to decrease as the population increases. Thus we look at how $(dP/dt)/P$ changes as P changes. This time we will use a two-sided estimate for dP/dt. For example:

$$\frac{dP}{dt} \text{ at } 1830 \approx \frac{\text{Population in 1840} - \text{Population 1820}}{20}.$$

Table 9.11 contains values of $(dP/dt)/P$ computed this way for some years between 1790 and 1940. Comparing the last two columns suggests that the values in the last column may be an approximately linear function of P. To get a better picture of how $(dP/dt)/P$ varies with P, we plot $(dP/dt)/P$ versus P for all the years from 1800 to 1930 and see whether the points lie on a straight line. Figure 9.44 shows the *scatterplot* of the points together with the line that best fits the points. You can see that the line fits quite well. Note that since the horizontal axis represents population, not time, the points are clumped to the left and not equally spaced. The equation for the line is

$$\frac{1}{P}\frac{dP}{dt} = 0.0318 - 0.000170P.$$

TABLE 9.11 *Some values for* $\frac{1}{P}\frac{dP}{dt}$.

Year	P	$\left(P(t+10) - P(t-10)\right)/(20P)$
1860	31.4	0.0245
1890	62.9	0.0205
1910	92.0	0.0161
1930	122.8	0.0106

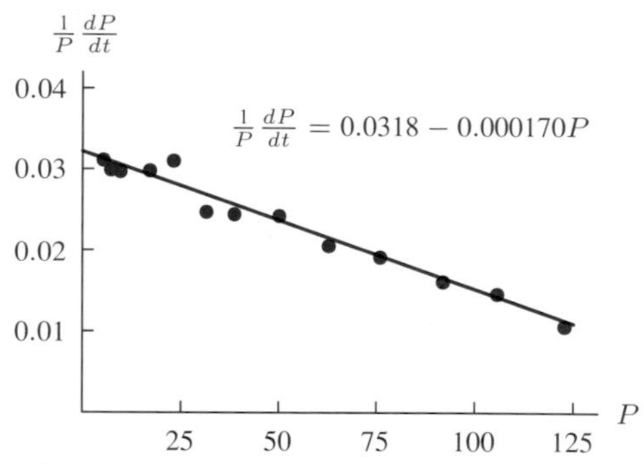

Figure 9.44: Scatterplot for US census data of $\frac{1}{P}\frac{dP}{dt}$ versus P.

Our new model is therefore that P satisfies the differential equation

$$\frac{dP}{dt} = 0.0318P - 0.000170P^2$$

This is known as a *logistic equation*. Before solving the equation analytically, we look at its slope field in Figure 9.45, plotted with the solution with $P(0) = 3.9$ superimposed. (Notice that $t = 0$ in 1790.) The most striking difference in this model compared with the exponential model is that it predicts that the US population will level off somewhere below 200 million. The population will continue to grow until $dP/dt = 0$, which occurs when

$$0 = 0.0318P - 0.000170P^2$$

so

$$P = 0 \quad \text{or} \quad P \approx 187 \text{ million}.$$

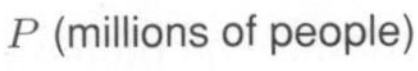
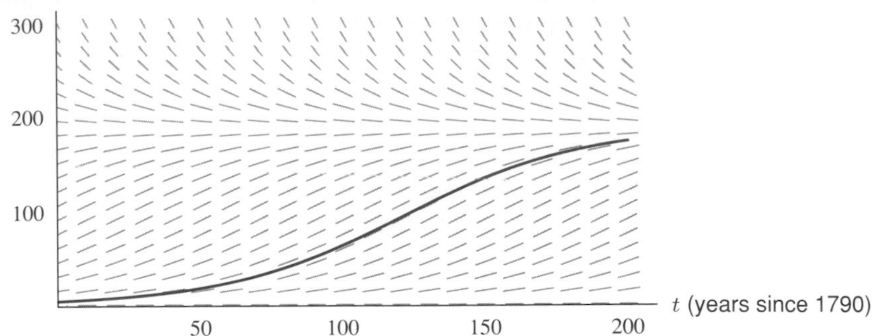

Figure 9.45: Solution to $\frac{dP}{dt} = 0.0318P - 0.000170P^2$ with $P(0) = 3.9$.

Looking at the S-shape of the solution curve in Figure 9.45, we see that initially the population grows faster and faster and then slows down as the limiting value of 187 is approached; the fastest growth rate appears to be about half-way to the limiting value.

Later on in this section we derive the formula for the solution to the logistic equation. For now, you can check by substitution that the function

$$P = \frac{187}{1 + 47e^{-0.0318t}}$$

is a solution to our logistic equation modeling the US population. (The numbers 187 and the 47 are not exact values, but have been rounded.) The values predicted by this equation for P agree very well with the actual populations up to 1940. See Table 9.12. The largest deviation from 1700 to 1940 is an error of about 3% in 1840 and 1870 (the Civil War accounts for the second one). All other errors are less than 2%.

Of course, the final test is how well our model, based on data from 1790 to 1840, predicts the population in the "future," 1950 to 1990. Table 9.12 contains predicted and actual data for this period also.

TABLE 9.12 *Predicted versus actual US population 1790–1980 (logistic model)*

Year	Actual	Predicted	Year	Actual	Predicted	Year	Actual	Predicted
1790	3.9	3.9	1860	31.4	30.8	1930	122.8	120.8
1800	5.3	5.3	1870	38.6	39.9	1940	131.7	133.7
1810	7.2	7.2	1880	50.2	50.7	1950	150.7	145.0
1820	9.6	9.8	1890	62.9	63.3	1960	179.3	154.4
1830	12.9	13.2	1900	76.0	77.2	1970	203.3	162.1
1840	17.1	17.7	1910	92.0	91.9	1980	226.5	168.2
1850	23.2	23.5	1920	105.7	106.7	1990	248.7	172.9

The fit between predicted and actual population values is obviously not good from 1950 on. Despite World War II, which undoubtedly depressed population growth between 1942 and 1945, in the last half of the 1940s the US population surged, wiping out in five years a deficit caused by 15 years of depression and war. The 1950s saw a population growth of 28 million, leaving our logistic model in the dust. This surge in population is referred to as the baby boom. All one can say is that based on 150 years of data, what happened in the US in the 20 years after World War II was completely without precedent. The baby boom could well end up being for the United States one of the most important sociological events of the twentieth century, and its consequences will be felt for many years to come.

Once again we have reached a point where our model is no longer useful. This should not lead you to believe that a reasonable mathematical model cannot be found; rather it should serve to point out that no model is perfect and that when one model fails, we seek a better one. Just as we abandoned the exponential model in favor of the logistic model for the US population, we could look further. (See Problems 8 and 9 on page 538.)

The Logistic Model

The logistic model we used to model the US population from 1790 to 1940 assumed that the relative growth rate of the population was a linearly decreasing function of P:

$$\frac{1}{P}\frac{dP}{dt} = k - aP.$$

(We took $k = 0.0318$ and $a = 0.000170$). For small P, we have approximately

$$\frac{1}{P}\frac{dP}{dt} \approx k.$$

The solution to this equation is an exponential function. This is why an exponential model fit the US population well during the years 1790-1860 when the population was relatively small. In the logistic model, as P increases, the relative growth rate decreases to zero; it reaches zero when P is given by

$$k - aP = 0.$$

Solving for P, we get

$$P = \frac{k}{a}.$$

This is the limiting value of the population, which we call L:

$$L = \frac{k}{a}.$$

The value L is called the *carrying capacity* of the environment, and represents the largest population the environment can support. Writing $a = k/L$, the logistic equation becomes

$$\frac{1}{P}\frac{dP}{dt} = k - \frac{k}{L}P$$
$$\frac{dP}{dt} = kP - \frac{k}{L}P^2$$

or

$$\frac{dP}{dt} = kP\left(1 - \frac{P}{L}\right).$$

This is the general *logistic differential equation*, first proposed as a model for population growth by the Belgian mathematician P. F. Verhulst in the 1830s.

Qualitative Solution to the Logistic Equation

Figure 9.46 shows the slope field and characteristic *sigmoid*, or S-shaped, solution curve for the logistic model. Notice that for each fixed value of P, that is, along each horizontal line, the slopes are all the same because dP/dt depends only on P and not on t. The slopes are small near $P = 0$ and near $P = L$; they are steepest around $P = L/2$. For $P > L$, the slopes are negative, meaning that if the population is above the carrying capacity, the population will decrease.

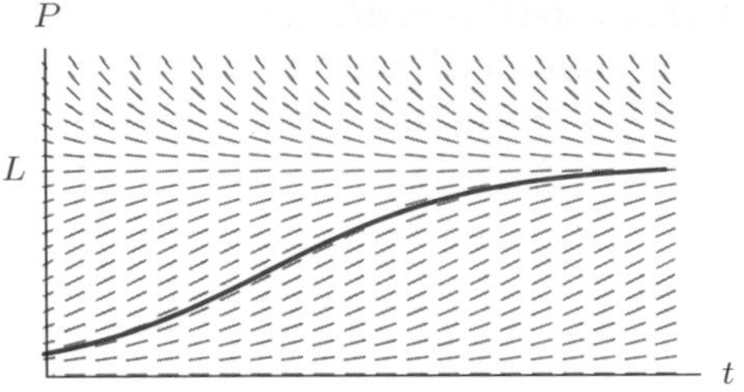

Figure 9.46: Slope field for $\frac{dP}{dt} = kP(1 - \frac{P}{L})$

We can locate precisely the inflection point where the slopes are greatest using the graph of dP/dt against P in Figure 9.47. The graph is a parabola because dP/dt is a quadratic function of P. The horizontal intercepts are at $P = 0$ and $P = L$, so the maximum, where the slope is greatest, is at $P = L/2$.

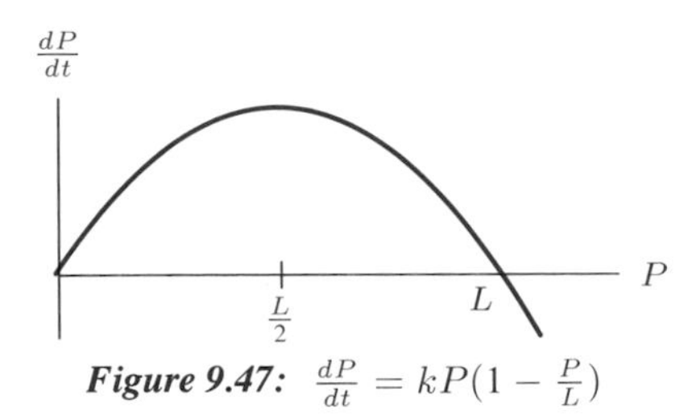

Figure 9.47: $\frac{dP}{dt} = kP(1 - \frac{P}{L})$

The graph in Figure 9.47 also tells us that for $0 < P < L/2$, the slope dP/dt is positive and increasing, so the graph of P against t is concave up. (See Figure 9.48.) For $L/2 < P < L$, the slope dP/dt is positive and decreasing, so the graph of P against t is concave down. For $P > L$, the slope dP/dt is negative, so the graph of P against t is decreasing.

If $P = 0$ or $P = L$, there is an equilibrium solution (not a very interesting one if $P = 0$). Figure 9.49 shows that $P = 0$ is an unstable equilibrium because solutions which start near 0 move away. However, $P = L$ is a stable equilibrium. For small P, the solution curve is similar to an exponential.

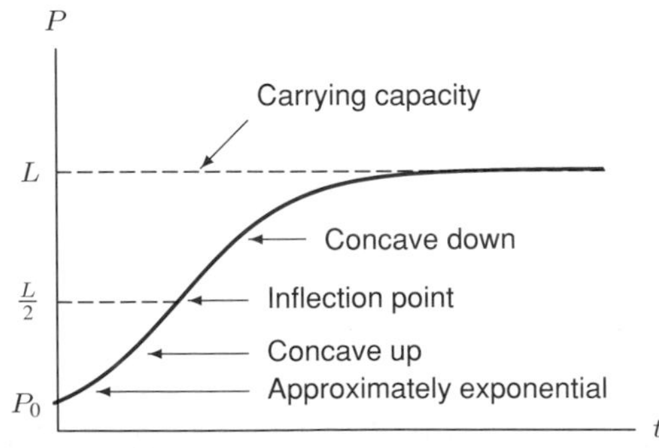

Figure 9.48: Logistic growth with inflection point

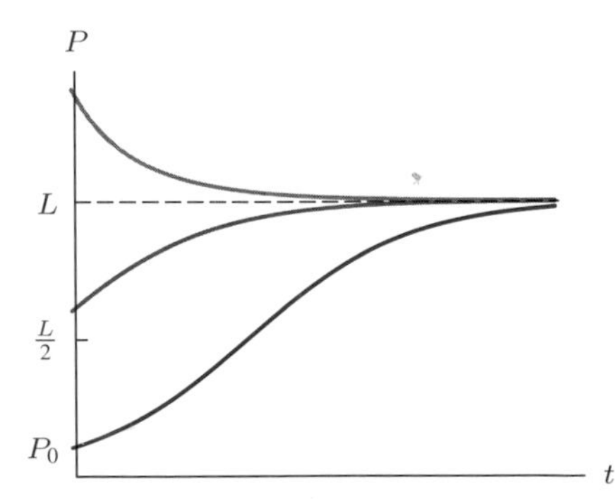

Figure 9.49: Solutions to the logistic equation

The Analytic Solution to the Logistic Equation

We have already obtained a lot of information about logistic growth without finding a formula for the solution. However, the equation can be solved analytically by separating variables:

$$\frac{dP}{dt} = kP\left(1 - \frac{P}{L}\right) = kP\left(\frac{L-P}{L}\right)$$

giving

$$\int \frac{dP}{P(L-P)} = \int \frac{k}{L}\,dt.$$

We can integrate the left side using the integral tables (Formula 26), or by rewriting

$$\frac{1}{P(L-P)} = \frac{1}{L}\left(\frac{1}{P} + \frac{1}{L-P}\right).$$

Thus the equation has become

$$\int \frac{1}{L}\left(\frac{1}{P} + \frac{1}{L-P}\right)dP = \int \frac{k}{L}\,dt.$$

Canceling the constant L, we get

$$\int \left(\frac{1}{P} + \frac{1}{L-P}\right)dP = \int k\,dt$$

which can be integrated to give

$$\ln|P| - \ln|L-P| = kt + C.$$

Multiplying through by (-1) and using the fact that $\ln M - \ln N = \ln(M/N)$, we have

$$\ln\left|\frac{L-P}{P}\right| = -kt - C.$$

Exponentiating both sides gives

$$\left|\frac{L-P}{P}\right| = e^{-kt-C} = e^{-C}e^{-kt},$$

so

$$\frac{L-P}{P} = Ae^{-kt} \quad \text{where} \quad A = \pm e^{-C}.$$

We find A by substituting $P = P_0$ when $t = 0$, which gives

$$\frac{L-P_0}{P_0} = Ae^0 = A.$$

Thus

$$\frac{L-P}{P} = Ae^{-kt} \quad \text{where} \quad A = \frac{L-P_0}{P_0}.$$

Since $(L-P)/P = (L/P) - 1$, we have

$$\frac{L}{P} = 1 + Ae^{-kt}$$

giving

$$P = \frac{L}{1+Ae^{-kt}} \quad \text{where} \quad A = \frac{L-P_0}{P_0}.$$

This is the formula for the logistic curve. In many cases, it is easier to argue graphically than to use this formula. But if, for example, you want to calculate the exact time at which the population reaches a certain value, this formula would be quite useful.

Problems for Section 9.7

1. Assuming that Switzerland's population is growing exponentially at a continuous rate of 0.2% a year and that its 1988 population was 6.6 million, write an expression for the population as a function of time in years. (Let $t = 0$ in 1988.)

2. Consider the logistic model
$$\frac{dP}{dt} = 3P - 3P^2.$$
   (a) On the slope field in Figure 9.50, sketch three solution curves showing different types of behavior.
   (b) Is there a stable value of the population? If so, what is it?
   (c) Describe the meaning of the shape of the solution curves for the population: Where is P increasing? Decreasing? What happens in the long run? Are there any inflection points? Where? What do they mean for the population?

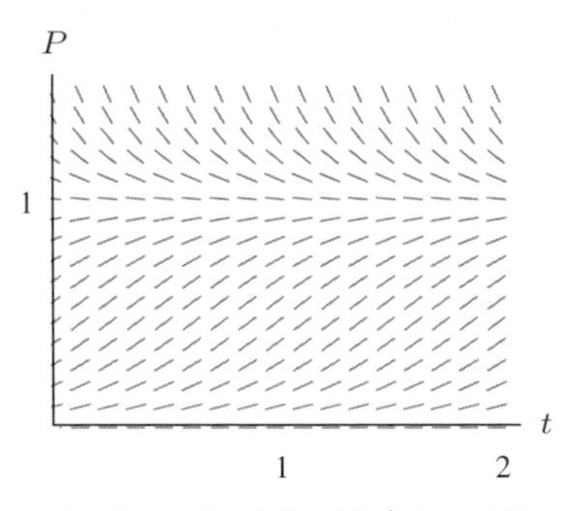

Figure 9.50: Slope field for $dP/dt = 3P - 3P^2$.

   (d) Sketch a graph of dP/dt against P. Where is dP/dt positive? Negative? Zero? Maximum? How do your observations about dP/dt explain the shapes of your solution curves?

3. Many organ pipes in old European churches are made of tin. In cold climates such pipes can be affected with *tin pest*, when the tin becomes brittle and crumbles into a grey powder. This transformation can appear to take place very suddenly because the presence of the grey powder encourages the reaction to proceed. At the start, when there is little grey powder, the reaction proceeds slowly. Similarly, toward the end, when there is little metallic tin left, the reaction is also slow. In between, however, when there is plenty of both metallic tin and powder, the reaction can be alarmingly fast.

   Suppose that the rate of the reaction is proportional to the product of the amount of tin left and the quantity of grey powder, p, present at time t. Assume also that when metallic tin is converted to grey powder, its weight does not change.

   (a) Write a differential equation for p. Let the total quantity of metallic tin present originally be B.
   (b) Sketch a graph of the solution $p = f(t)$ if there is a small quantity of powder initially. How much metallic tin has crumbled when it is crumbling fastest?
   (c) Suppose there is no grey powder initially. (For example, suppose the tin is completely new.) What does this model predict will happen? How do you reconcile this with the fact that many organ pipes do get tin pest?

4. The growth of a certain animal population is governed by the equation

$$\frac{1000}{P}\frac{dP}{dt} = 100 - P,$$

where $P(t)$ is the number of individuals in the colony at time t. The initial population is known to be 200 individuals. Sketch a graph of $P(t)$. Will there ever be more than 200 individuals in the colony? Will there ever be fewer than 100 individuals? Explain.

5. It is of considerable interest to policymakers to model the spread of information through a population. For example, various agricultural ministries use models to help them understand the spread of technical innovations or new seed types through their countries. Two models, based on how the information is spread, are given below. Assume the population is of a constant size M.
   (a) If the information is spread by mass media (TV, radio, newspapers), the rate at which information is spread is believed to be proportional to the number of people not having the information at that time. Write a differential equation for the number of people having the information by time t. Sketch a solution assuming that no one (except the mass media) has the information initially.
   (b) If the information is spread by word of mouth, the rate of spread of information is believed to be proportional to the product of the number of people who know and the number who don't. Write a differential equation for the number of people having the information by time t. Sketch the solution for the cases in which
      (i) no one
      (ii) 5% of the population
      (iii) 75% of the population

   knows initially. In each case, when is the information spreading fastest?

6. In the 1930s, the Soviet ecologist G. F. Gause[10] studied the population growth of yeast. Fit a logistic curve, $dP/dt = kP(1 - P/L)$, to his data below using the method outlined below.

Time (hours)	0	10	18	23	34	42	47
Yeast population	0.37	8.87	10.66	12.50	13.27	12.87	12.70

   (a) Plot the data and use it to estimate (by eye) the carrying capacity, L.
   (b) Use the first two pieces of data in the table, and your value for L, to estimate k.
   (c) Using your values for k and L, sketch the solution curve

$$P = \frac{L}{1 + Ae^{-kt}} \quad \text{where} \quad A = \frac{L - P_0}{P_0}$$

   on the same axes as the data points.

7. The population data from another experiment on yeast by the ecologist G. F. Gause is given.

Time (hours)	0	13	32	56	77	101	125
Yeast population	1.00	1.70	2.73	4.87	5.67	5.80	5.83

   (a) Do you think the population is growing exponentially or logistically? Give reasons for your answer.
   (b) Estimate the value of k (for either model) from the first two pieces of data. If you chose a logistic model in part (a), estimate the carrying capacity, L, from the data.
   (c) Sketch the data and the approximate growth curve given by the parameters you estimated.

[10]Data adapted from G. F. Gause, *The Struggle for Existence* (New York: Hafner Publishing Company, 1969).

8. (a) In the text we fitted a logistic model to the US population from 1790–1940. In this problem, we try to fit a logistic equation to the US population all the way from 1790 to 1990. No logistic equation fits the data exactly over this entire period, but we can use the method of page 531 to find an equation that does reasonably well throughout.

   To fit a logistic equation to the data in Table 9.13, we estimate the relative growth rate, $(dP/dt)/P$, and plot it against P. To approximate $(dP/dt)/P$, calculate $(\Delta P/\Delta t)/P$ from the data for six fairly spread-out points. Draw a reasonable line (by eye) through your points, and thus estimate k and a for the equation $(dP/dt)/P = k - aP$.

TABLE 9.13 *US population in millions, 1790–1990*

Year	Population	Year	Population	Year	Population
1790	3.9	1860	31.4	1930	122.8
1800	5.3	1870	38.6	1940	131.7
1810	7.2	1880	50.2	1950	150.7
1820	9.6	1890	62.9	1960	179.0
1830	12.9	1900	76.0	1970	205.0
1840	17.1	1910	92.0	1980	226.5
1850	23.1	1920	105.7	1990	248.7

   (b) What does this model predict about the US population in the long run?

9. (a) In Problem 8 we saw that a logistic model cannot be made to fit the US population very closely, because the points on the graph of $(dP/dt)/P$ against P are not exactly on a line. In this problem we'll try another model. This time we'll assume that $(dP/dt)/P$ is a linear function of t, so you should plot $(dP/dt)/P$ against t (using the same approximate values you calculated in Problem 8). Put a line through the points by eye, and estimate a and b to fit the equation

$$\frac{1}{P}\frac{dP}{dt} = a - bt.$$

   (b) When, if ever, does this model predict that the US population will be at its maximum?
   (c) Solve the differential equation and sketch its solution.

10. An alternative method of finding the analytic solution to the logistic equation

$$\frac{dP}{dt} = kP\left(1 - \frac{P}{L}\right)$$

uses the substitution $P = 1/u$.

   (a) Show that

$$\frac{dP}{dt} = -\frac{1}{u^2}\frac{du}{dt}$$

   (b) Rewrite the logistic equation in terms of u and t, and solve for u in terms of t.
   (c) Using your answer to part (b), find P as a function of t.

11. On page 527, we used one-sided estimates for dP/dt to fit an exponential model to the US population from 1790–1860. In this problem we will use two-sided estimates for the years 1800–1850. Suppose P is the US population in millions.

   (a) Estimate dP/dt by the symmetric difference quotient $\left(P(t+10) - P(t-10)\right)/20$. Then compute $(dP/dt)/P$ for each of these years and average them to estimate k for the exponential model $dP/dt = kP$. Compare your value of k with the estimate $k \approx 3.47\%$ obtained on page 528 by using $\left(P(t+10) - P(t)\right)/10$ for dP/dt.

(b) Using your answer to part (a), compute k as the continuous rate of change that leads to the observed 10-year percentage change. Then compare your k with the value $k \approx 2.98\%$ obtained on page 528.

12. This section suggested two ways of estimating dP/dt: the one-sided $\left(P(t+h) - P(t)\right)/h$, and the two-sided $\left(P(t+h) - P(t-h)\right)/(2h)$. Let $f(x) = x^3$. Consider the approximations $f'(2) \approx \left(f(2+h) - f(2)\right)/h$ and $f'(2) \approx \left(f(2+h) - f(2-h)\right)/2h$ for $h = 0.1, 0.01, 0.001$. Which is the better approximation? Is there a pattern to the errors in the approximation as you decrease h? If so, describe it.

13. Show that if $f(x) = x^2$, then $(f(x+h) - f(x-h))/(2h) = f'(x)$ for any value of h. This shows that if $f(x) = x^2$, the two-sided estimate for the derivative of $f(x)$ is exactly equal to the derivative.

14. Estimate the US population in 1870 from the 1860 population of 31.4 million, assuming that the population increased at the same percentage rate during the 1860s as it did in the decades previous to 1860 (about 34.7% each decade). Compare your estimate with the actual population of 38.6 million. Find some estimate of the number of people who died in the Civil War. Does the number of deaths in the Civil War explain the shortfall in the actual population in 1870? What else might influence the shortfall?

15. Another way to estimate the limiting population L for the logistic differential equation is to note that the derivative dP/dt is largest when $P = L/2$. Compute dP/dt for the US census data for 1790–1940 using the two-sided difference quotient $\left(P(t+10) - P(t-10)\right)/20$. Estimate L by doubling the population when dP/dt is largest. Compare this estimate with the estimate $L = 187$ given in the text.

Any population, P, for which we can ignore immigration, satisfies

$$\frac{dP}{dt} = \text{Birth rate} - \text{Death rate}.$$

For organisms which need a partner for reproduction but rely on a chance encounter for meeting a mate, the birth rate is proportional to the square of the population. Thus, the population of such an organism satisfies a differential equation of the form

$$\frac{dP}{dt} = aP^2 - bP \quad \text{with } a,\, b > 0.$$

Problems 16–18 investigate the solutions to such an equation.

16. Consider the equation

$$\frac{dP}{dt} = 0.02P^2 - 0.08P.$$

(a) Sketch the slope field for this differential equation for $0 \le t \le 50$, $0 \le P \le 8$.

(b) Use your slope field to sketch the general shape of the solutions to the differential equation satisfying the following initial conditions:

(i) $P(0) = 1$ (ii) $P(0) = 3$ (iii) $P(0) = 4$ (iv) $P(0) = 5$

(c) Are there any equilibrium values of the population? If so, are they stable?

17. Consider the equation

$$\frac{dP}{dt} = P^2 - 6P.$$

(a) Sketch a graph of dP/dt against P for positive P.
(b) Use the graph you drew in part (a) to sketch the approximate shape of the solution curve with $P(0) = 5$. To do this, consider the following question. For $0 < P < 6$, is dP/dt positive or negative? What does this tell you about the graph of P against t? As you move along the solution curve with $P(0) = 5$, how does the value of dP/dt change? What does this tell you about the concavity of the graph of P against t?
(c) Use the graph you drew in part (a) to sketch the solution curve with $P(0) = 8$.
(d) Describe the qualitative differences in the behavior of populations with initial value less than 6 and initial value more than 6. Why do you think $P = 6$ is called the *threshold population*?

18. Consider a population satisfying

$$\frac{dP}{dt} = aP^2 - bP \quad \text{with constants } a,\, b > 0.$$

(a) Sketch a graph of dP/dt against P.
(b) Use this graph to sketch the shape of solution curves with various initial values. Use your graph from part (a) to decide where dP/dt is positive or negative, and where it is increasing or decreasing. What does this tell you about the graph of P against t?
(c) Why is $P = b/a$ called the threshold population? What happens if $P(0) = b/a$? What happens in the long-run if $P(0) > b/a$? What if $P(0) < b/a$?

For Problems 19–21, we look at the effects of *harvesting* a population which is growing logistically. Harvesting could be, for example, fishing or logging. An important question is what level of harvesting leads to a *sustainable yield*. In other words, how much can be harvested without having the population depleted in the long run?

19. When there is no fishing, suppose a population of fish is governed by the differential equation

$$\frac{dP}{dt} = 2P - 0.01P^2,$$

where P is the number of fish at time t in years. Suppose in addition that the fish are removed by fishermen at a continuous rate of 75 fish/year.

(a) Explain why the fish population must satisfy the differential equation

$$\frac{dP}{dt} = 2P - 0.01P^2 - 75.$$

(b) Sketch the slope field for the differential equation in part (a).
(c) Use the slope field to sketch the general shape of solutions to the differential equation satisfying the following initial conditions:

(i) $P(0) = 40$ (ii) $P(0) = 50$ (iii) $P(0) = 60$
(iv) $P(0) = 150$ (v) $P(0) = 170$

(d) There are two equilibrium populations. What are they? Are they stable?

20. In this problem we will investigate the solutions to the differential equation

$$\frac{dP}{dt} = 2P - 0.01P^2 - 75$$

by looking at the graph of dP/dt against P. Suppose P is a population of fish at time, t, in years.

(a) In terms of fish, what is the meaning of the -75 term? Why is it negative? What are its units?

(b) First we will work without the -75 term. Sketch a graph of dP/dt against P if

$$\frac{dP}{dt} = 2P - 0.01P^2$$

(See Figure 9.47 in the text on page 534.) Mark on your graph the equilibrium values of P. Now, sketch P against t for various different initial values. (See Figure 9.49 in the text on page 534.)

(c) Now sketch dP/dt against P if

$$\frac{dP}{dt} = 2P - 0.01P^2 - 75.$$

Find and label the intercepts.

(d) Use the graph you drew in part (c) to sketch graphs of P against t, with the initial values:

(i) $P(0) = 40$ (ii) $P(0) = 60$ (iii) $P(0) = 160$ (iv) $P(0) = 140$

(e) Using the graphs you drew in part (d), decide what the equilibrium values of the populations are and whether or not they are stable.

21. In this problem we will look at the effect of different levels of fishing on a fish population. If fishing takes place at a continuous rate of H fish/year and t is time in years, the fish population P satisfies the differential equation

$$\frac{dP}{dt} = 2P - 0.01P^2 - H.$$

(a) For each of the values $H = 75$, $H = 100$, $H = 200$, plot a graph of dP/dt against P, and use it to sketch graphs of P against t for various different initial values.

(b) For which of the three values of H that you considered in part (a) is there an initial condition such that the fish population does not die out eventually?

(c) Looking at your answer to part (b), decide for what range of values of H there is an initial value for P such that the population does not die out eventually.

(d) What policy would you recommend to ensure long-term survival of the fish population?

9.8 SYSTEMS OF DIFFERENTIAL EQUATIONS

In the preceding section we modeled the growth of a single population over time. We now consider the growth of two populations which interact, such as a population of sick people infecting the healthy people around them. This will involve not just one differential equation, but a system of two or more.

Diseases and Epidemics

The progress of a disease through a population can be modeled using differential equations. Such models can be used, for example, to predict when an outbreak of a disease will become so severe that it is called an *epidemic*,[11] and to decide what level of vaccination will be necessary to prevent an epidemic. Let's consider a specific example.

Flu in a British Boarding School

In January 1978, 763 students returned to a boys' boarding school after their winter vacation. A week later, one boy developed the flu, followed immediately by two more. By the end of the month, nearly half the boys were sick. Most of the school had been affected by the time the epidemic was over in mid-February.[12]

Being able to predict how many people will get sick, and when, is an important step toward controlling an epidemic, and this is one of the responsibilities of Britain's Communicable Disease Surveillance Centre and the US's Center for Disease Control and Prevention.

The S-I-R model

We will apply one of the most commonly used models for an epidemic, called the S-I-R model, to the boarding school flu example. Imagine the population of the school divided into three groups:

S = the number of *susceptibles*, the people who are not yet sick but who could become sick

I = the number of *infecteds*, the people who are currently sick

R = the number of *recovered*, or *removed*, the people who have been sick and can no longer infect others or be reinfected.

In this model, the number of susceptibles decreases with time, as people become infected. We will assume that the rate people become infected is proportional to the number of contacts between susceptible and infected people. We expect the number of contacts between the two groups to be proportional to both S and I. (If S doubles, we expect the number of contacts to double; similarly, if I doubles, we expect the number of contacts to double.) Thus we assume that the number of contacts is proportional to the product, SI. In other words, we assume that for some constant $a > 0$:

$$\frac{dS}{dt} = -\left(\begin{array}{c}\text{Rate susceptibles}\\ \text{get sick}\end{array}\right) = -aSI.$$

(The negative sign is used because S is decreasing.)

The number of infecteds is changing in two ways: newly sick people are being added to the infected group, and others are being removed. The newly sick people are exactly those people leaving the susceptible group and so accrue at a rate of aSI (with a positive sign this time). People leave the infected group either because they recover, or (in the case of a fatal disease) because they die, or because they are physically removed from the rest of the group and can no longer infect

[11]Exactly when a disease should be called an epidemic is not always clear. The medical profession generally classifies a disease an epidemic when the frequency is higher than usually expected — leaving open the question of what is usually expected. See, for example, *Epidemiology in Medicine* by C. H. Hennekens and J. Buring (Boston: Little, Brown, 1987)

[12]Data from the Communicable Disease Surveillance Centre (UK); reported in "Influenza in a Boarding School," *British Medical Journal* March 4, 1978, and by J. D. Murray in *Mathematical Biology* (New York: Springer Verlag, 1990)

others. We assume that people are removed at a rate proportional to the number sick, or bI, where b is a positive constant. Thus,

$$\frac{dI}{dt} = \text{Rate susceptibles get sick} - \text{Rate infecteds get removed} = aSI - bI$$

Assuming that those who have recovered from the disease are no longer susceptible, the recovered group increases at the rate of bI, so

$$\frac{dR}{dt} = bI.$$

We are assuming that having the flu confers immunity on a person, that is, that the person cannot get the flu again. (This is true for a given strain of flu, at least in the short run.)

In analyzing the flu, we can use the fact that the total population $S + I + R$ is not changing. (The total population, the total number of boys in the school, did not change during the epidemic.) Thus, once we know S and I, we can calculate R. So we will restrict our attention to the two equations

$$\frac{dS}{dt} = -aSI$$
$$\frac{dI}{dt} = aSI - bI$$

The Constants *a* and *b*

The constant a measures how infectious the disease is — that is, how quickly it is transmitted from the infecteds to the susceptibles. In the case of the flu, we know from medical accounts that the epidemic started with one sick boy, with two more becoming sick roughly a day later. Thus, when $I = 1$ and $S = 762$, we have $dS/dt \approx -2$, enabling us to roughly approximate a:[13]

$$a = -\frac{dS/dt}{SI} = \frac{2}{(762)(1)} = 0.0026.$$

The constant b represents the rate at which infected people are removed from the infected population. In this case of the flu, boys were generally taken off to the infirmary within one or two days of becoming sick. About half the infected population was removed each day, so we take $b \approx 0.5$. Thus, our equations are:

$$\frac{dS}{dt} = -0.0026SI$$
$$\frac{dI}{dt} = 0.0026SI - 0.5I$$

The Phase Plane

While it is possible to solve these equations for S and I analytically, this is often not done because we can get a good idea of the progress of the disease from rough graphs. You might expect that we would look for graphs of S and I against t, and eventually we will. However, we first look at a graph of I against S. If we plot a point (S, I) representing the number of susceptibles and the number of infecteds at any moment in time, then, as the numbers of susceptibles and infected changes, the

[13]The values of a and b are close to those obtained by J. D. Murray in *Mathematical Biology* (New York: Springer Verlag, 1990).

point will move. The SI-plane on which the point moves is called the *phase plane*. The path along which the point moves is called the *phase trajectory*, or *orbit*, of the point.

To find the phase trajectory, we need a differential equation relating S and I directly. Thinking of I as a function of S, and S as a function of t, we use the chain rule to get

$$\frac{dI}{dt} = \frac{dI}{dS} \cdot \frac{dS}{dt}$$

giving

$$\frac{dI}{dS} = \frac{dI/dt}{dS/dt}.$$

Substituting for dI/dt and dS/dt, we get

$$\frac{dI}{dS} = \frac{0.0026SI - 0.5I}{-0.0026SI}.$$

Assuming I is not zero, this equation simplifies to approximately

$$\frac{dI}{dS} = -1 + \frac{192}{S}.$$

We now have a differential equation whose slope field is shown in Figure 9.51. The initial condition is $S_0 = 762$, $I_0 = 1$ (1 boy sick, 762 susceptible), and the trajectory starting from this point is shown in Figure 9.52.

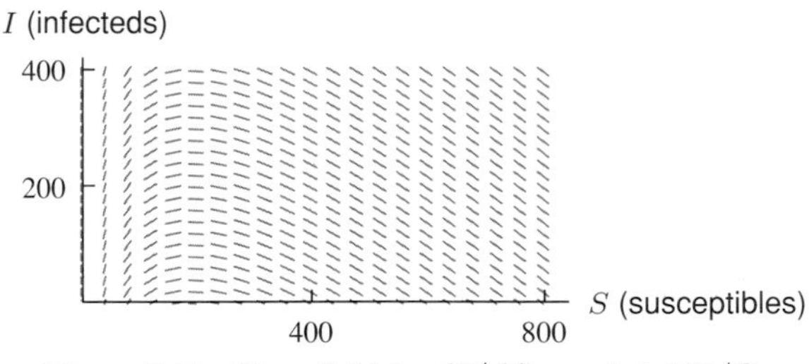

Figure 9.51: Slope field for $dI/dS = -1 + 192/S$

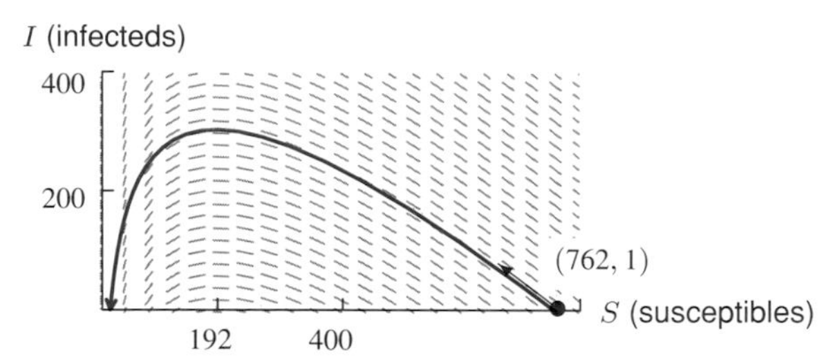

Figure 9.52: Trajectory for $S_0 = 762$, $I_0 = 1$

There's one important difference between the original system of differential equations and the single equation $dI/dS = -1 + 192/S$, namely that the original equations involve time, t. In the SI-phase plane, the passage of time is represented by the direction that a point moves on the trajectory. The disease starts at the point $S_0 = 762$, $I_0 = 1$, and as time passes, more people become infected and fewer are susceptible. In other words, S decreases and I increases. Thus, we move along the curve in the direction shown by the arrow in Figure 9.52.

What does the SI-Phase Plane Tell Us?

To learn how the disease progresses, look at the shape of the curve in Figure 9.52. The value of I increases to about 300 (the maximum number infected and infectious at any one time); then I decreases to zero. From the graph, we can see that this peak value of I occurs when $S \approx 200$. The peak value occurs where $dI/dS = 0$, or when

$$\frac{dI}{dS} = -1 + \frac{192}{S} = 0,$$

giving

$$S = 192.$$

Notice that the peak value for I always occurs at the same value of S, namely $S = 192$, for any trajectory that starts with S above 192. The graph shows that if a trajectory starts with $S_0 > 192$, then I first increases and then decreases to zero. On the other hand, if $S_0 < 192$, then I decreases right away.

> For this example, the value $S_0 = 192$ is called a *threshold value*. If S_0 is around or below 192, there is no epidemic. If S_0 is significantly greater than 192, an epidemic occurs.[14]

The phase diagram makes clear that the maximum value of I is about 300. Another question answered by the phase plane diagram is the total number of students who are expected to get sick during the epidemic. (This is not the maximum value reached by I, which gives the maximum number infected at any one time.) The point at which the trajectory crosses the S-axis represents the time when the epidemic has passed (since $I = 0$). Thus, the S-intercept shows how many boys never get the flu and, hence, how many did get it.

How Many People Should Be Vaccinated?

Faced with an outbreak of the flu or, as happened on several US campuses in the 1980s, of the measles, many institutions consider a vaccination program. How many students must be vaccinated in order to control an outbreak? To answer this, we can think of vaccination as removing people from the S category (without increasing I), which amounts to moving the initial point on the trajectory to the left, parallel to the S-axis. To avoid an epidemic, the initial value of S_0 should be around or below the threshold value. Therefore, for the boarding school, all but about 192 students need to be vaccinated to avoid an epidemic.

Graphs of S and I against t

To find out exactly when I reaches its maximum, we need numerical methods. A modification of Euler's method can be used to generate the solution curves of S and I against t in Figure 9.53. Notice that, as you would expect, the number of susceptibles drops throughout the disease as healthy people get sick. The number of infecteds peaks after about 6 days and then drops, and the epidemic has run its course in 20 days.

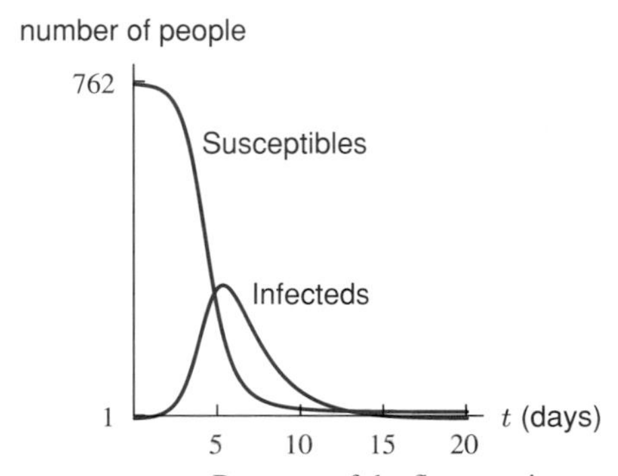

Figure 9.53: Progress of the flu over time.

[14]Here we are using J. D. Murray's definition of an epidemic as an outbreak in which the number of infecteds increases from the initial value, I_0. See *Mathematical Biology* (New York: Springer Verlag, 1990)

Analytical Solution for the SI-Phase Trajectory

The differential equation

$$\frac{dI}{dS} = -1 + \frac{192}{S}$$

can be integrated directly giving

$$I = -S + 192 \ln S + C$$

Using $S_0 = 762$, $I_0 = 1$, gives $1 = -762 + 192 \ln 762 + C$, so we get $C = 763 - 192 \ln(762)$. Putting this value for C into the equation for I, we get:

$$I = -S + 192 \ln S - 192 \ln(762) + 763$$

$$I = -S + 192 \ln\left(\frac{S}{762}\right) + 763.$$

This is the equation of the solution curve in Figure 9.52.

Two Interacting Populations: Predator-Prey

We now consider two populations which interact. For example, they may compete for food, one may prey on the other, or they may enjoy a symbiotic relationship in which each helps the other.

As an example, we will model a predator-prey system using what are called the Lotka-Volterra equations. One of the predator-prey systems about which we have long-term data is that of the Canadian lynx and the hare. Since both animals were of interest to fur trappers, records of the Hudson Bay Company throw some light on their populations through much of the last century. These records show that both populations oscillate up and down, quite regularly, with a period of about ten years. In addition, the maximum values of the two populations did not occur at the same time; the hare was a quarter of a cycle ahead of the lynx. You will see that this behavior is predicted by Lotka-Volterra equations.

Robins and Worms

Let's look at a simplified and idealized case in which robins are the predators and worms the prey. Suppose there are r thousand robins and w million worms. If there were no robins, the worms would increase exponentially according to the equation

$$\frac{dw}{dt} = aw \quad \text{where } a \text{ is a constant and } a > 0.$$

We will assume that if the robins were alone, and so had no worms for food, they would decrease according to the equation[15]

$$\frac{dr}{dt} = -br \quad \text{where } b \text{ is a constant and } b > 0.$$

Now we model for the effect of the two populations on one another. Clearly, the presence of the robins is bad for the worms, so

$$\frac{dw}{dt} = aw - \text{Effect of robins on worms.}$$

[15] You might criticize this assumption because it predicts that the number of robins will decay exponentially, rather than die out in finite time.

On the other hand, the robins do better with the worms around, so

$$\frac{dr}{dt} = -br + \text{Effect of worms on robins.}$$

How exactly do the two populations interact? Let's assume the effect of one population on the other will be proportional to the number of "encounters." (An encounter is when a robin meets and eats a worm.) The number of encounters is likely to be proportional to the product of the populations because the more there are of either population, the more encounters there will be. So we will assume

$$\frac{dw}{dt} = aw - cwr \quad \text{and} \quad \frac{dr}{dt} = -br + kwr,$$

where c and k are positive constants.

To analyze this system of equations, let's look at the specific example when $a = b = c = k = 1$:

$$\frac{dw}{dt} = w - wr \quad \text{and} \quad \frac{dr}{dt} = -r + wr.$$

To visualize the solutions to these equations, we look for trajectories in the phase plane. First we use the chain rule,

$$\frac{dr}{dw} = \frac{dr/dt}{dw/dt},$$

to obtain

$$\frac{dr}{dw} = \frac{-r + wr}{w - wr}.$$

The Slope Field and Equilibrium Points

You can get an idea of what solutions of this equation look like from the slope field in Figure 9.54. Probably the first thing that strikes you is that something special is going on at the point $(1, 1)$. You will also see that the solution curves appear to be closed curves centered at $(1, 1)$.

Let's look at both of these observations in more detail. At the point $(1, 1)$ there is no slope drawn because at this point the rate of change of the worm population with respect to time is zero:

$$\frac{dw}{dt} = 1 - (1)(1) = 0.$$

The rate of change of the robin population with respect to time is also zero:

$$\frac{dr}{dt} = -1 + (1)(1) = 0.$$

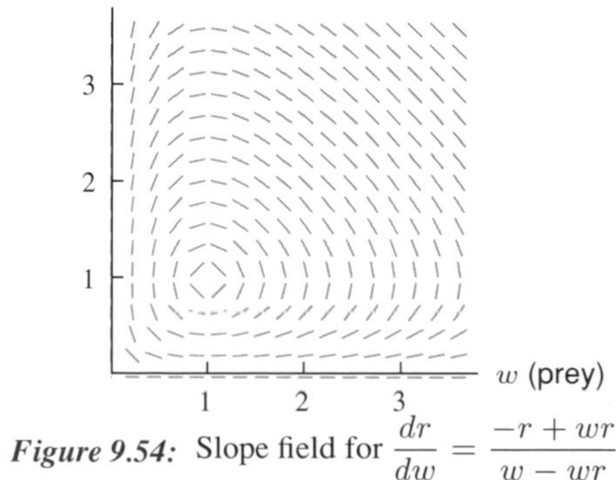

Figure 9.54: Slope field for $\dfrac{dr}{dw} = \dfrac{-r + wr}{w - wr}$

Thus dr/dw is undefined. In terms of worms and robins, this means that if at some moment $w = 1$ and $r = 1$ (i.e., there are 1 million worms and 1 thousand robins), then w and r remain constant forever. The point $w = 1$, $r = 1$ is therefore an equilibrium solution. The slope field suggests that there are no other equilibrium points except the origin.

> At an **equilibrium point**, both w and r are constant, so
> $$\frac{dw}{dt} = 0 \quad \text{and} \quad \frac{dr}{dt} = 0.$$

Therefore, we look for equilibrium points by solving

$$\frac{dw}{dt} = w - wr = 0 \quad \text{and} \quad \frac{dr}{dt} = -r + rw = 0,$$

which has $w = 0$, $r = 0$ and $w = 1$, $r = 1$ as the only solutions.

Trajectories in the *wr*-Phase Plane

Let's look at the trajectories in the phase plane. Remember that a point on one of the curves represents a pair of populations (w, r) which exist at the same time t (though t is not shown on the graph). At a short time later, the pair of populations will be represented by a nearby point. Thus as time passes, the point traces out the curve. The direction is marked on the curve by an arrow. (See Figure 9.55.)

How do we figure out which way to move on the trajectory? Approximating the solution numerically shows that the trajectory is traversed counterclockwise. Alternatively, look at the original pair of differential equations. Because they involve time, they tell us how w and r change as time passes. Imagine, for example, that you are at the point P_0 in Figure 9.56, where $w > 1$ and $r = 1$; then

$$\frac{dr}{dt} = -r + wr = -1 + w > 0.$$

Therefore, r is increasing, so the point is moving counterclockwise around the closed curve.

Now let's think about whether the solution curves are really closed curves (i.e., do they come back and meet themselves?). To see why the solution curves are closed, notice that the slope field is symmetric about the line $w = r$. You can confirm this by observing that interchanging w and r does not alter the differential equation for dr/dw. This means that if you start at point P on the line $w = r$ and travel once around the point $(1, 1)$, you have to arrive back at the same point P. The reason is that the second half of your path is the reflection of the first half in the line $w = r$. (See Figure 9.55.) If you did not end up at P again, the second half of your path would have had a different shape from the first half.

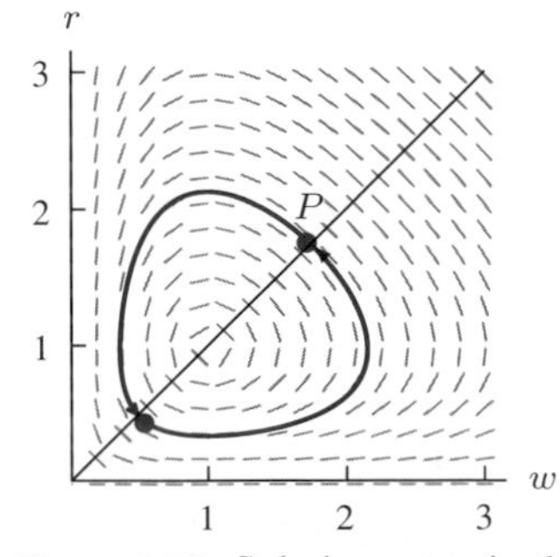

Figure 9.55: Solution curve is closed

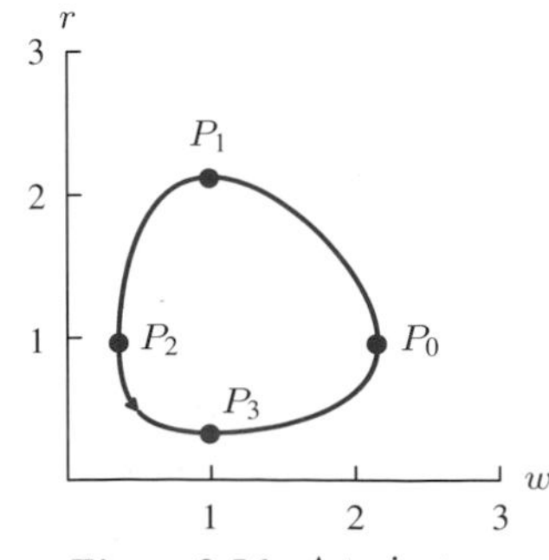

Figure 9.56: A trajectory

The Populations as Functions of Time

We will now see what the shape of trajectories tell us about how the populations vary with time. Suppose we start at $t = 0$ at the point P_0 in Figure 9.56. Then suppose we move to P_1 at time t_1, to P_2 at time t_2, to P_3 at time t_3, and so on. At time t_4 we will be back at P_0, and the whole cycle will repeat. Thus *the fact that the trajectory is a closed curve means that both populations will oscillate up and down through the same values.* Both populations oscillate with the same period, and the worms (the prey) are at their maximum a quarter of a cycle before the robins. (See Figure 9.57.)

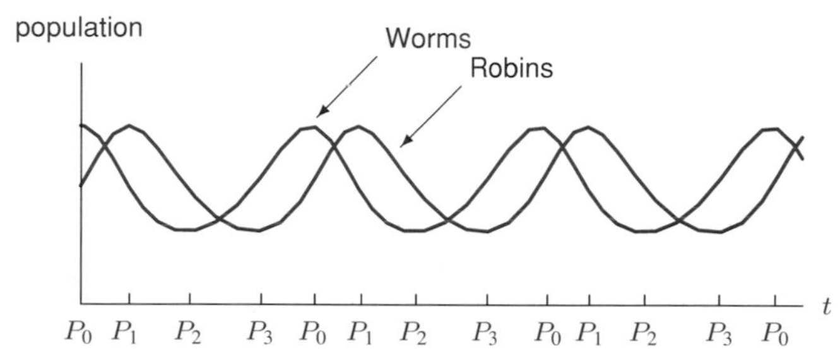

Figure 9.57: Populations of robins (in thousands) and worms (in millions) over time

Problems for Section 9.8

1. Show that if S, I, and R satisfy the differential equations on page 542, the total population, $S + I + R$, is a constant.

For Problems 2–5, suppose x and y are the populations of two different species. Describe in words how each population changes with time.

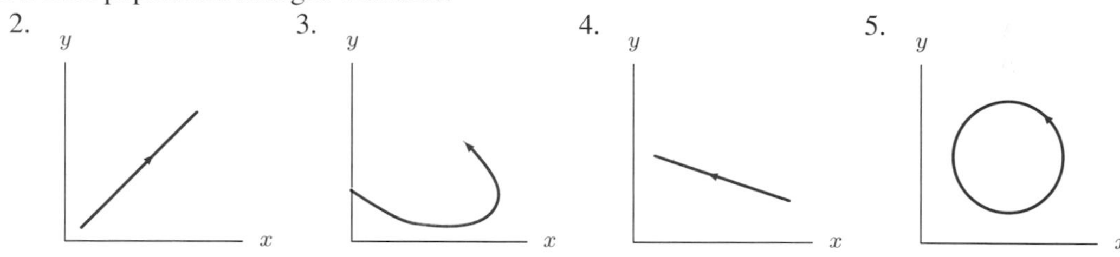

For Problems 6–12 let w be the number of worms (in millions) and r the number of robins (in thousands) living on an island. Suppose w and r satisfy the differential equations

$$\frac{dw}{dt} = w - wr$$
$$\frac{dr}{dt} = -r + wr.$$

6. Explain why this might be a reasonable way to model the interaction between the two populations. Why have the signs been chosen this way?
7. Solve these equations in the two special cases when there are no robins and when there are no worms living on the island.

For Problems 8–12, you may use the slope field in Figure 9.58.

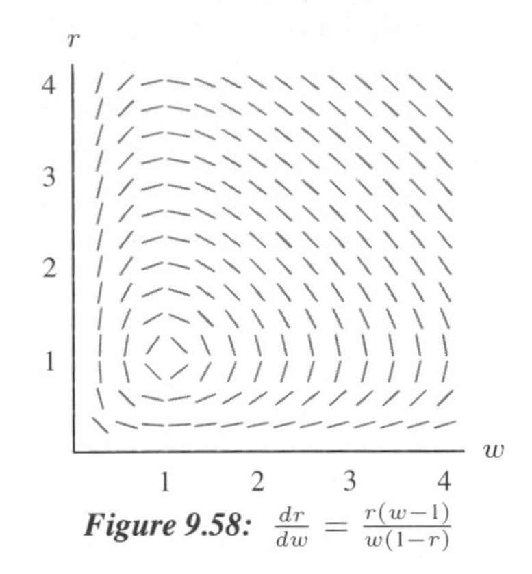

Figure 9.58: $\frac{dr}{dw} = \frac{r(w-1)}{w(1-r)}$

8. Describe and explain the symmetry you observe in the slope field. What consequences does this symmetry have for the solution curves?

9. Assume $w = 2$ and $r = 2$ when $t = 0$. Do the numbers of robins and worms increase or decrease at first? What happens in the long run?

10. For the case discussed in Problem 9, estimate the maximum and the minimum values of the robin population. How many worms are there at the time when the robin population reaches its maximum?

11. On the same axes, graph w and r (the worm and the robin populations) against time. Use initial values of 1.5 for w and 1 for r. You may do this without units for t.

12. People on our island like robins so much that they decide to import 200 robins all the way from England, to increase the initial population to $r = 2.2$ when $t = 0$. Does this make sense? Why or why not?

For Problems 13–16, consider a conflict between two armies of x and y soldiers, respectively. During World War I, F. W. Lanchester assumed that if both armies are fighting a conventional battle within sight of one another, the rate at which soldiers in one army are put out of action (killed or wounded) is proportional to the amount of fire the other army can concentrate on them, which is in turn proportional to the number of soldiers in the opposing army. Thus Lanchester assumed that if there are no reinforcements and t represents time since the start of the battle, then x and y obey the differential equations

$$\frac{dx}{dt} = -ay$$
$$\frac{dy}{dt} = -bx \quad a,\, b > 0.$$

13. Near the end of World War II a fierce battle took place between US and Japanese troops over the island of Iwo Jima, off the coast of Japan. Applying Lanchester's analysis to this battle, with x representing the number of US troops and y the number of Japanese troops, it has been estimated[16] that $a = 0.05$ and $b = 0.01$.

    (a) Using these values for a and b, and ignoring reinforcements, write a differential equation

[16]See Martin Braun, *Differential Equations and their Applications*, 2nd ed. (New York: Springer and Verlag, 1975)

involving dy/dx and sketch its slope field.

(b) Assuming that the initial strength of the US forces was 54,000 and that of the Japanese was 21,500, draw the trajectory which describes the battle. What outcome is predicted? (That is, which side do the differential equations predict will win?)

(c) Would knowing that the US in fact had 19,000 reinforcements, while the Japanese had none, alter the outcome predicted?

14. (a) For two armies of strengths x and y fighting a conventional battle governed by Lanchester's differential equations, write a differential equation involving dy/dx and the constants of attrition a and b.

(b) Solve the differential equation and hence show that the equation of the phase trajectory is

$$ay^2 - bx^2 = C.$$

for some constant C. This equation is called *Lanchester's square law*. The value of C depends on the initial sizes of the two armies.

15. Consider the battle of Iwo Jima, described in Problem 13. Take $a = 0.05$, $b = 0.01$ and assume the initial strength of the US troops to be 54,000 and that of the Japanese troops to be 21,500. (Again, ignore reinforcements.)

(a) Using Lanchester's square law derived in Problem 14, find the equation of the trajectory describing the battle.

(b) Assuming that the Japanese fought without surrendering until they had all been killed, as was the case, how many US troops does this model predict would be left when the battle ended?

16. We will apply Lanchester's model to the Battle of Trafalgar (1805), when a fleet of 40 British ships expected to face a combined French and Spanish fleet of 46 ships. Suppose that there were x British ships and y opposing ships at time t. We will assume that the ships are all identical so that constants in the differential equations in Lanchester's model are equal:

$$\frac{dx}{dt} = -ay$$
$$\frac{dy}{dt} = -ax.$$

(a) Write a differential equation involving dy/dx, and solve it using the initial sizes of the two fleets.

(b) If the battle were fought until all the British ships were put out of action, how many French/Spanish ships does this model predict would be left at the end of the battle?

Admiral Nelson, who was in command of the British fleet, did not in fact send his 40 ships against the 46 French and Spanish ships. Instead he split the battle into two parts, sending 32 of his ships against 23 of the French/Spanish ships and his other 8 ships against their other 23.

(c) Analyze each of these two sub-battles using Lanchester's model. Find the solution trajectory for each sub-battle. Which side is predicted to win each one? How many ships from each fleet are expected to be left at the end?

(d) Suppose, as in fact happened, that the remaining ships from each sub-battle then fought each other. Which side is predicted to win with how many ships remaining?

17. Two companies share the market for a new technology. They have no competition except each other. Let $A(t)$ be the net worth of one company and $B(t)$ the net worth of the other at time t. Assume that net worth cannot be negative, because a company goes out of business if its net

worth is zero. Suppose A and B satisfy the differential equations

$$A' = 2A - AB$$
$$B' = B - AB.$$

(a) What do these equations predict about the net worth of each company if the other were not present? What effect do the companies have on each other?
(b) Are there any equilibrium points? If so, what are they?
(c) Sketch a slope field for these equations (using a computer or calculator), and hence describe the different possible long-run behaviors.

The differential equations in Problems 18–20 describe the rates of growth of two populations x and y (both measured in thousands) of species A and B, respectively. For each set:
(a) Describe in words what happens to the population of each species in the absence of the other.
(b) Describe in words how the species interact with one another. Give reasons why the populations might behave as described by the equations. Suggest species that might interact as described by the equations.

18. $$\frac{dx}{dt} = 0.01x - 0.05xy$$
$$\frac{dy}{dt} = -0.2y + 0.08xy$$

19. $$\frac{dx}{dt} = 0.01x - 0.05xy$$
$$\frac{dy}{dt} = 0.2y - 0.08xy$$

20. $$\frac{dx}{dt} = 0.2x$$
$$\frac{dy}{dt} = 0.4xy - 0.1y$$

21. (a) Find the equilibrium points for the following system of equations:

$$\frac{dx}{dt} = 15x - 3xy$$
$$\frac{dy}{dt} = -14y + 7xy.$$

(b) Explain why $x = 2$, $y = 0$ is not an equilibrium point for this system.

22. (a) Find the equilibrium points for the following system of equations

$$\frac{dx}{dt} = x - 0.001x^2 - 0.005xy,$$
$$\frac{dy}{dt} = 0.02y - 3\frac{y}{x}.$$

(b) Explain why $x = 0$, $y = 0$ is not an equilibrium point for this system.

The systems of differential equations in Problems 23–25 model the interaction of two populations x and y. In each case, answer the following two questions:
(a) What kinds of interaction (symbiosis,[17] competition, predator-prey) do the equations describe?
(b) What happens in the long run? (For one of the systems, your answer will depend on the initial populations.) Use a calculator or computer to draw slope fields.

23. $$\frac{1}{x}\frac{dx}{dt} = y - 1$$
$$\frac{1}{y}\frac{dy}{dt} = x - 1$$

24. $$\frac{1}{x}\frac{dx}{dt} = 1 - \frac{x}{2} - \frac{y}{2}$$
$$\frac{1}{y}\frac{dy}{dt} = 1 - x - y$$

25. $$\frac{1}{x}\frac{dx}{dt} = y - 1 - 0.05x$$
$$\frac{1}{y}\frac{dy}{dt} = 1 - x - 0.05y$$

[17] Symbiosis takes place when the interaction of two species benefits both. An example is the pollination of plants by insects.

9.9 ANALYZING THE PHASE PLANE

In the previous section we analyzed a system of differential equations using a slope field. In this section we will again analyze a system of differential equations, using curves called *nullclines*. The problem we will consider is of two similar species in competition for food and space. In practice, one of the two species usually becomes extinct, and the other takes over completely. Biologists say that if two species occupy similar *niches*, or ways of living, then one will drive the other to extinction. This phenomenon is called the *Principle of Competitive Exclusion*. We will see how differential equations predict this in a particular case.

Competitive Exclusion: Citrus Tree Parasites

The citrus farmers of Southern California are interested in controlling the insects that live on their trees. Some of these insects can be controlled by parasites which live on the trees too. Scientists are therefore interested in understanding under what circumstances these parasites flourish or die out. One such parasite was introduced accidentally from the Mediterranean; later, other parasites were introduced from China and India; in each case the previous parasite became extinct over part of its habitat. In 1963 a lab experiment was carried out to determine which one of a pair of species became extinct when they were in competition with each other. The data on one pair of species, called *A. fisheri* and *A. melinus*, with populations P_1 and P_2 respectively, is given in Table 9.14 and shows that *A. melinus* (P_2) became extinct after 8 generations.[18]

TABLE 9.14 *Population (in thousands) of two species of parasite as a function of time*

Generation number	1	2	3	4	5	6	7	8
Population P_1 (thousands)	0.193	1.093	1.834	5.819	13.705	16.965	18.381	16.234
Population P_2 (thousands)	0.083	0.229	0.282	0.378	0.737	0.507	0.13	0

Data from the same experimenters indicates that, when alone, each population grows logistically. In fact, their data suggests that, when alone, the population of P_1 might grow according to the equation

$$\frac{dP_1}{dt} = 0.05P_1\left(1 - \frac{P_1}{20}\right),$$

and when alone, the population of P_2 might grow according to the equation

$$\frac{dP_2}{dt} = 0.09P_2\left(1 - \frac{P_2}{15}\right).$$

Now suppose both parasites are present. Each tends to reduce the growth rate of the other, so each differential equation is modified by subtracting a term on the right. The experimental data shows that together P_1 and P_2 can be well described by the equations

$$\frac{dP_1}{dt} = 0.05P_1\left(1 - \frac{P_1}{20}\right) - 0.002P_1P_2$$

$$\frac{dP_2}{dt} = 0.09P_2\left(1 - \frac{P_2}{15}\right) - 0.15P_1P_2.$$

The fact that P_2 dies out with time is reflected in these equations: the coefficient of P_1P_2 is so much larger in the equation for P_2 than in the equation for P_1. This indicates that the interaction has a much more devastating effect upon the growth of P_2 than on the growth of P_1.

[18]Data adapted from Paul DeBach and Ragnhild Sundby, "Competitive Displacement between Ecological Homologues," *Hilgardia 34*:17 (1963).

The Phase Plane and Nullclines

We consider the phase plane with the P_1 axis horizontal and the P_2 axis vertical. To find the trajectories in the P_1P_2 phase plane, we could draw a slope field as in the previous section. Instead, we will use a method which is frequently used to give a good qualitative picture of the behavior of the trajectories even without a calculator or computer. We will find the *nullclines* or curves along which $dP_1/dt = 0$ or $dP_2/dt = 0$. At points where $dP_2/dt = 0$, the population P_2 is momentarily constant, so only population P_1 is changing with time. Therefore, at this point the trajectory is horizontal. (See Figure 9.59.) Similarly, at points where $dP_1/dt = 0$, the population P_1 is momentarily constant and population P_2 is the only one changing, so the trajectory is vertical there. A point where both $dP_1/dt = 0$ and $dP_2/dt = 0$ is called an *equilibrium point* because P_1 and P_2 both remain constant if they reach these values.

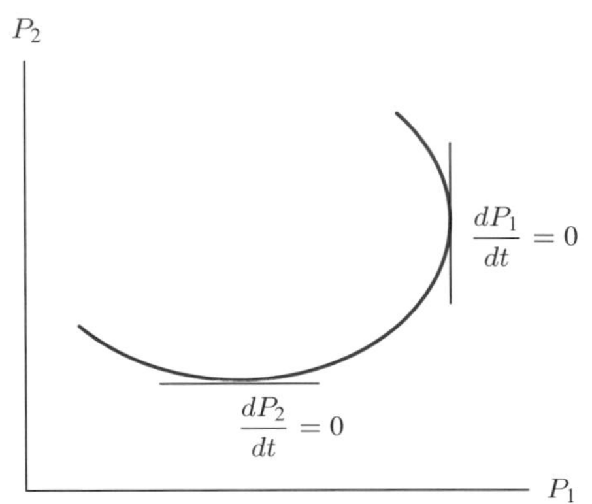

Figure 9.59: What happens to a trajectory at points where $dP_1/dt = 0$ or $dP_2/dt = 0$

On the P_1P_2 phase plane:

- If $\dfrac{dP_1}{dt} = 0$, the trajectory is vertical
- If $\dfrac{dP_2}{dt} = 0$, the trajectory is horizontal
- If $\dfrac{dP_1}{dt} = \dfrac{dP_2}{dt} = 0$, there is an equilibrium point

Using Nullclines to Analyze the Parasite Populations

In order to see where $dP_1/dt = 0$ or $dP_2/dt = 0$, rewrite our differential equations with the right side factored:

$$\frac{dP_1}{dt} = 0.05P_1\left(1 - \frac{P_1}{20}\right) - 0.002P_1P_2 = 0.001P_1(50 - 2.5P_1 - 2P_2)$$

$$\frac{dP_2}{dt} = 0.09P_2\left(1 - \frac{P_2}{15}\right) - 0.15P_1P_2 = 0.001P_2(90 - 150P_1 - 6P_2).$$

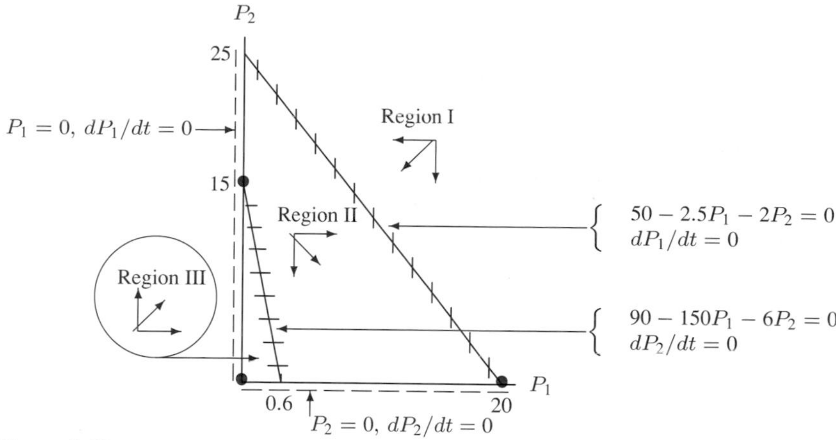

Figure 9.60: Analyzing three regions in the phase plane using nullclines (axes distorted) with equilibrium points represented by dots

Thus $dP_1/dt = 0$ where $P_1 = 0$ or where $50 - 2.5P_1 - 2P_2 = 0$. Graphing these equations in the phase plane gives two lines, which are nullclines. Since the trajectory is vertical where $dP_1/dt = 0$, in Figure 9.60 we draw small vertical line segments on these nullclines to represent the direction of the trajectories as they cross the nullcline. Similarly $dP_2/dt = 0$ where $P_2 = 0$ or where $90 - 150P_1 - 6P_2 = 0$. These equations are graphed in Figure 9.60 with small horizontal line segments on them. The equilibrium points are where both $dP_1/dt = 0$ and $dP_2/dt = 0$, namely the points $P_1 = 0, P_2 = 0$ (meaning that both species die out), $P_1 = 0, P_2 = 15$ (where P_1 is extinct), and $P_1 = 20, P_2 = 0$ (where P_2 is extinct).

What Happens in the Regions Between the Nullclines?

Nullclines are useful because they divide the plane into regions in which the signs of dP_1/dt and dP_2/dt remain the same. Thus, in each region, the direction of every trajectory remains roughly the same.

In region I, for example, we might try the point $P_1 = 20, P_2 = 25$. Then

$$\frac{dP_1}{dt} = 0.001(20)(50 - 2.5(20) - 2(25)) < 0$$

$$\frac{dP_2}{dt} = 0.001(25)(90 - 150(20) - 6(25)) < 0$$

Now $dP_1/dt < 0$, so P_1 is decreasing, which can be represented by an arrow in the direction $\leftarrow$. Also, $dP_2/dt < 0$, so P_2 is decreasing, as represented by the arrow $\downarrow$. Combining these directions, we know that the trajectories in this region go approximately in the diagonal direction shown: $\swarrow$ (See region I in Figure 9.60.)

In region II, try, for example, $P_1 = 1, P_2 = 1$. Then we have

$$\frac{dP_1}{dt} = 0.001(1)(50 - 2.5 - 2) > 0$$

$$\frac{dP_2}{dt} = 0.001(1)(90 - 150 - 6) < 0$$

So here, P_1 is increasing while P_2 is decreasing. (See region II in Figure 9.60.)

In region III, try $P_1 = 0.1, P_2 = 0.1$:

$$\frac{dP_1}{dt} = 0.001(0.1)(50 - 2.5(0.1) - 2(0.1)) > 0$$
$$\frac{dP_2}{dt} = 0.001(0.1)(90 - 150(0.1) - 6(0.1)) > 0$$

So here, both P_1 and P_2 are increasing. (See region III in Figure 9.60.)

Notice that the behavior of the populations in each region makes good biological sense. In region I both populations are so large that overpopulation is a problem, so both populations decrease. In region III both populations are so small that they are effectively not in competition, so both grow. In region II competition between the species does come into play. The fact that P_1 increases while P_2 decreases in region II means that P_1 wins.

Solution Trajectories

Suppose the system starts with some of each population. This means that the initial point of the trajectory is not on one of the axes, and so it is in region I, II, or III. Then the point moves on a trajectory like one of those computed numerically and shown in Figure 9.61. Notice that *all* these trajectories tend toward the point $P_1 = 20, P_2 = 0$, corresponding to a population of 20,000 for P_1 and extinction for P_2. Consequently, this model predicts that no matter what the initial populations are, provided $P_1 \neq 0$, the population of P_2 will be excluded by P_1, and P_1 will tend to a constant value. This makes biological sense: in the absence of P_2, we would expect P_1 to settle down to the carrying capacity of the niche, which is 20,000.

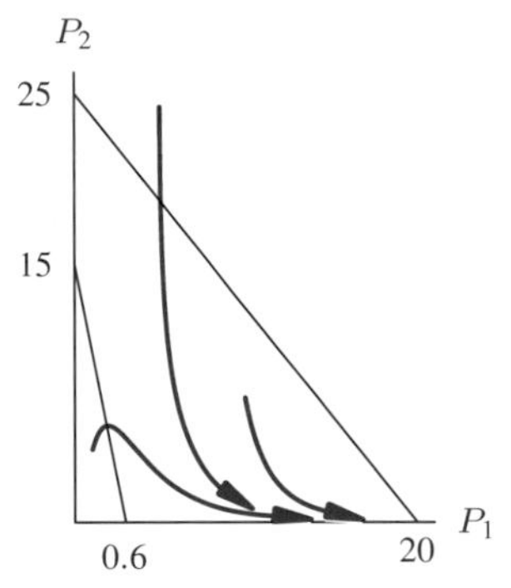

Figure 9.61: Trajectories showing exclusion of population P_2 (not to scale)

Problems for Section 9.9

1. The equations describing the flu epidemic in a boarding school are

$$\frac{dS}{dt} = -0.0026SI$$
$$\frac{dI}{dt} = 0.0026SI - 0.5I.$$

(a) Find the nullclines and equilibrium points in the SI phase plane.
(b) Find the direction of the trajectories in each region.
(c) Sketch some typical trajectories and describe their behavior in words.

2. Use the idea of nullclines dividing the plane into sectors to analyze the equations describing the interactions of robins and worms.

$$\frac{dw}{dt} = w - wr$$
$$\frac{dr}{dt} = -r + rw.$$

For Problems 3–5 analyze the phase plane of the differential equations for $x,\, y \geq 0$. Show the nullclines and equilibrium points, and sketch the direction of the trajectories in each region.

3. $$\frac{dx}{dt} = x(2 - x - y) \qquad \frac{dy}{dt} = y(1 - x - y)$$

4. $$\frac{dx}{dt} = x\left(1 - x - \frac{y}{3}\right) \qquad \frac{dy}{dt} = y\left(1 - y - \frac{x}{2}\right)$$

5. $$\frac{dx}{dt} = x\left(1 - \frac{x}{2} - y\right) \qquad \frac{dy}{dt} = y\left(1 - \frac{y}{3} - x\right)$$

In their study of parasites on California citrus trees, scientists studied three different species with populations P_1, P_2, and P_3 (in thousands). In the text (page 553) we analyzed the interaction between P_1 and P_2. Problems 6 and 7 below concern the competition between P_2 and P_3, and P_1 and P_3, respectively.

6. Scientists found that the populations P_2 and P_3 satisfied the differential equations

$$\frac{dP_2}{dt} = 0.09P_2\left(1 - \frac{P_2}{15}\right) - 0.45P_2P_3$$
$$\frac{dP_3}{dt} = 0.06P_3\left(1 - \frac{P_3}{10}\right) - 0.001P_2P_3.$$

(a) Sketch the phase plane showing nullclines and equilibrium points.
(b) Find the direction of the trajectories in each of the regions.
(c) Sketch some trajectories. What happens in the long run? Which parasite, if either, becomes extinct?

7. The populations P_1 and P_3 obey the equations

$$\frac{dP_1}{dt} = 0.05P_1\left(1 - \frac{P_1}{20}\right) - 0.5P_1P_3$$
$$\frac{dP_3}{dt} = 0.06P_3\left(1 - \frac{P_3}{10}\right) - 0.01P_1P_3.$$

(a) Analyze the phase plane, showing nullclines and equilibrium points, and find the direction of the trajectories in each region.
(b) What happens in the long run? Which parasite, if either, becomes extinct?

8. Naturalists have often noticed that when two closely related species invade an area, often the species that arrives earlier excludes the other. Thus either one of the two species can end up excluding the other, depending on which gets established first. By doing a phase plane analysis, decide which one of the following pairs of equations might describe this situation.

(a) $$\frac{dP_1}{dt} = 0.1P_1\left(1 - \frac{P_1}{16}\right) - 0.013P_1P_2$$
$$\frac{dP_2}{dt} = 0.2P_2\left(1 - \frac{P_2}{20}\right) - 0.05P_1P_2$$

(b) $$\frac{dP_1}{dt} = 0.5P_1\left(1 - \frac{P_1}{4}\right) - 0.025P_1P_2$$
$$\frac{dP_2}{dt} = 0.1P_2\left(1 - \frac{P_2}{8}\right) - 0.006P_1P_2$$

(c) $$\frac{dP_1}{dt} = 0.06P_1\left(1 - \frac{P_1}{16}\right) - 0.008P_1P_2$$
$$\frac{dP_2}{dt} = 0.12P_2\left(1 - \frac{P_2}{12}\right) - 0.007P_1P_2$$

9. In the 1930s, the Soviet ecologist G. F. Gause performed a series of experiments on competition among two yeasts with populations P_1 and P_2, respectively. By performing population studies at low density in large volumes he determined what he called the *coefficients of geometric increase* (and we would call continuous exponential growth rates). These coefficients described the growth of each yeast alone:

$$\frac{1}{P_1}\frac{dP_1}{dt} = 0.2$$
$$\frac{1}{P_2}\frac{dP_2}{dt} = 0.06$$

where P_1 and P_2 are measured in units that Gause established.

He also determined that, in his units, the carrying capacity of P_1 was 13 and the carrying capacity of P_2 was 6. He then observed that one P_2 occupies the niche space of 3 P_1 and that one P_1 occupied the niche space of 0.4 P_2. This led him to the following differential equations to describe the interaction of P_1 and P_2:

$$\frac{dP_1}{dt} = 0.2P_1\left(\frac{13 - (P_1 + 3P_2)}{13}\right)$$
$$\frac{dP_2}{dt} = 0.06P_2\left(\frac{6 - (P_2 + 0.4P_1)}{6}\right)$$

When both yeasts were growing together, Gause recorded the data in Table 9.15.

TABLE 9.15 *Gause's yeast populations*

Time (hours)	6	16	24	29	48	53
P_1	0.375	3.99	4.69	6.15	7.27	8.30
P_2	0.29	0.98	1.47	1.46	1.71	1.84

(a) Carry out a phase plane analysis of Gause's equations.
(b) Mark the data points on the phase plane and describe what would have happened had Gause continued the experiment.

10. In the 1930's L. F. Richardson proposed that an arms race between two countries could be modeled by a system of differential equations. One arms race which can be reasonably well described by differential equations is the US-Soviet Union arms race between 1945 and 1960. If $\$x$ represents the annual Soviet expenditures on armaments (in billions of dollars) and $\$y$ represents the corresponding US expenditures, it has been suggested[19] that x and y obey the following differential equations

$$\frac{dx}{dt} = -0.45x + 10.5,$$
$$\frac{dy}{dt} = 8.2x - 0.8y - 142.$$

[19] R. Taagepera, G. M. Schiffler, R. T. Perkins and D. L. Wagner *Soviet-American and Israeli-Arab Arms Races and the Richardson Model* (General Systems, XX, 1975)

(a) Find the nullclines and equilibrium points for these differential equations. Which direction do the trajectories go in each region?
(b) Sketch some typical trajectories in the phase plane.
(c) What do these differential equations predict will be the long term outcome of the US-Soviet arms race?
(d) Discuss these predictions in the light of the actual expenditures in Table 9.16.

TABLE 9.16 *Arms budgets of the United States and the Soviet Union for the years 1945–1960 (billions of dollars)*

	USSR	USA		USSR	USA
1945	14	97	1953	25.7	71.4
1946	14	80	1954	23.9	61.6
1947	15	29	1955	25.5	58.3
1948	20	20	1956	23.2	59.4
1949	20	22	1957	23.0	61.4
1950	21	23	1958	22.3	61.4
1951	22.7	49.6	1959	22.3	61.7
1952	26.0	69.6	1960	22.1	59.6

9.10 SECOND-ORDER DIFFERENTIAL EQUATIONS: OSCILLATIONS

A Second-Order Differential Equation

When a body moves freely under gravity, we know that

$$\frac{d^2s}{dt^2} = -g,$$

where s is the height of the body above ground at time t and g is the acceleration due to gravity. To solve this equation, we first integrate to get the velocity, $v = ds/dt$:

$$\frac{ds}{dt} = -gt + v_0,$$

where v_0 is the initial velocity. Then we integrate again, giving

$$s = -\frac{1}{2}gt^2 + v_0t + s_0,$$

where s_0 is the initial height.

Recall that the differential equation $d^2s/dt^2 = -g$ is called *second order* because the equation contains a second derivative but no higher derivatives. Since getting the solution involves integrating twice, the solution involves *two* arbitrary constants, here v_0 and s_0. Thus, the general solution to a second-order differential equation will be a family of functions with two parameters. Finding values for the two constants corresponds to picking a particular function out of this family.

A Mass on a Spring

Not every second-order differential equation can be solved simply by integrating twice. Consider the following situation: A mass m is attached to the end of a spring hanging from the ceiling. (We will assume that the mass of the spring itself is negligible in comparison with the mass m.) (See Figure 9.62.)

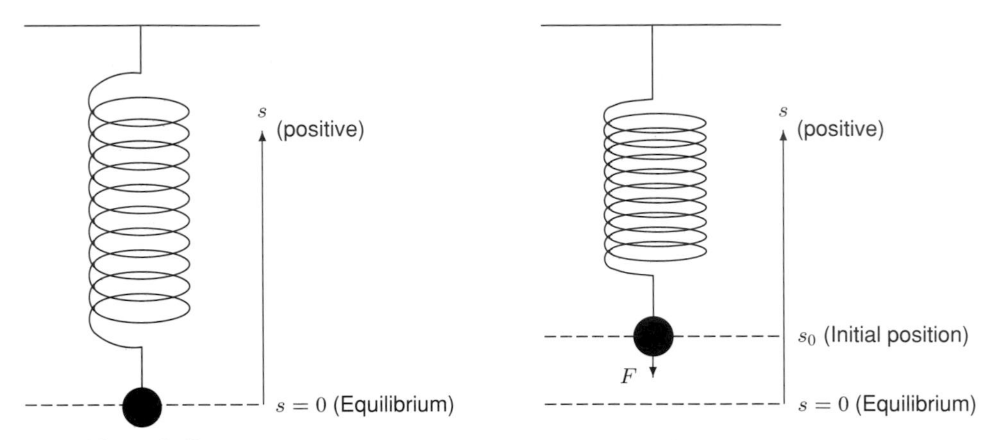

Figure 9.62: Spring and mass in equilibrium position and after upward displacement

When the system is left undisturbed, no net force acts on the mass. The force of gravity is balanced by the force the spring exerts on the mass, and the spring is said to be in the *equilibrium position.* If you pull down on the mass, you will feel a net force pulling upwards. If instead, you push upward on the mass, the opposite happens: the net force pushes the mass down.[20]

What happens if you push the mass upward and then release it? You'd expect the mass to oscillate up and down around the equilibrium position.

Springs and Hooke's Law

In order to figure out how the mass moves, we need to know the exact relationship between its displacement, s, from the equilibrium position and the net force, F, exerted on the mass. (See Figure 9.62.) We would expect that the further the mass is from the equilibrium position, and the more the spring is stretched or compressed, the larger the force. In fact, provided that the displacement is not large enough to deform the spring permanently, experiments show that the net force, F, is approximately proportional to the displacement, s:

$$F = -ks,$$

where k is the *spring constant* ($k > 0$) and the negative sign means that the net force is in the opposite direction to the displacement. The value of k depends on the physical properties of the particular spring. This relationship is known as *Hooke's Law.* Suppose you push the mass upward some distance and then release it. After you let go the net force causes the mass to accelerate toward the equilibrium position. By Newton's Second Law of Motion we have

$$\text{Force} = \text{Mass} \times \text{Acceleration}.$$

[20]Pulling down on the mass stretches the spring, increasing the tension, so the net force is upward. Pushing up the spring decreases tension in the spring, so the net force is downward.

Since acceleration is d^2s/dt^2, using Hooke's law that $F = -ks$, we have

$$-ks = m\frac{d^2s}{dt^2},$$

which is equivalent to the following differential equation.

Equation for Oscillations of a Mass on a Spring

$$\frac{d^2s}{dt^2} = -\frac{k}{m}s$$

Thus the motion of the mass is described by a second-order differential equation. Since we expect the mass to oscillate, we might guess that the solution to this equation will involve the trigonometric functions – and this turns out to be true.

Solving the Differential Equation by Guess-and-Check

Guessing is the tried-and-true method for solving differential equations. It may surprise you to learn that there is no systematic method for solving most differential equations analytically, so guesswork is often extremely important. Before looking for a general solution to the equation

$$\frac{d^2s}{dt^2} = -\frac{k}{m}s,$$

we'll consider the special case $k/m = 1$. In Example 1, you will see how to guess specific solutions, check that they satisfy the equation, and then build the general solution.

Example 1 Find the general solution to the equation

$$\frac{d^2s}{dt^2} = -s.$$

Solution We want to find functions whose second derivative is the negative of the original function. We are already familiar with two functions that have this property: $s(t) = \cos t$ and $s(t) = \sin t$. We check that they are solutions by substituting:

$$\frac{d^2}{dt^2}(\cos t) = \frac{d}{dt}(-\sin t) = -\cos t,$$

and

$$\frac{d^2}{dt^2}(\sin t) = \frac{d}{dt}(\cos t) = -\sin t.$$

The remarkable thing is that starting from these two particular solutions, we can build up *all* the solutions to our equation. Here's how: If C is a constant, then $C\sin t$ and $C\cos t$ are also solutions to the differential equation. (Why?) Also, $\sin t + \cos t$ is a solution to the differential equation. (Why?) In fact, given two constants C_1 and C_2, the function

$$s(t) = C_1\cos t + C_2\sin t$$

satisfies the differential equation, since

$$\begin{aligned}\frac{d^2}{dt^2}(C_1 \cos t + C_2 \sin t) &= \frac{d}{dt}(-C_1 \sin t + C_2 \cos t)\\ &= -C_1 \cos t - C_2 \sin t\\ &= -(C_1 \cos t + C_2 \sin t).\end{aligned}$$

It can be shown (though we will not do it) that $s(t) = C_1 \cos t + C_2 \sin t$ is the most general form of the solution. As expected, it contains two constants, C_1 and C_2.

If the differential equation represents a physical problem, then C_1 and C_2 are often determined by certain conditions of that physical problem. For example if the mass is pushed up to an initial displacement of s_0 and then released, then C_1 and C_2 can be computed, as shown in the next example.

Example 2 Find the solution to

$$\frac{d^2 s}{dt^2} = -s$$

if the mass were displaced by a distance of s_0 and then released.

Solution The position of the mass is given by the equation

$$s(t) = C_1 \cos t + C_2 \sin t.$$

We also know that the initial position is s_0; thus,

$$s(0) = C_1 \cos 0 + C_2 \sin 0 = C_1 \cdot 1 + C_2 \cdot 0 = s_0,$$

so $C_1 = s_0$, the initial displacement. What is C_2? To find it, we use the fact that at $t = 0$, when the spring has just been released, its velocity is 0. The velocity is the derivative of the displacement; thus,

$$\left.\frac{ds}{dt}\right|_{t=0} = \left.(-C_1 \sin t + C_2 \cos t)\right|_{t=0} = -C_1 \cdot 0 + C_2 \cdot 1 = 0,$$

so $C_2 = 0$. Therefore the solution is $s = s_0 \cos t$.

Solution to the General Spring Equation

Having found the general solution to the equation $\frac{d^2 s}{dt^2} = -s$, let us return to the more general equation for the motion of a spring, namely

$$\frac{d^2 s}{dt^2} = -\frac{k}{m} s.$$

To solve this we let $\omega = \sqrt{\frac{k}{m}}$ (why will be clear in a moment); then $\frac{d^2 s}{dt^2} = -\omega^2 s$ or $\frac{d^2 s}{dt^2} + \omega^2 s = 0$. Unfortunately, this equation no longer has $\sin t$ and $\cos t$ as solutions. For example,

$$\frac{d^2}{dt^2}(\sin t) = -\sin t \neq -\omega^2 \sin t.$$

However, we can adapt our guess as follows: we know that the functions $\sin t$ and $\cos t$ are *almost* solutions; we just have to deal with the extra ω^2 factor somehow. Ideally, we would like to be able to obtain a factor of ω^2 in the process of differentiating the sine or cosine function twice. This would be accomplished if, each time we differentiated, we obtained a factor of ω. The chain rule suggests an answer: try the function $\sin \omega t$. Checking this:

$$\frac{d^2}{dt^2}(\sin \omega t) = \frac{d}{dt}(\omega \cos \omega t) = -\omega^2 \sin \omega t$$

we see that $\sin \omega t$ is a solution, and you can probably guess that $\cos \omega t$ is a solution, too. As before, we can combine the sine and the cosine to find the general solution.

The general solution to the equation

$$\frac{d^2 s}{dt^2} + \omega^2 s = 0$$

is of the form

$$s(t) = C_1 \cos \omega t + C_2 \sin \omega t,$$

where C_1 and C_2 are arbitrary constants. (We assume that $\omega > 0$.) The period of this oscillation is

$$T = \frac{2\pi}{\omega}.$$

Such oscillations are called **simple harmonic motion**.

The solution to our original equation, $\frac{d^2 s}{dt^2} + \frac{k}{m}s = 0$, is thus $s = C_1 \cos \sqrt{\frac{k}{m}}t + C_2 \sin \sqrt{\frac{k}{m}}t$.

Initial-Value and Boundary-Value Problems

A problem in which the initial position and the initial velocity are used to determine the particular solution is called an *initial-value problem.* (See Example 2.) It is sometimes possible to find a specific solution from a general solution given the position at two known times, as in the following example. Such a problem is known as a *boundary-value problem.*

Example 3 Find a solution to the differential equation satisfying each set of conditions below

$$\frac{d^2 s}{dt^2} + 4s = 0.$$

(a) The boundary conditions $s(0) = 0$, $s(\pi/4) = 20$.

(b) The initial conditions $s(0) = 1$, $s'(0) = -6$.

Solution Since $\omega^2 = 4$, $\omega = 2$, the general solution to the differential equation is

$$s(t) = C_1 \cos 2t + C_2 \sin 2t.$$

(a) Substituting the boundary condition $s(0) = 0$ into the general solution gives

$$s(0) = C_1 \cos(2 \cdot 0) + C_2 \sin(2 \cdot 0) = C_1 \cdot 1 + C_2 \cdot 0 = C_1 = 0.$$

Thus $s(t)$ must have the form $s(t) = C_2 \sin 2t$. The second condition yields the value of C_2:

$$s\left(\frac{\pi}{4}\right) = C_2 \sin\left(2 \cdot \frac{\pi}{4}\right) = C_2 = 20.$$

Therefore, the solution satisfying the boundary conditions is

$$s(t) = 20 \sin 2t.$$

(b) For the initial-value problem, we start from the same general solution: $s(t) = C_1 \cos 2t + C_2 \sin 2t$. Substituting 0 for t once again, we find

$$s(0) = C_1 \cos(2 \cdot 0) + C_2 \sin(2 \cdot 0) = C_1 = 1.$$

Differentiating $s(t) = \cos 2t + C_2 \sin 2t$ gives

$$s'(t) = -2 \sin 2t + 2C_2 \cos 2t,$$

and applying the second initial condition gives us C_2:

$$s'(0) = -2 \sin(2 \cdot 0) + 2C_2 \cos(2 \cdot 0) = 2C_2 = -6,$$

so $C_2 = -3$ and our solution is

$$s(t) = \cos 2t - 3 \sin 2t.$$

What do the Graphs of Our Solutions Look Like?

Since the general solution of the equation $d^2s/dt^2 + \omega^2 s = 0$ is of the form

$$s(t) = C_1 \cos \omega t + C_2 \sin \omega t$$

it would be useful to know what the graph of such a sum of sines and cosines looks like. We'll start with a straightforward example, $s(t) = \cos t + \sin t$, which is graphed in Figure 9.63.

Interestingly, the graph in Figure 9.63 looks precisely like another sine function, and in fact it is one. If you measure the graph carefully, you will find that its period is 2π, the same as $\sin t$ and $\cos t$, but the amplitude appears to be larger than 1, and the graph is shifted along the t axis. (This is known as a phase shift or phase angle.) If you measure carefully, you will find that the amplitude of this graph is approximately 1.414. (In fact, it's $\sqrt{2}$.)

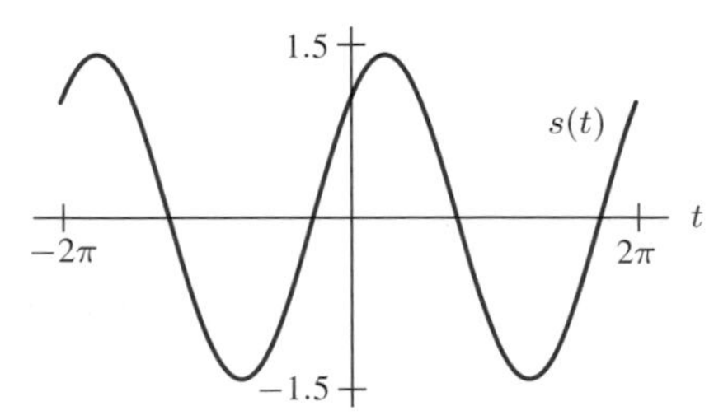

Figure 9.63: Graph of the sum of the sine and cosine: $s(t) = \cos t + \sin t$

If we were to plot $C_1 \cos t + C_2 \sin t$ for a variety of C_1 and C_2, we would find that the resulting graph is always a sine function with period 2π, though the amplitude and phase shift can vary. For example, the graphs of $s = 4\cos t + 3\sin t$ and $s = 6\cos t - 8\sin t$ are in Figure 9.64.

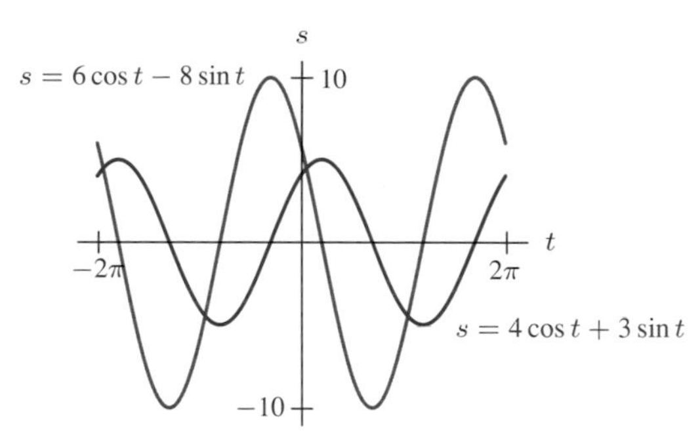

Figure 9.64: Graphs of two sums of sine and cosine functions

Both graphs have a period of 2π, but for $C_1 = 4$, $C_2 = 3$, the amplitude is 5, and for $C_1 = 6$, $C_2 = -8$, the amplitude is 10; their phase shifts also differ. These graphs suggest that we should be able to write the sum of a sine and a cosine as one single sine function.[21] It turns out that a sum of sine and cosine functions of the form

$$C_1 \cos \omega t + C_2 \sin \omega t$$

can always be written in the form

$$A \sin(\omega t + \varphi)$$

for some amplitude A and phase shift φ, which tells us how the oscillation has been shifted horizontally. How do we find the constants A and φ in terms of C_1 and C_2? You may already suspect the formula for A from the three functions we have graphed so far: with $C_1 = C_2 = 1$, $A = \sqrt{2}$; with $C_1 = 4$ and $C_2 = 3$, $A = 5$; and with $C_1 = 6$ and $C_2 = -8$, $A = 10$. As you will see in Problem 15 on page 568, it turns out that the following relations hold:

If $C_1 \cos \omega t + C_2 \sin \omega t = A \sin(\omega t + \varphi)$, then the *amplitude*, A, is

$$A = \sqrt{C_1^2 + C_2^2}.$$

We choose the *phase* (or *phase angle*) φ so that $-\pi < \varphi \le \pi$, and φ must satisfy

$$\tan \varphi = \frac{C_1}{C_2}.$$

We choose φ so that if $C_1 > 0$, φ is positive, and if $C_1 < 0$, φ is negative.

The reason that it is often useful to write the solution to the differential equation as a single sine function $A \sin(\omega t + \varphi)$, as opposed to the sum of a sine and a cosine, is that its amplitude A and the phase shift φ are easier to recognize in this form.

[21] The sum of a sine and a cosine can also be written as a single cosine function.

Example 4 Write the function $s(t) = \cos t - \sin t$ as a single sine function. Draw its graph.

Solution We want to find A and φ such that

$$\cos t - \sin t = A \sin(t + \varphi).$$

We know that $A = \sqrt{1^2 + (-1)^2} = \sqrt{2}$. Also, $\tan \varphi = 1/(-1) = -1$, so $\varphi = -\pi/4$ or $\varphi = 3\pi/4$. Since $C_1 = 1 > 0$, we take $\varphi = 3\pi/4$, giving

$$s(t) = \sqrt{2} \sin\left(t + \frac{3\pi}{4}\right)$$

as our solution. The graph of $s(t)$ is in Figure 9.65.

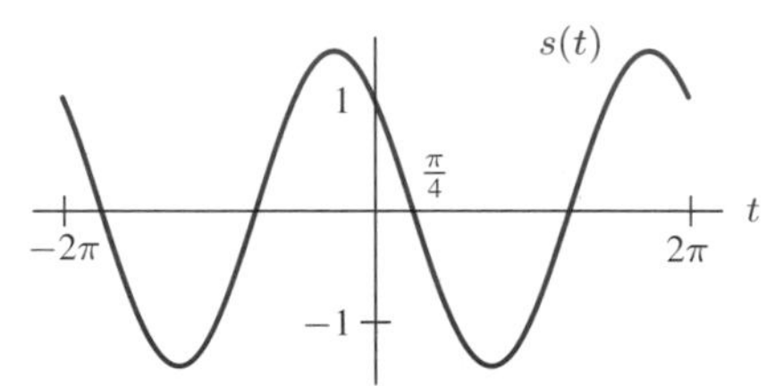

Figure 9.65: Graph of the function $s(t) = \sqrt{2}\sin(t + \frac{3\pi}{4})$

Warning: Phase Shift Is Not the Same as Horizontal Translation

Let's look at $s = \sin(3t + \pi/2)$. If we rewrite this as $s = \sin(3(t + \pi/6))$, you can see from Figure 9.66 that the graph of this function is the graph of $s = \sin 3t$ shifted to the left a distance of $\pi/6$. (Remember that replacing x by $x - 2$ shifts a graph by 2 to the right.) But $\pi/6$ is *not* the phase shift; the phase shift[22] is $\pi/2$. From the point of view of a scientist, the important question is often not the distance the curve has shifted, but the relation between the distance shifted and the length of a cycle. (A cycle is one complete oscillation.) When one is at its maximum, what is the other doing? When one is at its minimum, where is the other?

In this case, $s = \sin(3t + \pi/2)$ is at its maximum when $s = \sin 3t$ is zero. In fact, $s = \sin 3t$ has its maximum a quarter of a cycle later than $s = \sin(3t + \pi/2)$. Thus we say that $s = \sin(3t + \pi/2)$ is a quarter of a cycle ahead of $s = \sin 3t$. Since for $s = \sin t$, with period 2π, a quarter of a cycle is $2\pi/4 = \pi/2$, we say that $s = \sin(3t + \pi/2)$ is $\pi/2$ ahead of $s = \sin 3t$. In other words, the *phase difference* between $s = \sin(3t + \pi/2)$ and $s = \sin 3t$ is $\pi/2$.

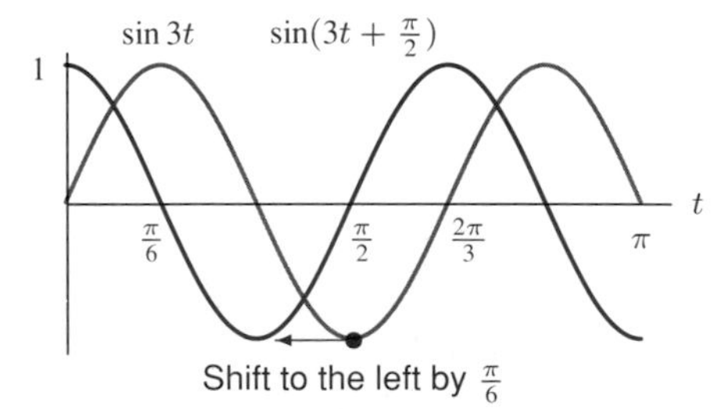

Figure 9.66: Phase shift is $\frac{\pi}{2}$; horizontal translation is $\frac{\pi}{6}$

[22]This definition of phase shift is the one usually used in the sciences.

Problems for Section 9.10

1. Check by differentiation that $y = 2\cos t + 3\sin t$ is a solution to $y'' + y = 0$.
2. Check that $y = A\cos t + B\sin t$ is a solution to $y'' + y = 0$ for any constants A and B.
3. What values of α and A make $y = A\cos\alpha t$ a solution to $y'' + 5y = 0$ such that $y'(1) = 3$?
4. If $y = A\cos\omega t + B\sin\omega t$ is a solution to $y'' + 16y = 0$ subject to the boundary conditions $y(0) = 2$ and $y(\pi/8) = 3$, find the constants A, B and ω.

The functions in Problems 5–7 describe the motion of a mass on a spring satisfying the differential equation $y'' = -9y$, where y is the displacement of the mass from the equilibrium position at time t, with upwards as positive. In each case, describe in words how the motion starts when $t = 0$. For example, is the mass at the highest point, the lowest point, or in the middle? Is it moving up or down or is it at rest?

5. $y = 2\cos 3t$ 6. $y = -0.5\sin 3t$ 7. $y = -\cos 3t$

8. Consider the equation $y'' + 4y = 0$.
   (a) Find a general solution to this equation.
   (b) For each set of initial conditions below, find a solution which satisfies them:
   (i) $y(0) = 5, y'(0) = 0$. (ii) $y(0) = 0, y'(0) = 10$. (iii) $y(0) = 5, y'(0) = 5$.
   (c) Sketch a graph of each of the three solutions you found in part (b).
9. Each graph in Figure 9.67 represents a solution to one of the differential equations:
   (a) $x'' + x = 0$, (b) $x'' + 4x = 0$, (c) $x'' + 16x = 0$.
   Assuming the t-scales on the four graphs are the same, which graph represents a solution to which equation? Find an equation for each graph.

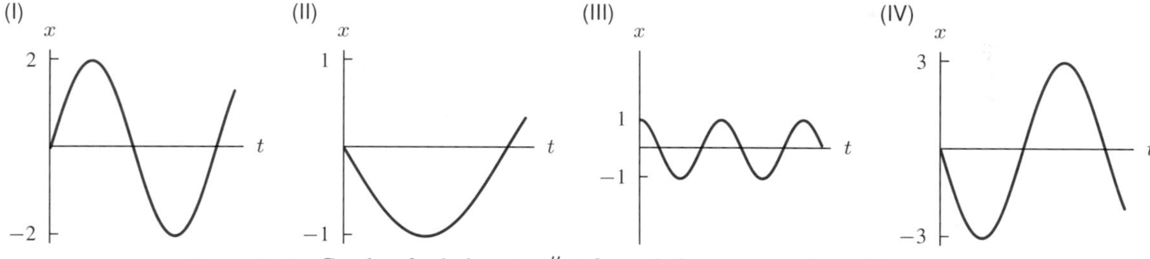

Figure 9.67: Graphs of solutions to $x'' + kx = 0$, for various values of k

10. The following differential equations represent oscillating springs.

(i)	$s'' + 4s = 0$	$s(0) = 5$,	$s'(0) = 0$
(ii)	$4s'' + s = 0$	$s(0) = 10$,	$s'(0) = 0$
(iii)	$s'' + 6s = 0$	$s(0) = 4$,	$s'(0) = 0$
(iv)	$6s'' + s = 0$	$s(0) = 20$,	$s'(0) = 0$

Which differential equation represents:

(a) The spring oscillating most quickly (with the shortest period)?
(b) The spring oscillating with the largest amplitude?
(c) The spring oscillating most slowly (with the longest period)?
(d) The spring with largest maximum velocity?

11. A pendulum of length l makes an angle of x (radians) with the vertical (see Figure 9.68). When x is small, it can be shown that, approximately:
$$\frac{d^2x}{dt^2} = -\frac{g}{l}x,$$
where g is the acceleration due to gravity.

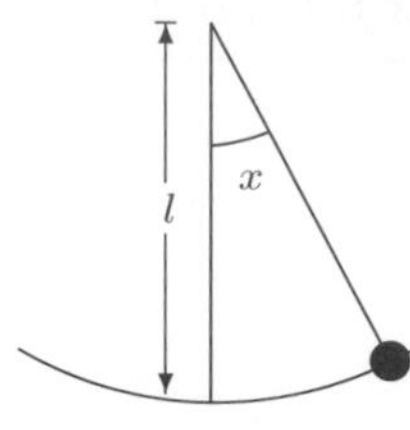

Figure 9.68

    (a) Solve this equation assuming that $x(0) = 0$ and $x'(0) = v_0$.
    (b) Solve this equation assuming that the pendulum is let go from the position where $x = x_0$. ("Let go" means that the velocity of the pendulum is zero when $x = x_0$. Measure t from the moment when the pendulum is let go.)

12. Look at the pendulum motion in Problem 11. What effect does it have on x as a function of time if:

    (a) x_0 is increased? (b) l is increased?

13. Find the amplitude of $3\sin 2t + 7\cos 2t$.

14. Write $7\sin\omega t + 24\cos\omega t$ in the form $A\sin(\omega t + \varphi)$.

15. (a) Expand $A\sin(\omega t + \varphi)$ using the trigonometric identity $\sin(x + y) = \sin x\cos y + \cos x\sin y$.
    (b) Assume $A > 0$. If $A\sin(\omega t + \varphi) = C_1\cos\omega t + C_2\sin\omega t$, show that we must have
$$A = \sqrt{C_1^2 + C_2^2} \quad \text{and} \quad \tan\varphi = C_1/C_2.$$

Problems 16–18 show how a second order differential equation can be used to describe an electric circuit. A charged capacitor connected to an inductor (as shown in Figure 9.69) will cause a current to flow through the inductor until the capacitor is fully discharged. The current in the inductor will, in turn, charge up the capacitor until the capacitor is fully charged again. We will let $Q(t)$ be the amount of charge on the capacitor at time t. We can then find the current, the rate at which charge is moving in the circuit, by looking at the rate of change of Q. Thus we have the relation between the current I and the charge Q:
$$I = \frac{dQ}{dt}.$$
The current can be positive or negative; its sign tells us which way the current is flowing. Q can also be positive or negative; its sign tells us whether the charge on the capacitor is positive or negative. If we assume the circuit resistance is zero, then it can be shown that the charge Q and the current I in the circuit satisfy the differential equation
$$L\frac{dI}{dt} + \frac{Q}{C} = 0.$$
so
$$L\frac{d^2Q}{dt^2} + \frac{Q}{C} = 0$$
where C, the capacitance, and L, the inductance, are constants depending on the particular capacitor and inductor. The unit of charge is the coulomb, the unit of capacitance the farad, the unit of inductance the henry, the unit of current is the ampere, and time is measured in seconds. This equation is similar to the differential equations we have seen for springs based on Hooke's law.

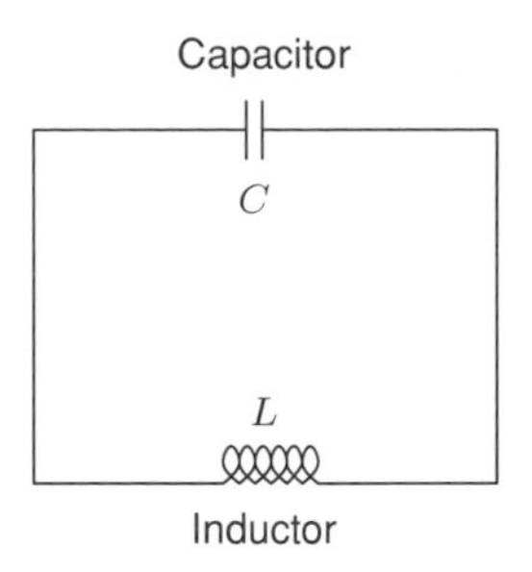

Figure 9.69

16. If $L = 36$ henry and $C = 9$ farad, find a formula for $Q(t)$ if
    (a) $Q(0) = 0, I(0) = 2$.
    (b) $Q(0) = 6, I(0) = 0$.
17. Suppose we wanted to set up our circuit so that $Q(0) = 0$, $Q'(0) = I(0) = 4$, and the maximum possible charge is $2\sqrt{2}$ coulombs. What should the capacitance of our capacitor be if we have an inductor with inductance 10 henry?
18. What happens to the charge and current as t goes to infinity? What does it mean that the charge and current are sometimes positive and sometimes negative?

9.11 DAMPED OSCILLATIONS AND NUMERICAL METHODS

A Spring with Friction

The differential equation $d^2s/dt^2 = -(k/m)s$, which we used to describe the motion of a spring, disregards friction. But there is friction in every real system. For a mass on a spring, the frictional force from air resistance increases with the velocity of the mass. The frictional force is often approximately proportional to velocity, and so we introduce a *damping term* of the form $c(ds/dt)$, where c is a constant called the *damping coefficient* and ds/dt is the velocity of the mass. Remember that without damping, the differential equation was obtained from

$$\text{Force} = \text{Mass} \times \text{Acceleration},$$

or

$$F = m\frac{d^2s}{dt^2}.$$

With damping, the force $-ks$ is replaced by $-ks - c(ds/dt)$, where c is positive and the $c(ds/dt)$ term is subtracted because the frictional force is in the direction opposite to the motion. The new differential equation is therefore

$$-ks - c\frac{ds}{dt} = m\frac{d^2s}{dt^2}$$

which is equivalent to the following differential equation:

Equation for Damped Oscillations of a Spring

$$\frac{d^2s}{dt^2} + \frac{c}{m}\frac{ds}{dt} + \frac{k}{m}s = 0.$$

What Are the Solutions to the Equation for Damped Motion?

Let's first try to get an intuitive sense of what the solution to the equation

$$\frac{d^2s}{dt^2} + \frac{c}{m}\frac{ds}{dt} + \frac{k}{m}s = 0$$

will look like.

Without damping, a mass on a spring oscillates, and the graph of its displacement against time is a sine curve. With some (but not too much) damping, the mass will still oscillate, but the amplitude of the oscillations will decrease with time. With a great deal of damping, the oscillation may be damped out entirely.

A Numerical Example

To understand the solutions to the damped equation, we first look at a numerical example. Suppose, for example, that $k/m = 4.04$ and $c/m = 0.4$:

$$\frac{d^2s}{dt^2} + 0.4\frac{ds}{dt} + 4.04s = 0.$$

Suppose the mass passes through the point $s = 0.8415$ meters at time $t = 0$ seconds with velocity $ds/dt = 0.9123$ m/sec. We can solve the above equation numerically using an adaptation of Euler's method described on page 572, to approximate the solution of a second-order differential equation and get the oscillation in Figure 9.70.

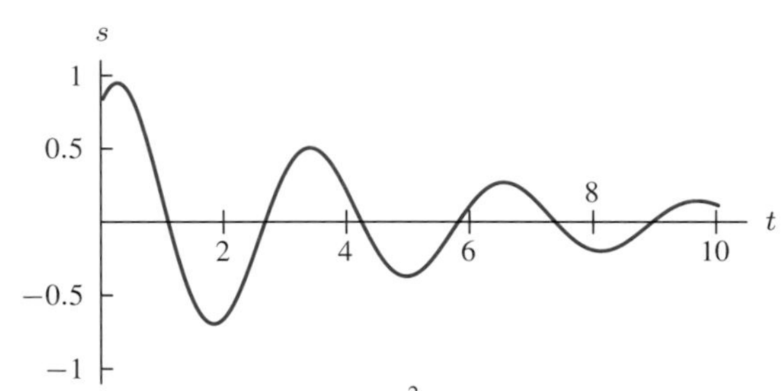

Figure 9.70: Solution to $\frac{d^2s}{dt^2} + 0.4\frac{ds}{dt} + 4.04s = 0$

If you measure the distance between the points where the curve crosses the t-axis, you will see that they occur approximately every $1.57 \approx \pi/2$ units, and so the position function s looks like a sine function with period approximately π and a decaying amplitude. To get a period of π, the solution must involve $\sin 2t$ or $\cos 2t$.

Let's draw the function that touches the peaks of the curve as in Figure 9.71. The dotted curve represents the decay of the oscillations; if we take a sine function with the appropriate period, amplitude, and phase and multiply it by this decay function, we would get the function for the position s.

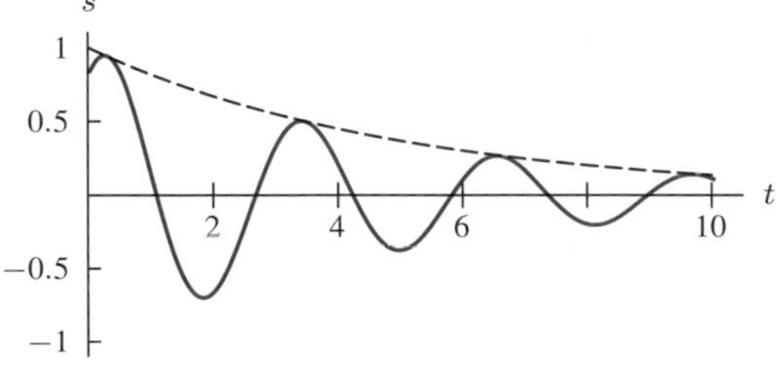

Figure 9.71: Solution with decay curve superimposed

TABLE 9.17 *Values of s at successive peaks*

t	s
0.29	0.9445
3.43	0.5039
6.57	0.2688
9.71	0.1434

What kind of function is the decay function? The approximate values of s at the peaks (calculated using Euler's method) are given in Table 9.17. Notice that the peaks are equally spaced and that the ratio of the values is approximately constant: $0.5039/0.9445 \approx 0.53$, $0.2688/0.5039 \approx 0.53$, etc. Hence, the decay curve seems to be exponential, and the solution to the original differential equation is probably the product of a sine or cosine and a decaying exponential. Problem 1 on page 574 shows that the solution can be written both in the form

$$s(t) = (0.53)^{t/\pi}(A\sin(2t+\varphi)),$$

as well as in the form

$$s(t) = e^{-0.20t}(C_1\cos 2t + C_2\sin 2t).$$

As you will see in Section 9.12, this is the form of the solution if the damping is not too large; the form of the solution will be different if the damping is large.

Numerical Solutions to Second-Order Differential Equations

Many second-order differential equations cannot be solved analytically. Therefore, it is useful to be able to solve such equations numerically, using an extension of Euler's method. We will now see how to generate numerical solutions such as those in Table 9.17 on page 570. This is done by first writing the second-order equation as a pair of first-order equations. To illustrate the process, let's apply the technique to a differential equation that we have already solved:

$$\frac{d^2s}{dt^2} = -s,$$

with initial conditions that at $t = 0$, $s = 0$ and $ds/dt = 1$. Using the methods of Section 9.10 we can show that the exact solution to this equation is

$$s = \sin t, \quad \text{so} \quad v = \frac{ds}{dt} = \cos t.$$

Now we'll solve the same equation numerically, using the phase plane.

Rewriting the Second-Order Equation as a Pair of First-Order Equations:

We let $v = ds/dt$ be a new variable. Then our original equation $d^2s/dt^2 = -s$ can be rewritten as $dv/dt = -s$, so the system is:

$$\frac{ds}{dt} = v$$
$$\frac{dv}{dt} = -s.$$

As in Section 9.8, consider the sv phase plane. Then, using the chain rule we have

$$\frac{dv}{ds} = \frac{dv/dt}{ds/dt} = -\frac{s}{v}.$$

The slope field for the system of differential equations is the slope field for $dv/ds = -s/v$, shown in Figure 9.72. If we plot the exact solution, $s = \sin t$, $v = \cos t$, in the phase plane, the trajectory will be a unit circle, because

$$s^2 + v^2 = \sin^2 t + \cos^2 t = 1.$$

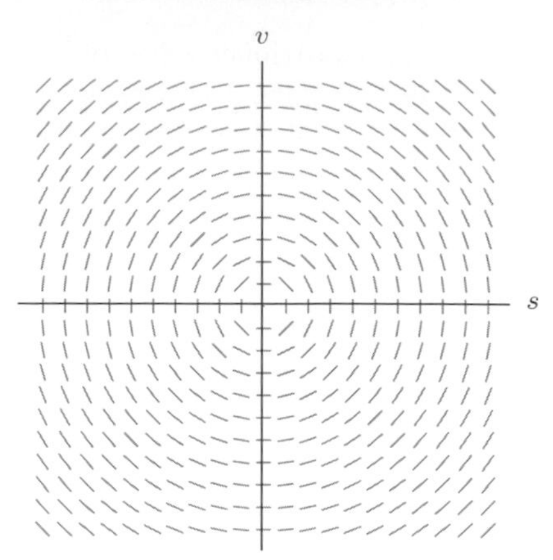

Figure 9.72: Slope field for the system $\frac{ds}{dt} = v$, $\frac{dv}{dt} = -s$

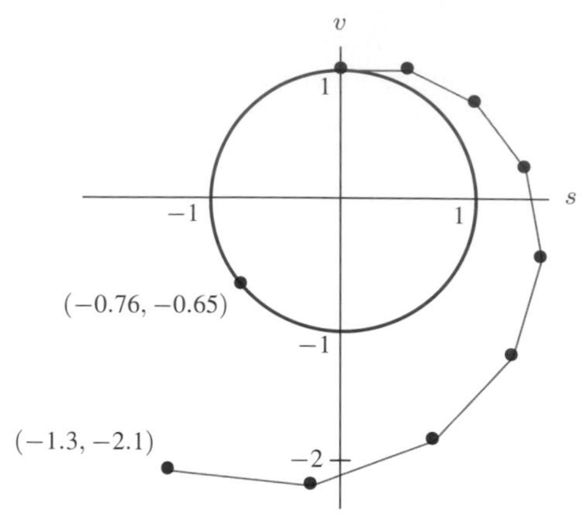

Figure 9.73: Euler's method trajectory compared with true trajectory

Using Euler's Method on a System of Differential Equations

Suppose that we want to approximate the solution using Euler's method with 8 steps and $\Delta t = 0.5$, starting at the point $(s, v) = (0, 1)$ corresponding to the initial condition. For each step, we calculate ds/dt and dv/dt at a particular point in the phase plane and use these slopes to move to a new point in the phase plane, as illustrated in Figure 9.73. Each step we take is an approximation, in the sense that it assumes that the slope field doesn't change over the step in the phase plane. As you can see in Figure 9.73, the Euler's method trajectory ends up deviating further and further from the true circular trajectory. The true solution to the differential equation is $s = \sin t$, and so at $t = 4$, $s = \sin 4 \approx -0.76$ and $v = \cos 4 \approx -0.65$. Euler's method with 8 steps and $\Delta t = 0.5$ gives $s = -1.3$ and $v = -2.1$. One way of improving the accuracy, of course, is to take smaller steps, that is, a smaller value for Δt.

In the next example we use Euler's method on a differential equation whose solution we do not already know.

Example 1 Given the differential equation

$$\frac{d^2y}{dt^2} + 2\frac{dy}{dt} + 2y = 0$$

with initial conditions that at $t = 0$, $y = 2$ and $y' = 0$, estimate the value of y and y' at $t = 1$ by using Euler's method with $n = 10$ steps ($\Delta t = 0.1$).

Solution First we must rewrite the differential equation as a system of first-order differential equations. To do this, let

$$\frac{dy}{dt} = v,$$

and consider the yv−phase plane. Then $d^2y/dt^2 = -2(dy/dt) - 2y$, so the second equation in the system is

$$\frac{dv}{dt} = -2v - 2y.$$

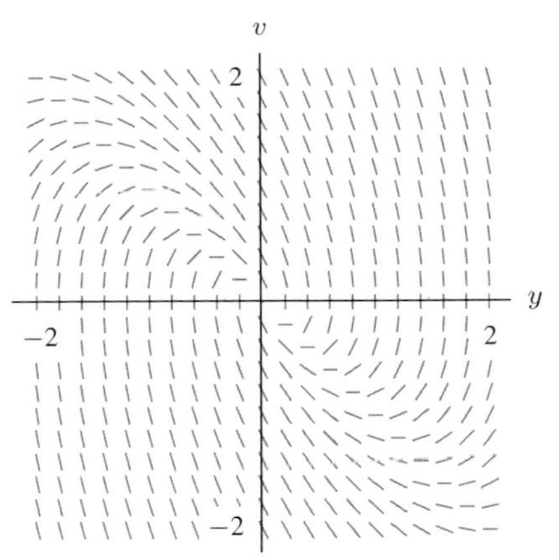

Figure 9.74: Slope field for the system $\frac{dy}{dt} = v$, $\frac{dv}{dt} = -2v - 2y$

To draw the slope field for this system of equations in Figure 9.74, we write $dv/dy = (dv/dt)/(dy/dt) = (-2v - 2y)/v$. Using Euler's method starting at $y_0 = 2$ and $v_0 = 0$, we calculate

$$\frac{dy}{dt} = v_0 = 0,$$
$$\frac{dv}{dt} = -2v_0 - 2y_0 = -4.$$

Therefore, the values of y_1 and v_1 are given by

$$y_1 = y_0 + \frac{dy}{dt}\Delta t = 2 + 0(0.1) = 2,$$
$$v_1 = v_0 + \frac{dv}{dt}\Delta t = 0 - 4(0.1) = -0.4.$$

For the next step, we calculate the derivatives at $y_1 = 2$, $v_1 = -0.4$, giving

$$\frac{dy}{dt} = v_1 = -0.4,$$
$$\frac{dv}{dt} = -2v_1 - 2y_1 = 0.8 - 4 = -3.2.$$

Therefore,

$$y_2 = y_1 + \frac{dy}{dt}\Delta t = 2 - 0.4(0.1) = 1.96,$$
$$v_2 = v_1 + \frac{dv}{dt}\Delta t = -0.4 - 3.2(0.1) = -0.72.$$

Table 9.18 lists the values of y, v, dy/dt, and dv/dt at each step of the process, and Figure 9.75 shows the Euler trajectory. Hence, at $t = 1$, $y \approx 0.99$ and $v \approx -1.33$.

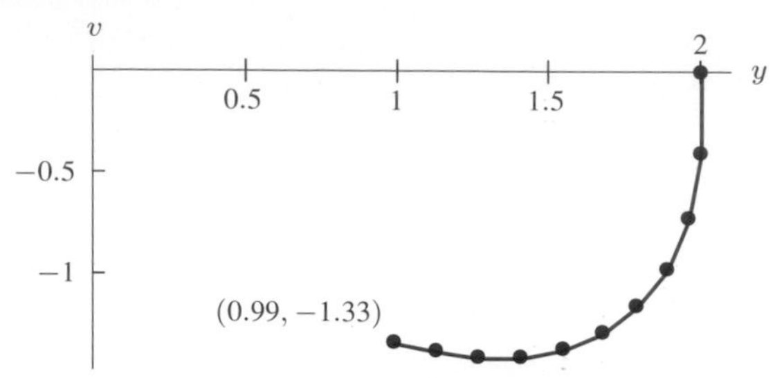

Figure 9.75: Euler approximation for the system $\frac{dy}{dt} = v$, $\frac{dv}{dt} = -2v - 2y$

TABLE 9.18 *Values from Euler's method*

i	t_i	y_i	v_i	dy/dt	dv/dt
0	0.0	2.00	0.00	0.00	−4.00
1	0.1	2.00	−0.40	−0.40	−3.20
2	0.2	1.96	−0.72	−0.72	−2.48
3	0.3	1.89	−0.97	−0.97	−1.84
4	0.4	1.79	−1.15	−1.15	−1.28
5	0.5	1.68	−1.28	−1.28	−0.79
6	0.6	1.55	−1.36	−1.36	−0.38
7	0.7	1.41	−1.40	−1.40	−0.03
8	0.8	1.27	−1.40	−1.40	0.25
9	0.9	1.13	−1.37	−1.37	0.48
10	1.0	0.99	−1.33	−1.33	0.66

We will find the exact solution to this differential equation in Section 9.12, but for now let us borrow the results to check the accuracy of the numerical solution. At $t = 1$, the true value is

$$y = \frac{2}{e}(\sin 1 + \cos 1) = 1.016\ldots \qquad \text{and} \qquad v = \left.\frac{dy}{dt}\right|_{t=1} = -\frac{4}{e}\sin 1 = -1.238\ldots$$

With only 10 steps, Euler's method gives solutions correct to within 3% in y and 7% in v.

Problems for Section 9.11

1. Assume that the position function, s, for which data is given in Table 9.17 on page 570 can be expressed in the form

$$s(t) = a^t \left(A \sin(2t + \varphi)\right)$$

where a^t is a decaying exponential.

(a) Using the data in Table 9.17 and the fact that the distance between successive peaks is approximately π, explain why we write $a^t = (0.53)^{t/\pi}$.

(b) Explain why $s(t)$ can also be written in the form

$$s(t) = e^{-0.20t}(C_1 \cos 2t + C_2 \sin 2t).$$

On page 571 we saw how to rewrite a second order differential equation as a system of first order differential equations. In Problems 2-4, we will rewrite a system of two first order differential equations as a second order differential equation, and thereby solve the system.

2. Consider the system

$$\frac{dx}{dt} = y$$
$$\frac{dy}{dt} = -x.$$

   (a) Differentiate the second equation with respect to t and substitute for dx/dt to obtain a second order differential equation for y.
   (b) Solve the equation you obtained for y as a function of t; hence find x as a function of t.

3. (a) Rewrite the system

$$\frac{ds}{dt} = u$$
$$\frac{du}{dt} = -4s$$

   as a second-order differential equation for u by differentiating the second equation and substituting for s.
   (b) Solve for u and hence find s as a function of t.

4. Consider the system

$$\frac{dx}{dt} = -y$$
$$\frac{dy}{dt} = x \qquad x(0) = 1,\ y(0) = 0.$$

   (a) Write a differential equation involving dy/dx and sketch the slope field and the trajectory starting at the point $x(0) = 1,\ y(0) = 0$.
   (b) Using the method of Problem 2, find x and y as functions of t.

5. Show the first two steps of the Euler's method calculation for $ds/dt = v$, $dv/dt = -s$, starting at $s = 0$, $v = 1$, and using $\Delta t = 0.5$.

6. Continue the calculation of Problem 5 to show that after 8 steps you reach $s = -1.3$, $v = -2.1$.

7. Write a spreadsheet, calculator or computer program to generate the data in Table 9.18, page 574.

8. (a) Write the differential equation $y'' = y$ as a system of two first order differential equations.
   (b) Starting at the point $t = 0$, $y = 1$, $y' = 1$, use Euler's method with $\Delta t = 0.25$ to approximate $y(1)$.
   (c) Check that $y = e^t$ is the exact solution to this differential equation which satisfies the initial conditions. Compare the true value $y(1)$ to your approximation.

9. Using $I(0) = 1$, $S(0) = 762$, and $\Delta t = 1$, carry out five steps of Euler's method for the system of equations on page 543:

$$\frac{dS}{dt} = -0.0026SI$$
$$\frac{dI}{dt} = 0.0026SI - 0.5I.$$

10. How does the computation in Problem 9 compare to the graph in Figure 9.53 on page 545? Is I over- or underestimated? Explain why.
11. Consider the differential equations used to describe the interaction of robins and worms on page 547:
$$\frac{dw}{dt} = w - wr$$
$$\frac{dr}{dt} = -r + wr.$$
Carry out three steps of Euler's method, using $w(0) = 2$, $r(0) = 2$, $\Delta t = 0.1$; you should get $r(0.3) = 2.5$, $w(0.3) = 1.4$.
12. In Problem 11, why would the choice $\Delta t = 1$ be inappropriate?

9.12 LINEAR SECOND-ORDER DIFFERENTIAL EQUATIONS

More General Second-Order Differential Equations

The equations we have been studying,
$$\frac{d^2s}{dt^2} + \omega^2 s = 0 \quad \text{and} \quad \frac{d^2s}{dt^2} + \frac{c}{m}\frac{ds}{dt} + \frac{k}{m}s = 0,$$
are important examples of the class of second-order differential equations of the form
$$\frac{d^2s}{dt^2} + b\frac{ds}{dt} + cs = 0.$$
Such equations are called *linear second-order differential equations*. We call b and c the *coefficients* of the equation; they may be either constants or functions of t. If b and c are constants, we will say we have a *linear second-order differential equation with constant coefficients*. This section explains an analytic method of solving any such equation.

The Principle of Superposition

As we have seen for the spring equation, if $f_1(t)$ and $f_2(t)$ satisfy a differential equation of the form
$$\frac{d^2y}{dt^2} + b\frac{dy}{dt} + cy = 0,$$
where b and c are constants, then the *principle of superposition* says that, for any constants C_1 and C_2, the function
$$y(t) = C_1 f_1(t) + C_2 f_2(t)$$
is a solution as well. It can be shown that provided $f_1(t)$ is not a multiple of $f_2(t)$, the general solution is of this form.

The Characteristic Equation

We will now look at a powerful method of solving the differential equation
$$\frac{d^2y}{dt^2} + b\frac{dy}{dt} + cy = 0.$$

The method is again a form of guess-and-check. This time we ask what kind of function might satisfy a differential equation which says that the second derivative d^2y/dt^2 is a sum of multiples of dy/dt and y. One possibility is the exponential function, so we try to find a solution of the form:

$$y = Ce^{rt},$$

where r may be a complex number.[23]

In order to find r, we substitute into the differential equation:

$$\frac{d^2y}{dt^2} + b\frac{dy}{dt} + cy = r^2Ce^{rt} + b \cdot rCe^{rt} + c \cdot Ce^{rt} = Ce^{rt}(r^2 + br + c) = 0.$$

We can divide this equation through by $y = Ce^{rt}$ so long as $C \neq 0$, because the exponential function is never zero. If $C = 0$, then $y = 0$, which is not a very interesting solution (though it is a solution). So we assume $C \neq 0$. Then $y = Ce^{rt}$ will be a solution to the differential equation if

$$r^2 + br + c = 0.$$

This quadratic equation is called the *characteristic equation* of the differential equation. Its solutions are

$$r = -\frac{1}{2}b \pm \frac{1}{2}\sqrt{b^2 - 4c}.$$

There are three different types of solution to the differential equation, depending on whether the solutions to the characteristic equation are real or complex or repeated, which is determined by $b^2 - 4c$.

The Overdamped Case: $b^2 - 4c > 0$

There are two real solutions r_1 and r_2 to the characteristic equation, and so the following two functions satisfy the differential equation:

$$C_1e^{r_1t} \quad \text{and} \quad C_2e^{r_2t}.$$

The sum of these two solutions is the general solution to the differential equation:

The Overdamped Case

If $b^2 - 4c > 0$, the general solution to:

$$\frac{d^2y}{dt^2} + b\frac{dy}{dt} + cy = 0$$

is

$$y(t) = C_1e^{r_1t} + C_2e^{r_2t}$$

where r_1 and r_2 are the solutions to the characteristic equation.

A physical system satisfying a differential equation of this type is called *overdamped* because it occurs when there is a lot of friction in the system. For example, a spring moving in a thick fluid such as oil or molasses is overdamped: it will not oscillate.

[23]See Appendix D on complex numbers.

Example 1 Suppose a spring is placed in oil, where it satisfies the differential equation

$$\frac{d^2s}{dt^2} + 3\frac{ds}{dt} + 2s = 0.$$

Solve this equation with the initial conditions $s = -0.5$ and $\frac{ds}{dt} = 3$ when $t = 0$.

Solution The characteristic equation is

$$r^2 + 3r + 2 = 0,$$

with solutions $r = -1$ and $r = -2$, and so the general form of the solution to the differential equation is

$$s(t) = C_1e^{-t} + C_2e^{-2t}.$$

We use the initial conditions to find C_1 and C_2. At $t = 0$, we have

$$s = C_1e^{-0} + C_2e^{-2(0)} = C_1 + C_2 = -0.5.$$

Furthermore, since $ds/dt = -C_1e^{-t} - 2C_2e^{-2t}$, we have

$$\left.\frac{ds}{dt}\right|_{t=0} = -C_1e^{-0} - 2C_2e^{-2(0)} = -C_1 - 2C_2 = 3.$$

Solving these equations simultaneously, we find $C_1 = 2$ and $C_2 = -2.5$, so that the solution is

$$s(t) = 2e^{-t} - 2.5e^{-2t}.$$

The graph of this function is in Figure 9.76. The behavior is what you might expect: the mass is so slowed by the oil that it passes through the equilibrium point only once (when $t \approx 1/4$) and for all practical purposes, it comes to rest after a short time. The motion has been "damped out" by the oil.

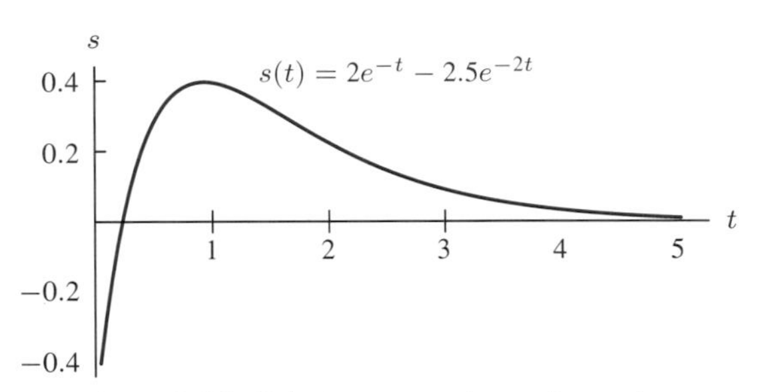

Figure 9.76: Solution to overdamped equation $\frac{d^2s}{dt^2} + 3\frac{ds}{dt} + 2s = 0$

The Critically Damped Case: $b^2 - 4c = 0$

In the critically damped case, the characteristic equation has only one solution, $r = -b/2$. By substitution, you can check that both $y = e^{-bt/2}$ and $y = te^{-bt/2}$ are solutions.

The Critically Damped Case

$$\frac{d^2y}{dt^2} + b\frac{dy}{dt} + \frac{b^2}{4}y = 0$$

has general solution

$$y(t) = (C_1 t + C_2)e^{-bt/2}.$$

The Underdamped Case: $b^2 - 4c < 0$

Example 2 An object of mass $m = 10$ kg is attached to a spring with spring constant $k = 20$ kg/sec^2, and the object experiences a frictional force proportional to the velocity, with constant of proportionality $c = 20$ kg/sec. At time $t = 0$, the object is released from rest 2 meters above the equilibrium position. Write the differential equation that describes the motion.

Solution The differential equation that describes the motion can be obtained from the following general expression for the damped motion of a spring:

$$\underbrace{m\frac{d^2s}{dt^2}}_{\text{Mass Acceleration}} = \underbrace{F}_{\text{Net Force}} = -\underbrace{ks}_{\text{Spring Force}} - \underbrace{c\frac{ds}{dt}}_{\text{Frictional Force}}.$$

Substituting $m = 10$ kg, $k = 20$ kg/sec^2, and $c = 20$ kg/sec, we obtain the differential equation for the motion:

$$\frac{d^2s}{dt^2} + 2\frac{ds}{dt} + 2s = 0.$$

At $t = 0$, the object is at rest 2 meters above equilibrium, so the initial conditions are $s(0) = 2$ and $s'(0) = 0$, where s is in meters and t in seconds.

Notice that this is the same differential equation as in Example 1 except that the coefficient of ds/dt has decreased from 3 to 2, which means that the frictional force has been reduced. This time, the roots of the characteristic equation have imaginary parts which lead to oscillations.

Example 3 Solve the differential equation

$$\frac{d^2s}{dt^2} + 2\frac{ds}{dt} + 2s = 0,$$

subject to $s(0) = 2$, $s'(0) = 0$.

Solution The characteristic equation is

$$r^2 + 2r + 2 = 0 \quad \text{giving} \quad r = -1 \pm i.$$

The solution to the differential equation is

$$s(t) = A_1 e^{(-1+i)t} + A_2 e^{(-1-i)t},$$

where A_1 and A_2 are arbitrary complex numbers. The initial condition $s(0) = 2$ gives

$$2 = A_1 e^{(-1+i)\cdot 0} + A_2 e^{(-1-i)\cdot 0} = A_1 + A_2.$$

Also,

$$s'(t) = A_1(-1+i)e^{(-1+i)t} + A_2(-1-i)e^{(-1-i)t},$$

so $s'(0) = 0$ gives

$$0 = A_1(-1+i) + A_2(-1-i).$$

Solving the simultaneous equations for A_1 and A_2 gives (after some algebra):

$$A_1 = 1 - i \quad \text{and} \quad A_2 = 1 + i.$$

The solution is therefore

$$s(t) = (1-i)e^{(-1+i)t} + (1+i)e^{(-1-i)t} = (1-i)e^{-t}e^{it} + (1+i)e^{-t}e^{-it}.$$

Using Euler's formula, $e^{it} = \cos t + i \sin t$ and $e^{-it} = \cos t - i \sin t$, we get:

$$s(t) = (1-i)e^{-t}(\cos t + i \sin t) + (1+i)e^{-t}(\cos t - i \sin t).$$

Multiplying out and simplifying, all the complex terms drop out, giving

$$\begin{aligned} s(t) &= e^{-t}\cos t + ie^{-t}\sin t - ie^{-t}\cos t + e^{-t}\sin t \\ &\quad + e^{-t}\cos t - ie^{-t}\sin t + ie^{-t}\cos t + e^{-t}\sin t \\ &= 2e^{-t}\cos t + 2e^{-t}\sin t. \end{aligned}$$

The $\cos t$ and $\sin t$ terms tell us that the solution oscillates. However the oscillations are damped extremely quickly by the factor e^{-t}. (See Figure 9.77.) Alternatively, you can see the behavior of the solution by rewriting it in the form

$$s(t) = 2e^{-t}(\cos t + \sin t) = 2\sqrt{2}e^{-t}\sin\left(t + \frac{\pi}{4}\right).$$

Physically the most interesting feature of the solution is that the period of the oscillations does not change as the amplitude decreases. This is why a spring-driven clock can keep accurate time even as it is running down.

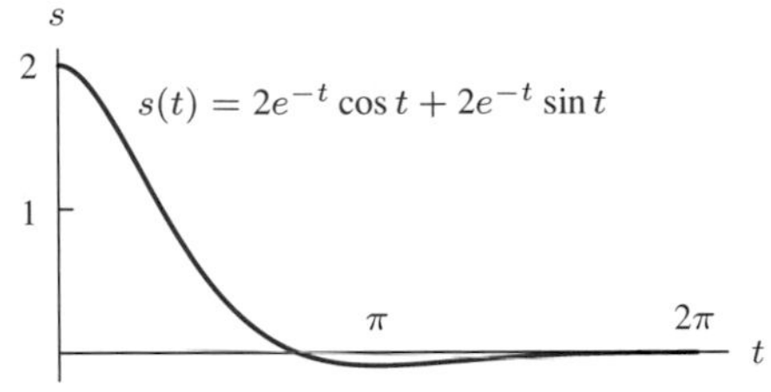

Figure 9.77: Solution to underdamped equation $\frac{d^2s}{dt^2} + 2\frac{ds}{dt} + 2s = 0$

In Example 3, notice that although the coefficients A_1 and A_2 are complex, the solution, $s(t)$, is real. (You might expect this, since $s(t)$ represents a real displacement.) In general, provided the coefficients b and c in the original differential equation and the initial values are real, the solution will always be real too. The complex coefficients A_1 and A_2 will always be complex conjugates.

The Underdamped Case

If $b^2 - 4c < 0$, to solve

$$\frac{d^2y}{dt^2} + b\frac{dy}{dt} + cy = 0,$$

- find the solutions $r = \alpha \pm i\beta$ to the characteristic equation $r^2 + br + c = 0$.
- The general solution to the differential equation is, for some real C_1 and C_2,

$$y = C_1 e^{\alpha t} \cos \beta t + C_2 e^{\alpha t} \sin \beta t.$$

Such oscillations are called **damped simple harmonic motion.**

Example 4 Find the general solution of the equations
(a) $y'' = 9y$ (b) $y'' = -9y$.

Solution (a) The characteristic equation is $r^2 - 9 = 0$, so $r = \pm 3$. Thus the general solution is

$$y = C_1 e^{3t} + C_2 e^{-3t}.$$

(b) The characteristic equation is $r^2 + 9 = 0$, so $r = \pm 3i$. Thus $y = e^{3it}$ and $y = e^{-3it}$ are solutions, and the general solution is

$$y = C_1 \cos 3t + C_2 \sin 3t.$$

Notice that we have seen the solution to this equation in Section 9.10; it's the equation of undamped simple harmonic motion.

Problems for Section 9.12

For Problems 1–8, find the general solution to the given differential equation.

1. $y'' + 4y' + 3y = 0$
2. $y'' + 4y' + 4y = 0$
3. $y'' + 4y' + 5y = 0$
4. $s'' - 7s = 0$
5. $s'' + 7s = 0$
6. $4z'' + 8z' + 3z = 0$
7. $\dfrac{d^2x}{dt^2} + 4\dfrac{dx}{dt} + 8x = 0$
8. $\dfrac{d^2p}{dt^2} + \dfrac{dp}{dt} + p = 0$

For Problems 9–12, solve the initial value problem.

9. $y'' + 6y' + 5y = 0, \quad y(0) = 1, \quad y'(0) = 0$
10. $y'' + 6y' + 5y = 0, \quad y(0) = 5, \quad y'(0) = 5$
11. $y'' + 6y' + 10y = 0, \quad y(0) = 0, \quad y'(0) = 2$
12. $y'' + 6y' + 10y = 0, \quad y(0) = 0, \quad y'(0) = 0$

For Problems 13–14, solve the boundary value problem.

13. $p'' + 2p' + 2p = 0, \quad p(0) = 0, \quad p(\pi/2) = 20$
14. $p'' + 4p' + 5p = 0, \quad p(0) = 1, \quad p(\pi/2) = 5$
15. Match the graphs of solutions in Figure 9.78 with the differential equations below.
(a) $x'' + 4x = 0$ (b) $x'' - 4x = 0$ (c) $x'' - 0.2x' + 1.01x = 0$
(d) $x'' + 0.2x' + 1.01x = 0$.

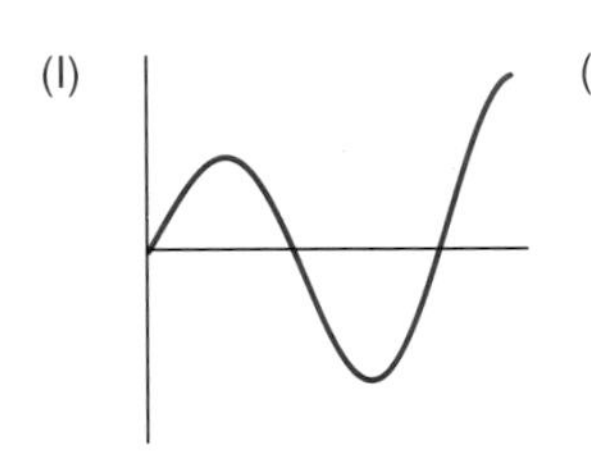

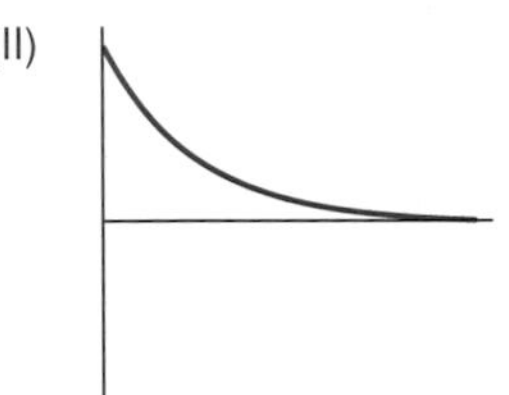

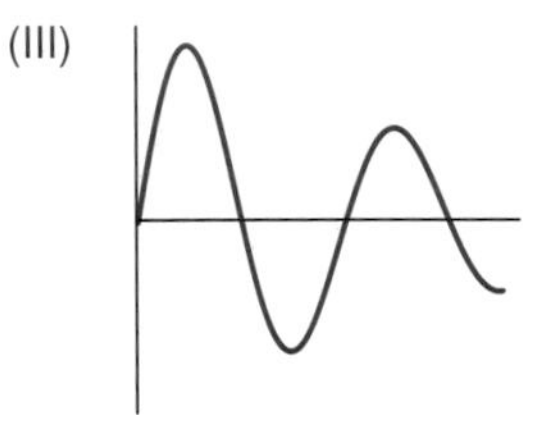

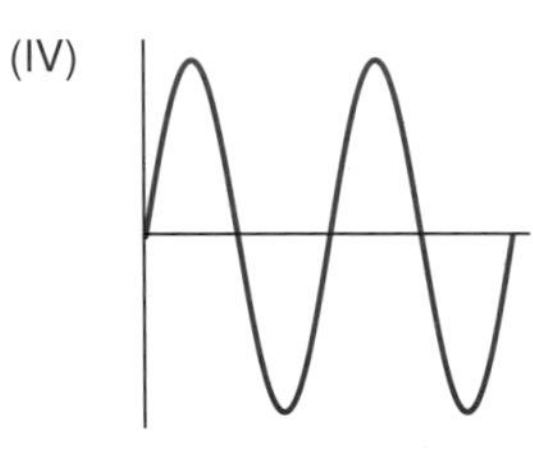

Figure 9.78

Each of the differential equations below represents the position of a 1 gram mass oscillating on the end of a damped spring. For Problems 16–20 below, pick the differential equation representing the system which answers the question.

(i) $s'' + s' + 4s = 0$ (iii) $s'' + 3s' + 3s = 0$
(ii) $s'' + 2s' + 5s = 0$ (iv) $s'' + 0.5s' + 2s = 0$

16. Which spring has the largest coefficient of damping?
17. Which spring exerts the smallest restoring force for a given displacement?
18. In which system does the mass experience the frictional force of smallest magnitude for a given velocity?
19. Which oscillation has the longest period?
20. Which spring is the stiffest? [Hint: You need to determine what it means for a spring to be stiff. Think of an industrial strength spring and a slinky.]

For each of the differential equations in Problems 21–23, find the values of c that make the general solution: (a) overdamped, (b) underdamped, (c) critically damped.

21. $s'' + 4s' + cs = 0.$ 22. $s'' + 2\sqrt{2}s' + cs = 0.$ 23. $s'' + 6s' + cs = 0.$

24. If $y = e^{2t}$ is a solution to the differential equation

$$\frac{d^2y}{dt^2} - 5\frac{dy}{dt} + ky = 0,$$

find the value of the constant k and the general solution to this equation.

25. Find a solution to the equation
$$\frac{d^2z}{dt^2} + \frac{dz}{dt} - 2z = 0,$$
which satisfies $z(0) = 3$ and does not tend to infinity as $t \to \infty$.

26. Assuming $b,\ c > 0$, explain how you know that the solutions of an underdamped differential equation must go to 0 as $t \to \infty$.

27. Consider an overdamped differential equation with $b,\ c > 0$.
    (a) Show that both roots of the characteristic equation are negative.
    (b) Show that any solution to the differential equation goes to 0 as $t \to \infty$.

28. Could the graph in Figure 9.79 show the position of a mass oscillating at the end of an overdamped spring? Why or why not?

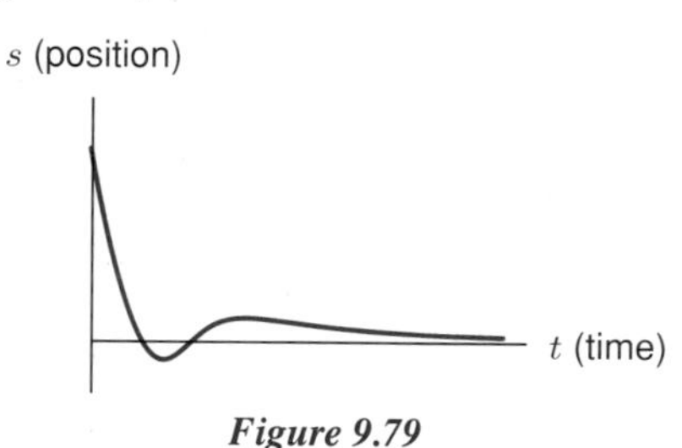

Figure 9.79

29. Consider the system of differential equations
$$\frac{dx}{dt} = -y \qquad \frac{dy}{dt} = -x$$
    (a) Convert this system to a second order differential equation in y by differentiating the second equation with respect to t and substituting for x from the first equation.
    (b) Solve the equation you obtained for y as a function of t; hence find x as a function of t.

Recall the discussion of electric circuits in Section 9.10 on page 568. Just as a spring can have a damping force which affects its motion, so can a circuit. Problems 30–33 illustrate this situation. The damping force is caused by a resistor shown in Figure 9.80. Just as the damping force for a mass on a spring

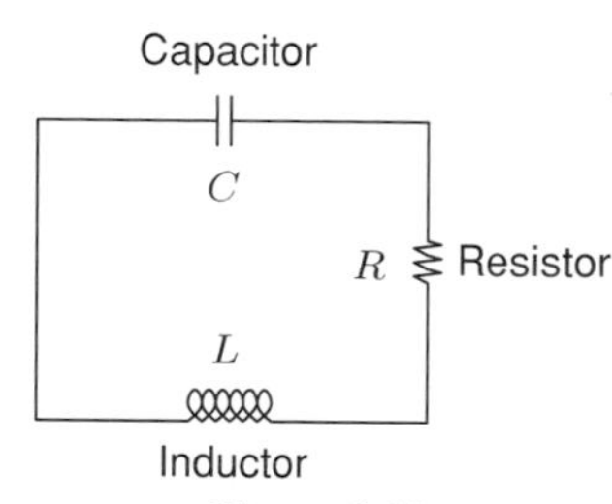

Figure 9.80

is proportional to the velocity of the mass, the damping force in the circuit is proportional to the current. We get an extra term $RI = R(dQ/dt)$ in the differential equation. The resistance, R, is a constant that depends on the resistor, and is measured in ohms. The equation for the charge Q on a capacitor in a circuit with inductance L, capacitance C, and resistance R satisfies the differential equation
$$L\frac{d^2Q}{dt^2} + R\frac{dQ}{dt} + \frac{1}{C}Q = 0.$$

30. If $L = 1$ henry, $R = 2$ ohms, and $C = 4$ farads, find a formula for the charge when
    (a) $Q(0) = 0$, $Q'(0) = 2$.
    (b) $Q(0) = 2$, $Q'(0) = 0$.
31. If $L = 1$ henry, $R = 1$ ohm, and $C = 4$ farads, find a formula for the charge when
    (a) $Q(0) = 0$, $Q'(0) = 2$.
    (b) $Q(0) = 2$, $Q'(0) = 0$.
    (c) How did reducing the resistance affect the charge? Compare with your solution to Problem 30.
32. If $L = 8$ henry, $R = 2$ ohm, and $C = 4$ farads, find a formula for the charge when
    (a) $Q(0) = 0$, $Q'(0) = 2$.
    (b) $Q(0) = 2$, $Q'(0) = 0$.
    (c) How did increasing the inductance affect the charge? Compare with your solution to Problem 30.
33. Given any positive values for R, L and C, what happens to the charge as t goes to infinity?

REVIEW PROBLEMS FOR CHAPTER NINE

For the differential equations in Problems 1–15, find the equation of a solution which passes through the given point.

1. $\dfrac{dP}{dt} = 0.03P + 400\,, \quad P(0) = 0$

2. $\dfrac{dy}{dt} = y(10 - y)\,, \quad y(0) = 1$

3. $\dfrac{df}{dx} = \sqrt{xf(x)}\,, \quad f(1) = 1$

4. $\dfrac{dy}{dx} = \dfrac{y(3 - x)}{x(0.5y - 4)}\,, \quad y(1) = 5$

5. $\dfrac{dy}{dx} = e^{x-y}\,, \quad y(0) = 1$

6. $\dfrac{dk}{dt} = (1 + \ln t)k\,, \quad k(1) = 1$

7. $2\sin x - y^2\dfrac{dy}{dx} = 0\,, \quad y(0) = 3$

8. $1 + y^2 - \dfrac{dy}{dx} = 0\,, \quad y(0) = 0$

9. $\dfrac{dy}{dx} + xy^2 = 0\,, \quad y(1) = 1$

10. $\dfrac{dy}{dx} = \dfrac{y(100 - x)}{x(20 - y)}\,, \quad (1, 20)$

11. $e^{-\cos\theta}\dfrac{dz}{d\theta} = \sqrt{1 - z^2}\sin\theta, \quad z(0) = \frac{1}{2}$.

12. $\dfrac{dy}{dt} = 2^y \sin^3 t, \quad y(0) = 0$.

13. $(1 + t^2)y\dfrac{dy}{dt} = 1 - y, \quad y(1) = 0$.

14. $\dfrac{dy}{dx} = e^{x+y}, \quad y = 0$ where $x = 1$.

15. $\dfrac{dy}{dx} = \dfrac{0.2y(18 + 0.1x)}{x(100 + 0.5y)}\,, \quad (10, 10)$

Solve the differential equations in Problems 16–19. Assume $x \geq 0$, $y \geq 0$.

16. $\dfrac{dy}{dt} = \dfrac{y \ln y}{t^2}$.

17. $(y\sqrt{x^3 + 1})\dfrac{dy}{dx} + x^2y^2 + x^2 = 0$.

18. $\dfrac{dQ}{dt} + t^2Q^2 + Q^2 - 4t^2 - 4 = 0$.

19. $\dfrac{x\,dx}{y\,dy} = e^{(x/a)^2}\ln y \quad$ (a is constant)

For Problems 20–21, sketch solution curves with a variety of initial values for the differential equations. You do not need to find an equation for the solution.

20. $\dfrac{dy}{dt} = \alpha - y,$
where α is a positive constant.

21. $\dfrac{dw}{dt} = (w-3)(w-7)$

22. Consider the initial value problem

$$y' = 5 - y, \quad y(0) = 1.$$

(a) Use Euler's method with five steps to estimate $y(1)$.
(b) Sketch the slope field for this differential equation in the first quadrant, and use it to decide if your estimate is an over- or underestimate.
(c) Find the exact solution to the differential equation and hence find $y(1)$ exactly.
(d) Without doing the calculation, roughly what would you expect the approximation for $y(1)$ to be with ten steps?

23. Solve the differential equation

$$\frac{dy}{dx} = \frac{1}{(\cos x)(\cos y)}$$

by Euler's method, for the solution passing through $(0,0)$. Find y when $x = \frac{1}{2}$.

(a) Find the solution from Euler's method when $\Delta x = \frac{1}{2}$ (1 step),
(b) $\Delta x = \frac{1}{4}$ (2 steps),
(c) $\Delta x = \frac{1}{8}$ (4 steps).
(d) Solve the differential equation by separation of variables for y passing through $(0,0)$ and find $y(\frac{1}{2})$. Compare with your results in parts (a)–(c).

Find a general solution to the differential equations in Problems 24–27.

24. $y'' + 6y' + 8y = 0$

25. $9z'' + z = 0$

26. $9z'' - z = 0$

27. $x'' + 2x' + 10x = 0$

For each of the differential equations in Problems 28–29, find the values of b that make the general solution: (a) overdamped, (b) underdamped, (c) critically damped.

28. $s'' + bs' + 5s = 0.$

29. $s'' + bs' - 16s = 0.$

30. Explain why $s'' + bs' - s = 0$ cannot have a critically damped solution for any real number b.

31. Suppose that the rate at which a drug leaves the bloodstream and passes into the urine is proportional to the quantity of the drug in the blood at that time. If an initial dose of Q_0 is injected directly into the blood, only 20% is left in the blood after 3 hours.

(a) Write and solve a differential equation for the quantity, Q, of the drug in the blood at time, t, measured in hours.
(b) How much of this drug is in a patient's body after 6 hours if the patient is given 100 mg initially?

32. If a body at temperature T is placed in surroundings of a different temperature, then it will either gain or lose heat so as to acquire the temperature of the surroundings according to the differential equation $dT/dt = -k(T-A)$, where A is the the temperature of the surroundings. Here k is a positive constant that depends on the physical nature of the body.

(a) Why is k positive?
(b) Suppose the units in which time is measured are changed, say from days to hours, or from hours to minutes. Then k will change; how?
(c) Suppose the body in question is a cup of hot coffee. Will k be larger or smaller if the cup itself is styrofoam or thin china?
(d) Suppose the coffee comes out of the coffee machine at 170 degrees, the room temperature is 70, and the constant k is 0.14 when time is measured in minutes. How long must you wait to drink the coffee if you can't stand it above 120 degrees? How soon must you drink it, if you hate coffee below 90 degrees?

33. A bank account earns 10% annual interest, compounded continuously. Money is deposited in a continuous cash flow at a rate of \$1200 per year into the account.

(a) Write a differential equation that describes the rate at which the balance $B = f(t)$ is changing.
(b) Solve the differential equation given an initial balance $B_0 = 0$.
(c) Find the balance after 5 years.

34. (Continuation of Problem 33.) Now suppose the money is deposited once a month (instead of continuously) but still at a rate of \$1200 per year.

(a) Write down the sum that gives the balance after 5 years, assuming the first deposit is made one month from today, and today is $t = 0$.
(b) The sum you wrote in part (a) is a Riemann sum approximation to the integral

$$\int_0^5 1200e^{0.1t}dt.$$

Determine whether it is a left sum or right sum, and determine what Δt and N are. Then use your calculator to evaluate the sum.
(c) Compare your answer in part (b) to your answer to Problem 33(c).

35. A species of bird that eats only the rare poha berry is introduced to the Hawaiian Islands. As a result, the maximum number of birds that can be supported is 10,000. Hence, we choose to model the rate of growth of the bird's population, dP/dt, as proportional to the number, P, of birds currently on the islands, times the remaining capacity for birds, or $10{,}000-P$. At time $t = 0$ years, there are 10 birds on the islands, and their population is increasing at the rate $dP/dt = 100$ birds per year. Assume nothing disastrous happens to the birds.

(a) What differential equation describes the rate of growth of the bird population? Solve for the constant of proportionality from the given information.
(b) What is the maximum rate of growth of the bird population, and at what population does it occur?
(c) Sketch P, the solution to the differential equation, as a function of t.
(d) Find a formula for P as a function of t.

36. The following equations describe the rates of growth of an insect and bird population in a particular region, where x is the insect population in millions at time t and y is the bird population in thousands:

$$\frac{dx}{dt} = 3x - 0.02xy$$
$$\frac{dy}{dt} = -10y + 0.001xy$$

(a) Describe in words the growth of each population in the absence of the other, and describe in words their interaction.
(b) Find the two points (x, y) at which the populations are in equilibrium.
(c) Suppose that when the populations are at the nontrivial equilibrium, 10 thousand additional birds are suddenly introduced. Let A be the point in the phase plane representing these populations. Get a differential equation in terms of just x and y (i.e., eliminate t), and find an equation for the particular solution passing through the point A.
(d) Show that the following points lie on the trajectory in the phase plane that passes through point A: (i) B (9646.91, 150) (ii) C (10,000, 140.43) (iii) D (10,361.60, 150)
(e) Sketch this trajectory in the phase plane, with x on the horizontal axis, y on the vertical. Show the equilibrium point relative to this trajectory.
(f) In what order will the points A, B, C, D be traversed? [Hint: Find dy/dt, dx/dt at each point.]
(g) On another graph, sketch x and y versus time, t. Use the same initial value as in part (c). You do not need to indicate actual numerical values on the t-axis.
(h) Find dy/dx at points A and C. Find dx/dy at points B and D. [Hint: $dx/dy = 1/(dy/dx)$. Notice that the points A and C are the maximum and minimum values of y, respectively, and B and D are the maximum and minimum values of x, respectively.]

37. In this problem we adapt Lanchester's model for a conventional battle (described before Problem 13 on page 550), to the case in which one or both of the armies is a guerrilla force. We assume that the rate at which a guerrilla force is put out of action is proportional to the product of the strengths of the two armies.

(a) Give a justification for the assumption that the rate at which a guerrilla force is put out of action is proportional to the product of the strengths of the two armies.
(b) Write the differential equations which describe a conflict between a guerrilla army of strength x and a conventional army of strength y, assuming all the constants of proportionality are 1.
(c) Find a differential equation involving dy/dx and solve it to find equations of phase trajectories.
(d) If $C > 0$, which side wins? If $C < 0$, which side wins? What if $C = 0$?
(e) Use your solution to part (d) to divide the phase plane into regions according to which side wins.

38. To model a conflict between two guerrilla armies, we assume that the rate that each one is put out of action is proportional to the product of the strengths of the two armies.

(a) Write the differential equations which describe a conflict between two guerrilla armies, of strengths x and y, respectively.
(b) Find a differential equation involving dy/dx and solve to find equations of phase trajectories.
(c) If $C > 0$, which side wins? If $C < 0$, which side wins? What if $C = 0$?
(d) Use your solution to part (b) to divide the phase plane into regions according to which side wins.

39. When two countries are in an arms race, the rate at which each one spends money on arms is determined by its own current level of spending and by its opponent's level of spending. We expect that the more a country is already spending on armaments, the less willing it will be to increase its military expenditures. On the other hand, the more a country's opponent spends on armaments, the more rapidly the country will arm. If $\$x$ billion is the country's yearly expenditure on arms, and $\$y$ billion is its opponent's, then the *Richardson arms race model* proposes that x and y are determined by differential equations. Suppose that for some particular arms race the equations are

$$\frac{dx}{dt} = -0.2x + 0.15y + 20$$
$$\frac{dy}{dt} = 0.1x - 0.2y + 40.$$

(a) Explain the signs of the three terms on the right side of the equations for dx/dt.
(b) Find the nullclines and equilibrium points for this system.
(c) Analyze the direction of the trajectories in each region.
(d) Are the equilibrium points stable or unstable?
(e) What does this model predict will happen if both countries disarm?
(f) What does this model predict will happen in the case of unilateral disarmament (one country disarms, and the other country does not)?
(g) What does the model predict will happen in the long run?

40. Juliet is in love with Romeo, who happens (in our version of this story) to be a fickle lover. The more Juliet loves him, the more he begins to dislike her. When she hates him, his feelings for her warm up. On the other hand, her love for him grows when he loves her and withers when he hates her. A model for their ill-fated romance is

$$\frac{dj}{dt} = Ar, \qquad \frac{dr}{dt} = -Bj,$$

where A and B are positive constants, $r(t)$ represents Romeo's love for Juliet at time t, and $j(t)$ represents Juliet's love for Romeo at time t. (Negative love is hate.)

(a) The constant on the right-hand side of Juliet's equation (the one including dj/dt) has a positive sign, whereas the constant in Romeo's equation is negative. Explain why these signs follow from the story.
(b) Derive a second-order differential equation for $r(t)$ and solve it. (Your equation should involve r and its derivatives, but not j and its derivatives.)
(c) Express $r(t)$ and $j(t)$ as functions of t, given $r(0) = 1$ and $j(0) = 0$. Your answer will contain A and B.
(d) As you may have discovered, the outcome of the relationship is a never-ending cycle of love and hate. Find what fraction of the time they both love one another.

CHAPTER TEN

APPROXIMATIONS

In this chapter we will see how to approximate functions by simpler functions. One way to think about this is through an analogy with numbers. Numbers can be described by exact expressions, such as $\frac{1}{2}$, $\sqrt{2}$, or π, or they can be given in decimal form, which may not be exact. For example,

$$\begin{aligned}\frac{1}{2} &= 0.5\\ \sqrt{2} &= 1.4142\ldots\\ \pi &= 3.14159\ldots.\end{aligned}$$

The first equation is exact, but the other two are not. In the last two cases the dots indicate that there are more digits in the decimal expansion, and that the decimal on the right side of the equation is only an approximation.

Just as rational numbers are the simplest sorts of numbers, so are rational functions; in particular, polynomials are the simplest sorts of functions because they can be evaluated by "hand," unlike the transcendental functions. In a Taylor approximation, we construct something like a decimal expansion that enables us to approximate functions by polynomials. Fourier approximations use trigonometric functions instead of polynomials. Taylor approximations are generally good approximations to the function locally (that is, near a specific point), whereas Fourier approximations are generally good approximations over an interval.

10.1 TAYLOR POLYNOMIALS

In this section, we approximate a function by polynomials. To get more accuracy, we take a higher-degree polynomial. The approximation is usually good near one particular point, but often not so good further away from that point.

Linear Approximations

We have already seen how to approximate a function using a tangent line. The tangent line at a point is the best linear approximation to the function at that point; it is the line through that point which lies closest to the graph of the function there. Figure 10.1 shows the graph of a function and its tangent line. Notice that the tangent line and the curve have the same slope at the point of tangency. This means they are both climbing up or down at the same rate there. As in Chapters 2 and 4, we have the following equation.

Tangent Line Approximation of $f(x)$ for x near a

$$f(x) \approx f(a) + f'(a)(x - a).$$

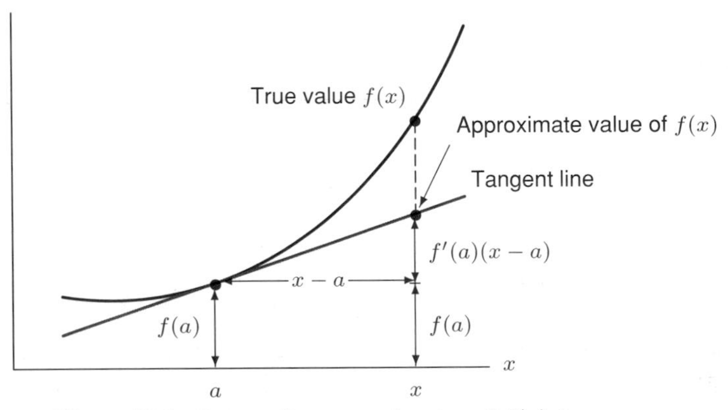

Figure 10.1: Tangent line approximation of $f(x)$ for x near a

Example 1 Approximate $f(x) = \cos x$, with x in radians, by its tangent line at $x = 0$.

Solution The tangent line at $x = 0$ is just the horizontal line $y = 1$, as shown in Figure 10.2, so

$$f(x) = \cos x \approx 1, \quad \text{for } x \text{ near } 0.$$

If we take $x = 0.05$, then

$$f(x) = \cos(0.05) = 0.999,$$

which is quite close to the approximation $\cos x \approx 1$. Similarly, if $x = -0.1$, then

$$f(x) = \cos(-0.1) = 0.995 \approx 1.$$

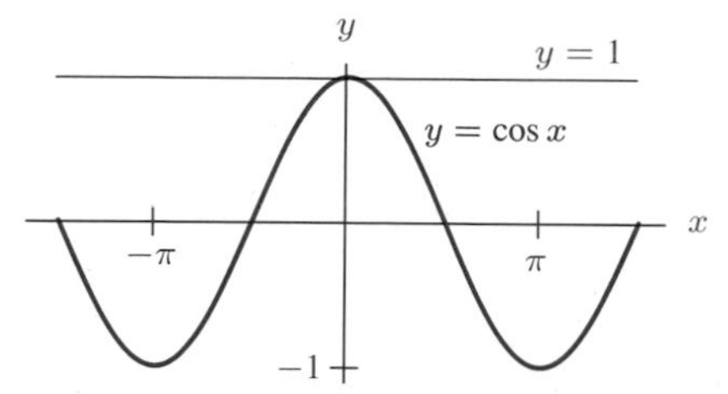

Figure 10.2: Graph of $\cos x$ and its tangent line at $x = 0$

However, if $x = 0.4$, then

$$f(x) = \cos(0.4) = 0.921,$$

so the approximation $\cos x \approx 1$ is less accurate. The further a point x is away from the original point, i.e., 0, the worse the corresponding approximation is likely to be.

Notice that if $a = 0$, we have the following approximation:

> **Tangent Line Approximation of $f(x)$ for x near 0**
>
> $$f(x) \approx f(0) + f'(0)x.$$
>
> This approximation is also called the *Taylor polynomial* of degree 1, or $P_1(x)$.

Why Do This?

At this point, you are undoubtedly wondering why we should be interested in approximating the values for $\cos x$ at all. After all, you can obtain very accurate approximations simply by punching the right keys on a calculator. On the other hand, have you ever wondered how a calculator actually generates the values for the cosine function? What would happen if you were to face some new function whose values cannot be found directly from a calculator? Some of the most effective methods for calculating such values involve approximating one function by another, simpler function, such as by a tangent line approximation.

Quadratic Approximations

Suppose we want a more accurate way of approximating $f(x) = \cos x$ for x near 0. The problem with using a linear approximation, as we did above, is that the graph of f usually bends away from its tangent line. Consequently, it makes sense to approximate the function not by a line, but by a curve which bends in the same way as the original curve. We will require that at $x = 0$ the graphs of the original function f and the approximating function have the same slope, $f'(0)$, and that they bend at the same rate — in other words, that they have the same second derivative, $f''(0)$. The simplest type of function we might use for this approximation is a quadratic polynomial

$$P_2(x) = C_0 + C_1x + C_2x^2,$$

where we must determine the values for C_0, C_1, and C_2. At $x = 0$, we want the two functions to agree, so $P_2(0) = f(0)$; we want the two functions to have the same derivative at this point, so $P_2'(0) = f'(0)$; and we also want the two functions to have the same second derivative, which gives $P_2''(0) = f''(0)$. These three conditions will determine C_0, C_1, and C_2.

Example 2 Find the quadratic approximation to $f(x) = \cos x$ for x near 0.

Solution We find C_0, C_1, C_2 using $P_2(0) = f(0)$, $P_2'(0) = f'(0)$, $P_2''(0) = f''(0)$.
Since

$$\begin{aligned} P_2(x) &= C_0 + C_1x + C_2x^2 \\ P_2'(x) &= C_1 + 2C_2x \\ P_2''(x) &= 2C_2 \end{aligned} \quad \text{and} \quad \begin{aligned} f(x) &= \cos x \\ f'(x) &= -\sin x \\ f''(x) &= -\cos x, \end{aligned}$$

we have

$$\begin{aligned} C_0 &= P_2(0) = f(0) = \cos 0 = 1 \\ C_1 &= P_2'(0) = f'(0) = -\sin 0 = 0 \\ 2C_2 &= P_2''(0) = f''(0) = -\cos 0 = -1 \end{aligned} \quad \text{so} \quad \begin{aligned} C_0 &= 1 \\ C_1 &= 0 \\ C_2 &= -\tfrac{1}{2} \end{aligned}$$

Consequently, the quadratic approximation is

$$\cos x \approx P_2(x) = 1 + 0 \cdot x - \frac{1}{2}x^2 = 1 - \frac{x^2}{2}, \quad \text{for } x \text{ near } 0.$$

From Figure 10.3, you can see that the quadratic approximation $\cos x \approx P_2(x)$ is better than the linear approximation $\cos x \approx P_1(x)$ for x near 0. Let's compare the accuracy of the two approximations numerically. At $x = 0.4$, $\cos(0.4) = 0.921$ and $P_2(0.4) = 0.920$, so the quadratic approximation is a significant improvement over the linear approximation. (The error is about 0.001 instead of 0.08.) In addition, the quadratic approximation is much better than the linear approximation for x values near 0: $\cos(0.1) = 0.995004$ and $P_2(0.1) = 0.995$.

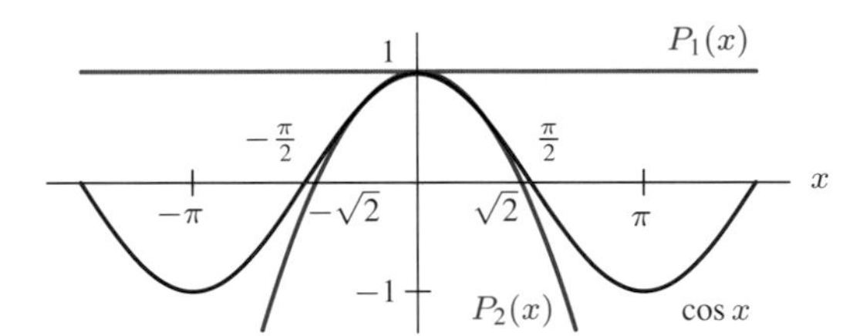

Figure 10.3: Graph of $\cos x$ and its linear, $P_1(x)$, and quadratic, $P_2(x)$, approximations for x near 0

In general, suppose we have a function $f(x)$ that is defined, along with its first two derivatives, in an interval about $x = 0$. We can construct a quadratic polynomial approximating $f(x)$ for x near 0 by the same steps we used above. Assume that the quadratic polynomial is

$$P_2(x) = C_0 + C_1x + C_2x^2.$$

We choose C_0, C_1, and C_2 so that the original function $f(x)$ and the quadratic polynomial $P_2(x)$ have the same value and the same first and second derivatives at $x = 0$, and obtain what is called a Taylor polynomial of degree 2.

Taylor Polynomial of Degree 2 Approximating $f(x)$ for x near 0

$$f(x) \approx P_2(x) = f(0) + f'(0)x + \frac{f''(0)}{2}x^2$$

Higher-Degree Polynomials

As a rule, over a small interval around $x = 0$, the quadratic approximation to a function will be a better approximation than the linear (tangent line) approximation. However, Figure 10.3 shows that even though we have matched up the function and the quadratic in terms of their values, their slopes, and their concavity at the point $x = 0$, they usually still bend away from one another for large x. We can fix this somewhat by using an approximating polynomial of higher degree. Suppose that we want to approximate a function $f(x)$ for x near 0 by a polynomial of degree n:

$$f(x) \approx P_n(x) = C_0 + C_1x + C_2x^2 + \cdots + C_{n-1}x^{n-1} + C_nx^n.$$

We now need to find the values of the constants: $C_0,\ C_1,\ C_2, \ldots,\ C_n$. To do this, we require that the function $f(x)$ and all of its first n derivatives agree with those of the polynomial $P_n(x)$ at the point $x = 0$. Notice that the higher-order derivatives of a function contribute more subtle information about its shape than the first two derivatives do. (For instance, the third derivative measures how fast the concavity changes.) The more derivatives there are that agree at $x = 0$, the longer the function and the polynomial are likely to remain close to each other.

To see how to find the constants, let's take as an example

$$f(x) \approx P_3(x) = C_0 + C_1x + C_2x^2 + C_3x^3.$$

Substituting $x = 0$ gives

$$f(0) = P_3(0) = C_0.$$

Differentiating $P_3(x)$ yields

$$P_3'(x) = C_1 + 2C_2x + 3C_3x^2,$$

so substituting $x = 0$ shows that

$$f'(0) = P_3'(0) = C_1.$$

Differentiating and substituting again, we get

$$P_3''(x) = 2 \cdot 1C_2 + 3 \cdot 2 \cdot 1C_3x,$$

which gives

$$f''(0) = P_3''(0) = 2C_2,$$

so that

$$C_2 = \frac{f''(0)}{2}.$$

The third derivative, denoted by P_3''', is

$$P_3'''(x) = 3 \cdot 2 \cdot 1C_3,$$

so that

$$f'''(0) = P_3'''(0) = 3 \cdot 2 \cdot 1 C_3,$$

and so

$$C_3 = \frac{f'''(0)}{3 \cdot 2 \cdot 1}.$$

You can imagine a similar calculation starting with $P_4(x)$, using the fourth derivative $f^{(4)}$, which would give

$$C_4 = \frac{f^{(4)}(0)}{4 \cdot 3 \cdot 2 \cdot 1},$$

and so on. Using factorial notation,[1] we write these expressions as

$$C_3 = \frac{f'''(0)}{3!}, \quad C_4 = \frac{f^{(4)}(0)}{4!}.$$

In general, for any integer k between 1 and n

$$C_k = \frac{f^{(k)}(0)}{k!},$$

where $f^{(k)}$ means the kth derivative of f.

Taylor Polynomial of Degree *n* Approximating *f*(*x*) for *x* near *0*

$$\begin{aligned} f(x) &\approx P_n(x) \\ &= f(0) + f'(0)x + \frac{f''(0)}{2!}x^2 + \frac{f'''(0)}{3!}x^3 + \frac{f^{(4)}(0)}{4!}x^4 + \cdots + \frac{f^{(n)}(0)}{n!}x^n \end{aligned}$$

We call this a Taylor polynomial centered at $x = 0$, or a Taylor polynomial about $x = 0$.

Example 3 Construct the Taylor polynomial of degree 7 approximating the function $f(x) = \sin x$ for x near 0. Compare the value of the Taylor approximation with the true value of f at $x = \pi/3$.

Solution We have

$$\begin{aligned} f(x) &= \sin x \quad \text{giving} & f(0) &= 0 \\ f'(x) &= \cos x & f'(0) &= 1 \\ f''(x) &= -\sin x & f''(0) &= 0 \\ f'''(x) &= -\cos x & f'''(0) &= -1 \\ f^{(4)}(x) &= \sin x & f^{(4)}(0) &= 0 \\ f^{(5)}(x) &= \cos x & f^{(5)}(0) &= 1 \\ f^{(6)}(x) &= -\sin x & f^{(6)}(0) &= 0 \\ f^{(7)}(x) &= -\cos x & f^{(7)}(0) &= -1. \end{aligned}$$

[1] We define 3! (read *three factorial*) by $3! = 3 \cdot 2 \cdot 1 = 6$, $4! = 4 \cdot 3 \cdot 2 \cdot 1 = 24$, and $k! = k(k-1) \cdots 2 \cdot 1$.

Using these values, we see that the Taylor polynomial approximation of degree 7 is

$$\begin{aligned}\sin x \approx P_7(x) &= 0 + x + 0\cdot\frac{x^2}{2!} - \frac{x^3}{3!} + 0\cdot\frac{x^4}{4!} + \frac{x^5}{5!} + 0\cdot\frac{x^6}{6!} - \frac{x^7}{7!}\\ &= x - \frac{x^3}{3!} + \frac{x^5}{5!} - \frac{x^7}{7!}, \quad \text{for } x \text{ near } 0.\end{aligned}$$

In Figure 10.4 we show the graphs of the sine function and the approximating polynomial of degree 7 for x near 0. They are indistinguishable where x is close to 0. However, as we look at values of x further away from 0 in either direction, the two graphs diverge. To see the accuracy of this approximation numerically, we see how well it approximates $\sin(\pi/3) = \sqrt{3}/2 \approx 0.8660254$. When we substitute $\pi/3 \approx 1.0471976$ into the polynomial approximation, we obtain $P_7(\pi/3) \approx 0.8660213$, which is extremely accurate—to about five parts in a million.

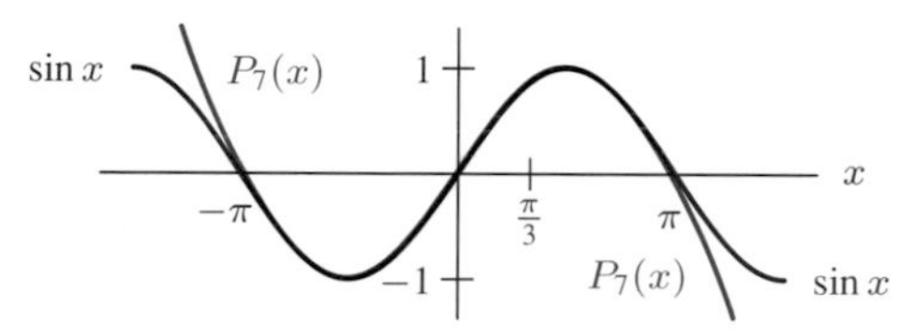

Figure 10.4: Graph of $\sin x$ and its seventh degree Taylor polynomial, $P_7(x)$, for x near 0

Example 4 Show that the Taylor polynomial of degree 8 approximating $f(x) = \cos x$ for x near 0 is

$$\cos x \approx P_8(x) = 1 - \frac{x^2}{2!} + \frac{x^4}{4!} - \frac{x^6}{6!} + \frac{x^8}{8!}.$$

Solution We can find the coefficients by the method of the preceding example. Figure 10.5 shows that $P_8(x)$ is close to the cosine function for more x values than the quadratic approximation $P_2(x) = 1 - x^2/2$ that we found in Example 2 on page 592.

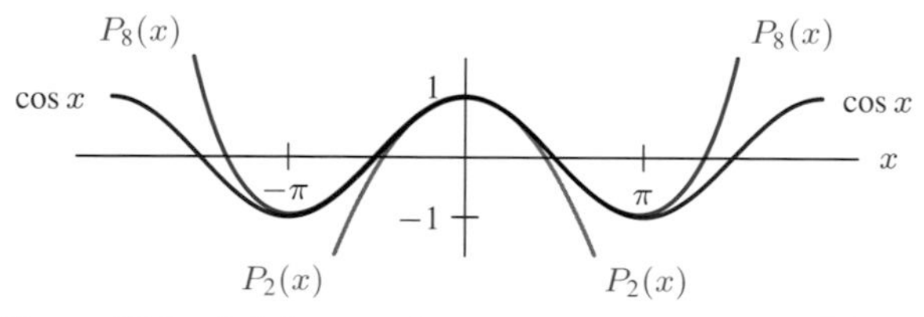

Figure 10.5: $P_8(x)$ approximates $\cos x$ better than $P_2(x)$ for x near 0

Example 5 Construct the Taylor polynomial of degree 10 about $x = 0$ for the function $f(x) = e^x$.

Solution We have $f(0) = 1$. Since the derivative of e^x is equal to e^x, all the higher-order derivatives will be equal to e^x. Consequently, for any $k = 1, 2, \ldots, 10$, $f^{(k)}(x) = e^x$ and $f^{(k)}(0) = e^0 = 1$. Therefore, the Taylor polynomial approximation of degree 10 is given by

$$e^x \approx P_{10}(x) = 1 + x + \frac{x^2}{2!} + \frac{x^3}{3!} + \frac{x^4}{4!} + \cdots + \frac{x^{10}}{10!}, \quad \text{for } x \text{ near } 0.$$

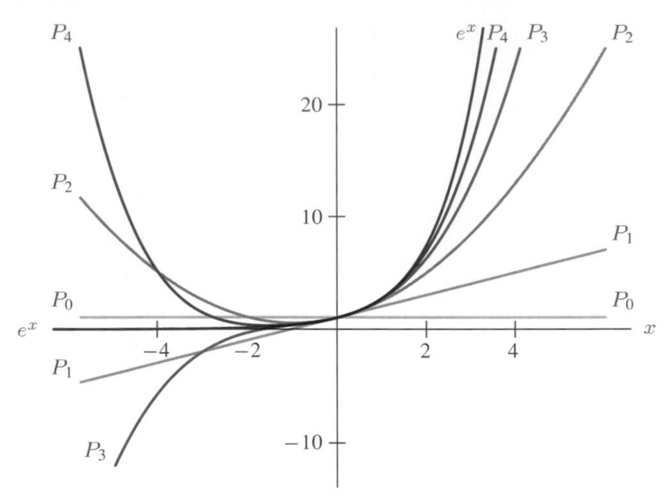

Figure 10.6: For x near 0, e^x is more closely approximated by higher-degree Taylor polynomials

To verify the accuracy of this approximation, suppose we use it to approximate $e = e^1$ which is 2.718281828 correct to nine decimal places. If we substitute $x = 1$ into the approximating polynomial, we obtain $P_{10}(1) = 2.718281801$. Thus, this tenth-degree polynomial yields the first seven decimal places for e. For large values of x, however, the accuracy must diminish because e^x grows much faster than any polynomial as $x \to \infty$. Figure 10.6 shows graphs of $f(x) = e^x$ and the Taylor polynomials of degree $n = 0, 1, 2, 3, 4$. Notice that each successive approximation remains close to the exponential curve for a larger interval of x values.

Example 6 Construct the Taylor polynomial of degree 5 approximating $f(x) = \dfrac{1}{1-x}$ for x near 0.

Solution You can check, by differentiating, that $f(0) = 1$, $f'(0) = 1$, $f''(0) = 2$, $f'''(0) = 3!$, $f^{(4)}(0) = 4!$, and $f^{(5)}(0) = 5!$, so

$$\frac{1}{1-x} \approx P_5(x) = 1 + x + x^2 + x^3 + x^4 + x^5, \quad \text{for } x \text{ near } 0.$$

Example 7 Construct the Taylor polynomial of degree 5 about $x = 0$ for the function $f(x) = \ln x$.

Solution As soon as we notice that $f(0)$ is not defined, we realize there is no such polynomial.

Moving Away from $x = 0$

The graph of $f(x) = \ln x$ in Figure 10.7 shows why there is no Taylor polynomial about $x = 0$ for $\ln x$. After all, the function is not defined for $x = 0$ or for $x < 0$. Can we find another Taylor polynomial for $f(x) = \ln x$? Although there is no tangent line at $x = 0$, the graph of $\ln x$ certainly does have tangent lines for $x > 0$.

Thus, rather than constructing a polynomial approximation for $f(x)$ about $x = 0$, let's construct a polynomial centered about some other point, $x = a$. Before constructing a polynomial approximating

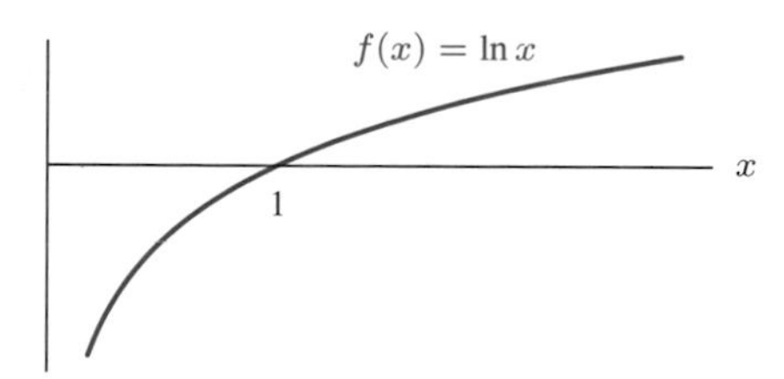

Figure 10.7: The natural logarithm is not defined for values less than or equal to 0

$f(x)$ for x near a, let's look at the equation of the tangent line at $x = a$. Recall that since the tangent line goes through the point $(a, f(a))$ and has slope $f'(a)$, its equation is

$$y = f(a) + f'(a)(x - a).$$

This gives the approximation

$$f(x) \approx f(a) + f'(a)(x - a) \quad \text{for } x \text{ near } a.$$

The $f'(a)(x - a)$ term is a correction term which approximates how much $f(x)$ moves away from $f(a)$ as x moves away from a.

The approximating polynomial $P_n(x)$ centered at $x = a$ will be set up as $f(a)$ plus correction terms which depend on the derivatives of $f(x)$ and which are zero for $x = a$. This is achieved by writing the polynomial in powers of $(x - a)$ instead of powers of x:

$$f(x) \approx P_n(x) = C_0 + C_1(x - a) + C_2(x - a)^2 + \cdots + C_n(x - a)^n.$$

If we require the derivatives of the approximating polynomial $P_n(x)$ and the original function $f(x)$ to agree at $x = a$, we get the following result.

Taylor Polynomial of Degree *n* Approximating *f*(*x*) for *x* near *a*

$$\begin{aligned} f(x) &\approx P_n(x) \\ &= f(a) + f'(a)(x - a) + \frac{f''(a)}{2!}(x - a)^2 + \cdots\cdots + \frac{f^{(n)}(a)}{n!}(x - a)^n \end{aligned}$$

We call this a Taylor polynomial centered at $x = a$, or a Taylor polynomial about $x = a$.

You can derive these coefficients in the same way as for $a = 0$. (See Problem 35, page 599.)

Example 8 Construct the Taylor polynomial of degree 4 approximating the function $f(x) = \ln x$ for x near 1.

Solution We have

$$\begin{aligned} f(x) &= \ln x & \quad \text{so} \quad f(1) &= \ln(1) = 0 \\ f'(x) &= 1/x & f'(1) &= 1 \\ f''(x) &= -1/x^2 & f''(1) &= -1 \\ f'''(x) &= 2/x^3 & f'''(1) &= 2 \\ f^{(4)}(x) &= -6/x^4 & f^{(4)}(1) &= -6 \end{aligned}$$

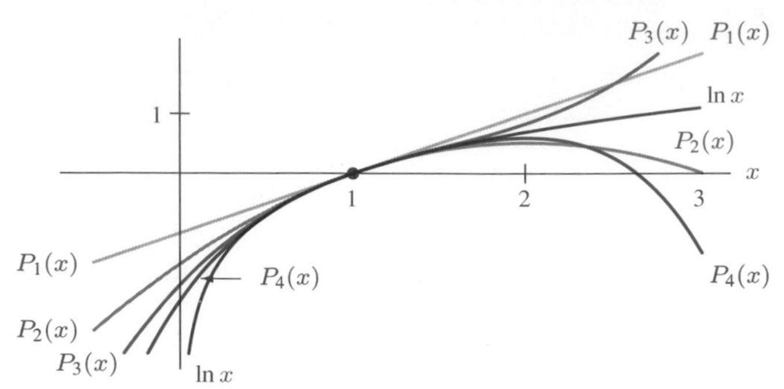

Figure 10.8: Taylor polynomials approximate $\ln x$ closely for x near 1 but not necessarily further away

so that the Taylor polynomial is

$$\ln x \approx P_4(x) = 0 + (x-1) - \frac{(x-1)^2}{2!} + 2\frac{(x-1)^3}{3!} - 6\frac{(x-1)^4}{4!}$$

$$= (x-1) - \frac{(x-1)^2}{2} + \frac{(x-1)^3}{3} - \frac{(x-1)^4}{4}, \quad \text{for } x \text{ near } 1.$$

We show $\ln x$ and several Taylor polynomials in Figure 10.8. You will notice that $P_4(x)$ stays reasonably close to $\ln x$ for x near 1, but bends away as x gets further from 1. Also, note that the Taylor polynomials are defined for $x \le 0$, but $\ln x$ is not.

The moral of the story is that you can be sure that the Taylor polynomials are a good approximation only when you are "close to home" — for x near a. Further away, they may be good or they may not be.

Problems for Section 10.1

For Problems 1–10, find the Taylor polynomials of degree n approximating the given functions for x near 0.

1. $\cos x, \quad n = 2, 4, 6$
2. $\sqrt{1+x}, \quad n = 2, 3, 4$
3. $\dfrac{1}{1-x}, \quad n = 3, 5, 7$
4. $\dfrac{1}{1+x}, \quad n = 4, 6, 8$
5. $\tan x, \quad n = 3, 4$
6. $\arctan x, \quad n = 3, 4$
7. $\ln(1+x), \quad n = 5, 7, 9$
8. $\sqrt[3]{1-x}, \quad n = 2, 3, 4$
9. $\dfrac{1}{\sqrt{1+x}}, \quad n = 2, 3, 4$
10. $(1+x)^p, \quad n = 2, 3, 4$ (p is a constant.)
11. Show how the results of Problems 2, 3, 4, 8, and 9 are special cases of the result of Problem 10.

For Problems 12–22, find the Taylor polynomial of degree n for x near the given point a.

12. $e^x, \quad a = 1, \quad n = 4$
13. $\cos x, \quad a = \pi/2, \quad n = 4$
14. $\sin x, \quad a = \pi/2, \quad n = 4$
15. $\cos x, \quad a = \pi/4, \quad n = 3$
16. $\sin x, \quad a = -\pi/4, \quad n = 3$
17. $\sqrt{1-x}, \quad a = 0, \quad n = 3$
18. $\sqrt{1+x}, \quad a = 1, \quad n = 3$
19. $\dfrac{1}{1+x}, \quad a = 2, \quad n = 4$

20. $\dfrac{1}{1-x}, \quad a = 2, \quad n = 4$

21. $\ln x, \quad a = 1, \quad n = 6$
[Hint: Compare with Example 8, page 597.]

22. $\ln x, \quad a = 2, \quad n = 4$

For Problems 23–26, suppose $P_2(x) = a + bx + cx^2$ is the second degree Taylor polynomial for the function f about $x = 0$. What can you say about the signs of a, b, c if f has the graph given below?

23.

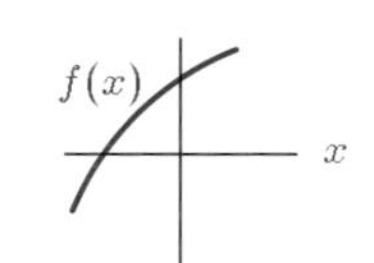

24.

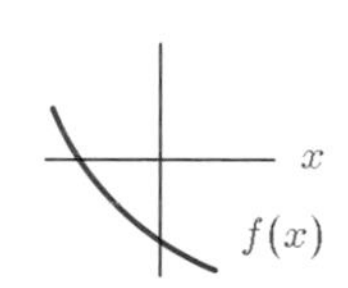

25.

26.

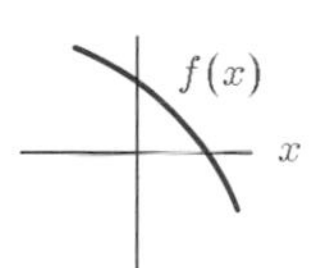

27. Show how you can use the Taylor approximation $\sin x \approx x - \dfrac{x^3}{3!}$, for x near 0, to explain why $\lim\limits_{x\to 0} \dfrac{\sin x}{x} = 1$.

28. Use the fourth-degree Taylor approximation $\cos x \approx 1 - \dfrac{x^2}{2!} + \dfrac{x^4}{4!}$, for x near 0, to evaluate $\lim\limits_{x\to 0} \dfrac{1-\cos x}{x^2}$.

29. Use a fourth degree Taylor approximation for e^h, for h near 0, to evaluate the following limits. Would your answer be different if you used a Taylor polynomial of higher degree?
   (a) $\lim\limits_{h\to 0} \dfrac{e^h - 1 - h}{h^2}$
   (b) $\lim\limits_{h\to 0} \dfrac{e^h - 1 - h - \frac{h^2}{2}}{h^3}$

30. When we model the motion of a pendulum, we replace the differential equation

$$\frac{d^2\theta}{dt^2} = -\frac{g}{l}\sin\theta \quad \text{by} \quad \frac{d^2\theta}{dt^2} = -\frac{g}{l}\theta$$

where θ is the angle between the pendulum and the vertical. Explain why, and under what circumstances, it is reasonable to make this replacement.

31. Find the second-degree Taylor polynomial for $f(x) = 4x^2 - 7x + 2$ about $x = 0$. What do you notice?

32. Find the third-degree Taylor polynomial for $f(x) = x^3 + 7x^2 - 5x + 1$ about $x = 0$. What do you notice?

33. (a) Based on your observations in Problems 31–32, make a conjecture about Taylor approximations in the case when f is itself a polynomial.
   (b) Show that your conjecture is true.

34. (a) Find the Taylor polynomial approximation of degree 4 about $x = 0$ for the function $f(x) = e^{x^2}$.
   (b) Compare this result to the Taylor polynomial approximation of degree 2 for the function $f(x) = e^x$ about $x = 0$. What do you notice?
   (c) Use your observation in part (b) to write out the Taylor polynomial approximation of degree 20 to the function in part (a).
   (d) What is the Taylor polynomial approximation of degree 5 for the function $f(x) = e^{-2x}$?

35. Derive the formulas given in the box on page 597 for the coefficients of the Taylor polynomial approximating a function f for x near a.

10.2 TAYLOR SERIES

We have just seen how to approximate a function near a point by Taylor polynomials. Now we will define a Taylor series, which can be thought of as a Taylor polynomial that goes on forever, and look at where such a series serves as a good approximation to the function.

In the preceding section, we calculated Taylor polynomials centered at $x = 0$ such as those given below for $\cos x$:

$$\cos x \approx P_0(x) = 1$$
$$\cos x \approx P_2(x) = 1 - \frac{x^2}{2!}$$
$$\cos x \approx P_4(x) = 1 - \frac{x^2}{2!} + \frac{x^4}{4!}$$
$$\cos x \approx P_6(x) = 1 - \frac{x^2}{2!} + \frac{x^4}{4!} - \frac{x^6}{6!}$$
$$\cos x \approx P_8(x) = 1 - \frac{x^2}{2!} + \frac{x^4}{4!} - \frac{x^6}{6!} + \frac{x^8}{8!}$$

Here we have a sequence of polynomials, $P_0(x)$, $P_2(x)$, $P_4(x)$, $P_6(x)$, $P_8(x)$, ..., each of which is a better approximation to $\cos x$ than the last, at least for x near 0. Notice that when you go to a higher-degree polynomial (say from P_6 to P_8), you add more terms ($x^8/8!$, for example), but the terms of lower degree don't change. Thus each polynomial includes all the previous ones. We will write the *Taylor series* for $\cos x$:

$$T(x) = 1 - \frac{x^2}{2!} + \frac{x^4}{4!} - \frac{x^6}{6!} + \frac{x^8}{8!} - \cdots$$

to represent the whole sequence of polynomials.

What Does $\cdots$ Mean? The three dots indicate that the terms in the series go on forever. This is the same idea as in $\pi = 3.1415\ldots$ where the three dots mean that the digits go on forever after the 5.

$$1 - \frac{x^2}{2!} + \frac{x^4}{4!} \quad \text{is a finite polynomial of degree 4,}$$

whereas

$$1 - \frac{x^2}{2!} + \frac{x^4}{4!} - \cdots \quad \text{is an infinite series, and continues forever.}$$

Convergence of Taylor Series for cos *x*

We have already seen that the Taylor polynomials centered at $x = 0$ are good approximations to $\cos x$ for x near 0. (See Figure 10.9.) Amazingly enough, for any value of x, if you take a Taylor polynomial centered at $x = 0$ of high enough degree, its graph is indistinguishable from the graph of the cosine near that point.

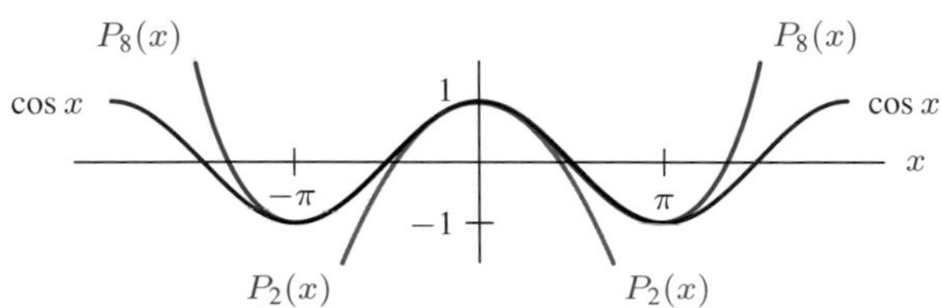

Figure 10.9: Graph of $\cos x$ and two Taylor polynomials for x near 0

Let's see what happens numerically. Suppose we take $x = \pi/2$. The successive Taylor approximations to $\cos(\pi/2) = 0$ are

$$\begin{aligned}
P_2(\pi/2) &= 1 - (\pi/2)^2/2! &&= -0.23370\\
P_4(\pi/2) &= 1 - (\pi/2)^2/2! + (\pi/2)^4/4! &&= 0.01997\\
P_6(\pi/2) &= \cdots &&= -0.00089\\
P_8(\pi/2) &= \cdots &&= 0.00002
\end{aligned}$$

and we see that the approximations converge to the true value very rapidly. If we take a value of x somewhat further away from 0, say $x = \pi$, then $\cos\pi = -1$ and

$$\begin{aligned}
P_2(\pi) &= 1 - (\pi)^2/2! &&= -3.93480\\
P_4(\pi) &= \cdots &&= 0.12391\\
P_6(\pi) &= \cdots &&= -1.21135\\
P_8(\pi) &= \cdots &&= -0.97602\\
P_{10}(\pi) &= \cdots &&= -1.00183\\
P_{12}(\pi) &= \cdots &&= -0.99990\\
P_{14}(\pi) &= \cdots &&= -1.000004
\end{aligned}$$

and we see that the rate of convergence is somewhat slower. If x were taken still further away from 0, then we would need still more terms to obtain an accurate approximation of $\cos x$. However it turns out that no matter how large x is, either positive or negative, the approximations $P_n(x)$ will converge to $\cos x$ as n gets larger and larger.

Thus, we feel justified in replacing the "$\approx$" by "$=$", and writing

$$\cos x = T(x) = 1 - \frac{x^2}{2!} + \frac{x^4}{4!} - \frac{x^6}{6!} + \frac{x^8}{8!} - \cdots \qquad \text{for all } x.$$

We say that the Taylor series *converges* to $\cos x$ for all x, because for all x, the numbers

$$P_1(x),\ P_2(x),\ P_3(x), \ldots,\ P_n(x), \ldots$$

converge to $\cos x$ as $n \to \infty$.

Taylor Series for sin x, cos x, e^x

Similarly, we can define the Taylor series for $\sin x$ and e^x. It turns out that they, too, converge for all x. Thus, we can write the following.

For all x,

$$\begin{aligned}
\sin x &= x - \frac{x^3}{3!} + \frac{x^5}{5!} - \frac{x^7}{7!} + \frac{x^9}{9!} - \cdots\\
\cos x &= 1 - \frac{x^2}{2!} + \frac{x^4}{4!} - \frac{x^6}{6!} + \frac{x^8}{8!} - \cdots\\
e^x &= 1 + x + \frac{x^2}{2!} + \frac{x^3}{3!} + \frac{x^4}{4!} + \cdots
\end{aligned}$$

These series are also called *Taylor expansions* of the functions $\sin x$, $\cos x$, and e^x.

Taylor Series in General

The series for $\sin x$, $\cos x$, and e^x are all examples of Taylor series. Any function f, all of whose derivatives exist at 0, has a Taylor series. However, for many functions f the series does not converge to $f(x)$ for all values of x. Assuming that the series does converge to $f(x)$, we have the following general formula.

Taylor Series for $f(x)$ about $x = 0$

$$f(x) = f(0) + f'(0)x + \frac{f''(0)}{2!}x^2 + \frac{f'''(0)}{3!}x^3 + \cdots$$

In addition, just as we have Taylor polynomials centered at points other than 0, so (provided all the derivatives exist at $x = a$) we can have a Taylor series centered at $x = a$. Assuming the series converges to $f(x)$, we have the following result.

Taylor Series for $f(x)$ about $x = a$

$$f(x) = f(a) + f'(a)(x-a) + \frac{f''(a)}{2!}(x-a)^2 + \frac{f'''(a)}{3!}(x-a)^3 + \cdots$$

Again, for many functions f, this Taylor series may converge to $f(x)$ only for x at, or near, a.

Intervals of Convergence

Let us look again at the Taylor polynomials for $\ln x$ about $x = 1$ that we derived on page 597.

Graphical Viewpoint

What happens if we attempt to improve the approximation to $\ln x$ by using higher-degree Taylor polynomials? Figure 10.10 suggests that for $0 < x < 2$, the polynomials fit the curve well, though outside that interval they are not close at all. In fact for $0 < x < 2$, the higher the degree of the polynomial, the better it fits the curve. Thus we say that, for $0 < x < 2$, the polynomials

$$P_5(x), P_6(x), P_7(x), \ldots, P_n(x), \ldots$$

converge to $\ln x$ as $n \to \infty$. For such x values, a higher-degree polynomial will, in general, give a better approximation.

However, when $x > 2$, the polynomials move away from the curve and the approximations get worse as the degree of the polynomial increases. For such x values, we say the polynomials *diverge* from the original function.

Thus, the Taylor polynomial approximations about $x = 1$ are effective only as approximations to $\ln x$ for values of x between 0 and 2; beyond that, they definitely should not be used. We say the *interval of convergence* of the Taylor polynomials is $0 < x < 2$. As x gets near 0 or 2, the polynomials converge very slowly. This means you might have to take a polynomial of very high degree to get an accurate value for $\ln x$. At the endpoints of the interval of convergence, in this case $x = 0$ and $x = 2$, the series may or may not converge. We will not consider the question. For us, an interval of convergence is always an open interval.

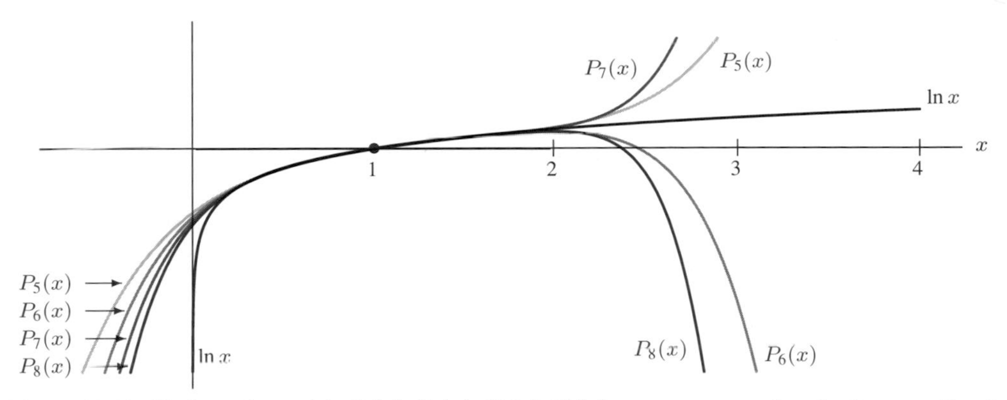

Figure 10.10: Taylor polynomials $P_5(x), P_6(x), P_7(x), P_8(x), \ldots$ converge to $\ln x$ for $0 < x < 2$ and diverge outside that interval

Numerical Viewpoint

To illustrate the interval of convergence numerically, consider the successive approximations obtained from the Taylor polynomial

$$\ln x \approx (x-1) - \frac{(x-1)^2}{2} + \frac{(x-1)^3}{3} - \frac{(x-1)^4}{4} + \cdots + (-1)^{n-1}\frac{(x-1)^n}{n} \quad \text{for } x \text{ near } 1.$$

If $x = 1.4$, with $\ln 1.4 = 0.33647$, we find that

$$P_2(1.4) = 0.32 \qquad P_6(1.4) = 0.33630$$
$$P_4(1.4) = 0.33493 \qquad P_8(1.4) = 0.33645$$

so we see that the convergence is quite rapid.

In Table 10.1 we show the results of using $x = 1.9$ and $x = 2.3$ in the Taylor series for $\ln x$. Notice that for $x = 1.9$, which is inside the interval of convergence but close to an endpoint, the approximations converge, though rather slowly. For $x = 2.3$, which is outside the interval of convergence of the Taylor series, the approximations diverge: the larger the value for n, the less accurate the "approximations" become. In fact, the contribution of the twenty-fifth term is about 28; the contribution of the hundredth term is about $-2{,}500{,}000{,}000$. Thus, outside the interval of convergence, the "approximations" will oscillate ever more wildly away from the desired value. They are not approximations at all!

TABLE 10.1 *Approximations for* $\ln 1.9 = 0.64185$ *and* $\ln 2.3 = 0.83291$

n	$P_n(1.9)$	n	$P_n(2.3)$
2	0.495	2	0.455
5	0.69021	5	1.21589
8	0.61802	8	0.28817
11	0.65473	11	1.71710
14	0.63440	14	−0.70701

Thus, when we define the Taylor series for $\ln x$ about $x = 1$, we say that it converges for $0 < x < 2$, or that its interval of convergence is $0 < x < 2$, and write

$$\ln x = (x-1) - \frac{(x-1)^2}{2} + \frac{(x-1)^3}{3} - \frac{(x-1)^4}{4} + \cdots \qquad \text{for } 0 < x < 2.$$

For $x < 0$ or $x > 2$, the series does not converge. Notice that the interval of convergence is centered at $x = 1$. For most functions f that you will ever meet, a Taylor series about $x = a$ either converges to $f(x)$ for all x, or has an interval of convergence centered at $x = a$.

Example 1 Find the Taylor series for $\ln(1+x)$ about $x = 0$, and its interval of convergence.

Solution Taking derivatives of $\ln(1+x)$ and substituting $x = 0$ leads to the Taylor series

$$\ln(1+x) = x - \frac{x^2}{2} + \frac{x^3}{3} - \frac{x^4}{4} + \cdots.$$

Notice that this is the same series that you get by substituting $(1+x)$ for x in the series for $\ln x$:

$$\ln x = (x-1) - \frac{(x-1)^2}{2} + \frac{(x-1)^3}{3} - \frac{(x-1)^4}{4} + \cdots \qquad \text{for } 0 < x < 2.$$

Since the series for $\ln x$ about $x = 1$ converges for $0 < x < 2$, you shouldn't be surprised to see in Figure 10.11 that the interval of convergence for the Taylor series for $\ln(1+x)$ about $x = 0$ is $-1 < x < 1$. Thus we write

$$\ln(1+x) = x - \frac{x^2}{2} + \frac{x^3}{3} - \frac{x^4}{4} + \cdots \qquad \text{for } -1 < x < 1.$$

By the way, it should be clear that the series cannot possibly converge to $\ln(1+x)$ for $x \leq -1$ since $\ln(1+x)$ is not defined there.

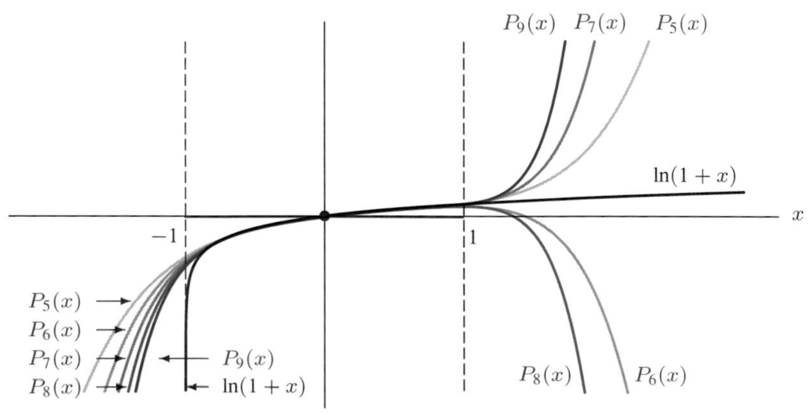

Figure 10.11: Interval of convergence for the Taylor series for $\ln(1+x)$ is $-1 < x < 1$

The Binomial Series

We will find the Taylor series about $x = 0$ for the function $f(x) = (1+x)^p$, with p a constant, not necessarily an integer. Taking derivatives:

$$\begin{aligned} f(x) &= (1+x)^p & \text{so} \qquad f(0) &= 1 \\ f'(x) &= p(1+x)^{p-1} & f'(0) &= p \\ f''(x) &= p(p-1)(1+x)^{p-2} & f''(0) &= p(p-1) \\ f'''(x) &= p(p-1)(p-2)(1+x)^{p-3} & f'''(0) &= p(p-1)(p-2) \end{aligned}$$

Thus the third-degree Taylor polynomial for x near 0 is

$$(1+x)^p \approx P_3(x) = 1 + px + \frac{p(p-1)}{2!}x^2 + \frac{p(p-1)(p-2)}{3!}x^3.$$

For a specific value of p, graphing $P_3(x)$, $P_4(x)$, and so on, suggests that the Taylor polynomials converge to $f(x)$ for $-1 < x < 1$. (See Problems 27–29, page 607.) Thus the Taylor series for $f(x) = (1+x)^p$ is as follows.

The Binomial Series

$$(1+x)^p = 1 + px + \frac{p(p-1)}{2!}x^2 + \frac{p(p-1)(p-2)}{3!}x^3 + \cdots \qquad \text{for } -1 < x < 1.$$

As you will see in the next example, this series gives the binomial theorem of algebra when p is a positive integer.

Example 2 Use the binomial series with $p = 3$ to expand $(1+x)^3$.

Solution The series is

$$(1+x)^3 = 1 + 3x + \frac{3\cdot 2}{2!}x^2 + \frac{3\cdot 2\cdot 1}{3!}x^3 + \frac{3\cdot 2\cdot 1\cdot 0}{4!}x^4 + \cdots.$$

The term in x^4 and all terms beyond it turn out to be zero, because each coefficient contains a factor of 0. Simplifying gives

$$(1+x)^3 = 1 + 3x + 3x^2 + x^3,$$

which is the usual expansion of $(1+x)^3$ we get from the binomial theorem of algebra. (It was Newton who discovered that the binomial series can be used for noninteger exponents.)

Example 3 Find the Taylor series about $x = 0$ for $\frac{1}{1+x}$.

Solution Since $\frac{1}{1+x} = (1+x)^{-1}$, let $p = -1$. Then

$$\begin{aligned} \frac{1}{1+x} = (1+x)^{-1} &= 1 + (-1)x + \frac{(-1)(-2)}{2!}x^2 + \frac{(-1)(-2)(-3)}{3!}x^3 + \cdots \\ &= 1 - x + x^2 - x^3 + \cdots \qquad \text{for } -1 < x < 1. \end{aligned}$$

This series is a special case of the binomial series and so converges for $-1 < x < 1$. It is also an example of a *geometric series*, which we will study in more detail in Section 10.4.

Another View of Euler's Formula

If you read Appendix D on complex numbers, you may have found the expression $e^{i\theta}$ quite mysterious. What does it mean to raise a number to a complex power? We can now answer this question by another method. Suppose we consider the expression $\cos\theta + i\sin\theta$, with $\cos\theta$ and $\sin\theta$ replaced by their Taylor series:

$$\cos\theta + i\sin\theta = \left(1 - \frac{\theta^2}{2!} + \frac{\theta^4}{4!} - \frac{\theta^6}{6!} + \cdots\right) + i\left(\theta - \frac{\theta^3}{3!} + \frac{\theta^5}{5!} - \cdots\right)$$

Reordering terms, we have

$$\cos\theta + i\sin\theta = 1 + i\theta - \frac{\theta^2}{2!} - \frac{i\theta^3}{3!} + \frac{\theta^4}{4!} + \frac{i\theta^5}{5!} - \frac{\theta^6}{6!} - \cdots$$

Using the fact that $i^2 = -1$, $i^3 = -i$, $i^4 = 1$, $i^5 = i, \cdots$, we can rewrite the series as

$$\cos\theta + i\sin\theta = 1 + i\theta + \frac{(i\theta)^2}{2!} + \frac{(i\theta)^3}{3!} + \frac{(i\theta)^4}{4!} + \frac{(i\theta)^5}{5!} + \frac{(i\theta)^6}{6!} + \cdots$$

Amazingly enough, this series is the Taylor series for e^x with $i\theta$ substituted for x. We define $e^{i\theta}$ to be the sum of this series, that is, we define

$$e^{i\theta} = 1 + i\theta + \frac{(i\theta)^2}{2!} + \frac{(i\theta)^3}{3!} + \frac{(i\theta)^4}{4!} + \frac{(i\theta)^5}{5!} + \frac{(i\theta)^6}{6!} + \cdots$$

Therefore, we have shown that

$$\cos\theta + i\sin\theta = e^{i\theta}.$$

Problems for Section 10.2

Which of the series in Problems 1– 5 are Taylor series?

1. $x - x^3 + x^6 - x^{10} + x^{15} - \cdots$
2. $\frac{1}{x} + \frac{1}{x^2} + \frac{1}{x^3} + \frac{1}{x^4} + \cdots$
3. $1 + x + (x-1)^2 + (x-2)^3 + (x-3)^4 + \cdots$
4. $x^7 + x + 2$
5. $\pi + (x-\pi) + 2!(x-\pi)^2 + 3!(x-\pi)^3 + \cdots$

For Problems 6–11, find the Taylor series for the given function about 0.

6. $\frac{1}{1-x}$
7. $\sqrt{1+x}$
8. $\arctan z$
9. $\ln(1-t)$
10. $\frac{1}{\sqrt{1+x}}$
11. $\sqrt[3]{1-y}$

For Problems 12–18, find the Taylor series for the function about the given point a.

12. $1/x$, $a = 1$
13. $1/x$, $a = -1$
14. $1/x$, $a = 2$
15. $\sin x$, $a = \pi/4$
16. $\cos\theta$, $a = \pi/4$
17. $\sin\theta$, $a = -\pi/4$
18. $\tan x$, $a = \pi/4$

Use Taylor series to calculate the limits in Problems 19–24.

19. $\lim_{\theta \to 0} \frac{\theta - \sin\theta}{\theta^3}$

20. $\lim_{\alpha \to 0} \frac{\tan\alpha}{\alpha}$

21. $\lim_{x \to 0} \frac{\sqrt{1+x} - 1}{x}$

22. $\lim_{h \to 0} \left(\frac{h}{\sqrt{1+h} - 1} \right)$

23. $\lim_{x \to \pi} \left(\frac{\sin x}{x - \pi} \right)$

24. $\lim_{\theta \to \pi/2} \frac{\cos\theta}{(\theta - \pi/2)}$

25. (a) Find the Taylor series for $f(x) = \ln(1+2x)$ about $x = 0$ by taking derivatives.
    (b) Compare your result in part (a) to the series for $\ln(1+x)$. How could you have obtained your answer to part (a) from the series for $\ln(1+x)$?
    (c) What do you expect the interval of convergence for the series for $\ln(1+2x)$ to be?

26. (a) Find the Taylor series for $f(x) = \sin(x^2)$ about $x = 0$ by taking derivatives.
    (b) Compare your result in part (a) to the series for $\sin x$. How could you have obtained your answer to part (a) from the series for $\sin x$?

27. By graphing the function $f(x) = \frac{1}{1-x}$ and several of its Taylor polynomials, estimate the interval of convergence of the series you found in Problem 6.

28. By graphing the function $f(x) = \sqrt{1+x}$ and several of its Taylor polynomials, estimate the interval of convergence of the series you found in Problem 7.

29. By graphing the function $f(x) = \frac{1}{\sqrt{1+x}}$ and several of its Taylor polynomials, estimate the interval of convergence of the series you found in Problem 10.

30. Suppose you know that all the derivatives of some function f exist at 0, and that the function has Taylor series

$$x + \frac{x^2}{2} + \frac{x^3}{3} + \frac{x^4}{4} + \cdots + \frac{x^n}{n} + \cdots.$$

Find $f'(0)$, $f''(0)$, $f'''(0)$, and $f^{(10)}(0)$.

31. Suppose that you are told that the Taylor series of $f(x) = x^2 e^{x^2}$ is

$$x^2 + x^4 + \frac{x^6}{2!} + \frac{x^8}{3!} + \frac{x^{10}}{4!} + \cdots.$$

Find $\left.\frac{d}{dx}(x^2 e^{x^2})\right|_{x=0}$ and $\left.\frac{d^6}{dx^6}(x^2 e^{x^2})\right|_{x=0}$.

10.3 FINDING AND USING TAYLOR SERIES

Finding a Taylor series for a function means finding the coefficients. Assuming the function has all its derivatives, finding the coefficients can always be done, in theory at least, by differentiation. That is how we derived the four most important Taylor series, those for the functions e^x, $\sin x$, $\cos x$, and $(1+x)^p$.

For many functions, however, computing Taylor series coefficients by differentiation can be a very laborious business. The purpose of this section is to give you easier ways of finding Taylor series, at least if the series you want is closely related to a series that you already know.

New Series by Substitution

Suppose you want to find the Taylor series for e^{-x^2} about $x = 0$. You could find the coefficients by differentiation. Differentiating e^{-x^2} by the chain rule gives $-2xe^{-x^2}$, and differentiating again gives $-2e^{-x^2} + 4x^2e^{-x^2}$. Now, every time you differentiate you will need the product rule, and the number of terms will grow. Finding the tenth or twentieth derivative of e^{-x^2}, and thus the series for e^{-x^2} up to the terms in x^{10} or x^{20} by this method, will be tiresome (at least without a computer or calculator that can differentiate).

Fortunately, there's a quicker way. Recall that

$$e^y = 1 + y + \frac{y^2}{2!} + \frac{y^3}{3!} + \frac{y^4}{4!} + \cdots, \quad \text{for all } y.$$

Substituting $y = -x^2$ tells us that

$$e^{-x^2} = 1 + (-x^2) + \frac{(-x^2)^2}{2!} + \frac{(-x^2)^3}{3!} + \frac{(-x^2)^4}{4!} + \cdots, \quad \text{for all } x.$$

Simplifying shows that

$$e^{-x^2} = 1 - x^2 + \frac{x^4}{2!} - \frac{x^6}{3!} + \frac{x^8}{4!} + \cdots, \quad \text{for all } x,$$

which is the Taylor series for e^{-x^2}. Using this method, it is easy to find the series up to the terms in x^{10} or x^{20}.

Example 1 Find the Taylor series about $x = 0$ for $f(x) = \dfrac{1}{1+x^2}$.

Solution The binomial theorem tells us that

$$\frac{1}{1+y} = (1+y)^{-1} = 1 - y + y^2 - y^3 + y^4 + \cdots, \quad \text{for } -1 < y < 1.$$

Substituting $y = x^2$ gives

$$\frac{1}{1+x^2} = 1 - x^2 + x^4 - x^6 + x^8 + \cdots \quad \text{for } -1 < x < 1.$$

which is the Taylor series for $\dfrac{1}{1+x^2}$.

These examples show that you can get new series from old ones by substitution. Proving that the new series you get is the same one you would have obtained by direct calculation of the derivatives is more difficult, and is done in more advanced texts.

In Example 1, we made the substitution $y = x^2$. We can also substitute an entire series into another one, as in the next example.

Example 2 Find the Taylor series about $\theta = 0$ for $g(\theta) = e^{\sin\theta}$.

Solution For all y and θ, we know that

$$e^y = 1 + y + \frac{y^2}{2!} + \frac{y^3}{3!} + \frac{y^4}{4!} + \cdots$$

and

$$\sin\theta = \theta - \frac{\theta^3}{3!} + \frac{\theta^5}{5!} - \cdots.$$

Let's substitute the series for $\sin\theta$ for y:

$$e^{\sin\theta} = 1 + \left(\theta - \frac{\theta^3}{3!} + \frac{\theta^5}{5!} - \cdots\right) + \frac{1}{2!}\left(\theta - \frac{\theta^3}{3!} + \frac{\theta^5}{5!} - \cdots\right)^2 + \frac{1}{3!}\left(\theta - \frac{\theta^3}{3!} + \frac{\theta^5}{5!} - \cdots\right)^3 + \cdots.$$

To simplify, we multiply out and collect terms. The only constant term is the 1, and there's only one θ term. The only θ^2 term is the first term you get by multiplying out the square, and it is $\theta^2/2!$. There are two contributors to the θ^3 term: the $-\theta^3/3!$ from within the first parentheses, and the first term you get from multiplying out the cube, which is $\theta^3/3!$. Thus the series starts

$$\begin{aligned} e^{\sin\theta} &= 1 + \theta + \frac{\theta^2}{2!} + \left(-\frac{\theta^3}{3!} + \frac{\theta^3}{3!}\right) + \cdots \\ &= 1 + \theta + \frac{\theta^2}{2!} + 0\cdot\theta^3 + \cdots \quad \text{for all } \theta. \end{aligned}$$

New Series by Integration

Just as we can get new series by substitution, we can also get new series by integration. Here again, proving that the new series obtained this way is actually the Taylor series we want, and that it has the same interval of convergence as the original series, are important questions that we will not deal with.

Example 3 Find the Taylor series about $x = 0$ for $\arctan x$ from the series for $\dfrac{1}{1+x^2}$.

Solution We know that $\dfrac{d(\arctan x)}{dx} = \dfrac{1}{1+x^2}$, so, using the series derived in Example 1, we have

$$\frac{d(\arctan x)}{dx} = \frac{1}{1+x^2} = 1 - x^2 + x^4 - x^6 + x^8 - \cdots \quad \text{for } -1 < x < 1.$$

Antidifferentiating term by term (which turns out to be legal) gives

$$\arctan x = C + x - \frac{x^3}{3} + \frac{x^5}{5} - \frac{x^7}{7} + \frac{x^9}{9} - \cdots \quad \text{for } -1 < x < 1.$$

where C is the constant of integration. The fact that $\arctan 0 = 0$ tells us that we must have $C = 0$, so

$$\arctan x = x - \frac{x^3}{3} + \frac{x^5}{5} - \frac{x^7}{7} + \frac{x^9}{9} - \cdots \quad \text{for } -1 < x < 1.$$

Applications of Taylor Series

Example 4 Use a series to estimate the numerical value of π.

Solution Since $\arctan 1 = \pi/4$, we use the series for $\arctan x$ that we worked out in Example 3. It is shown in more advanced courses that the series does converge to $\pi/4$ at $x = 1$; we will assume this fact. We therefore substitute $x = 1$ into the series for $\arctan x$, getting

$$\pi = 4\arctan 1 = 4\left(1 - \frac{1}{3} + \frac{1}{5} - \frac{1}{7} + \frac{1}{9} - \cdots\right).$$

Table 10.2 shows the value of the sum, S_n, obtained by summing the terms from 1 through n. As you see, the values of S_n do seem to converge to $\pi = 3.141\ldots$. Unfortunately, though, this series converges very slowly, meaning that you have to take a large number of terms to get an accurate estimate for π. Thus, this way of calculating π is not particularly practical (a better one is given in Problem 19, page 615). However, the expression for π given by this series is surprising and elegant.

TABLE 10.2 *Approximating π using the series for* $\arctan x$

n	7	9	25	100	500	1000	10,000
S_n	2.895	3.340	3.218	3.122	3.138	3.140	3.141

One of the most basic questions we can ask about two functions is which one is larger. The first few terms of the Taylor series for the two functions can often be used to answer this question over a small interval. If the constant terms of the two series are the same, compare the linear terms; if the linear terms are the same, compare the quadratic terms, and so on.

Example 5 By looking at the Taylor series, decide which of the following functions is largest, and which is smallest, for a small positive θ.

(a) $1 + \sin\theta$ (b) e^θ (c) $\dfrac{1}{\sqrt{1-2\theta}}$

Solution The Taylor expansion about $\theta = 0$ for $\sin\theta$ is

$$\sin\theta = \theta - \frac{\theta^3}{3!} + \frac{\theta^5}{5!} - \frac{\theta^7}{7!} + \cdots.$$

So

$$1 + \sin\theta = 1 + \theta - \frac{\theta^3}{3!} + \frac{\theta^5}{5!} - \frac{\theta^7}{7!} + \cdots.$$

The Taylor expansion about $\theta = 0$ for e^θ is

$$e^\theta = 1 + \theta + \frac{\theta^2}{2!} + \frac{\theta^3}{3!} + \frac{\theta^4}{4!} + \cdots.$$

The Taylor expansion about $\theta = 0$ for $1/\sqrt{1+\theta}$ is

$$\begin{aligned}\frac{1}{\sqrt{1+\theta}} &= (1+\theta)^{-1/2} = 1 - \frac{1}{2}\theta + \frac{(-\frac{1}{2})(-\frac{3}{2})}{2!}\theta^2 + \frac{(-\frac{1}{2})(-\frac{3}{2})(-\frac{5}{2})}{3!}\theta^3 + \cdots \\ &= 1 - \frac{1}{2}\theta + \frac{3}{8}\theta^2 - \frac{5}{16}\theta^3 + \cdots.\end{aligned}$$

So, substituting -2θ for θ:

$$\frac{1}{\sqrt{1-2\theta}} = 1 - \frac{1}{2}(-2\theta) + \frac{3}{8}(-2\theta)^2 - \frac{5}{16}(-2\theta)^3 + \cdots$$
$$= 1 + \theta + \frac{3}{2}\theta^2 + \frac{5}{2}\theta^3 + \cdots.$$

For θ near 0, we can neglect the higher-order terms in these expansions. Keeping the constant, linear, and second-degree terms, we are left with three approximations, valid for θ near 0:

$$1 + \sin\theta \approx 1 + \theta$$
$$e^\theta \approx 1 + \theta + \frac{\theta^2}{2}$$
$$\frac{1}{\sqrt{1-2\theta}} \approx 1 + \theta + \frac{3}{2}\theta^2.$$

Since

$$1 + \theta < 1 + \theta + \frac{1}{2}\theta^2 < 1 + \theta + \frac{3}{2}\theta^2,$$

we conclude that, for small positive θ,

$$1 + \sin\theta < e^\theta < \frac{1}{\sqrt{1-2\theta}}.$$

Example 6 Two electrical charges of equal magnitude and opposite signs located near one another are called an electrical dipole. Suppose the charges Q and $-Q$ are a distance r apart. (See Figure 10.12.) The electric field, E, at the point P is given by

$$E = \frac{Q}{R^2} - \frac{Q}{(R+r)^2}.$$

Use series to investigate the behavior of the electric field far away from the dipole. Show that when R is large in comparison to r, the electric field is approximately proportional to $1/R^3$.

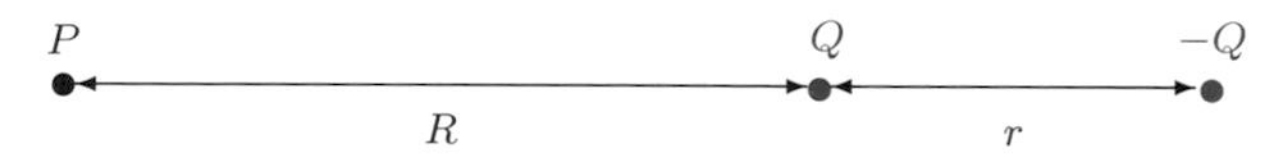

Figure 10.12: A dipole

Solution In order to use a series approximation, we need to choose a variable whose value we know to be small. Although we know that r is much smaller than R, we do not know that r is itself small. The quantity r/R is, however, very small — much smaller than 1. Hence we expand $1/(R+r)^2$ in terms of r/R so that we can safely use only the first few terms:

$$\frac{1}{(R+r)^2} = \frac{1}{R^2(1+r/R)^2} = \frac{1}{R^2}\left(1+\frac{r}{R}\right)^{-2}$$
$$= \frac{1}{R^2}\left(1 + (-2)\left(\frac{r}{R}\right) + \frac{(-2)(-3)}{2!}\left(\frac{r}{R}\right)^2 + \frac{(-2)(-3)(-4)}{3!}\left(\frac{r}{R}\right)^3 + \cdots\right)$$
$$= \frac{1}{R^2}\left(1 - 2\frac{r}{R} + 3\frac{r^2}{R^2} - 4\frac{r^3}{R^3} + \cdots\right).$$

So

$$E = \frac{Q}{R^2} - \frac{Q}{(R+r)^2} = Q\left[\frac{1}{R^2} - \frac{1}{R^2}\left(1 - 2\frac{r}{R} + 3\frac{r^2}{R^2} - 4\frac{r^3}{R^3} + \cdots\right)\right]$$
$$= \frac{Q}{R^2}\left(2\frac{r}{R} - 3\frac{r^2}{R^2} + 4\frac{r^3}{R^3} - \cdots\right).$$

Since r/R is smaller than 1, the binomial expansion for $(1+r/R)^{-2}$ will converge. We are interested in the electric field far away from the dipole. The quantity r/R is small there, and $(r/R)^2$ and higher powers are much smaller still. Thus, we approximate by disregarding all except the first term, giving

$$E \approx \frac{Q}{R^2}\left(\frac{2r}{R}\right) \quad \text{so} \quad E \approx \frac{2Qr}{R^3}.$$

Since Q and r are constants, this means that E is nearly proportional to $1/R^3$.

In the previous example, we say that E is *expanded in terms of* r/R, meaning that the independent variable in the expansion is r/R.

Series Solutions to Differential Equations

A Taylor series can often be used to approximate the solution to a differential equation with an initial condition. The idea is to substitute the series into the differential equation and to find the coefficients in the series one by one.

Example 7 Find the Taylor polynomial of degree 4 about $x = 0$ of the function $y(x)$ that satisfies each of the following initial-value problems.

(a) $\frac{dy}{dx} = y \qquad y(0) = 1.$ (b) $\frac{dy}{dx} = y + \frac{1}{1+x} \qquad y(0) = 1.$

Solution (a) We express both sides of the differential equation $dy/dx = y$ as Taylor series about $x = 0$. The Taylor series for y has the form

$$y(x) = C_0 + C_1x + C_2x^2 + C_3x^3 + C_4x^4 + C_5x^5 + \cdots.$$

Since $C_0 = y(0)$, the initial condition $y(0) = 1$ gives $C_0 = 1$, so

$$y(x) = 1 + C_1x + C_2x^2 + C_3x^3 + C_4x^4 + C_5x^5 + \cdots.$$

Taking the derivative of the series for y term by term (which is legal), we get

$$\frac{dy}{dx} = C_1 + 2C_2x + 3C_3x^2 + 4C_4x^3 + 5C_5x^4 + \cdots.$$

Since the differential equation tells us that $dy/dx = y$, we must have

$$C_1 + 2C_2x + 3C_3x^2 + 4C_4x^3 + 5C_5x^4 + \cdots = 1 + C_1x + C_2x^2 + C_3x^3 + C_4x^4 + \cdots.$$

In order for the series on each side of this equation to be equal, the coefficients of corresponding powers of x must be equal. Therefore, we equate coefficients of corresponding powers of x:

Constant terms	$C_1 = 1$	
Coefficients of x	$2C_2 = C_1 = 1,$	so $C_2 = 1/2$
Coefficients of x^2	$3C_3 = C_2 = 1/2,$	so $C_3 = 1/6$
Coefficients of x^3	$4C_4 = C_3 = 1/6,$	so $C_4 = 1/24.$

We have found the approximation

$$y(x) \approx 1 + x + \frac{x^2}{2} + \frac{x^3}{6} + \frac{x^4}{24} \quad \text{for } x \text{ near } 0.$$

In this case, we can see that the coefficients can be written as

$$C_2 = \frac{1}{2} = \frac{1}{2!}, \quad C_3 = \frac{1}{6} = \frac{1}{3!}, \quad C_4 = \frac{1}{24} = \frac{1}{4!},$$

so we recognize the series as the Taylor series for e^x, which is the solution we expect:

$$y(x) = 1 + x + \frac{x^2}{2!} + \frac{x^3}{3!} + \frac{x^4}{4!} + \cdots = e^x.$$

(b) In the case of the equation

$$\frac{dy}{dx} = y + \frac{1}{1+x} \qquad y(0) = 1,$$

we do not already know a solution, so using a series will be the only method we currently have of approximating the solution by a formula. As before, express both sides of the differential equation as Taylor series about $x = 0$. As in part (a), $y(0) = C_0 = 1$, so

$$y(x) = 1 + C_1x + C_2x^2 + C_3x^3 + C_4x^4 + C_5x^5 + \cdots$$

and

$$\frac{dy}{dx} = C_1 + 2C_2x + 3C_3x^2 + 4C_4x^3 + 5C_5x^4 + \cdots.$$

We also need the series for $1/(1+x)$, which is a binomial series:

$$\frac{1}{1+x} = 1 - x + x^2 - x^3 + x^4 - x^5 + \cdots.$$

Thus, substituting into the series:

$$\begin{aligned} &C_1 + 2C_2x + 3C_3x^2 + 4C_4x^3 + 5C_5x^4 + \cdots \\ &\quad = (1 + C_1x + C_2x^2 + C_3x^3 + C_4x^4 + \cdots) + (1 - x + x^2 - x^3 + x^4 - \cdots) \\ &\quad = 2 + (C_1 - 1)x + (C_2 + 1)x^2 + (C_3 - 1)x^3 + (C_4 + 1)x^4 + \cdots \end{aligned}$$

Equating coefficients gives

Constant terms	$C_1 = 2$	
Coefficients of x	$2C_2 = C_1 - 1 = 1,$	so $C_2 = 1/2$
Coefficients of x^2	$3C_3 = C_2 + 1 = 3/2,$	so $C_3 = 1/2$
Coefficients of x^3	$4C_4 = C_3 - 1 = -1/2,$	so $C_4 = -1/8.$

We have shown that the solution is approximated by

$$y(x) \approx 1 + 2x + \frac{x^2}{2} + \frac{x^3}{2} - \frac{x^4}{8} \quad \text{for } x \text{ near } 0.$$

Problems for Section 10.3

Find the first four nonzero terms of the Taylor series about 0 for the functions in Problems 1–12.

1. e^{-x}
2. $\sqrt{1-2x}$
3. $\cos(\theta^2)$
4. $\ln(1-2y)$
5. $\dfrac{t}{1+t}$
6. $\dfrac{1}{\sqrt{1-z^2}}$
7. $\arcsin x$
8. $\dfrac{z}{e^{z^2}}$
9. $\phi^3\cos(\phi^2)$
10. $\sqrt{1+\sin\theta}$
11. $e^t\cos t$
12. $\sqrt{(1+t)}\sin t$

For Problems 13–14, expand the quantity about 0 in terms of the variable given. Give the first four nonzero terms.

13. $\dfrac{1}{2+x}$ in terms of $\dfrac{x}{2}$
14. $\dfrac{a}{\sqrt{a^2+x^2}}$ in terms of $\dfrac{x}{a}$, where $a>0$.
15. For values of y near 0, put the following functions in increasing order, using their Taylor expansions.
    (a) $\ln(1+y^2)$ (b) $\sin(y^2)$ (c) $1-\cos y$
16. By looking at the Taylor series, decide which of the following functions is largest, and which is smallest, for small positive θ.
    (a) $1+\sin\theta$ (b) $\cos\theta$ (c) $\dfrac{1}{1-\theta^2}$
17. Figure 10.13 shows the graph of the four functions below for values of x near 0. Use Taylor series to match graphs and formulas.
    (a) $\dfrac{1}{1-x^2}$ (b) $(1+x)^{1/4}$ (c) $\sqrt{1+\dfrac{x}{2}}$ (d) $\dfrac{1}{\sqrt{1-x}}$

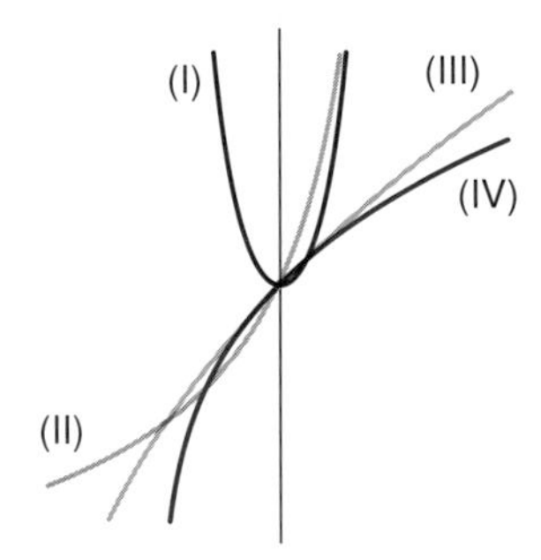

Figure 10.13

18. Consider the two functions $y=e^{-x^2}$ and $y=1/(1+x^2)$.
    (a) Write the Taylor expansions for the two functions about $x=0$. What is similar about the two series? What is different?
    (b) Looking at the series, which function do you predict will be greater over the interval $(-1,1)$? Graph both and see.
    (c) Are these functions even or odd? How might you see this by looking at the series expansions?
    (d) By looking at the coefficients, explain why it is reasonable that the series for $y=e^{-x^2}$ converges for all values of x, but the series for $y=1/(1+x^2)$ converges only on $(-1,1)$.

19. Machin's formula says $\pi/4 = 4\arctan(1/5) - \arctan(1/239)$.
    (a) Verify that $\pi/4$ and the quantity $4\arctan(1/5) - \arctan(1/239)$ agree to as many decimal places as your calculator shows.
    (b) Use the Taylor polynomial approximation of degree 5 to the arctangent function to approximate the value of π. (Note: In 1873 William Shanks used this approach to calculate π to 707 decimal places. Unfortunately, in 1946 it was found that he made an error in the 528th place.)
    (c) Can you explain why the two series for arctangent converge so rapidly here while the series used in Example 4 on page 610 converges so slowly?
20. One of Einstein's most amazing predictions was that light traveling from distant stars would bend around the sun on the way to earth. His calculations involved solving for ϕ in the equation

$$\sin\phi + b(1 + \cos^2\phi + \cos\phi) = 0$$

where b is a very small positive constant.
    (a) Explain why the equation has a solution for ϕ which is near 0.
    (b) Expand the left-hand side of the equation in Taylor series about $\phi = 0$, disregarding terms of order ϕ^2 and higher. Solve for ϕ. (Your answer will involve b.)
21. An electric dipole on the x axis consists of a charge Q at $x = 1$ and a charge $-Q$ at $x = -1$. The electric field, E, at the point $x = R$ on the x-axis is given (for $R > 1$) by

$$E = \frac{kQ}{(R-1)^2} - \frac{kQ}{(R+1)^2}$$

where k is a positive constant whose value depends on the units. Expand E as a series in $1/R$, giving the first two nonzero terms.
22. A hydrogen atom consists of an electron, of mass m, orbiting a proton, of mass M, where m is much smaller than M. The *reduced mass*, μ, of the hydrogen atom is defined by

$$\mu = \frac{mM}{m+M}.$$

    (a) Show that $\mu \approx m$.
    (b) To get a more accurate approximation for μ, express μ as m times a series in m/M.
    (c) The approximation $\mu \approx m$ is obtained by disregarding all but the constant term in the series. The first-order correction is obtained by including the linear term but no higher terms. If $m \approx M/1836$, by what percentage does including the first-order correction change the estimate $\mu \approx m$?
23. The electric potential, V, at a distance R along the axis perpendicular to the center of a charged disc with radius a and constant charge density σ, is given by

$$V = 2\pi\sigma(\sqrt{R^2 + a^2} - R).$$

Show that, for large R,

$$V \approx \frac{\pi a^2 \sigma}{R}.$$

24. Studying resonance in electric circuits leads us to consider the expression

$$\left(\omega L - \frac{1}{\omega C}\right)^2,$$

where ω is the variable and L and C are constants.

(a) Find ω_0, the value of ω making the expression zero.

(b) Because, in practice, ω fluctuates about ω_0, we are interested in the behavior of this expression for values of ω near ω_0. Let $\omega = \omega_0 + \Delta\omega$ and expand the expression in terms of $\Delta\omega$ up to the first nonzero term. Give your answer in terms of ω_0 and L but not C.

25. The electric force, F, between two atoms depends on the distance r separating them. See Figure 10.14. A positive F represents a repulsive force; a negative F represents an attractive force.

(a) Why does $r = a$ represent an equilibrium position?

(b) Qualitatively, what happens to the force if the atoms start with $r = a$ and are pulled slightly further apart?

(c) Qualitatively, what happens to the force if the atoms start with $r = a$ and are pushed slightly closer together?

(d) Write the Taylor series for F around $r = a$.

(e) By discarding all except the first nonzero term in this series, describe how the force between the atoms depends on the displacement from the equilibrium, when that displacement is small.

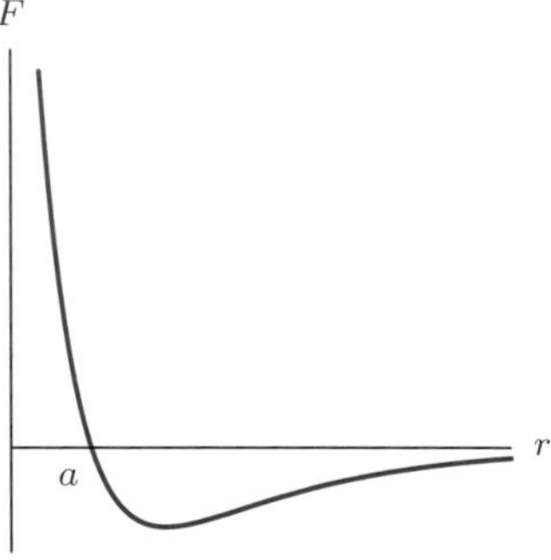

Figure 10.14

26. When a body is near the surface of the earth, we usually assume that the force due to gravity on it is a constant mg, where m is the mass of the body and g is the acceleration due to gravity at sea level. For a body at a distance h above the surface of the earth, a more accurate expression for the force F is

$$F = \frac{mgR^2}{(R+h)^2}$$

where R is the radius of the earth. We will consider the situation in which the body is close to the surface of the earth so that h is much smaller than R.

(a) Show that $F \approx mg$.

(b) Express F as mg multiplied by a series in h/R.

(c) The first-order correction to the approximation $F \approx mg$ is obtained by taking the linear term in the series but no higher terms. How far above the surface of the earth can you go before the first-order correction changes the estimate $F \approx mg$ by more than 10%? (Assume $R = 6400$ km.)

In Problems 27–31, find a Taylor polynomial of degree at least four which is a solution to the given initial value problem. Is the answer a Taylor series you have seen before?

27. $\dfrac{dy}{dx} = \dfrac{1}{1-x}, \quad y(0) = 1.$

28. $\dfrac{d^2 f}{dx^2} = -f, \quad f(0) = 1, f'(0) = 0.$

29. $\dfrac{d^2 f}{d\theta^2} = -f, \quad f(0) = 0, f'(0) = 1.$

30. $\dfrac{d^2 y}{dx^2} = xy, \quad y(0) = 1, y'(0) = -1$

31. $\dfrac{dy}{dt} = 1 - yt, \quad y(0) = 2.$

10.4 GEOMETRIC SERIES

In the last two sections we expressed a given function by a Taylor series. Notice that there we started with the function and then derived the series. In this section we will work in the opposite direction, starting with a series and finding its sum, the function. This can't be done for just any old series, but for some particular types of series it can.

Repeated Drug Dosage

Suppose you have an ear infection and are told to take antibiotic tablets regularly for several days. Since the drug is being excreted by the body in between doses, how do we calculate the quantity of the drug remaining in the body at any particular time?

To be specific, let's suppose the drug is ampicillin (a common antibiotic) taken in 250 mg doses four times a day (that is, every six hours). It is known that at the end of six hours, about 4% of the drug is still in the body. What quantity of the drug is in the body right after the tenth tablet? The fortieth?

Let's suppose Q_n represents the quantity, in milligrams, of ampicillin in the blood right after the nth tablet. Then

$$\begin{aligned}
Q_1 &= 250 \\
Q_2 &= \underbrace{250(0.04)}_{\text{Remnants of first tablet}} + \underbrace{250}_{\text{New tablet}} \\
Q_3 &= Q_2(0.04) + 250 = \big(250(0.04) + 250\big)(0.04) + 250 \\
&= \underbrace{250(0.04)^2 + 250(0.04)}_{\text{Remnants of first and second tablets}} + \underbrace{250}_{\text{New tablet}} \\
Q_4 &= Q_3(0.04) + 250 = \big(250(0.04)^2 + 250(0.04) + 250\big)(0.04) + 250 \\
&= \underbrace{250(0.04)^3 + 250(0.04)^2 + 250(0.04)}_{\text{Remnants of first, second, and third tablets}} + \underbrace{250}_{\text{New tablet}}
\end{aligned}$$

Looking at the pattern that is emerging, you can probably guess that

$$\begin{aligned}
Q_5 &= 250(0.04)^4 + 250(0.04)^3 + 250(0.04)^2 + 250(0.04) + 250 \\
Q_{10} &= 250(0.04)^9 + 250(0.04)^8 + \cdots + 250(0.04) + 250.
\end{aligned}$$

Notice that there are 10 terms in this sum — one for every tablet — but that the highest power of 0.04 is the ninth, because no tablet has been in the body for more than 9 six-hour time periods. (Check this: do you see why?) Now suppose we actually want to find the numerical value of Q_{10}. It seems that we have to add terms — and if we want the value of Q_{40}, we would be faced with adding 40 terms:

$$Q_{40} = 250(0.04)^{39} + 250(0.04)^{38} + \cdots + 250(0.04) + 250.$$

Fortunately, there's a better way. We can put the sum into what is called *closed-form*. Let's start with Q_{10}.

$$Q_{10} = 250(0.04)^9 + 250(0.04)^8 + 250(0.04)^7 + \cdots + 250(0.04)^2 + 250(0.04) + 250.$$

Notice the remarkable fact that if you subtract $(0.04)Q_{10}$ from Q_{10}, a great many terms (all but two, in fact) drop out. Multiplying by 0.04, we get

$$(0.04)Q_{10} = 250(0.04)^{10} + 250(0.04)^9 + 250(0.04)^8 + \cdots + 250(0.04)^3 + 250(0.04)^2 + 250(0.04).$$

Subtracting gives

$$Q_{10} - (0.04)Q_{10} = 250 - 250(0.04)^{10}.$$

Factoring Q_{10} on the left and solving for Q_{10}:

$$\begin{aligned} Q_{10}(1 - 0.04) &= 250\left(1 - (0.04)^{10}\right) \\ Q_{10} &= \frac{250\left(1 - (0.04)^{10}\right)}{1 - 0.04}. \end{aligned}$$

This is the closed-form expression for Q_{10}, and it is easy to evaluate on a calculator, giving $Q_{10} = 260.42$ (to two decimal places). Similarly, Q_{40} is given in closed-form by

$$Q_{40} = \frac{250\left(1 - (0.04)^{40}\right)}{1 - 0.04}.$$

Evaluating this on a calculator shows $Q_{40} = 260.42$, the same (to two decimal places) as the value of Q_{10}. Thus after ten tablets, the value of Q_n appears to have stabilized at just over 260 mg.

Looking at the closed-forms for Q_{10} and Q_{40}, you can see that Q_n must be given by

$$Q_n = \frac{250\left(1 - (0.04)^{n}\right)}{1 - 0.04}.$$

What Happens as $n \to \infty$?

What does this closed-form for Q_n predict about the long-run level of ampicillin in the body right after a tablet is taken, assuming that 250 mg continue to be taken every six hours? As $n \to \infty$, the quantity $(0.04)^n \to 0$, so in the long run,

$$Q_n = \frac{250\left(1 - (0.04)^{n}\right)}{1 - 0.04} \to \frac{250(1 - 0)}{1 - 0.04} = 260.42.$$

Alternative Method of Calculating Long-Run Ampicillin Level

There's another way of deriving this limiting ampicillin level. Suppose that in the long run the ampicillin levels off to Q mg right after each tablet is taken. Six hours later, right before the next dose, there will be less ampicillin in the body. However, if stability has been reached, the amount of ampicillin that has been excreted must be exactly 250 mg because taking one more tablet must raise the level back to Q mg. Thus

$$\begin{aligned} \text{Amount of ampicillin excreted} &= \text{Original quantity} - \text{Final quantity} \\ &= Q - (0.04)Q \\ &= 250. \end{aligned}$$

Solving for Q gives, as before,

$$Q = \frac{250}{1 - 0.04} = 260.42.$$

The Geometric Series in General

In the previous example we encountered sums of the form $a + ax + ax^2 + \cdots + ax^8 + ax^9$ (with $a = 250$ and $x = 0.04$). Such a sum is called a finite *geometric series*. In general, a geometric series is defined as one in which each term is a constant multiple of the one before. (In our example, the constant multiple was 0.04.)

A **finite geometric series** has the form

$$a + ax + ax^2 + \cdots + ax^{n-2} + ax^{n-1}$$

An **infinite geometric series** has the form

$$a + ax + ax^2 + \cdots + ax^{n-2} + ax^{n-1} + ax^n + \cdots$$

The $\cdots$ at the end of the second series tells you that the series is going on forever — in other words, that it's infinite.

Sum of a Finite Geometric Series

The same remarkable trick that enabled us to find the closed-form for Q_{10} above can be used to find the sum of any finite geometric series. Suppose we define S_n to be the sum of the first n terms, which means up to the term in x^{n-1}:

$$S_n = a + ax + ax^2 + \cdots + ax^{n-2} + ax^{n-1}.$$

Multiply S_n by x:

$$xS_n = ax + ax^2 + ax^3 + \cdots + ax^{n-1} + ax^n.$$

Now subtract xS_n from S_n, which cancels out all terms except for two, giving

$$S_n - xS_n = a - ax^n$$
$$(1 - x)S_n = a(1 - x^n).$$

Thus, provided $x \neq 1$, we can solve for S_n as follows.

The **sum of a finite geometric series** is given by

$$S_n = a + ax + ax^2 + \cdots + ax^{n-1} = \frac{a(1 - x^n)}{1 - x}$$

You may find it helpful to remember that the value of n which appears in the closed-form for S_n is the number of terms in the sum S_n.

If $x = 1$ then $S_n = n \cdot a$.

Sum of an Infinite Geometric Series

On the face of it, it's not clear how we should find the sum of an infinite series, since we can't simply add up an infinite number of terms. In the ampicillin example we found the sum Q_n and then let $n \to \infty$. We do the same here. Suppose we want to find the sum S of the series $a + ax + ax^2 + \cdots + ax^{n-1} + \cdots$. First we consider the *partial sum*, S_n,

$$S_n = a + ax + ax^2 + \cdots + ax^{n-1} = \frac{a(1 - x^n)}{1 - x}$$

and we look at what happens as $n \to \infty$. What does happen? It depends on the value of x. If $|x| < 1$, then $x^n \to 0$ as $n \to \infty$, and so

$$S_n = \frac{a(1 - x^n)}{1 - x} \to \frac{a(1 - 0)}{1 - x} = \frac{a}{1 - x}.$$

Thus if $|x| < 1$, we say the infinite geometric series *converges*.

For $|x| < 1$, the **sum of the infinite geometric series** is given by

$$S = a + ax + ax^2 + \cdots + ax^n + \cdots = \frac{a}{1-x}$$

If, on the other hand, $|x| > 1$, then x^n has no limit as $n \to \infty$, and we say that the series doesn't converge. This corresponds to the fact that when $|x| > 1$, the terms in the series get larger and larger, so adding up infinitely many of them couldn't possibly give a finite sum.

What happens when $x = \pm 1$? When $x = 1$, the formula for S_n doesn't apply, but the series doesn't converge. (Why?) If $x = -1$, the quantity x^n has no limit as $n \to \infty$, so again the series doesn't converge. (Why?)

Relationship between Geometric and Taylor Series

You may have recognized that the geometric series is a Taylor series. For example, the fact that

$$1 + x + x^2 + x^3 + \cdots = \frac{1}{1-x}$$

suggests that this geometric series is the Taylor expansion of $f(x) = 1/(1-x)$ for x near 0. You can check that the geometric series really is the Taylor series by taking derivatives. Alternatively, this geometric series can also be obtained from the binomial series:

$$(1+x)^p = 1 + px + \frac{p(p-1)}{2!}x^2 + \frac{p(p-1)(p-2)}{3!}x^3 + \cdots.$$

First we substitute $p = -1$:

$$\frac{1}{1+x} = (1+x)^{-1} = 1 + (-1)x + \frac{(-1)(-2)}{2!}x^2 + \frac{(-1)(-2)(-3)}{3!}x^3 + \cdots$$
$$= 1 - x + x^2 - x^3 + \cdots.$$

Now replace x by $-x$:

$$\frac{1}{1-x} = (1-x)^{-1} = 1 - (-x) + (-x)^2 - (-x)^3 + \cdots$$
$$= 1 + x + x^2 + x^3 + \cdots.$$

Example 1 Find the sum of the series

(a) $1 + \frac{1}{2} + \frac{1}{4} + \frac{1}{8} + \cdots$

(b) $1 + 2 + 4 + 8 + \cdots$

Solution (a) This series may be written

$$1 + \frac{1}{2} + \left(\frac{1}{2}\right)^2 + \left(\frac{1}{2}\right)^3 + \cdots$$

which we can identify as a geometric series with $a = 1$ and $x = \frac{1}{2}$, so $S = \dfrac{1}{1-(1/2)} = 2$. Let's check this by finding the partial sums:

$$
\begin{aligned}
S_1 &= 1 \\
S_2 &= 1 + \frac{1}{2} = \frac{3}{2} = 2 - \frac{1}{2} \\
S_3 &= 1 + \frac{1}{2} + \frac{1}{4} = \frac{7}{4} = 2 - \frac{1}{4} \\
S_4 &= 1 + \frac{1}{2} + \frac{1}{4} + \frac{1}{8} = \frac{15}{8} = 2 - \frac{1}{8} \\
S_5 &= 1 + \frac{1}{2} + \frac{1}{4} + \frac{1}{8} + \frac{1}{16} = \frac{31}{16} = 2 - \frac{1}{16}.
\end{aligned}
$$

Clearly the partial sums are creeping up on the sum $S = 2$, so $S_n \to 2$ as $n \to \infty$.

(b) In contrast, the partial sums of this series grow uncontrollably, so the whole series has no sum:

$$
\begin{aligned}
S_1 &= 1 \\
S_2 &= 1 + 2 = 3 \\
S_3 &= 1 + 2 + 4 = 7 \\
S_4 &= 1 + 2 + 4 + 8 = 15 \\
S_5 &= 1 + 2 + 4 + 8 + 16 = 31.
\end{aligned}
$$

Regular Deposits into a Savings Account

People who save money often do so by putting some fixed amount aside regularly, perhaps every week or every month. To be specific, suppose \$1000 is deposited every year in a savings account earning 5% a year, compounded annually. What is the balance B in dollars in the savings account right after the nth deposit?

As before, let's start by making a chart:

$$
\begin{aligned}
B_1 &= 1000 \\
B_2 &= B_1(1.05) + 1000 = \underbrace{1000(1.05)}_{\text{Original deposit}} + \underbrace{1000}_{\text{New deposit}} \\
B_3 &= B_2(1.05) + 1000 = \underbrace{1000(1.05)^2 + 1000(1.05)}_{\text{First two deposits}} + \underbrace{1000}_{\text{New deposit}} \\
B_4 &= B_3(1.05) + 1000 = \underbrace{1000(1.05)^3 + 1000(1.05)^2 + 1000(1.05)}_{\text{First three deposits}} + \underbrace{1000}_{\text{New deposit}}
\end{aligned}
$$

Observing the pattern, we see

$$B_n = 1000(1.05)^{n-1} + 1000(1.05)^{n-2} + \cdots + 1000(1.05) + 1000.$$

So B_n is a finite geometric series with $a = 1000$ and $x = 1.05$. Thus in closed-form,

$$B_n = \frac{1000\left(1 - (1.05)^n\right)}{1 - 1.05}.$$

Rewriting this so both the numerator and denominator of the fraction are positive gives

$$B_n = \frac{1000\left((1.05)^n - 1\right)}{1.05 - 1}.$$

What Happens as $n \to \infty$?

Common sense tells you that if you keep depositing $1000 in an account and it keeps earning interest, your balance will tend toward infinity. This is what the formula for B_n shows also: $(1.05)^n \to \infty$ as $n \to \infty$, so B_n has no limit. (Alternatively, observe that the infinite geometric series of which B_n is a partial sum has $x > 1$ and so does not converge.)

Present Value of Series of Payments

When basketball player Patrick Ewing was signed by the New York Knicks, he was given a contract for $30 million: $3 million a year for ten years. Of course, since much of the money was to be paid in the future, the team's owners did not have to have all $30 million available on the day of the signing. How much money would the owners have to deposit in a bank account on the day of the signing in order to cover all the payments? Assuming the account was earning interest, the owners would have to deposit much less than $30 million. This smaller amount is called the *present value* of $30 million. We will calculate the present value of Ewing's contract on the day he signed.

Definition of Present Value

> The **present value**, $\$P$, of a future payment, $\$B$, is the amount which would have to be deposited in a bank account today to have exactly $\$B$ in the account at the relevant time in the future.

If the interest is compounded n times a year for t years at an annual rate of r, then from Section 8.4, page 446, we know the following:

$$B = P\left(1 + \frac{r}{n}\right)^{nt}, \quad \text{or equivalently,} \quad P = \frac{B}{(1 + r/n)^{nt}}.$$

If interest is compounded continuously at an annual rate of r, we have the following result:

$$B = Pe^{rt}, \quad \text{or equivalently,} \quad P = \frac{B}{e^{rt}} = Be^{-rt}.$$

Calculating the Present Value of Ewing's Contract

The present value of Ewing's contract represents what it was worth on the day it was signed. Let's suppose that he will receive his money in 10 payments of $3 million each, the first payment to be made on the day the contract was signed. We will calculate the present value of the contract, assuming that interest is compounded annually at a rate of 5% per year, throughout the period of the contract.

Since the first payment is made the day the contract is signed:

$$\text{Present value of first payment, in millions of dollars } = 3.$$

Since the second payment is made a year in the future:

$$\text{Present value of second payment, in millions of dollars } = \frac{3}{(1+0.05)^1} = \frac{3}{1.05}.$$

The third payment is made two years in the future:

$$\text{Present value of third payment, in millions of dollars } = \frac{3}{(1.05)^2}.$$

Similarly,

$$\text{Present value of tenth payment, in millions of dollars } = \frac{3}{(1.05)^9}$$

Thus, in millions of dollars,

$$\text{Total present value} = 3 + \frac{3}{1.05} + \frac{3}{(1.05)^2} + \cdots + \frac{3}{(1.05)^9}$$

This is a finite geometric series with $x = 1/1.05$ and sum:

$$\text{Total present value of contract in millions of dollars} = \frac{3\left(1-(1/1.05)^{10}\right)}{1-1/1.05}$$

Evaluating this expression shows that the total present value of the contract is about $24.3 million.

Example 2 What would be the present value of a contract which paid $3 million a year forever? Assume that payments are made once a year, starting the day the contract was signed, and continuing forever. This time assume interest is 5% a year, compounded continuously.

Solution As before:

$$\text{Present value of first payment, in millions of dollars } = 3.$$

Since the second payment is made a year in the future, with continuous compounding,

$$\text{Present value of second payment, in millions of dollars } = 3e^{-0.05}.$$

Similarly, since the next payment is two years in the future,

$$\text{Present value of third payment, in millions of dollars } = 3e^{-0.05(2)}.$$

Thus, continuing forever,

$$\text{Total present value} = 3 + 3e^{-0.05} + 3e^{-0.05(2)} + 3e^{-0.05(3)} + \cdots.$$

Since $e^{-0.05(n)} = \left(e^{-0.05}\right)^n$ for any n, we can write

$$\text{Total present value} = 3 + 3e^{-0.05} + 3\left(e^{-0.05}\right)^2 + 3\left(e^{-0.05}\right)^3 + \cdots.$$

This is an infinite geometric series with $x = e^{-0.05}$ and sum:

$$\text{Total present value, in millions of dollars} = \frac{3}{1-e^{-0.05}} = 61.5.$$

Thus it would have cost the New York Knicks about three times as much — but only three times — to pay Ewing or his heirs $3 million a year forever.

Problems for Section 10.4

In Problems 1–10, decide which of the following are geometric series. For those which are, give the first term and the ratio between successive terms. For those which are not, explain why not.

1. $2+1+\frac{1}{2}+\frac{1}{4}+\frac{1}{8}+\cdots$

2. $1-\frac{1}{2}+\frac{1}{4}-\frac{1}{8}+\frac{1}{16}+\cdots$

3. $1+\frac{1}{2}+\frac{1}{3}+\frac{1}{4}+\frac{1}{5}+\cdots$

4. $5-10+20-40+80-\cdots$

5. $1-x+x^2-x^3+x^4-\cdots$

6. $1+x+2x^2+3x^3+4x^4+\cdots$

7. $y^2+y^3+y^4+y^5+\cdots$

8. $1-y^2+y^4-y^6+\cdots$

9. $1+2z+(2z)^2+(2z)^3+\cdots$

10. $3+3z+6z^2+9z^3+12z^4+\cdots$

11. Find the sum of the series in Problem 5.

12. Find the sum of the series in Problem 7.

13. Find the sum of the series in Problem 8.

14. Find the sum of the series in Problem 9.

Find the sum of the series in Problems 15–18.

15. $3+\frac{3}{2}+\frac{3}{4}+\frac{3}{8}+\cdots+\frac{3}{2^{10}}$

16. $-2+1-\frac{1}{2}+\frac{1}{4}-\frac{1}{8}+\frac{1}{16}-\cdots$

17. $\sum_{n=4}^{\infty}\left(\frac{1}{3}\right)^n$

18. $\sum_{n=0}^{\infty}\frac{3^n+5}{4^n}$

19. A repeating decimal can always be expressed as a fraction. This problem shows how writing a repeating decimal as a geometric series enables you to find the fraction. Consider the decimal $0.232323\ldots$.
   (a) Use the fact that $0.232323\ldots=0.23+0.0023+0.000023+\cdots$ to write $0.232323\ldots$ as a geometric series.
   (b) Use the formula for the sum of a geometric series to show that $0.232323\ldots=23/99$.

20. Figure 10.15 shows the quantity of the drug atenolol in the blood as a function of time, with the first dose at time $t=0$. Atenolol is taken in 50 mg doses once a day to lower blood pressure.

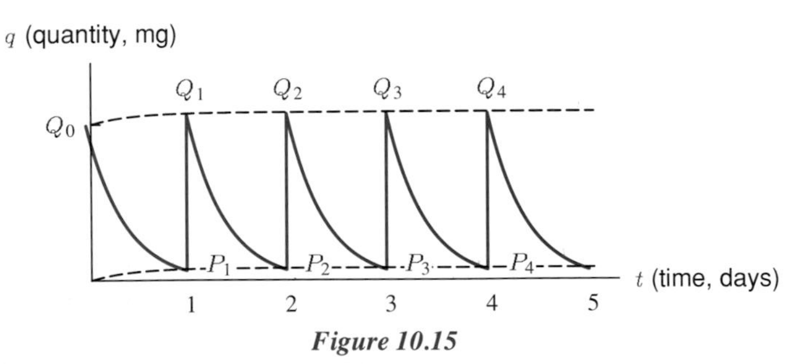

Figure 10.15

   (a) If the half-life of atenolol in the blood is 6.3 hours, what percentage of the atenolol present at the start of a 24-hour period is still there at the end?
   (b) Find expressions for the quantities $Q_0, Q_1, Q_2, Q_3, \ldots$, and Q_n shown in Figure 10.15. Write the expression for Q_n in closed-form.
   (c) Find expressions for the quantities $P_1, P_2, P_3, \ldots$, and P_n shown in Figure 10.15. Write the expression for P_n in closed-form.

21. On page 617, you saw how to compute the quantity Q_n mg of ampicillin in the body right after the nth tablet of 250 mg, taken once every six hours.

    (a) Do a similar calculation for P_n, the quantity of ampicillin (in mg) in the body right *before* the nth tablet is taken.
    (b) Express P_n in closed form.
    (c) What is $\lim_{n\to\infty} P_n$? Is this limit the same as $\lim_{n\to\infty} Q_n$? Explain in practical terms why your answer makes sense.

22. Draw a graph like that in Figure 10.15 for 250 mg of ampicillin taken every 6 hours, starting at time $t = 0$. Put on the graph the values of $Q_1, Q_2, Q_3, \ldots$ introduced in the text on page 617 and the values of $P_1, P_2, P_3, \ldots$ calculated in Problem 21.

23. A ball is dropped from a height of 10 feet and bounces. Each bounce is $3/4$ of the height of the bounce before. Thus after the ball hits the floor for the first time, the ball rises to a height of $10(3/4) = 7.5$ feet, and after it hits the floor for the second time, it rises to a height of $7.5(3/4) = 10(3/4)^2 = 5.625$ feet.

    (a) Find an expression for the height to which the ball rises after it hits the floor for the nth time.
    (b) Find an expression for the total vertical distance the ball has traveled when it hits the floor for the first, second, third, and fourth times.
    (c) Find an expression for the total vertical distance the ball has traveled when it hits the floor for the nth time. Express your answer in closed-form.

24. You might think that the ball in Problem 23 keeps bouncing forever since it takes infinitely many bounces. This is not true!

    (a) Show that a ball dropped from a height of h feet reaches the ground in $\frac{1}{4}\sqrt{h}$ seconds.
    (b) Show that the ball in Problem 23 stops bouncing after

$$\frac{1}{4}\sqrt{10} + \frac{1}{2}\sqrt{10}\sqrt{\frac{3}{4}}\left(\frac{1}{1-\sqrt{3/4}}\right)$$

    seconds, or approximately 11 seconds.

25. Consider Patrick Ewing's contract described on page 622. Determine the present value of the contract if the interest rate is 7% per year, compounded continuously, for the entire 10-year period of the contract.

26. Consider Patrick Ewing's contract described on page 622. Determine the present value of the contract if the interest rate is 5% per year, compounded twice a year, and he is to receive 20 payments of \$1.5 million, one every six months. Compare this with the present value of ten \$3 million payments, one per year, if the interest rate is 5% per year, compounded annually.

27. Around January 1, 1993, Barbra Streisand signed a contract with Sony Corporation for \$2 million a year for 10 years. Suppose the first payment was made on the day of signing and that all other payments are made on the first day of the year. Suppose also that all payments are made into a bank account earning 4% a year, compounded annually.

    (a) How much money will be in the account on the night of December 31, 1999?
    (b) How much money will be in the account on the day the last payment is made?
    (c) What was the present value of the contract on the day it was signed?

28. One way of valuing a company is to calculate the present value of all its future earnings. Suppose a farm expects to sell \$1000 worth of Christmas trees once a year forever, with the

first sale in the immediate future. What is the present value of this Christmas tree business? Assume that the interest rate is 4% per year, compounded continuously.

29. Before World War I, the British government issued what are called *consols*, which pay the owner or his heirs a fixed amount of money every year forever. (Cartoonists of the time described aristocrats living off such payments as "pickled in consols.") What should a person expect to pay for a consol which pays £10 a year forever? Assume the first payment is one year from the date of purchase and that interest remains 4% per year, compounded annually. (£ denotes pounds, the British unit of currency.)

Problems 30–32 are about *bonds*, which are issued by a government to raise money. An individual who buys a \$1000 bond gives the government \$1000 and in return receives a fixed sum of money, called the *coupon*, every six months or every year for the life of the bond. At the time of the last coupon, the individual also gets the \$1000, or *principal*, back.

30. What is the present value of a \$1000 bond which pays \$50 a year for 10 years, starting one year from now? Assume interest rate is 6% per year, compounded annually.
31. What is the present value of a \$1000 bond which pays \$50 a year for 10 years, starting one year from now? Assume the interest rate is 4% per year, compounded annually.
32. (a) What is the present value of a \$1000 bond which pays \$50 a year for 10 years, starting one year from now? Assume the interest rate is 5% per year, compounded annually.
    (b) Since \$50 is 5% of \$1000, this bond is often called a 5% bond. What does your answer to part (a) tell you about the relationship between the principal and the present value of this bond when the interest rate is 5%?
    (c) If the interest rate is more than 5% per year, compounded annually, which is larger: the principal or the value of the bond? Why do you think the bond is then described as *trading at discount*?
    (d) If the interest rate is less than 5% per year, compounded annually, why is the bond described as *trading at a premium*?
33. This problem illustrates how banks create credit and can thereby lend out more money than has been deposited. Suppose that initially \$100 is deposited in a bank. Experience has shown bankers that on average only 8% of the money deposited is withdrawn by the owner at any time. Consequently, bankers feel free to lend out 92% of their deposits. Thus \$92 of the original \$100 is loaned out to other customers (to start a business, for example). This \$92 will become someone else's income and, sooner or later, will be redeposited in the bank. Then 92% of \$92, or $\$92(0.92) = \$84.64$, is loaned out again and eventually redeposited. Of the \$84.64, the bank again loans out 92%, and so on.
    (a) Find the total amount of money deposited in the bank.
    (b) The total amount of money deposited divided by the original deposit is called the *credit multiplier*. Calculate the credit multiplier for this example and explain what this number tells us.
34. In a number of different games (e.g., tennis, volleyball, pickup basketball), winning requires a lead of two points. That is, if the score is tied you have to score two points in a row to win the game. Suppose your probability of scoring the next point is always p. (Your opponent's probability of scoring the next point then is always $1 - p$.)
    (a) What is the probability that you win the next two points?
    (b) What is the probability that you and your opponent split the next two points, that is, that neither of you wins both points?
    (c) What is the probability that you split the next two points but you win the two after that?

(d) What is the probability that you either win the next two points or split the next two and then win the next two after that?

(e) Give a formula for your probability w of winning a tied game.

(f) Compute your probability of winning a tied game when $p = 0.5$; when $p = 0.6$; when $p = 0.7$; when $p = 0.4$. Comment on your answers.

35. In some games you can score a point only if it is your turn; for example, in basketball you have to have the ball to score and in volleyball you have to be serving. Suppose your probability of scoring a point when it is your turn is p, and your opponent's probability of scoring a point when it is his turn is q.

(a) Find a formula for the probability S that you are the first to score a point, assuming it is your turn and players alternate turns until a point is scored.

(b) In both volleyball and "winners" basketball, if you score a point, the next turn is yours. Using your answers to part (a) and to Problem 34, compute the probability of winning a tied game if you need two points in a row to win.

(i) Assume $p = 0.5$ and $q = 0.5$ and it is your turn.

(ii) Assume $p = 0.6$ and $q = 0.5$ and it is your turn.

10.5 THE ERROR IN TAYLOR APPROXIMATIONS

In order to use an approximation intelligently, we need to be able to estimate the size of the error, which is the difference between the exact answer (which we usually do not know) and the approximate value. In general, we cannot find the exact value of the error; if we could, we would not need the approximation. However, we can often find a maximum possible value for the error. This gives us a feeling for how good the approximation is.

We now consider how we can assess the size of the error involved in using $P_n(x)$, the nth degree Taylor polynomial, to approximate $f(x)$ for values of x near a. The error is the difference

$$E = f(x) - P_n(x)$$

If E is positive, the approximation is smaller than the true value. If E is negative, the approximation is too large. Sometimes we are only interested in the magnitude of the error, $|E|$.

Graphical Approach: Error in $P_0(x)$

We begin with the simplest case, that of approximating $f(x)$ by the constant function $P_0(x)$ for values of x close to a. Since $P_0(x)$ is the Taylor polynomial with only the constant term,

$$P_0(x) = f(a).$$

Suppose we consider one particular value of x, say $x = b$, which is close to $x = a$. We want to estimate how far the graph of the function is from the horizontal line $y = f(a)$ where $x = b$, as shown in Figure 10.16. That is, we want to *bound* the difference

$$E = f(b) - P_0(b) = f(b) - f(a)$$

for any given value of b. Figure 10.16 makes it clear that if the derivative at a is positive, the error is positive for $b > a$ and negative for $b < a$, at least if b is near enough to a. If the derivative were negative, the situation would be reversed. We can bound the error as shown by the geometric argument used in the following theorem.

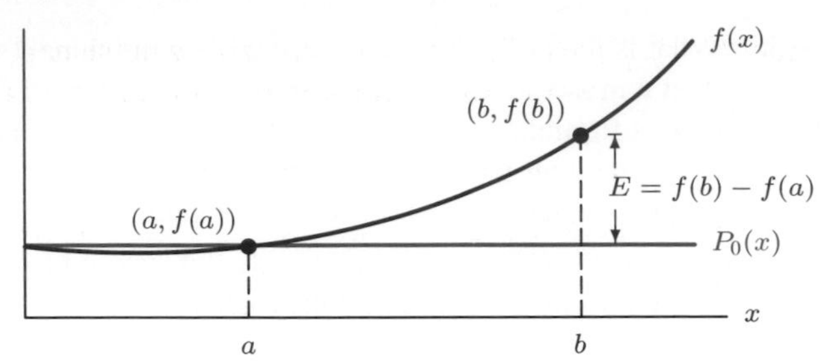

Figure 10.16: The error, E, in $P_0(x)$ at $x = b$

The Mean Value Theorem

Consider the graph shown in Figure 10.17 for some continuous and differentiable function $f(x)$ on the interval $[a, b]$. Join the points on the curve where $x = a$ and $x = b$ with a line and observe that the slope of this secant line AB is given by

$$m = \frac{f(b) - f(a)}{b - a}.$$

Consider now the tangent line drawn to the curve at each point between $x = a$ and $x = b$. In general, these lines will have different slopes. For the curve shown in Figure 10.17, the tangent line at $x = a$ is flatter than the secant line from A to B. Similarly, the tangent line at $x = b$ is steeper than the secant line. However, there is at least one point between a and b where the slope of the tangent line to the curve is precisely the same as the slope of the secant line. Suppose this occurs at $x = c$. Then we must have

$$f'(c) = m = \frac{f(b) - f(a)}{b - a}.$$

The fact that such a point $x = c$ exists is called the Mean Value Theorem.

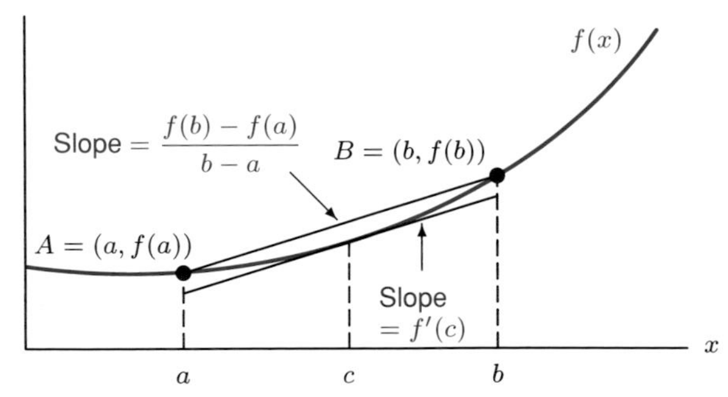

Figure 10.17: Estimating $f(b) - f(a)$

Estimating the Error in $P_0(x)$ Using the Mean Value Theorem

We can rewrite the condition satisfied by c so that

$$E = f(b) - f(a) = f'(c)(b - a).$$

Since we usually do not know where the point c is, we cannot get an exact value for $f'(c)$. However, if we know the derivative $f'(x)$, then we can find its maximum magnitude on the interval $[a, b]$ and so find a maximum possible value for the error in the approximation:

> **Bounding the Error in $P_0(x)$**
>
> If $|f'(c)| \le M$ for all c in $[a, b]$, then
>
> $$|E| = |f(x) - f(a)| \le M|x - a| \quad \text{for all } x \text{ in } [a, b].$$

Symbolic Approach: Error in $P_1(x)$

Let's try to bound the error in the first-degree Taylor polynomial approximation to a function $f(x)$. For simplicity, we will assume the Taylor polynomial is centered at $x = 0$. We have

$$f(x) \approx P_1(x) = f(0) + f'(0)x.$$

Notice that the error of $P_0(x)$ depends on the magnitude of the first derivative, $f'(x)$. For $P_1(x)$, the error will depend on how straight or curved the graph of f is for values of x near 0. The more curved f is, the worse the approximation. Thus we would expect that the size of the second derivative $f''(x)$ will be important. If we know how big $f''(x)$ is, we should be able to tell how good or bad the approximation can be. Suppose we know that $f''(x)$ is bounded above by M and bounded below by L for values of x near 0. That is,

$$L \le f''(x) \le M \quad \text{for values of } x \text{ near } 0.$$

First consider the upper bound. Then

$$f''(x) \le M \quad \text{for all nonnegative } x \text{ near } 0, \quad \text{say} \quad 0 \le x \le d.$$

Then it follows that

$$\int_0^x f''(x)\,dx \le \int_0^x M\,dx, \quad \text{for} \quad 0 \le x \le d.$$

Since integrating the second derivative gives the first derivative, the Fundamental Theorem of Calculus tells us that $\int_0^x f''(x)\,dx = f'(x) - f'(0)$. Integrating both sides of the inequality gives us

$$f'(x) - f'(0) \le M(x - 0)$$

so

$$f'(x) \le f'(0) + Mx.$$

Integrating again gives

$$\int_0^x f'(x)\,dx \le \int_0^x (f'(0) + Mx)\,dx.$$

On the left side, use the Fundamental Theorem. On the right side, observe that $f'(0)$ is just a constant, so $\int_0^x f'(0)\,dx = f'(0)(x - 0)$. We now have

$$f(x) - f(0) \le f'(0)x + \frac{M}{2}x^2.$$

Thus we have shown that if $f''(x) \leq M$ for $0 \leq x \leq d$, then

$$f(x) \leq f(0) + f'(0)x + \frac{M}{2}x^2.$$

Now we consider the lower bound. By reversing the inequalities, it can be shown that if $L \leq f''(x)$ for $0 \leq x \leq d$, then

$$f(0) + f'(0)x + \frac{L}{2}x^2 \leq f(x).$$

If we subtract $P_1(x) = f(0) + f'(0)x$ from both sides of these last two inequalities, we get a bound on the error $E = f(x) - P_1(x)$:

Bounding the Error in $P_1(x)$

If $\quad L \leq f''(x) \leq M \quad$ for $\quad 0 \leq x \leq d, \quad$ then

$$\frac{L}{2}x^2 \leq f(x) - P_1(x) \leq \frac{M}{2}x^2 \quad \text{for} \quad 0 \leq x \leq d.$$

Error in $P_2(x)$

Suppose we now want to bound the error in the second-degree Taylor polynomial approximation. By analogy with the error estimate for $P_1(x)$, we expect that the size of $f'''(x)$ will control the size of the error in $P_2(x)$. Thus we assume that

$$f'''(x) \leq M \quad \text{for} \quad 0 \leq x \leq d$$

and try to find a bound on $E = f(x) - P_2(x)$. Integrating as before, we get

$$\begin{aligned} \int_0^x f'''(x)\,dx &\leq \int_0^x M\,dx \\ f''(x) - f''(0) &\leq M(x-0) \\ f''(x) &\leq f''(0) + Mx. \end{aligned}$$

Integrating again:

$$\begin{aligned} \int_0^x f''(x)\,dx &\leq \int_0^x (f''(0) + Mx)\,dx \\ f'(x) - f'(0) &\leq f''(0)x + \frac{M}{2}x^2 \\ f'(x) &\leq f'(0) + f''(0)x + \frac{M}{2}x^2 \end{aligned}$$

and again:

$$\begin{aligned} \int_0^x f'(x)\,dx &\leq \int_0^x \left(f'(0) + f''(0)x + \frac{M}{2}x^2\right)\,dx \\ f(x) - f(0) &\leq f'(0)x + \frac{f''(0)}{2}x^2 + \frac{M}{3\cdot 2}x^3. \end{aligned}$$

We conclude that if $f'''(x) \le M$ for $0 \le x \le d$, then

$$f(x) \le f(0) + f'(0)x + \frac{f''(0)}{2}x^2 + \frac{M}{3!}x^3.$$

Similarly, if $L \le f'''(x)$ for $0 \le x \le d$, then

$$f(0) + f'(0)x + \frac{f''(0)}{2!}x^2 + \frac{L}{3!}x^3 \le f(x).$$

Thus, the error in the approximation by $P_2(x)$ is bounded as follows:

Bounding the Error in $P_2(x)$

If $\quad L \le f'''(x) \le M \quad$ for $\quad 0 \le x \le d, \quad$ then

$$\frac{L}{3!}x^3 \le f(x) - P_2(x) \le \frac{M}{3!}x^3 \quad \text{for} \quad 0 \le x \le d.$$

Error in $P_n(x)$

Needless to say, the error in higher-degree Taylor polynomials follows the same pattern. In general, if $L \le f^{(n+1)}(x) \le M$ for $0 \le x \le d$, then

$$\frac{L}{(n+1)!}x^{n+1} \le f(x) - P_n(x) \le \frac{M}{(n+1)!}x^{n+1}.$$

General Error Formula

When x is to the left of 0, so $-d \le x \le 0$, and when the Taylor series is centered at $a \ne 0$, similar calculations give us the following statement:

Error Bound for Taylor Polynomial Approximations

If $\quad |f^{(n+1)}(x)| \le M \quad$ for $\quad |x-a| \le d, \quad$ then

$$|E_n| = |f(x) - P_n(x)| \le \frac{M}{(n+1)!}|x-a|^{n+1} \quad \text{for} \quad |x-a| \le d.$$

Example 1 Give a bound on the error when e^x is approximated by its fourth-degree Taylor polynomial for $-0.5 \le x \le 0.5$.

Solution Let $f(x) = e^x$. Then the fifth derivative $f^{(5)}(x) = e^x$. Since e^x is increasing,

$$|f^{(5)}(x)| \leq e^{0.5} = \sqrt{e} < \sqrt{3} < 2 \quad \text{for} \quad -0.5 \leq x \leq 0.5.$$

Thus

$$|f(x) - P_4(x)| \leq \frac{2}{5!}|x|^5.$$

This means, for example, that on $-0.5 \leq x \leq 0.5$, the approximation

$$e^x \approx 1 + x + \frac{x^2}{2!} + \frac{x^3}{3!} + \frac{x^4}{4!}$$

has an error of at most $\frac{2}{120}(0.5)^5 < 0.0006$.

Importance of the Error Bound for Taylor Polynomials

To be honest, we do not often use the error bound for Taylor polynomials to bound the error in a particular numerical approximation. The reason for this is that if we know the function $f(x)$ so well that we can find a bound M on the $(n+1)^{\text{st}}$ derivative $f^{(n+1)}(x)$, we can probably compute $f(x)$ as accurately as we want anyway. For example, why would we want a bound on a Taylor polynomial approximation to $e^{0.5}$, when we can get $e^{0.5}$ directly off our calculator?

On the other hand, the error formula allows us to see how the accuracy in an approximation changes as the value of x changes. Observe that the error for a Taylor polynomial of degree n depends on the $(n+1)^{\text{st}}$ power of x. That means, for example, for a Taylor polynomial of degree n centered at 0, if you decrease x by a factor of 2, the error bound decreases by a factor of 2^{n+1}.

Example 2 Compare the errors in the approximations

$$e^{0.1} \approx 1 + 0.1 + \frac{1}{2!}(0.1)^2$$
$$e^{0.05} \approx 1 + (0.05) + \frac{1}{2!}(0.05)^2$$

Solution We are approximating e^x by its second-degree Taylor polynomial, first at $x = 0.1$, and then at $x = 0.05$. Since we have decreased x by a factor of 2, we should expect that the error has decreased by a factor of $2^3 = 8$. Let's see what actually happens. Here are the approximate values of the two errors:

$$e^{0.1} - \left(1 + 0.1 + \frac{1}{2!}(0.1)^2\right) = 1.105171 - 1.105000 = 0.000171$$
$$e^{0.05} - \left(1 + 0.05 + \frac{1}{2!}(0.05)^2\right) = 1.051271 - 1.051250 = 0.000021$$

Thus the error has decreased by a factor of $(0.000171)/(0.000021) = 8.1$, which is about what we expected.

Convergence of Taylor Series

The most important thing about the error bound is that in some cases it allows us to show that the Taylor series for a function does converge to that function. Since the error represents the difference between the Taylor polynomial and the function, showing that the series converges to the function for a particular value of x means showing that the error in the nth degree Taylor approximation goes to 0 as $n \to \infty$ for that value of x.

Showing $1 + x + \frac{x^2}{2!} + \frac{x^3}{3!} + \cdots$ Converges to e^x

When we write the equation

$$e^x = 1 + x + \frac{x^2}{2!} + \frac{x^3}{3!} + \cdots,$$

the equal sign and the "$\cdots$" mean that as we look at more and more terms of the series on the right, we get closer and closer to e^x. Since

$$E_n = e^x - P_n(x) = e^x - \left(1 + x + \frac{x^2}{2!} + \cdots + \frac{x^n}{n!}\right),$$

we can write

$$e^x = 1 + x + \frac{x^2}{2!} + \cdots + \frac{x^n}{n!} + E_n.$$

Thus, if the Taylor series converges to e^x, we must have $E_n \to 0$ as $n \to \infty$. We assumed this to be true when we defined Taylor series in Section 10.2; we will now prove it.

Showing $E_n \to 0$ as $n \to \infty$

The error bound is the key to the proof. Start by choosing any number M larger than e^x. (Remember that x is fixed so there must be such an M.) Since $f(x) = e^x$, the $(n+1)^{\text{st}}$ derivative $f^{(n+1)}(x)$ is also e^x, no matter what n is. Thus $|f^{(n+1)}(x)| \leq M$ for all n. The important observation is:

For $f(x) = e^x$, the *same* M bounds all the higher derivatives $f^{(n+1)}(x)$.

Thus

$$|E_n| = |e^x - P_n(x)| \leq \frac{M|x|^{n+1}}{(n+1)!} \quad \text{for every } n.$$

To show that the errors go to zero, we must show that for a fixed x and a fixed number M,

$$\frac{M}{(n+1)!}|x|^{n+1} \to 0 \quad \text{as} \quad n \to \infty.$$

Since M is fixed, we need only to show that

$$\frac{1}{(n+1)!}|x|^{n+1} \to 0 \quad \text{as} \quad n \to \infty.$$

To see why this is true, think of what happens when n is much larger than x. Suppose, for example, that $x = 17.3$. Let's look at the value of

$$\frac{1}{(n+1)!}(17.3)^{n+1}$$

for n at least twice as big as 17.3, that is $n = 35$, or $n = 36$, or $n = 37, \ldots$. We get the values

$$\frac{1}{35!}(17.3)^{35},$$

$$\frac{1}{36!}(17.3)^{36} = \frac{17.3}{36} \cdot \frac{1}{35!}(17.3)^{35},$$

$$\frac{1}{37!}(17.3)^{37} = \frac{17.3}{37} \cdot \frac{17.3}{36} \cdot \frac{1}{35!}(17.3)^{35}, \quad \ldots$$

The $n = 38$ term will be $17.3/38$ times the $n = 37$ term and so on. Each time we increase n by 1, the value of $\frac{1}{(n+1)!}(17.3)^{n+1}$ is multiplied by a number less than $\frac{1}{2}$. No matter what the value of $\frac{1}{35!}(17.3)^{35}$ is, if you keep on halving it, the result will get closer and closer to zero. Thus $\frac{1}{(n+1)!}(17.3)^{n+1}$ goes to 0 as n goes to infinity. In general, for fixed x

$$\frac{M}{(n+1)!}|x|^{n+1} \to 0 \quad \text{as} \quad n \to \infty.$$

Therefore, the Taylor series for e^x does converge to e^x.

Problems for Section 10.5

1. Suppose you approximate $f(t) = e^t$ by a Taylor polynomial of degree 0 about $t = 0$ on the interval $[0, 0.5]$.
   (a) Is the approximation an overestimate or an underestimate?
   (b) Estimate the magnitude of the largest possible error. Check your answer graphically on a computer or calculator.
2. Repeat Problem 1 using the second-degree Taylor approximation, $P_2(t)$, to e^t.
3. Consider the error in using the approximation $\sin\theta \approx \theta$ on the interval $[-1, 1]$.
   (a) Where is the approximation an overestimate, and where is it an underestimate?
   (b) Estimate the magnitude of the largest possible error. Check your answer graphically on a computer or calculator.
4. Repeat Problem 3 for the approximation $\sin\theta \approx \theta - \theta^3/3!$.

Use the methods of this section to show how you can estimate the magnitude of the error in approximating the quantities in Problems 5–8 using a third-degree Taylor polynomial about $x = 0$.

5. $\tan 1$ 6. $0.5^{1/3}$ 7. $\ln(1.5)$ 8. $1/\sqrt{3}$

9. Give a bound for the maximum possible error for the nth degree Taylor polynomial about $x = 0$ approximating $\cos x$ on the interval $[0, 1]$. What is the bound for $\sin x$?
10. What degree Taylor polynomial about $x = 0$ do you need to calculate $\cos 1$ to four decimal places? To six decimal places? Justify your answer using the results of Problem 9.
11. Show that the Taylor series about 0 for $\sin x$ converges to $\sin x$ for every x.
12. Show that the Taylor series about 0 for $\cos x$ converges to $\cos x$ for every x.
13. (a) Using a calculator, make a table of the values to four decimal places of $\sin x$ for
$$x = -0.5, -0.4, \ldots, -0.1, 0, 0.1, \ldots, 0.4, 0.5.$$
   (b) Add to your table the values of the error $E_1 = \sin x - x$ for these x values.
   (c) Using a calculator or computer, draw a graph of the quantity $E_1 = \sin x - x$ showing that
$$|E_1| < 0.03 \quad \text{for} \quad -0.5 \le x \le 0.5.$$

14. In this problem, you will investigate the error in the nth degree Taylor approximation to e^x for various values of n.

    (a) Let $E_1 = e^x - P_1(x) = e^x - (1 + x)$. Using a calculator or computer, graph E_1 for $-0.1 \leq x \leq 0.1$. What shape is the graph of E_1? Use the graph to confirm that

    $$|E_1| \leq x^2 \quad \text{for} \quad -0.1 \leq x \leq 0.1.$$

    (b) Let $E_2 = e^x - P_2(x) = e^x - (1 + x + x^2/2)$. Choose a suitable range and graph E_2 for $-0.1 \leq x \leq 0.1$. What shape is the graph of E_2? Use the graph to confirm that

    $$|E_2| \leq x^3 \quad \text{for} \quad -0.1 \leq x \leq 0.1.$$

    (c) Explain why the graphs of E_1 and E_2 have the shapes they do.

15. Graph the error

    $$E_0 = \cos x - P_0(x) = \cos x - 1$$

    for $|x| \leq 0.1$. Explain the shape of the graph, using information given by the Taylor expansion of $\cos x$, and find a bound for $|E_0|$ for $|x| \leq 0.1$.

10.6 FOURIER SERIES

We have seen how to approximate a function by a Taylor polynomial of fixed degree. Such a polynomial is usually very close to the true value of the function near one point (the point at which the Taylor polynomial is centered), but not necessarily at all close anywhere else. In other words, Taylor polynomials are good approximations of a function *locally*, but not necessarily *globally*. In this section, we take another approach: we approximate the function by trigonometric functions, called *Fourier approximations*. The resulting approximation may not be as close to the original function at some points as the Taylor polynomial, but in general, it is closer over a larger interval. In other words, a Fourier approximation can be a better approximation globally. In addition, unlike Taylor approximations, Fourier approximations are periodic, so they are useful for approximating periodic functions.

Many processes in nature are periodic or repeating, so it makes sense to approximate them by periodic functions. For example, sound waves are made up of periodic oscillations of air molecules. Heartbeats, the movement of the lungs, and the electrical current that powers our homes are all periodic phenomena. Two of the simplest periodic functions are the square wave of Figure 10.18 and the triangular wave of Figure 10.19. Electrical engineers use the square wave as the model for the flow of electricity as a switch is repeatedly flicked on and off.

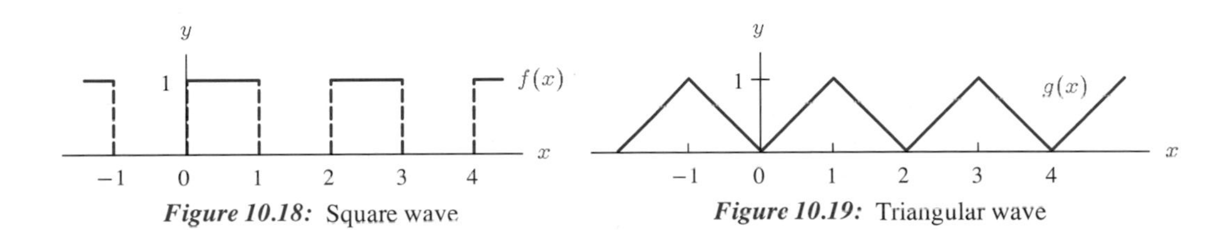

Figure 10.18: Square wave

Figure 10.19: Triangular wave

Fourier Polynomials

While we can express the square wave and the triangular wave by the formulas

$$f(x) = \begin{cases} \vdots & \vdots \\ 0 & -1 \le x < 0 \\ 1 & 0 \le x < 1 \\ 0 & 1 \le x < 2 \\ 1 & 2 \le x < 3 \\ 0 & 3 \le x < 4 \\ \vdots & \vdots \end{cases} \qquad g(x) = \begin{cases} \vdots & \vdots \\ -x & -1 \le x < 0 \\ x & 0 \le x < 1 \\ 2-x & 1 \le x < 2 \\ x-2 & 2 \le x < 3 \\ 4-x & 3 \le x < 4 \\ \vdots & \vdots \end{cases}$$

these formulas are not particularly easy to work with. Worse, the functions are not differentiable at various points. Here we will approximate such a function by a differentiable, periodic function.

Since the sine and cosine are the simplest periodic functions, they are the building blocks we will use. Because they repeat every 2π, we will assume that the function f we want to approximate repeats every 2π. (Later, we will deal with the case where f has some other period.) We will start by considering the square wave in Figure 10.20. Because of the periodicity of all the functions concerned, we only have to consider what happens in the course of a single period; the same behavior repeats in any other period.

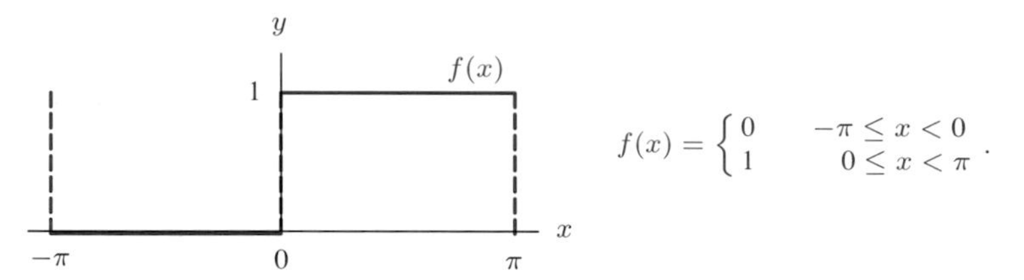

Figure 10.20: Square wave on $[-\pi, \pi]$

We will attempt to approximate f with a sum of trigonometric functions of the form

$$\begin{aligned} f(x) &\approx F_n(x) \\ &= a_0 + a_1 \cos x + a_2 \cos 2x + a_3 \cos 3x + \cdots + a_n \cos nx \\ &\qquad + b_1 \sin x + b_2 \sin 2x + b_3 \sin 3x + \cdots + b_n \sin nx \\ &= a_0 + \sum_{k=1}^{n} a_k \cos kx + \sum_{k=1}^{n} b_k \sin kx. \end{aligned}$$

$F_n(x)$ is known as a *Fourier polynomial of degree* n, named after the French mathematician Joseph Fourier, who was one of the first to investigate it.[2] The coefficients a_k and b_k are the *Fourier coefficients*. Since each of the component functions $\cos kx$ and $\sin kx$, $k = 1, 2, \ldots, n$, repeats every 2π, $F_n(x)$ must repeat every 2π and so is a potentially good match for $f(x)$, which also repeats every 2π. The problem is to determine the values for the Fourier coefficients to achieve a close match between $f(x)$ and $F_n(x)$. We choose the following values:

[2]The Fourier polynomials are not polynomials in the usual sense of the word.

The Fourier Coefficients for a Periodic Function f of Period 2π

$$a_0 = \frac{1}{2\pi}\int_{-\pi}^{\pi} f(x)\,dx,$$
$$a_k = \frac{1}{\pi}\int_{-\pi}^{\pi} f(x)\cos kx\,dx \quad \text{for } k > 0,$$
$$b_k = \frac{1}{\pi}\int_{-\pi}^{\pi} f(x)\sin kx\,dx \quad \text{for } k > 0.$$

Notice that a_0 is just the average value of f over the interval $[-\pi, \pi]$.

For an informal justification of these formulas, see the appendix to this section.

Example 1 Construct successive Fourier polynomials for the square wave function f, with period 2π, given by

$$f(x) = \begin{cases} 0 & -\pi \le x < 0 \\ 1 & 0 \le x < \pi. \end{cases}$$

Solution Since a_0 is the average value of f on $[-\pi, \pi]$, we suspect from the graph of f that $a_0 = \frac{1}{2}$. We can verify this analytically:

$$a_0 = \frac{1}{2\pi}\int_{-\pi}^{\pi} f(x)\,dx = \frac{1}{2\pi}\int_{-\pi}^{0} 0\,dx + \frac{1}{2\pi}\int_{0}^{\pi} 1\,dx = 0 + \frac{1}{2\pi}(\pi) = \frac{1}{2}.$$

Furthermore,

$$a_1 = \frac{1}{\pi}\int_{-\pi}^{\pi} f(x)\cos x\,dx = \frac{1}{\pi}\int_{0}^{\pi} 1\cos x\,dx = 0$$

and

$$b_1 = \frac{1}{\pi}\int_{-\pi}^{\pi} f(x)\sin x\,dx = \frac{1}{\pi}\int_{0}^{\pi} 1\sin x\,dx = \frac{2}{\pi}.$$

Therefore, the Fourier polynomial of degree 1 is given by

$$f(x) \approx F_1(x) = \frac{1}{2} + \frac{2}{\pi}\sin x$$

and the graphs of the function and the first Fourier approximation are shown in Figure 10.21.

We next construct the Fourier polynomial of degree 2. The coefficients a_0, a_1, b_1 are the same as before. In addition,

$$a_2 = \frac{1}{\pi}\int_{-\pi}^{\pi} f(x)\cos 2x\,dx = \frac{1}{\pi}\int_{0}^{\pi} 1\cos 2x\,dx = 0$$

and

$$b_2 = \frac{1}{\pi}\int_{-\pi}^{\pi} f(x)\sin 2x\,dx = \frac{1}{\pi}\int_{0}^{\pi} 1\sin 2x\,dx = 0.$$

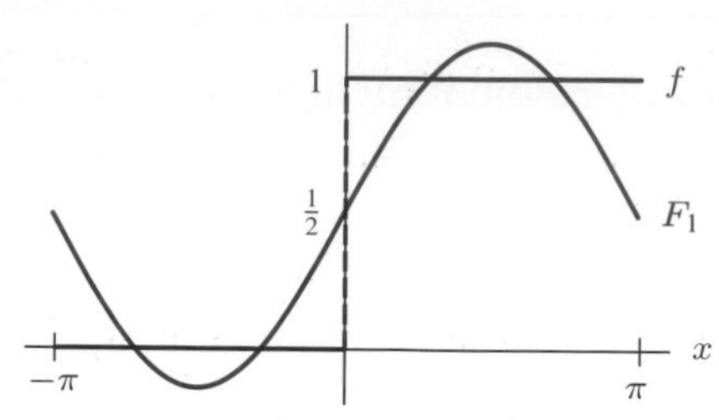

Figure 10.21: First Fourier approximation to the square wave

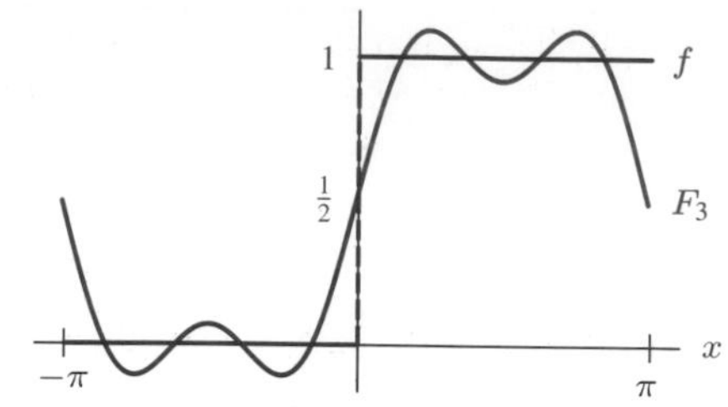

Figure 10.22: Third Fourier approximation to the square wave

Since $a_2 = b_2 = 0$, the Fourier polynomial of degree 2 is identical to the Fourier polynomial of degree 1. Let's look at the Fourier polynomial of degree 3:

$$a_3 = \frac{1}{\pi}\int_{-\pi}^{\pi} f(x)\cos 3x\,dx = \frac{1}{\pi}\int_{0}^{\pi} 1\cos 3x\,dx = 0$$

and

$$b_3 = \frac{1}{\pi}\int_{-\pi}^{\pi} f(x)\sin 3x\,dx = \frac{1}{\pi}\int_{0}^{\pi} 1\sin 3x\,dx = \frac{2}{3\pi}$$

and so the approximation is given by

$$f(x) \approx F_3(x) = \frac{1}{2} + \frac{2}{\pi}\sin x + \frac{2}{3\pi}\sin 3x.$$

The corresponding graphs are shown in Figure 10.22. This approximation is much more accurate than $F_1(x) = \frac{1}{2} + \frac{2}{\pi}\sin x$, as comparing Figure 10.22 to Figure 10.21 shows.

Without going through the details, we calculate the coefficients for higher-degree Fourier approximations:

$$F_5(x) = \frac{1}{2} + \frac{2}{\pi}\sin x + \frac{2}{3\pi}\sin 3x + \frac{2}{5\pi}\sin 5x$$
$$F_7(x) = \frac{1}{2} + \frac{2}{\pi}\sin x + \frac{2}{3\pi}\sin 3x + \frac{2}{5\pi}\sin 5x + \frac{2}{7\pi}\sin 7x.$$

Figure 10.23 shows that higher-degree approximations match the steplike nature of the square wave function more and more closely.

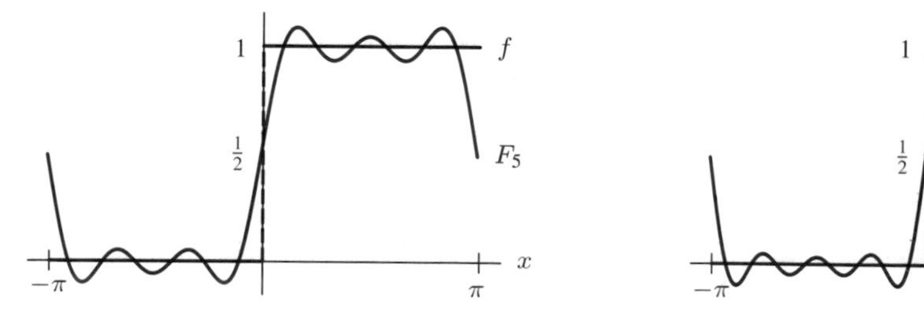

Figure 10.23: Fifth and seventh Fourier approximations to the square wave

We could have used a Taylor series to approximate the square wave, provided we did not center the series at a point of discontinuity. Since the square wave is a constant function on each interval, all its derivatives are zero, and so its Taylor series approximations are the constant functions: 0 or 1, depending on where your Taylor series is centered. They approximate the square wave perfectly on each piece, but they do not do a good job over the whole interval of length 2π. That is what Fourier polynomials succeed in doing: they approximate a curve fairly well everywhere, rather than just near a particular point. The Fourier approximations above look a lot like square waves, so they approximate well *globally*; but they may not give good values near points of discontinuity. (For example, near $x = 0$, they all give values near $1/2$, which are incorrect.) Thus Fourier polynomials may not be good *local* approximations.

Taylor polynomials give good *local* approximations to a function;
Fourier polynomials give good *global* approximations to a function.

Fourier Series

As with Taylor polynomials, the higher the degree of the Fourier approximation, the more accurate it is. Therefore, we carry this procedure on indefinitely and call the infinite sequence of approximations a Fourier series.

The Fourier Series for f on $[-\pi, \pi]$

$$f(x) = a_0 + a_1 \cos x + a_2 \cos 2x + a_3 \cos 3x + \cdots$$
$$+ b_1 \sin x + b_2 \sin 2x + b_3 \sin 3x + \cdots$$

where a_k and b_k are the Fourier coefficients.

Thus, the Fourier series for the square wave is

$$f(x) = \frac{1}{2} + \frac{2}{\pi} \sin x + \frac{2}{3\pi} \sin 3x + \frac{2}{5\pi} \sin 5x + \frac{2}{7\pi} \sin 7x + \cdots.$$

Harmonics

Let us start with a function $f(x)$ that is periodic with period 2π, expanded in a Fourier series.

$$f(x) = a_0 + a_1 \cos x + a_2 \cos 2x + a_3 \cos 3x + \cdots$$
$$+ b_1 \sin x + b_2 \sin 2x + b_3 \sin 3x + \cdots$$

The function

$$a_k \cos kx + b_k \sin kx$$

is referred to as the kth *harmonic* of f, and it is customary to say that the Fourier series expresses f in terms of its harmonics. The first harmonic, $a_1 \cos x + b_1 \sin x$ is sometimes called the *fundamental harmonic* of f.

Example 2 Find a_0 and the first four harmonics of the *pulse train* function $f(x)$ of period 2π (see Figure 10.24) where

$$f(x) = \begin{cases} 1 & 0 \le x < \pi/2 \\ 0 & \pi/2 \le x < 2\pi. \end{cases}$$

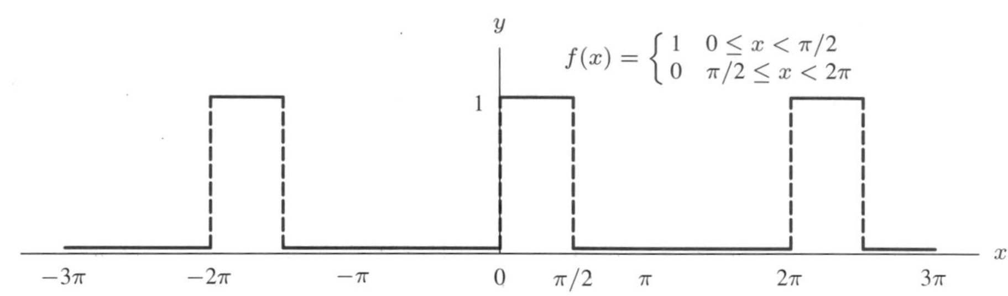

Figure 10.24: A train of pulses with period 2π

Solution First, a_0 is the average value of the function, so

$$a_0 = \frac{1}{2\pi}\int_{-\pi}^{\pi} f(x)\,dx = \frac{1}{2\pi}\int_0^{\pi/2} 1\,dx = \frac{1}{4}.$$

Next, we compute a_k and b_k, $k = 1, 2, 3,$ and 4. The formulas

$$a_k = \frac{1}{\pi}\int_{-\pi}^{\pi} f(x)\cos kx\,dx = \frac{1}{\pi}\int_0^{\pi/2} \cos kx\,dx$$

$$b_k = \frac{1}{\pi}\int_{-\pi}^{\pi} f(x)\sin kx\,dx = \frac{1}{\pi}\int_0^{\pi/2} \sin kx\,dx$$

lead to the harmonics

$$a_1\cos x + b_1 \sin x = \frac{1}{\pi}\cos x + \frac{1}{\pi}\sin x$$

$$a_2\cos 2x + b_2\sin 2x = \frac{1}{\pi}\sin 2x$$

$$a_3\cos 3x + b_3\sin 3x = -\frac{1}{3\pi}\cos 3x + \frac{1}{3\pi}\sin 3x$$

$$a_4\cos 4x + b_4\sin 4x = 0.$$

Figure 10.25 shows the graph of the sum of a_0 and these harmonics, which is the fourth Fourier approximation of f.

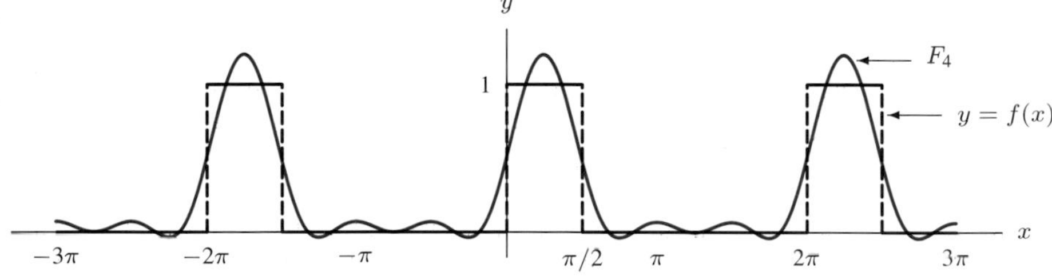

Figure 10.25: Fourth Fourier approximation to pulse train f equals the sum of a_0 and the first four harmonics

Energy and the Energy Theorem

The quantity $A_k = \sqrt{a_k^2 + b_k^2}$ is called the amplitude of the kth harmonic. The square of the amplitude has a useful interpretation. Adopting terminology from the study of periodic waves, we define the *energy* E of a periodic function f of period 2π to be the number

$$E = \frac{1}{\pi}\int_{-\pi}^{\pi} [f(x)]^2\,dx.$$

Problem 18 on page 648 verifies that for all positive integers k,

$$\frac{1}{\pi}\int_{-\pi}^{\pi} (a_k \cos kx + b_k \sin kx)^2\,dx = a_k^2 + b_k^2 = A_k^2$$

which shows that the kth harmonic of f has energy A_k^2. The energy of the constant term a_0 of the Fourier series is $\frac{1}{\pi}\int_{-\pi}^{\pi} a_0^2\,dx = 2a_0^2$, so we make the definition

$$A_0 = \sqrt{2}a_0.$$

It turns out that for all reasonable periodic functions f, the energy of f equals the sum of the energy of its harmonics:

The Energy Theorem for a Periodic Function f of Period 2π

$$E = \frac{1}{\pi}\int_{-\pi}^{\pi} [f(x)]^2\,dx = A_0^2 + A_1^2 + A_2^2 + \cdots$$

where $A_0 = \sqrt{2}a_0$ and $A_k = \sqrt{a_k^2 + b_k^2}$ (for all integers $k \geq 1$).

The graph of A_k^2 against k is called the *energy spectrum* of f. It shows how the energy of f is distributed among its harmonics.

Example 3

(a) Graph the energy spectrum of the square wave of Example 1.

(b) What fraction of the energy of the square wave is contained in the constant term and first three harmonics of its Fourier series?

Solution (a) We know from Example 1 that $a_0 = 1/2$, $a_k = 0$ for $k \geq 1$, $b_k = 0$ for k even, and $b_k = 2/(k\pi)$ for k odd. Thus

$$A_0^2 = 2a_0^2 = \frac{1}{2}$$

$$A_k^2 = 0 \quad \text{if } k \text{ is even}, \quad k \geq 1,$$

$$A_k^2 = \left(\frac{2}{k\pi}\right)^2 = \frac{4}{k^2\pi^2} \quad \text{if } k \text{ is odd}, \quad k \geq 1.$$

The energy spectrum is graphed in Figure 10.26. Notice that it is customary to represent the energy A_k^2 of the kth harmonic by a vertical line of length A_k^2. The graph shows that the constant term and first harmonic carry most of the energy of f.

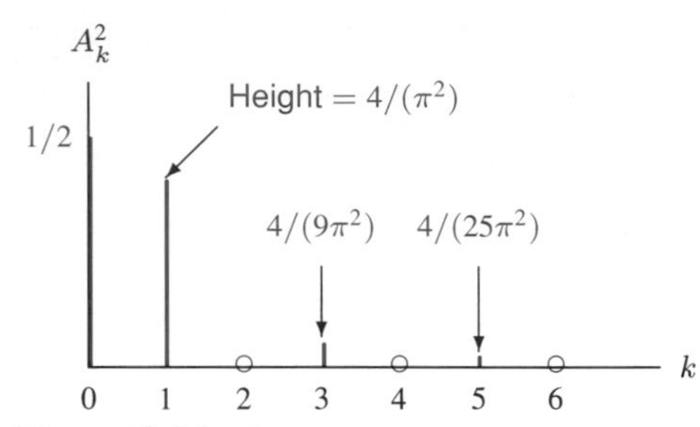

Figure 10.26: The energy spectrum of a square wave

(b) The energy of the square wave $f(x)$ is

$$E = \frac{1}{\pi}\int_{-\pi}^{\pi} [f(x)]^2\,dx = \frac{1}{\pi}\int_0^{\pi} 1\,dx = 1.$$

The energy in the constant term and the first three harmonics of the Fourier series is

$$A_0^2 + A_1^2 + A_2^2 + A_3^2 = \frac{1}{2} + \frac{4}{\pi^2} + 0 + \frac{4}{9\pi^2} = 0.950.$$

The fraction of energy carried by the constant term and the first three harmonics is

$$0.95/1 = 0.95, \text{ or } 95\%.$$

Musical Instruments

You may have wondered why different musical instruments sound different, even when playing the same note. A first step might be to graph the periodic variations in air pressure that form the sound

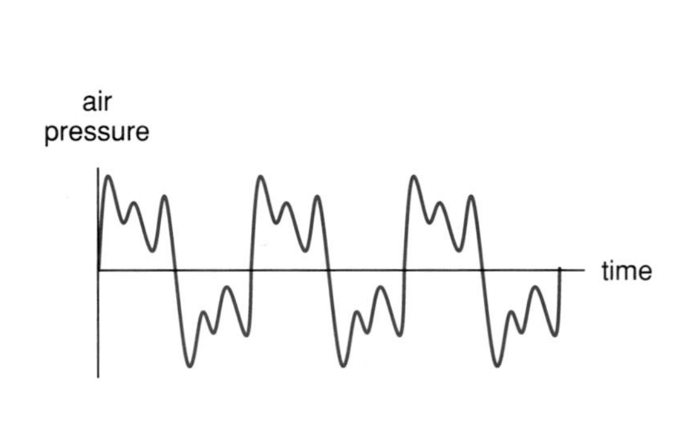

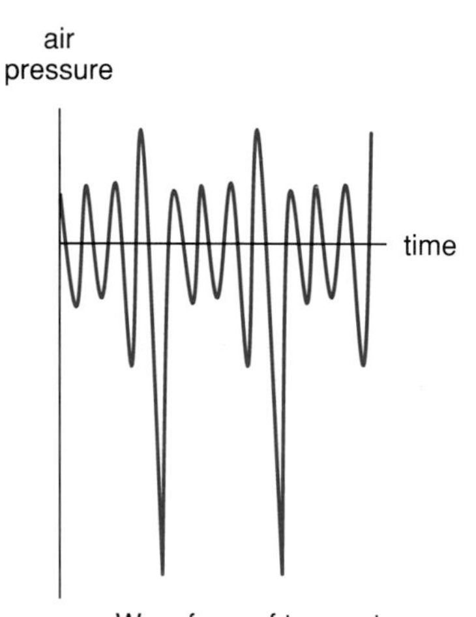

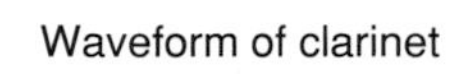

Waveform of clarinet

Waveform of trumpet

Figure 10.27: Sound waves of a clarinet and trumpet

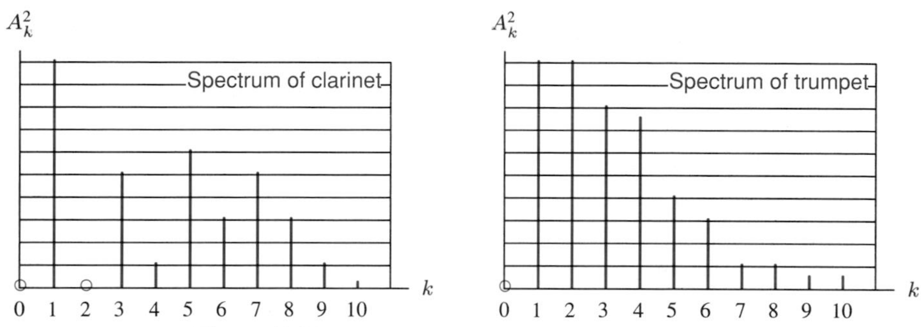

Figure 10.28: Energy spectra of a clarinet and trumpet

waves they produce. This has been done for clarinet and trumpet in Figure 10.27.[3] However, it is more revealing to graph the energy spectra of these functions, as in Figure 10.28 The most striking difference is the relative weakness of the second, fourth, and sixth harmonics for the clarinet, with the second harmonic completely absent. The trumpet sounds the second harmonic with as much energy as it does the fundamental.

What Do We Do if Our Function Does Not Have Period 2π?

We can easily adapt what we have done above by changing variables. Suppose we have a function $f(x)$ that is periodic with period b. If we define $x = bt/2\pi$, then x varies over the interval $[0, b]$ while t varies over the interval $[0, 2\pi]$; similarly, x varies over $[-b/2, b/2]$ while t varies over $[-\pi, \pi]$. Thus, if we substitute $x = bt/2\pi$ into f and define a new function g by

$$g(t) = f\left(\frac{bt}{2\pi}\right) = f(x),$$

then g has period 2π if f has period b. Thus we can find the Fourier series for g as before, giving

$$\begin{aligned} g(t) = a_0 &+ a_1 \cos t + a_2 \cos 2t + a_3 \cos 3t + \cdots \\ &+ b_1 \sin t + b_2 \sin 2t + b_3 \sin 3t + \cdots . \end{aligned}$$

Substituting $t = 2\pi x/b$ allows us to convert back:

$$\begin{aligned} f(x) = g\left(\frac{2\pi x}{b}\right) &= a_0 + a_1 \cos\left(\frac{2\pi x}{b}\right) + a_2 \cos\left(\frac{4\pi x}{b}\right) + a_3 \cos\left(\frac{6\pi x}{b}\right) + \cdots \\ &+ b_1 \sin\left(\frac{2\pi x}{b}\right) + b_2 \sin\left(\frac{4\pi x}{b}\right) + b_3 \sin\left(\frac{6\pi x}{b}\right) + \cdots \end{aligned}$$

Note that the terms in the Fourier series of a periodic function of period b are not $\cos kx$ and $\sin kx$ but instead turn out to be $\cos(2\pi kx/b)$ and $\sin(2\pi kx/b)$. In addition, for a periodic function of period 2π, integrals over $[-\pi, \pi]$ can be replaced by integrals over any interval of length 2π. Thus the integrals for a_k and b_k can be evaluated over any interval of length 2π, not just $[-\pi, \pi]$.

Example 4 Find the fifth-degree Fourier polynomial of the square wave $f(x)$ graphed in Figure 10.29.

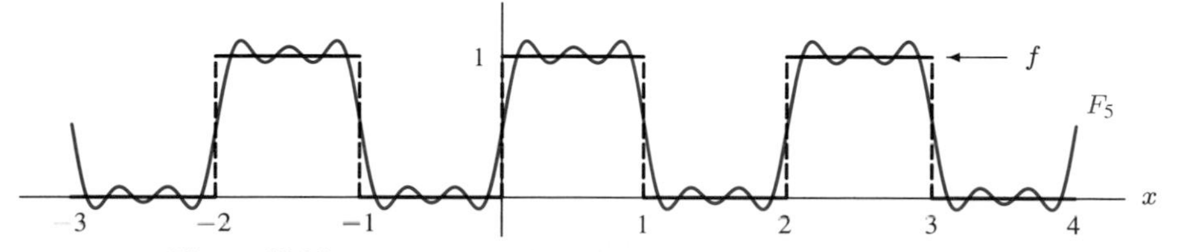

Figure 10.29: Square wave f and its fifth Fourier approximation F_5

[3]Adapted from C.A. Culver, *Musical Acoustics* (New York: McGraw-Hill, 1956), pages 204, 220.

Solution Since f has period $b = 2$, we let $x = bt/2\pi = t/\pi$ and consider the function

$$g(t) = f\left(\frac{t}{\pi}\right) = f(x)$$

which has period 2π. Then

$$g(t) = \begin{cases} 0 & -\pi \le t < 0 \\ 1 & 0 \le t < \pi \end{cases}$$

so $g(t)$ is the square wave of Example 1. We have, therefore,

$$g(t) \approx \frac{1}{2} + \frac{2}{\pi}\sin t + \frac{2}{3\pi}\sin 3t + \frac{2}{5\pi}\sin 5t.$$

Hence, substituting $t = \pi x$

$$f(x) = g(t) = g(\pi x) \approx \frac{1}{2} + \frac{2}{\pi}\sin(\pi x) + \frac{2}{3\pi}\sin(3\pi x) + \frac{2}{5\pi}\sin(5\pi x).$$

Graphs of $f(x)$ and its fifth Fourier approximation are in Figure 10.29.

Seasonal Variation in the Incidence of Measles

Example 5 Fourier approximations have been used to analyze the seasonal variation in the incidence of diseases. One study[4] done in Baltimore, Maryland, for the years 1901–1931, studied $I(t)$, the average number of cases of measles per 10,000 susceptible children in the tth month of the year. The data points in Figure 10.30 show $f(t) = \log I(t)$. The curve in Figure 10.30 shows the second Fourier approximation of $f(t)$. Figure 10.31 contains the graphs of the first and second harmonics of $f(t)$, plotted separately as deviations about a_0, the average logarithmic incidence rate. Describe what these two harmonics tell you about incidence of measles.

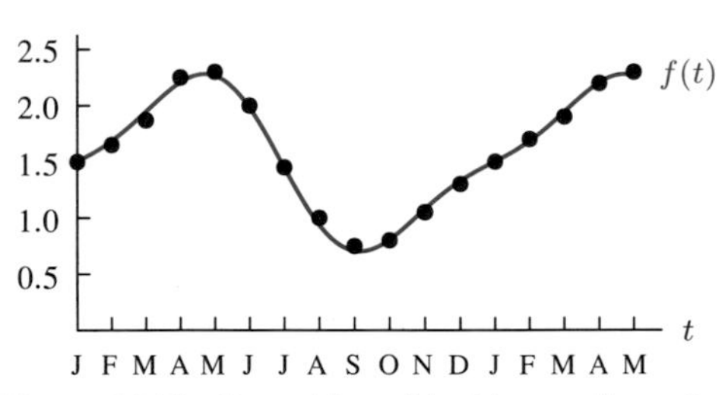

Figure 10.30: Logarithm of incidence of measles per month (dots) and second Fourier approximation (curve)

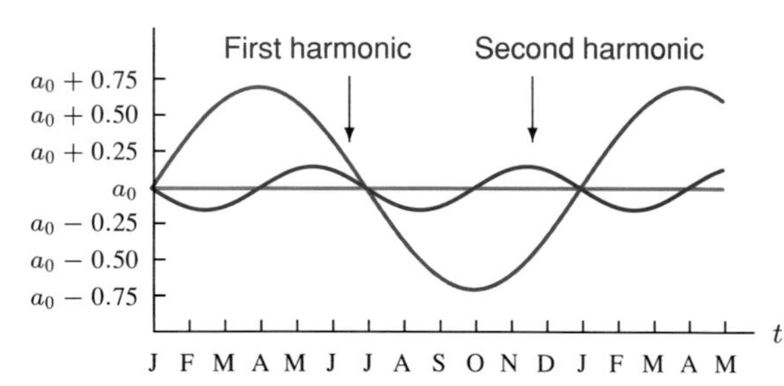

Figure 10.31: First and second harmonics of $f(t)$ plotted as deviations from average log incidence rate, a_0

Solution Taking the log of $I(t)$ has the effect of reducing the amplitude of the oscillations. However, since the log of a function increases when the function increases, and decreases when it decreases, oscillations in $f(t)$ correspond to oscillations in $I(t)$.

Figure 10.31 shows that the first harmonic in the Fourier series has a period of one year (the same period as the original function); the second harmonic has a period of six months. Reading off

[4]From C. I. Bliss and D. L. Blevins, *The Analysis of Seasonal Variation in Measles* (Am. J. Hyg. 70, 1959), reported by Edward Batschelet, *Introduction to Mathematics for the Life Sciences* (Springer-Verlag, Berlin, 1979).

Figure 10.31 shows that the first harmonic is approximately a sine with amplitude about 0.7; the second harmonic is approximately the negative of a sine with amplitude about 0.15. Thus, for t in months ($t = 0$ in January)

$$\log I(t) = f(t) \approx a_0 + 0.7 \sin\left(\frac{\pi}{6}t\right) - 0.15 \sin\left(\frac{\pi}{3}t\right)$$

where $\pi/6$ and $\pi/3$ are introduced to make the periods 12 and 6 months, respectively. We can read a_0 off the original graph of f: it is the average value of approximately 1.5. Thus

$$f(t) \approx 1.5 + 0.7 \sin\left(\frac{\pi}{6}t\right) - 0.15 \sin\left(\frac{\pi}{3}t\right).$$

Figure 10.30 shows that the second Fourier approximation of $f(t)$ is quite good. The harmonics of $f(t)$ beyond the second must be rather insignificant. This suggests that the variation in incidence in measles comes from two sources, one with a yearly cycle that is reflected in the first harmonic and one with a half-yearly cycle reflected in the second harmonic. At this point the mathematics can tell us no more; we must turn to the epidemiologists for further explanation.

Appendix: Justification of the Formulas for the Fourier Coefficients

Recall that the coefficients in a Taylor series (which is a good approximation locally) are found by differentiation. In contrast, the coefficients in a Fourier series (which is a good approximation globally) are found by integration.

We will use a $\sum$ sign with ∞ at the top to denote a series that does not have a last term:

$$f(x) = a_0 + \sum_{k=1}^{\infty} a_k \cos kx + \sum_{k=1}^{\infty} b_k \sin kx.$$

Consider the integral

$$\int_{-\pi}^{\pi} f(x)\,dx = \int_{-\pi}^{\pi} \left[a_0 + \sum_{k=1}^{\infty} a_k \cos kx + \sum_{k=1}^{\infty} b_k \sin kx \right] dx.$$

Splitting the integral into separate terms, and assuming we can interchange integration and summation, we get

$$\begin{aligned}\int_{-\pi}^{\pi} f(x)\,dx &= \int_{-\pi}^{\pi} a_0\,dx + \int_{-\pi}^{\pi} \sum_{k=1}^{\infty} a_k \cos kx\,dx + \int_{-\pi}^{\pi} \sum_{k=1}^{\infty} b_k \sin kx\,dx \\ &= \int_{-\pi}^{\pi} a_0\,dx + \sum_{k=1}^{\infty} \int_{-\pi}^{\pi} a_k \cos kx\,dx + \sum_{k=1}^{\infty} \int_{-\pi}^{\pi} b_k \sin kx\,dx.\end{aligned}$$

But for $k \geq 1$, thinking of the integral as an area shows that

$$\int_{-\pi}^{\pi} \sin kx\,dx = 0 \quad \text{and} \quad \int_{-\pi}^{\pi} \cos kx\,dx = 0,$$

so all terms drop out except the first, giving

$$\int_{-\pi}^{\pi} f(x)\,dx = \int_{-\pi}^{\pi} a_0\,dx = a_0 x\Big|_{-\pi}^{\pi} = 2\pi a_0$$

and so we get the following result:

$$\boxed{a_0 = \frac{1}{2\pi}\int_{-\pi}^{\pi} f(x)\,dx}$$

Thus a_0 is the average value of f on the interval $[-\pi, \pi]$.

To determine the values of any of the other a_k, $k = 1, 2, ..., n$, we use a rather clever method which depends on the fact that for all integers k and m

$$\int_{-\pi}^{\pi} \sin kx \cos mx\,dx = 0,$$

and that, provided $k \neq m$,

$$\int_{-\pi}^{\pi} \cos kx \cos mx\,dx = 0.$$

(See Problems 25–29 on page 650.) In addition, provided $m \neq 0$, we have

$$\int_{-\pi}^{\pi} \cos^2 mx\,dx = \pi.$$

To use this, we multiply through the expression for the Fourier series by $\cos mx$, where m is any positive integer, and integrate term by term. (The justification for this step is omitted.) We then have

$$f(x)\cos mx = a_0 \cos mx + \sum_{k=1}^{\infty} a_k \cos kx \cos mx + \sum_{k=1}^{\infty} b_k \sin kx \cos mx.$$

When we integrate this between $-\pi$ and π, we obtain

$$\begin{aligned}\int_{-\pi}^{\pi} f(x)\cos mx\,dx &= \int_{-\pi}^{\pi}\left(a_0\cos mx + \sum_{k=1}^{\infty} a_k \cos kx\cos mx + \sum_{k=1}^{\infty} b_k \sin kx \cos mx\right)dx\\ &= a_0\int_{-\pi}^{\pi}\cos mx\,dx + \sum_{k=1}^{\infty}\left(a_k\int_{-\pi}^{\pi}\cos kx\cos mx\,dx\right)\\ &\quad + \sum_{k=1}^{\infty}\left(b_k\int_{-\pi}^{\pi}\sin kx\cos mx\,dx\right).\end{aligned}$$

Provided $m \neq 0$, we have $\int_{-\pi}^{\pi}\cos mx\,dx = 0$. Since the integral $\int_{-\pi}^{\pi}\sin kx\cos mx\,dx = 0$, all the terms in the third sum are zero. Since $\int_{-\pi}^{\pi}\cos kx\cos mx\,dx = 0$ provided $k \neq m$, all the terms in the second sum are zero except where $k = m$. Thus the right-hand side reduces to one term:

$$\int_{-\pi}^{\pi} f(x)\cos mx\,dx = a_m\int_{-\pi}^{\pi}\cos mx\cos mx\,dx = \pi a_m,$$

which leads, for each value of $m = 1, 2, 3 \ldots$ to the following formula

$$a_m = \frac{1}{\pi} \int_{-\pi}^{\pi} f(x) \cos mx \, dx$$

Using a similar line of argument, we multiply through by $\sin mx$ instead of $\cos mx$ and eventually obtain, for each value of $m = 1, 2, 3 \ldots$ the following result.

$$b_m = \frac{1}{\pi} \int_{-\pi}^{\pi} f(x) \sin mx \, dx$$

Problems for Section 10.6

Which of the series in Problems 1– 4 are Fourier series?

1. $\frac{1}{2} - \sin x + \sin 2x - \sin 3x + \cdots$
2. $1 + \cos x + \cos^2 x + \cos^3 x + \cos^4 x + \cdots$
3. $\sin x + \sin(x + 1) + \sin(x + 2) + \cdots$
4. $\cos x + \sin x - \cos 2x - \frac{\sin 2x}{2} + \cos 3x + \frac{\sin 3x}{3} - \cdots$
5. (a) For $-2\pi \leq x \leq 2\pi$, use a graphing calculator to sketch:
   i) $y = \sin x + \frac{1}{3} \sin 3x$
   ii) $y = \sin x + \frac{1}{3} \sin 3x + \frac{1}{5} \sin 5x$
   (b) Each of the functions in part (a) is a Fourier approximation to a function whose graph is a square wave. What term would you add to the right-hand side of the second function in part (a) to get a better approximation to the square wave?
   (c) What is the equation of the square wave function? Is this function continuous?
6. Construct the first three Fourier approximations to the square wave function
   $$f(x) = \begin{cases} -1 & -\pi \leq x < 0 \\ 1 & 0 \leq x < \pi. \end{cases}$$
   Use a calculator or computer to draw the graph of each approximation.
7. Repeat Problem 6 with the function
   $$f(x) = \begin{cases} -x & -\pi \leq x < 0 \\ x & 0 \leq x < \pi. \end{cases}$$

For Problems 8–10, find the nth Fourier polynomial for the given functions, assuming them to be periodic with period 2π. Graph the first three approximations with the original function.

8. $g(x) = x, \quad -\pi < x \leq \pi.$
9. $f(x) = x^2, \quad -\pi < x \leq \pi.$
10. $h(x) = \begin{cases} 0 & -\pi < x \leq 0 \\ x & 0 < x \leq \pi. \end{cases}$

11. If the Fourier coefficients of f are a_k and b_k, and the Fourier coefficients of g are c_k and d_k, and if A and B are real, show that the Fourier coefficients of $Af + Bg$ are $Aa_k + Bc_k$ and $Ab_k + Bd_k$.

12. Use the results of Problem 11 to find the Fourier coefficients of the function h in Problem 10 from those of f and g in Problems 7 and 8.

13. Suppose we have a periodic function f with period 1 defined by $f(x) = x$ for $0 \le x < 1$. Find the fourth degree Fourier polynomial for f and graph it on the interval $0 \le x < 1$. [Hint: Remember that since the period is not 2π, you will have to start by doing a substitution. Notice that the terms in the sum are not $\sin(nx)$ and $\cos(nx)$, but instead turn out to be $\sin(2\pi nx)$ and $\cos(2\pi nx)$.]

14. Suppose f has period 2 and $f(x) = x$ for $0 \le x < 2$. Find the fourth-degree Fourier polynomial and graph it on $0 \le x < 2$. [Hint: See Problem 13.]

15. (a) Find and graph the third Fourier approximation of the square wave $g(x)$ of period 2π such that

$$g(x) = \begin{cases} 0 & -\pi \le x < \pi/2 \\ 1 & -\pi/2 \le x < \pi/2 \\ 0 & \pi/2 \le x < \pi \end{cases}$$

(b) How does the result of part (a) differ from that of the square wave in Example 1?

16. Suppose that a spacecraft near Neptune has measured a quantity A and sent it to earth in the form of a periodic signal $A \cos t$ of amplitude A. On its way to earth, the signal picks up periodic noise, containing only second and higher harmonics. Suppose that the signal $h(t)$ actually received on earth is graphed below in Figure 10.32. Determine the signal that the spacecraft originally sent and hence the value A of the measurement.

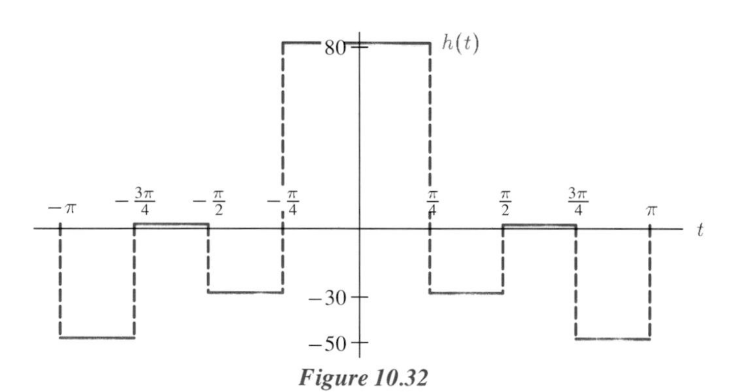

Figure 10.32

17. What fraction of the energy of the function in Problem 7 is contained in the constant term and first three harmonics of its Fourier series?

18. Show that for positive integers k, the periodic function $f(x) = a_k \cos kx + b_k \sin kx$ of period 2π has energy $a_k^2 + b_k^2$.

19. Figures 10.33 and 10.34 show the waveforms and energy spectra for notes produced by flute and bassoon.[5] Describe the principal differences between the two spectra.

[5] Adapted from C.A. Culver, *Musical Acoustics* (New York: McGraw-Hill, 1956), pp. 200, 213.

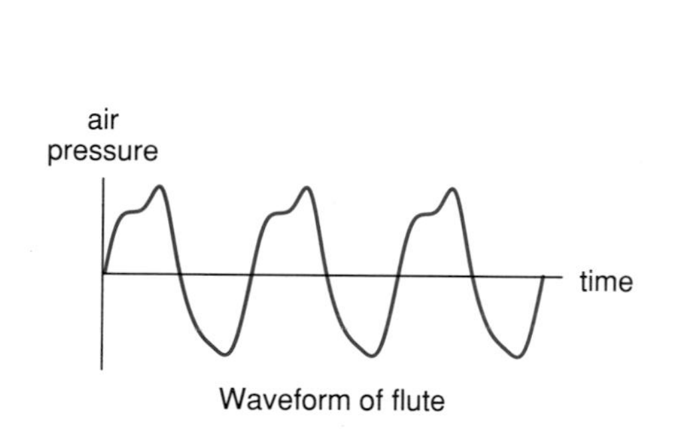

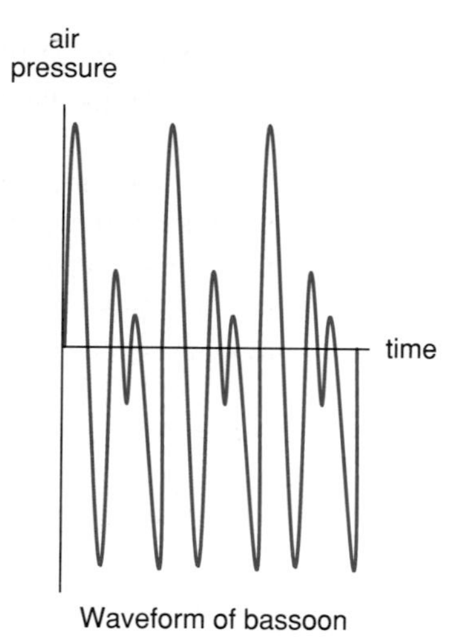

Figure 10.33

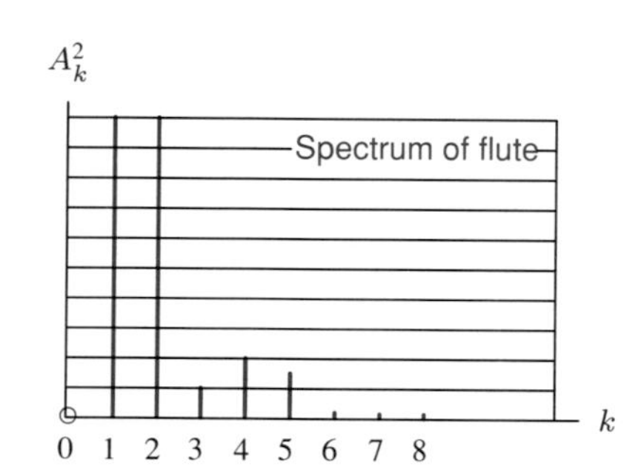

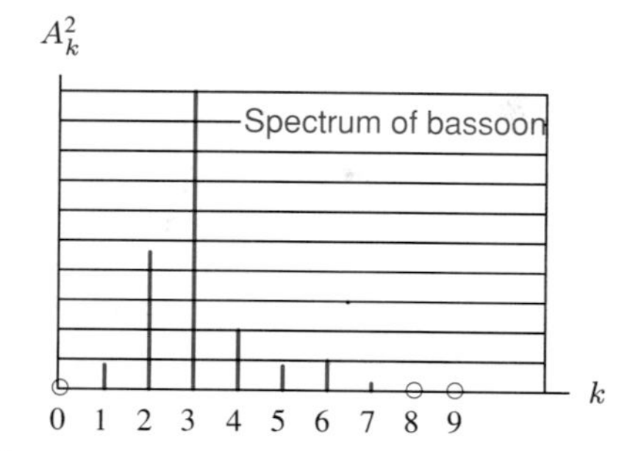

Figure 10.34

In Problems 20–23, the pulse train of width c, where $0 < c < \pi$, is the periodic function f of period 2π such that

$$f(x) = \begin{cases} 0 & -\pi \le x < -c/2 \\ 1 & -c/2 \le x < c/2 \\ 0 & c/2 \le x < \pi \end{cases}$$

20. Suppose that f is the pulse train of width 2.
    (a) What fraction of the energy of f is contained in the constant term of its Fourier series? In the constant term and the first harmonic together?
    (b) How many terms of the Fourier series of f are needed to capture 90% of the energy of f?
    (c) Graph f and its third Fourier approximation on the interval $[-3\pi, 3\pi]$.
21. Suppose that f is the pulse train of width 1.
    (a) What fraction of the energy of f is contained in the constant term of its Fourier series? In the constant term and the first harmonic together?
    (b) Find a formula for the energy of the kth harmonic of f. Use it to sketch the energy spectrum of f.
    (c) How many terms of the Fourier series of f are needed to capture 90% of the energy of f?
    (d) Graph f and its fifth Fourier approximation on the interval $[-3\pi, 3\pi]$.

22. Suppose that f is the pulse train of width 0.4.
    (a) What fraction of the energy of f is contained in the constant term of its Fourier series? In the constant term and the first harmonic together?
    (b) Find a formula for the energy of the kth harmonic of f. Use it to sketch the energy spectrum of f.
    (c) What fraction of the energy of f is contained in the constant term and the first five harmonics of f? (You need the constant term and the first thirteen harmonics to capture 90% of the energy of f.)
    (d) Graph f and its fifth Fourier approximation on the interval $[-3\pi, 3\pi]$.
23. After working Problems 20, 21, 22, write a paragraph about the approximation of pulse trains by Fourier polynomials. Explain how the energy spectrum of a pulse train of width c will change as c gets closer and closer to 0 and how this affects the number of terms required for an accurate approximation.
24. Given the graph of f in Figure 10.35, find the first and second Fourier approximations numerically.

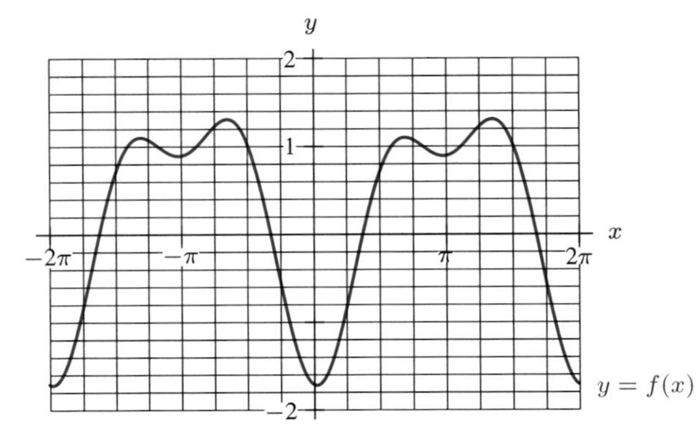

Figure 10.35

For Problems 25–29 use the table of integrals in Chapter 7, page 366, to show that the following statements are true for any positive integers k and m. (That is, $k > 0$ and $m > 0$.)

25. $\displaystyle\int_{-\pi}^{\pi} \sin kx \cos mx \, dx = 0.$
26. $\displaystyle\int_{-\pi}^{\pi} \cos kx \cos mx \, dx = 0, \quad \text{if } k \neq m.$
27. $\displaystyle\int_{-\pi}^{\pi} \cos^2 mx \, dx = \pi.$
28. $\displaystyle\int_{-\pi}^{\pi} \sin kx \sin mx \, dx = 0, \quad \text{if } k \neq m.$
29. $\displaystyle\int_{-\pi}^{\pi} \sin^2 mx \, dx = \pi.$

REVIEW PROBLEMS FOR CHAPTER TEN

In Problems 1–4, find the first four nonzero terms of the Taylor series about the origin of the given functions.

1. $\sin t^2$ **2.** $\theta^2 \cos \theta^2$ **3.** $\dfrac{1}{1-4z^2}$ **4.** $\dfrac{1}{\sqrt{4-x}}$

For Problems 5–6, expand the quantity in a Taylor series around the origin in terms of the variable given. Give the first four nonzero terms.

5. $\dfrac{a}{a+b}$ in terms of $\dfrac{b}{a}$

6. $\sqrt{R-r}$ in terms of $\dfrac{r}{R}$, for $R > 0$

For Problems 7–9, find the second-degree Taylor polynomial about the given point.

7. $\sin x, \quad x = -\pi/4$

8. $e^x, \quad x = 1$

9. $\ln x, \quad x = 2$

10. Find the third-degree Taylor polynomial for $f(x) = x^3 + 7x^2 - 5x + 1$ at $x = 1$.

11. (a) Find $\lim\limits_{\theta \to 0} \dfrac{\sin 2\theta}{\theta}$. Explain your reasoning.

(b) Use series to explain why $f(\theta) = \dfrac{\sin 2\theta}{\theta}$ looks like a parabola near $\theta = 0$. What is the equation of the parabola?

12. Suppose x is positive but very small. Arrange the following expressions in increasing order:

$$x, \quad \sin x, \quad \ln(1+x), \quad 1 - \cos x, \quad e^x - 1, \quad \arctan x, \quad x\sqrt{1-x}.$$

13. (a) Find the Taylor series for $f(t) = te^t$ about $t = 0$.

(b) Using your answer to part (a), find a Taylor series expansion about $x = 0$ for

$$\int_0^x te^t\, dt.$$

(c) Using your answer to part (b), show that

$$\frac{1}{2} + \frac{1}{3} + \frac{1}{4(2!)} + \frac{1}{5(3!)} + \frac{1}{6(4!)} + \cdots = 1.$$

14. Padé approximants are rational functions used to approximate more complicated functions. In this problem, you will derive the Padé approximant to the exponential function.

(a) Let $f(x) = (1 + ax)/(1 + bx)$, where a and b are constants. Write down the first three terms of the Taylor series for $f(x)$ about $x = 0$.

(b) By equating the first three terms of the Taylor series about $x = 0$ for $f(x)$ and for e^x, find a and b so that $f(x)$ approximates e^x as closely as possible near $x = 0$.

15. The theory of relativity predicts that when an object moves at speeds close to the speed of light, the object appears heavier. The apparent, or relativistic, mass, m, of the object when it is moving at speed v is given by the formula

$$m = \frac{m_0}{\sqrt{1 - v^2/c^2}}$$

where c is the speed of light and m_0 is the mass of the object when it is at rest.

(a) Use the formula for m to decide what values of v are possible.

(b) Sketch a rough graph of m against v, labeling intercepts and asymptotes.

(c) Write the first three nonzero terms of the Taylor series for m in terms of v.

(d) For what values of v do you expect the series to converge?

16. The *gravitational field* at a point in space is the gravitational force that would be exerted on a unit mass placed there. We will assume that the gravitational field strength at a distance d away from a mass M is

$$\frac{GM}{d^2}$$

where G is constant. In this problem you will investigate the gravitational field strength, F, exerted by a system consisting of a large mass M and a small mass m, with a distance r between them. (See Figure 10.35.)

Figure 10.35

(a) Write an expression for the gravitational field strength, F, at the point P.
(b) Assuming r is small in comparison to R, expand F in a series in r/R.
(c) By discarding terms in $(r/R)^2$ and higher powers, explain why the series allows you to look at the field as resulting from a single particle of mass $M+m$, plus a correction term. What is the position of the particle of mass $M+m$? Why is the sign of the correction term what it is?

17. The potential energy, V, of two gas molecules separated by a distance r is given by

$$V = -V_0\left(2\left(\frac{r_0}{r}\right)^6 - \left(\frac{r_0}{r}\right)^{12}\right),$$

where V_0 and r_0 are positive constants.

(a) Show that if $r = r_0$, then V takes on its minimum value, $-V_0$.
(b) Write V as a series in $(r - r_0)$ up through the quadratic term.
(c) For r near r_0, show that the difference between V and its minimum value is approximately proportional to $(r-r_0)^2$. In other words, show that $V-(-V_0) = V+V_0$ is approximately proportional to $(r - r_0)^2$.
(d) The force, F, between the molecules is given by $F = -dV/dr$. What is F when $r = r_0$? For r near r_0, show that F is approximately proportional to $(r - r_0)$.

18. Use the fact that the Taylor series of $g(x) = \sin(x^2)$ is

$$x^2 - \frac{x^6}{3!} + \frac{x^{10}}{5!} - \frac{x^{14}}{7!} + \cdots$$

to find $g''(0)$, $g'''(0)$, and $g^{(10)}(0)$. (There is an easy way and a hard way to do this!)

19. Consider the Taylor expansion

$$f(x) = \frac{1}{1+x} = 1 - x + x^2 - x^3 + x^4 - \cdots$$

By plotting several Taylor polynomials and the function $f(x) = 1/(1+x)$, confirm that the interval of convergence of this series is $-1 < x < 1$.

20. Using the fact that, for example, $\ln 4 = -\ln(\frac{1}{4})$, develop an algorithm for approximating $\ln x$ for $x > 2$ by Taylor polynomials. (Remember that the Taylor series for $\ln x$ about $x = 1$ diverges for $x > 2$.)

21. Suppose all the derivatives of g exist at $x = 0$ and that g has a critical point at $x = 0$.

(a) Write the nth Taylor polynomial for g at $x = 0$.
(b) What does the Second Derivative test for local maxima and minima say?
(c) Use the Taylor polynomial to explain why the Second Derivative test works.

22. (Continuation of Problem 21) You may remember that the Second Derivative test tells us nothing when the second derivative is zero at the critical point. In this problem you will investigate that special case.

Assume g has the same properties as in Problem 21, and that, in addition, $g''(0) = 0$. What does the Taylor polynomial tell you about whether g has a local maximum or minimum at $x = 0$?

23. Cephalexin is an antibiotic with a half-life in the body of 0.9 hours, taken in tablets of 250 mg every six hours.

(a) What percentage of the cephalexin in the body at the start of a six-hour period is still there at the end (assuming no tablets are taken during that time)?
(b) Write an expression for Q_1, Q_2, Q_3, Q_4, where Q_n mg, is the amount of cephalexin in the body right after the nth tablet is taken.
(c) Express Q_3, Q_4 in closed-form and evaluate them.
(d) Write an expression for Q_n and put it in closed-form.
(e) If the patient keeps taking the tablets, use your answer to part (d) to find the quantity of cephalexin in the body in the long run, right after taking a pill.

24. In the nineteenth century, the railroads issued 100-year bonds. Consider a \$100 bond which paid \$5 a year, starting a year after it was sold. Assume interest rates are 4% per year, compounded annually.

(a) What was such a bond worth on the day it was sold?
(b) Suppose that instead of maturing in 100 years, the bond was to have paid \$5 a year forever. This time the principal, \$100, is never repaid. How much would such a bond be worth on the day of its sale?

25. Finding the *yield to maturity* of a bond means finding the interest rate which leads to a particular present value for a bond of given coupons, principal, and term. What is the yield to maturity of a \$1000 bond which pays \$50 a year for 10 years, starting one year from now, and has a present value, or price, of \$950? Assume interest is compounded annually. [Hint: You will get an equation which cannot be solved algebraically.]

26. In an old puzzle, there are two trains, each moving at 10 km/hr toward one another. Initially the trains are 30 kilometers apart. At the same moment a fly, whose velocity is 20 km/hr, starts at one train and flies till it meets the other, then turns around and flies back till it meets the first train, and so on.

(a) How far has the fly traveled the first time it turns around? The second time? The third time? The fourth time?
(b) How far has the fly traveled by the nth time it turns around? Write your answer in closed-form.
(c) Use you answer to part (b) to decide how far the fly has traveled by the time the trains meet in the middle and squash it.
(d) How long does it take the trains to meet in the middle? Use this to answer part (c) without summing a series.

27. Use the methods of this section to give a bound on the magnitude of the error if we approximate $\sqrt{2}$ using the Taylor approximation of degree three for $\sqrt{1+x}$ about $x = 0$.

28. Suppose you wanted to approximate π using a Taylor polynomial. You could use either the arctangent or the arcsine to give you an approximation. In this problem, you will compare the two methods.

(a) Using the fact that $\arctan 1 = \pi/4$ and $d(\arctan x)/dx = 1/(1+x^2)$, approximate the value of π using the third-degree Taylor polynomial of $4\arctan x$ about $x = 0$.

(b) Using the fact that $\arcsin 1 = \pi/2$ and $d(\arcsin x)/dx = 1/\sqrt{1-x^2}$, approximate the value of π using the third-degree Taylor polynomial of $2\arcsin x$ about $x = 0$.

(c) Estimate the maximum error of the approximation you found in part (a).

(d) Looking at the error formula for the arcsine approximation, explain why using the Taylor polynomials for the arctangent seems like a better idea than using the Taylor polynomial for the arcsine.

29. Find a Fourier polynomial of degree three for $f(x) = e^{2\pi x}$, for $0 \le x < 1$.

30. Use the Fourier polynomials for the square wave

$$f(x) = \begin{cases} -1 & -\pi < x \le 0 \\ 1 & 0 < x \le \pi \end{cases}$$

to explain why

$$1 - \frac{1}{3} + \frac{1}{5} - \frac{1}{7} + \cdots + (-1)^{2n+1}\frac{1}{2n+1}$$

must approach $\frac{\pi}{4}$ as $n \to \infty$.

31. Find the third Fourier approximation of the triangular wave graphed in Figure 10.19 on page 635, and graph it on $-2 \le x < 5$. (You may find the Fourier coefficients by numerical integration if you wish.)

32. Suppose that $f(x)$ is a differentiable periodic function of period 2π. Assume the Fourier series of f is differentiable term by term.

(a) If the Fourier coefficients of f are a_k and b_k, show that the Fourier coefficients of its derivative f' are kb_k and $-ka_k$.

(b) How are the amplitudes of the harmonics of f and f' related?

(c) How are the energy spectra of f and f' related?

33. Suppose that f is a periodic function of period 2π and that g is a horizontal shift of f, say $g(x) = f(x+c)$. Show that f and g have the same energy.

APPENDICES

Appendix A ROOTS AND ACCURACY

It is often necessary to find the zeros of a polynomial or the points of intersection of two curves. So far, you have probably used algebraic methods, such as the quadratic formula, to solve such problems. Unfortunately, however, mathematicians' search for formulas for the solutions to equations, such as the quadratic formula, has not been all that successful. The formulas for the solutions to third- and fourth-degree equations are so complicated that you'd never want to use them. Early in the nineteenth century, it was proved that there is no algebraic formula for the solutions to equations of degree 5 and higher. Most nonpolynomial equations cannot be solved using a formula either.

However, we can still find roots of equations, provided we use approximation methods, not formulas. In this section we will discuss three ways to find roots: algebraic, graphical, and numerical. Of these, only the algebraic method gives exact solutions.

First, let's get some terminology straight. Given the equation $x^2 = 4$, we call $x = -2$ and $x = 2$ the *roots*, or *solutions of the equation*. If we are given the function $f(x) = x^2 - 4$, then -2 and 2 are called the *zeros of the function*; that is, the zeros of the function f are the roots of the equation $f(x) = 0$.

The Algebraic Viewpoint: Roots by Factoring

If the product of two numbers is zero, then one or the other or both must be zero, that is, if $AB = 0$, then $A = 0$ or $B = 0$. This observation lies behind finding roots by factoring. You may have spent a lot of time factoring polynomials. Here you will also factor expressions involving trigonometric and exponential functions.

Example 1 Find the roots of $x^2 - 7x = 8$.

Solution Rewrite the equation as $x^2 - 7x - 8 = 0$. Then factor the left side: $(x+1)(x-8) = 0$. By our observation about products, either $x + 1 = 0$ or $x - 8 = 0$, so the roots are $x = -1$ and $x = 8$.

Example 2 Find the roots of $\dfrac{1}{x} - \dfrac{x}{(x+2)} = 0$.

Solution Rewrite the left side with a common denominator:

$$\frac{x + 2 - x^2}{x(x+2)} = 0.$$

Whenever a fraction is zero, the numerator must be zero. Therefore we must have

$$x + 2 - x^2 = (-1)(x^2 - x - 2) = (-1)(x-2)(x+1) = 0.$$

We conclude that $x - 2 = 0$ or $x + 1 = 0$, so 2 and -1 are the roots. They can be checked by substitution.

Example 3 Find the roots of $e^{-x}\sin x - e^{-x}\cos x = 0$.

Solution Factor the left side: $e^{-x}(\sin x - \cos x) = 0$. The factor e^{-x} is never zero; it is impossible to raise e to a power and get zero. Therefore, the only possibility is that $\sin x - \cos x = 0$. This equation is equivalent to $\sin x = \cos x$. If we divide both sides by $\cos x$, we get

$$\frac{\sin x}{\cos x} = \frac{\cos x}{\cos x} \quad \text{so} \quad \tan x = 1.$$

The roots of this equation are

$$\ldots, \frac{-7\pi}{4}, \frac{-3\pi}{4}, \frac{\pi}{4}, \frac{5\pi}{4}, \frac{9\pi}{4}, \frac{13\pi}{4}, \ldots.$$

Warning: Using factoring to solve an equation only works when one side of the equation is 0. It is not true that if, say, $AB = 7$ then $A = 7$ or $B = 7$. For example, you *cannot* solve $x^2 - 4x = 2$ by factoring $x(x-4) = 2$ and then assuming that either x or $x - 4$ equals 2.

The problem with factoring is that factors are not easy to find. For example, the left side of the quadratic equation $x^2 - 4x - 2 = 0$ does not factor, at least not into "nice" factors with integer coefficients. For the general quadratic equation:

$$ax^2 + bx + c = 0$$

there is the quadratic formula for the roots:

$$x = \frac{-b \pm \sqrt{b^2 - 4ac}}{2a}.$$

Thus the roots of $x^2 - 4x - 2 = 0$ are $(4 \pm \sqrt{24})/2$, or $2 + \sqrt{6}$ and $2 - \sqrt{6}$.

Notice that in each of these examples, we have found the roots exactly.

The Graphical Viewpoint: Roots by Zooming

To find the roots of an equation $f(x) = 0$, it helps to draw the graph of f. The roots of the equation, that is the zeros of f, are *the values of x where the graph of f crosses the x-axis*. Even a very rough sketch of the graph can be useful in determining how many zeros there are and their approximate values. If you have a computer or graphing calculator, then finding solutions by graphing is the easiest method, especially if you use the zoom feature. However, a graph can never tell you the exact value of a root, only an approximate one.

Example 4 Find the roots of $x^3 - 4x - 2 = 0$.

Solution Attempting to factor the left side with integer coefficients will convince you it cannot be done, so we cannot easily find the roots by algebra. We know the graph of $f(x) = x^3 - 4x - 2$ will have the usual cubic shape; see Figure A.1. There are clearly three roots: one between $x = -2$ and $x = -1$, another between $x = -1$ and $x = 0$, and a third between $x = 2$ and $x = 3$. Zooming in on the largest root with a graphing calculator or computer shows that it lies in the following interval:

$$2.213 < x < 2.215.$$

Thus, the root is $x = 2.21$, accurate to two decimal places. Zooming in on the other two roots shows them to be $x = -1.68$ and $x = -0.54$, accurate to two decimal places.

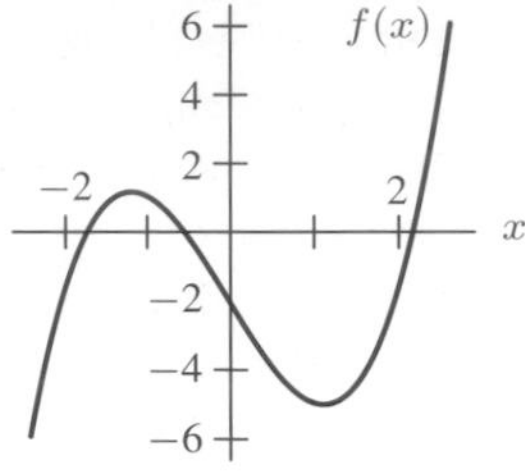

Figure A.1: The cubic $f(x) = x^3 - 4x - 2$

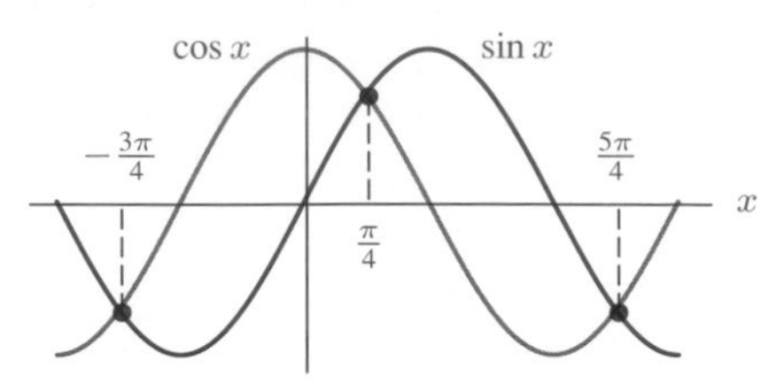

Figure A.2: Finding roots of $\sin x - \cos x = 0$

Useful trick: Suppose you want to solve the equation $\sin x - \cos x = 0$ graphically. Instead of graphing $f(x) = \sin x - \cos x$ and looking for zeros, you may find it easier to rewrite the equation as $\sin x = \cos x$ and graph $g(x) = \sin x$ and $h(x) = \cos x$. (After all, you already know what these two graphs look like. See Figure A.2.) The roots of the original equation are then precisely the x coordinates of the points of intersection of the graphs of $g(x)$ and $h(x)$.

Example 5 Find the roots of $2\sin x - x = 0$.

Solution Rewrite the equation as $2\sin x = x$, and graph both sides. Since $g(x) = 2\sin x$ is always between -2 and 2, there are no roots of $2\sin x = x$ for $x > 2$ or for $x < -2$. Thus, we need only consider the graphs between -2 and 2 (or between $-\pi$ and π, which makes graphing the sine function easier). Figure A.3 shows the graphs. There are three points of intersection: one appears to be at $x = 0$, one between $x = \pi/2$ and $x = \pi$, and one between $x = -\pi/2$ and $x = -\pi$. You can tell that $x = 0$ is the exact value of one root because it satisfies the original equation exactly. Zooming in shows that there is a second root $x \approx 1.9$, and the third root is $x \approx -1.9$ by symmetry.

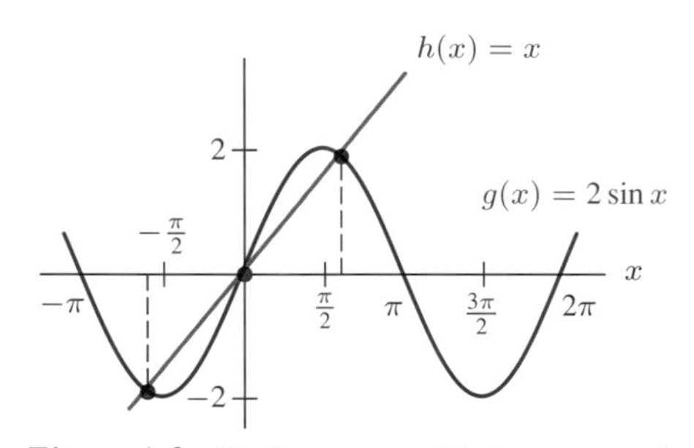

Figure A.3: Finding roots of $2\sin x - x = 0$

The Numerical Viewpoint: Roots by Bisection

We now look at a numerical method of approximating the solutions to an equation. This method depends on the idea that if the value of a function $f(x)$ changes sign in an interval, and if we believe there is no break in the graph of the function there, then there is a root of the equation $f(x) = 0$ in that interval.

Let's go back to the problem of finding the root of $f(x) = x^3 - 4x - 2 = 0$ between 2 and 3. To locate the root, we close in on it by evaluating the function at the midpoint of the interval, $x = 2.5$. Since $f(2) = -2$, $f(2.5) = 3.625$, and $f(3) = 13$, the function changes sign between $x = 2$ and $x = 2.5$, so the root is between these points. Now we look at $x = 2.25$.

Since $f(2.25) = 0.39$, the function is negative at $x = 2$ and positive at $x = 2.25$, so there is a root between 2 and 2.25. Now we look at 2.125. We find $f(2.125) = -0.90$, so there is a root between 2.125 and 2.25, ... and so on. (You may want to round the decimals as you work.) See Figure A.4. The intervals containing the root are listed in Table A.1 and show that the root is $x = 2.21$ to two decimal places.

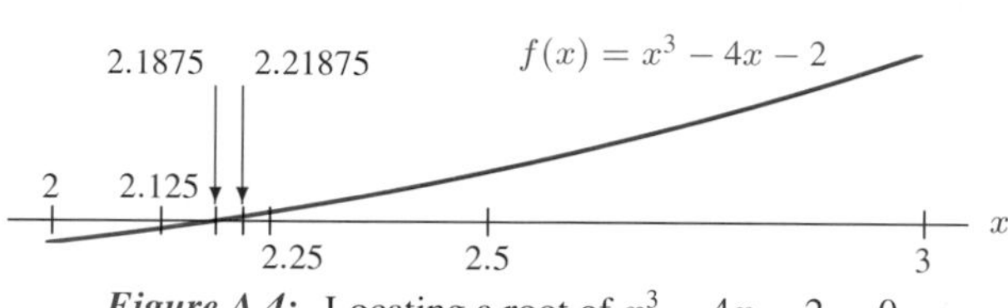

Figure A.4: Locating a root of $x^3 - 4x - 2 = 0$

TABLE A.1 *Intervals containing root to $x^3 - 4x - 2 = 0$ (Note that $[2, 3]$ means $2 \le x \le 3$)*

$[2, 3]$	
$[2, 2.5]$	
$[2, 2.25]$	
$[2.125, 2.25]$	
$[2.1875, 2.25]$	So $x = 2.2$ rounded to one decimal place
$[2.1875, 2.21875]$	
$[2.203125, 2.21875]$	
$[2.2109375, 2.21875]$	
$[2.2109375, 2.2148438]$	So $x = 2.21$ rounded to two decimal places

This method of finding roots is called the **Bisection Method**:

- To solve an equation $f(x) = 0$ using the bisection method, we need two starting values for x, say, $x = a$ and $x = b$, such that $f(a)$ and $f(b)$ have opposite signs and f is continuous on $[a, b]$.
- Evaluate f at the midpoint of the interval $[a, b]$, and decide in which half-interval the root lies.
- Repeat, using the new half-interval instead of $[a, b]$.

There are some problems with the bisection method:

- The function may not change signs near the root. For example, $f(x) = x^2 - 2x + 1 = 0$ has a root at $x = 1$, but $f(x)$ is never negative because $f(x) = (x - 1)^2$, and a square cannot be negative. (See Figure A.5.)

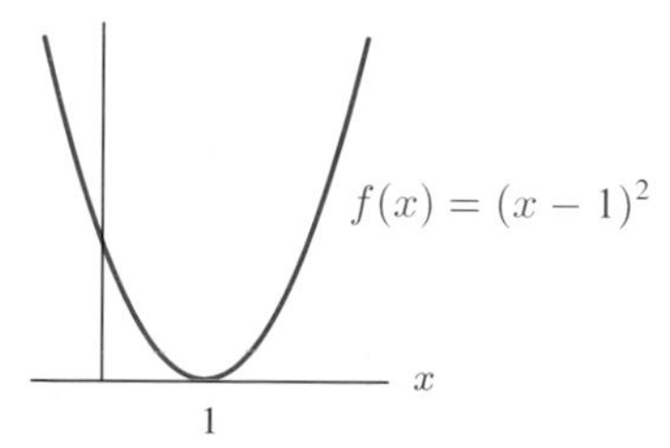

Figure A.5: f doesn't change sign at the root

- The function f must be continuous between the starting values $x = a$ and $x = b$.
- If there is more than one root between the starting values $x = a$ and $x = b$, the method will find only one of the roots. For example, if we had tried to solve $x^3 - 4x - 2 = 0$ starting at $x = -12$ and $x = 10$, the bisection method would zero in on the root between $x = -2$ and $x = -1$, not the root between $x = 2$ and $x = 3$ that we found earlier. (Try it! Then see what happens if you use $x = -10$ instead of $x = -12$.)
- The bisection method is slow and not very efficient. Applying bisection three times in a row only traps the root in an interval $(\frac{1}{2})^3 = \frac{1}{8}$ as large as the starting interval. Thus, if we initially know that a root is between, say, 2 and 3, then we would need to apply the bisection method at least four times to know the first digit after the decimal point.

There are much more powerful methods available for finding roots, such as Newton's method, which is more complicated, but which avoids some of these difficulties.

Example 6 Find all the roots of $xe^x = 5$ to at least one decimal place.

Solution If we rewrite the equation as $e^x = 5/x$ and graph both sides, as in Figure A.6, it is clear that there is exactly one root, and it is somewhere between 1 and 2. Table A.2 shows the intervals obtained by the bisection method. After five iterations, we have the root trapped between 1.3125 and 1.34375, so we can say the root is $x = 1.3$ to one decimal place.

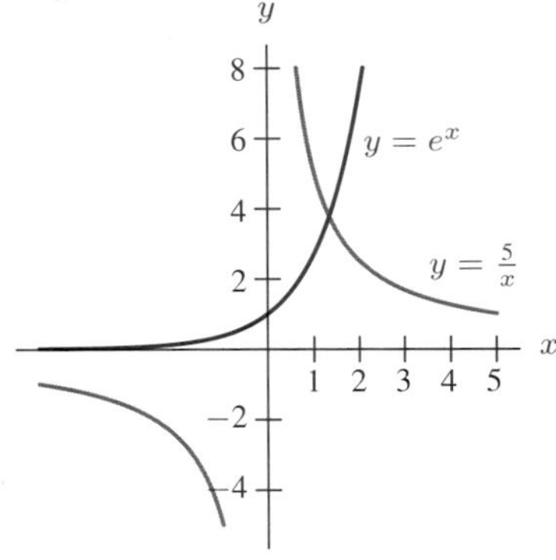

Figure A.6: Intersection of $y = e^x$ and $y = 5/x$

TABLE A.2 *Bisection method for* $f(x) = xe^x - 5 = 0$ *(Note that* $[1, 2]$ *means the interval* $1 \le x \le 2$*)*

Interval Containing Root
[1, 2]
[1, 1.5]
[1.25, 1.5]
[1.25, 1.375]
[1.3125, 1.375]
[1.3125, 1.34375]

Iteration

Both zooming in and bisection as discussed here are examples of *iterative* methods, in which a sequence of steps is repeated over and over again, using the results of one step as the input for the next. We can use the method to locate a root to any degree of accuracy. In bisection, each iteration traps the root in an interval that is half the length of the previous one. Each time you zoom in on a calculator, you trap the root in a smaller interval; how much smaller depends on the settings on the calculator.

Accuracy and Error

In the previous discussion, we used the phrase "accurate to 2 decimal places." For an iterative process where we get closer and closer estimates for some quantity, we take a commonsense approach to accuracy: we watch the numbers carefully, and when a digit stays the same for a few iterations, we assume it has stabilized and is correct, especially if the digits to the right of that digit also stay the same. For example, suppose 2.21429 and 2.21431 are two successive estimates for a zero of $f(x) = x^3 - 4x - 2$. Since these two estimates agree to the third digit after the decimal point, we probably have at least 3 decimal places correct.

There is a problem with this, however. Suppose we are finding a root whose true value is 1, and the estimates are converging to the value from below — say, 0.985, 0.991, 0.997 and so on. In this case, not even the first decimal place is "correct," even though the difference between the estimates and the actual answer is very small — much less than 0.1. To avoid this difficulty, we say that an estimate a for some quantity r is *accurate to p decimal places* if the error, which is the absolute value of the difference between a and r, or $|r - a|$, is as follows:

Accuracy to p decimal places	means	Error less than
$p = 1$		0.05
2		0.005
3		0.0005
⋮		⋮
n		$0.\underbrace{000\ldots0}_{n}5$

This is the same as saying that r must lie in an interval of length twice the maximum error, centered on a. For example, if a is accurate to 1 decimal place, r must lie in the following interval:

$a - 0.05$ ——— a ——— $a + 0.05$

Since both the graphing calculator and the bisection method give us an interval in which the root is trapped, this definition of decimal accuracy is a natural one for these processes.

Example 7 Suppose the numbers $\sqrt{10}$, $22/7$, and 3.14 are given as approximations to $\pi = 3.1415\ldots$. To how many decimal places is each approximation accurate?

Solution Using $\sqrt{10} = 3.1622\ldots$,

$$|\sqrt{10} - \pi| = |3.1622\ldots - 3.1415\ldots| = 0.0206\ldots < 0.05,$$

so $\sqrt{10}$ is accurate to one decimal place. Similarly, using $22/7 = 3.1428\ldots$,

$$\left|\frac{22}{7} - \pi\right| = |3.1428\ldots - 3.1415\ldots| = 0.0013\ldots < 0.005,$$

so 22/7 is accurate to two decimal places. Finally,

$$|3.14 - 3.1415\ldots| = 0.0015\ldots < 0.005,$$

so 3.14 is accurate to two decimal places.

Warning:

- Saying that an approximation is accurate to, say, 2 decimal places does *not* guarantee that its first two decimal places are "correct," that is, that the two digits of the approximation are the same as the corresponding two digits in the true value. For example, an approximate value of 5.997 is accurate to 2 decimal places if the true value is 6.001, but neither of the 9s in the approximation agrees with the 0s in the true value (nor does the digit 5 agree with the digit 6).
- The number of decimal places of accuracy refers to the number of digits that have stabilized in the root, r. It does *not* refer to the number of digits of $f(r)$ that are zero. For example, Table A.1 on page 659 shows that $x = 2.2$ is a root of $f(x) = x^3 - 4x - 2 = 0$, accurate to one decimal place. Yet, $f(2.2) = -0.152$, so $f(2.2)$ does not have one zero after the decimal point. Similarly, $x = 2.21$ is the root accurate to two decimal places, but $f(2.21) = -0.046$ does not have two zeros after the decimal point.

Example 8 Is $x = 2.2143$ a zero of $f(x) = x^3 - 4x - 2$ accurate to four decimal places?

Solution We want to know whether r, the exact value of the zero, lies in the interval

$$2.2143 - 0.00005 < r < 2.2143 + 0.00005$$

which is the same as

$$2.21425 < r < 2.21435.$$

Since $f(2.21425) < 0$ and $f(2.21435) > 0$, the zero does lie in this interval, and so $r = 2.2143$ is accurate to four decimal places.

How to Write a Decimal Answer

The graphing calculator and bisection method naturally give an interval for a root or zero. Other numerical techniques, however, do not give a pair of numbers bounding the true value, but rather a single number near the true value. What should you do if you want to give a single number, rather than an interval, for an answer?

When giving a single number as an answer and interpreting it, be careful about giving rounded answers. For example, suppose you have computed a value to be 0.84, and you know the true value is between 0.81 and 0.87. It would be wrong to round 0.84 to 0.8 and say that the answer is 0.8 accurate to one decimal place; the true value could be 0.86, which is not within 0.05 of 0.8. The right thing to say is that the answer is 0.84 accurate to one decimal place. Similarly, to give an answer accurate to, say, 2 decimal places, you may have to show 3 or more decimal places in your answer.

Problems for Appendix A

1. Use a calculator or computer graph of $f(x) = 13 - 20x - x^2 - 3x^4$ to determine:
   (a) The range of this function;
   (b) The number of zeros of this function.

For Problems 2–12, determine the roots or points of intersection to an accuracy of one decimal place.

2. (a) The root of $x^3 - 3x + 1 = 0$ between 0 and 1
   (b) The root of $x^3 - 3x + 1 = 0$ between 1 and 2
   (c) The third root of $x^3 - 3x + 1 = 0$
3. The root of $x^4 - 5x^3 + 2x - 5 = 0$ between -2 and -1
4. The root of $x^5 + x^2 - 9x - 3 = 0$ between -2 and -1
5. The largest real root of $2x^3 - 4x^2 - 3x + 1 = 0$
6. All real roots of $x^4 - x - 2 = 0$
7. All real roots of $x^5 - 2x^2 + 4 = 0$
8. The first positive root of $x \sin x - \cos x = 0$
9. The first positive point of intersection between $y = 2x$ and $y = \tan x$
10. The first positive point of intersection between $y = 1/2^x$ and $y = \sin x$
11. The point of intersection between $y = e^{-x}$ and $y = \ln x$
12. All roots of $\cos t = t^2$
13. Estimate all real zeros of the following polynomials, accurate to 2 decimal places:
    (a) $f(x) = x^3 - 2x^2 - x + 3$
    (b) $f(x) = x^3 - x^2 - 2x + 2$
14. Find the largest zero of

$$f(x) = 10xe^{-x} - 1$$

to two decimal places, using the bisection method. Make sure to demonstrate that your approximation is as good as you claim.

15. (a) Find the smallest positive value of x where the graphs of $f(x) = \sin x$ and $g(x) = 2^{-x}$ intersect.
    (b) Repeat with $f(x) = \sin 2x$ and $g(x) = 2^{-x}$.
16. Use a graphing calculator to sketch $y = 2\cos x$ and $y = x^3 + x^2 + 1$ on the same set of axes. Find the positive zero of $f(x) = 2\cos x - x^3 - x^2 - 1$. A friend claims there is one more real zero. Is your friend correct? Explain.
17. Use the table below to investigate the zeros of the function

$$f(\theta) = (\sin 3\,\theta)(\cos 4\,\theta) + 0.8$$

in the interval $0 \le \theta \le 1.8$.

θ	0	0.2	0.4	0.6	0.8	1.0	1.2	1.4	1.6	1.8
$f(\theta)$	0.80	1.19	0.77	0.08	0.13	0.71	0.76	0.12	−0.19	0.33

   (a) Decide how many zeros the function has in the interval $0 \le \theta \le 1.8$.
   (b) Locate each zero, or a small interval containing each zero.
   (c) Are you sure you have found all the zeros in the interval $0 \le \theta \le 1.8$? Graph the function on a calculator or computer to decide.

18. (a) Use the accompanying table to locate approximate solution(s) to

$$(\sin 3x)(\cos 4x) = \frac{x^3}{\pi^3}$$

in the interval $1.07 \le x \le 1.15$. Give an interval of length 0.01 in which each solution lies.

(b) Make an estimate for each solution accurate to two decimal places.

x	x^3/π^3	$(\sin 3x)(\cos 4x)$
1.07	0.0395	0.0286
1.08	0.0406	0.0376
1.09	0.0418	0.0442
1.10	0.0429	0.0485
1.11	0.0441	0.0504
1.12	0.0453	0.0499
1.13	0.0465	0.0470
1.14	0.0478	0.0417
1.15	0.0491	0.0340

19. (a) With your calculator in radian mode, take the arctangent of 1 and multiply that number by 4. Now, take the arctangent of the result and multiply *it* by 4. Continue this process 10 times or so and record each result as in the accompanying table. At each step, you get $4\arctan$ of the result of the previous step.

1
3.14159...
5.05050...
5.50129...
⋮

(b) Your table allows you to find a solution of the equation

$$4\arctan x = x.$$

Why? What is that solution?

(c) What does your table in part (a) have to do with Figure A.7?
[Hint: The coordinates of P_0 are $(1, 1)$. Find the coordinates of $P_1, P_2, P_3, \ldots$]

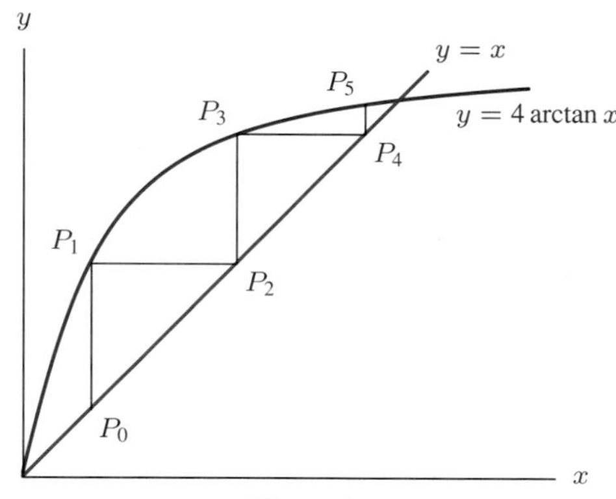

Figure A.7

(d) In part (a), what happens if you start with an initial guess of 10? Of -10? What types of behavior do you observe? (I.e., for which initial guesses is the sequence increasing, and for which is it decreasing; does the sequence approach a limit?) Explain your answers graphically, as in part (c).

20. Using radians, apply the iteration method of Problem 19 to the equation

$$\cos x = x.$$

Represent your results graphically, as in Figure A.7.

Appendix B CONTINUITY AND BOUNDS

Continuity of a Function on an Interval

How much do we know about a function if we only know its values at a few points? Can we find its zeros, at least approximately? The answer must be yes, at least for some functions, since we can find zeros with a calculator, which graphs by plotting a finite number of points.

When we found the roots of the cubic equation $f(x) = x^3 - 4x - 2 = 0$ by bisection, we knew that there is a root somewhere between $x = 2$ and $x = 3$ because the function changes sign between these points: $f(2) = -2$ and $f(3) = 13$. (See Figure B.8.)

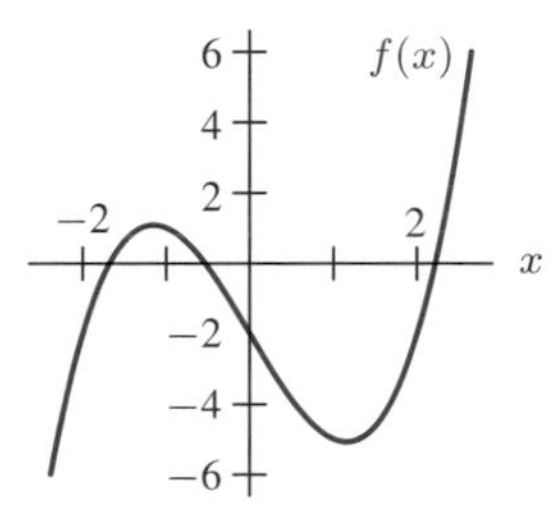

Figure B.8: The graph of $f(x) = x^3 - 4x - 2$

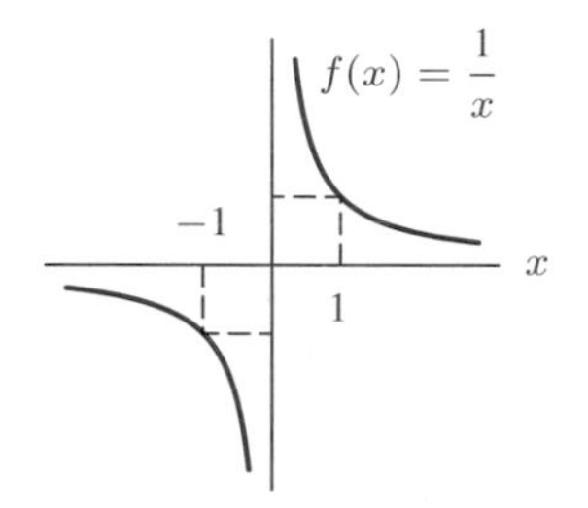

Figure B.9: No zero although $f(-1)$ and $f(1)$ have opposite signs

In general, if we find an interval on which $f(x)$ changes sign, can we be sure that $f(x)$ has a zero there? To be certain, we need to know that the graph of the function has no breaks or jumps in it. Otherwise the graph could "jump" across the x-axis, changing sign but not creating a zero. For example, $f(x) = 1/x$ has opposite signs at $x = -1$ and $x = 1$ but no zeros at all because of the break at $x = 0$. (See Figure B.9.) To avoid this problem, we consider only intervals where our functions are continuous.

What Does Continuity Mean Graphically?

A function is said to be *continuous* on an interval if its graph has no breaks, jumps, or holes in that interval. A continuous function has a graph which can be drawn without lifting the pencil from the paper.

Example: Figure B.9 shows that $f(x) = 1/x$ is not continuous on any interval containing the origin because of the break at $x = 0$, where f is not defined.

Example: The function $g(\theta) = \sin\theta/\theta$ is graphed in Figure B.10. Since g is not defined at $\theta = 0$, the graph has a hole there. This function is not continuous on any interval containing the origin.

Example: Suppose $p(x)$ is the price of mailing a first-class letter weighing x ounces. It costs 29¢ up to the first ounce, and 52¢ between the first and second, so the graph (in Figure B.11) is a series of steps. We say that this function is not continuous at any positive integer because the graph jumps at these points.

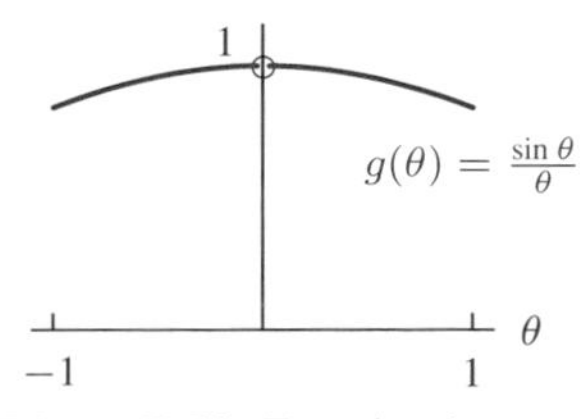

Figure B.10: Function is not continuous on interval containing the origin

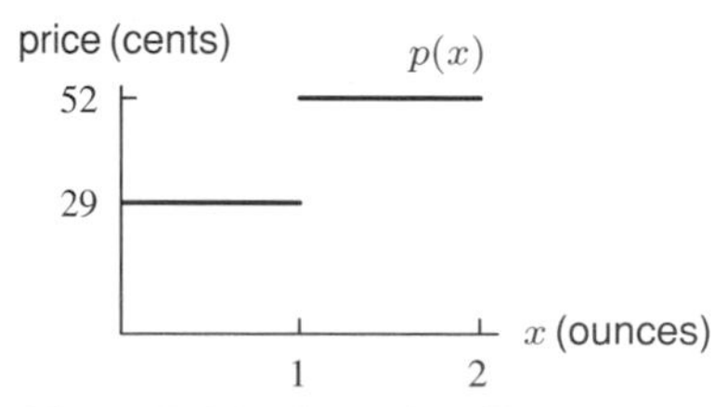

Figure B.11: Cost of mailing a letter

What Does Continuity Mean Numerically?

A function is continuous if nearby values of the independent variable give nearby values of the function. In practical work, continuity is important because it means that small errors in the independent variable lead to relatively small errors in the value of the function.

Example: Suppose $f(x) = x^2$ and that we want to compute $f(\pi)$. Knowing f is continuous tells us that taking $\pi = 3.14$ gives a relatively good approximation to $f(\pi)$, and that we can get more accuracy in $f(\pi)$ by taking more digits of accuracy for π.

Example: If $p(x)$ is the cost of mailing a first class letter weighing x ounces, then $p(0.99) = p(1) =$ 29¢, whereas $p(1.01) =$ 52¢, because as soon as you get over 1 ounce, the price jumps up to 52¢. So a small difference in the weight of a letter can lead to a significant difference in the cost of mailing it.

Which Functions are Continuous?

Requiring a function to be continuous is not asking very much, as any function whose graph is a single curve over any interval on which it is defined is continuous. For example, polynomials are continuous everywhere (meaning for $-\infty < x < \infty$), and rational functions are continuous on any interval in which their denominators are not zero. Exponential functions are continuous everywhere, as are the sine and cosine. All the new functions created by composing continuous functions are continuous too.

Example 1 What do the values in Table B.3 tell you about the zeros of $f(x) = \cos x - 2x^2$?

TABLE B.3

x	$f(x)$
0	1.00
0.2	0.90
0.4	0.60
0.6	0.11
0.8	−0.58
1.0	−1.46

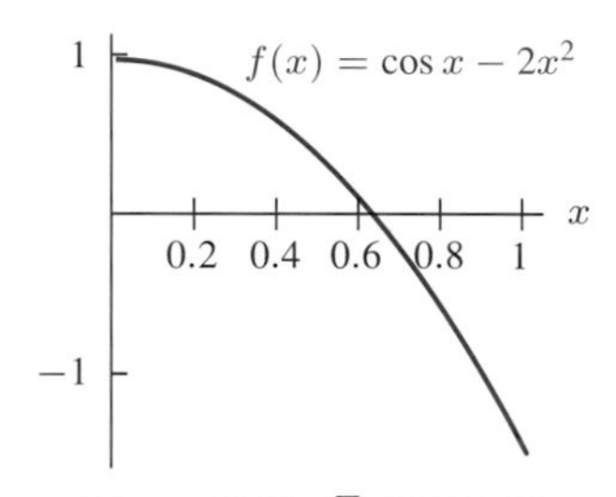

Figure B.12: Zeros occur where graph of continuous function crosses horizontal axis

Solution You can conclude that $f(x)$ has at least one zero in the interval $0.6 < x < 0.8$ since $f(x)$ is continuous and changes from positive to negative on that interval. From the graph of $f(x)$ in Figure B.12, it seems that there is in fact only one zero in the interval $0 \leq x \leq 1$, but one cannot be sure of this just from the given table of values.

Bounds of a Function

Knowing how big or how small a function gets can sometimes be useful, especially when you can't easily find exact values of the function. You can say, for example, that $\sin x$ always stays between -1 and 1 and that $2 \sin x + 10$ always stays between 8 and 12. But 2^x is not confined between any two numbers, because 2^x will exceed any number you can name if x is large enough. We say that $\sin x$ and $2 \sin x + 10$ are *bounded* functions, and that 2^x is an *unbounded* function.

A function f is **bounded** on an interval if there are numbers L and U such that

$$L \leq f(x) \leq U$$

for all x in the interval. Otherwise, f is **unbounded** on the interval.

We say that L is a **lower bound** for f on the interval, and that U is an **upper bound** for f on the interval.

Example 2 Use Figures B.13 and B.14 to decide which of the following functions are bounded.

(a) x^3 on $-\infty < x < \infty$; on $0 \leq x \leq 100$.

(b) $2/x$ on $0 < x < \infty$; on $1 \leq x < \infty$.

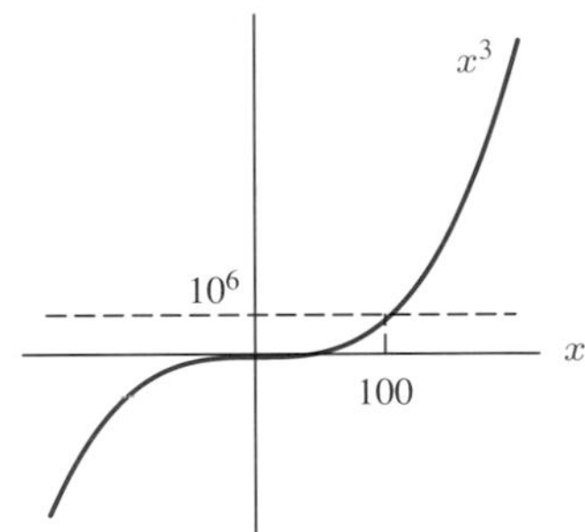

Figure B.13: Is x^3 bounded?

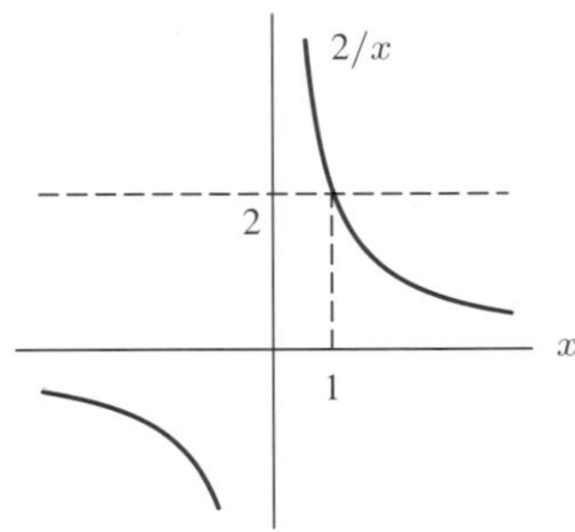

Figure B.14: Is $2/x$ bounded?

Solution (a) The graph of x^3 in Figure B.13 shows that x^3 will exceed any number, no matter how large, if x is big enough, so x^3 does not have an upper bound on $-\infty < x < \infty$. Therefore, x^3 is unbounded on $-\infty < x < \infty$. But on the interval $0 \leq x \leq 100$, x^3 stays between 0 (a lower bound) and $100^3 = 1{,}000{,}000$ (an upper bound). Therefore, x^3 is bounded on the interval $0 \leq x \leq 100$. Notice that upper and lower bounds, when they exist, are not unique. For example, -100 is another lower bound and 2,000,000 another upper bound for x^3 on $0 \leq x \leq 100$.

(b) $2/x$ is unbounded on $0 < x < \infty$ since it has no upper bound there. But $0 \leq 2/x \leq 2$ for $1 \leq x < \infty$, so $2/x$ is bounded on $1 \leq x < \infty$. (See Figure B.14.)

Best Possible Bounds

Consider a group of people whose height in feet, h, ranges from 5 feet to 6 feet. Then 5 feet is a lower bound for the people in the group and 6 feet is an upper bound:

$$5 \leq h \leq 6.$$

But the people in this group are also all between 4 feet and 7 feet, so it is also true that

$$4 \leq h \leq 7.$$

Thus, there are many lower bounds and many upper bounds. However, the 5 and the 6 are considered the best bounds because they are the closest together of all the possible pairs of bounds.

> The **best possible bounds** for a function, f, over an interval are numbers A and B such that, for all x in the interval,
> $$A \leq f(x) \leq B$$
> and where A and B are as close together as possible. A is called the **greatest lower bound** and B is the **least upper bound**.

What Do Bounds Mean Graphically?

Upper and lower bounds can be visualized on a graph by realizing that they can be represented by horizontal lines. See Figure B.15.

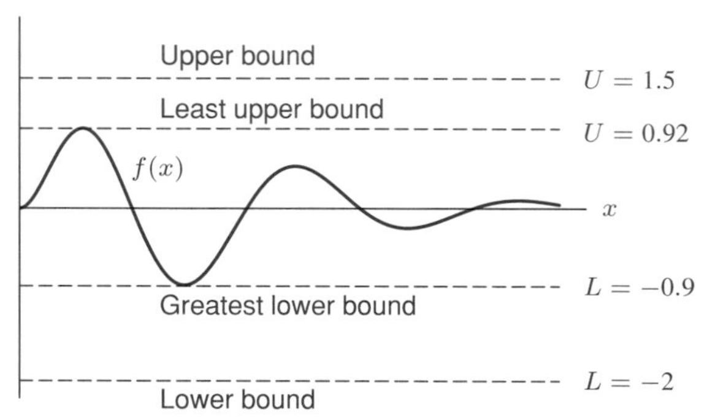

Figure B.15: Upper and lower bounds for the function f

Problems for Appendix B

Are the functions in Problems 1–4 continuous on the interval $[-1, 1]$?

1. $f(x) = |x|$
2. $g(x) = \dfrac{|x|}{x}$
3. $h(\theta) = \theta \sin\theta$
4. $f(t) = \dfrac{\sin t}{t^2}$

For Problems 5–7 draw a graph to decide if the function is bounded on the interval given. Give the best possible upper and lower bounds for any function which is bounded.

5. $f(x) = 4x - x^2$ on $[-1, 4]$
6. $h(\theta) = 5 + 3\sin\theta$ on $[-2\pi, 2\pi]$
7. $f(t) = \dfrac{\sin t}{t^2}$ on $[-10, 10]$

8. Sketch the graphs of three different functions $f(x)$ that are continuous on $0 \leq x \leq 1$ and that have the table of values given. The first function is to have exactly one zero in $[0, 1]$, the second is to have at least two zeros in the interval $[0.6, 0.8]$, and the third is to have at least two zeros in the interval $[0, 0.6]$.

x	0	0.2	0.4	0.6	0.8	1.0
$f(x)$	1.00	0.90	0.60	0.11	−0.58	−1.46

Appendix C POLAR COORDINATES

A point P with Cartesian coordinates (x, y) can be referred to by its *polar coordinates*, r and θ. The number r is the distance between P and the origin, and θ is the angle between the positive x-axis and the line joining P to the origin (with the convention that counterclockwise is positive). Figure C.16 shows the connection between Cartesian and polar coordinates:

Relation between Cartesian and Polar Coordinates

$$x = r\cos\theta \qquad r = \sqrt{x^2 + y^2}$$
$$y = r\sin\theta \qquad \tan\theta = \frac{y}{x}$$

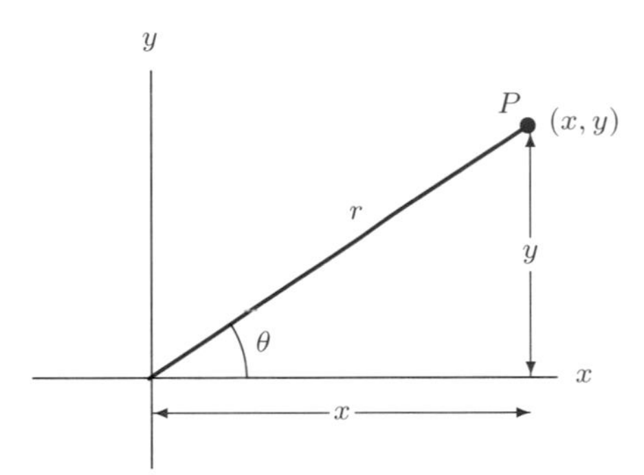

Figure C.16: Cartesian and polar coordinates

Problems for Appendix C

For Problems 1–7, give Cartesian coordinates for the points with the following polar coordinates (r, θ). The angles are measured in radians.

1. $(1, 0)$
2. $(0, 1)$
3. $(2, \pi)$
4. $(\sqrt{2}, 5\pi/4)$
5. $(5, -\pi/6)$
6. $(3, \pi/2)$
7. $(1, 1)$

For Problems 8–15, give polar coordinates for the points with the following Cartesian coordinates. Choose $0 \le \theta < 2\pi$.

8. $(1, 0)$
9. $(0, 2)$
10. $(1, 1)$
11. $(-1, 1)$
12. $(-3, -3)$
13. $(0.2, -0.2)$
14. $(3, 4)$
15. $(-3, 1)$

16. Every point in the plane can be represented by some pair of polar coordinates, but are the polar coordinates (r, θ) uniquely determined by the Cartesian coordinates (x, y)? In other words, for each pair of Cartesian coordinates, is there one and only one pair of polar coordinates for that point? Why or why not?

Appendix D COMPLEX NUMBERS

The quadratic equation

$$x^2 - 2x + 2 = 0$$

is not satisfied by any real number x. If you try applying the quadratic formula, you get

$$x = \frac{2 \pm \sqrt{4-8}}{2} = 1 \pm \frac{\sqrt{-4}}{2}.$$

Apparently, you need to take a square root of -4. But -4 doesn't have a square root, at least, not one which is a real number. Let's give it a square root.

We define the imaginary number i to be a number such that

$$i^2 = -1.$$

Using this i, we see that $(2i)^2 = -4$, so

$$x = 1 \pm \frac{\sqrt{-4}}{2} = 1 \pm \frac{2i}{2} = 1 \pm i.$$

This solves our quadratic equation. The numbers $1 + i$ and $1 - i$ are examples of complex numbers.

A **complex number** is defined as any number that can be written in the form

$$z = a + bi,$$

where a and b are real numbers and $i = \sqrt{-1}$.
The *real part* of z is the number a; the *imaginary part* is the number bi.

Calling the number i imaginary makes it sound like i doesn't exist in the same way that real numbers exist. In some cases, it is useful to make such a distinction between real and imaginary numbers. For example, if we measure mass or position, we want our answers to be real numbers. But the imaginary numbers are just as legitimate mathematically as the real numbers are.

As an analogy, consider the distinction between positive and negative numbers. Originally, people thought of numbers only as tools to count with; their concept of "five" or "ten" was not far removed from "five arrows" or "ten stones." They were unaware that negative numbers existed at all. When negative numbers were introduced, they were viewed only as a device for solving equations like $x + 2 = 1$. They were considered "non-numbers," or, in Latin, "negative numbers." Thus, even though people started to use negative numbers, they did not view them as existing in the same way that positive numbers did. An early mathematician might have reasoned: "The number 5 exists because I can have 5 coins in my hand. But how can I have -5 coins in my hand?" Today we have an answer: "I have -5 coins" means I owe somebody 5 coins. We have realized that negative numbers are just as real as positive ones, and that in some cases, negative numbers can have physical meaning, even though they are useless for measuring length or keeping baseball scores. As we will see, complex numbers can have physical meaning as well. For example, complex numbers are used in studying wave motion in electric circuits.

Algebra of Complex Numbers

Numbers such as 0, 1, $\frac{1}{2}$, π, and $\sqrt{2}$ are called *purely real*, because they contain no imaginary components. Numbers such as i, $2i$, and $\sqrt{2}i$ are called *purely imaginary*, because they contain only the number i multiplied by a nonzero real coefficient.

Two complex numbers are called *conjugates* if their real parts are equal and if their imaginary parts are opposites. The complex conjugate of the complex number $z = a + bi$ is denoted $\overline{z}$, so

$$\overline{z} = a - bi.$$

(Note that z is real if and only if $z = \overline{z}$.) Complex conjugates have the following remarkable property: if $f(x)$ is any polynomial with real coefficients ($x^3 + 1$, say) and $f(z) = 0$, then $f(\overline{z}) = 0$. This means that if z is the solution to a polynomial equation with real coefficients, then so is $\overline{z}$.

- Adding two complex numbers is done by adding real and imaginary parts separately: $(a + bi) + (c + di) = (a + c) + (b + d)i.$
- Subtracting is similar: $(a + bi) - (c + di) = (a - c) + (b - d)i.$
- Multiplication works just like for polynomials, using $i^2 = -1$:

$$\begin{aligned}(a + bi)(c + di) &= a(c + di) + bi(c + di) = ac + adi + bci + bdi^2 \\ &= ac + adi + bci - bd = (ac - bd) + (ad + bc)i.\end{aligned}$$

- Powers of i: We know that $i^2 = -1$; then, $i^3 = i \cdot i^2 = -i$, and $i^4 = (i^2)^2 = (-1)^2 = 1$. Then $i^5 = i \cdot i^4 = i$, and so on. Thus we have

$$(bi)^n = b^n i^n = \begin{cases} b^n i & \text{for } n = 1, 5, 9, 13, \ldots \\ -b^n & \text{for } n = 2, 6, 10, 14, \ldots \\ -b^n i & \text{for } n = 3, 7, 11, 15, \ldots \\ b^n & \text{for } n = 4, 8, 12, 16, \ldots \end{cases}$$

- The product of a number and its conjugate is always real and nonnegative:

$$z \cdot \overline{z} = (a + bi)(a - bi) = a^2 - abi + abi - b^2 i^2 = a^2 + b^2.$$

- Dividing is done by multiplying the denominator by its conjugate, thereby making the denominator real:

$$\frac{a + bi}{c + di} = \frac{a + bi}{c + di} \cdot \frac{c - di}{c - di} = \frac{ac - adi + bci - bdi^2}{c^2 + d^2} = \frac{ac + bd}{c^2 + d^2} + \frac{bc - ad}{c^2 + d^2} i.$$

Example 1 Compute $(2 + 7i)(4 - 6i) - i$.

Solution $(2 + 7i)(4 - 6i) - i = 8 + 28i - 12i - 42i^2 - i = 8 + 15i + 42 = 50 + 15i.$

Example 2 Compute $\dfrac{2 + 7i}{4 - 6i}$.

Solution $\dfrac{2 + 7i}{4 - 6i} = \dfrac{2 + 7i}{4 - 6i} \cdot \dfrac{4 + 6i}{4 + 6i} = \dfrac{8 + 12i + 28i + 42i^2}{4^2 + 6^2} = \dfrac{-34 + 40i}{52} = \dfrac{-17}{26} + \dfrac{10}{13} i.$

The Complex Plane and Polar Coordinates

It is often useful to picture a complex number $z = x + iy$ in the plane, with x along the horizontal axis and y along the vertical. The xy-plane is then called the *complex plane*. Figure D.17 shows the complex numbers $-2i$, $1 + i$, and $-2 + 3i$.

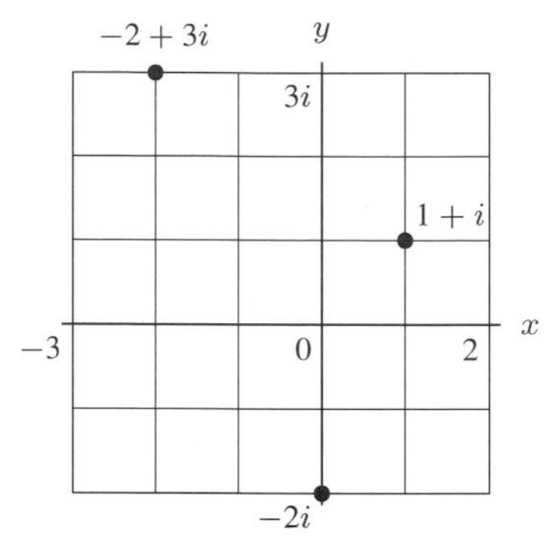

Figure D.17: Points in the complex plane

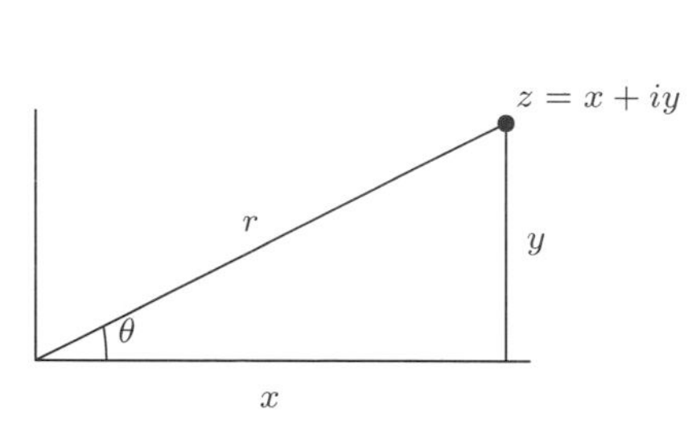

Figure D.18: The point $z = x + iy$ in the complex plane, showing polar coordinates

The triangle in Figure D.18 shows that a complex number can be written using polar coordinates as follows:

$$z = x + iy = r\cos\theta + ir\sin\theta.$$

Example 3 Express $z = -2i$ and $z = -2 + 3i$ using polar coordinates. (See Figure D.17.)

Solution For $z = -2i$, the distance of z from the origin is 2. Thus $r = 2$. Also, one value for θ is $\theta = 3\pi/2$. Thus, using polar coordinates, $-2i = 2\cos(3\pi/2) + i\,2(\sin 3\pi/2)$.

For $z = -2+3i$, we have $x = -2$, $y = 3$, so $r = \sqrt{(-2)^2 + 3^2} \approx 3.61$ and $\tan\theta = 3/(-2)$, so, for example, $\theta \approx 2.16$. Thus using polar coordinates, $-2 + 3i \approx 3.61\cos(2.16) + i\,3.61\sin(2.16)$.

Example 4 Consider the point with polar coordinates $r = 5$ and $\theta = 3\pi/4$. What complex number does this point represent?

Solution Since $x = r\cos\theta$ and $y = r\sin\theta$ we see that $x = 5\cos 3\pi/4 = -5/\sqrt{2}$, and $y = 5\sin 3\pi/4 = 5/\sqrt{2}$, so $z = -5/\sqrt{2} + i\,5/\sqrt{2}$.

Euler's Formula

Consider the complex number z lying on the unit circle in Figure D.19. Writing z in polar coordinates, and using the fact that $r = 1$, we have

$$z = f(\theta) = \cos\theta + i\sin\theta.$$

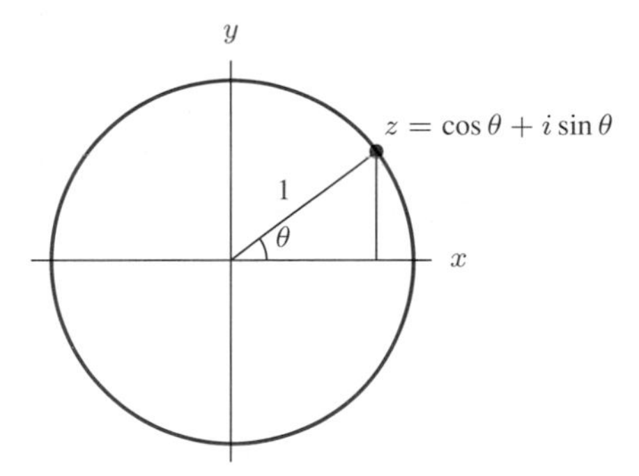

Figure D.19: Complex number represented by a point on the unit circle

It turns out that there is a particularly beautiful and compact way of rewriting $f(\theta)$ using complex exponentials. Let's take the derivative of f, treating i like any other constant but using the fact that $i^2 = -1$:

$$f'(\theta) = -\sin\theta + i\cos\theta = i\cos\theta + i^2\sin\theta.$$

Factoring out an i gives

$$f'(\theta) = i(\cos\theta + i\sin\theta) = i \cdot f(\theta).$$

As you know from Chapter 9, page 501, the only function whose derivative is proportional to the function itself is the exponential function. In other words, we know that if

$$g'(x) = k \cdot g(x), \quad \text{then} \quad g(x) = Ce^{kx}$$

for some constant C. In our case, we have

$$f'(\theta) = i \cdot f(\theta) \quad \text{so} \quad f(\theta) = Ce^{i\theta}$$

for some constant C. To find C we substitute $\theta = 0$. Now $f(0) = Ce^{i \cdot 0} = C$, and since $f(0) = \cos 0 + i\sin 0 = 1$, we must have $C = 1$. Therefore $f(\theta) = e^{i\theta}$. Thus we have *Euler's formula:*

$$e^{i\theta} = \cos\theta + i\sin\theta.$$

This elegant and surprising relationship was discovered by the Swiss mathematician Leonhard Euler in the eighteenth century, and it is particularly useful in solving second-order differential equations. Another derivation of Euler's formula (using Taylor series) is given in Chapter 10 on page 606. For now, it allows us to write the complex number represented by the point with polar coordinates (r, θ) in the following form:

$$z = r(\cos\theta + i\sin\theta) = re^{i\theta}.$$

Similarly, since $\cos(-\theta) = \cos\theta$ and $\sin(-\theta) = -\sin\theta$, we have

$$re^{-i\theta} = r\left(\cos(-\theta) + i\,\sin(-\theta)\right) = r(\cos\theta - i\,\sin\theta).$$

Example 5 Evaluate $e^{i\pi}$.

Solution Using Euler's formula, $e^{i\pi} = \cos\pi + i\sin\pi = -1$.

Example 6 Express the complex number represented by the point $r = 8, \theta = 3\pi/4$ in Cartesian form and polar form, $z = re^{i\theta}$.

Solution Using Cartesian coordinates, the complex number is

$$z = 8\left(\cos\left(\frac{3\pi}{4}\right) + i\sin\left(\frac{3\pi}{4}\right)\right) = \frac{-8}{\sqrt{2}} + i\frac{8}{\sqrt{2}}.$$

Since $e^{i\theta} = \cos\theta + i\sin\theta$, using polar coordinates we have

$$z = 8e^{i\,3\pi/4}.$$

Among its many benefits, the polar form of complex numbers makes finding powers and roots of complex numbers much easier. Using the polar form, $z = re^{i\theta}$, for a complex number, we may find any power of z as follows:

$$z^p = (re^{i\theta})^p = r^p e^{ip\theta}.$$

To find roots, we let p be a fraction, as in the following example.

Example 7 Find a cube root of the complex number represented by the point with polar coordinates $(8, 3\pi/4)$.

Solution In Example 6, we saw that this complex number could be written as $z = 8e^{i3\pi/4}$. Thus,

$$\begin{aligned}\sqrt[3]{z} = \left(8e^{i\,3\pi/4}\right)^{1/3} = 8^{1/3}e^{i(3\pi/4)\cdot(1/3)} = 2e^{\pi i/4} &= 2\left(\cos(\pi/4) + i\sin(\pi/4)\right)\\ &= 2\left(1/\sqrt{2} + i/\sqrt{2}\right) = \sqrt{2}(1+i).\end{aligned}$$

Problems for Appendix D

For Problems 1–8, express the given complex number in polar form, $z = re^{i\theta}$.

1. $2i$
2. -5
3. $1 + i$
4. $-3 - 4i$
5. 0
6. $-i$
7. $-1 + 3i$
8. $5 - 12i$

For Problems 9–18, perform the indicated calculations. Give your answer in Cartesian form, $z = x + iy$.

9. $(2 + 3i) + (-5 - 7i)$
10. $(2 + 3i)(5 + 7i)$
11. $(2 + 3i)^2$
12. $(1 + i)^2 + (1 + i)$
13. $(0.5 - i)(1 - i/4)$
14. $(2i)^3 - (2i)^2 + 2i - 1$
15. $(e^{i\pi/3})^2$
16. $\sqrt{e^{i\pi/3}}$
17. $(5e^{i7\pi/6})^3$
18. $\sqrt[4]{10e^{i\pi/2}}$

By writing the complex numbers in polar form, $z = re^{i\theta}$, find a value for the quantities in Problems 19–28. Give your answer in Cartesian form, $z = x + iy$.

19. $\sqrt{i}$
20. $\sqrt{-i}$
21. $\sqrt[3]{i}$
22. $\sqrt{7i}$
23. $(1+i)^{100}$
24. $(1+i)^{2/3}$
25. $(-4+4i)^{2/3}$
26. $(\sqrt{3}+i)^{1/2}$
27. $(\sqrt{3}+i)^{-1/2}$
28. $(\sqrt{5}+2i)^{\sqrt{2}}$

Solve the simultaneous equations in Problems 29–30 for A_1 and A_2.

29. $A_1 + A_2 = 2$
    $(1-i)A_1 + (1+i)A_2 = 3$
30. $A_1 + A_2 = 2$
    $(i-1)A_1 + (1+i)A_2 = 0$

31. Let $z_1 = -3 - i\sqrt{3}$ and $z_2 = -1 + i\sqrt{3}$.
    (a) Find $z_1 z_2$ and z_1/z_2. Give your answer in Cartesian form, $z = x + iy$.
    (b) Put z_1 and z_2 into polar form, $z = re^{i\theta}$. Find $z_1 z_2$ and z_1/z_2 using the polar form, and verify that you get the same answer as in part (a).
32. If the roots of the equation $x^2 + 2bx + c = 0$ are the complex numbers $p \pm iq$, find expressions for p and q in terms of a and b.

Are the statements in Problems 33–38 true or false? Explain your answer.

33. Every nonnegative real number has a real square root.
34. For any complex number z, the product $z \cdot \bar{z}$ is a real number.
35. The square of any complex number is a real number.
36. If f is a polynomial, and $f(z) = i$, then $f(\bar{z}) = i$.
37. Every nonzero complex number z can be written in the form $z = e^w$, where w is another complex number.
38. If $z = x + iy$, where x and y are positive, then $z^2 = a + ib$ has a and b positive.

For Problems 39–43, use Euler's formula to derive the following relationships. (Note that if a, b, c, d are real numbers, $a + bi = c + di$ means that $a = c$ and $b = d$.)

39. $\sin^2\theta + \cos^2\theta = 1$
40. $\sin 2\theta = 2\sin\theta\cos\theta$
41. $\cos 2\theta = \cos^2\theta - \sin^2\theta$
42. $\dfrac{d}{d\theta}\sin\theta = \cos\theta$
43. $\dfrac{d^2}{d\theta^2}\cos\theta = -\cos\theta$

INDEX